Wind Power

Wind Power

Edited by

S. M. Muyeen

Department of Electrical Engineering
The Petroleum Institute
Abu Dhabi, U.A.E.

Wind Power

Edited by S. M. Muyeen

CBS, Edition 2016

Published by InTech

Janeza Trdine 9, 51000 Rijeka, Croatia

Technical Editor Teodora Smiljanic

Cover Designer Dino Smrekar

Additional hard copies can be obtained from orders@intechopen.com

Wind Power, Edited by S. M. Muyeen
p. cm.
ISBN 978-953-7619-81-7

Dedicated to My Parents

Contents

Preface XI

1. Introduction 001
   S.M.Muyeen

Part A Wind Turbine Generator System

2. An Analytical Analysis of a Wind Power Generation System Including Synchronous Generator with Permanent Magnets, Active Rectifier and Voltage Source Inverter 023
   Sergey A. Kharitonov, Member. IEEE

3. Speed Sensorless Vector Control of Permanent Magnet Wind Power Generator – The Redundant Drive Concept 073
   Tero Halkosaari

4. Wind Turbine Modelling of a Fully-Fed Induction Machine 093
   Umashankar S, Dr. Kothari D P and Mangayarkarasi P

5. Doubly-fed Induction Generator Drives for Wind Power Plants 113
   Balduino Rabelo and Wilfried Hofmann

6. Control Methods for Variable Speed Wind Energy Converters 141
   Sourkounis, Constantinos and Ni, Bingchang

7. Small Wind Turbine Power Controllers 165
   Dr. Horizon Gitano-Briggs

8. Analysis and Investigation of the Inverter for Energy Transfer from Small Wind Power Plant to Common Grid 189
   Romuald Luczkowski and Roman Muszynski

9. Control Strategies for Variable-speed Fixed-pitch Wind Turbines 209
Bunlung Neammanee, Somporn Sirisumrannukul and Somchai Chatratana

10. Real-time Physical Simulation of Wind Energy Conversion Systems 233
Iulian Munteanu, Antoneta Iuliana Bratcu, Seddik Bacha and Daniel Roye

Part B The Variability of Wind Power

11. Variability and Predictability of Large-Scale Wind Energy in the Netherlands 259
A.J. Brand, M. Gibescu and W.W. de Boer

12. Variability of Wind and Wind Power 289
Joaquin Mur-Amada and Ángel Bayod-Rújula

13. Impact of Real Case Transmission Systems Constraints on Wind Power Operation 321
François Vallée – Olivier Deblecker – Jacques Lobry

14. Wind Power at Sea as Observed from Space 341
W. Timothy Liu, Wenqing Tang, and Xiaosu Xie

Part C The Grid Integration Issues

15. Methods and Models for Computer Aided Design of Wind Power Systems for EMC and Power Quality 353
Vladimir Belov, Peter Leisner, Nikolay Paldyaev, Alexey Shamaev and Ilja Belov

16. Design of Robust Power System Stabilizer in an interconnected Power System with Wind Power Penetrations 379
Sarjiya, Cuk Supriyadi A.N and Tumiran

17. Wind Power Impact on Power System Dynamic Performance 395
Emmanuel S. Karapidakis

18. Wind Power: Integrating Wind Turbine Generators (WTG's) with Energy Storage 415
Septimus van der Linden

19. Optimization of Spinning Reserve in Stand-alone Wind-Diesel Power Systems 437
Fernando Olsina and Carlos Larisson

20. Power Characteristics of Compound Microgrid Composed from PEFC and Wind Power Generation 465
Shin'ya Obara

21. Large Scale Integration of Wind Power in Thermal Power Systems 479
Lisa Göransson and Filip Johnsson

22. The Future Energy Mix Paradigm: How to Embed Large Amounts of Wind Generation While Preserving the Robustness and Quality of the Power Systems? 499
Ana Estanqueiro

Part D The Environmental Issues

23. Environmental Impact of Modern Wind Power under LCA Methodology 521
Eduardo Martínez Cámara, Emilio Jiménez Macías, Julio Blanco Fernández and Mercedes Pérez de la Parte

24. Wind-Solar Driven Natural Electric Hybrid Ventilators 537
N.A.Ahmed

Preface

In electrical power generation system the present trend is to maximize the renewable energy penetration ratio as much as possible. Among the renewable energy sources such as wind, solar, biogas/biomass, tidal, geothermal, etc., wind energy has the huge potential to play an important role in energy market along with conventional energy sources. Therefore, reasonably the research on wind power is progressing rapidly. Its roots include many engineering and science disciplines, from mechanical, electrical and electronics to computer, and aerospace engineering. Each of the fields is unique, awesome, and has its own beauty. The joint effort from different fields make one technology successful.

This book is the result of inspirations and contributions from many researchers of different fields. A wide verity of research results are merged together to make this book useful for students and researchers who will take contribution for further development of the existing technology. I hope you will enjoy the book, so that my effort to bringing it together for you will be successful. In my capacity, as the Editor of this book, I would like to thanks and appreciate the chapter authors, who ensured the quality of the material as well as submitting their best works. Most of the results presented in to the book have already been published on international journals and appreciated in many international conferences.

A large number of individuals and organizations have assisted the authors in a variety of ways in the preparation of this work. We have made extensive use of the material from Global Wind Energy Council (GWEC), and American Wind Energy Association (AWEA) and record our special thanks to these organizations for making documents available to us free of charge and sanctioning the permission to use some of the material therein. I would like to acknowledge Dr. Vedran Kordic and others who gave significant efforts to chose some best works on wind power at the present development stage of this technology that eventually ensured the quality of the book. I wish to take this opportunity to express my gratitude to Prof. Junji Tamura for the encouragement he gave me and the confidence he showed in me during the early part of my carrier at Kitami Institute of Technology, Japan.

In Chapter 1 of the book, I discussed on each of the chapter from where the readers may give a general overview and essence of the chapters. Based on the topics, 24 chapters of the book are described under 4 parts. In Part A, the recent trend of wind turbine generator systems are presented where both large and small scale wind generators including their control strategies are described. Different types of machines used in wind energy conversion systems including the control strategy are presented in Part A. In Part B, the variability of wind power is described in detailed. relationship between wind power variability and its forecast, influence of the wind variability on the grid, geographical diversity on wind power, stochastic wind generation models, Wind Power at Sea as Observed from Space are the key content of Part B. Part C presents the grid integration issues. Wind power impact on power system dynamic performance, power quality, fault ride through, future growth of wind power are discussed in Part C. Finally, some environmental issues related to wind energy conversion systems are presented in Part D. In a brief, theoretical analysis, experimental validation, computer simulation as well as real time physical simulation are presented throughout the book.

June 10, 2010

Editor

S.M.Muyeen
Department of Electrical Engineering
The Petroleum Institute
Abu Dhabi, U.A.E.

1

Introduction

S. M. Muyeen
Department of Electrical Engineering
The Petroleum Institute
Abu Dhabi, U.A.E.

The conventional energy sources such as oil, natural gas, coal, or nuclear are finite and generate pollution. Alternatively, the renewable energy sources like wind, fuel cell, solar, biogas/biomass, tidal, geothermal, etc. are clean and abundantly available in nature. Among those the wind energy has the huge potential of becoming a major source of renewable energy for this modern world. Wind power is a clean, emissions-free power generation technology. Like all renewable sources it is based on capturing the energy from natural forces and has none of the polluting effects associated with 'conventional' fuels. In 2008, 27 GW wind power has been installed all over the world, bringing world-wide installed capacity to 120.8 GW (GWEC publication, 2008). This is an increase of 36% compared with the 2007 market, and represents an overall increase in the global installed capacity of about 28.8%. From this scenario, it is clear that wind power is going to dominate the renewable as the conventional energy market in sooner future. Wind energy is the only power generation technology that can deliver the necessary cuts in CO2 emissions from the power sector in the critical period up to 2020, when greenhouse gas emissions must peak and begin to decline if we are to have any hope of avoiding the worst impacts of climate change . The 120 .8 GW of global wind capacity installed by the end of 2008 will produce 260 TWh of electricity and save 158 million tons of CO2 every year (GWEC publication, 2008).

2. Background of Wind Power

In the 1970s, concern for the limited fossil fuel resources and their impact on the environment awakened. Due to this growing concern, interest revived in using renewable energy sources to meet the constantly rising world electricity demand. In addition, the oil crises of 1973 and 1979 led to the awareness that the amount of energy imported should be decreased so as to become less dependent on oil ex-porting countries. The Gulf-War (1990-1991) confirmed this concern.

The increasing concerns over environmental issues and the depletion of fossil fuel demanded the search for more sustainable electrical sources. One technology for generating electricity from renewable resources is to use wind turbines that convert the energy contained by the wind into electricity. The wind is a vast, worldwide renewable source of energy. Since ancient times, humans have har-nessed the power of the wind. The earliest known use of wind power is the sail-boat. Boats propelled by wind energy sailed up the

Nile against the current as early as 5000 B.C. By A.D. 1000, the Vikings had explored and conquered the North Atlantic. The wind was also the driving force behind the voyages of discovery of the Verenigde Oost-Indische Compagnie (VOC) between 1602 and 1799. Windmills have been providing useful mechanical power for at least the last thousand years, and wind turbines have generated electricity since 1888.

3. Current Installation Status of Wind Power Worldwide

Wind energy has become a mainstream energy source and an important player in the world's energy markets, with the 2008 market for turbine installations worth about € 36 .5 billion . The wind industry also creates many new jobs; over 400,000 people are now employed in this industry, and that number is expected to be in the millions in the near future .The following sub-sections will provide a condensed overview of wind energy status around the world until the year 2008. The prediction of future growth of wind energy until 2020 is also presented. This section is written in light of the data from Global Wind Energy Council (GWEC), European Wind Energy Association (EWEA), and Ameriical Wind Energy Association (AWEA).

Figure 1 shows the total wind power installation capacities throughout Europe at the end of 2008 (GWEC publication, 2008). Global cumulative installed capacity from 1996-2008, global annual installed capacity from 1996-2008, and annual installed capacity by region from 2003-2008 scenarios are shown in Figs. 2 to 4 (GWEC publication, 2008). The installation scenario for world top 10 countries in 2008 is shown in Fig. 5 (GWEC publication, 2008). The regional analyses are given in light of Global Wind Energy Council (GWEC) and American Wind Energy Association (AWEA) reports as shown below.

3.1 United States

In North America, the US market broke all previous records with new installations of 8.5 GW, reaching a total installed capacity of over 25 GW. In 2008 the US was the number one market both in terms of new capacity and in terms of total wind generation capacity. Fig. 5 shows that the US passed Germany to become the number one market in wind power. The massive growth in the US wind market in 2008 increased the country's total power generating capacity by half . The new wind projects completed in 2008 accounted for about 42% of the entire new power producing capacity added in the US last year, and created 35,000 new jobs, bringing the total employed in the sector up to 85,000 (GWEC publication, 2008).

The U.S. wind industry installed over 1,600 MW of new wind capacity in the third quarter of the year, bringing the wind power capacity installed so far in 2009 to over 5,800 MW and the total installed capacity in the U.S. to over 31,100 MW overall. Over 5,000 MW more are under construction for completion this year or next year (AWEA publication, 2009).

The following map (Fig. 6) shows the installed megawatts (MW) for each state of the United States, as of end June 2009. Wind power generating capacity existing and under construction at the 3rd quarter of 2009 is shown in Table 1 (AWEA resources, 2009). The top ten ranking of the wind power generating states are shown in Fig. 7 (AWEA publication, 2009).

In 2008, the US Department of Energy released a groundbreaking report, finding that wind power could provide 20% of US electricity by 2030 (GWEC publication, 2008).

		End 2007	New 2008	Total end 2008
AFRICA & MIDDLE EAST	Egypt	310	55	365
	Morocco	124	10	134
	Iran	67	17	85
	Tunisia	20	34	54
	Other[1]	17	14	31
	Total	539	130	669
ASIA	China	5,910	6,300	12,210
	India	7,845	1,800	9,645
	Japan	1,538	346	1,880
	Taiwan	281	81	358
	South Korea	193	43	236
	Philippines	25	8	33
	Other[2]	5	1	6
	Total	15,795	8,579	24,368
EUROPE	Germany	22,247	1,665	23,903
	Spain	15,145	1,609	16,754
	Italy	2,726	1,010	3,736
	France	2,454	950	3,404
	UK	2,406	836	3,241
	Denmark	3,125	77	3,180
	Portugal	2,150	712	2,862
	Netherlands	1,747	500	2,225
	Sweden	788	236	1,021
	Ireland	795	208	1,002
	Austria	982	14	995
	Greece	871	114	985
	Poland	276	196	472
	Norway	326	102	428
	Turkey	147	286	433
	Rest of Europe[3]	955	362	1,305
	Total Europe	57,139	8,877	65,946
	of which EU-27[4]	*56,531*	*8,484*	*64,948*
LATIN AMERICA & CARIBBEAN	Brazil	247	94	341
	Mexico	87	0	87
	Costa Rica	70	0	70
	Caribbean	55	0	55
	Argentina	29	2	31
	Other[5]	45	0	45
	Total	533	95	629
NORTH AMERICA	USA	16,824	8,358	25,170
	Canada	1,846	526	2,372
	Total	18,670	8,884	27,542
PACIFIC REGION	Australia	824	482	1,306
	New Zealand	322	4	326
	Pacific Islands	12	0	12
	Total	1,158	486	1,644
	World total	93,835	27,051	120,798

1 South Africa, Cape Verde, Israel, Lebanon, Nigeria, Jordan;
2 Thailand, Bangladesh, Indonesia, Sri Lanka;
3 Belgium, Bulgaria, Croatia, Cyprus, Czech Republic, Estonia, Faroe Islands, Finland, Hungary, Latvia, Lithuania, Luxembourg, Romania, Russia, Slovakia, Switzerland, Ukraine;
4 Austria, Belgium, Bulgaria, Cyprus, Czech Republic, Denmark, Estonia, Finland, France, Germany, Greece, Hungary, Ireland, Italy, Latvia, Lithuania, Luxembourg, Malta, Netherlands, Poland, Portugal, Romania, Slovakia, Slovenia, Spain, Sweden, UK;
5 Colombia, Chile, Cuba;

Please note: project decommissioning of 89 MW and rounding affect the final sums

Source: GWEC

Fig. 1. Regional distribution of global installed wind power capacity in MW (Source: Global Wind Energy Council, GWEC)

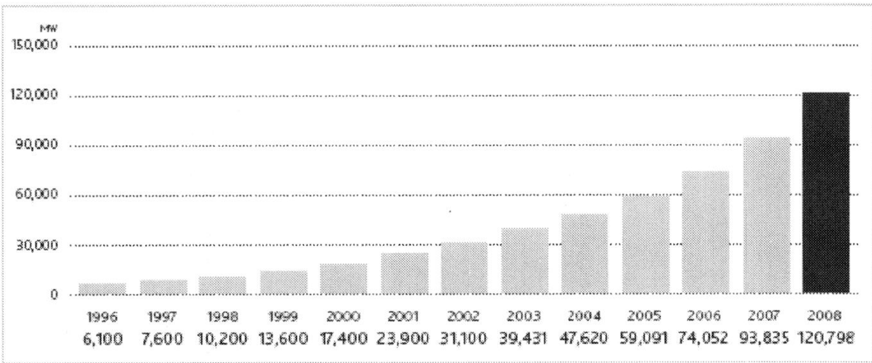

Fig. 2. Global cumulative installed capacity 1996-2008 (Source: Global Wind Energy Council, GWEC)

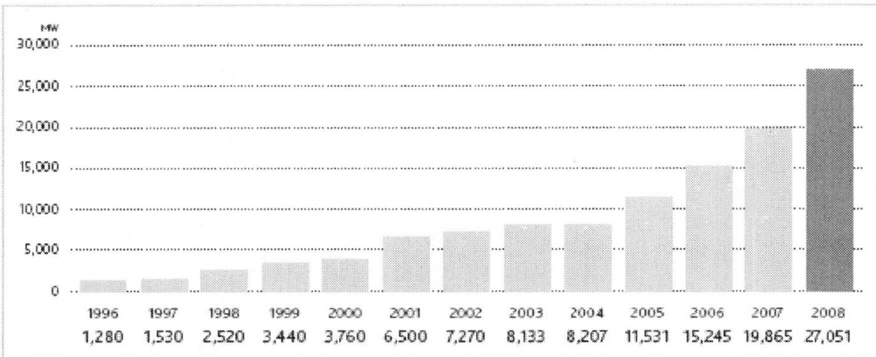

Fig. 3. Global annual installed capacity 1996-2008 (Source: Global Wind Energy Council, GWEC)

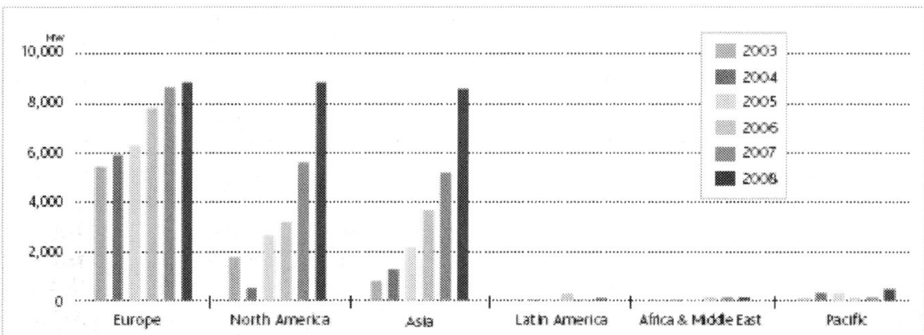

Fig. 4. Annual installed capacity by region 2003-2008 (Source: Global Wind Energy Council, GWEC)

TOP 10 TOTAL INSTALLED CAPACITY 2008

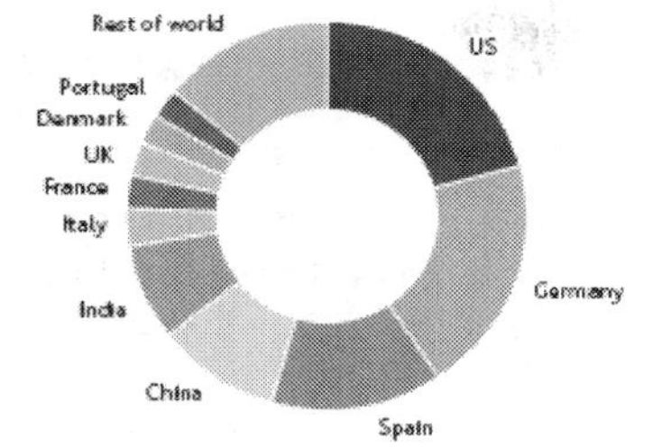

	MW	%
USA	25,170	20.8
Germany	23,903	19.8
Spain	16,754	13.9
China	12,210	10.1
India	9,645	8.0
Italy	3,736	3.1
France	3,404	2.8
UK	3,241	2.7
Denmark	3,180	2.6
Portugal	2,862	2.4
Rest of world	16,693	13.8
Total top 10	**104,104**	**86.2**
World total	**120,798**	**100.0**

TOP 10 NEW CAPACITY 2008

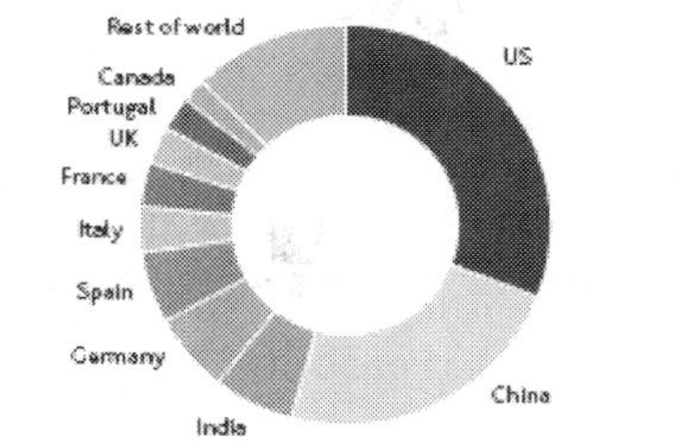

	MW	%
USA	8,358	30.9
China	6,300	23.3
India	1,800	6.7
Germany	1,665	6.2
Spain	1,609	5.9
Italy	1,010	3.7
France	950	3.5
UK	836	3.1
Portugal	712	2.6
Canada	526	1.9
Rest of world	3,285	12.2
Total top 10	**23,766**	**87.8**
World total	**27,051**	**100.0**

Fig. 5. Installation scenario for top 10 countries in 2008 (Source: Global Wind Energy Council, GWEC)

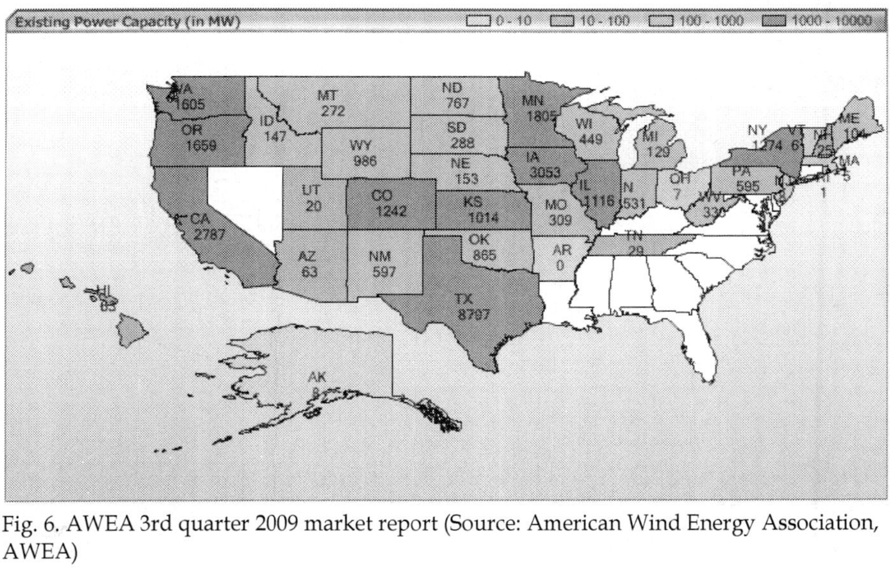

Fig. 6. AWEA 3rd quarter 2009 market report (Source: American Wind Energy Association, AWEA)

Existing	Under Construction
16818.78	3506.38

Table 1. National Total Power Capacities from Wind Energy (MW) in U.S (Source: American Wind Energy Association, AWEA)

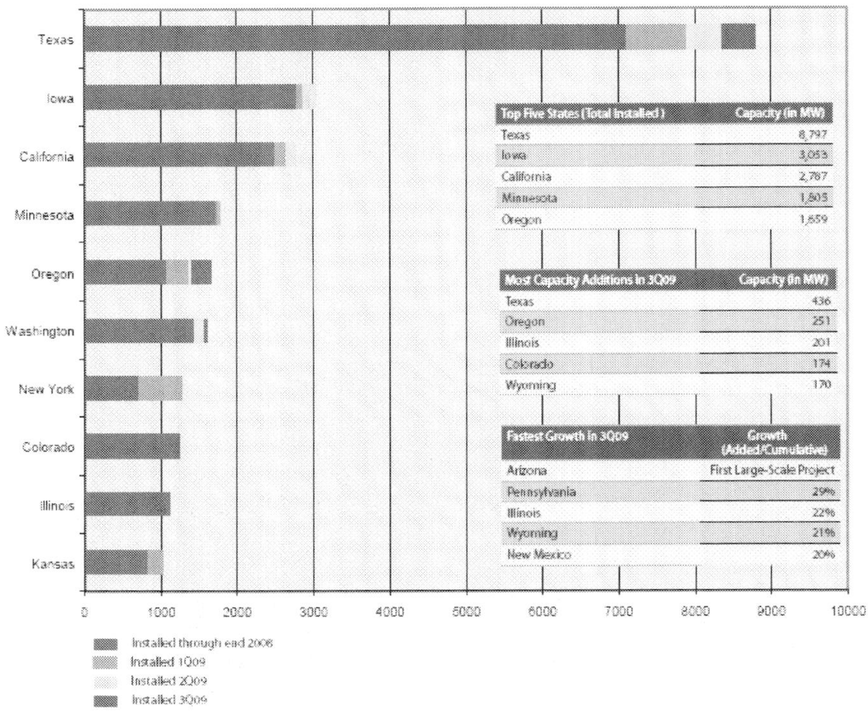

Fig. 7. Wind Project Installations by Top Ten States (Source: American Wind Energy Association, AWEA)

3.2 Canada

Canada in 2008 surpassed the 2 GW mark for installed wind energy capacity, ending the year with 2 .4 GW. Canada's wind farms now produce enough power to meet almost 1% of Canada's total electricity demand. 2008 was Canada's second best ever year for new wind energy installations with ten new wind farms coming online, representing 526 MW of installed wind energy capacity. Included in this total were the first wind farms in the provinces of New Brunswick, Newfoundland and Labrador. In British Columbia, the only Canadian province without a wind farm, construction began on the first wind farm with completion expected in early 2009 (GWEC publication, 2008).

3.3 Asia

The growth in Asian markets has been breathtaking, as nearly a third of the 8 .6 GW installed in 2008 was installed in Asia (GWEC publication, 2008).
China continued its spectacular growth in 2008, once again doubling its installed capacity by adding about 6 .3 GW, to reach a total of 12 .2 GW. The prospects for future growth in the Chinese market are very good. In response to the financial crisis, the Chinese government has identified the development of wind energy as one of the key economic growth areas, and in 2009, new installed capacity is expected to nearly double again. At this rate, China is on its way to overtake Germany and Spain to reach second place in terms of total wind power capacity in 2010. This means that China would have met its 2020 target of 30 GW ten years ahead of time.
The growing wind power market in China has also encouraged domestic production of wind turbines and components, and the Chinese manufacturing industry is becoming increasingly mature, stretching over the whole supply chain. According to the Chinese Renewable Energy Industry Association (CREIA), the supply is starting to not only satisfy domestic demand, but also meet international needs, especially for components. In 2009, Chinese companies are set to start entering the UK and Japanese markets. There are also ambitions for exploring the US market in the coming years.
In 2008, the newly-established National Energy Bureau established wind energy as a priority for diversifying China's energy mix away from coal, and it implemented the 10 GW-Size Wind Base Program (Wind Base) in order to achieve this. The bureau selected six locations for Wind Base projects; Xinjiang, Inner Mongolia, Gansu, Hebei, and Jiangsu. The objective is to install of 10 GW or more of new wind generating capacity by 2020 on each of these Wind Base sites.
India is continuing its steady growth, with 1,800 MW of wind energy capacity added in 2008, bringing the total up to 9 .6 GW. The leading wind producing state in India is Tamil Nadu, which hosts over 4 GW of installed capacity, followed by Maharashtra with 1 .8 GW and Gujarat with 1 .4 GW.
Other Asian countries with new capacity additions in 2008 include Japan (346 MW, taking the total to 1 .9 GW), Taiwan (81 MW for a total of 358 MW) and South Korea (43 MW for a total of 236 MW) (GWEC publication, 2008).

3.4 Europe

Although Europe was home to only one third of the world's new installed capacity in 2008, the European market continues its steady growth, and wind power is now the fastest growing power generation technology in the EU. Indeed, more than 35% of all new energy installations in 2008 were wind power, which meant that renewable energy accounted for more than half of all new power generation capacity in the EU considering the fact of 4.2 GW of PV installation (GWEC publication, 2008).
Overall, almost 8 .9 .GW of new wind turbines brought European wind power generation capacity up to nearly 66 GW. There is now clear diversification of the European market, relying less and less on the traditional wind markets of Germany, Spain and Denmark. 2008 saw a much more balanced expansion, with a 'second wave' led by Italy, France and the UK. Ten of the EU's 27 member states now have more than 1 GW of wind power capacity.

In 2008 the European wind turbine market was worth € 11 billion. The entire wind fleet will produce 142 TWh of electricity, or about 4 .2% of EU demand in an average wind year. This will save about 100m tons of CO_2 each year.
While at the global level, Germany has been surpassed by the US, it continues to be Europe's leading market, both in terms of new and total installed capacity. Over 1 .6 GW of new capacity was installed in 2008, bringing the total up to nearly 24 GW.
Wind energy is continuing to play an important role in Germany's energy mix. In 2008, 40 .4 TWh of wind power were generated, representing 7 .5% of the country's net electricity consumption. In economic terms too, wind power has become a serious player in Germany, and the sector now employs close to 100,000 people.
Spain is Europe's second largest market, and has seen growth in line with previous years (with the exception of 2007, when regulatory change brought about a higher than usual amount of new wind capacity). In 2008, 1 .6 GW of new generating equipment was added to the Spanish wind fleet, bringing the total up to 16 .8 GW. This development confirms Spain as a steadily growing market, which at this rate is likely to reach the government's 2010 target of 20 GW of installed wind capacity. In 2008, wind energy generated more than 31,000 GWh, covering more than 11% of the country's electricity demand.
One noteworthy newcomer among the growing European markets in 2008 was Italy, which experienced a significant leap in wind power capacity. Over 1,000 MW of new wind turbines came on line in 2008, bringing total installed capacity up to 3 .7 GW. At the end of 2008, the Italian government passed an important decree that resolves many of the main problems related to the value of green certificates. This measure is designed to avoid speculative fluctuations in the price of green certificates that negatively affected the Italian market in the past.
France is also continuing to see strong growth, after progressing steadily in recent years. At the end of 2008, the total installed capacity stood at 3 .4 GW, representing an annual growth rate of 38%.

3.5 Latin America
The Latin American market, despite the tremendous wind resources in the region, saw only slow growth in 2008. The only country installing substantial new capacity was brazil, which added 94 MW of wind energy across five wind farms, mostly located in Ceará in the north east of the country.

3.6 Australia
After several years of stagnation in Australia's wind market, the speed of development picked up again in 2008, with 482 MW of new installations, a 58% leap in terms of total installed capacity. Australia is now home to 50 wind farms, with a total capacity of 1 .3 GW. Six additional projects totalling 555 MW are currently under construction and expected to become operational in 2009.

3.7 Africa and Middle East
In North Africa, the expansion of wind power continues in Egypt, Morocco and Tunisia, with 55 MW, 10 MW and 34 MW of new capacity installed respectively. In the Middle East, Iran installed 17 MW of new capacity. The total for Africa and the Middle East now stands at 669 MW.

3.8 United Kingdom

In 2008, the UK installed 3,240 MW of wind energy capacity, and there are another 8, 827 MW of projects either under construction or awaiting planning permission. After failing to pick up the pace of development in the 1990s and struggling to reach 1 GW of installed capacity, a clearly revitalized and reenergized UK wind sector has delivered over 2 GW since 2006, and continues to attract interest from developers and investors. The UK government published a Renewable Energy Strategy in June 2008 which proposes 14 GW of onshore and 14 GW of offshore wind by 2020. This would increase the current installed capacity by eight times in 12 years (GWEC publication, 2008).

3.9 Market Forecast for 2009-2013

GWEC predicts that in 2013, five years from now, global wind generating capacity will stand at 332 GW, up from 120 GW at the end of 2008. During 2013, 56 .3 GW of new capacity will be added to the global total, more than double the annual market in 2008 (GWEC publication, 2008).

The annual growth rates during this period will average 22 .4% in terms of total installed capacity, and 15 .8% for the annual market. These rates are modest compared to past developments: in the last ten years, we have seen an average increase of 28 .2% for total capacity and 28 .3% for annual capacity.

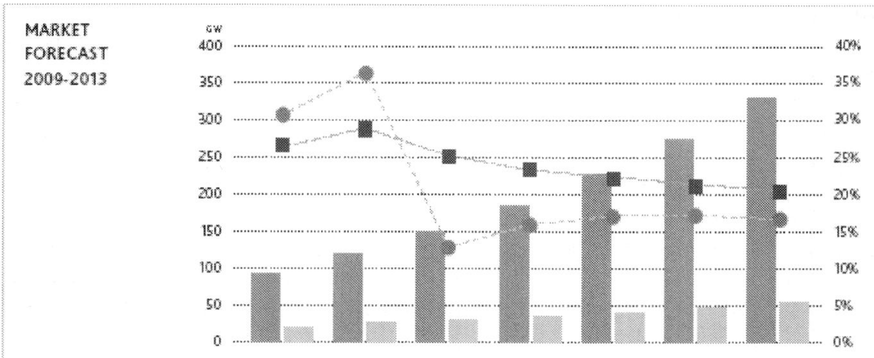

	2007	2008	2009	2010	2011	2012	2013
Annual installed capacity (GW)	19.9	27.1	30.4	35.2	41.2	48.2	56.3
Cumulative installed capacity (GW)	93.8	120.8	151.2	186.4	227.6	275.8	332.1
Annual install capacity growth rate	30.7%	36.2%	12.5%	15.7%	17.0%	17.0%	16.8%
Cumulative capacity growth rate	26.7%	28.7%	25.2%	23.3%	22.1%	21.2%	20.4%

Fig. 8. Market forecast for 2009-2013 (Source: Global Wind Energy Council, GWEC)

4. Technology and Trend

4.1 Wind Energy Technology

GWEC published an excellent report on wind energy technology in light of past and present wind turbine development (GWEC publication, 2008)

Since the 1980s, when the first commercial wind turbines were deployed, their installed capacity, efficiency and visual design have all improved enormously.
Although many different pathways towards the ideal turbine design have been explored, significant consolidation has taken place over the past decade. The vast majority of commercial turbines now operate on a horizontal axis with three evenly spaced blades. These are attached to a rotor from which power is transferred through a gearbox to a generator. The gearbox and generator are contained within a housing called a nacelle. Some turbine designs avoid a gearbox by using direct drive. The electricity is then transmitted down the tower to a transformer and eventually into the grid network.
Wind turbines can operate across a wide range of wind speeds - from 3-4 metres per second up to about 25 m/s, which translates into 90 km/h (56 mph), and would be the equivalent of gale force 9 or 10.
The majority of current turbine models make best use of the constant variations in the wind by changing the angle of the blades through 'pitch control', by turning or "yawing" the entire rotor as wind direction shifts and by operating at variable speed. Operation at variable speed enables the turbine to adapt to varying wind speeds and increases its ability to harmonise with the operation of the electricity grid. Sophisticated control systems enable fine tuning of the turbine's performance and electricity output.
Modern wind technology is able to operate effectively at a wide range of sites – with low and high wind speeds, in the desert and in freezing arctic climates. Clusters of turbines collected into wind farms operate with high availability, are generally well integrated with the environment and accepted by the public. Using lightweight materials to reduce their bulk, modern turbine designs are sleek, streamlined and elegant.
The main design drivers for current wind technology are:

- reliability
- grid compatibility
- acoustic performance (noise reduction)
- maximum efficiency and aerodynamic performance
- high productivity for low wind speeds
- offshore expansion

Wind turbines have also grown larger and taller. The generators in the largest modern turbines are 100 times the size of those in 1980. Over the same period, their rotor diameters have increased eight-fold. The average capacity of turbines installed around the world during 2007 was 1,492 kW, whilst the largest turbine currently in operation is the Enercon E126, with a rotor diameter of 126 metres and a power capacity of 6 MW.
The main driver for larger capacity machines has been the offshore market, where placing turbines on the seabed demands the optimum use of each foundation. Fixing large foundations in the sea bed, collecting the electricity and transmitting it to the shore all increase the costs of offshore development over those on land. Although the offshore wind farms installed so far have used turbines in the capacity range up to 3.6 MW, a range of designs of 5 MW and above are now being deployed and are expected to become the 'standard' in the coming years.
For turbines used on land, however, the past few years have seen a levelling of turbine size in the 1.5 to 3 MW range. This has enabled series production of many thousands of turbines of the same design, enabling teething problems to be ironed out and reliability increased.

Ongoing innovations in turbine design include the use of different combinations of composite materials to manufacture blades, especially to ensure that their weight is kept to a minimum, variations in the drive train system to reduce loads and increase reliability, and improved control systems, partly to ensure better compatibility with the grid network.

4.2 Variability of Wind Power

In 2008 the article variability of wind power is reported by GWEC (Angelika et. Al, 2008). The interested readers may have a look at EWEA report (EWEA Publication, 2009). Wind power is often described as an "intermittent" energy source, and therefore unreliable. Iin fact, at power system level, wind energy does not start and stop at irregular intervals, so the term "intermittent" is misleading. The output of aggregated wind capacity is variable, just as the power system itself is inherently variable.

Since wind power production is dependent on the wind, the output of a turbine and wind farm varies over time, under the influence of meteorological fluctuations. Variations occur on all time scales: by seconds, minutes, hours, days, months, seasons and years. Understanding and predicting these variations is essential for successfully integrating wind power into the power system and to use it most efficiently.

Electricity flows - both supply and demand - are inherently variable, as power systems are influenced by a large number of planned and unplanned factors, but they have been designed to cope effectively with these variations through their configuration, control systems and interconnection.

Changing weather makes people switch their heating, cooling and lighting on and off, millions of consumers expect instant power for TVs and computers. On the supply side, when a large power station, especially, if it is a nuclear reactor, goes offline, whether by accident or planned shutdown, it does so instantaneously, causing an immediate loss of many hundreds of megawatts. By contrast, wind energy does not suddenly trip off the system. Variations are smoother because there are hundreds or thousands of units rather than a few large power stations, making it easier for the system operator to predict and manage changes in supply. Especially in large, interconnected grids, there is little overall impact if the wind stops blowing in one particular place.

Predictability is key in managing wind power's variability, and significant advances have been made in improving forecasting methods. Today, wind power prediction is quite accurate for aggregated wind farms and large areas. Using increasingly sophisticated weather forecasts, wind power generation models and statistical analysis, it is possible to predict generation from five minute to hourly intervals over timescales up to 72 hours in advance, and for seasonal and annual periods. Using current tools, the forecast error for a single wind farm is between 10 and 20% of the power output for a forecast horizon of 36 hours. For regionally aggregated wind farms the forecast error is in the order of 10% for a day ahead and less than 5% for 1-4 hours in advance.

The effects of geographical distribution can also be significant. Whereas a single wind farm can experience power swings from hour to hour of up to 60% of its capacity, monitoring by the German ISET research institute has shown that the maximum hourly variation across 350 MW of aggregated wind farms in Germany does not exceed 20%. Across a larger area, such as the Nordel system covering four countries (Finland, Sweden, Norway and Eastern Denmark), the greatest hourly variations would be less than 10%, according to studies [1]

4.3 Storage Option

Storage option is described in GWEC publication in 2008 (Angelika et al., 2008). There is increasing interest in both large scale storage implemented at transmission level, and in smaller scale dedicated storage embedded in distribution networks. The range of storage technologies is potentially wide.

For large-scale storage, pumped hydro accumulation storage (PAC) is the most common and best known technology, which can also be done underground. Another technology option available for large scale is compressed air energy storage (CAES).

On a decentralized scale storage options include flywheels, batteries, possibly in combination with electric vehicles, fuel cells, electrolysis and super-capacitors. Furthermore, an attractive solution consists of the installation of heat boilers at selected combined heat and power locations (CHP) in order to increase the operational flexibility of these units.

However, it has to be pointed out that storage leads to energy losses, and is not necessarily an efficient option for managing wind farm output. If a country does not have favourable geographical conditions for hydro reservoirs, storage is not an attractive solution because of the poor economics at moderate wind power penetration levels (up to 20%). In any case, the use of storage to balance variations at wind plant level is neither necessary nor economic.

4.4 Grid Infrastructures

GWEC and Greenpeace International also reported on grid infrastructure in its 2008 report (Angelika et al., 2008). The specific nature of wind power as a distributed and variable generation source requires specific infrastructure investments and the implementation of new technology and grid management concepts. High levels of wind energy in system can impact on grid stability, congestion management, transmission efficiency, and transmission adequacy.

In many parts of the world, substantial upgrades of grid infrastructure will be required to allow for the levels of grid integration proposed in this report. Significant improvements can be achieved by network optimisation and other 'soft' measures, but an increase in transmission capacity and construction of new transmission lines will also be needed. At the same time, adequate and fair procedures for grid access for wind power need to be developed and implemented, even in areas where grid capacity is limited.

However, the expansion of wind power is not the only driver. Extensions and reinforcements are needed to accommodate whichever power generation technology is chosen to meet a rapidly growing electricity demand. The IEA estimates that by 2030, over 1.8 trillion USD will have to be invested in transmission and distribution networks in the OECD alone.

[1] Holttinen, H. (2004): The impact of large scale wind power on the Nordic electricity system

In the present situation wind power is disadvantaged in relation to conventional sources, whose infrastructure has been largely developed under national vertically integrated monopolies which were able to finance grid network improvements through state subsidies and levies on electricity bills. But while a more liberalised market has closed off those options in some countries, numerous distortions continue to disadvantage renewable generators in the power market - from discriminatory connection charges to potential abuse of their dominant power by incumbent utilities.

4.5 Grid Integration Issues

A grid code covers all material technical aspects relating to connections to, and the operation and use of, a country's electricity transmission system. They lay down rules which define the ways in which generating stations connecting to the system must operate in order to maintain grid stability (Angelika et al., 2008).

Technical requirements within grid codes vary from system to system, but the typical requirements for generators normally concern tolerance, control of active and reactive power, protective devices and power quality. Specific requirements for wind power generation are changing as penetration increases and as wind power is assuming more and more power plant capabilities, i.e. assuming active control and delivering grid support services (Angelika et al., 2008).

In response to increasing demands from the network operators, for example to stay connected to the system during a fault event, the most recent wind turbine designs have been substantially improved. The majority of MW-size turbines being installed today are capable of meeting the most severe grid code requirements, with advanced features including fault-ride-through capability. This enables them to assist in keeping the power system stable when disruptions occur. Modern wind farms are moving towards becoming wind energy power plants that can be actively controlled (Angelika et al., 2008).

Grid codes developed in many countries are more or less similar. The rest of the system is written in light of the grid code reports developed in different countries (Zavadil R. et al., 2005), (FERC Report, 2005), (Tsili M. et al., 2009), (P. Gardner et al., 2009), (Qiao W. et al., 2009), (ING Report, 2007), (E. On Netz Resources, 2006), and (Bharat Singh, S.N. Singh, 2009).

4.5.1Transient Fault Ride Through

The fault ride-through (FRT) requirement is imposed on a wind power generator so that it remains stable and connected to the network during network faults. In the past, the common practice was to disconnect the wind turbine generator unit during network disturbance. However, disconnection from the grid may worsen a critical grid situation and can threaten the security standards when wind penetration is high. In Germany, wind generating plants are expected to acquit themselves during a low-voltage disturbance as summarized in a voltage versus time curve shown in Fig. 9. Wind turbines are required to stay on the grid within areas 1 and 2 (E. On Netz Resources, 2006), and (Bharat Singh, S.N. Singh, 2009).

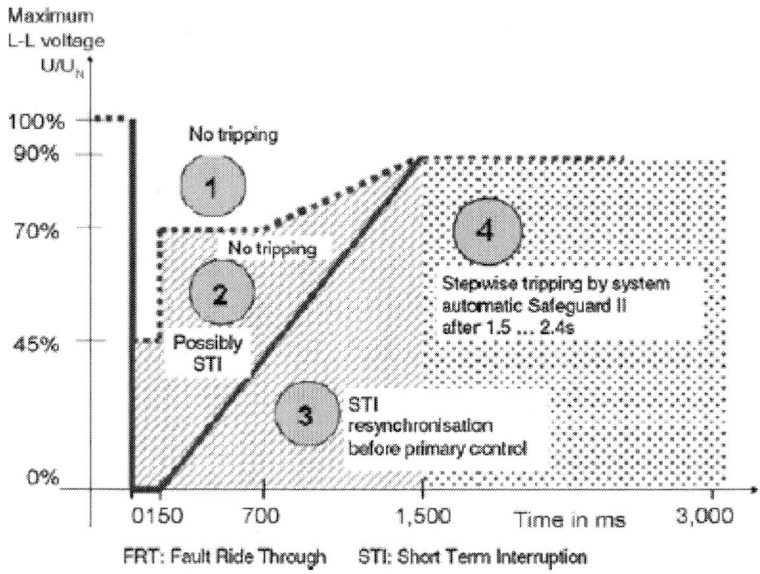

Fig. 9. Low voltage ride-through standard set by E.On Netz..

4.5.2 Active Power Control

Active power control means to control the system frequency by changing the real power into the grid. This is a requirement for generating units to be able to deliver power and remain connected to the network even if the system frequency deviates from the specified one. It gives the flexibility to ensure power quality, avoid large voltage steps and inrush currents during startup and shutdown of wind turbines. It has also an influence on voltage and transient stability augmentation.

4.5.3 Power Quality Ensure

The fluctuating nature of wind arises the issue of power quality such as flicker, voltage fluctuation, etc. At the present, wind generator uses the frequency converter for its variable speed operation. However, it introduces harmonics into the power system and there is also high possibility of resonant effect due to the reactance of wind turbine generator system electrical unit. Therefore most grid code will request wind power plant to maintain voltage fluctuations, flickers and harmonic current/voltage in the desired range.

4.5.4 Frequency Control

For secure grid operation the frequency of the power system should be maintained constant to its rated value. However, when the power imbalance occurs between supply and demand, the frequency varies from its rated value, which is undesarable. Frequency control is a requirement for generating units to be able to increase or decrease power output with falling or rising frequency. The acceptable frequency deviation varies in different countries. As for example, according to E.On, wind turbines have to stay connected in the grid within the frequency range of 47.5-51.5 Hz. Outside this range, disconnection without any time delay is necessary.

4.5.5 Voltage Control

Voltage control requirement defines that the wind power station should be capable of automatic regulating its terminal voltage according to the given set point. The dynamic performance of voltage control is normally defined as rise time, overshoot and settling time. Voltage control has the relation with leading/lagging reactive power at the grid connection point of wind power station. Among the different grid codes, this requirement can be classified as constant reactive power mode or constant power factor mode.

5. Wind Power Explained in this Book

The chapters of the book are written in light of recent development of wind power industry by many renowned authors in the wind energy arena. The chapters of the book are described under 4 parts.

In part A, the recent trend of wind turbine generator systems are presented where both large and small scale wind generators including their control strategies are described.

Chapter 2 discusses about active rectifier and voltage source inverter based topology for permanent magnet synchronous wind generator that connects the grid. The active rectifier and voltage source inverter based topology are well studied at a constant frequency and voltage of the generator. However, this chapter considered the factors to operate the system at variable speed operation of wind generator. Changing the frequency and the voltage of the synchronous generator, as well as dependence of the generated power on shaft speed of the wind turbine are discussed. The detailed mathematical modeling of the overall system are presented in detail including choice of control algorithm and structural circuits.

The permanent magnet synchronous generator (PMSG) is becoming popular in wind power industry using a full capacity frequency converter for grid interconnection. PMSG is recognized as a promising technology for using as wind generator, both in direct-drive system and the system using a simple single-stage gearbox. One of the major advantages is the high power density of this type of machine. The transverse flux design of it makes it possible to fish out some other benefits. The use of PMSG seems to be more prospective in wind power application from now on because the price for the rare-earth magnets decreases remarkably, in the last few years. In chapter 3, the author proposed new types of permanent magnet machine for wind power application considering its reliability issue. The reliability of the PM wind power generators can be increased by using multiple stator modules, or stator segments, independent of each other. These segments can be considered, for example, as independent stator parallel windings which are each fed by an own frequency converter. If one of the frequency converters fails, other ones can continue the operation while the failed units are changed. In this chapter, the redundant drive concept is introduced based on the modular drive concept. Then, the speed sensorless vector control theory applied to wind power generators is discussed. Simulations as well as experimental results are presented from which the effectiveness of the proposed control strategy can be verified.

Few types of variable speed wind turbine (VSWT) generator systems are commercially available, nowadays. Doubly fed induction generator (DFIG) is holding the major market share in wind power industry. Chapter 4 describes the detailed modeling and controls strategies of VSWT driven DFIG, which might be pretty much effective for the students or researchers to build the simulation model and learn the characteristics of DFIG. The simulation model is developed in Matlab Simulink environment. The chapter includes both dynamic and transient analyses.

Doubly-fed Induction Generator (DFIG) is most commonly used wind generator, nowadays, in wind energy conversion system. In Chapter 5, the authors present theoretical basics for doubly-fed induction generator that can be considered for the electrical power generation using wind power. The chapter includes the space vector theory, mathematical modeling of DFIG, electromechanical energy conversion process, active and reactive power flow issues, filter, topological overview, and steady state analysis. The chapter is written in an easy way for the students and researchers who are interested about doubly-fed machine for wind power application.

In Chapter 6, the theoretical fundamental of wind energy converter controls is presented. The energy conversion chains in light of fixed and variable speed wind turbine generator systems are described. The control path and basic control structure of variable speed wind turbine driven synchronous generator are described in detailed. Process control and operation management for variable speed wind energy converters are presented in detail. The conventional and the proposed methods are also compared therein.

Chapter 7 focuses on the control of a small size wind turbine generator system. Larger wind turbine systems have complex control systems such as advanced power maximizing control, blade pitch angle control, gearing control, etc. Compared to that small wind turbine unit used in the past were not designed sophistically. However, the level of sophistication of theses small wind turbine generator systems has been steadily improving nowadays. In this chapter, the authors investigated the controls associated with small wind turbine generator systems, culminating in a detailed description of the Peak Power Tracking controller utilizing a microcontroller running an impedance matching DC-DC converter between the turbine and the load.

Chapter 8 discusses on the small scale wind energy conversion system suitable for home in remote areas. The problem of transfer of energy from small wind turbine generator system to the power grid is presented. The scheme of that small wind turbine generator system, theoretical and simulation analyses of the proposed system along with the design and control are demonstrated throughout the chapter. The laboratory plant description, cost and reliability analyses of the converter design are also presented in the chapter.

Chapter 9 describes the variable speed fixed pitch wind turbine design and operations. Wind turbine nonlinear model, linearization around a set of equilibrium points, open loop characteristics to analyze the dynamic behavior are presented. To describe the control objectives for variable speed fixed pitch wind turbines, three important issues are emphasized. First is maximization of extracted energy, second, is limitation of extracted energy with active stall with rotational speed control, and the third is the control of maximum power point tracking and stall regulation at the overlapping region. The MPPT algorithms are implemented on a low cost DSC board and tested with a developed wind turbine simulator. In general, the hardware as well as software implementation for the control algorithm are explained throughout the chapter.

The real time simulation helps to validate the control system, control algorithm of a system. In Chapter 10, authors discussed about real-time physical simulation for wind energy conversion system along with its control units. The authors demonstrated the way to build physical simulator as well as the implementation and assessment techniques. This chapter is pretty useful for the readers who are interested about physical simulation of horizontal-axis wind turbines (HAWT), though the procedure will not differ so much for vertical-axis wind turbine (VAWT).

In Part B, the variability of wind power is described.

Wind energy in Netherlands by 2020 is reported in chapter 11. This chapter presents in a national context energy balancing requirements due to the variability and the limited predictability of wind energy in the thermal energy system of the Netherlands. The chapter includes different scenarios for wind and other energy capacity in the Netherlands, balancing energy reduction options in the context of some other countries in Europe, wind modeling aspects of wind energy integration, impact of extra variability due to wind, impact of limited wind predictability, and options to reduce extra balancing energy requirements in terms of short term forecast, aggregation, pumped storage, compressed air energy storage and so on.

Wind varies in space and in time and therefore the wind power is a fluctuating quality. Chapter 12 discusses about the variability of wind and wind power. A significant amount of this renewable electricity is likely to come from wind, and the variability of this power needs to be managed. The chapter discusses, relationship between wind power variability and its forecast, influence of the wind variability on the grid, geographical diversity on wind power, estimated cost of wind power variability, classification of wind and wind power oscillations,, turbulence characterization, equivalent wind of turbine clusters, spectrum and coherence estimated from weather station, and so many useful topic related to wind and wind power variability.

In chapter 13, the impact of bulk wind power penetration to the transmission system is analyzed by the authors. Stochastic electrical production from wind is expected to play a vital role in power systems, in the near future. Due to that reason, the authors proposed and developed stochastic wind generation models into an HLI Monte Carlo simulation tool that will assist system planners and transmission system operators (TSOs) to qualitatively assess the system impact of wind production. This will be helpful to take the managerial decision for integrating large scale wind farm to the existing grid. The methodology used to efficiently introduce wind generation in HLI simulation tool is presented. Then the hypotheses based on real observations are given in order to introduce wind generation into an economic dispatch with classical parks and transmission constraints. Finally, the reliability and reinforcement analysis for transmission system is computed using Ray Billinton Test System (RBTS).

Chapter 14 focuses on a very interesting area of wind energy conversion system which has significant importance on the future growth of wind power industry at the sea. As well as the onshore trend, offshore wind farms have also been continuing its growth rapidly. Some leading countries in wind energy arena are focusing more on offshore technology. The main reasons for adopting offshore are lack of the suitable onshore sites and much better wind conditions of offshore sites (wind is much stronger and more constant). However, the wind information at the sea came from merchant ships or Numerical Weather Prediction (NWP). NWP hugely depends on numerical models as well as statistical data. Chapter 14 discussed about the space-based microwave sensors which give the wind information with sufficient temporal and spatial sampling in clear and cloudy condition during day and night time. Both active and passive filters are discussed in detailed to measure wind speed and direction.

In part C, the grid integration issues are described.

Integration of power converter with variable speed wind generator causes high levels of low and high frequency emission which worsen the quality of consumed electric power, increase power losses as well as adversely affect reliability of connected appliances. Chapter 15 is focused on a simulation-based spectral technique for power quality and EMC design of wind power systems including a power source or synchronous generator (G), an AC/DC/AC converter and electronic equipment with power supplies connected to a power distribution network. Mathematical models of single- and three-phase devices in WPS are obtained as a particular case of multi-phase B-element concept. The computational experiments demonstrated the efficiency of the developed mathematical model in designing of the power quality and EMC in the wind power system.

In chapter 16, a robust power system stabilizer is proposed to stabilize wind farm interconnected power system. The power output fluctuations from wind power generation causes low frequency oscillation, which deteriorate the power system stability. To overcome this problem, the fixed structure robust PSS design by the $H\infty$ loop shaping technique is proposed by the author. The normalized coprime factor is used to model system uncertainties. To optimize the control parameters, the performance and robust stability conditions in the $H\infty$ loop shaping technique are formulated as the objective function. As a result, the proposed PSSs are very robust against various uncertainties. The configuration of PSS is a conventional second-order lead-lag compensator. To tune the PSS parameters, the concept of enhancement of system robust stability margin is formulated as the optimization problem. The genetic algorithm (GA) is applied to solve the problem and achieve the PSS parameters. The author validates the robustness of the proposed PSS by simulation analyses considering a two areas four machines power system model that interconnects wind farm.

The integration of a bulk amount of wind power in isolated electrical network needs special attention considering the reliability and security aspects. In chapter 17, the impact of large scale wind power generation on the dynamic performance especially of islands power systems are discussed. During the dynamic behavior analysis, modeling of the system is emphasized for examining the impact of higher wind penetration up to 40%. Create Island of Greece is taken as the base model system for the analysis. It is reported that spinning reserves of the conventional units have a great influence on the stability of isolated network when high wind power penetration is considered. The authors found that wind turbine affect the transient stability of the isolated network though it may not be the main obstacle for secure and reliable operation of the power system. By taking preventive actions, improving control technique the overall stability of the isolated network can be maintained even if the high wind power penetration is considered.

In Chapter 18, integration of wind farm with bulk energy storage system has been discussed which is one of the major challenge of wind energy conversion system. The author proposed 100 MW class Compressed Air Energy Storage (CAES) as the bulk energy storage system which might be suitable for large scale wind farm application. Stored energy integration into the generation-grid system is also explained. This covers a wide field in every aspect of generation- transmission and distribution. Converting the stored energy back to electricity readily provides three primary functions: Energy Management (hours of duration) load leveling or peak period needs; Bridging Power (seconds or minutes duration) assuring continuity of service, contingency reserves or UPS (Uninterruptible Power Supply); and Power Quality & Reliability (milliseconds or seconds duration) in support of manufacturing facilities, voltage and frequency controls. The chapter broadly discussed principal of CAES system, storage concept, application, benefits, future prospects, projects in development, and economics of CAES system.

Chapter 19 focuses on the stand-alone wind-diesel power system emphasizing to determine the optimum spinning reserve for managing unforeseen power unbalances. The problem of keeping the power balance is not so easy in stand-alone wind-diesel power systems, since these systems are additionally subjected to random power fluctuations originated in the stochastic and intermittent nature of the wind resource. Therefore, the authors proposed a novel method for determining the optimal amount of spinning reserve that should be carried in autonomous hybrid wind-diesel generation systems. The optimal spinning reserve is determined by comparing the cost of its provision with the economic benefits it delivers in terms of supply reliability. Global search methods like Particle Swarm Optimization (PSO) are proposed for finding the optimal scheduling policy and spinning reserve requirement that minimizes the sum of the expected operation costs and the expected costs of the energy not served.

Chapter 20 discussed about a micro-grid system composed of fuel cell and wind turbine generator system. The author considered Proton Exchange Membrane (PEM) type fuel cell installed at few locations of the micro-grid system. The home equipped with the fuel cell is also connected to the city gas network considering the issue of hydrogen gas. Wind generator is directly connected to the micro-grid system and the author investigated its influence on the micro-grid system. Adjustment of the production of electricity of each fuel cell connected to the micro-grid may operate some fuel cell with a partial load with low efficiency. Therefore, the number of operations of fuel cells is controlled to follow fluctuations in the electricity demand. The wind power to the micro-grid causes grid power instability because of the supply and demand difference, especially when the load is small compared to the electricity generation from wind turbine. When wind power equipment is connected to the micro-grid with load fluctuation, the operating point of the fuel cell system may shift and power generation efficiency may improve.

Chapter 21 presents variations in wind power generation and its impact on the thermal plants. The impact of load variation on thermal plant is another feature of this chapter. The integration of different types of storage such as pumped hydro power, compressed air energy storage (CAES), flow batteries, and sodium sulphur batteries are also discussed to moderate variations from wind power. The authors proposed to use Plug-in Hybrid Electric Vehicle (PHEV) to handle the demand side management by choosing appropriate charging strategies. The authors chose the data from the power system of western Denmark to illustrate various aspects influencing the ability of a power system to accommodate wind power.

Wind power penetration to the power grid is increasing rapidly throughout the world. This requires additional attention from different transmission system operators, regulatory official agencies along with the wind park developers to secure the power quality and other grid interfacing issues. In chapter 22, the author attempts to emphasize on such issues, which is a timely reporting at the present development of wind power. This chapter discussed on technical barriers and solutions of high wind penetration to power grid. Detail of large scale wind power integrations issues are discussed as well.

In Part D, the environmental issues are presented.

In chapter 23, the authors analyzed about environmental impact due to the manufacturing process of the wind turbine and the disposal process at the end of the wind turbine life cycle. The Life Cycle Assessment (LCA) model is developed with the purpose of determining and quantifying the related emissions and the impact of wind energy production technology. The model can even be used to define the energy payback time. The authors reported on the significant impact of wind turbine blades from the viewpoint of their non-recycling status. Some good conclusions are also given about the environmental impact and energy payback time of wind turbines and other conventional power plants. The LCA methodology is explained in terms of method and scope, system boundary, functional unit, data collection, key assumptions, and analyses scenarios. The results on environmental impact, cumulative energy demand, and recycling process are also described inside the chapter.

In chapter 24, a super sense is applied by designing wind-solar based hybrid ventilators for domestic, commercial, and industrial buildings. It is necessary to provide an optimum or at least a satisfactory environment to where we are living. Current ventilation devices are using conventional electric power or solar or wind power to drive the turbines. The author and his team at University of North South Wales (UNSW) have been working in the area of ventilation research, design and performance studies for over a decade. The turbine ventilator relies entirely of the prevailing wind conditions with no facility to extract energy from the sun. The solar ventilator is at the complete mercy of ambient solar radiation conditions and cannot extract energy from the wind. In this chapter, the author proposes wind-solar driven natural hybrid electrical ventilators which is environment friendly and cost-effective also. The horizontal axis ventilator might be a solution to the marginal performance of a turbine ventilator at low wind speeds. Testing of the horizontal axis ventilator found significantly improved performance at low wind speed conditions.

6. Conclusion

Wind energy is playing a vital role in the world's energy markets nowadays, considering it's striking growth rate in the last few years. The wind turbine and generator technology has reached to a matured stage. The developments and improvements of the power electronic devices added an extra pace in its overall growth and therefore hundreds of MW level wind farms are available these days. However, the high penetration of wind power to the electrical network needs further consideration of the existing grid infrastructures. Grid integration issues of wind farms are the most important challenge for the future growth of this technology, which must be handled carefully.

7. Acknowledgement

Special thanks and appreciations to Global Wind Energy Council (GWEC), and American Wind Energy Association (AWEA) for providing necessary permission to use their material free of cost.

8. References

GWEC Publications (2008). Global Wind 2008 Report, The Global Wind Energy Council, 2008

AWEA Resources (2009). AWEA Third Quarter 2009 Market Report, The American Wind Energy Association, 2009

AWEA Resources (2009). Projects (US), The American Wind Energy Association, 2009

GWEC Publication (2009). About Wind-Technology, The Global Wind Energy Council, 2009

Angelika Pullen, Steve Sawyer, Sven Teske, Crispin Aubrey (2008). Global Wind Energy Outlook, The Global Wind Energy Council and Greenpeace International, 2008

EWEA Publication (2009). Wind Energy The FACTS, The European Wind Energy Association, 2009, Vol. 2

Zavadil R., Miller N., Ellis A., Muljadi E. (2005). Making connections, IEEE Power Energy Mag., Vol.3, No.6, pp.20

FERC Report (2005). Interconnection for Wind Energy, Federal Energy Regulatory Commission (FERC), United States of America, Docket No. RM05-4-000 - Order No. 661, Issued June 2

FERC Report (2005). Interconnection for Wind Energy, Federal Energy Regulatory Commission (FERC), United States of America, Docket No. RM05-4-001; Order No. 661-A, Issued December 12

Tsili M., Papathanassiou S. (2009). A review of grid code technical requirements for wind farms, IET-Renewable Power Generation, Vol.3, No.3, pp.308-332.

P. Gardner, M. Tremblay, D. Price (2009). Technical requirements for high-penetration wind : what system operators need, and what wind technology can deliver, Integration of wide-scale renewable resources into the power delivery system, CIGRE/IEEE PES Joint Symposium

Qiao W. Harley R.G. (2008). Grid connection requirements and solutions for DFIG wind turbines, Energy 2030 Conference, IEEE.

ING Report (2007). Wind Farm Power Station Grid Code Provisions, WFPS1, Ireland National Grid, Grid Code Version 2, pp.213-216, January

E.On Netz Resources (2006). E.On Netz, Grid Code, High- and Extra-High Voltage, available at www.eon-netz.com/

Bharat Singh, S.N. Singh (2009). Wind Power Interconnection into the Power System: A Review of Grid Code Requirements, The Electricity Journal, Vol.22, No.5, pp. 54-63

Part A
Wind Turbine Generator System

2

An Analytical Analysis of a Wind Power Generation System Including Synchronous Generator with Permanent Magnets, Active Rectifier and Voltage Source Inverter

Sergey A. Kharitonov, Member. IEEE
Novosibirsk State Technical University
Rusia

1. Introduction

For the high-power Wind Power Installation (WPI) with a variable speed wind turbine the system of transformation of mechanical energy into electric energy of the alternating current, constructed under the scheme "the synchronous generator with constant magnets - the active rectifier - the voltage inverter" (fig. 1) is perspective. Then the system is called the Wind Power Generation System - WPGS.

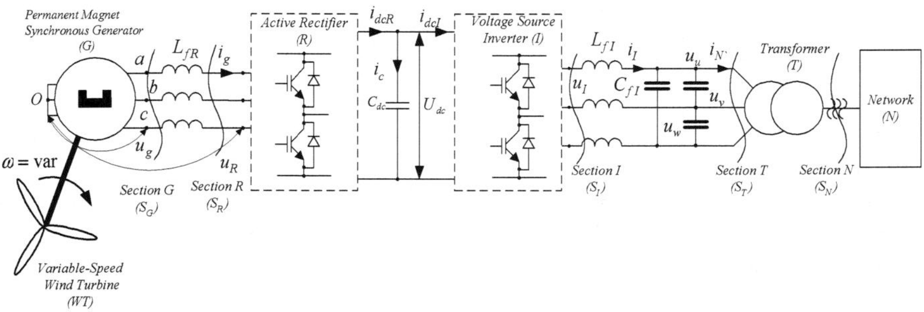

Fig. 1.

WPGS of this type implements the full set of options required from the generation for high-power WPI, namely this: a generating mode at work on the nonlinear, asymmetric and non-stationary loads, electric starter startup mode of the wind turbine, in-phase and parallel operation of an electric network and other WPI.

In this paper we attempt to identify the main energy characteristics of system in sections S_G, S_R, S_I, S_T, S_N (fig. 1) when working on high-power electric network. Processes in the active rectifier and the voltage inverter are well studied at a constant frequency and voltage of the generator. The peculiarity of this study is the consideration of factors that arise when working within the system to generate WPI with a variable rotation speed of the wind turbine. These factors include: changing the frequency and the voltage of the synchronous generator (G), as well as dependence of the generated power on shaft speed of the wind turbine.

Presence in the system of the active rectifier (R) modifies the functional and energy potential of WPGS. The active rectifier with a PWM, which frequency is much higher than the voltage frequency of a synchronous generator (G), allows for a number of modes, significantly affecting the power consumption of G in the WPGS.

As a result of the conducted research where as an example it is accepted that the active rectifier and the inverter are based on the classical scheme of the two-level voltage inverter, the analytical description of WPGS system is obtained at a variable speed of rotation of the wind turbine, the basic expressions for currents, voltage and capacities of the synchronous generator, the voltage inverter are defined, algorithms of the management are offered by the active rectifier, and also the modular principle of construction of the voltage inverter and WPGS system as a whole are considered.

2. Basic assumptions. Mathematical model of the system.

For a generality of results of the analysis in the scheme elements which not always are obligatory are entered:

L_{fR} - inductance of the cable connecting the generator and the active rectifier;

T - the matching transformer;

C_{fI} - capacity of the output filter, for the smoothing of pulsations on a transformer input.

The active rectifier and the voltage inverter are controlled by a high-frequency PWM, and their frequencies ω_{cR} and ω_{cI}, correspondingly, are significantly higher than frequencies of the fundamental harmonic voltage of synchronous generator (ω) and the electric voltage network (Ω). The multiplicities of frequencies are constant, i.e. $\omega_{cR}/\omega = a_R = const$ and $\omega_{cI}/\Omega = a_I = const$.

The electrical network has a capacity much bigger than the power of WPI.

Let's assume also that the synchronous generator does not contain soothing contours and its magnetic system is linear.

The generating mode is a subject to consideration. In this case, the active rectifier (R) entrusted with the tasks of forming a given voltage in the DC link U_{dc} and reactive power control on the frequency ω in sections S_G and S_R, while the voltage inverter (I) is tasked to ensure the specified quality and quantity of the generated current in the electric network .

Let's consider that the capacity C_{dc} in a DC link is big, the voltage regulator U_{dc} in a control system works with the maximum speed and is non-static then in the established mode it is possible to accept that $U_{dc} = const$. In this case in sections S_G, S_R it is possible to consider the electromagnetic processes irrespective of processes in sections S_I, S_N and S_T.

It is convenient to study WPGS in the rotating coordinate systems. In this case in section S_G the coordinate system rotates synchronously with the frequency of the generator voltage (ω), and in sections S_I, S_N respectively, with the frequency of the mains voltage (Ω).

Taking into account the accepted assumptions *the mathematical model of SG* in rotating system of co-ordinates, under condition of axis orientation d on a longitudinal axis of the synchronous generator will look like:

$$u_R = -r_\Sigma i_G - \frac{d}{dt}\psi_\Sigma - \omega\psi_\Sigma, \qquad \psi_\Sigma = L_\Sigma i_G - \psi_0 \tag{1}$$

where: $\psi_\Sigma = \begin{bmatrix} \Psi_{\Psi d} & \Psi_{\Sigma q} \end{bmatrix}^t$, $\Psi_{\Sigma d} = \Psi_d + L_{fR} i_{Gd}$, $\Psi_{\Sigma q} = \Psi_q + L_{fR} i_{Gq}$, Ψ_d, Ψ_q - the magnetic flux of generator in the longitudinal and transverse axes, $u_R = \begin{bmatrix} u_{Rd} & u_{Rq} \end{bmatrix}^t$, $i_g = \begin{bmatrix} i_{Gd} & i_{Gq} \end{bmatrix}^t$ - vectors of the active rectifier voltages and currents of the generator; $\psi_o = \begin{bmatrix} \Psi_o & 0 \end{bmatrix}^t$, $\Psi_o = const$ - the magnetic flux created by permanent magnets; $r_\Sigma = diag\{ r_\Sigma, r_\Sigma \}$, $r_\Sigma = r_s + r_{LfR}$, r_s, r_{LfR} - the active resistances of stator phase windings of the generator and cable connecting the generator and active rectifier; $L_\Sigma = diag\{ L_{\Sigma d}, L_{\Sigma q} \}$, $L_{\Sigma d} = L_d + L_{fR}$, $L_{\Sigma q} = L_q + L_{fR}$, L_d, L_q - inductance of the generator in the longitudinal and transverse axes; $\omega = \begin{bmatrix} 0 & -\omega \\ \omega & 0 \end{bmatrix}$, ω - circular frequency of the electromotive force (EMF) of SG ($\omega = \text{var}$).

Selecting the generator currents as variables, after simple transformations, we obtain from equation (1):

$$u_R = -r_\Sigma i_G - L_\Sigma \frac{d}{dt} i_G - \omega L_\Sigma i_G + e \quad e = \begin{bmatrix} 0 & E_o \end{bmatrix}^t, \tag{2}$$

here $E_o = \omega \Psi_o$ - EMF-load of the generator ($E_o = \text{var}$).

Neglecting the active resistance it is possible to write down parity (2) in the scalar form

$$u_{Rd} = -i_{Gd} \cdot r_s - L_{d\Sigma} \frac{di_{Gd}}{dt} + \omega L_{q\Sigma} i_{Gq} \approx -L_{d\Sigma} \frac{di_{Gd}}{dt} + \omega L_{q\Sigma} i_{Gq},$$

$$u_{Rq} = -i_{Gq} \cdot r_s - L_{q\Sigma} \frac{di_{Gq}}{dt} - \omega L_{d\Sigma} i_{Gd} + E_o \approx -L_{q\Sigma} \frac{di_{Gq}}{dt} - \omega L_{d\Sigma} i_{Gd} + E_o \tag{3}$$

In any section S active (P_S), reactive (Q_S) and apparent (S_S) powers will be defined by means of the following parities:

$$P_S = \frac{3}{2}(u, i) = \frac{3}{2}\left(u_d i_d + u_q i_q\right), Q_S = \frac{3}{2}[u, i] = \frac{3}{2}\left(u_d i_q - u_q i_d\right), S_S = \left[P_S^2 + Q_S^2 \right]^{\frac{1}{2}}. \tag{4}$$

The mathematical description of the active rectifier and the inverter will be obtained by means of switching functions. We will consider that transistors and diodes are ideal keys. R and I are realized on base of the voltage inverter schematically presented in fig. 2.

The phase voltage on alternating current clips is defined by means of parity:

$$u_m = U_{dc} \left(2F_m - \sum_{\substack{k=1 \\ k \neq m}}^{3} F_k \right) \Big/ 3; \quad m = 1, 2, 3, \tag{5}$$

where F_m - the switching functions of transistors VT_m of the inverter which are defined by means of a following parity $F_m = \begin{cases} 1, & VT_m - \textit{is switched on;} \\ 0, & VT_m - \textit{is switched off.} \end{cases}$

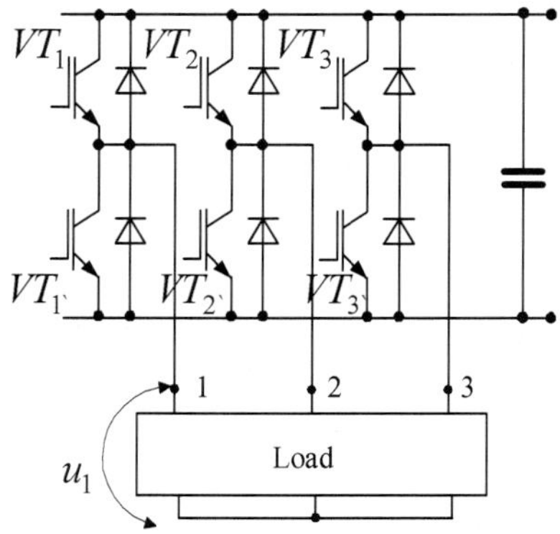

Fig. 2.

They also can be defined by means of a series:

$$F_m = M\sin\theta_m/2 + \sum_{j=1}^{\infty} F_{msj}\sin(j\cdot a_k\vartheta) + F_{mcj}\cos(j\cdot a_k\vartheta);$$

$$F_{msj} = \frac{(-1)^j}{j\pi}[(-1)^j - \cos(j\pi\cdot M\sin\theta_m)];\quad F_{mcj} = \frac{(-1)^j}{j\pi}\sin(j\pi\cdot M\sin\theta_m);$$

$$M = \begin{cases} u_c/u_{sc} & \text{SPWM;} \\ 2u_c/\sqrt{3}u_{sc} & \text{SVPWM;} \end{cases}\quad M_{\max} = \begin{cases} 1 & \text{SPWM;} \\ 2/\sqrt{3} & \text{SVPWM;} \end{cases}\quad \theta_m = \vartheta - (m-1)\frac{2\pi}{3} + \phi_c;$$

where: $\vartheta = \omega t$, $a_k = \omega_c/\omega$, ω_c, ω - the cyclic frequencies of PWM and an of operating signal, accordingly, M, $M_{\max}$ - the depth (index) of modulation and its maximum value, u_c - amplitude of the control input wave, u_{sc} - the amplitude of saw tooth carrier wave, φ_c - the phase of the control input wave, is defined by the chosen algorithm of control.

After a number of transformations we will obtain for SPWM

$$u_m = U_{dc}M\sin(\theta_m)/2 + \frac{U_{dc}}{\pi}\sum_{\substack{p=-\infty \\ p\neq 0}}^{\infty}\sum_{k=-\infty}^{\infty}(-1)^{p+3k}/p\cdot J_{3k+1}(p\pi M)\cdot\sin[a_k p\vartheta + (3k+1)\theta_m];^{1} \quad (6)$$

where $J_{3k+1}(....)$ - Bessel functions of the first kind of an order $3k+1$.

Obviously, expression for the fundamental component will look like: $u_{m(1)} = U_{dc}M\sin(\theta_m)/2$.

[1] In this parity and further the high-frequency harmonics are defined for SPWM.

If we introduce the operator of rotation $a = \exp(i \cdot 2\pi/3)$, $i = \sqrt{-1}$, u_m, $m = 1,2,3$ the three voltages can be written in the orthogonal coordinate system:

$$u_{\alpha\beta} = u_\alpha + i \cdot u_\beta = \frac{2}{3}(u_1 + a \cdot u_2 + a^2 \cdot u_3).$$

Using parity (6), we will obtain:

$$u_\alpha = \frac{U_{dc}}{2} M \sin(\theta_1) + \frac{U_{dc}}{\pi} \sum_{\substack{p=-\infty \\ p \neq 0}}^{\infty} \sum_{k=-\infty}^{\infty} \frac{(-1)^{p+3k}}{p} \cdot J_{3k+1}(p\pi \cdot M) \sin[ap\vartheta + (3k+1)\theta_1]; \quad (7)$$

$$u_\beta = -\frac{U_{dc}}{2} M \cos(\theta_1) - \frac{U_{dc}}{\pi} \sum_{\substack{p=-\infty \\ p \neq 0}}^{\infty} \sum_{k=-\infty}^{\infty} \frac{(-1)^{p+3k}}{p} \cdot J_{3k+1}(p\pi \cdot M) \cos[ap\vartheta + (3k+1)\theta_1]. \quad (8)$$

We will determine the current in a direct current link (i_{dc}) by means of parity: $i_{dc} = \sum_{m=1}^{3} i_{gm} F_m$, where i_{gm} - the instant value of phase currents of the generator. The average value of current i_{dc} from the condition of equality of the active power in AC and DC circuits of inverter is: $I_{dco} = 3MI_{(1)} \cos\varphi/4$, here φ- an angle shift between the fundamental harmonic of phase voltage and an inverter current, $I_{(1)}$ - the amplitude of inverter current, I_{dco} - the mean value of a current in a DC link.

In the analysis of electromagnetic processes in sections S_I, S_T, S_N we will assume that $\omega_{kI} >> \Omega$. In a first approximation it allows to neglect the effect of capacitors C_{fI}. When considering the transformer (*T*), we assume that its magnetic system is unsaturated, active losses and the magnetization current are zero and its influence on the processes we take into account with the total leakage inductance ($L_{\sigma\Sigma T}$) and transformation factor (k_T): $k_T = w_1/w_2$; $L_{\sigma\Sigma T} = L_{\sigma 1T} + L`_{\sigma 2T} = L_{\sigma 1T} + L_{\sigma 2T} \cdot k_T^2$; where w_1, w_2, - the number of turns of primary and secondary windings of the transformer, $L_{\sigma 1T}$, $L_{\sigma 2T}$ - the leakage inductance of primary and secondary windings, respectively.

We will express the voltage of an electric network through the voltage on a primary winding of the transformer using the relation: $u_{N`} = u_N \cdot k_T$. The equivalent inductance in the output circuit of inverter: $L_I = L_f + L_{\sigma\Sigma T}$.

Taking into account the accepted assumptions WPGS can be presented in the form of two equivalent circuits, for example, in phase coordinates fig. 3 and fig. 4.

In these figures dependent sources of voltage which reflect the voltages R and I in alternating current clips are presented in the form of rhombs. These voltages are defined by relation (5).

3. Basic energy indicators in the chain of "synchronous generator - active rectifier"

The power quality parameters of electromagnetic processes in WPGS determine the technical efficiency of converting mechanical energy of a shaft rotating with a variable speed

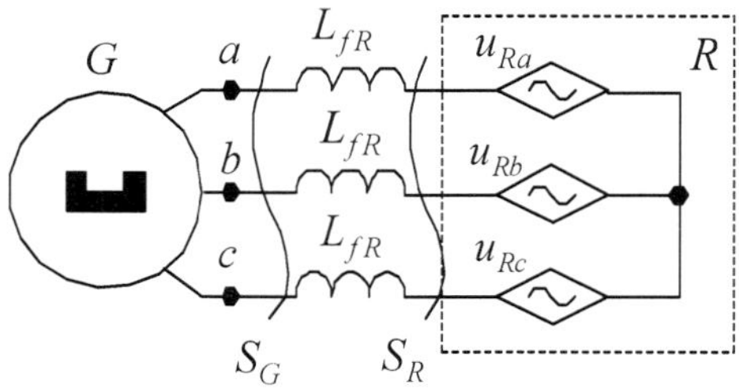

Fig. 3.

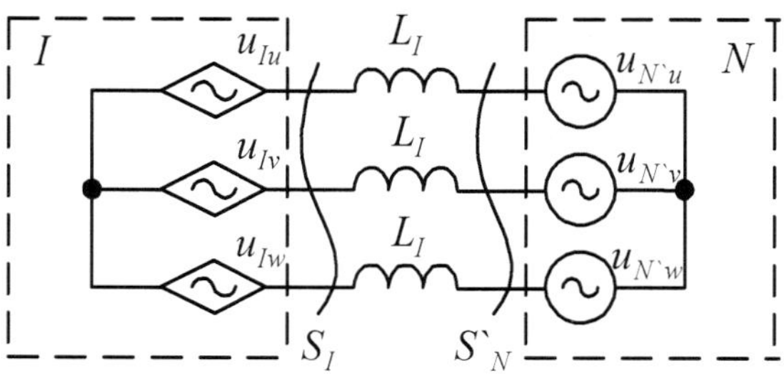

Fig. 4.

wind turbine into electrical energy by means of a synchronous generator and the voltage inverter, and the degree of influence of WPGS on the electric network through quality indicators of the electricity generated.

The basic indicators include: efficiency, power factor, and factors of harmonics and distortions of currents and voltages. For calculation of these indicators the definition of the operating values of currents and voltages in system elements is necessary, therefore they also are indirect power indicators.

For generality of the results, we introduce the relative units, as a basic value we choose the following:

$$U_{dc} = \sqrt{3}U_{N`}; \quad E_б = \frac{U_{dc}}{\sqrt{3}} = U_{N`}; \quad \omega_б = \frac{E_б}{\Psi_o}; \quad X_б = \omega_б\left(L_f + L_d\right) = \omega_б L_d\left(1+q\right);$$

$$I_б = I_{кз} = \frac{E_б}{X_б}; S_б = \frac{3}{2}\cdot E_б I_б; \omega^* = \frac{\omega}{\omega_б}, E_o^* = \frac{E_o}{E_б} = \omega^*, u^* = \frac{u}{E_б}, i^* = \frac{i}{I_б}, a_R = \frac{\omega_{cR}}{\omega}, \tag{9}$$

where $U_{N`}$ - the amplitude value of voltage of electrical network, referred to the primary winding, ω_{cR} the cyclic frequency PWM of an active rectifier.

Denote:

$$q = \frac{L_f}{L_d}; \; k_x = \frac{L_{aq}}{L_{ad}}; \sigma = \frac{L_\sigma}{L_{ad}}; k_L = \frac{L_q}{L_d} = \frac{\sigma + k_x}{\sigma + 1}. \tag{10}$$

Taking into account (9) and (10) we obtain:

$$X_{fR}^{*}=\frac{X_{fR}}{X_{б}}=\frac{\omega^{*}q}{1+q},\ X_{d}^{*}=\frac{X_{d}}{X_{б}}=\frac{\omega^{*}}{1+q},\ X_{q}^{*}=\frac{X_{q}}{X_{б}}=\frac{\omega^{*}}{1+q}k_{L},$$

$$X_{d\Sigma}^{*}=X_{d}^{*}+X_{fR}^{*}=\omega^{*},\ X_{q\Sigma}^{*}=X_{q}^{*}+X_{fR}^{*}=\omega^{*}\frac{k_{L}+q}{1+q}.$$

Considering relative units and the entered designations of the equation (3) will become:

$$\begin{aligned} u_{Rd}^{*}&=-X_{d\Sigma}^{*}\frac{di_{Gd}^{*}}{d\vartheta}+X_{q\Sigma}^{*}i_{Gq}^{*}=-\omega^{*}\frac{di_{Gd}^{*}}{d\vartheta}+\omega^{*}\frac{k_{L}+q}{1+q}\cdot i_{Gq}^{*};\\ u_{Rq}^{*}&=-X_{q\Sigma}^{*}\frac{di_{Gq}^{*}}{d\vartheta}-X_{d\Sigma}^{*}i_{Gd}^{*}+\omega^{*}=-\omega^{*}\frac{k_{L}+q}{1+q}\cdot\frac{di_{Gq}^{*}}{d\vartheta}-\omega^{*}i_{Gd}^{*}+\omega^{*}, \end{aligned} \tag{11}$$

where $\vartheta=\omega t$.

Accordingly, we will define the power for the basic harmonics in sections S_G also S_R by means of expressions:

$$\begin{cases} P_{SGo}^{*}=P_{SRo}^{*}=P_{SNo}^{*}=\omega^{*}i_{gqo}^{*},\\ Q_{SGo}^{*}=u_{Gdo}^{*}i_{Gqo}^{*}-u_{Gqo}^{*}i_{Gdo}^{*};\quad Q_{SRo}^{*}=u_{Rdo}^{*}i_{Gqo}^{*}-u_{Rqo}^{*}i_{Gdo}^{*};\\ S_{SGo}^{*}=[(P_{SNo}^{*})^{2}+(Q_{SGo}^{*})^{2}]^{\frac{1}{2}};\quad S_{SRo}^{*}=[(P_{SNo}^{*})^{2}+(Q_{SRo}^{*})^{2}]^{\frac{1}{2}}; \end{cases} \tag{12}$$

here it is considered that at the accepted assumptions the active power is identical and equal in all sections to the power generated in an electric network (P_{SNo}^{*}).

Taking into account the higher harmonics value of active power will not change, and to calculate inactive and a total power it is necessary to apply the following relations:

$$P_{SG}^{*}=P_{SR}^{*}=P_{SGo}^{*};\ S_{SG}^{*}=U_{G,rms}^{*}I_{G,rms}^{*};\quad S_{SR}^{*}=U_{R,rms}^{*}I_{G,rms}^{*};$$

$$Q_{SG}^{*}=\sqrt{(S_{SG}^{*})^{2}-(P_{SG}^{*})^{2}};\quad Q_{SR}^{*}=\sqrt{(S_{SR}^{*})^{2}-(P_{SR}^{*})^{2}}$$

We will define the power factor in sections S_R also S_R by means of parities:

$$\chi_{G}=P_{SG}^{*}/S_{SG}^{*}=\nu_{iG}\nu_{uG}\cos\varphi_{SG};\quad \chi_{R}=P_{SR}^{*}/S_{SR}^{*}=\nu_{iG}\nu_{uR}\cos\varphi_{SR}, \tag{13}$$

where: $\nu_{iSG}=I_{G(1),rms}^{*}/I_{G,rms}^{*}$, $\nu_{uSG}=U_{G(1),rms}^{*}/U_{G,rms}^{*}$, $\nu_{uSR}=U_{R(1),rms}^{*}/U_{R,rms}^{*}$, ν_{iS}, ν_{uS}, φ_{S} - a fundamental factors of current and voltage, and also a shift angle between the basic harmonics of current and voltage accordingly in sections S_G and S_R, $I_{G(1),rms}^{*}$, $I_{G,rms}^{*}$, $U_{G(1),rms}^{*}$, $U_{G,rms}^{*}$, $U_{R(1),rms}^{*}$, $U_{R,rms}^{*}$ - root-mean-square - RMS of the basic harmonics and full values of a current and a voltage in corresponding sections.

Assuming that the EMF-load of generator (e_{Gm}^{*}) and the control voltage of an active rectifier (u_{Rcm}^{*}) varies according to the law:

$$e^*_{Gm} = \omega^* \cos\left[\vartheta - (m-1)\frac{2\pi}{3}\right]; \; u^*_{Rcm} = M\sin(\theta_m);$$

$$\theta_m = \vartheta - (m-1)\frac{2\pi}{3} + \frac{\pi}{2} - \phi_{Rc}; \; \vartheta = \omega t; \; m = 1,2,3\,(a,b,c);$$

and taking into account the relation (7, 8), we obtain expressions for the quantities u^*_{Rd} and u^*_{Rq} : $u^*_{Rdq} = u_{Rd} + i \cdot u_{Rq} = u^*_{R\alpha\beta}\exp[-i \cdot \gamma(\vartheta)]$, where $\gamma(\vartheta) = \vartheta - \pi/2$.

At the analysis in «*d q*» co-ordinates it is convenient to present the three control input waves of the active rectifier in the form of two orthogonal projections on *d* and *q* axes, then $M_d = M\sin\varphi_{Rc}$; $M_q = M\cos\varphi_{Rc}$, it is obvious that $M = \sqrt{M_d^2 + M_q^2}$.

After the transformations we will obtain an expression for voltage of the active rectifier in «*d q*» co-ordinates:

$$u^*_{Rd} = u^*_{Rdo} + \Delta u^*_{Rd}; \quad u^*_{Rq} = u^*_{Rqo} + \Delta u^*_{Rq}; \quad u^*_{Rdo} = \sqrt{3}\,M_d/2; \quad u^*_{Rqo} = \sqrt{3}\,M_q/2;$$

$$\Delta u^*_{Rd} = \frac{\sqrt{3}}{\pi}\sum_{\substack{p=-\infty \\ p\neq 0}}^{\infty}\sum_{k=-\infty}^{\infty}\frac{(-1)^{p+k}}{p}\cdot J_{3k+1}(p\pi \cdot M)\cos\left[a_R p\vartheta + 3k\vartheta + (3k+1)(\pi/2 - \varphi_{Rc})\right]; \tag{14}$$

$$\Delta u^*_{Rq} = \frac{\sqrt{3}}{\pi}\sum_{\substack{p=-\infty \\ p\neq 0}}^{\infty}\sum_{k=-\infty}^{\infty}\frac{(-1)^{p+k}}{p}\cdot J_{3k+1}(p\pi \cdot M)\sin\left[a_R p\vartheta + 3k\vartheta + (3k+1)(\pi/2 - \varphi_{Rc})\right],$$

here u^*_{Rdo}, u^*_{Rqo} - the orthogonal components in *d* and *q* coordinates of the basic harmonic of voltage of the active rectifier; Δu^*_{Rd}, Δu^*_{Rq} - the orthogonal components in *d* and *q* co-ordinates of the high-frequency harmonics of voltage of the active rectifier.

In the steady operating mode for a particular value of generator voltage frequency (ω^*) with the help of relations (11) and (14) we can determine an analytical expression for the generator currents. To do this in (14) we will allocate sinus and cosine components ($U^*_{Rds\,pk}$, $U^*_{Rdc\,pk}$, $U^*_{Rqs\,pk}$, $U^*_{Rqc\,pk}$) of the harmonics with frequencies $\nu_{pk} = a_R p\omega + 3k\omega$:

$$\begin{aligned} U^*_{Rds\,pk} &= -g_{kp}\sin\left[(3k+1)(\pi/2 - \phi_{Rc})\right]; \; U^*_{Rqs\,pk} = g_{kp}\cos\left[(3k+1)(\pi/2 - \phi_{Rc})\right]; \\ U^*_{Rdc\,pk} &= g_{kp}\cos\left[(3k+1)(\pi/2 - \phi_{Rc})\right]; \; U^*_{Rqc\,pk} = g_{kp}\sin\left[(3k+1)(\pi/2 - \phi_{Rc})\right], \end{aligned} \tag{15}$$

here $g_{kp} = \dfrac{\sqrt{3}}{\pi}\dfrac{(-1)^{p+k}}{p}\cdot J_{3k+1}(p\pi \cdot M)$.

The equation for the generator current can be represented as a sum of components from the fundamental (i^*_{Gdo}, i^*_{Gqo}) and the high-frequency (Δi^*_{Gd}, Δi^*_{Gq}) harmonics

$$i^*_{Gd} = i^*_{Gdo} + \Delta i^*_{Gd}; \quad i^*_{Gq} = i^*_{Gqo} + \Delta i^*_{Gq}\,.$$

Then, using equations (11) and (15), we obtain

$$i^*_{Gqo} = \frac{u^*_{Rdo}}{X^*_{q\Sigma}} = \frac{1+q}{\omega^*(k_L+q)} \cdot u^*_{Rdo} = \frac{\sqrt{3}}{2}\frac{1+q}{\omega^*(k_L+q)} M_d, \quad i^*_{Gdo} = 1 - \frac{u^*_{Rqo}}{\omega^*} = 1 - \frac{\sqrt{3}}{2}\frac{1}{\omega^*} M_q.$$

$$\Delta i^*_{Gd}(\vartheta) = \sum_{\substack{p=-\infty \\ p\neq 0}}^{\infty} \sum_{k=-\infty}^{\infty} \left[I^*_{Gds\,pk} \sin(a_R p + 3k)\vartheta + I^*_{Gdc\,pk} \cos(a_R p + 3k)\vartheta \right];$$

$$\Delta i^*_{Gq}(\vartheta) = \sum_{\substack{p=-\infty \\ p\neq 0}}^{\infty} \sum_{k=-\infty}^{\infty} \left[I^*_{Gqs\,pk} \sin(a_R p + 3k)\vartheta + I^*_{Gqc\,pk} \cos(a_R p + 3k)\vartheta \right]; \tag{16}$$

where:

$$I^*_{Gds\,pk} = gd_{kp}[U_{Rqspk} - U_{Rdcpk}(a_R p + 3k)]; I^*_{Gdc\,pk} = gd_{kp}[U_{Rqcpk} - U_{Rdspk}(a_R p + 3k)];$$

$$I^*_{Gqs\,pk} = gq_{kp}[U_{Rdspk} + U_{Rqcpk}(a_R p + 3k)]; I^*_{Gqc\,pk} = gq_{kp}[U_{Rdcpk} + U_{Rqspk}(a_R p + 3k)],$$

here $gd_{kp} = 1/[(a_R p + 3k)^2 - 1]X_{d\Sigma}$, $gq_{kp} = -1/[(a_R p + 3k)^2 - 1]X_{q\Sigma}$.

Voltage of the synchronous generator is defined as follows:

$$u^*_{Gd} = \Delta u^*_{Gd} + u^*_{Gdo}; \quad u^*_{Gq} = \Delta u^*_{Gq} + u^*_{Gqo}$$

$$u^*_{Gdo} = u^*_{Rdo} - X^*_{Rf} i^*_{Gqo} = \frac{\sqrt{3}}{2}\frac{k_L}{k_L+q} M \sin(\varphi_{Rc});$$

$$u^*_{Gqo} = u^*_{Rqo} + X^*_{Rf} i^*_{Gdo} = \frac{1}{1+q}\left[\omega^* q + \frac{\sqrt{3}}{2} M \cos(\varphi_{Rc})\right]; \tag{17}$$

$$\Delta u^*_{Gd} = \Delta u^*_{Rd} + X^*_{Rf}\frac{di^*_{Gd}}{d\vartheta} - X^*_{Rf}\Delta i^*_{Gq}; \quad \Delta u^*_{Gq} = \Delta u^*_{Rq} + X^*_{Rf}\frac{d\Delta i^*_{Gq}}{d\vartheta} + X^*_{Rf}\Delta i^*_{Gd}.$$

here u^*_{Gdo}, u^*_{Gqo} - the orthogonal components in the d and q coordinates of the fundamental harmonic of generator voltage, $\Delta u^*_{Gd}, \Delta u^*_{Gq}$ - the orthogonal components in the d and q coordinates of the high-frequency harmonics of generator voltage.

From (16a) and (17) can easily be obtained the following useful relations:

$$u^*_{Gdo} = k_L u^*_{Rdo}/(k_L + q), \quad u^*_{Gqo} = (u^*_{Rqo} + q\omega^*)/(1+q). \tag{18}$$

RMS of the active rectifier voltage ($U^*_{R,rms}$) and its fundamental factor can be determined if we use the following properties of switching functions F_m [1]:

$$\frac{1}{2\pi}\cdot\int_0^{2\pi}(F_j)^2 d\vartheta = \frac{1}{2}; \quad \frac{1}{2\pi}\cdot\int_0^{2\pi} F_j \cdot F_i d\vartheta \approx \frac{1}{2}\cdot(1-0.5M); \quad i,j = 1,2,3; (i \approx j).$$

Then:

$$U^*_{R,rms} = \sqrt{M/2}; \quad \nu_{uR} = \frac{U_{R(1),rms}}{U_{R,rms}} = \sqrt{3M}/2.$$

We define RMS of the generator current through the equation:

$$I^*_{G,rms} = \sqrt{(i^*_{Gdo})^2 + (i^*_{Gqo})^2 + (\Delta I^*_G)^2} \Big/ \sqrt{2}, \tag{19}$$

where

$$\Delta I^*_G \approx \left(\frac{\sqrt{6}}{\pi \cdot \omega^*}\right) \cdot \left(J_1(\pi \cdot M)^2 \frac{a_R^2 + 1}{(a_R+1)^2 (a_R-1)^2}\right)^{\frac{1}{2}}. \tag{19a}$$

Then the fundamental factor of the generator current can be estimated using the relation:

$$\nu_{iG} = \sqrt{\frac{(i^*_{Gdo})^2 + (i^*_{Gqo})^2}{(i^*_{Gdo})^2 + (i^*_{Gqo})^2 + (\Delta I^*_{Gdq})^2}}.$$

We will determine RMS of the generator voltage considering that, $k_L \to 1$, then:

$$U^*_{G,rms} = \frac{1}{\sqrt{2}} \sqrt{\left(\omega^* \cos\theta\right)^2 + \frac{1}{(1+q)^2} \cdot \left[M - \left(\frac{\sqrt{3}}{2} M \cos\phi_c\right)^2\right]},$$

where θ - the angle between the main harmonics of the EMF and the generator voltage. In accordance with the vector diagram (fig. 5) θ is given by: $\theta = arctg\, u^*_{Gdo} / u^*_{Gqo}$.

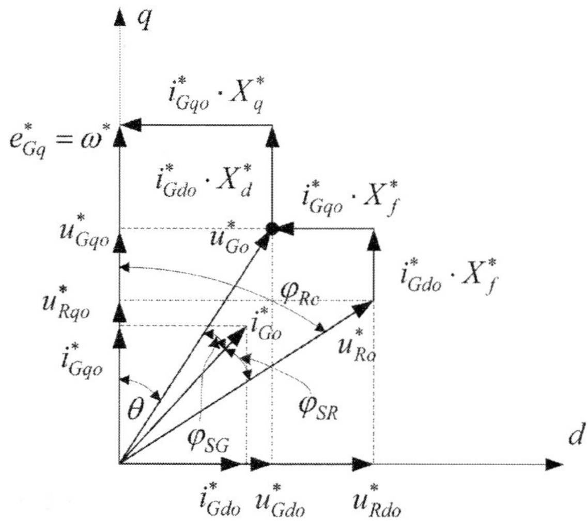

Fig. 5.

In fig.6 as an illustration of the possibilities suggested by the mathematical model of the system the calculated dependences of some power indicators as change of frequency of rotation of the wind turbine for a mode $i^*_{Gdo} = 0$, i.e. when a phase of a current and EMF of the generator coincide are presented. In this case the active filter increases the voltage $u^*_{Ro} > e^*_G$.

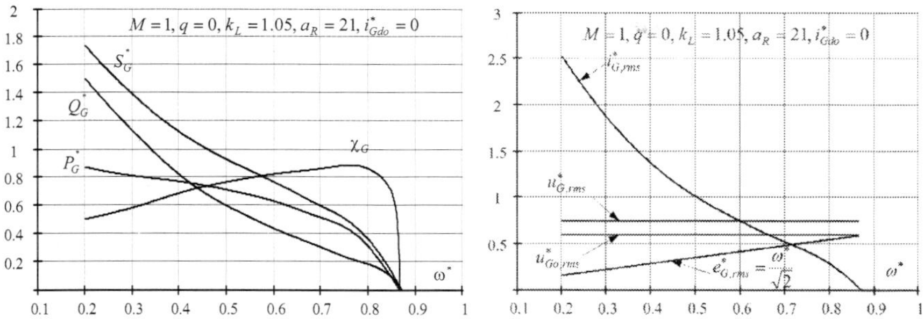

(a) (b)

Fig. 6.

Dependence of the fundamental factor of the generator current on the modulation depth and speed of the wind turbine for different multiplicities of frequencies is presented in fig. 7. It follows from these graphs, in engineering calculations, and when $a_R > 15 \div 20$ and $\omega^* = 0.4 \div 0.8$ you can take $\nu_{iR} \approx 1$. This means that the active power generated by the system is determined by the fundamental harmonics of current and voltage.

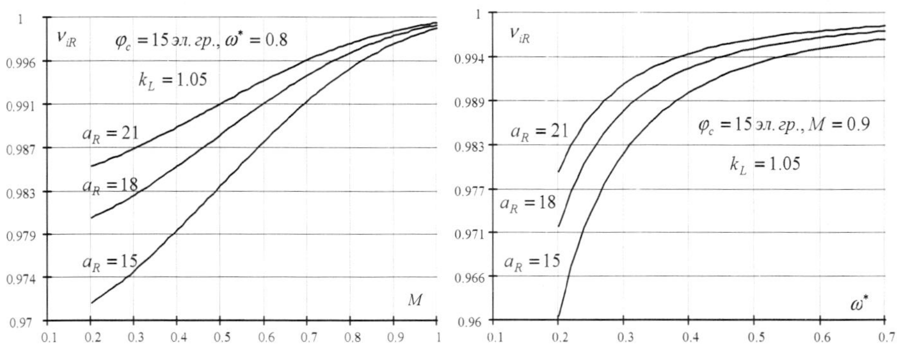

(a) (b)

Fig. 7.

Let's consider the character of change of generated power, and also currents and voltages using (11), (12) and (16a,). For the real synchronous generators it is characteristic $L_d \approx L_q$, i.e. $k_L \to 1$; besides this, usually, the inequality $L_f < L_d$ takes place, so $q < 1$.

For WPI with a variable speed wind turbine the generated active power (P^*_{WTo}) in the working range ($\omega^* \in \{\omega^*_{WT\min}, \omega^*_{WT\max}\}$) is determined by the speed of the wind and can be calculated for the known characteristics of the turbine using the relation:

$$P^*_{WTo} = \gamma \cdot (\omega^*)^3 . \tag{20}$$

where γ - a constant coefficient determined by the design of wind turbine; $\{\omega^*_{WT\min}, \omega^*_{WT\max}\}$ - the working range, also characterized by the value $D_{WT} = \omega^*_{WT\max} / \omega^*_{WT\min}$; $\omega^*_{WT\min}, \omega^*_{WT\max}$ - the minimum and maximum operating speed of WT, respectively.

Obviously, the active power generated by the system (P^*_{Ro}) should satisfy the inequality:

$$P^*_{Ro} \geq P^*_{WTo} .$$

The orthogonal components of voltages in the sections S_R and S_G for a given active power are determined, as it follows from (12), according to the expression:

$$u^*_{Rqo} = \omega^* \cdot \left(1 - \frac{P^*_{Ro}}{u^*_{Rdo}}\right) \frac{k_L + q}{k_L - 1}; \quad u^*_{Gqo} = \omega^* \cdot \left(1 - \frac{P^*_{Ro}}{u^*_{Gdo}(1+q)}\right) \frac{k_L}{k_L - 1};$$

On the other hand, the active power generated by system

$$P^*_{Ro} = e^*_{Gq} i^*_{Gqo} = \omega^* i^*_{Gqo} = \omega^* \frac{u^*_{Rdo}}{X^*_{q\Sigma}} = \omega^* \frac{u^*_{Gdo}}{X^*_q} = \frac{1+q}{k_L + q} u^*_{Rdo} = \frac{1+q}{k_L} u^*_{Gdo} . \tag{21}$$

At $k_L \to 1$ $P^*_{Ro} \approx u^*_{Rdo} = (1+q) u^*_{Gdo}$.

Fig. 8 shows the dependence u^*_{Rqo} of u^*_{Rdo} for different values P^*_{Ro} . Constancy of the active power is carried out on the sites of characteristics between the points of «a» and «b» outside of these points the modulation index (M) is limited and the active power decreases.

The total power and power factor in the section S_R are defined by the relations:

$$S^*_{Ro} = [(Q^*_{Go})^2 + (P^*_{Go})^2]^{\frac{1}{2}} , \ \cos\varphi_R = P^*_{Ro} / S^*_{Ro} ,$$

where the reactive power Q^*_{Ro} is determined by the ratio

$$Q^*_{Ro} = \frac{1+q}{\omega^*(k_L + q)} (u^*_{Rdo})^2 + \frac{1}{\omega^*} (u^*_{Rqo})^2 - u^*_{Rqo} . \tag{22}$$

On fig. 9, 10 the dependences of S^*_R and $\cos(\varphi_R)$ on ω^* are presented. As can be seen from figure 9 at a certain frequencies, there is a minimum total power that takes place at zero values of the inactive power. Denote the frequency at which there is a minimum of full power as ω^*_0 , while its value is determined using the relation

$$\omega^*_0 = u^*_{Rqo} + \frac{(u^*_{Rdo})^2}{u^*_{Rqo}} \cdot \frac{(1+q)}{(1+k_L)} = u^*_{Gqo} + \frac{(u^*_{Gdo})^2}{u^*_{Gqo}} \cdot \frac{1}{k_L}$$

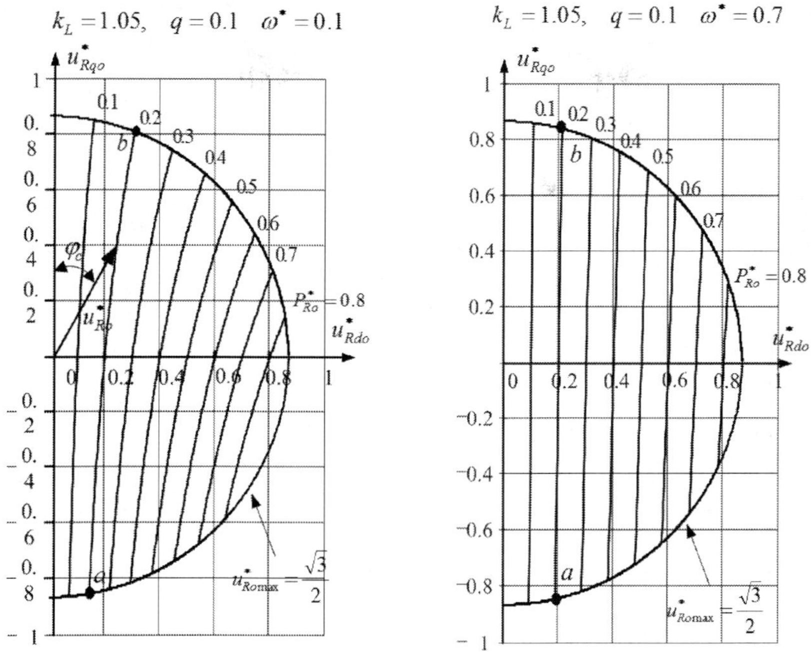
$k_L = 1.05, \quad q = 0.1 \quad \omega^* = 0.1$
u^*_{Rqo}
u^*_{Rdo}
$P^*_{Ro} = 0.8$
u^*_{Ro}
φ_o
$u^*_{Romax} = \frac{\sqrt{3}}{2}$
a
b
$k_L = 1.05, \quad q = 0.1 \quad \omega^* = 0.7$

a) b)

Fig. 8.

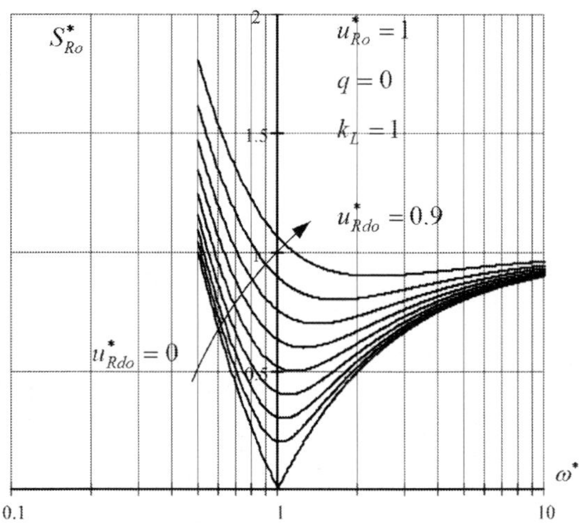
S^*_{Ro}
$u^*_{Ro} = 1$
$q = 0$
$k_L = 1$
$u^*_{Rdo} = 0.9$
$u^*_{Rdo} = 0$
ω^*
0.1
1
10

Fig. 9.

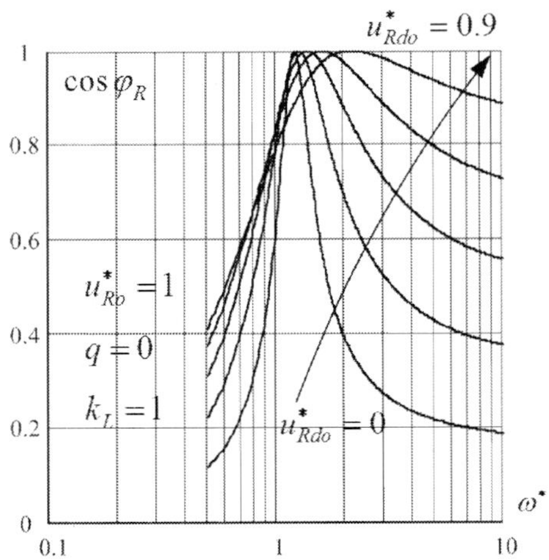

Fig. 10.

As appears from (22), the orthogonal components u^*_{Rdo} also u^*_{Rqo} are connected by the equation

$$\left(\frac{u^*_{Rdo}}{\gamma_Q R_Q}\right)^2 + \left(\frac{u^*_{Rqo} - \dfrac{\omega^*}{2}}{R_Q}\right)^2 = 1, \tag{23}$$

where:

$$\gamma_Q = \sqrt{(k_L + q)/(1+q)}, \quad R_Q = \sqrt{\omega^* Q^*_{Ro} + (\omega^*/2)^2}\ . \tag{24}$$

Expression (23) is the equation of an ellipse with the major axis - $2a_d$ and the minor axis - $2b_q$ that we find from the following expressions:

$$a_d = \gamma_Q R_Q; \quad b_q = R_Q\ .$$

Thus the ellipse centre in «*d q*» co-ordinates is located in a point $\left(0, \omega^*/2\right)$.

Considering the known relation [8], equation (23) in polar coordinates takes the form:

$$\rho(\phi)^2 \cdot [1 - (\varepsilon\cos\phi)^2] - 2\rho_{ucr}(\phi)\rho_0 \left[\left(\frac{b_q}{a_d}\right)^2 \cos\phi\cos\phi_0 + \sin\phi\sin\phi_0\right] + \rho_0^2 \cdot [1 - (\varepsilon\cdot\cos\phi_0)^2] - b_q{}^2 = 0,$$

where co-ordinates of the centre of the ellipse ρ_0, ϕ_0 and the parameter ε are defined by means of expressions

$$\rho_0 = \frac{n^*}{2}; \quad \phi_0 = \frac{\pi}{2}; \quad \varepsilon^2 = 1 - \left(\frac{b_q}{a_d}\right)^2 = \frac{k_L - 1}{k_L + q}. \tag{25}$$

Considering (24) ÷ (25) we will obtain the following expression for the locus of voltage $u^*_{Ro} = \sqrt{(u^*_{Rdo})^2 + (u^*_{Rqo})^2}$ in section S^*_R

$$u^*_{Ro} = \left[\omega^* \sin\phi + \sqrt{\left(\omega^* \sin\phi\right)^2 + 4\omega^* Q^*_{Ro}\left(1-\varepsilon^2 \cos\phi\right)} \right] \Big/ \left[2\left(1-\varepsilon^2 \cos\phi\right)\right]; \quad \phi \in (0, 2\pi).$$

The dependence $u^*_{Ro}(\phi)$ for different values ω^* and the value of reactive power Q^*_{Ro} are shown in Figure 11. Here and below, a circle with a radius $u^*_{Ro\max} = \sqrt{3}/2$ is limiting mode with $M = 1$, i.e. outside this circle the modulation depth is limited, and therefore the ratios obtained above are valid only inside the circle. The negative value of the reactive power Q^*_{Ro} means that in the given section the current of the basic harmonic lags behind of a voltage phase.

Considering $P^*_{Ro} \approx u^*_{Rdo}$, from fig. 11b it follows that the maximum active power ($P^*_{Ro\max}$) which is defined by the maximum projection of locus on a «d» axis, essentially depends on size of the reactive power (Q^*_{Ro}), and at a negative value of Q^*_{Ro} $P^*_{Ro\max}$ decreases. Indeed from (23) and (24) we obtain:

$$P^*_{Ro\max} \approx \gamma_Q R_Q \approx \sqrt{\omega^* Q^*_{Ro} + (\omega^*/2)^2} \ .$$

By changing the coordinates u^*_{Rdo}, u^*_{Rqo} we obtain the possibility to control the active power generated and the reactive power consumed from the generator on the fundamental harmonic.

Using the relation $Q^*_{Ro} = tg\varphi_{SR} P^*_{Ro}$ and assuming $q = 0$ that equation (23) can be rewritten in the variables P^*_{Ro} and u^*_{Rqo}

$$\left(\frac{P^*_{Ro} - \frac{\omega^*}{2} tg\varphi_{SR} \cdot k_L}{\sqrt{k_L} \cdot \frac{\omega^*}{2} \sqrt{1 + k_L \left(tg\varphi_{SR}\right)^2}} \right)^2 + \left(\frac{u^*_{Rqo} - \frac{\omega^*}{2}}{\frac{\omega^*}{2} \sqrt{1 + k_L \left(tg\varphi_{SR}\right)^2}} \right)^2 = 1 \qquad (26)$$

$$a_d = \frac{\omega^*}{2} \sqrt{k_L} \sqrt{1 + k_L \left(tg\varphi_{SR}\right)^2}; \quad b_q = \frac{\omega^*}{2} \sqrt{1 + k_L \left(tg\varphi_{SR}\right)^2} .$$

Thus the ellipse centre in «d q» co-ordinates is located in a point $\left(\omega^* tg\varphi_{SR} \cdot k_L \cdot (1+q)/2, \omega^*/2\right)$

$$\rho_0 = \frac{n^*}{2} \sqrt{1 + (tg\varphi_{SR} \cdot k_L)^2}; \quad \phi_0 = arctg\left(\frac{1}{tg\varphi_{SR} \cdot k_L} \right); \varepsilon^2 = 1 - \left(\frac{b}{a}\right)^2 = \left(1 - \frac{1}{k_L (1+q)^2}\right).$$

The maximum active power generated for the each set of parameters is determined by the point on the graph, as shown, for example, in fig.12a. From the relation (26) we obtain

$$P^*_{Ro\max} = a_d + \omega^* k_L tg\varphi_{SR}/2 = \omega^*/2[\sqrt{k_L} tg\varphi_{SR} + \sqrt{1 + k_L (tg\varphi_{SR})^2}] .$$

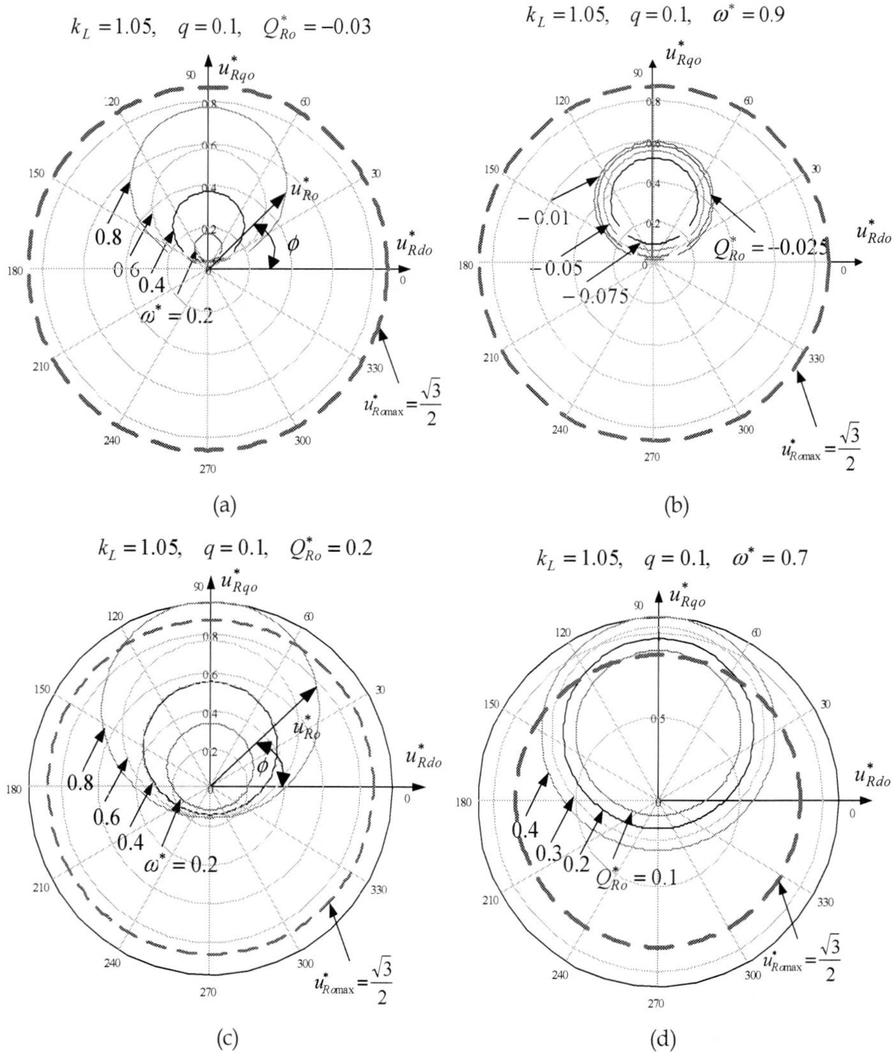

Fig. 11.

The graph of dependence $P^*_{Ro\max}$ on the frequency of rotation ω^* and φ_{SR} for the various k_L is presented in fig. 13.

The above reasoning and results of calculations allow drawing a conclusion that thanks to possibility of independent regulation by means of the active rectifier of orthogonal components of the resultant voltage vector u_R, modes with various $\cos(\varphi_R)$ values at change ω^* are possible in the system.

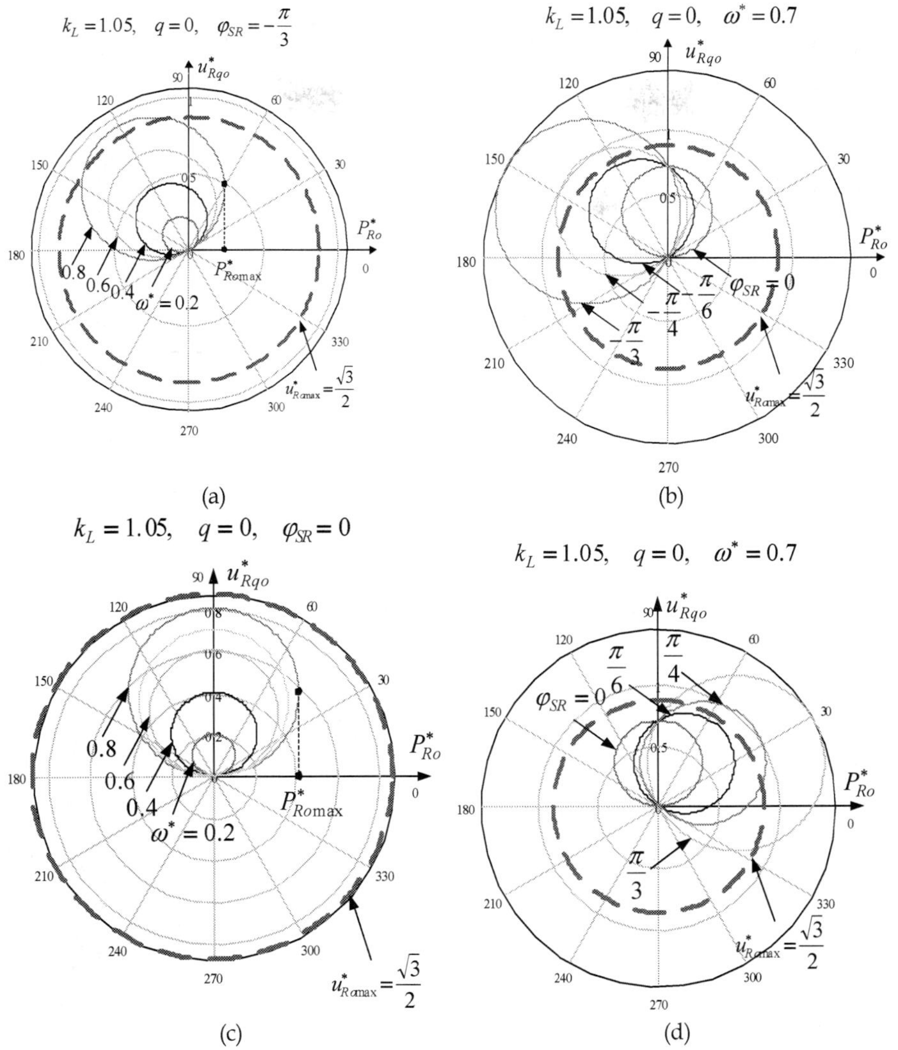

Fig. 12.

Phases of the fundamental harmonics of current and voltage of the generator coincide.

The vector diagram for the fundamental harmonics of current and voltage in the given mode is shown in fig. 14. As appears from the given diagram

$$tg(\theta) = i^*_{do} / i^*_{qo} = u^*_{gdo} / u_{gqo} \,.$$

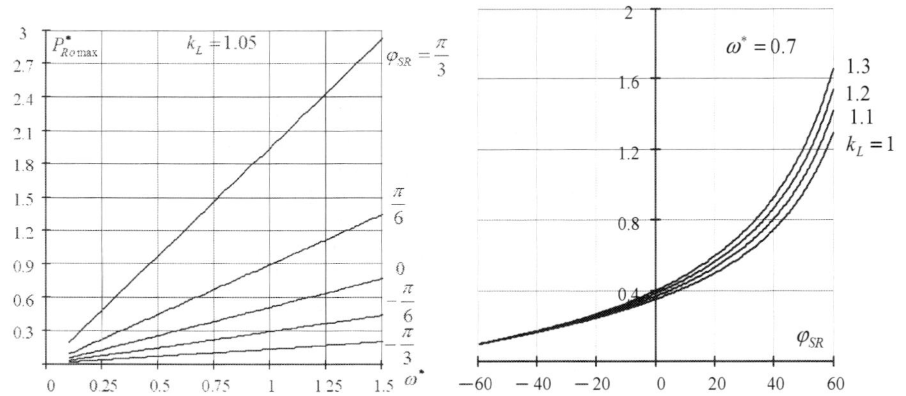

(a) (b)

Fig. 13.

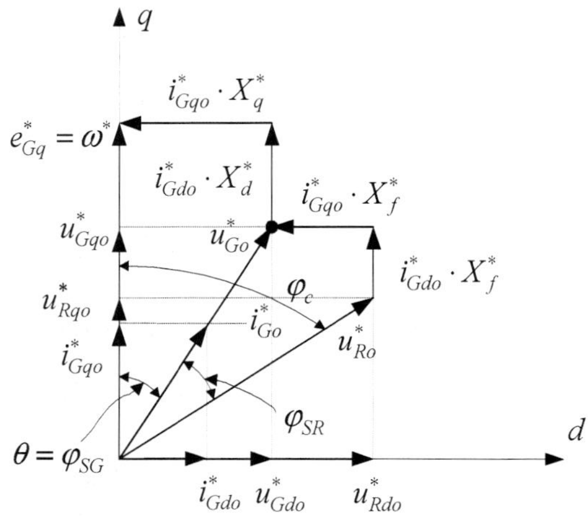

Fig. 14.

This ratio provides a functional link between the orthogonal components $u^*_{do} = \sqrt{3}M_d/2$ and $u^*_{qo} = \sqrt{3}M_q/2$. For this we use the relations (16) and (17), as a result we obtain:

$$\frac{\omega^* - u^*_{Rqo}}{u^*_{Rdo}} \cdot \frac{k_L + q}{1+q} = \frac{\gamma_q u^*_{Rdo}}{\gamma_d u^*_{Rqo} + (1-\gamma_d)\omega^*}, \tag{27}$$

where $\gamma_q = X_q / X_{q\Sigma} = k_L/(k_L + q)$, $\gamma_d = X_d / X_{d\Sigma} = 1/(1+q)$.

Expression (27) represents the ellipse equation, in a canonical form this equation will become:

$$\frac{(u^*_{Rdo})^2}{(\gamma_u R)^2} + \frac{(u^*_{Rqo} + u^*_{0q})^2}{R^2} = 1, \quad (28)$$

where the semi-major axis $a_d = \gamma_u R$ and semi-minor axis $b_q = R$, values of γ_u , R and u^*_{0q} are determined by the relations:

$$\gamma_u = \frac{k_L + q}{1+q} \cdot \frac{1}{\sqrt{k_L}}; R = \frac{\omega^*}{2}(1+q); u^*_{0q} = \frac{\omega^*}{2}(q-1).$$

The equation (28) in polar co-ordinates looks like:

$$u^*_{Ro} = \sqrt{(u^*_{do})^2 + (u^*_{qo})^2} = \frac{\omega^*}{2} \frac{-(q-1)\sin\phi + \sqrt{(q-1)^2 \sin\phi^2 + 4q \cdot (1 - \varepsilon^2 \cos\varphi^2)}}{1 - {\varepsilon_{Ro}}^2 \cos\phi^2}, \phi \in (0, 2\pi), \quad (29)$$

where $\varepsilon^2_{Ro} = \begin{cases} 1 - \frac{(\gamma_u R)^2}{R^2} = 1 - (\gamma_u)^2 = 1 - \left(\frac{k_L + q}{1+q}\right)^2 \frac{1}{k_L}, \text{если } q > \sqrt{k_L}; \\ 1 - \frac{R^2}{(\gamma_u R)^2} = 1 - \frac{1}{(\gamma_u)^2} = 1 - \left(\frac{1+q}{k_L + q}\right)^2 k_L, \text{если } q < \sqrt{k_L}. \end{cases}$

The character of change u^*_{Ro} as the function of ϕ is shown in fig.15.

In figure 16 the loci are constructed in accordance with the relation (29) for the different values of ω^*, q and k_L . With the increase of q (fig.16a), with the same value $u^*_{Rdo} \equiv P^*_{Ro}$, the coordinate u^*_{Rqo} changes sign and increases the module.

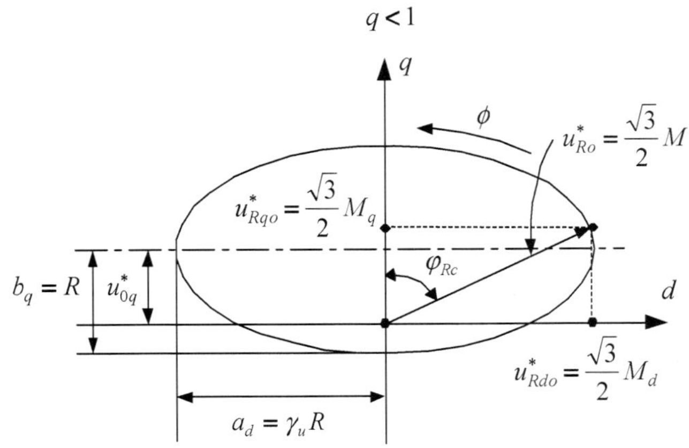

Fig. 15.

At $k_L \to 1$ (fig.16) the loci are circles, while it should be noted that when $k_L = 1 \div 1.1$ you can take $k_L = 1$. At a constant ω^* in this mode, the voltage module u^*_{Ro} increase should be accompanied by the growth of the both orthogonal components u^*_{do} and u^*_{qo} .

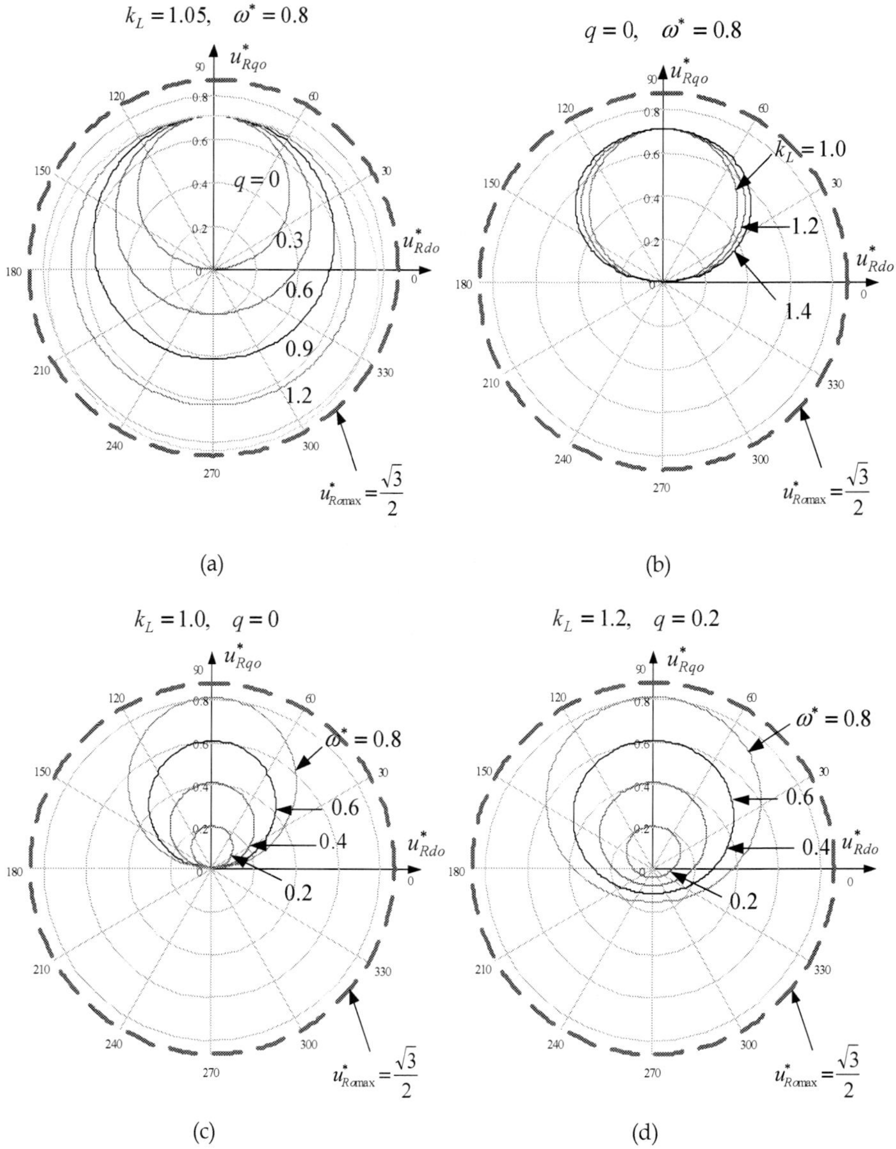

Fig. 16.

The loci of voltage and current of the generator in this mode are determined by the following relations:

$$\left(\frac{u^*_{Gdo}}{\sqrt{k_L}\,\frac{\omega^*}{2}}\right)^2+\left(\frac{u^*_{Gqo}-\frac{\omega^*}{2}}{\frac{\omega^*}{2}}\right)^2=1\,,\ u^*_{Go}=\frac{\omega^* k_L\sin\phi+}{k_L\sin\phi^2+\cos\phi^2}\,,$$

$$\left(\frac{i^*_{do}-\frac{1+q}{2}}{\frac{1+q}{2}}\right)^2+\left(\frac{i^*_{qo}}{\frac{1+q}{2}\frac{1}{\sqrt{k_L}}}\right)^2=1\,,\ i^*_{Go}=\frac{\cos\phi\cdot(q+1)}{k_L\sin\phi^2+\cos\phi^2}\,. \tag{30}$$

The three loci (u^*_{Go} , u^*_{Ro} , i^*_{Go}), built on the same graph, are shown in figure 17.

From the graph and the relation (30) it follows that the maximum values of a current in orthogonal axes are:

$$i^*_{do\max}=(1+q)/2;\quad i^*_{qo\max}=(1+q)/2\sqrt{k_L}\,. \tag{31}$$

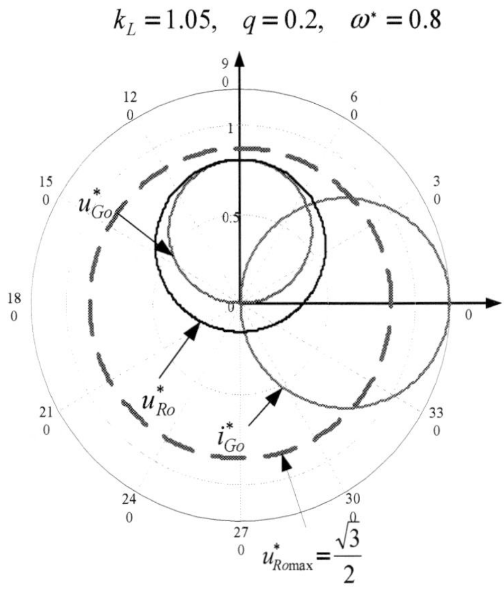

Fig. 17.

Taking into account, that

$$P^*_{Ro}=\frac{P_{Ro}\omega_6 L_d(1+q)}{E_6^2}=\frac{1+q}{k_L+q}u^*_{Rdo}=\frac{\sqrt{3}}{2}\frac{1+q}{k_L+q}M_d\,,$$

Equation (28) can be rewritten in the coordinates (P^*_{Ro}, u^*_{Rqo})

$$\frac{(P^*_{Ro})^2}{\left(R/\sqrt{k_L}\right)^2}+\frac{(u^*_{Rqo}+u^*_{0q})^2}{R^2}=1\,, \tag{32}$$

$$R=\frac{\omega^*}{2}(1+q);\; u^*_{0q}=\frac{\omega^*}{2}(q-1)\,.$$

$$\sqrt{(P^*_{Ro})^2+(u^*_{Rqo})^2}=\omega^*\cdot\frac{(1-q)\sin\phi+\{[(1+q)\sin\phi]^2+4qk_L(\cos\phi)^2\}^{\frac{1}{2}}}{2[(\sin\phi)^2+(\cos\phi)^2k_L]}\,.$$

The graph of u^*_{Rqo} change at a variation of P^*_{Ro} is shown in fig. 18.

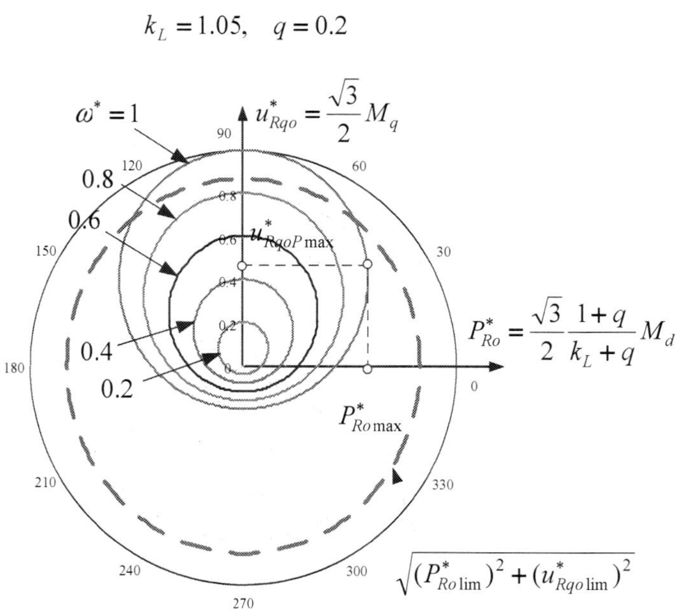

Fig. 18.

The limiting locus is determined by the condition

$$M_d^2+M_q^2\le 1\,,$$

whence we obtain

$$\frac{(P^*_{Ro\lim})^2}{\left(\frac{1+q}{k_L+q}\frac{\sqrt{3}}{2}\right)^2}+\frac{(u^*_{Rqo\lim})^2}{\left(\frac{\sqrt{3}}{2}\right)^2}=1\,. \tag{33}$$

$$\sqrt{(P^*_{Ro\lim})^2+(u^*_{Rqo\lim})^2}=\frac{\sqrt{3}}{2}\frac{1}{\sqrt{\left(\frac{k_L+q}{1+q}\right)^2(\cos\phi)^2+(\sin\phi)^2}}.$$

For a given active power P^*_{Ro} limiting the value of the modulation of control input wave on the transverse axis (q) is determined using the relation

$$M_{q\lim}=\sqrt{1-\left[\frac{2}{\sqrt{3}}\frac{(k_L+q)}{(1+q)}P^*_{Ro}\right]^2}.$$

As appears from fig.18, with increasing of the frequency of rotation of the shaft WT (ω^*) there is an increase in active power and for each the value ω^* holds the maximum value of active power $P^*_{Ro\max}$, which is defined by the relation:

$$P^*_{Ro\max}=\frac{P_{Ro\max}\omega_б L_d(1+q)}{E_б^2}=\frac{\omega^*}{2\sqrt{k_L}}(1+q)=\frac{\sqrt{3}}{2}\frac{1+q}{k_L+q}M_{dP\max}, \tag{34}$$

while taking into account the introduced relative units (9), the real value of power can be found from the expression:

$$P_{Ro\max}=E_б^2/\omega_б L_d\cdot\omega^*/2\sqrt{k_L}.$$

The maximum power is achieved with the following voltage $u^*_{Rqo}=u^*_{RqoP\max}$ and $u^*_{Rdo}=u^*_{RdoP\max}$ and also voltage $u^*_{Ro}=u^*_{RoP\max}$:

$$u^*_{RqoP\max}=-u^*_{0q}=\frac{\omega^*}{2}(1-q)=\frac{\sqrt{3}}{2}M_{qP\max},\quad u^*_{RdoP\max}=\frac{\omega^*}{2}\frac{k_L+q}{\sqrt{k_L}}=\frac{\sqrt{3}}{2}M_{dP\max},$$
$$u^*_{RoP\max}=\sqrt{(u^*_{RqoP\max})^2+(u^*_{RqoP\max})^2}=\frac{\omega^*}{2}\cdot\sqrt{(q-1)^2+\frac{(k_L+q)^2}{k_L}}. \tag{35}$$

At this point the generator voltage in accordance with (35) and (18) is determined by the relations:

$$u^*_{GqoP\max}=\omega^*/2,\quad u^*_{GdoP\max}=\omega^* k_L(1-q)/2(k_L+q),$$

$$u^*_{GoP\max}=\sqrt{(u^*_{GqoP\max})^2+(u^*_{GqoP\max})^2}=\frac{\omega^*}{2}\cdot\sqrt{1+\left[\frac{k_L(1-q)}{k_L+q}\right]^2}. \tag{36}$$

In accordance with (34) and (35) the parameters of control signals for the point with maximum capacity are determined by the following relations

$$M_{dP\max}=\frac{\omega^*}{\sqrt{3}}\cdot\frac{k_L+q}{\sqrt{k_L}};\quad M_{qP\max}=\frac{\omega^*}{\sqrt{3}}\cdot(1-q);$$

$$M_{P\max} = \frac{\omega^*}{\sqrt{3}} \cdot \sqrt{(q-1)^2 + \frac{(k_L+q)^2}{k_L}}; \quad \varphi_{RcP\max} = arctg\left[\frac{M_{dP\max}}{M_{qP\max}}\right] = arctg\left[\frac{k_L+q}{\sqrt{k_L}\cdot(1-q)}\right];$$

Dependence of the active power on the frequency of rotation (ω^*), values of the modulation depth of control input wave on the transverse axis (M_q) and parameters k_L and q is defined as follows:

$$P_{Ro}^* = \frac{1}{\sqrt{k_L}}\left\{\left(\frac{\omega^*}{2}\right)^2 - \left[\frac{\sqrt{3}}{2}M_q + (q-1)\frac{\omega^*}{2}\right]^2\right\}^{\frac{1}{2}}.$$

Graphs of the dependence of active power from the values ω^*, M_q are presented in fig.19.

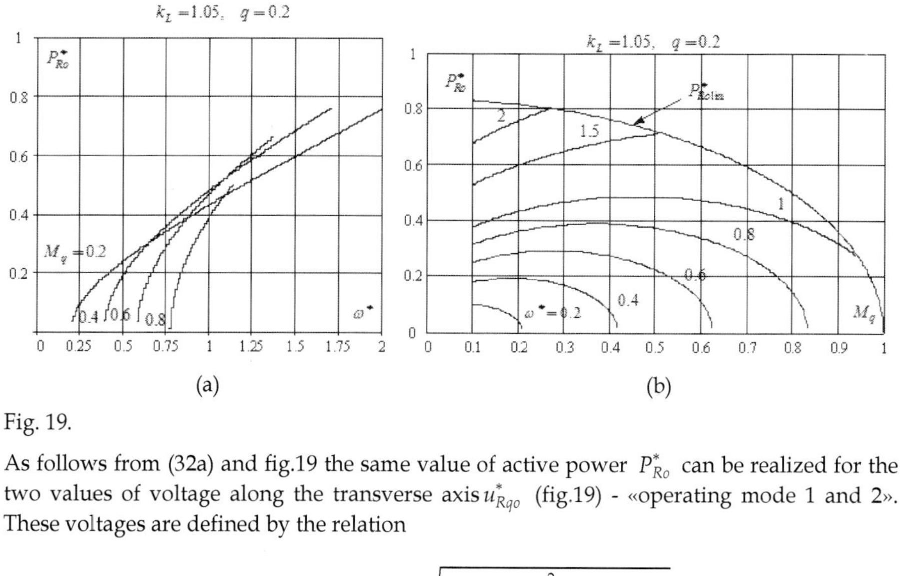

(a) (b)

Fig. 19.

As follows from (32a) and fig.19 the same value of active power P_{Ro}^* can be realized for the two values of voltage along the transverse axis u_{Rqo}^* (fig.19) - «operating mode 1 and 2». These voltages are defined by the relation

$$u_{Rqo1,2}^* = -\frac{\omega^*}{2}(q-1) \mp \sqrt{\left[\frac{\omega^*}{2}(q+1)\right]^2 - (\sqrt{k_L}P_{Ro}^*)^2}.$$

Considering (18) the generator voltage in these modes will be defined as follows:

$$u_{Gqo1,2}^* = \frac{\omega^*}{2} \mp \sqrt{\left(\frac{\omega^*}{2}\right)^2 - \left(\frac{\sqrt{k_L}P_{Ro}^*}{1+q}\right)^2}, \quad u_{Gdo}^* = \frac{k_L}{1+q}P_{Ro}^*,$$

$$u_{Go1,2}^* = \frac{\omega^*}{\sqrt{2}}\sqrt{1 + \left(\frac{\sqrt{2k_L}P_{Ro}^*}{(1+q)\cdot\omega^*}\right)^2 (k_L-1) \mp \sqrt{(1+q)^2 - \left(\frac{2}{\omega^*}\frac{\sqrt{k_L}P_{Ro}^*}{1+q}\right)^2}}. \qquad (37)$$

$k_L = 1.0, \quad q = 0, \quad \omega^* = 0.8$

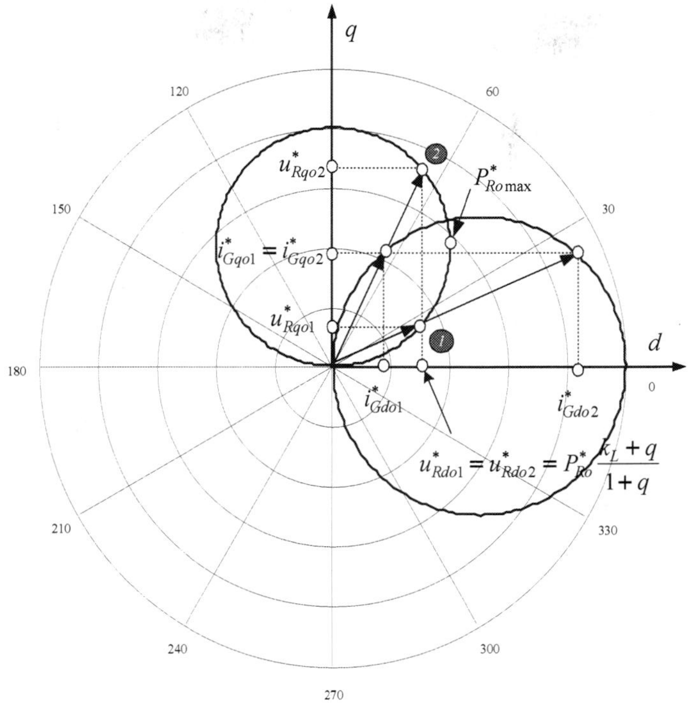

Fig. 20.

From fig.20 it is clear that the larger value of voltage u^* corresponds to the lower value of current i^*_{Go}, while $i^*_{Gqo1} = i^*_{Gqo2}$ that follows from (5a).

$$\begin{cases} i^*_{Gqo1} = i^*_{Gqo2} = u^*_{Rdo}/X^*_{q\Sigma} = P^*_{Ro}/\omega^*, \\ i^*_{Gdo1,2} = 1 - \dfrac{u^*_{Rqo1,2}}{\omega^*} = \dfrac{1+q}{2} \pm \sqrt{\left(\dfrac{1+q}{2}\right)^2 - \left(\dfrac{\sqrt{k_L}\,P^*_{Ro}}{\omega^*}\right)^2}. \end{cases}$$

At the point when the power is maximum for a given frequency of rotation ($P^*_{Ro} = P^*_{Ro\max}$), (the relation (34)), the orthogonal components of currents and voltages are determined by the relations (31), (37).

The full value of the generator current is:

$$i^*_{Go1,2} = \frac{1+q}{\sqrt{2}}\sqrt{1 + (1-k_L)\left(\frac{2P^*_{Ro}}{\omega^*(1+q)}\right)^2 \pm \sqrt{1 - \left(2\frac{\sqrt{k_L}\,P^*_{Ro}}{\omega^*(1+q)}\right)^2}}.$$

Graphs of the dependences u^*_{Rqo}, i^*_{Go} and u^*_{Go} from ω^* are presented in fig.30. In these graphs to the right of points «a, b, c, d» there is a limitation of the depth of modulation, and the proposed model becomes inadequate to the real modes.

The first mode on the graphs of fig.21 is characterized by the fact that the generator voltage does not change significantly with variation of ω^*. From fig.21d it follows that in mode 1 the system has a positive internal differential resistance and therefore may be potentially unstable at some disturbing effects.

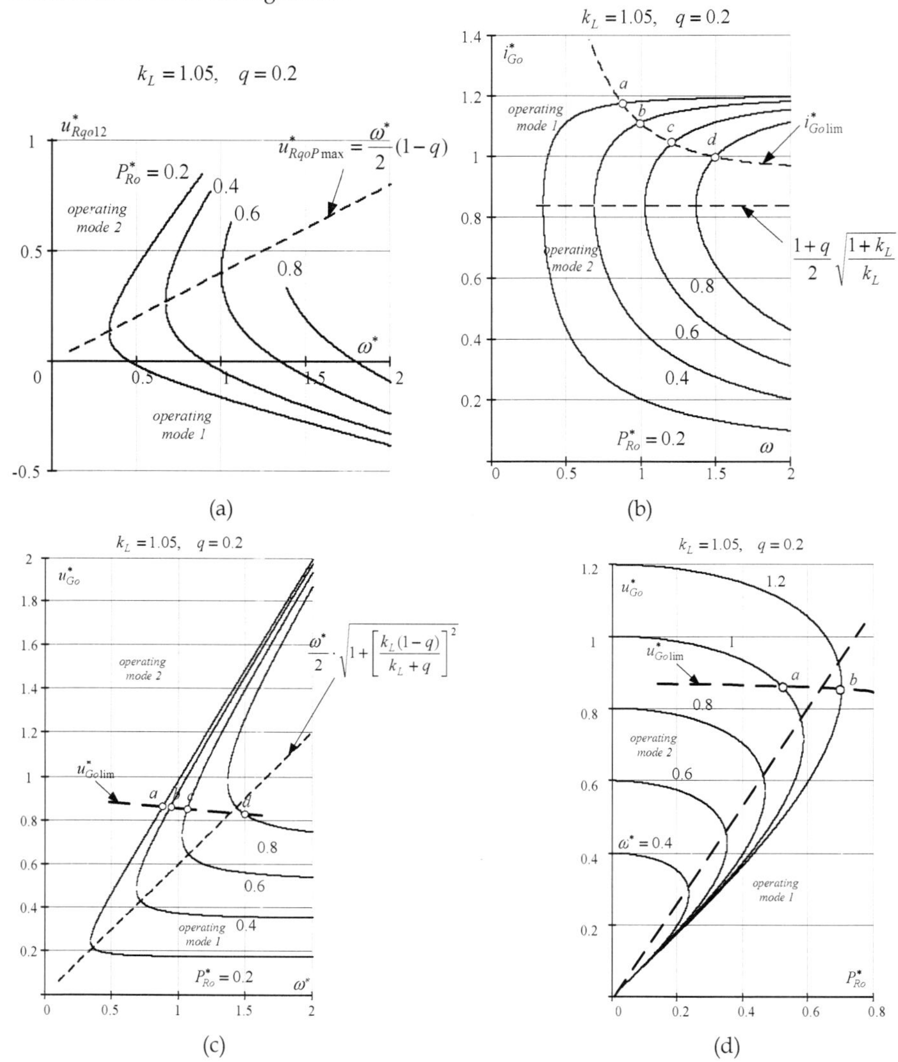

Fig. 21.

It should be noted that the mode when the phases of voltage and current of the fundamental harmonics are the same, essentially involves a change of u^*_{Go} and i^*_{Go} under the change of frequency of rotation of the shaft of WT (ω^*).
The equation (28) can be rewritten in the co-ordinates of orthogonal components of a control input wave (M_d, M_q):

$$\frac{(M_d)^2}{\left[\gamma_u \frac{\omega^*}{\sqrt{3}}(1+q)\right]^2} + \frac{\left[M_q + \frac{\omega^*}{\sqrt{3}}(q-1)\right]^2}{\left[\frac{\omega^*}{\sqrt{3}}(1+q)\right]^2} = 1, \tag{38}$$

Expression (38) defines a parametric relationship between the orthogonal components, which ensures the phase coincidence of the main of harmonics of the generator voltage and current. This ratio allows us to propose the following control algorithm, in which the active power is given. This solution is useful for the application of the system in the WPI.
In accordance with (16b), (21) and (38), we obtain:

- given the active power - P^*_{Ro}, when the condition $P^*_{Ro\max} \le P^*_{Ro\max} = \omega^*(1+q)/2\sqrt{k_L}$ must be respected;
- the longitudinal component of the modulation (M_d) of the control signal is determined by $M_d = 2(k_L + q)P^*_{Ro}/\sqrt{3}(1+q)$;
- the transverse component of the modulation (M_q) of the control input wave is determined by $M_{q1,2} = -\frac{\omega^*}{\sqrt{3}}(q-1) \mp \sqrt{\left[\frac{\omega^*}{\sqrt{3}}(q+1)\right]^2 - \left(\frac{2}{\sqrt{3}}\sqrt{k_L P^*_{Ro}}\right)^2}$ - here M_{q1} - corresponds to the mode 1, and M_{q2} - to the mode 2.

When $q > 0$, the fundamental harmonics of inverter current and voltage (section S_R) do not coincide in phase. The current phase is ahead of the voltage phase ($\varphi_{SR} > 0$). Fig. 22a shows the dependence of $\cos\varphi_{SR}$ on the parameters q and k_L . From figure 22a it follows that in the first mode there is a significant reduction of $\cos\varphi_{SR}$ with the increase of the parameter q . Dependence of $\cos\varphi_{SR}$ on the parameter k_L is ambiguous, namely, in the first mode $\cos\varphi_{SR}$ decreases with the increase of k_L , while in the second mode, on the contrary increases.
The phases of the fundamental harmonics of generator voltage and current (section S_G) are always the same ($\cos\varphi_{SG} = 1$), the power factor in section S_G will be determined according to (15a) by the relation $\chi_G = P^*_{SG}/S^*_{SG} = P^*_{Ro}/S^*_{SG} = \nu_{iG}\nu_{uG}$.
The RMS of the fundamental harmonic of generator voltage is determined by the following expression:

$$u^*_{Go,rms} = \frac{1}{\sqrt{2}} \frac{\omega^*}{(q^2-1)} \sqrt{q^2 - \left(\frac{\sqrt{3}M}{2\omega^*}\right)^2} \sqrt{\left(\frac{k_L(q+1)}{(k_L+q)}\right)^2 \left[\left(\frac{\sqrt{3}M_{1,2}}{2\omega^*}\right)^2 - 1\right] + \left[q^2 - \left(\frac{\sqrt{3}M_{1,2}}{2\omega^*}\right)^2\right]} . \tag{39}$$

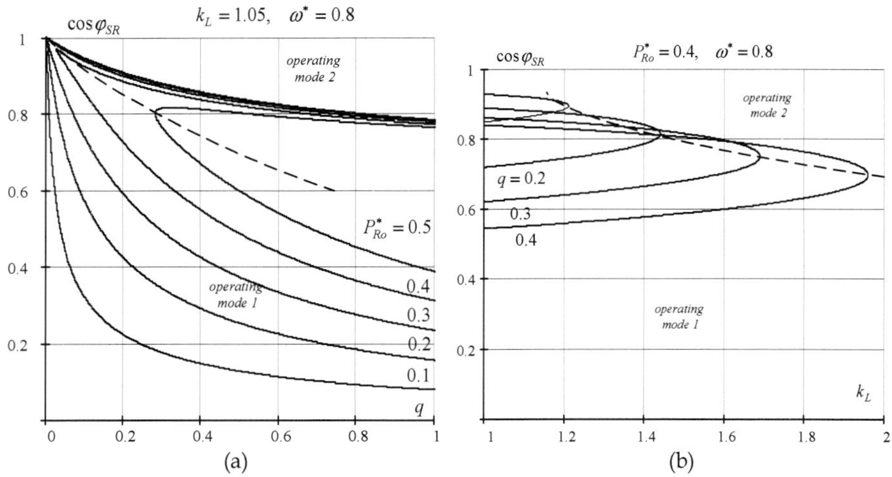

Fig. 22.

Complete RMS of the generator voltage is found from the ratio

$$u^*_{Go,\,rms} = \frac{1}{\sqrt{2}}\sqrt{(\omega^* \cos\theta)^2 + \frac{M_{1,2}}{(1+q)^2}\left(1 - \frac{3}{4}M_{1,2}\right)}. \tag{40}$$

where: θ - the angle shift between the fundamental harmonic of generator voltage and EMF of generator is defined as follows $\theta = arctg\left[\frac{k_L(1+q)}{k_L+q}\cdot\frac{M_d}{M_{q1,2}+\omega^*\frac{2}{\sqrt{3}}}\right]$.

The RMS of fundamental harmonic of the generator current

$$i^*_{Go,\,rms1,2} = \frac{1}{\sqrt{2}}\sqrt{1 - \frac{\sqrt{3}}{\omega^*}M_{q1,2} + \frac{1}{(\omega^*)^2}\left[\left(\frac{\sqrt{3}}{2}M_{q1,2}\right)^2 + \left(P^*_{Ro}\right)^2\right]} \tag{41}$$

Complete RMS of the generator current is found from the relation (19) and (41).
In fig.23 the distortion coefficients (ν_{uSG}) and harmonics (THD_{uSG}) of generator voltage as functions of the active power generated (P^*_{Ro}) and frequency of rotation (ω^*) are presented. As can be seen from fig.23 the best quality of generator voltage is characteristic for mode 2. In addition, analysis of the relations (39) and (40) suggests a reduction factor of harmonics THD_{uSG} with the increase of the parameter q. Similar conclusions can be drawn from the consideration of fig. 24, which shows the distortion coefficients (ν_{iSG}) and harmonics (THD_{iSG}) of current as a function of the active power generated (P^*_{Ro}) and frequency of rotation (ω^*). In the engineering calculations we can take $\nu_{iSG} \approx 1$.
Power factor of the generator according to (15a), taking into account that $\cos\varphi_{SG} = 1$, is determined using the relation: $\chi_G = \nu_{iG}\nu_{uG}$.

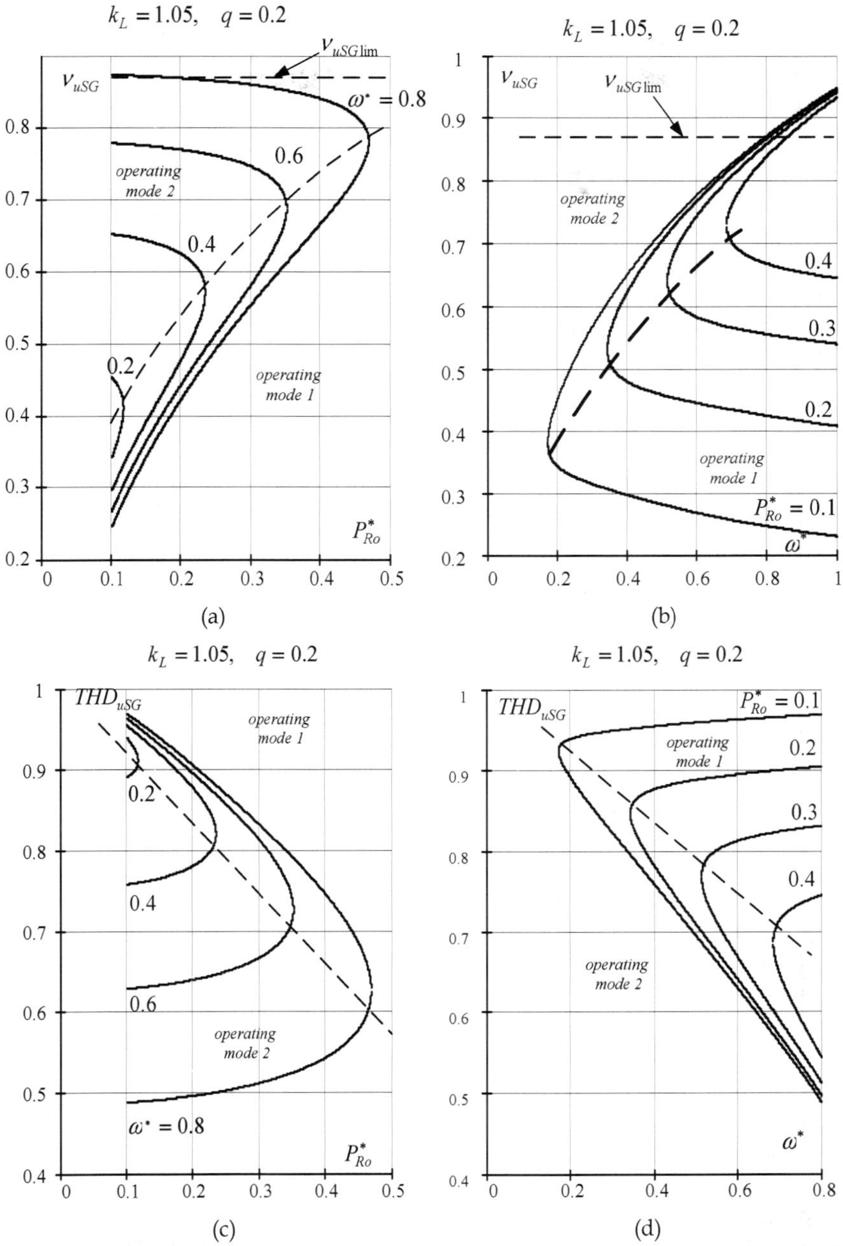

$k_L = 1.05, \quad q = 0.2$
v_{uSG}
$v_{uSG\,\mathrm{lim}}$
$\omega^* = 0.8$
0.6
0.4
0.2
operating mode 2
operating mode 1
P^*_{Ro}
(a)
$k_L = 1.05, \quad q = 0.2$
v_{uSG}
$v_{uSG\,\mathrm{lim}}$
operating mode 2
operating mode 1
0.4
0.3
0.2
$P^*_{Ro} = 0.1$
ω^*
(b)
$k_L = 1.05, \quad q = 0.2$
THD_{uSG}
operating mode 1
0.2
0.4
0.6
operating mode 2
$\omega^* = 0.8$
P^*_{Ro}
(c)
$k_L = 1.05, \quad q = 0.2$
THD_{uSG}
$P^*_{Ro} = 0.1$
0.2
0.3
0.4
operating mode 1
operating mode 2
ω^*
(d)

Fig. 23.

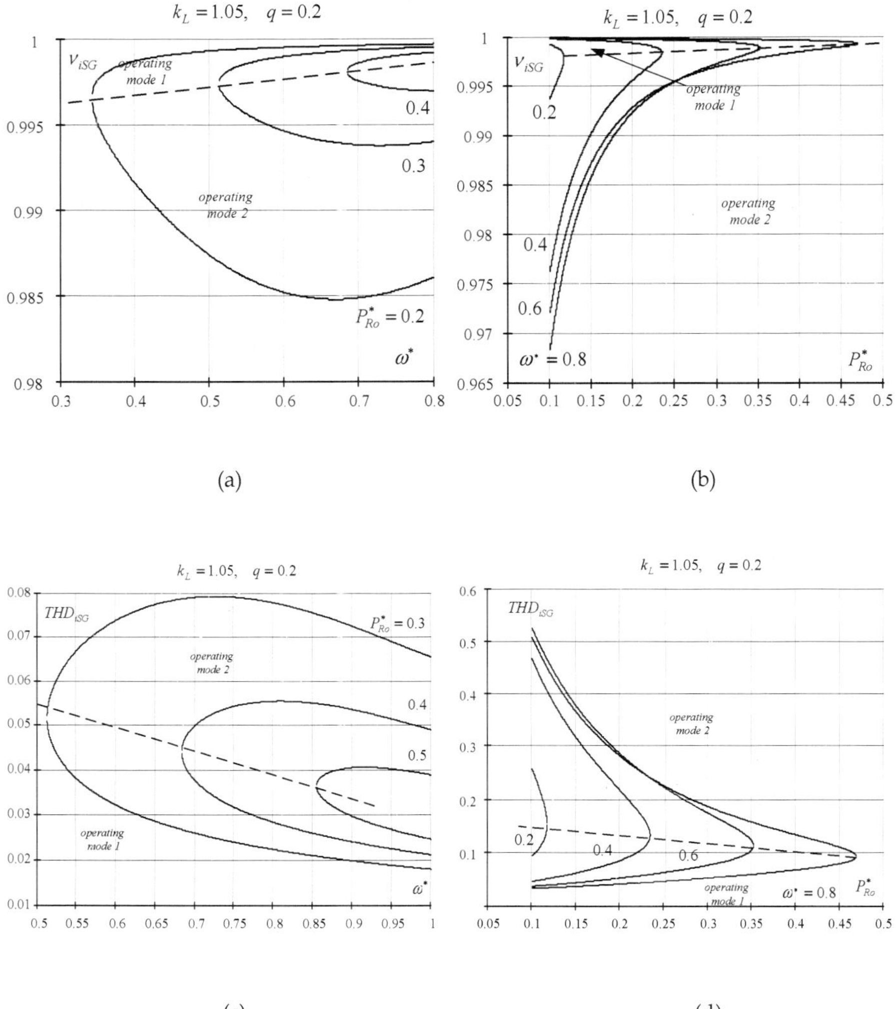

Fig. 24.

Dependence of χ_G on the generated active power (P^*_{Ro}) and the frequency of rotation (ω^*) is shown in fig.25. From this figure and the previous findings it can be taken: $\chi_G \approx \nu_{uG}$.
As expected, the power factor is higher in mode 2. In the mode 1 with a decrease in power χ_G is significantly reduced because of the need to reduce the modulation depth M in order to maintain $\cos\varphi_{SG} = 1$.

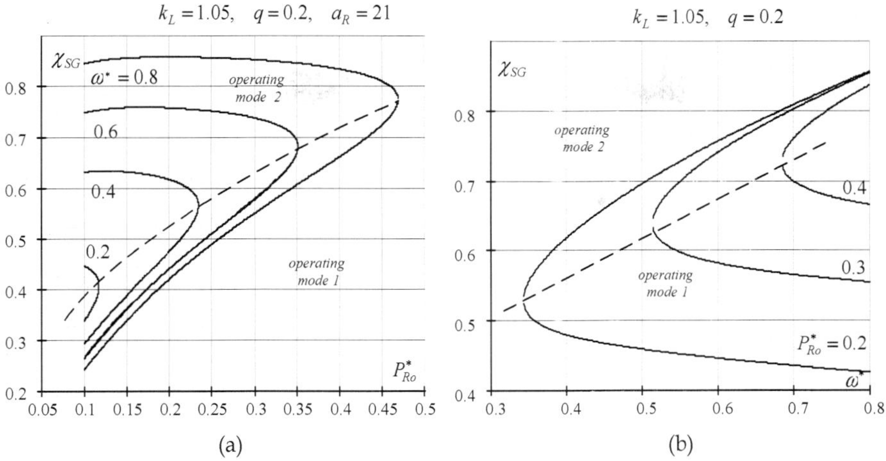

Fig. 25.

From (33), (34) and (35) the conditions can be found under which the maximum power $P^*_{Ro\max}$ (the point where modes 1 and 2 are the same, fig. 20) is attained at the beginning of the limitations of the modulation depth ($M = 1$), i. e. $P^*_{Ro\max} = P^*_{Ro\lim}$ holds; the value of active power, in the performance of this condition is denoted as $P^*_{Ro\max\lim}$.

$$P^*_{Ro\max\lim} = P^*_{Ro\max} = P^*_{Ro\lim} = \frac{\sqrt{3}}{2} \cdot \left\{ \left[\frac{1-q}{\sqrt{k_L}\,(1+q)} \right]^2 + \left(\frac{k_L + q}{1+q} \right)^2 \right\}^{-\frac{1}{2}}. \tag{42}$$

The power $P^*_{Ro\max\lim}$ can be reached at a frequency of rotation $\omega^* = \omega^*_{\max\lim}$:

$$\omega^*_{\max\lim} = \sqrt{3} \cdot \left\{ (1-q)^2 + \left((k_L + q) / \sqrt{k_L} \right)^2 \right\}^{-\frac{1}{2}}. \tag{43}$$

Taking into account the taken relative units we determine the real value of active power $P_{Ro\max\lim}$ from the ratio

$$P_{Ro\max\lim} = \frac{E_б^2}{\omega_б L_d} \cdot \frac{\sqrt{3}}{2} \cdot \left\{ \left(\frac{1-q}{\sqrt{k_L}} \right)^2 + (k_L + q)^2 \right\}^{-\frac{1}{2}}.$$

Figure 26 shows the active power $P_{Ro\max\lim} / E_б^2 / \omega_б L_d$ and frequency $\omega^*_{\max\lim}$ as a function of the parameter q at different values of k_L.

As can be seen from fig.26, the maximum possible active power ($P^*_{Ro\max\lim}$) and the corresponding speed of rotation in this mode ($\omega^*_{\max\lim}$) occur at q = 0, k_L = 1. From (42) and (43) we obtain:

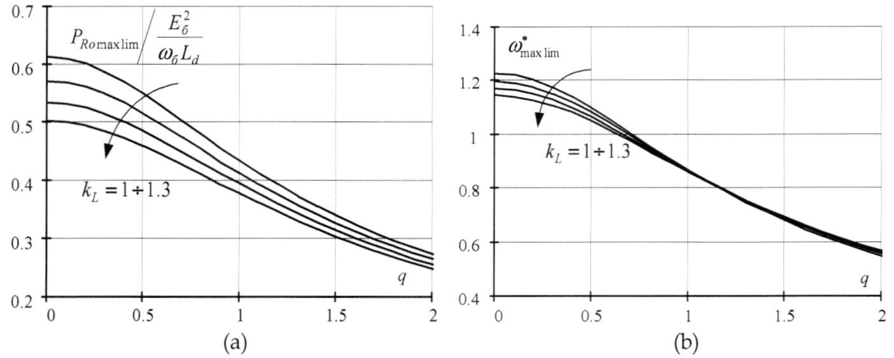

Fig. 26.

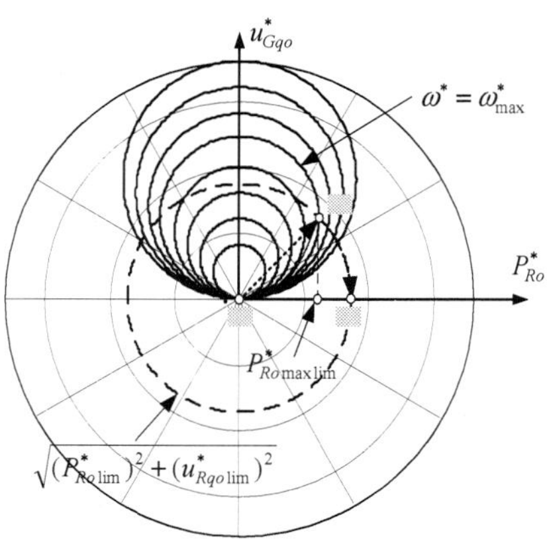

Fig. 27.

$$P^*_{Ro\max\lim} = \begin{cases} \sqrt{3}/(2\sqrt{2}) = 0.6124 - \text{SPWM}; \\ 1/\sqrt{2} = 0.7071 - \text{SVPWM}; \end{cases} \qquad \omega^*_{\max\lim} = \begin{cases} \sqrt{3/2} = 1.2247 - \text{SPWM}; \\ \sqrt{2} = 1.4142 - \text{SVPWM}. \end{cases} \tag{44}$$

When we select the power generation system PGS in the WPI, it is convenient to use fig.27. In this figure (for $q = 0$, $k_L = 1$) the trajectory « $a \rightarrow b$ »corresponds to the points of maximum power ($P^*_{Ro\max}$) for the different frequencies of rotation ω^*. At the point « b » $P^*_{Ro\max} = P^*_{Ro\max\lim}$ and $\omega^* = \omega^*_{\max\lim}$. With the further increase in frequency ω^* to keep the value $\cos\varphi_{SG} = 1$ without the restrictions of the modulation depth it should be the moving on a trajectory« $b \rightarrow c$ ». If $\omega^* \rightarrow \infty$ the point « c » is reached. The value of active power at the point« c »will be the maximum possible for $\cos\varphi_{SG} = 1$. The value of this power is

$$P^*_{Ro\max\lim}(\omega^* \rightarrow \infty) = \begin{cases} \sqrt{3}/2 = 0.866 - \text{SPWM}; \\ 1 - \text{SVPWM}. \end{cases}$$

At work of the PGS from WT an active power will change under the law (20). The mode of the maximum power with $\omega^* = \omega^*_{WT\max}$ is desirable by choosing from a condition $\omega^*_{WT\max} = \omega^*_{\max\lim}$.
In this case according to (20) and (44):

$$P^*_{WTo\max}(\omega^*_{WT\max}) = \gamma \cdot (\omega^*_{WT\max})^3 = P^*_{Ro\max\lim};$$
$$\gamma = P^*_{Ro\max\lim} / (\omega^*_{WT\max})^3.$$

When we change ω^* the operating point should move along the trajectory « $a \to b$ » fig.28, where the power varies according to the law (20), but $\cos\varphi_{SG} = 1$ will be retained. For such a trajectory the dependences of the amplitude values of generator voltage and current (u^*_{Go}, i^*_{Go}), the power factor (χ_{SG}) and the generated power $P^*_{WTo} = P^*_{Ro}$ as a function of the frequency of rotation are shown in fig.29. In fig.29a the movement trajectory "« $a \to b$ »" occurs in mode 1, in fig.29b, respectively, in mode 2.

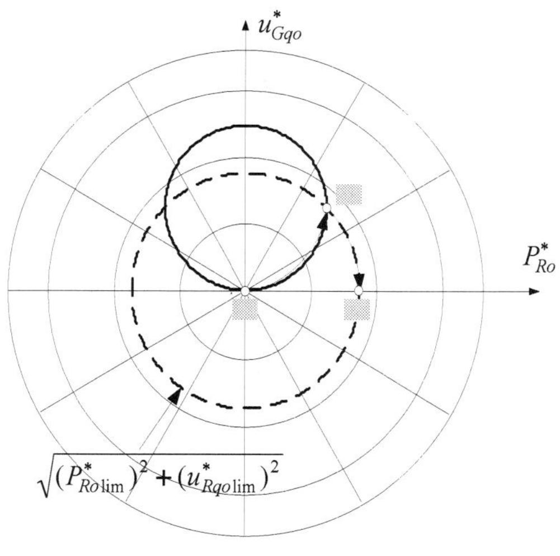

Fig. 28.

As follows from fig.29 at work in the 1st mode, despite the fact that $\cos\varphi_{SG} = 1$ the power factor of the generator has a little value when $\omega^* \to \omega^*_{WT\min}$ it is explained by sufficiently small value of the coefficient of distortion of the generator voltage at low frequencies. In addition, there is a large value of the generator current, so when $\omega^* \to \omega^*_{WT\min}$, $i^*_{Go} \to 1$ i.e. the current is close to the value of short-circuit current. When working in 2nd mode the power factor is much bigger, with the generator current is much smaller than in mode 1. In the 2nd mode, the generator voltage has increased, but it is less than the EMF-load of the generator current.
Note that if we want to save $\cos\varphi_{SG} = 1$ in the entire working range and $M \le 1$ at $\omega^* = \omega^*_{WT\max}$ as well as to choose the frequency of rotation of WT from the condition

$\omega^*_{WT\max} > \omega^*_{\max\lim}$ the working point of the trajectory at maximum frequency of rotation and maximum power generated will be in the 1st mode (fig.30) and, consequently, will have a low value of power factor.

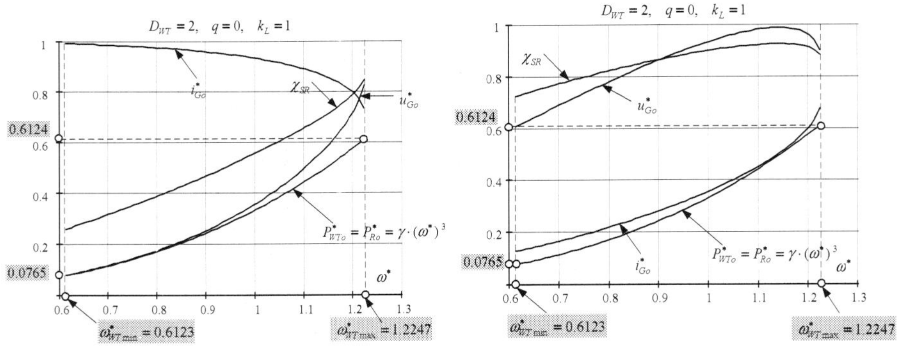

(a) (b)

Fig. 29.

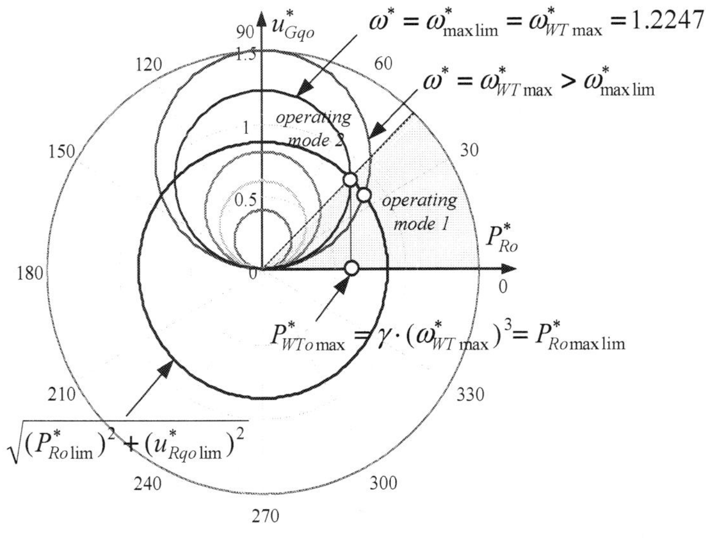

Fig. 30.

Taking into account the results obtained, we can conclude that the work with $\cos\varphi_{SG} = 1$ *in the 1st mode is not optimal for WPGS, because in the entire frequency range* $\omega^* \in \{\omega^*_{WT\min}, \omega^*_{WT\max}\}$ *there is a large value of the generator current (* $i^*_{Go} \to 1$ *at* $\omega^* \to \omega^*_{WT\min}$ *) and a low power factor (* χ_{SG} *). If condition* $\cos\varphi_{SG} = 1$ *remain in the range of frequencies* $\{\omega^*_{WT\min}, \omega^*_{WT\max}\}$ *for WPGS should be recommended the second mode, since in this case, the power factor of the generator in the working* *frequency range* $\omega^*_{w\min} < \omega^* < \omega^*_{w\max}$ *is large enough, with the generator current is much smaller than in the first mode, but there is an increase in generator voltage.*

Phases of the fundamental harmonics of current and voltage of the generator do not coincide.

In this mode, the angle can be $0 > \varphi_{SG}$ or $\varphi_{SG} > 0$. Vector diagram for the case $\varphi_{SG} < 0$ is shown in fig. 5. Basic relations for the determination of voltages, currents and power in the system are given in (14) ÷ (26). For these values of the angle φ_{SG}, as in the case of $\varphi_{SG} = 0$, the same value of power can be obtained in the two modes, corresponding to different values of the parameter M_q.

In the general case, when $q \geq 0, k_L \geq 1$ the active power P^*_{Ro} is related to M_q by the relation:

$$\left(\frac{P^*_{Ro} - P_0}{\gamma_P R_{uP}}\right)^2 + \left(\frac{u^*_{qo} - U_0}{R_{uP}}\right)^2 = 1\,, \tag{45}$$

where

$$\gamma_P = \sqrt{\frac{k_L + q}{1+q}};\quad R_{uP} = \frac{\omega^*}{2}\sqrt{1+\left(\frac{tg\varphi_{SG}}{\gamma_P}\right)^2};\quad U_0 = \frac{\omega^*}{2};\quad P_0 = \frac{\omega^*}{2}\frac{tg\varphi_{SG}}{\gamma_P^2}.$$

Whence:

$$u^*_{qo1,2} = U_0 \mp \sqrt{(R_{uP})^2 - \left(\frac{P^*_{Ro} - P_0}{\gamma_P}\right)^2};\quad u^*_{Ro} = \gamma_P^2 P^*_{Ro}; M_{q1,2} = \frac{2}{\sqrt{3}}u^*_{qo1,2};\quad M_d = \frac{2}{\sqrt{3}}\gamma_P^2 P^*_{Ro}.$$

Here the indices "1" and "2" correspond to the 1st and 2d modes in accordance with fig.20. Maximum power achievable at a given frequency of rotation (ω^*) is defined by the relation:

$$P^*_{Ro\max} = P_0 + \gamma_P R_{uP} \equiv \omega^*/2\,. \tag{46}$$

Relationships (97) make possible to determine the dependence of the currents and voltages in the system as a function of frequency of rotation for different values of the angle φ_{SG} and the parameters q and k_L. Major trends of these relationships can be seen on the graphs (22) ÷ (25). Let us consider the choice of mode of the system in WPI, while we assume that $q = 0, k_L = 1$. In this case, the equation (45) in polar coordinates will be:

$$\begin{aligned}
&\rho(\phi) = \omega^* \sec(\phi_{SG})\sin(\phi + \varphi_{SG});\\
&P^*_{Ro}(\phi) = U^*_{do} = \omega^* \sec(\varphi_{SG})\sin(\varphi + \varphi_{SG})\cos(\phi);\\
&u^*_{qo}(\phi) = \omega^* \sec(\varphi_{SG})\sin(\phi + \varphi_{SG})\sin(\phi).
\end{aligned} \tag{47}$$

Fig.31 shows the nature of the proposed change of the angle (φ_{SG}) of current shift (i^*_{Go}) on voltage (u^*_{Go}) and $\cos\varphi_{SG}$ on the frequency of rotation of the shaft of WT. The proposed

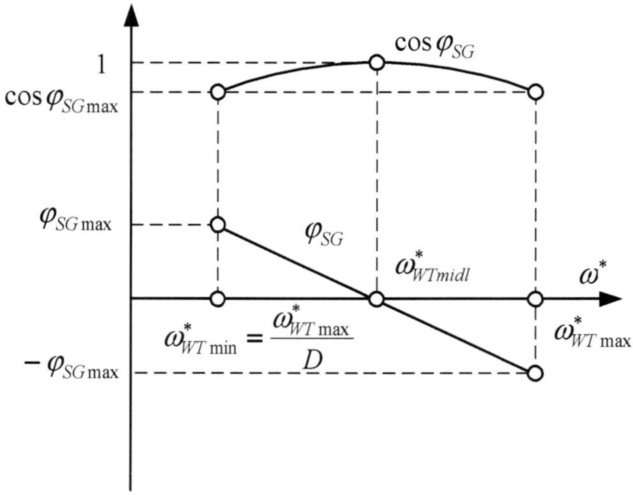
1
$\cos\varphi_{SG}$
$\cos\varphi_{SG\max}$
$\varphi_{SG\max}$
φ_{SG}
ω^*_{WTmidl}
ω^*
$\omega^*_{WT\min} = \frac{\omega^*_{WT\max}}{D}$
$\omega^*_{WT\max}$
$-\varphi_{SG\max}$

Fig. 31.

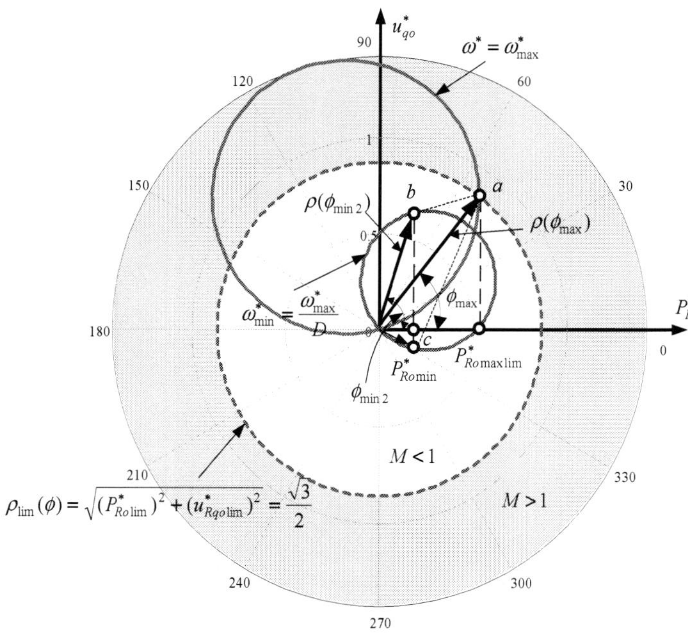
u^*_{qo}
$\omega^* = \omega^*_{\max}$
90
120
60
150
30
180
0
210
330
240
300
270
1
0.5
b
a
$\rho(\phi_{\min 2})$
$\rho(\phi_{\max})$
$\phi_{\max}$
$\omega^*_{\min} = \frac{\omega^*_{\max}}{D}$
P^*_{Ro}
c
$P^*_{Ro\min}$
$P^*_{Ro\max\lim}$
$\phi_{\min 2}$
$M < 1$
$M > 1$
$\rho_{\lim}(\phi) = \sqrt{(P^*_{Ro\lim})^2 + (u^*_{Rqo\lim})^2} = \frac{\sqrt{3}}{2}$

Fig. 32.

scenario allows us to work with $\omega^*_{WT\max} > \omega^*_{\max\lim}$ remaining in the second mode ($\omega^*_{\max\lim}$ is defined according to (44)). For this operating point with a maximum power of WT $P^*_{WTo\max}(\omega^*_{WT\max})$ is compatible with the maximal achievable power $P^*_{Ro\max}$ (46), (47). In addition, we require that the power $P^*_{Ro\max}$ corresponds to $M=1$ (fig.32), i.e. $P^*_{Ro\max} = P^*_{Ro\max\lim}$.

Angle $\phi_{\max}$ (fig.32) is determined from the equation $\dfrac{d\rho(\phi)}{d\phi} = 0$, $\phi_{\max} = \dfrac{\pi}{4} - \dfrac{\varphi_{SG}}{2}$.

We will find the frequency of rotation at which the equality $P^*_{Ro\max} = P^*_{Ro\max\lim}$ is realized, from the equation: $\rho(\phi_{\max}) = \sqrt{3}/2$, it follows that

$$\omega^*_{\max} = \frac{\sqrt{3}}{2}\left[\sec(\varphi_{SG})\sin\left(\frac{\pi}{4} + \frac{\varphi_{SG}}{2}\right)\right]^{-1}. \tag{48}$$

Based on (47) ÷ (48) $P^*_{Ro\max\lim} = \dfrac{\sqrt{3}}{2}\cos\left(\dfrac{\pi}{4} - \dfrac{\varphi_{SG}}{2}\right)$.

Then we require that

$$\omega^*_{WT\max} = \omega^*_{\max}; \quad P^*_{WTo\max} = P^*_{Ro\max\lim}.$$

Proceeding from the equation (20) $\gamma = P^*_{Ro\max\lim}/(\omega^*_{\max})^3$, $P^*_{WTo}(\omega^*) = \gamma(\omega^*)^3$.

In accordance with fig.31 we take $\varphi_{SG} = -\varphi_{SG\max}$ when $\omega^* = \omega^*_{\max}$. The law of change of φ_{SG} in the operating range $\omega^* \in \{\omega^*_{WT\min}, \omega^*_{WT\max}\}$ according to fig.31 will look as follows

$$\phi_{SG}(\omega^*) = \varphi_{SG\max}\left(1 - 2\frac{\omega^* - \omega^*_{\min}}{\omega^*_{\max} - \omega^*_{\min}}\right),$$

where $\omega^*_{\min} = \omega^*_{WT\min} = \omega^*_{WT\max}/D = \omega^*_{\max}/D$.

Then

$$\omega^*_{\max} = \frac{\sqrt{3}}{2}\left[\sec(\varphi_{SG})\sin\left(\frac{\pi}{4} - \frac{\varphi_{SG\max}}{2}\right)\right]^{-1}; \quad P^*_{Ro\max\lim} = \frac{\sqrt{3}}{2}\cos\left(\frac{\pi}{4} + \frac{\varphi_{SG\max}}{2}\right).$$

The minimum power at $\omega^* = \omega^*_{WT\min}$: $P^*_{Ro\min} = P^*_{Ro\max\lim}/D^3$.

The locus corresponding to the frequency of rotation $\omega^* = \omega^*_{WT\min}$ is: $\rho(\phi) = \omega^*_{WT\min}\sec(\varphi_{SG\max})\sin(\phi + \varphi_{SG\max})$.

The angle $\phi = \phi_{\min}$ at $\omega^* = \omega^*_{WT\min}$ is determined from the equation

$$\omega^*_{WT\min}\sec(\varphi_{SG\max})\sin(\phi_{\min} + \varphi_{SG\max})\cos\phi_{\min} = P^*_{Ro\max\lim}/D^3$$

$$\phi_{\min 1} = \frac{1}{2}\arcsin\left[2\frac{P^*_{Ro\max\lim}}{D^2\omega^*_{WT\max}\sec(\varphi_{SG\max})} - \sin(\varphi_{SG\max})\right] - \varphi_{SG\max}; \tag{49}$$

$$\phi_{\min 2} = \pi/2 - (\phi_{\min 1} + \varphi_{SG\max}).$$

In the relation (49) angles $\phi_{\min 1}$ and $\phi_{\min 2}$ correspond to the 1st and 2d modes.

When the rotation frequency $\omega^*_{WT\min} \leftrightarrow \omega^*_{WT\max}$ changes the two trajectories are possible (fig.32), namely, « $a \leftrightarrow c$ » and « $a \leftrightarrow b$ » with the first corresponding to the system in the 1st mode, and the second - in the 2nd mode.

As already noted, the first mode is characterized by the low value of power factor and the big value of current. For this reason, the second trajectory is desirable, i.e. work in the second mode. In this case:

$$M_d(\omega^*) = 2P^*_{WTo}(\omega^*)/\sqrt{3}; \quad M_q(\omega^*) = \omega^*/2 + \sqrt{(\omega^*/2)^2 - [P^*_{WTo}(\omega^*)]^2 + \omega^* tg[\varphi_{SG}(\omega^*)]P^*_{WTo}(\omega^*)}.$$

For $\varphi_{SG\max} = \pi/12$, $D = 2$ and $D = 3$ the result of calculation in fig. 33, when $\omega^* \in \{\omega^*_{WT\min}, \omega^*_{WT\max}\}$.

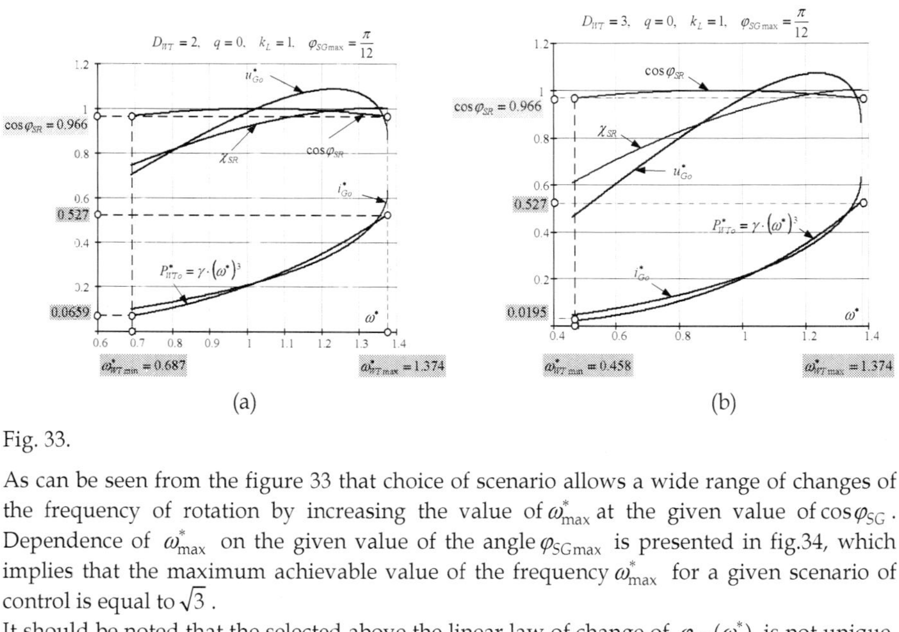

Fig. 33.

As can be seen from the figure 33 that choice of scenario allows a wide range of changes of the frequency of rotation by increasing the value of $\omega^*_{\max}$ at the given value of $\cos\varphi_{SG}$. Dependence of $\omega^*_{\max}$ on the given value of the angle $\varphi_{SG\max}$ is presented in fig.34, which implies that the maximum achievable value of the frequency $\omega^*_{\max}$ for a given scenario of control is equal to $\sqrt{3}$.

It should be noted that the selected above the linear law of change of $\varphi_{SG}(\omega^*)$ is not unique. In that case, if for the area of installing of WPI the prevailing wind speed is known, then the frequency of rotation of the shaft of WT is calculated and at an obtained frequency the point with $\cos\varphi_{SG} = 1$ is selected. The law of changes the function $\varphi_{SG}(\omega^*)$ can be optimized according to the change in the winds, with equality $\cos\varphi_{SG}$ at the extreme points of the operating range $\{\omega^*_{WT\min}, \omega^*_{WT\max}\}$ is not obligatory.

Thus, the scenario of the WPGS system working according to the given law of change of $\cos\varphi_{SG}$ with change of ω^*_{WT} allows to increase the maximum operating frequency of rotation while maintaining the 2-second mode, which is characterized by relatively high value of power factor.

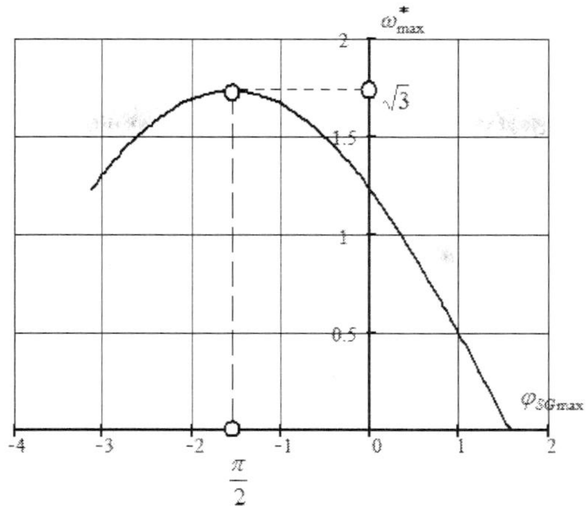

Fig. 34.

4. Basic power indicators in the circuit "voltage inverter - electrical network"

The schematic diagram of the circuit "voltage inverter - electrical network» taking into account the accepted assumptions is shown in fig.35. The estimated mathematical model of the electrical circuit is shown in fig. 4.

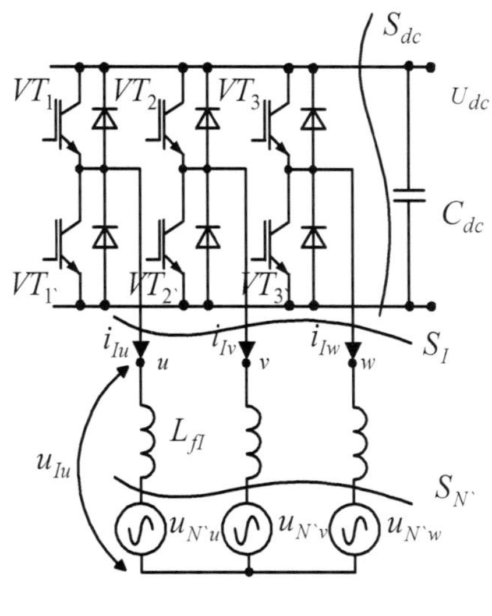

Fig. 35.

Voltage of the electrical network changes according to the law:

$$u_{N'm} = U_{N'} \cos[\upsilon - (m-1)2\pi/3]; \; \upsilon = \Omega t; \quad m = 1,2,3\,(u,v,w).$$

Change laws of the inverter control signals are $u_{Icm} = u_c \cos(\theta_m)$, where $\theta_m = \upsilon - (m-1)2\pi/3 + \varphi_{Ic}$;
Taking into account the accepted assumptions the mathematical model of an electric circuit in rotating system of coordinates, under condition of orientation on an axis of voltage of an electric network q, will look like:

$$\boldsymbol{u}_I - \boldsymbol{u}_{N`} = \boldsymbol{r}_I \boldsymbol{i}_I + L_I \frac{d}{dt} \boldsymbol{i}_I + L_I \boldsymbol{\Omega} \cdot \boldsymbol{i}_I , \tag{50}$$

where: $\boldsymbol{u}_I = \begin{bmatrix} u_{Id} & u_{Iq} \end{bmatrix}^t$, $\boldsymbol{u}_{N`} = \begin{bmatrix} 0 & U_{N`} \end{bmatrix}^t$, $\boldsymbol{i}_I = \begin{bmatrix} i_{Id} & i_{Iq} \end{bmatrix}^t$ - vectors of the inverter voltage and the mains voltage, vector of the inverter currents; $U_{N`}$ - the peak value of network voltage; $\boldsymbol{r}_I = diag\{ r_I, r_I \}$, r_I - the resistance of inductance of power filter and of the transformer windings; L_I - the equivalent inductance of the power filter and the transformer leakage inductance; $\boldsymbol{\Omega} = \begin{bmatrix} 0 & -\Omega \\ \Omega & 0 \end{bmatrix}$, Ω - circular frequency of the network voltage.
Neglecting the active resistance the ratio (50) can be written in a scalar form

$$u_{Id} \approx L_I \frac{di_{gd}}{dt} - \Omega \cdot L_I i_{Iq}, \quad u_{Iq} - U_{N`} \approx L_I \frac{di_{Iq}}{dt} + \Omega \cdot L_{d\Sigma} i_{Id}. \tag{51}$$

A mathematical model of the inverter will be determined by the relations (5) ÷ (8). In these relationships we take: $U_{dc} = \sqrt{3} \cdot U_{N`} \cdot \delta_{Udc}$, where $\sqrt{3} \cdot U_{N`}$ - is the minimal possible voltage in a direct current link with SPWM, δ_{Udc} - is excess of the minimal possible voltage of a link of a direct current.
As before, in order to preserve the universality of the results of the analysis, we introduce the following relative units: $E_б = U_{N`}$; $\omega_б = \Omega$; $X_б = \omega_б L_I$; $I_б = I_{кз} = E_б / X_б$; $S_б = 3E_б I_б / 2$; $a_I = \omega_{cI} / \Omega$; where ω_{cI} - a cyclic frequency of the PWM inverter.
Taking into account relative units the equation (51) will become:

$$u_{Id}^* = \frac{di_{Id}^*}{d\upsilon} - i_{Iq}^*, \quad u_{Iq}^* - 1 = \frac{di_{Iq}^*}{d\upsilon} + i_{Id}^*,$$

where $\upsilon = \Omega t$.
The voltages u_{Id}^* and u_{Iq}^* are determined by the relations

$$u_{Id}^* = u_{Ido}^* + \Delta u_{Id}^*; \quad u_{RIq}^* = u_{Iqo}^* + \Delta u_{Iq}^*;$$

$$u_{Ido}^* = \frac{\sqrt{3}}{2} \delta_{Udc} M \sin(-\varphi_{Rc}) = -\frac{\sqrt{3}}{2} \delta_{Udc} M_d; u_{Iqo}^* = \frac{\sqrt{3}}{2} \delta_{Udc} M \cos(-\varphi_{Rc}) = \frac{\sqrt{3}}{2} \delta_{Udc} M_q;$$

here u_{Ido}^*, u_{Iqo}^* - the orthogonal components in the d and q coordinates of the fundamental harmonic of inverter voltage; Δu_{Id}^*, Δu_{Iq}^* - the orthogonal components in the d and q coordinates of the high-frequency harmonics of inverter voltage.

We will define the high-frequency harmonics for SPWM from the relations (14).
The equation for the inverter current can be represented as the sum of the fundamental (i^*_{Ido} , i^*_{Iqo}) and the high frequency (Δi^*_{Id} , Δi^*_{Iq}) harmonics $i^*_{Id} = i^*_{Ido} + \Delta i^*_{Id}$; $i^*_{Iq} = i^*_{Iqo} + \Delta i^*_{Iq}$.
The fundamental harmonic of the inverter current is determined by the relation

$$i^*_{Iqo} = -u^*_{Ido}; \quad i^*_{Id} = u^*_{Iqo} - 1.$$

The high-frequency harmonics of the inverter current for SPWM are determined from the relations (16).
We assume such a control law of inverter, when the WPI in electrical circuit generates only an active power. Then the vector diagram for the fundamental harmonic of current and voltage will have the form shown in fig.36.
Under such a control $u^*_{Iqo} = 1$; $i^*_{Io} = i^*_{Iqo} = -u^*_{Ido}$; $i^*_{Id} = 0$. Generated in the electrical network active power is:

$$P^*_{No} = i^*_{Iqo} = i^*_{Io} = -u^*_{Ido}. \tag{52}$$

Vector diagram for the orthogonal components (M_d, M_q) of the inverter control signal in «*d q*» coordinates is presented in fig.37
The quantities M_d, M_q and φ_{Ic} are determined by the relations:

$$M_q = 2/(\sqrt{3}\delta_{Udc}), \quad M_d = 2P^*_{No}/(\sqrt{3}\delta_{Udc}), \quad \varphi_{Ic} = arctg M_d / M_q = arctg P^*_{SNo}. \tag{53}$$

The linear range of work of the inverter is limited by a condition:

$$\sqrt{(M_d)^2 + (M_q)^2} \le \begin{cases} 1 - SPWM; \\ \dfrac{2}{\sqrt{3}} - SVPWM. \end{cases}$$

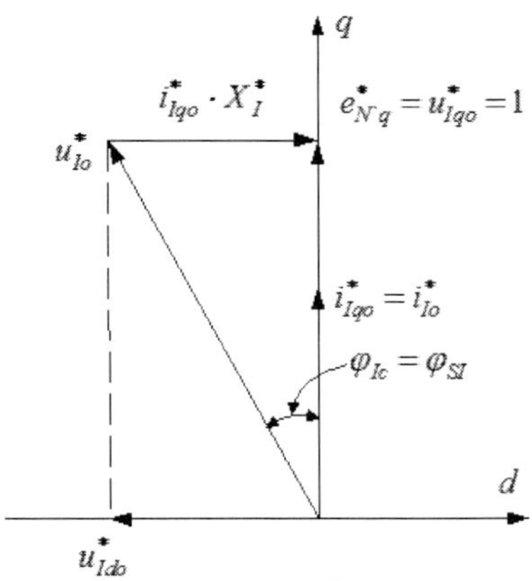

Fig. 36.

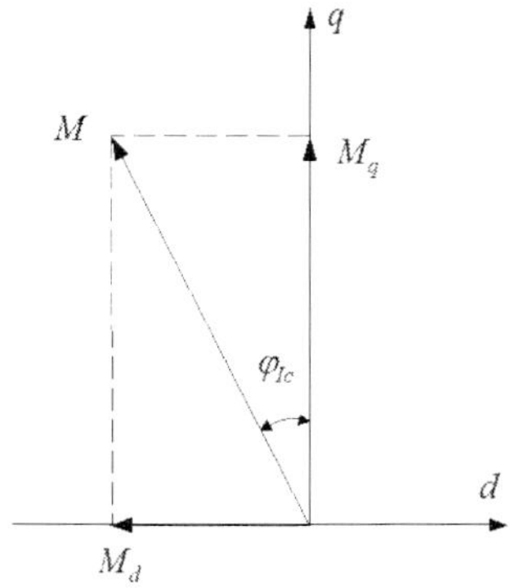

Fig. 37.

From (53) (in the case of equality), we obtain an expression for the maximum active power ($P^*_{No\max}$), which can be transferred to the electricity grid without distortion of the current.

$$P^*_{No\max} = \begin{cases} \sqrt{\left(\frac{\sqrt{3}}{2}\delta_{Udc}\right) - 1}, 1SPWM \\ \sqrt{(\delta_{Udc})^2 - 1}, -SVPWM. \end{cases}$$

The dependence of $P^*_{No\max}$ on the value of δ_{Udc} is shown in fig.38, which implies that the minimum value of $\delta_{Udc\min}$ at which the generation of active power begins is given by:

$$\delta_{Udc\min} = \begin{cases} \frac{2}{\sqrt{3}} - SPWM; \\ 1 - SVPWM. \end{cases}$$

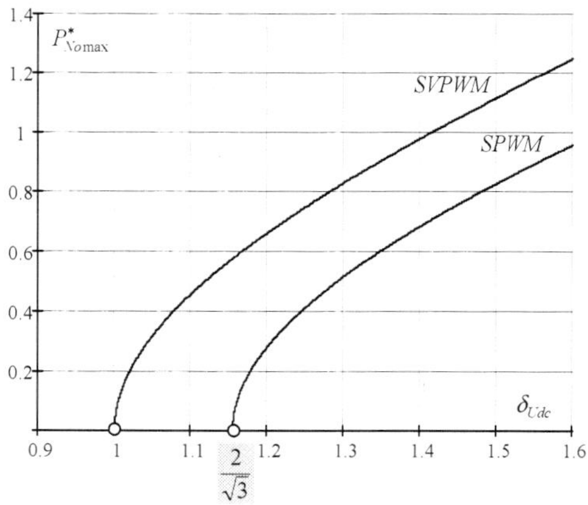

Fig. 38.

Dependence of the active power P^*_{No} from the relative values of voltage in the DC link $\delta_{Udc} = U_{dc} / \sqrt{3}U_N$, and the depth (index) of modulation M can be found from the following relation:

$$P^*_{No} = \sqrt{(\sqrt{3}\delta_{Udc}M / 2)^2 - 1}\,. \tag{54}$$

Taking into account the active losses in the output circuit of the generating system an expression for the fundamental harmonic current and the active power will look like

$$i^*_{Io} = P^*_{No} = \frac{1}{1+(\omega^*_R)^2}\cdot\left\{-\omega^*_R + \sqrt{[1+(\omega^*_R)^2]\left(\frac{2}{\sqrt{3}\delta_{Udc}M}\right)^2 - 1}\right\};$$

where $\omega^*_R = R/X_б$, R - the equivalent active resistance of the inverter phase.
Graph of this dependence (at $\omega^*_R = 0$) for SPWM and SVPWM is shown in fig.39, which implies that the adjustment range of active power decreases with decreasing of δ_{Udc}. It should be noted that when working on electrical network application of SVPWM can significantly increase the active power. As follows from fig.40 for each the value of δ_{Udc} there is a minimum value of modulation depth $M_{\min}$ below which the generating active power is equal to zero $M_{\min} = 2/(\sqrt{3}\delta_{Udc})$.
The dependence of $M_{\min}$ on δ_{Udc} is shown in fig.40.

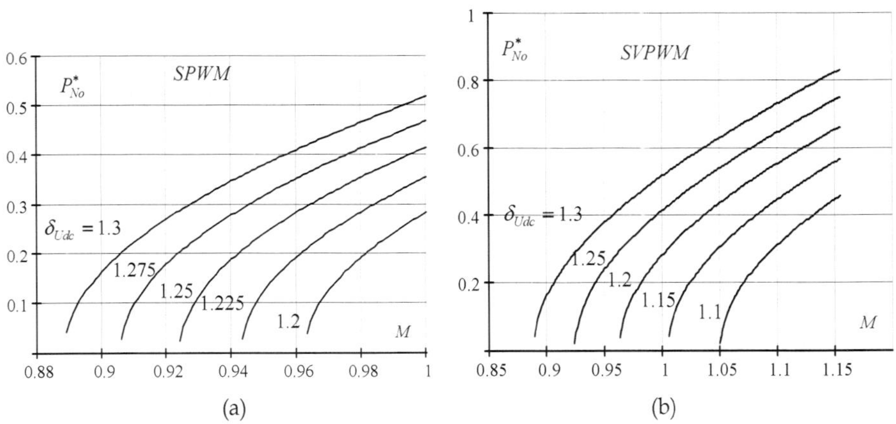

Fig. 39.

We determine how vary the coefficients of harmonics (THD_{iI}) and distortions (ν_{iI}) of the inverter current. In accordance with the relations (52) and (19) the effective value of the fundamental harmonic of inverter current ($i^*_{Io,rms}$), its fluctuating components ($\Delta i^*_{I,rms}$) and the total effective value ($i^*_{I,rms}$) are defined as follows:

$$i^*_{Io,rms} = P^*_{No}/\sqrt{2}\,,\ \Delta i^*_{I,rms} = \left(\frac{\sqrt{3}\delta_{Udc}}{\pi\cdot X^*_\Sigma}\right)\cdot\left(J_1(\pi\cdot M)^2\frac{a_I^2+1}{(a_I+1)^2(a_I-1)^2}\right)^{\frac{1}{2}},\ i^*_{I,rms} = \sqrt{(i^*_{Io,rms})^2 + (\Delta i^*_{I,rms})^2} \tag{55}$$

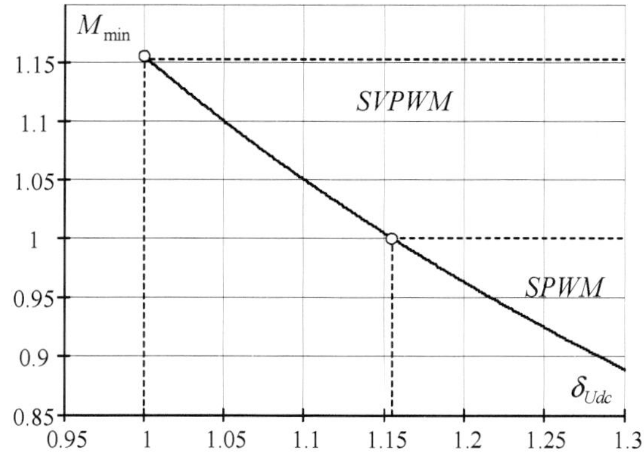

Fig. 40.

In fig.41 the dependence of the magnitudes THD_{iI} and ν_{iI} as a function of the depth of modulation M with $\delta_{Udc} = 1.3$ is presented. As follows from this figure, the qualitative indicators of a current are much worse with a decrease in modulation depth, while the value $THD_{iI} = 0.05$ is reached at $M \to 1$ and $\delta_{Udc} \geq 1.3$. For the Russian standards, the quality of the generated electric current in the WPI network must fulfill the condition $THD_{iI} \leq 0.05$. It should also be noted that THD_{iI} practically does not depend on the inductance L_I and is determined only by the multiplicity of frequencies a_I and the ratio of voltages δ_{Udc}.

Taking into account that the phase of the inverter current coincides with the phase of voltage of the electrical network, as well as a sinusoidal change of the voltage, taking into account the relations (13) we obtain the following expression for the power factor in the cross section S_N: $\chi_N = P^*_{No} / S^*_N = \nu_{iI}$.

We define the changes in THD_{iI} and ν_{iI} in the WPI, as function of the frequency of rotation of the shaft of WT. We assume

$$P^*_{WTo} = \gamma \cdot (\omega / \omega_{WT\max})^3 . \qquad (56)$$

We define the coefficient γ according to the condition $P^*_{WTo}(\omega_{WT\max}) = P^*_{No\max}$, then $\gamma = P^*_{No\max}$. Based on the (54) and (56) we obtain the dependence of modulation depth on the frequency of rotation of the shaft of WT

$$M = \frac{2}{\sqrt{3}\delta_{Udc}} \sqrt{\left[P^*_{No\max} \cdot \left(\frac{\omega}{\omega_{WT\max}} \right)^3 \right]^2 + 1} . \qquad (57)$$

In fig.42 the dependence of M on $\omega / \omega_{WT\max}$ for the two types of modulation (SPWM and SVPWM) is presented. It implies that the modulation depth varies slightly.

Knowing the dependence of M on $\omega / \omega_{WT\max}$, we can determine the changes of the qualitative characteristics of the generated energy as the function of the frequency of rotation of the shaft of WT. For this we use the relations (55) (57).

In fig.43 graphs of THD_{iI} and ν_{iI} on $\omega / \omega_{WT\max}$ for SPWM are presented.

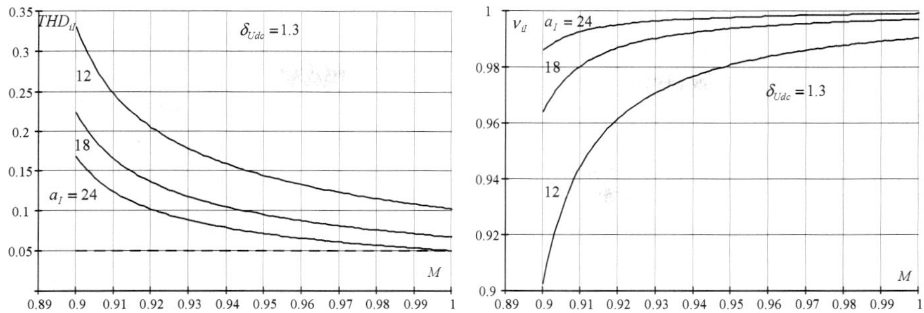

(a) (b)

Fig. 41.

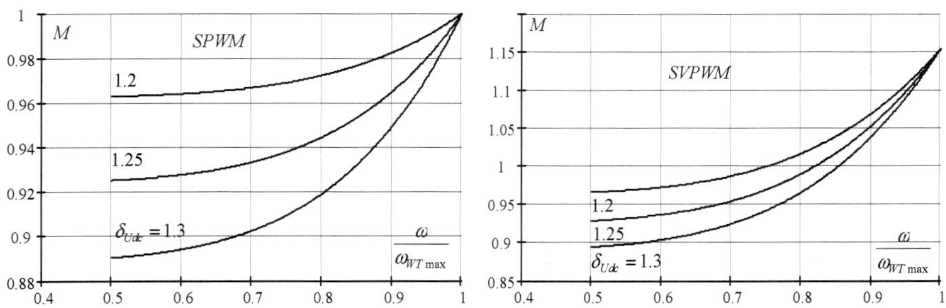

(a) (b)

Fig. 42.

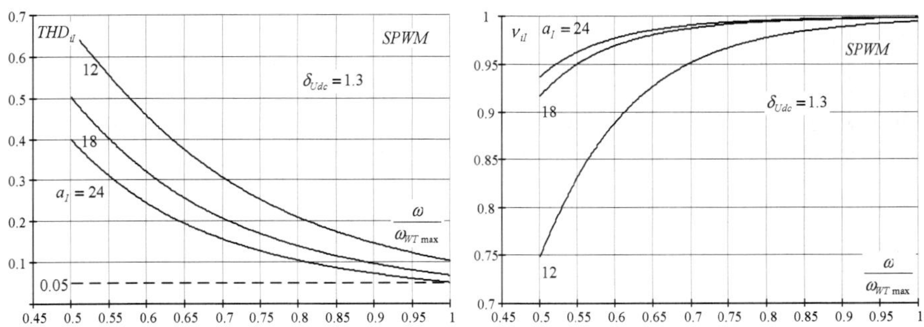

(a) (b)

Fig. 43.

Given that in the powerful WPI multiplicity of frequencies a_I is limited by the dynamic losses in the semiconductor switches, we can conclude that it is impossible to satisfy the requirements of the quality of the generated energy by increasing the *PWM* frequency, as in the case of *SPWM*, and when *SVPWM*.
There is a positive impact of increase in the parameter δ_{Udc} on the quality of electric power, but it leads to a significant increase in the DC link voltage and, consequently, to increase of the installed capacity of the electric power converter.
The solution of this problem can be modification of the voltage inverter circuit or reducing the range of power change implementing *SVPWM*.
When constructing the WPI of MW capacity and more one should be guided by the multilevel inverter circuits.
However, one of the ways of solving the problem may be the shunting "m" inverters connected to a DC voltage source controlled by the same modulating signal at the output of each phase, but the time of entry gates are shifted relative to each other in frequency ω and at an angle $2\pi/m$ that is carried out, for example, in bilateral sinusoidal PWM introducing of m sources of reference signals with a specified phase shift. This decision, in addition to improving the quality of energy, increases its level, providing a modular inverter and system as a whole. The modular principle of the considered WPGS also has the advantage that it can save about the same level of efficiency of large and small rotational velocities, which provided by the different numbers of modules in the function of the frequency of rotation of the shaft of the wind turbine. Fig.44 shows an example of parallel connection of m inverters.
In parallel connection, each inverter independently from the other forms voltage u_{Ii}, which value can be determined in accordance with the relation (6). This parallel connection is possible if the parallel channels have no common inductance, i.e. each channel operates on the electrical network, or at the entrance of a transformer there is a capacitive filter on which high-frequency ripple voltage is practically equal to zero.
Each of the phases of such a system can be represented as an equivalent circuit (fig.45). For this scheme the following relations are fair

$$i_I = \sum_{i=1}^{m} i_{Ii}\,,\ u_{eI} = \frac{1}{m}\sum_{i=1}^{m} u_{Ii};\qquad L_{Ie} = \frac{L_I}{m}\,,\ i_{dc} = \sum_{i=1}^{m} i_{dci}\,,$$

where u_{eI}, L_{Ie} - the equivalent internal voltage and inductance.
The total generated power of the system of m channels will be determined by the ratio $P_{N`o} = \sum_{i=1}^{m} P_{N`oi}$, where $P_{N`oi}$ - the active power of the *i*-th channel.
Fig.46 shows an example [2] of calculated the equivalent inverter voltage waveform (u_{Ie}) and locus of voltage u^*_{Idq} at $m = 3$. Locus structure is similar to the equivalent multi-level inverter. This conclusion is illustrated by the amplitude-frequency spectrum of the current i_I given at fig.47 for the three options $m = 1,3$ (at $a_I = 20$).
Increase in the number of channels leads to exclusion from the spectrum of the current of groups of combinational harmonics with frequencies $\nu = n\cdot\omega_{kI} \pm p\cdot\Omega;\quad n < m$.

Calculations show that a reduction in the harmonics of load current $THD_{iI(m)}$ when you turn on m parallel channels can be estimated by the ratio $THD_{iI(m)} = THD_{iI(1)} / m^2$, where $THD_{iI(1)}$ - the coefficient of harmonics with $m = 1$.

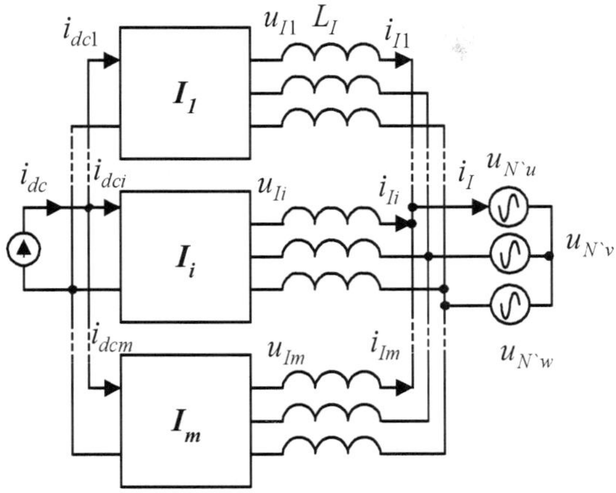

Fig. 44.

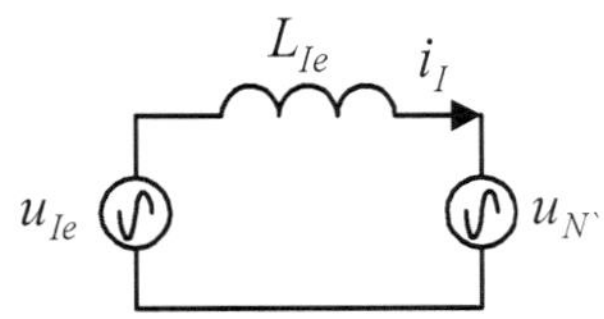

Fig. 45.

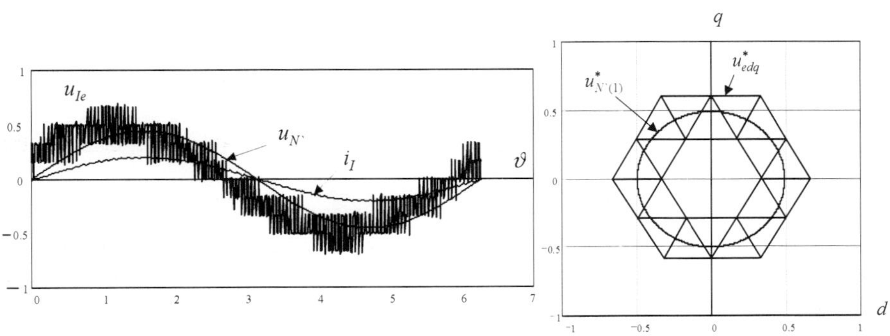

(a) (b)

Fig. 46.

Not identical distribution of the active power between the "two neighboring" channels can be evaluated using the following relation

$$\delta_{PN} = \frac{P^*_{No(i)} - P^*_{No(i+1)}}{P^*_{No(i)} + P^*_{No(i+1)}} * 100\% \approx \frac{1}{4}\left(\frac{\pi}{m \cdot a_I}\right)^2 100\% .$$

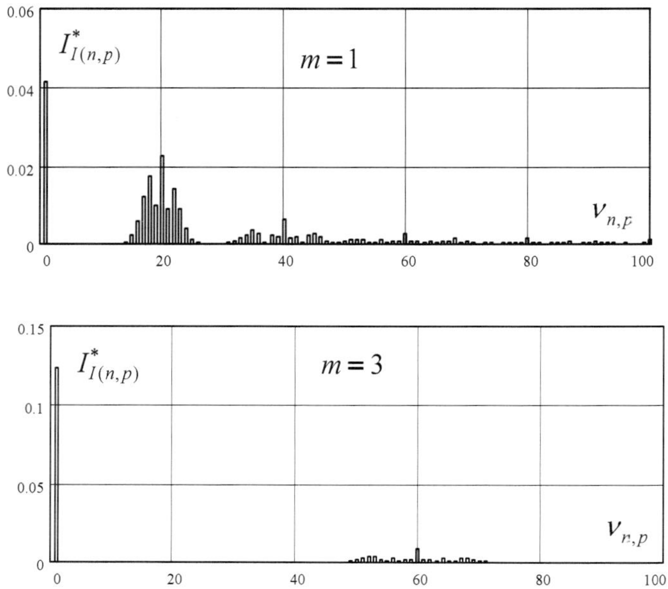

Fig. 47.

The dependence of δ_{PN} on the multiplicity of frequencies for different m is shown in fig.48, which implies that when a_I <8 and m = 2, imprecision in the distribution of active power does not exceed one percent. With the increasing m the error decreases.
Thus, with the reasonable accuracy it can be assumed that

$$P^*_{No(1)} = .. = P^*_{No(i)} = .. = P^*_{No(m)}, \quad P^*_{No} = \sum_{i=1}^{m} P^*_{No(i)} .$$

The analysis shows that the parallel connection of the channels leads to a decrease in the number of groups of combination harmonics in the amplitude-frequency spectrum i_{dc}, the amplitude of high-frequency harmonics decreases. The power factor ($\chi_{Sdc(m)}$) and the inactive power (Q_{Sdc}) in the section S_{dc} at the voltage inverter input (fig.44) subject to m parallel channels can be estimated using the relations of the form [2]:

$$\chi_{Sdc(m)} = \frac{P_{No}}{\sqrt{(P_{No})^2 + (Q_{Sdc})}} \approx \left[\sqrt{1 + \frac{1-(\chi_{Sdc(1)})^2}{(\chi_{Sdc(1)})^2} \cdot \frac{1}{(m)^2}} \right]^{-1}, Q_{Sdc} = P_{No} \cdot \sqrt{1/(\chi_{Sdc(m)})^2 - 1}, (58)$$

where $\chi_{Sdc(1)}$, $\chi_{Sdc(m)}$ - power ratios in the cross section S_{dc} of one channel and m channels respectively.

From (58) we can conclude that an increase in the number of parallel channels included in the input power increases the component of active power and virtually unchanged a reactive power. Thus, the parallel connection of m channels leads to an increase in the power generated by m times, to a decrease of ratio of harmonics of the generated current by factor of about m^2 while maintaining the multiplicity of frequencies a_I, the specific reactive power of capacitor (C_f) in the chain of dc decreases by m times. When we save the value of the coefficient of harmonics of generated current the shunting channels can reduce the multiplicity of frequencies a_I by approximately m^2 times.

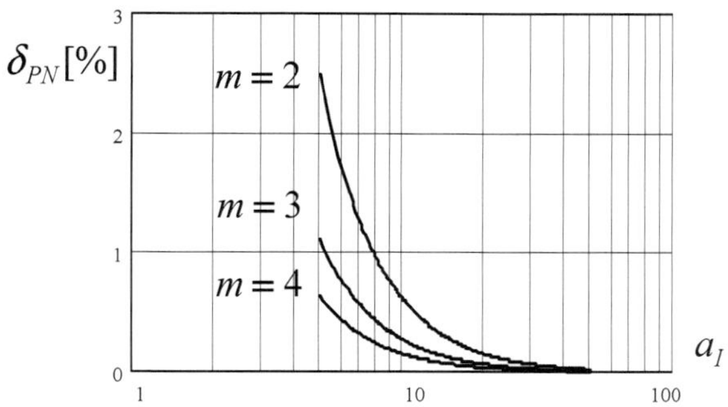

Fig. 48.

Thus, the analysis of energy data in the circuit "voltage inverter - electrical network" provides an opportunity for an analytical assessment of the main parameters and detects trends in integrated settings, and characteristics when changing operating modes and control algorithms. When using the voltage inverter of WPI on the electrical network of the power quality of output current varies considerably. So, for example, at a range of change of frequency of rotation of a shaft of the wind turbine $D \geq 2$ it is practically impossible to provide at real parameters $THD_{iI} \leq 0.05$ of the power scheme of the inverter in all range of frequencies of rotation. This problem is solved by using SVPWM, multi-level converter circuit, or using the modular principle of the converter. The modular construction principle allows increasing the efficiency of WPGS at low frequencies by turning off the multiple channels. This solution allows extending the working range of wind speeds of WPI. The principle of modular converter can be extended to WPGS; in this case the generation system is constructed, as shown in fig.49. Here, as an example, a system of 4-channel generation is presented. Fig.50 shows the change in power in one channel to generate at different ranges of speed of shaft of WT. It would start one channel, then two, three and four. Increase in the number of working modules reduces the current range of each module, which increases the efficiency of the system at low wind speeds; this increases the maximum capacity of the system, generated by maintaining its high quality. According to this principle the WPI «Raduga-1A" 1 MW was designed and built near Elista.

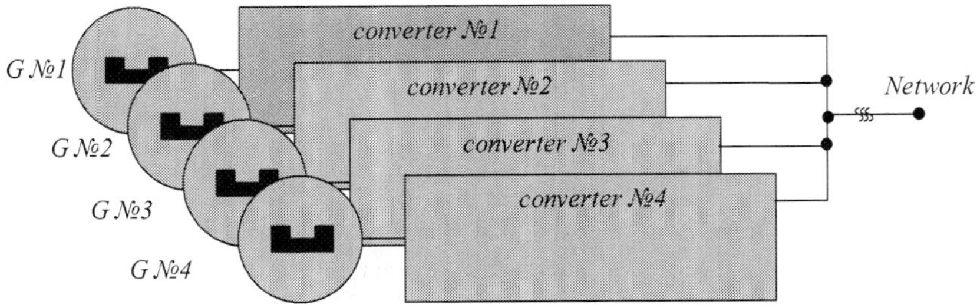

Fig. 49.

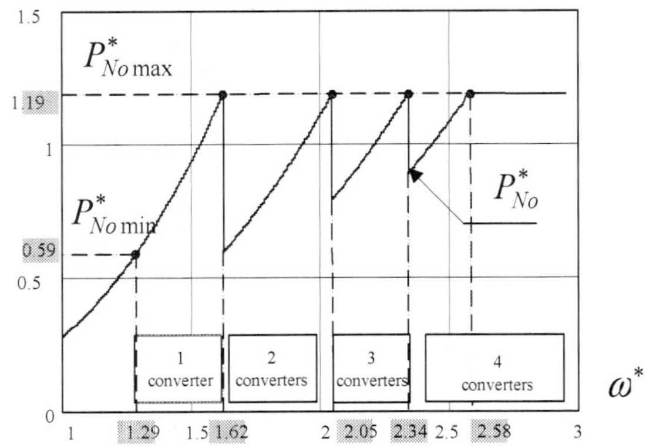

Fig. 50.

5. Conclusions

1. *A mathematical model for analysis of energy characteristics of electric power generation system consisting of a synchronous generator with excitation from permanent magnets, the active rectifier and the voltage inverter with PWM is considered.*
2. *Various algorithms to control the active rectifier and inverter for variable speed wind turbine shaft are analyzed.*
3. *Analytical relations for the calculation of currents, voltages and power generation in the system are obtained.*
4. *Recommendations on the choice of control algorithms and structural circuits of the generation electrical energy at a variable speed shaft WT are given.*

6. References

[1] Corn G., Corn T. Mathematics handbook (for science officers and engineers). - M. - the Science, 1974. in Russian

[2] Kharitonov S. A. Integrated parameters and characteristics of voltage inverters in structure of generating systems of an alternating current such as "variable speed - constant frequency" for wind-energetic installations / Scientific bulletin NSTU, Novosibirsk, 1999. 92-120 p., in Russian.

3

Speed Sensorless Vector Control of Permanent Magnet Wind Power Generator – The Redundant Drive Concept

Tero Halkosaari
Vacon Oyj
Finland

1. Introduction

Permanent Magnet motors (PM-motors) have become more and more popular, especially in low speed and high speed applications. The conventional motor drive, consisting of a standard speed induction motor with a gearbox, can be replaced of a very low speed, or a very high speed permanent magnet motor having no gearbox. The advantages are, for example, an increased efficiency and the reliability. Also total weight, noise and costs of the whole drive system are reduced.

Permanent magnet motors can be designed efficiently for very low running speeds by increasing the pole number. This makes permanent magnet machines very attractive for wind power generators, because in high power wind mills (> 1 MW), a wind turbine rotor is typically rotating about 10 to 20 rpm.

One important aspect of motor drives is the reliability of the drive system. Reliability comes even more important in installations, where the maintenance is difficult i.e. in wind mill nacelles. The reliability of the PM wind power generators can be increased by using multiple stator modules, or stator segments, independent of each other. These segments can be considered, for example, as independent stator parallel windings which are each fed by an own frequency converter. If one of the frequency converters fails, other ones can continue the operation while the failed units are changed. Hence, there is no long period total power interruption, because the wind turbine operation can continue anyway at a reduced power level.

This chapter describes the speed sensorless vector control of a variable speed multi-module PM wind power generator. First, the redundant drive concept is introduced based on the modular drive concept. Then, the speed sensorless vector control theory applied to wind power generators is discussed. Some emphasize is given also for the generator inverter overvoltage control, which is important in case of a grid lost. Theory and laboratory tests are shown for a 4x1 kW PM-generator. Some test results are shown also for the real size wind power generator, which is a 3.8 MW 17.5 rpm radial flux PM-generator consisting of 3 stator modules.

2. Drive system

Drive topologies used in high power wind generators are still mainly conventional solutions i.e. asynchronous induction generators with a gearbox coupling. These are limited speed range slip-ring induction generators directly connected to the power grid and having a resistive slip-control, variable speed double-fed slip-ring induction generators with an active slip-power recovery drive, or variable speed fully controlled induction generators with a 4-quadrant drive. However, nowadays development interests are directed towards variable speed direct drives which are externally excited synchronous generators, or now more often 4-quadrant drive permanent magnet synchronous generators. The advantage of a variable speed drive is that the power can be generated already at small wind speeds and the power production can be optimized according to the wind and the generator speed. This seems to be a trend also in smaller (20kW) wind power generators.

2.1 4-quadrant VSI drive topology

A direct drive variable speed PM wind generator system (Fig. 1) consists typically of a multi pole PM-generator, the generator side INverter Unit (INU), and the supply side Active Front End (AFE). In a Voltage Source Inverter (VSI) topology, drives are connected through the dc-link having a capacitor bank which is the momentary energy storage. LCL-filter is used in the supply side to suppress switching frequency harmonics generated by the AFE. Generator side inverter controls the generator torque according to the power reference of the generator based on the wind speed, while the active front end controls the dc-link voltage and the active and reactive power flow to the utility grid. The generator side inverter can also be used to rotate the generator in motoring mode, for example, in turbine service and installation situations.

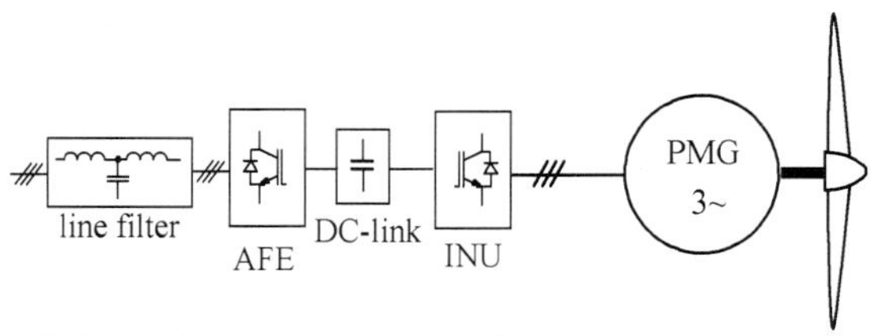

Fig. 1. A 4-quadrant VSI wind power generator drive.

2.2 Redundant PM-generator concept

A redundant PM-generator comprise multiple independent stator modules or segments preferably with a galvanic isolation from each other. Each module will be driven by an own inverter connected to a common dc-link and to the power grid with parallel connected AFEs. Alternatively each inverter can have an own dc-link and AFE (Fig. 2). Generator stator modules can consist of, for example, stator parallel winding groups carrying the same stator current. If winding groups are located in adjacent stator slots, there will be a voltage and a current phase shift between the winding groups. This phase shift can be compensated by the drive control and hence is not usually a problem. On the other hand, stator windings can be located in own groups in similar stator segments. For example, in (Albrich & Fuchs, 2002), it is proposed that instead of the conventional 16-pole stator winding, windings can be arranged in four 4-pole segments, in which case segments are independent from each other and can be run by an own inverter.

In all arrangements, if there is a galvanic isolation between the segment windings, there is no need to synchronize IGBT switching instants of different segment inverters, unless there is a strong magnetic or capacitive cross-coupling between the segment windings. The

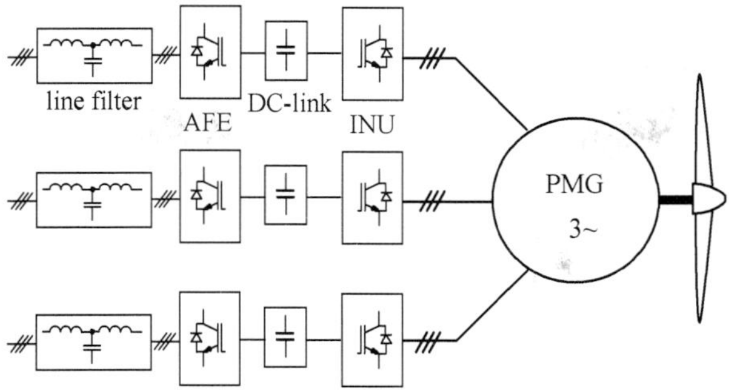

Fig. 2. Modular redundant PM wind generator drive.

cross-coupling can occur, for example, if segment windings are located in same stator slots. Strong cross-coupling means that the voltage or current transient in one drive segment will induce a transient current component to the other segment and the generator current control can fail due to the disturbance current component. Synchronization of IGBT switching instants is possible but requires more control hardware and makes the drive system more complicated. In case of drive segments are phase shifted, only partial synchronization is possible due to the phase shifted output voltages. Also if there is an own current control in each drive, a perfect synchronization is not possible due to the slight differences in segment's instantaneous voltages.

Fig. 2 modular system has several advantages. If one of the segment drive inverters fails, it can be separated from the system and the generator can continue on operation using the reduced power until the failed unit is changed. One main advantage is also that the system power can be increased by increasing the number of the stator segments while the generator power is increased. Hence there is necessarily no need to increase the power of individual inverters, just the amount of the inverters.

3. Control of a redundant PM-generator

3.1 Master-follower control concept with an encoder feedback

Optimally controlled redundant PM-generator requires the rotor speed and position information for each segment drive. Typically used position sensors are resolvers and serial data type (e.g. EnDat, SSI) absolute encoders. Typically, these encoders can deliver also the speed signal based on the separate incremental encoder pulses, sin/cos-signals or just the serial data transfer. Speed signal can also been derived from the resolver position signal with sufficient accuracy. Instead of absolute encoders, also incremental encoders can be used which are more common in industry. Then the rotor position can be calculated by summing incremental pulses, if the rotor start position is known. There are plenty of methods to identify the PM-generator start position (magnet pole position) either for standstill or rotating rotor. Also the reference-mark pulse (z-pulse) can be utilized to determine the rotor position because the z-pulse can be considered as a single step absolute encoder.

In high power wind generators, the cost of a speed and/or rotor position measurement utilizing the encoder or resolver is very small compared to the total costs of the whole wind turbine. Hence, it's mainly a question of the encoder and other system reliability, if the encoder is used for the generator control in order to have an accurate speed and shaft position measurement. In case of the encoder failure, the whole wind mill has to be stopped, if there are no backup devices. It is also difficult to avoid disturbances in encoder cables, because there is usually not much space to separate the power cables from the signal cables in a wind turbine nacelle. Anyway, some kind of speed measurement is required due to the safety but these can be robust low resolution devices, such as inductive sensors.

In a redundant drive concept, the simplest solution is to have multiple encoders, one for each of drive units. However, that it is not a mechanically robust and reliable solution, if there are more than two encoders. Another solution is to use one encoder and to repeat the encoder signal for each drive. Repeating encoder signals using commercial repeaters can be difficult depending on the type of encoder signals. A relatively simple solution (Fig. 3) is that one of the drives behaves as the repeater i.e. as a master drive which receives encoder signals and other command signals from the overriding system controller. The master drive sends further the processed speed and position information and other command signals to the other drives called followers. Hence, followers are able to track instantaneous currents of the master drive (Mäkinen & al, 2006).

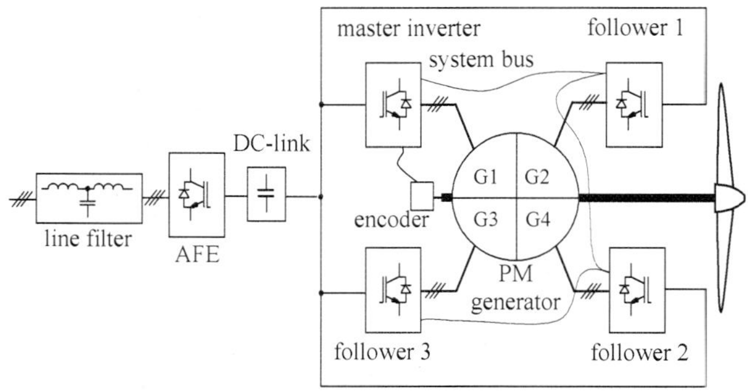

Fig. 3. Master-follower type control of 4-segment PM-generator with an encoder feedback.

Communication link between the master and follower drives can be done via the optical serial data link in order to avoid disturbances. The link can also be a galvanic bus, for example, a CAN-bus was tested in the drive test setup. The benefit of a galvanic bus is that the bus is always continuous from the first drive to the last drive and not depending on any auxiliary power like optical bus repeaters do. In case of the master drive inverter fails, one of the follower drives can be switched to be a master drive in order to continue the operation. In this master-follower configuration also the segment phase shift is easily implemented by shifting the follower current reference phase angle received from the master by the amount of each segment phase shift.

3.2 Speed sensorless control concept

In order to increase the drive system reliability, speed and position sensorless vector control can be used. In fact, the wind power generation is one of the applications which is very attractive for the sensorless vector control. First of all, there is no need to start the rotor at standstill or even to drive in a low speed area. Especially, driving the high torque at a low speed has been the most challenging and unreliable part of sensorless control methods. Typically, the wind generator is started and operated at speed above 15 % of the generator nominal speed. Hence, the flying start can be used to start the generator side inverter. The flying start is easily implemented, because the generator back e.m.f. voltage is already high enough to identify the rotor position and speed. Second, there is no need for a high torque at low speed region, because the generator power increases rapidly not until the speed is closer to the nominal speed of the generator. It is not usually a problem to identify the rotor position and the speed in high speed region whatever the torque is. In fact, even the speed estimation is not needed for the generator control because inverters are operating typically in torque control mode and the speed control loop is not used. However, the generator speed is usually the essential information for the overriding generator controller. Third, because each of the inverters is identifying independently the rotor speed and the position, there is no need for the communication between the units (Fig. 4). Only commands needed for each drive are low speed commands from the overriding control system which can be delivered to each drive using drive I/O-signals or industry standard buses e.g. Profibus.

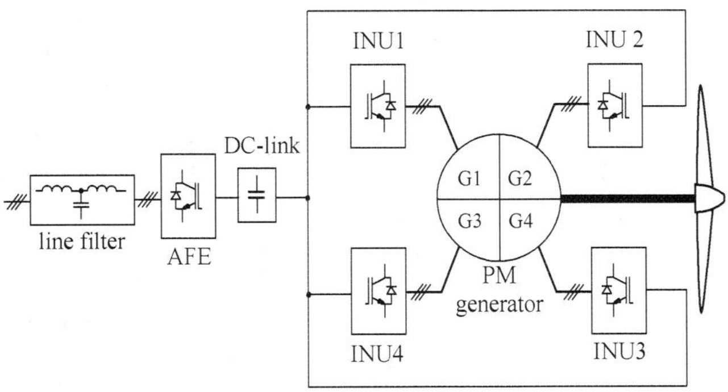

Fig. 4. Speed and position sensorless control of a 4-segment PM-generator.

3.3 Speed sensorless vector control

Speed sensorless vector control means that the generator current is controlled into a right angle with respect to the rotor magnet flux for the maximal torque production. Generator torque equation is defined as

$$T_e = \frac{3}{2} p(\underline{\psi}_s \times \underline{i}_s) = \frac{3}{2} p(\psi_m i_q + (L_d - L_q) i_d i_q), \quad (1)$$

where the first term is the torque produced by the magnet flux Ψ_m and the last term is the reluctance torque. According to equation (1), the right control angle means for smooth air gap machines ($L_d = L_q$) that the current phasor is leading the magnet flux phasor by 90° i.e. machine d-axis current is kept zero and the generator takes only the q-axis torque producing current (Fig. 5). For salient pole machines ($L_d < L_q$), some amount of additive reluctance torque can be produced by a negative d-axis current. This requires that the d-axis current is controlled as a function of q-axis current according to the Maximum Torque Per Ampere (MTPA) optimization. In order to divide the generator current into the d- and q-axis components, rotor magnet flux i.e. the rotor magnet pole position θ_r must be known all the time. There are a lot of different estimation methods for the rotor position and the speed. Main estimation techniques are such as; estimators based on the motor back e.m.f. voltage identification, estimators based on the Kalman-filters, estimators based on the state observers, estimators based on the stator flux identification such as the Model Reference Adaptive System (MRAS), and estimators based on the external signal injection into the motor. For simplicity, the MRAS (Andreescu, 1999) estimator is used, which is not a computationally intensive method.

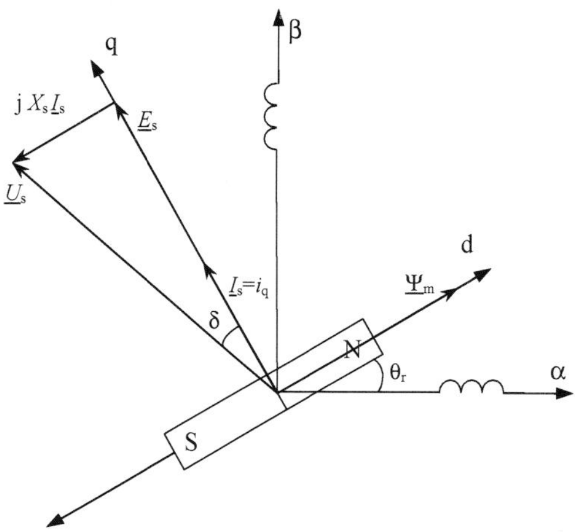

Fig. 5. Smooth air gap PM-motor voltage and current vector presentation in a rotor flux (d-q) frame of reference.

3.4 MRAS speed and position estimator

MRAS is based on the stator flux identification by using the motor mathematical model utilizing measured motor currents and voltages (Fig. 6). First, the stator flux is calculated based on motor voltages using the voltage model which is the reference model. On the other hand, the stator flux is calculated based on motor currents and the estimated rotor flux position using the current model which is the adaptive model. If there is an estimation error in the rotor flux position, the error is directly seen in a stator flux phase angle obtained from the current model. Hence, by comparing the stator flux vector phase angle of the current model to the stator flux vector phase angle of the voltage model, one can obtain the flux error angle. This error angle is used to correct the estimated rotor speed in a way that the error angle converges towards the zero. Correction mechanism i.e. the adaptation mechanism in this case is a simple PI-controller which input is the flux error angle and the output is the estimated speed. Estimated rotor flux angle is obtained by integrating the rotor speed. For simplicity, only a smooth air gap machine is considered i.e. it is assumed that the generator has $L_d = L_q$, which is a reasonable assumption for most of the generators with surface mounted magnets. However, the generator saliency can be included in the current model.

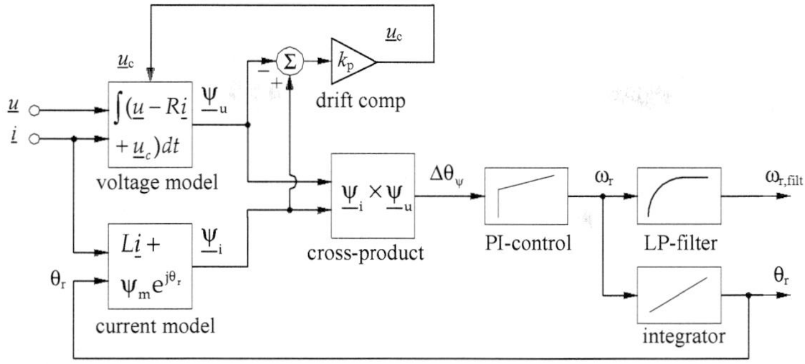

Fig. 6. MRAS estimator for the rotor speed and position estimation.

3.4.1 MRAS current model

Current model can be built utilizing the synchronous d-q frame of reference or the stationary α-β frame of reference. Stationary frame of reference is better in a way that it uses directly measured stator currents and voltages. Instead, the synchronous frame of reference uses converted d- and q-axis values, which are calculated based on the estimated rotor flux position which is not an exact value. Synchronous frame presentation is more simple in case of generator has salient poles but for smooth air gap machines there is not much computational difference. Current model in a stationary frame of reference is obtained from the PM-machine stator flux equations as

$$\psi_{i\alpha} = L_s i_\alpha + \psi_m \cos(\theta_r), \tag{2}$$

$$\psi_{i\beta} = L_s i_\beta + \psi_m \sin(\theta_r), \tag{3}$$

where θ_r is the estimated rotor flux position. In this case it's also assumed that there is no machine saliency i.e. $L_d = L_q = L_s$. In synchronous frame of reference, equations (2) - (3) are presented as

$$\psi_d = L_d i_d + \psi_m = L_s i_d + \psi_m, \tag{4}$$

$$\psi_q = L_q i_q = L_s i_q. \tag{5}$$

3.4.2 MRAS voltage model

PM-machine voltage equations in a stationary frame of reference are given as

$$u_\alpha = R i_\alpha + \frac{d\psi_\alpha}{dt}, \tag{6}$$

$$u_\beta = R i_\beta + \frac{d\psi_\beta}{dt}. \tag{7}$$

MRAS voltage model calculates stator flux components by integrating stator voltages. Using equations (6) and (7), flux components are obtained as

$$\psi_{u\alpha} = \int (u_\alpha - Ri_\alpha)dt \tag{8}$$

$$\psi_{u\beta} = \int (u_\beta - Ri_\beta)dt\,. \tag{9}$$

Equations (8) and (9) include the open integration, which means that the calculated stator flux vector tends to drift away from the origin due to the measurement and parameter errors. This can be avoided by replacing the integration algorithm by low-pass filters, or other algorithms tracking the behavior of integrator such as presented in (Hu & Wu, 1998). In this case, a corrective voltage term $\underline{u}_c$ (Andreescu, 1999) is added inside the integration giving

$$\psi_{u\alpha} = \int (u_\alpha - Ri_\alpha + u_{c\alpha})dt\,, \tag{10}$$

$$\psi_{u\beta} = \int (u_\beta - Ri_\beta + u_{c\beta})dt\,. \tag{11}$$

This correction term forces the calculated stator flux vector back to the origin. Correction voltage is obtained by comparing the current model flux components to the voltage model flux components at slower execution time level than the integration operation is performed according to

$$u_{c\alpha} = k_p(\psi_{i\alpha} - \psi_{u\alpha})\,, \tag{12}$$

$$u_{c\beta} = k_p(\psi_{i\beta} - \psi_{u\beta})\,. \tag{13}$$

Because the stator flux vector obtained from the current model is always centered at the origin, the voltage correction term forces the stator flux obtained from the voltage model also towards the origin.

3.4.3 Stator flux error angle

Stator flux error angle is obtained from the angle difference of the voltage model stator flux vector and the current model stator flux vector. This can be calculated, for example, by using the cross-product of these two stator flux vectors. The cross-product is given as

$$\underline{\psi}_u \times \underline{\psi}_i = \psi_u \psi_i \sin(\Delta\theta_\psi) = \psi_{u,\beta}\psi_{i,\alpha} - \psi_{u,\alpha}\psi_{i,\beta}\,. \tag{14}$$

The cross-product is zero, if estimated stator flux vectors are aligned. For small error angles, stator flux error angle is obtained from (14) as

$$\Delta\theta_\psi \approx \sin(\Delta\theta_\psi) = \frac{\psi_{u,\beta}\psi_{i,\alpha} - \psi_{u,\alpha}\psi_{i,\beta}}{\psi_u \psi_i}\,. \tag{15}$$

3.5 Rotor flux oriented vector control

A simplified control diagram for the overall closed loop control system is shown in Fig. 7. Torque control block defines the q-axis current reference $i_{q,\mathrm{ref}}$ for the q-axis current controller. Torque controller inputs comprise the torque reference and the estimated actual torque. The actual torque calculation is based on the equation (1) and the torque reference is calculated typically as a function of the wind speed. MTPA block defines the d-axis current reference $i_{d,ref}$ for the d-axis current controller as a function of the i_q-current. For smooth air gap machines, the d-axis current is referenced to zero, which gives the name '$i_d = 0$' control. In the field-weakening area, the field-weakening controller outputs an additional d-axis current reference. Current controllers ensure the fast tracking of current references by output voltage control using the Space Vector Modulator (SVM).

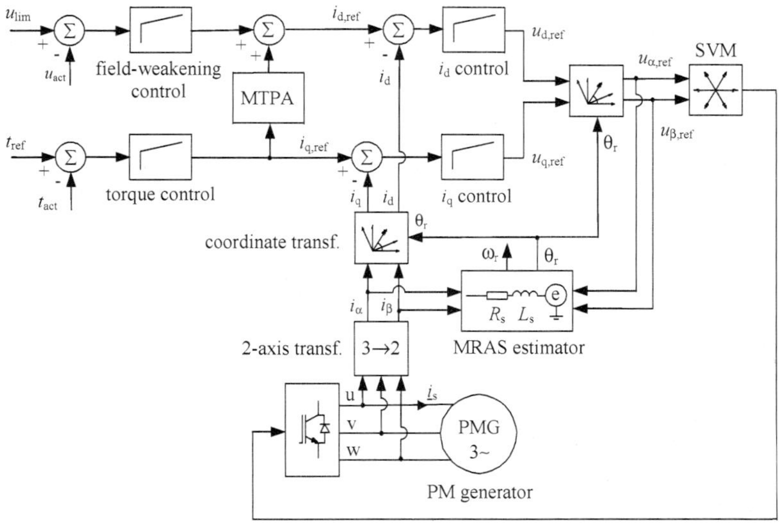

Fig. 7. A simplified block diagram of the speed sensorless control system

3.5.1 Field-weakening control

Field-weakening control is based on the reduction of the d-axis stator flux, which in turn reduces the stator voltage. According to the equation (4), stator flux can be reduced by injecting the negative d-axis current into the generator. Field-weakening controller inputs comprise the maximum motor voltage limit reference and the actual motor voltage. There is no need to measure the actual voltage because it can be calculated accurate enough based on voltage references. Due to the field-weakening controller, generator can operate also in the field-weakening area (constant power area), where the generator stator voltage is limited to the maximum voltage inverter is capable to produce. The extra i_d current needed depends on the generator design and is defined by

$$i_d = -\frac{\Psi_m}{L_d} + \frac{1}{L_d}\sqrt{\frac{u_{lim}^2}{\omega_r^2} - (L_q i_q)^2}\,, \tag{16}$$

where u_{lim} is the maximum voltage limit and ω_r is the rotor speed. Note, that in the equation (16), the stator resistance is neglected because it has only a small affect in field-weakening area. In practice, there is no need to calculate the field-weakening current based on the equation (16), because the PI-controller will automatically adjust the current in the level defined by the equation (16). Extra i_d-current component in a field-weakening operation means a higher generator current. Because the maximum thermal current limit of the inverter and the generator cannot be exceeded for a long period, the q-axis current must be limited in a field-weakening operation according to the required d-axis current and the current limit. Fig. 8 shows an example how the i_d- and the total current behave as a function of the i_q-current with a fixed speed and voltage limit.

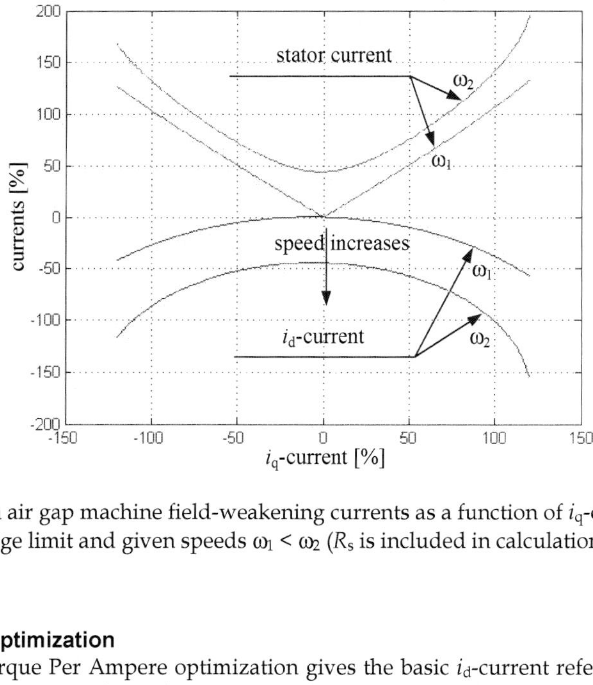

Fig. 8. Smooth air gap machine field-weakening currents as a function of i_q-current with a constant voltage limit and given speeds $\omega_1 < \omega_2$ (R_s is included in calculation).

3.5.2 MTPA optimization

Maximum Torque Per Ampere optimization gives the basic i_d-current reference. The target is that the generator produces the maximum torque using the minimum current. For smooth air gap machines, the optimal i_d-current reference equals to zero. Instead, for salient pole machines some extra torque can be produced by feeding the negative i_d-current into the generator. I_d-current needed, can be found by solving the torque global maximum with respect to the motor current using the motor torque equation (1). This gives the motor i_d-current reference as

$$i_d = \frac{\psi_m}{2(L_q - L_d)} - \sqrt{\frac{\psi_m^2}{4(L_q - L_d)^2} + i_q^2}\,. \tag{17}$$

Typically, for salient pole PM-machines L_d is lower than L_q, and the equation (17) results in a negative id-current reference (Fig. 9).

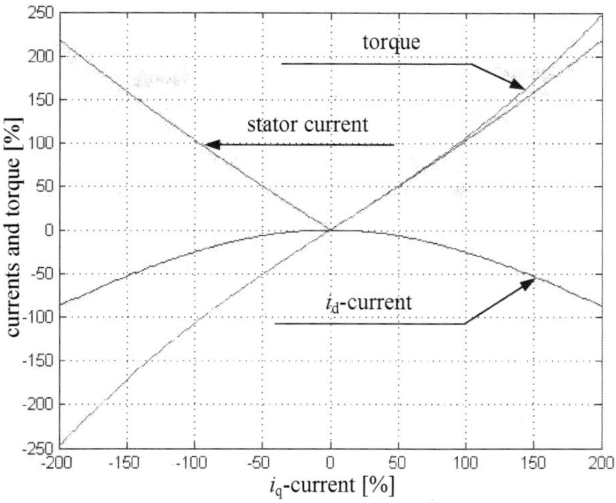

Fig. 9. Salient pole machine currents and torque according to the MTPA optimization (as an example $L_{d,pu}$-$L_{q,pu}$ = -0.27 p.u. and $L_{d,pu}/L_{q,pu}$ = 0.55).

3.5.3 Cos(φ) ctrl

PM-generator cos(φ) can be controlled by feeding the extra i_d-current to the machine. Negative i_d-current produces the reactive power, which means that the machine cos(φ) can be improved by a negative i_d-current injection. This means also that the MTPA-optimization and the field-weakening control tend to improve the generator power factor. Normally, there is no need for an additional cos(φ) control. MTPA-optimization for salient pole machines, and '$i_d = 0$' control for smooth air gap machines give already the optimal performance i.e. the maximum torque with a minimum current. Negative i_d-current advances the current vector angle and reduces the motor voltage (Fig. 10). Additional i_d-current component means also a higher stator current which increases the machine copper losses and the drive current rating. On the other hand, the lower stator voltage means the lower stator flux which in turn decreases machine iron losses. Motor cos(φ) can be improved up to the maximum cos(φ) = 1. For cos(φ) = 1 control, i_d-current can be calculated as a function of the i_q-current giving

$$i_d = -\frac{\psi_m}{2L_d} + \sqrt{\frac{\psi_m^2}{4L_d^{\,2}} - \frac{L_q}{L_d} i_q^2} \, . \tag{18}$$

It can be concluded from the equation (18) that there is a maximum i_q-current limit above the cos(φ) = 1 control is not any more possible. This condition is given as

$$\left| i_q \right| \leq \frac{\psi_m}{2\sqrt{L_d L_q}} \, . \tag{19}$$

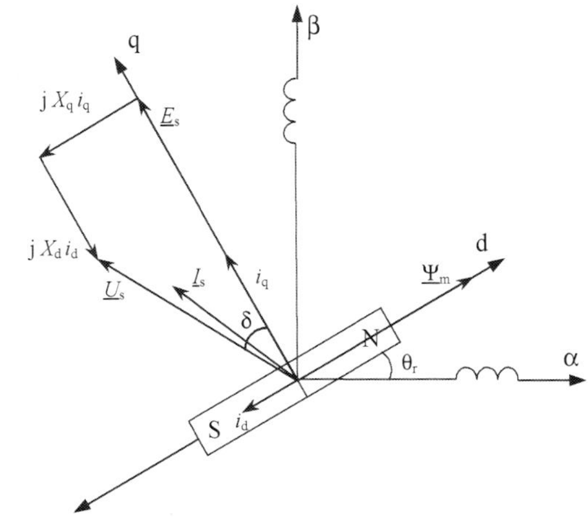

Fig. 10. PM-motor voltage and current vector presentation with i_d-current.

For smooth air gap machines $L_d = L_q = L_s$ and the equation (19) gives

$$|i_q| \leq \frac{\Psi_m}{2L_s} = \frac{E}{2X_s}, \tag{20}$$

where $E = \omega_r \Psi_m$ is the motor back e.m.f voltage and $X_s = \omega_r L_s$ is the stator reactance. Equation (20) means in p.u. values that, if the motor stator inductance is more than 0.5 p.u., $\cos(\varphi) = 1$ control must be restricted even when the motor operates below the nominal

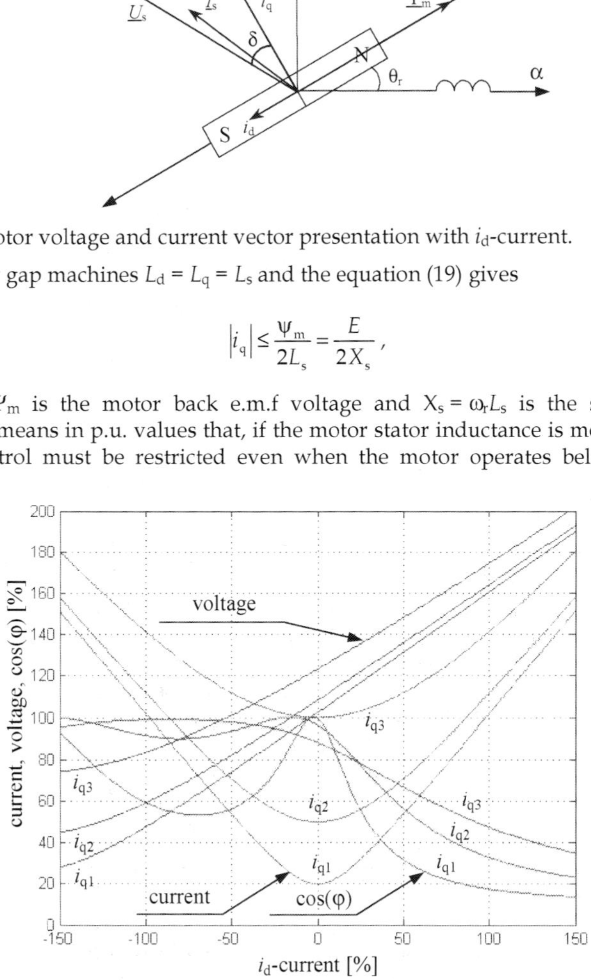

Fig. 11. Motor stator voltage, stator current and cos(φ) as a function of i_d-current for different load points i_{q1} = 0.2 p.u., i_{q2} = 0.5 p.u., i_{q3} = 1.0 p.u. (L_d=L_q= 0.59 p.u.).

torque point. Fig. 11 shows an example how the motor cos(φ) and the stator voltage behave as a function of negative i_d-current in three load points. Cos(φ) = 1 point can be found corresponding to the i_d-current according to the equation (18).

3.5.4 Pull-out torque

Pull-out torque defines the maximum torque generator is capable to produce. Trying to get out more torque, generator pulls out the synchronism. Pull-out torque is normally not an issue in windmill applications. Typically, the generator torque or the thermal current limit is the first torque limiter below the pull-out torque limit. Also, the generator operation area is not heavily in field-weakening area where the pull-out torque reduces approximately inverse proportional to the speed. Neglecting the machine stator resistance, pull-out torque is defines as

$$T_{max} = \frac{3p}{\omega_r}\left[\frac{U_{max}\psi_m}{L_d}\sin(\delta) + \frac{L_d - L_q}{2\omega_r L_q L_d}U_{max}^2\sin(2\delta)\right], \tag{21}$$

where ω_r is the rotor electrical angular speed, p is the pole pair number, U_{max} is the maximum voltage of the generator and δ is the power angle. The first term in equation (21) corresponds to the torque produced by the magnet flux and the second term the reluctance torque. It can be seen that the first term, torque produced by the magnet flux, is reduced inverse proportional to rotor speed when stator voltage is limited to the maximum value.

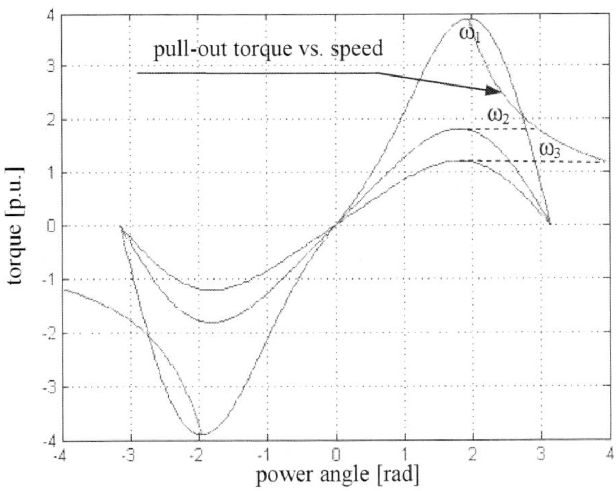

Fig. 12. Salient pole PM-machine torque characteristics at various speeds ω_1 = 1 p.u., ω_2 = 2 p.u., ω_3 = 3p.u. (as an example $L_{d,pu}$-$L_{q,pu}$ = -0.27 p.u. and $L_{d,pu}/L_{q,pu}$ = 0.55).

This corresponds to the operation in the field-weakening area, which is called also a constant power region. In the field-weakening area, the second term i.e. the reluctance torque vanishes more rapidly by inverse proportional to the square of rotor speed. For salient pole PM-machines the d-axis inductance L_d is usually smaller that the q-axis inductance L_q. This means that the reluctance torque, in addition to increasing the pull-out torque, increases also the power angle corresponding to the pull-out torque. Power angle corresponding to pull-out torque can be found by solving the zero of the derivative of the equation (21) with respect to the power angle. Unfortunately, the resulting equation is a transcendental equation which must be solved using iterative methods. Fig. 12 shows, as an example, the torque curve for a salient pole machine at various speeds in field-weakening region.

3.5.5 Dc-link overvoltage control

In a normal operation mode, the AFE drive takes care of the dc-link voltage control by regulating the active power flow into the utility grid. In case of sudden supply lost, active power cannot anymore generated into the grid. This means that all the power generated will be pushed in to the dc-link capacitors following that the dc-link voltage will start to rise. If no protection is available, such as an overvoltage control, brake choppers and resistors or crowbars, the dc-link voltage can increase such a high level that destroys the drive units. Need for the external hardware overvoltage protection depends a lot of whole drive system design: how big are the dc-link capacitors, what is the maximum generator back e.m.f. voltage, how much energy is stored in stator inductances etc. In addition to hardware protection, software protection can be used. Obviously in this kind of fault situation, the generator power production must be stopped as fast as possible. In practice it means that the generator control must limit the generative torque down to zero as fast as possible in order to limit the energy pushed to the dc-link. Time needed to bring the torque down to zero depends on the generator characteristics and the operation point before the fault happens. Generator current dynamics is obtained from the synchronous frame voltage equations as

$$L_\mathrm{d}\frac{di_\mathrm{d}}{dt} = u_\mathrm{d} - Ri_\mathrm{d} + \omega_\mathrm{r} L_\mathrm{q} i_\mathrm{q} \tag{22}$$

$$L_\mathrm{q}\frac{di_\mathrm{q}}{dt} = u_\mathrm{q} - Ri_\mathrm{q} - \omega_\mathrm{r}(L_\mathrm{d} i_\mathrm{d} + \psi_\mathrm{m})\,. \tag{23}$$

In a generative operation, the sign of the i_q-current and the torque is negative by assuming the positive speed direction. Hence, bringing the negative torque down to zero means a positive derivative of i_q-current i.e. $di_\mathrm{q}/dt > 0$. By neglecting the stator resistance, equation (23) gives

$$L_\mathrm{q}\frac{di_\mathrm{q}}{dt} = u_\mathrm{q} - \omega_\mathrm{r} L_\mathrm{d} i_\mathrm{d} - e\,, \tag{24}$$

where $e = \Psi_\mathrm{r}\omega_\mathrm{r}$ is the generator back e.m.f voltage. In addition, if it's assumed that the i_d-current equals to zero, which is true for '$i_\mathrm{d} = 0$' control in constant torque region, equation (24) gives

$$L_\mathrm{q}\frac{di_\mathrm{q}}{dt} = u_\mathrm{q} - e\,. \tag{25}$$

Generator back e.m.f voltage e can be assumed to be a constant value during a fast current transient. This means that in order to control the generative torque down to a zero as fast as possible, controller must increase the q-axis voltage u_q as high and fast as possible. However, higher the generator torque is, the higher is the generator speed and voltage in wind power generators. This means that in a worst case, in a full power operation point, the generator operates at the full voltage and the generator has a high back e.m.f voltage. Hence, there is no margin to increase the voltage any more. Then it is not the overvoltage or the current controller dynamics, which limits the current and torque derivative but rather the lack of the sufficient voltage. A simple calculation for an example generator with $\Delta i_q = 1$ p.u., $u_{qmax} = 1.15$ p.u., $e = 1$.p.u. and $L_q = 0.5$ p.u. yields the torque fall time 6.2 ms, which is much more than the typical current controller step-response time. In order to force the i_q-current faster towards to zero, i_d-current can be utilized and equation (24) can be written as

$$L_q \frac{di_q}{dt} = u_{q2} - e = (u_q - \omega_r L_d i_d) - e \,. \tag{26}$$

Hence, the effective voltage u_{q2} can be increased with the negative i_d-current which brings the current towards to zero much faster and reduces the energy pushed to the dc-link.

4. Test results

4.1 A 4 kW four segment motor

The redundant generator control principle was tested in a laboratory with a radial flux PM-motor comprising four independent stator segments. Each stator segment was fed with a Vacon NXP frequency converter. The generator was coupled to the asynchronous induction motor which was used to rotate the generator. Generator parameters for a one stator segment are: $R_s = 3\ \Omega$ (7.8 %), $L_s = 60$ mH (50.1 %), $E_n = 262$ V (78.8 %), n = 765 rpm, $f_n = 102$ Hz, $p = 8$, $I_n = 2.5$ A.

4.1.1 Closed loop control with an encoder

First, the closed loop control was tested with an EnDat absolute encoder feedback corresponding to the principle shown in Fig. 3. In this case, the encoder signal was wired only to one of the drives which was considered as the master drive. The master drive sends the necessary information such as the shaft position, torque reference, and the run command to the follower drives using the system bus, in this case a fiber optic bus. The generator is rotated by an asynchronous motor at 15 % speed while the inverters are started with a full 100 % generative torque command. Fig. 13 shows the measured U-phase currents for a master inverter and two of the followers. As expected, the currents of each drive are synchronized with respect to each other producing the optimal instantaneous total generator torque.

4.1.2 Sensorless control without the encoder

The test corresponding to the closed loop test with the encoder was repeated with the sensorless control corresponding to the principle shown in Fig. 4. Compared to the control with the encoder, the drive system is simpler: there is no encoder and no extra communication needed between the drives. Each drive is independently identifying the own rotor flux position and the speed. As in test with the encoder, currents are in phase synchronized with respect to each other (Fig. 14). It was also measured that the torque generated was the same w/o the encoder (Halkosaari, 2007). The main difference compared

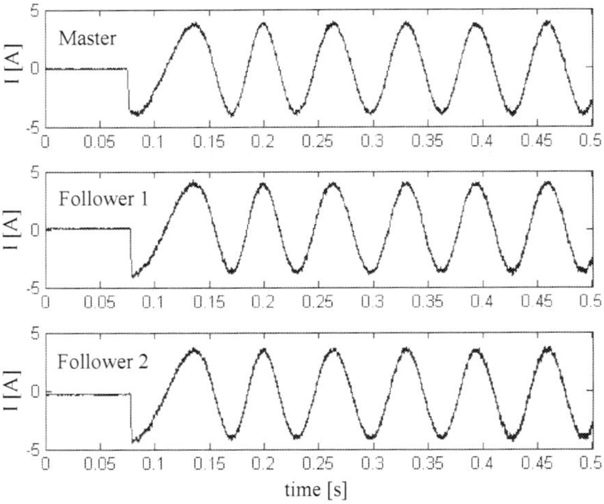

Fig. 13. Measured U-phase currents of the master and followers during starting the generator with 100 % torque and 15 % speed. Closed loop control with an absolute encoder.

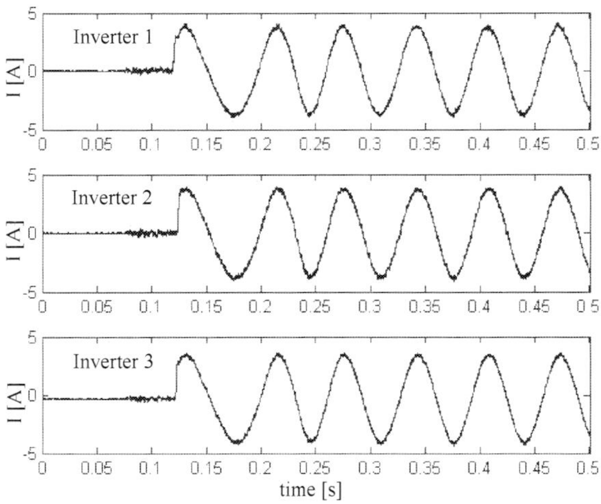

Fig. 14. Measured U-phase currents of inverter 1, 2, and 3, during starting the generator with 100 % torque and 15 % speed. Speed sensorless control.

to the closed loop control using the absolute encoder, is the start delay due to the flying start phase. This flying start phase is used to initialize the MRAS estimator and can be recognized in Fig. 14 as a 50 ms duration zero current phase at the beginning of the start.

Generator must be able to start also at higher speed than the 15 % shown. If generator voltages are not measured, starting at higher speed using the sensorless control is more difficult than at lower speeds due to the higher back e.m.f. voltage of the generator. The higher the generator back e.m.f. voltage is, the faster the generator currents increase until the steady state condition is reached. However, starting at full speed is usually not a problem due to the fast current control response time. Fig. 15 shows the load machine torque calculated by the load machine inverter corresponding to the total torque produced

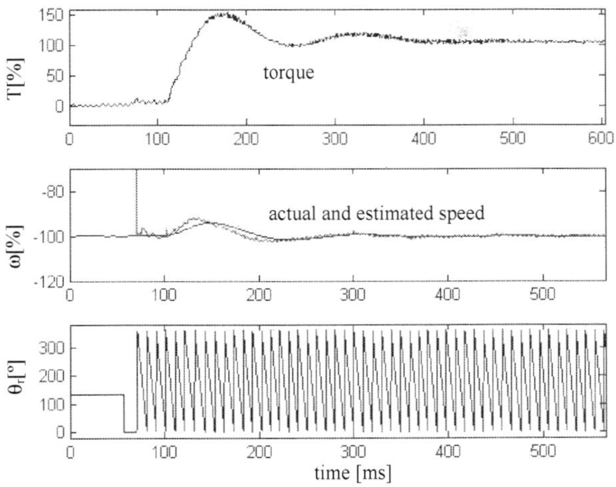

Fig. 15. Generator torque, estimated and measured speed, and the estimated rotor flux position θ_r. Generator is started at 100 % speed and torque using the sensorless control.

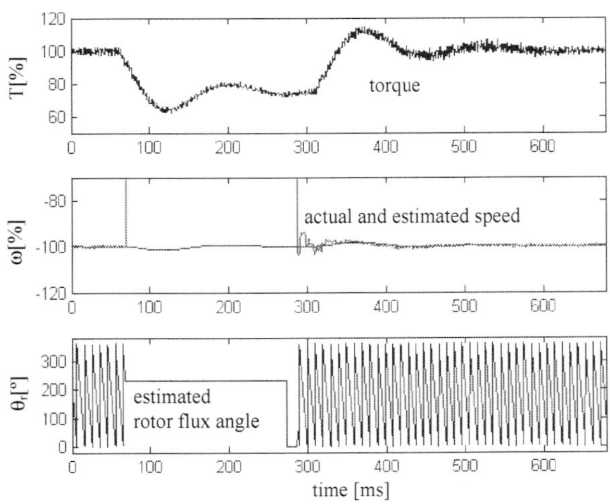

Fig. 16. Generator redundancy test: one of the four inverters is switched on fly off and on at 100 % speed and torque command.

by the generator when the generator is started at 100 % speed. In addition, the estimated rotor flux angle, the estimated generator speed and the measured actual speed are also shown in Fig. 15.
The redundancy check was made by switching one of the four inverters on and off with the full torque command while the other one inverters continue the operation. This could happen, for example, if one of the inverters for some reason fails and is switched back on the fly. Such a situation is shown in Fig. 16, which shows the generator torque, the estimated and measured speed, and the rotor flux position when a one inverter is switched on and off at nominal speed. It can be seen that the generator can continue on running without the interruption if one (or more) of the inverters are switched off. Naturally, the generator power is reduced if all of the inverters are not in use.

4.2 3.8 MW three segment generator

Some test results are shown also for a real size 3.8 MW direct drive wind power generator manufactured by The Switch. The generator comprises three stator segments (Pyrhönen & al, 2006) which are controlled using the sensorless control mode similar to the laboratory test generator described. The generator (Fig. 17) has about 6.6 m diameter and it weighs about 81000 kg. Generator nominal values are given as: U_n = 690 V, n = 17.5 rpm, f_n = 17.5 Hz, and I_n = 3500 A. At the moment, there are several of these or similar kind of generators in operation.

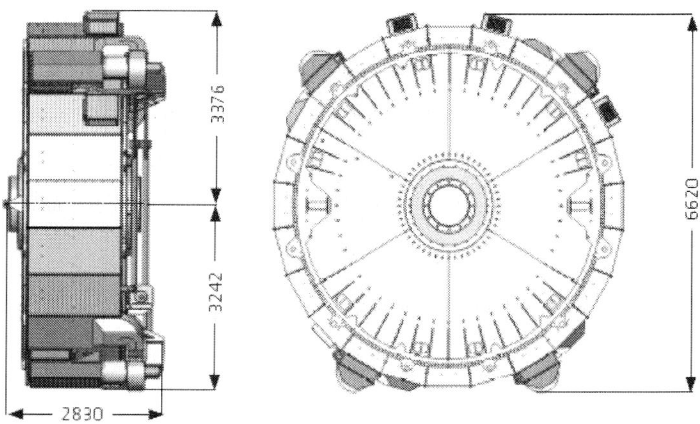

Fig. 17. 3.8 MW generator manufactured by The Switch: n_n = 17.5 rpm, f_n = 17.5 Hz, I_n = 3500 A, U_n = 690 V (Photo Courtesy of The Switch, Finland)

It is obviously expensive to build up test bench systems for this kind of large generators, which makes the small scale laboratory tests even more valuable. Normal laboratory tests are difficult due to the weight and size of the generator. Hence, most of the tests must be made during the factory installation or when the generator is in real operation. Some load tests can be made without the back-to-back connected generator-motor test benches due to the segment structure of the generator: one stator segment can be used to rotate the generator while the other segments are working in a generative mode. With this kind of test method, Fig. 18 shows a startup of one segment at the 20 % generative torque. In this case, all the signals shown in Fig. 18 are measured or calculated by the segment inverter.

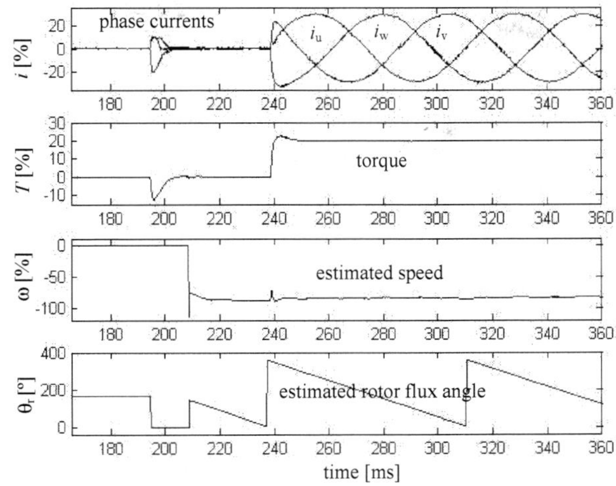

Fig. 18. Startup of one segment of the 3.8 MW generator at 20 % torque. Inverter measurements for a segment torque, phase currents, and the estimated speed and rotor flux angle.

5. Conclusions

The reliability of high power drives can be increased using a generator/motor structure comprising multiple independent stator modules or segments each having an own drive inverter. In case that one of the inverters fails, the operation can continue by other inverters assuming that there is no failure in generator windings. In order to increase the drive system power level, several smaller inverter units can be used instead of a one larger one.

Wind power generators can operate long periods at reduced power due to the lack of sufficient wind. Hence, it is important that the generator operates at high efficiency also when operating below the nominal load point. The optimal efficiency is obtained only when the generator control is functioning in an optimal way, which is ensured by using the optimal vector control and the MTPA optimization which keeps the minimum stator current and the optimal stator flux level.

Wind power generators are a fast increasing application area requiring high power levels. It is a suitable application for the speed sensorless vector control described. Also, a principle of the closed loop control using an absolute encoder feedback was shown for the comparison. The test results with a small laboratory test bench confirmed the validity of the control methods and the idea of the redundancy. At the moment, several large scale 3.8 MW permanent magnet generators comprising three stator segments are in operation in wind-mills. However, the applications of redundant drives are just not limited to the wind power production but the idea can be used everywhere requiring redundancy. For example, the same drive principle is used also in large 800 kW ski-lift motors comprising 8 stator segments and inverters.

6. References

Albrich, R. & Fuchs, E. (2002). Electric motor, *United States Patent*, Patent No.: 6,429,554, Aug. 2002.

Andreescu, G. D. (1999). Position and Speed Sensorless Control of PMSM Drives Based on Adaptive Observer, *in Proc. 8th European Conference on Power Electronics and Applications (EPE'99)*, 10 p., 1999.

Halkosaari, T.H. (2007). Speed Sensorless Vector Control of a Redundant Permanent Magnet Wind Power Generator, *in Proc. IEEE International Symposium on Industrial Electronics (ISIE 2007)*, pp. 2595-2600, 2007.

Hu, J. & Wu, B. (1998). New integration algorithms for estimating motor flux over a wide speed range, *IEEE Transaction on Power Electronics*, Vol. 13, No. 5, September 1998, pp. 969-977.

Mäkinen, J-P.; Haapamäki, P. & Halkosaari T. (2006). Control of parallel operation of frequency converters or inverters, *United States Patent*, Patent No.: 7,038,405, May 2006.

Pyrhönen, J.; Kurronen, P. & Parviainen, A. (2006). Permanent Magnet 3 MW Low-Speed Generator Development, *in Proc. XVII International Conference on Electrical Machine (ICEM2006)*, 2006.

4

Wind Turbine Modelling of a Fully-Fed Induction Machine

Umashankar S, Dr. Kothari D P and Mangayarkarasi P
VIT University & Anna University
Tamilnadu,
India

1. Introduction

This document provides detail about modelling Fully-Fed Induction Generator (FFIG) based wind farm. The generator torque model was specified either as a constant derived from nominal turbine power or pitch control block depends on the wind speed.
This model has been simulated under the normal operating condition and three different fault conditions. The performance of the model analysed based on the speed, torque, voltage, current, and power of generator, converter and grid.
This document contains the model design parameters and simulation results.

2. Wind farm model

The model in Fig.1. is the top level model of the fully-fed induction generator based wind farm with the following subsystems:

1. Induction generator and its torque subsystem
2. Converter Bridges
3. Machine bridge control
4. Network bridge control
5. Pitch Controller
6. Transmission line and grid

2.1 Induction generator
The Induction Generator block is a standard block in the SimPowerSystems Application Library, Machines. Its design is based on reference 3

2.2 Converter bridges
This is a standard block in SimPowerSystems, Power Electronics, Universal Bridge. The pulses for these two bridges given from subsystems MCB control and NWB control, which are modelled based on the vector control principles. Its design parameters are based on reference 12.

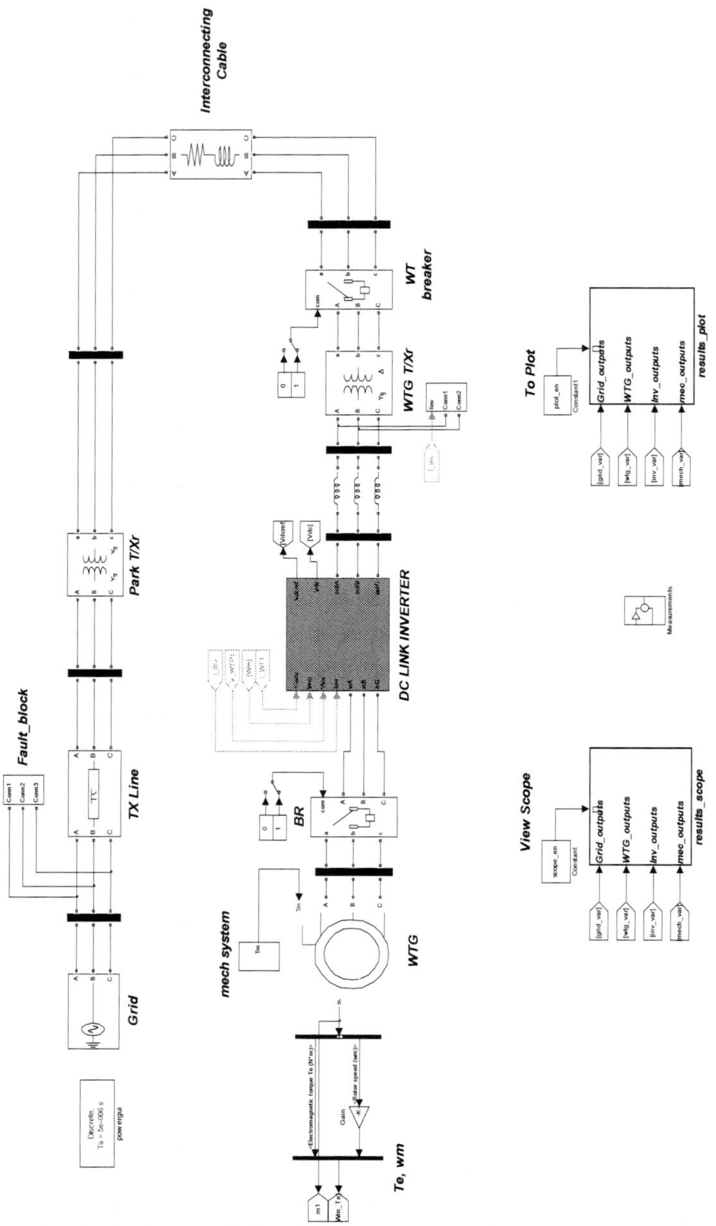

Fig. 1. Top Level Wind Farm Model

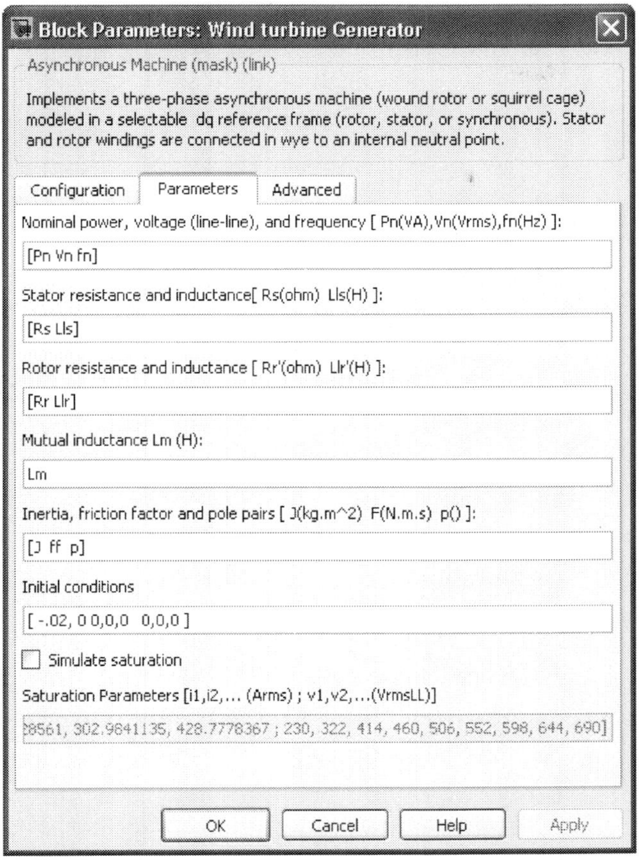

Fig. 2. Wind Turbine Generator mask dialogue

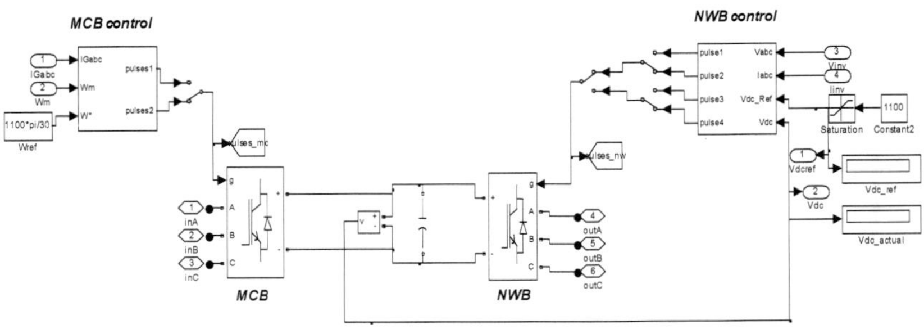

Fig. 3. DC link-Inverter with control blocks

2.3 Machine-bridge

This subsystem was modelled based on the vector control principles, in which the flux and torque controlled with the use of speed controller and hysteresis current controller. All the parameters are converted into two phase quantities and then the required flux and torque are calculated using abc to dq conversion. Here one provision is also available for checking the controller response for step change in speed.

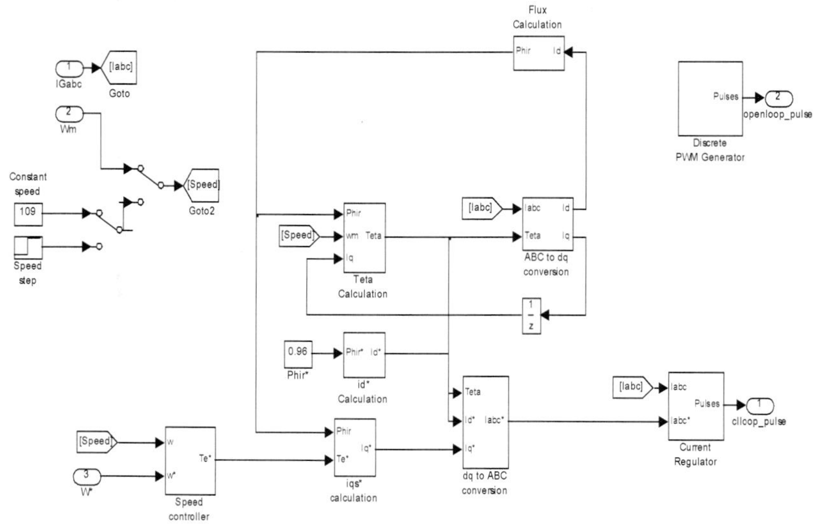

Fig. 4. Machine bridge control block

2.4 network-bridge

This subsystem was modelled based on the independent real and reactive power control using voltage and current controller. All the AC quantities are converted into two phase dqo and then the voltage and current references are estimated. Here one more provision is also available to give different type of pwm pulses, open loop as well as closed loop.

2.5 Transmission line and grid specifications

The transmission line is modelled as simple Π equivalent circuit block which is available with the standard block in SimPowerSystems, Elements. Its design parameters are specified in reference 8.

The Grid block is a standard block in the SimPowerSystems, Electrical Sources. Its design parameters are based on reference 8. The grid condition is based on the parameters under short-circuit section either by providing short-circuit power SCVA, Base Voltage Vrms, X/R ratio SCR or by Source Resistance Rs and inductance Ls.

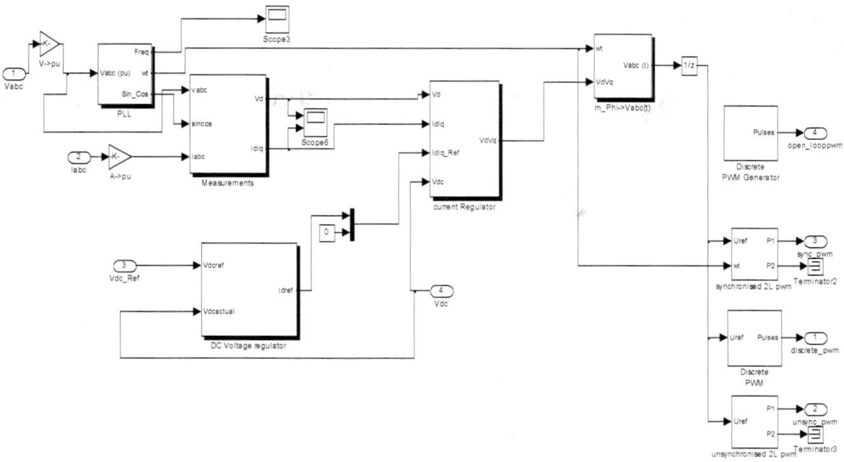

Fig. 5. Network bridge control block

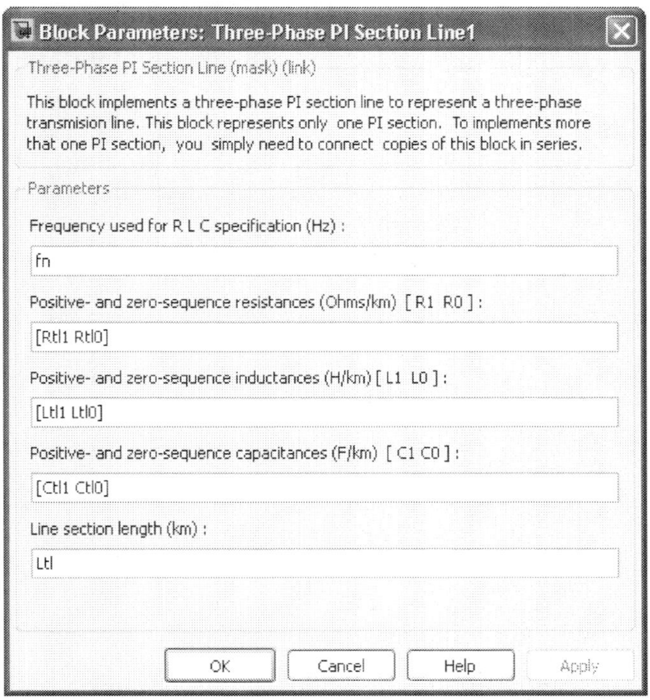

Fig. 6. Transmission Line Π equivalent mask dialogue

2.6 Pitch controller

The pitch controller block is a standard SimPowerSystems Pitch Controller Library block. We have made some modifications to enable or disable the controller based on the wind speed. Also the existing standard wind turbine block will give the per unit torque value. Then this is converted into actual value by manually adding the gain block which multiplies it with the base torque.

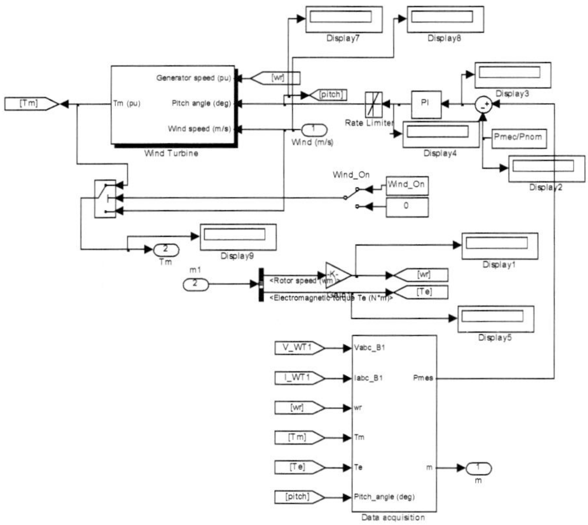

Fig. 7. Pitch controller block

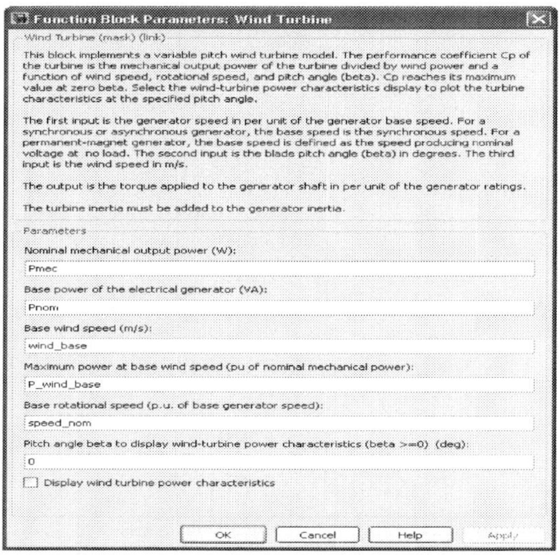

Fig. 8. Wind Turbine mask dialogue

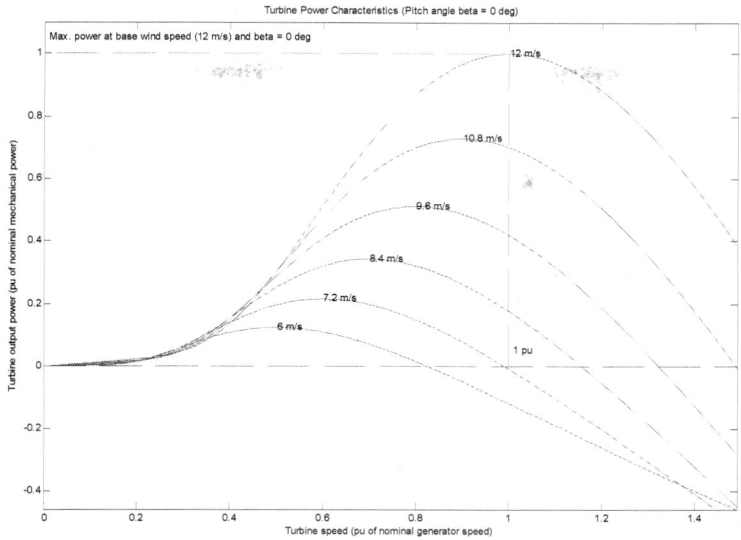

Fig. 9. Wind Turbine characteristic curves

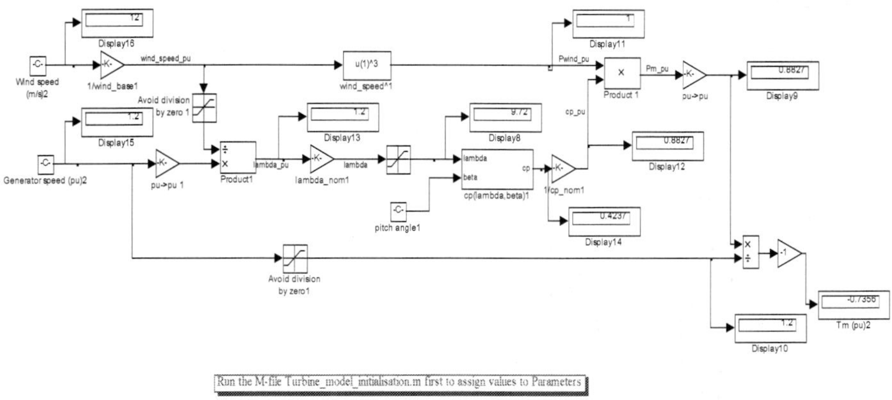

Fig. 10. Wind Turbine Block

3. Modelling notes and assumptions

The model is completely discrete; there are no continuous states.
The model is full parameterised; it has no hard coded parameter values.
All the model parameter values are set up in the initialisation file. The model could be used for a different wind turbine simply by changing the initialisation file; no change to the model is needed.

The stiffness of the shaft between the generator and wind turbine was chosen so that mechanical resonance occurred at 5Hz. This was done to demonstrate the effectiveness of active damping.
MATLAB mechanical dynamic model has been made discrete. The sample time chosen for the discrete model, 5X10-6 s, gave time domain results very close to the continuous model for 5Hz resonant frequency.
The model assumes that the converter DC link voltage is maintained constant during the simulations, thus the energy storage of the DC link due to the DC link capacitor has not been modeled.
The standard MATLAB wind turbine model has been used, which is analytic but doesn't completely agree with the data provided by client in reference 1.
The wind speed is assumed to be rated value and hence the torque and speed of the generator.
The wind turbine is considered to be delivering the power to grid and not to local loads.
All the initial conditions are assumed to be zero, unless otherwise specified externally and losses neglected to make the model simple and calculation easier.
As the objective of the modelling work was to get the correct transient response during a grid fault, generator and converter losses neglected or assumed to be zero.
The model is not analysed under Grid Fault Ride through Conditions, also not able to maintain the stability due to unavailable provision for voltage stabilization.

4. Results

Simulation results are provided for the following different cases

1. Normal operation
2. Grid Fault Analysis
   a. Three-Phase Line to Ground Fault
   b. Two-Phase Line to Line Fault

The following category figures are plotted for each scenario

1. Generator Voltage, Current, Real Power and Reactive Power
2. Generator Electromagnetic Torque, Speed
3. Mechanical Torque, Rotor and Wind Speed, Pitch
4. Inverter Voltage, current, PWM pulses, Vdc
5. Grid Voltage, Current, Real Power and Reactive Power

4.1 Normal operation

To analyse the model under normal operating condition, follow the steps in the below snapshot of the matlab command window. After the simulation, command window will prompt for user input to display the results and to save the figures. To start with run the main.m file in command window and follow the steps below:

4.1.1 Comments on the result

At rated torque and speed, the FFIG delivers the rated power to grid under normal operation. The negative sign in the generator power waveform denotes consumption and positive sign denotes generation of power. Thus the generator delivering rated power to grid by absorbing some reactive power.

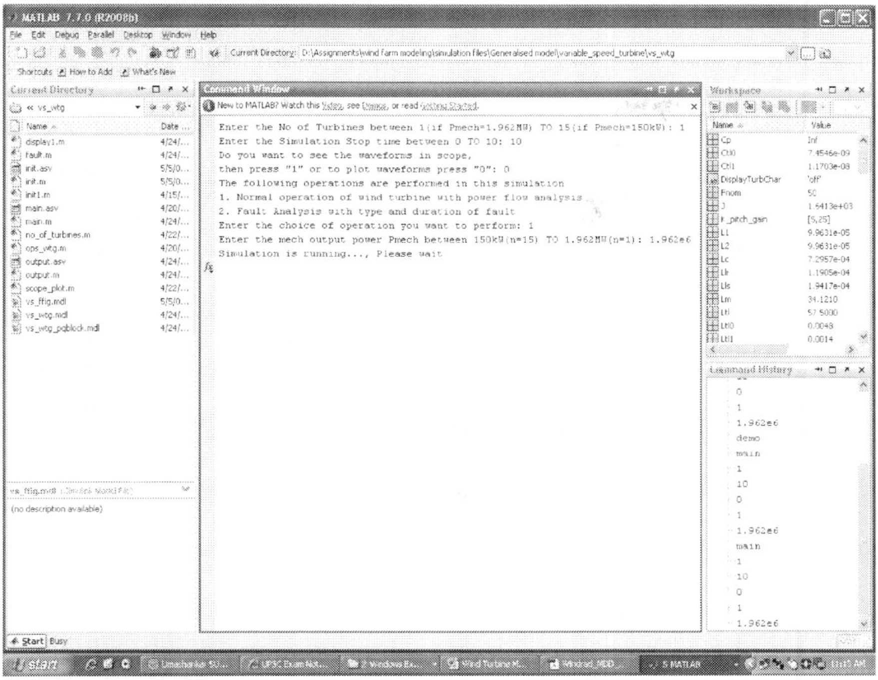

Fig. 11. Matlab command window as user interface, normal operation

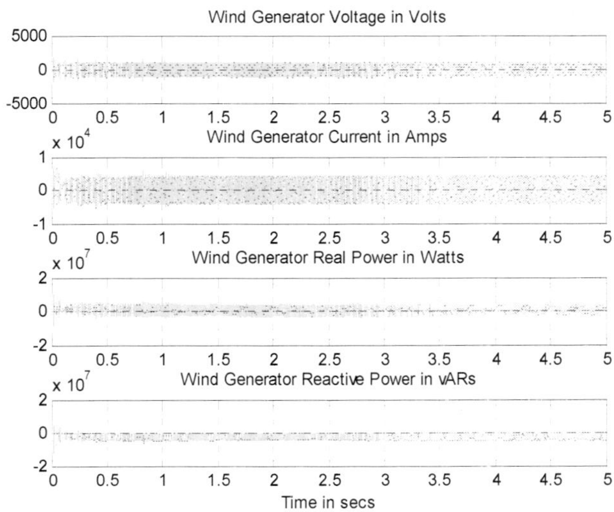

Fig. 12. Generator-side electrical outputs, normal operation

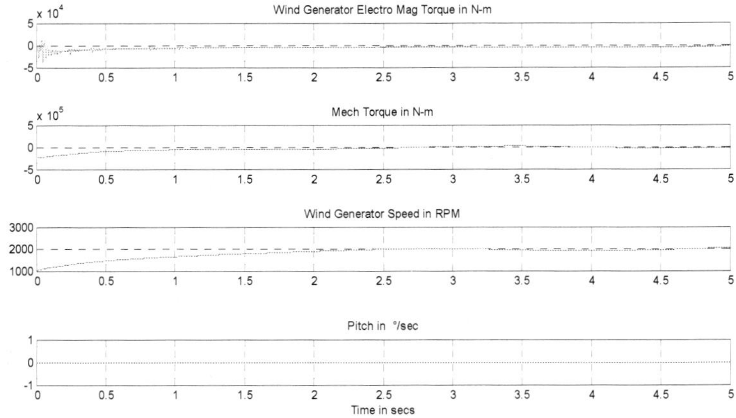

Fig. 13. Generator-side mechanical outputs, normal operation

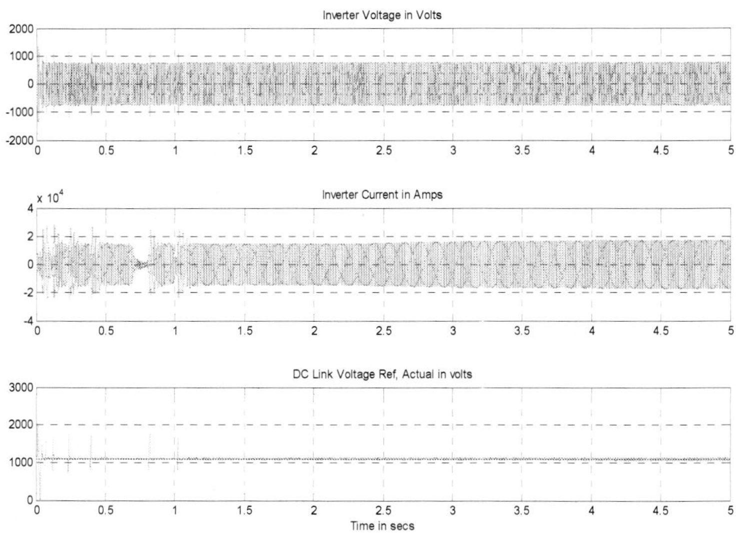

Fig. 14. Inverter Electrical outputs, normal operation

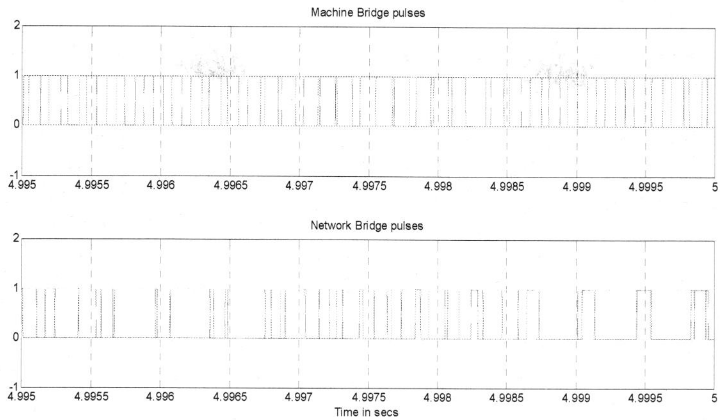

Fig. 15. PWM pulse outputs, normal operation

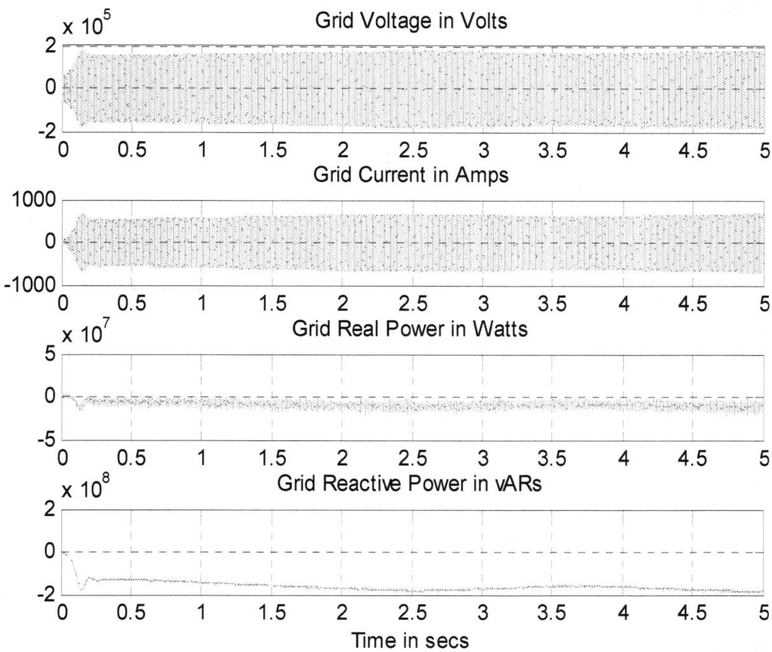

Fig. 16. Grid-side Electrical outputs, normal operation

The negative sign in torque denotes it as generating torque. The real power is delivered to grid with some loss of power in the transformer, cable and transmission line.
Fig.14. shows inverter output settles after one sec, before that there are some spikes occurring at regular intervals. This can be avoided by providing dv/dt filters in the output. The dc regulator maintains the dc voltage at its constant level as seen from fig.14.
The PWM pulse output shown in fig.15. for both machine and network bridges. The modulation is unsynchronised sine pwm and it's based on the closed loop bridge control.

4.2 Three-phase to ground grid fault

To analyse the model under three-phase to ground grid fault condition, follow the steps in the below snapshot of the matlab command window. After the simulation, command window will prompt for user input to display the results and to save the figures. To start with run the main.m file in command window and follow the steps below:

4.2.1 Comments on the result

At rated torque and speed, the FFIG delivers the rated power to grid under normal operation. At time t=2.5s, three-phase to ground fault is created at the grid side transmission line hence the grid current abruptly increased. The generator voltage, current will not become zero as like in fixed speed wind turbine and speed will increase above rated speed. Also the actual dc voltage becomes zero with that of reference.

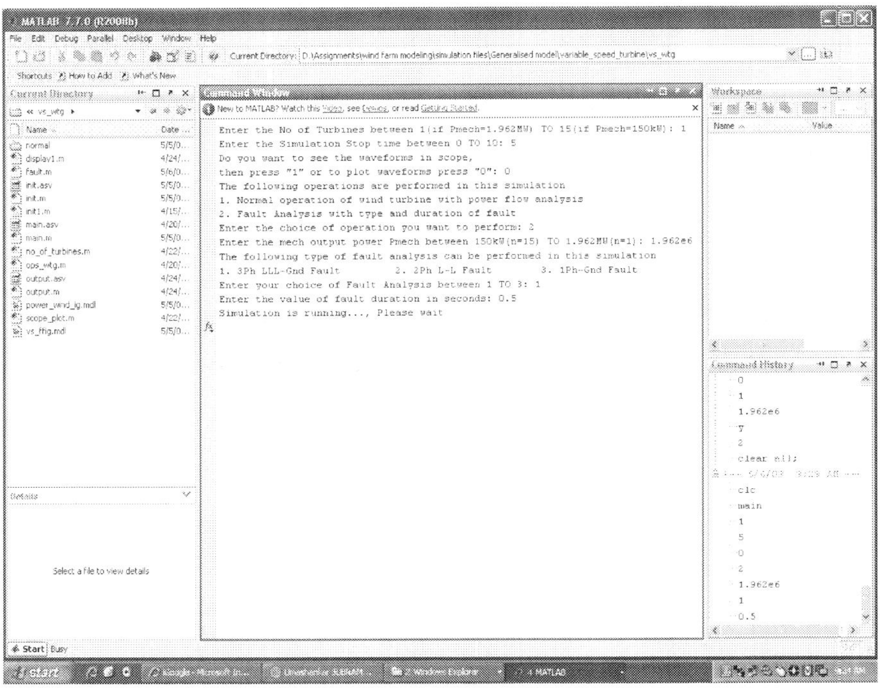

Fig. 17. Matlab command window as user interface, 3phase grid fault

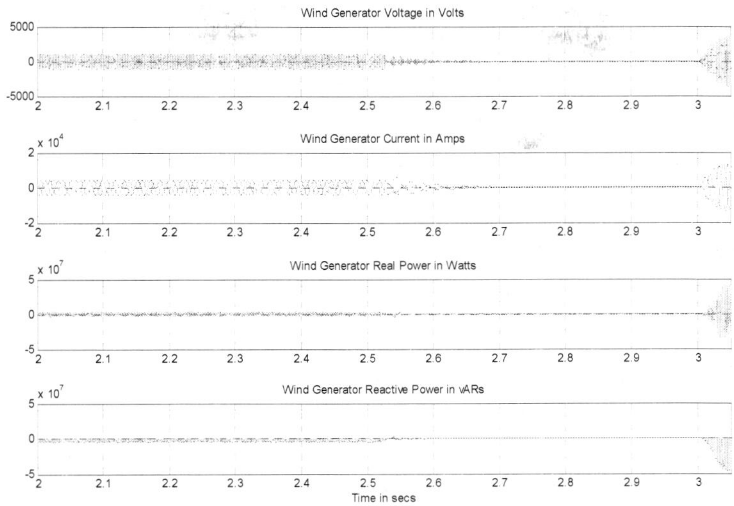

Fig. 18. Generator-side electrical outputs, 3phase grid fault

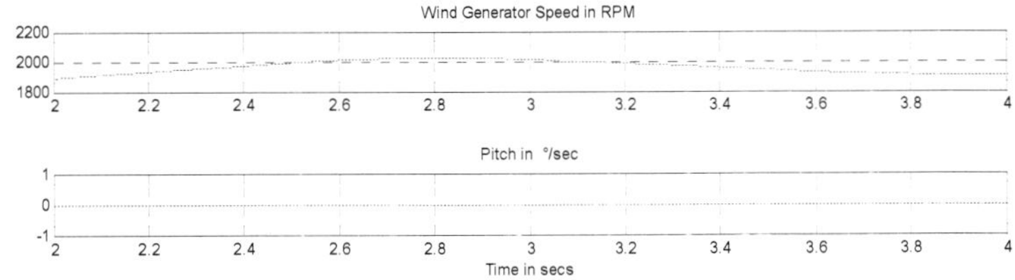

Fig. 19. Generator-side mechanical outputs, 3phase grid fault

It is evident from the fig.19. that there is a mechanical torque reversal during the grid fault and there are some oscillations in all the waveforms after the fault has been cleared. But the shaft between generator and wind turbine will ring at its natural frequency.
At time t=3s fault is cleared and all the quantities will slowly come to its previous condition of normal operation. Since electromagnetic torque becomes discontinuous during grid fault and initial conditions are changed after fault removal, it will take more time to settle in steady state. Implementation of GFRT will rectify this issue and stabilize the electrical parameters during fault and after fault removal.

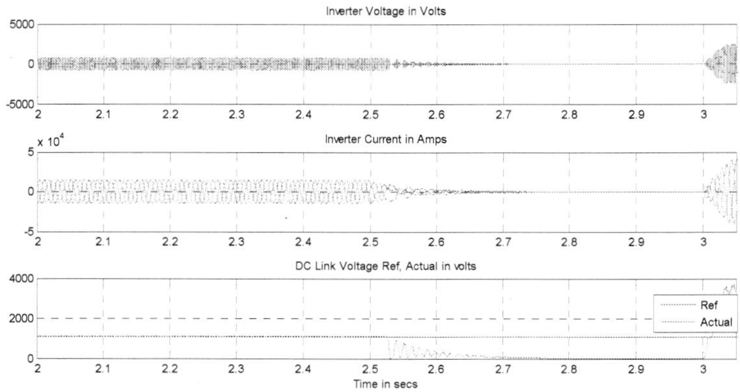

Fig. 20. Inverter Electrical outputs, 3phase grid fault

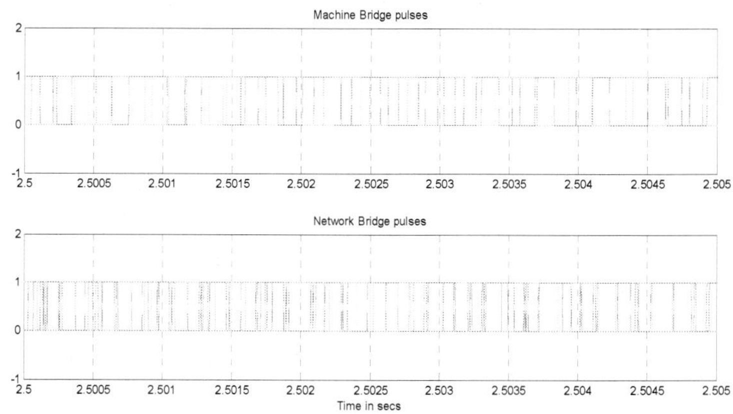

Fig. 21. PWM pulse outputs, 3phase grid fault

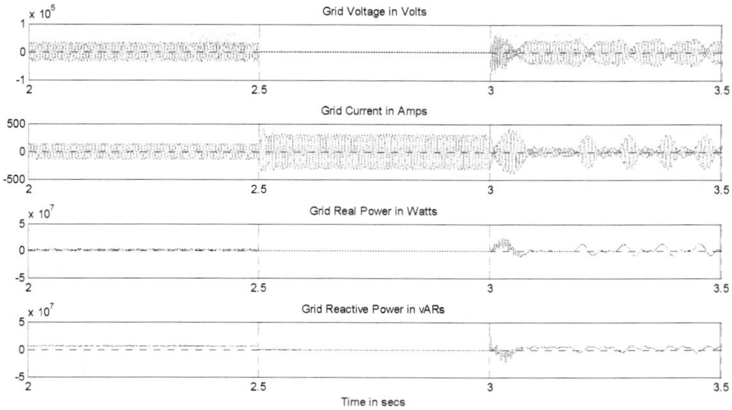

Fig. 22. Grid-side Electrical outputs, 3phase grid fault

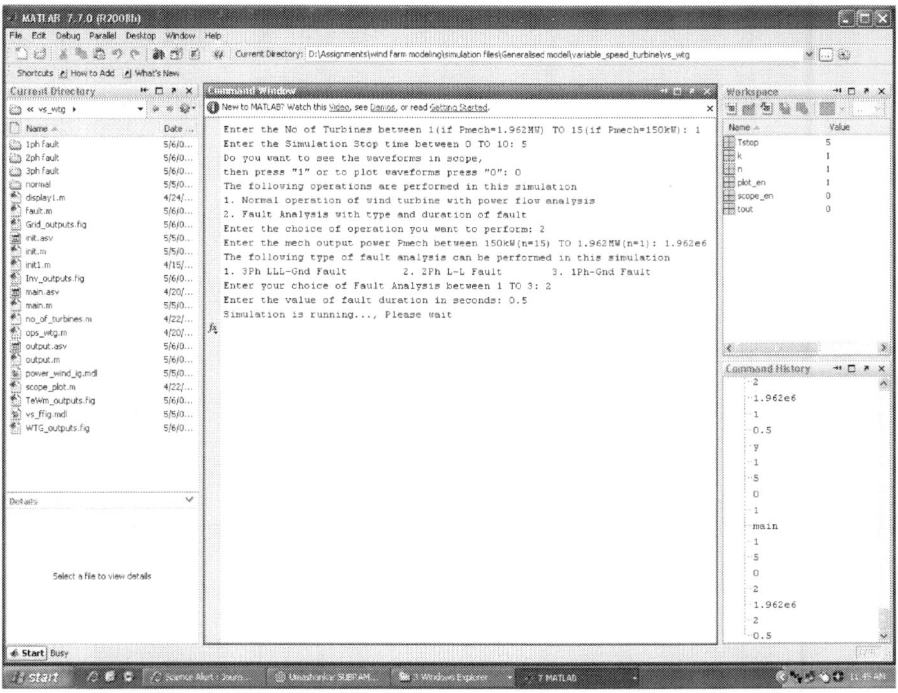

Fig. 23. Matlab command window as user interface, 2phase grid fault

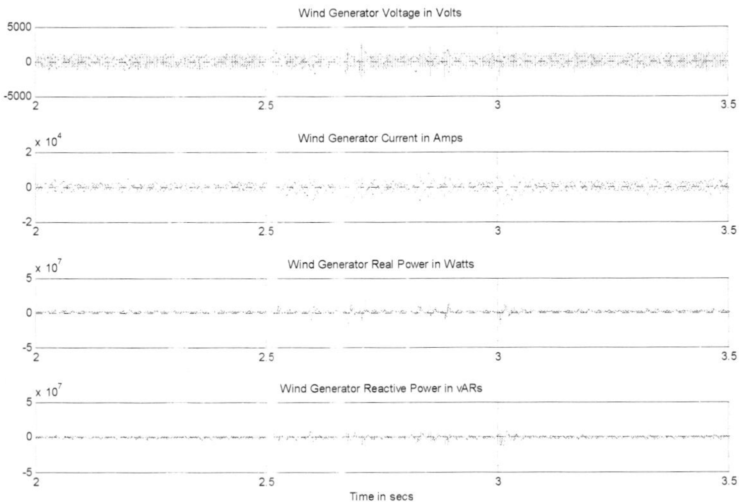

Fig. 24. Generator-side electrical outputs, 2phase grid fault

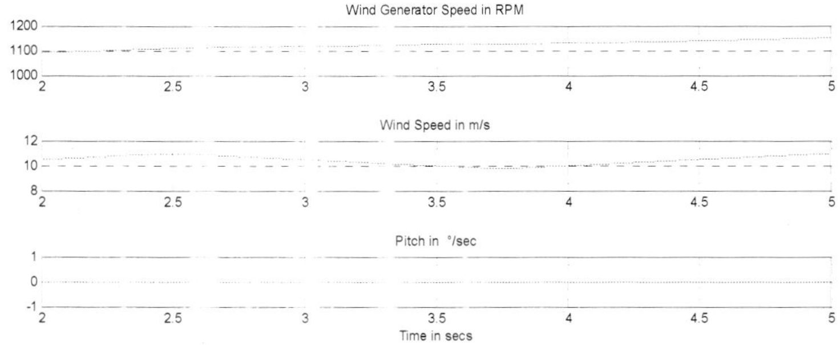

Fig. 25. Generator-side mechanical outputs, 2phase grid fault

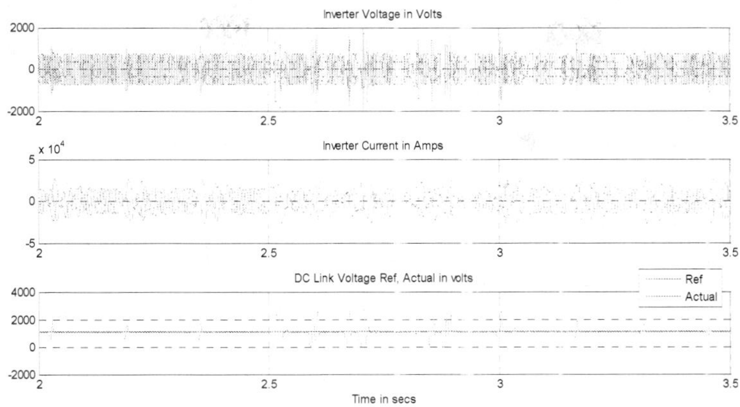

Fig. 26. Inverter Electrical outputs, 2phase grid fault

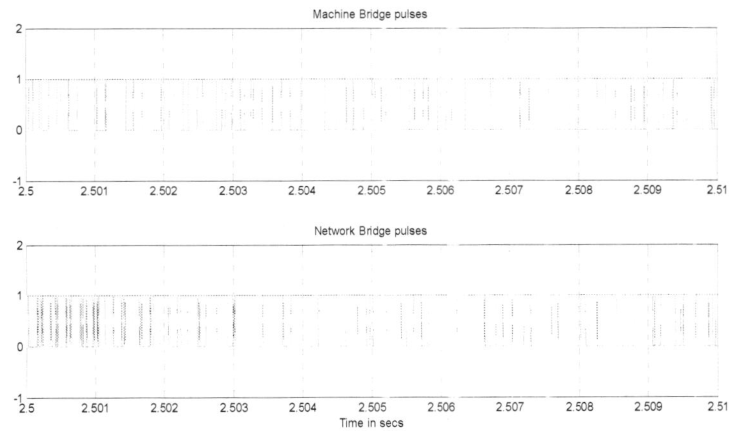

Fig. 27. PWM pulse outputs, 2phase grid fault

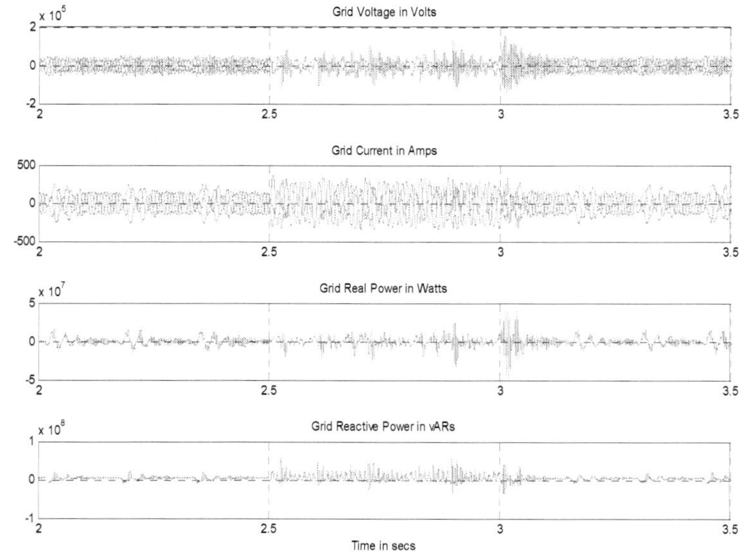

Fig. 28. Grid-side Electrical outputs, 2phase grid fault

4.3 Two-phase line to line grid fault

To analyse the model under two-phase line to line grid fault condition, follow the steps in the below snapshot of the matlab command window. After the simulation, command window will prompt for user input to display the results and to save the figures. To start with run the main.m file in command window and follow the steps below:

4.3.1 Comments on the result

At rated torque and speed, the FFIG delivers the rated power to grid under normal operation. At time t=2.5s, two-phase line to line fault is created at the grid side transmission line hence the grid current abruptly increased. The generator speed will increase above rated speed. Also the actual dc voltage maintained as that of reference.

It is evident from the fig.25. that there is a no mechanical torque reversal during the grid fault but the torque tends to reduce and the oscillations in all the waveforms during grid fault are not severe. Also the shaft between generator and wind turbine will ring at its natural frequency.

At time t=3s fault is cleared and all the quantities will slowly come to its previous condition of normal operation. Implementation of GFRT will rectify this issue and stabilize the electrical parameters during fault and after fault removal.

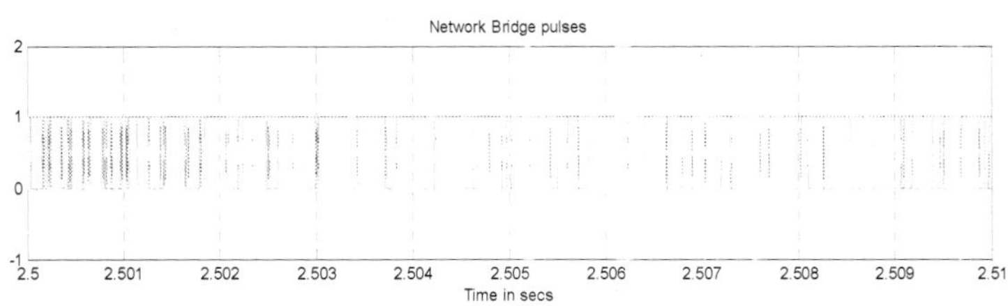

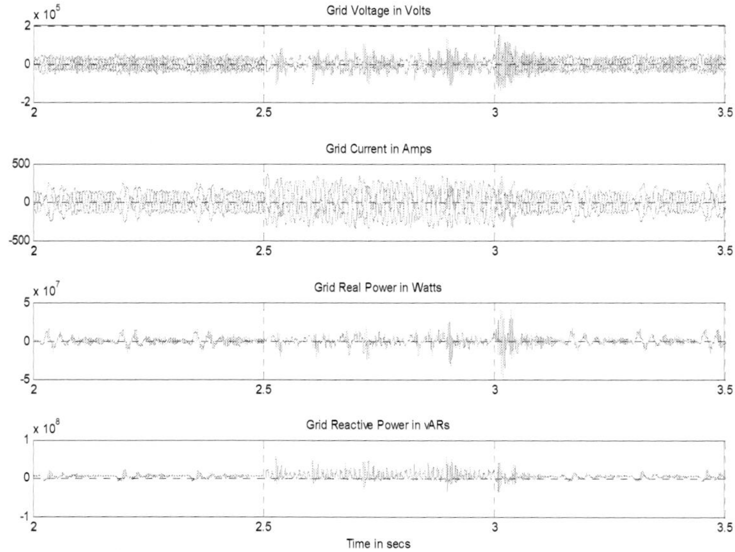

5. Conclusion and future work

The matlab model created for fully-fed induction generator based wind farm provides good performance under normal and transient (fault) operating conditions. It provides good results for different pwm techniques and fault conditions except the single-phase line to ground fault, which should be verified with help of practical hardware scaled down model or real time data from wind farms.

The present system also consist of machine and network bridge controllers in which the former maintains dc link energy by controlling torque, speed and the later provides controlled real, reactive power to grid.

The performance is achieved by considering assumptions mentioned in section (3), hence this matlab model should be implemented for Grid Fault ride-through and the performance should be evaluated by comparing the results. The system also requires a condition monitoring algorithm to check for faults & to protect the equipments from malfunctioning or damage and filters to arrest the spikes.

Also this model should be tested at different speeds above & below rated speed and performance of pitch controller should be verified as an extension to this work.

6. References

T. Ackermann (Editor), "Wind Power in Power Systems", John Wiley & Sons, Ltd, copyright 2005

V. Akhmatov, H. Knudsen, M. Bruntt, A.H. Nielsen, J.K. Pedersen, N.K. Poulsen, "A dynamic Stability Limit of Grid-Connected Induction Generators", IASTED, Mabella, Spain, September, 2000

M.B. Bana Sharifian, Y. Mohamadrezapour, M. Hosseinpour and S. Torabzade, "Maximum Power Control of Grid Connected Variable Speed Wind System through Back to Back Converters", Journal of Applied Sciences, Vol.8(23), 4416-4421, 2008

Divya, KC and Rao, Nagendra PS,"Study of Dynamic Behavior of Grid Connected Induction Generators", IEEE Power Engineering Society General Meeting, Denver, Colorado, USA, Vol.2, 2200 -2205, 6-10 June, 2004

P. Kundur, "Power System Stability and Control", EPRI, McGraw-Hill, Inc., Copyright 1994

Kim Johnsen, Bo Eliasson, "SIMULINK® Implementation of Wind Farm Model for use in Power System Studies", Nordic Wind Power Conference NWPC'04, Chalmers University of Technology, Göteborg, Sweden, 1-2 March 2004

Øyvind Rogne, Trond Gärtner," Stability for wind farm equipped with induction generators", Nordic PhD course on Wind Power, Smøla, Norway, June 5 - 11, 2005

Paulo Fischer de Toledo, Hailian Xie KTH, Kungl Tekniska Högskolan, "Wind Farm in Weak Grids compensated with STATCOM", Nordic PhD course on Wind Power, Smøla, Norway, June 5 - 11, 2005

J. Soens, J. Driesen, R. Belmans, "Generic Dynamic Wind Farm Model for Power System Simulations," Nordic Wind Power Conference NWPC'04, Chalmers University of Technology, Göteborg, Sweden, 1-2 March 2004

D. P. Kothari I. J. Nagrath, "Elecrtical Machines", 3rd Edition, Tata McGraw-Hill, NewDelhi, 2004

D. P. Kothari I. J. Nagrath, "Modern Power System Analysis", 3rd Edition, Tata McGraw-Hill, NewDelhi, 2003

5

Doubly-fed Induction Generator Drives for Wind Power Plants

Balduino Rabelo and Wilfried Hofmann
Dresden University of Technology
Germany

1. Introduction

This chapter presents the theoretical basics of the electric power generation using doubly-fed induction drives in wind turbines. This type of drive is the most utilized in wind power plants nowadays. This is mainly due to its fast dynamic response and reduced power electronics on the rotor side. This latter aspect is responsible for less harmonic pollution in the network and reduced acquisition costs.
To begin with, the mathematical background required for comprehending the formulation used on electrical drives theory is repeated. Then the electromechanical energy conversion process in an electrical generator is explained, followed by a steady-state analysis of the induction machine. Subsequently, the doubly-fed induction generator and the network side circuit are introduced. Finally, active and power flow issues are discussed. The feasibility of reactive power production is in accordance with the new requirements from the power companies on wind energy converters and the degrees of freedom in its production allow for optimisation of the power losses.

2. Basic definitions

2.1 Complex space vectors

The space vector theory can be applied to all 3-phase system variables where these latter assume a complex vector representation. Besides the reduction from a 3 to a 2-phase system the suitable choice of a non-stationary coordinates system as a reference enables us to work with DC quantities. The representation simplifies, to a great extent, the coupling equations between both stator and rotor, as well as the computation of the 3-phase power. As an example the current space vector $\underline{i}$ representing the 3-phase symmetrical currents

$$i_a(t) = \hat{I}\cos(\omega_S t) \tag{1}$$

$$i_b(t) = \hat{I}\cos(\omega_S t - \frac{2\pi}{3}) \tag{2}$$

$$i_c(t) = \hat{I}\cos(\omega_S t - \frac{4\pi}{3}), \tag{3}$$

where $\hat{I}$ is the current peak value and ω_S the angular frequency, can be determined as

$$\underline{i} = \frac{2}{3}\left[i_a(t) + \underline{a} i_b(t) + \underline{a}^2 i_c(t) \right] \tag{4}$$

where the operators $\underline{a} = e^{j\frac{2\pi}{3}}$ and $\underline{a}^2 = e^{j\frac{4\pi}{3}}$. The complex space vector can be related to the current instantaneous values by the expressions

$$i_a(t) = \Re\{\underline{i}\} + i_n \tag{5}$$

$$i_b(t) = \Re\{\underline{a}^2 \underline{i}\} + i_n \tag{6}$$

$$i_c(t) = \Re\{\underline{a}\underline{i}\} + i_n, \tag{7}$$

where i_n is the zero component of the current in a star connected system, as given by

$$i_n = \frac{1}{3}(i_a + i_b + i_c) \cdot \tag{8}$$

In a balanced symmetrical 3-phase system this term is not available.
The graphical representation of the vector addition of the single phase current components from expression (4) for i_a, $i_b > 0$ and $i_c < 0$ is depicted in figure 1. The $\alpha\beta$-coordinates frame is fixed on the stator with axis α on the same direction as that of the magnetic axis a. The dq-coordinates frame is rotating at an arbitrary angular speed

$$\omega_K = \frac{d\vartheta_K}{dt}, \tag{9}$$

where the phase displacement relative to the stator is a function of time

$$\vartheta_K(t) = \vartheta_0 + \int_0^t \omega_K dt \cdot \tag{10}$$

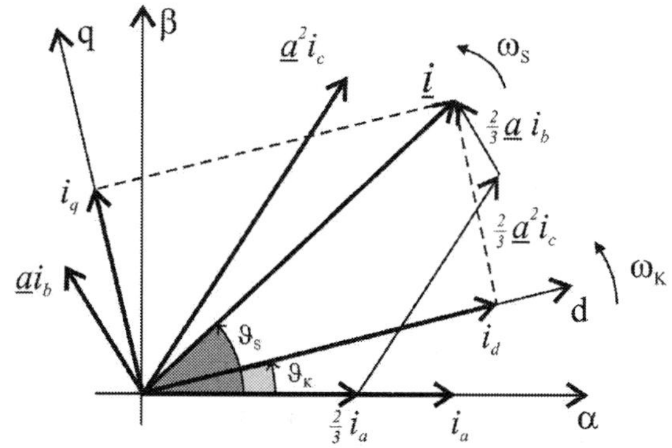

Fig. 1. Space vector in fixed and rotating coordinates

The current space vector $\underline{i}$ rotates at an angular speed

$$\omega_S = \frac{d\vartheta_S}{dt} \tag{11}$$

corresponding to the frequency of the 3-phase stator currents and can be represented as

$$\underline{i}^S = \underline{i}e^{j\vartheta_S}. \tag{12}$$

The complex exponential term performs the rotation of the space vector. Referring $\underline{i}$ to the rotating dq-frame require a coordinate transformation as follows

$$\underline{i}^K = \underline{i}e^{j(\vartheta_S - \vartheta_K)} = \underline{i}^S e^{-j\vartheta_K}. \tag{13}$$

The angular speed of the rotating coordinate system ω_K can be selected arbitrarily so that if $\omega_K = 0$ is chosen one gets a fixed coordinate system whose orientation will depend on the angle ϑ_0 in (10). If $\vartheta_0 = 0$, for example, one gets the $\alpha\beta$-coordinate system as a reference.
If ω_K is chosen to be equal the electrical rotor angular frequency

$$\omega = \frac{d\vartheta}{dt} = P_P\omega_m, \tag{14}$$

where P_P stands for the number of pole pairs and ω_m is the mechanical angular frequency of the rotor, the dq-coordinate system will be fixed on the rotor circuit.
If $\omega_K = \omega_S$ the difference angle $\vartheta_S - \vartheta_K$ is constant, i.e. the dq-coordinate frame rotates synchronously with the current space vector. Therefore, under steady state contitions the space vector $\underline{i}$ can be decomposed in 2 DC scalar components i_d and i_q as follows

$$i_d = \|\underline{i}\|\cos(\vartheta_S - \vartheta_K) \tag{15}$$

$$i_q = \|\underline{i}\|\sin(\vartheta_S - \vartheta_K), \tag{16}$$

where $\|\underline{i}\|$ is the bi-dimensional Euclidean norm of the space vector which gives the peak value of the phase current

$$\hat{I} = \|\underline{i}\| = \sqrt{i_d^2 + i_q^2}\,. \tag{17}$$

Therefore, i_d and i_q are the projections of $\underline{i}$ over the rotating coordinate system, as depicted in figure 1. Assuming the d-axis as real and the q-axis as imaginary a very useful complex notation for the space vectors can be derived

$$\underline{i} = i_d + ji_q\,. \tag{18}$$

2.1.1 Transformation operators.

Coordinate transformation is an operation used quite frequently in controlling modern electrical drives. The instantaneous 3-phase values are measured, sampled and given to the control algorithm at a rate ranging from a few to tens of kilohertz. If dq modeling is used, transformations have to be carried out using the same rate, which is a burden for the processor. Hence, instead of using transcendental functions, coordinate transformations are derived in matrix form having trigonometric functions as elements. These latter are computed as truncated series or as look-up tables

Assuming a 3-phase system a matrix operator $\mathbf{T}_0$ that maps the column-vector $\mathbf{i}_{abc}$ containing the instantaneous values of the 3-phase currents as elements onto the column-vector $\mathbf{i}_{dq0}$, whose elements are the $dq0$ components, as follows

$$\mathbf{i}_{dq0} = \mathbf{T}_0\mathbf{i}_{abc} \Rightarrow \begin{pmatrix} i_d \\ i_q \\ i_0 \end{pmatrix} = \mathbf{T}_0 \begin{pmatrix} i_a \\ i_b \\ i_c \end{pmatrix}. \tag{19}$$

After solving the equation above, one obtains the following matrix operator which defines the well-know Park transformation

$$\mathbf{T}_0 = \frac{2}{3}\begin{pmatrix} \cos(\omega_K t) & \cos(\omega_K t - \frac{2\pi}{3}) & \cos(\omega_K t - \frac{4\pi}{3}) \\ -\sin(\omega_K t) & -\sin(\omega_K t - \frac{2\pi}{3}) & -\sin(\omega_K t - \frac{4\pi}{3}) \\ 0.5 & 0.5 & 0.5 \end{pmatrix}. \tag{20}$$

The third line of the transformation matrix accommodates the zero-sequence component. The back-transformation matrix from the $dq0$ to the abc system is accomplished by using the inverse of (20)

$$\mathbf{i}_{abc} = \mathbf{T}_0^{-1}\mathbf{i}_{dq0}, \tag{21}$$

where

$$\mathbf{T}_0^{-1} = \frac{3}{2}\mathbf{T}_0^T = \begin{pmatrix} \cos(\omega_K t) & -\sin(\omega_K t) & \frac{1}{2} \\ \cos(\omega_K t - \frac{2\pi}{3}) & -\sin(\omega_K t - \frac{2\pi}{3}) & \frac{1}{2} \\ \cos(\omega_K t - \frac{4\pi}{3}) & -\sin(\omega_K t - \frac{4\pi}{3}) & \frac{1}{2} \end{pmatrix}. \tag{22}$$

Other useful operations to the electrical drives theory are the partial transformations $\mathbf{T}_1$, between a 3 to a 2-phase stationary system, and $\mathbf{T}_2$, between the stationary to a rotating coordinate system, defined as follows

$$\mathbf{i}_{\alpha\beta 0} = \mathbf{T}_1\mathbf{i}_{abc} \Rightarrow \mathbf{T}_1 = \frac{2}{3}\begin{pmatrix} 1 & -\frac{1}{2} & -\frac{1}{2} \\ 0 & \frac{\sqrt{3}}{2} & -\frac{\sqrt{3}}{2} \\ \frac{1}{2} & \frac{1}{2} & \frac{1}{2} \end{pmatrix} \tag{23}$$

$$\mathbf{i}_{dq0} = \mathbf{T}_2 \mathbf{i}_{\alpha\beta 0} \Rightarrow \mathbf{T}_2 = \begin{pmatrix} \cos(\omega_K t) & \sin(\omega_K t) & 0 \\ -\sin(\omega_K t) & \cos(\omega_K t) & 0 \\ 0 & 0 & 1 \end{pmatrix} \tag{24}$$

The respective back-transformations are found by inverting the matrices (23) and (24)

$$\mathbf{i}_{abc} = \mathbf{T}_1^{-1} \mathbf{i}_{\alpha\beta 0} \Rightarrow \mathbf{T}_1^{-1} = \begin{pmatrix} 1 & 0 & 1 \\ -\frac{1}{2} & \frac{\sqrt{3}}{2} & 1 \\ -\frac{1}{2} & -\frac{\sqrt{3}}{2} & 1 \end{pmatrix} \tag{25}$$

$$\mathbf{i}_{\alpha\beta 0} = \mathbf{T}_2^{-1} \mathbf{i}_{dq} \Rightarrow \mathbf{T}_2^{-1} = \begin{pmatrix} \cos(\omega_K t) & -\sin(\omega_K t) & 0 \\ \sin(\omega_K t) & \cos(\omega_K t) & 0 \\ 0 & 0 & 1 \end{pmatrix} \tag{26}$$

2.2 Electrical power computation

The computation of the electrical power is an extensive subject and the definitions according to the DIN 40110 standard *Wechselstromgrössen - Zweileiter Stromkreise* (1994) used throughout the text will be explained here. The matter is addressed in detail in the above mentioned norm and in other literature Späth (2000).

2.2.1 Non-sinusoidal wave forms.

Due to the rectangular voltage wave forms in power converters output and their effects on voltage drops and currents in connected circuits, it is necessary to have suitable mathematical tools in order to analyse these kind of phenomena. Thus, for any non-sinusoidal periodic wave form that submit to the Dirichlet conditions there exists a convergent Fourier series. Thus, the Fourier series representation of any periodic single phase voltage and current wave forms can be written as

$$u(t) = U_0 + \sum_{\mu=1}^{\infty} \hat{U}_{\mu} \cos(\mu\omega t + \alpha_{\mu}) \tag{27}$$

$$i(t) = I_0 + \sum_{\nu=1}^{\infty} \hat{I}_{\nu} \cos(\nu\omega t + \beta_{\nu}), \tag{28}$$

where U_0 and I_0 are the *DC* components and the summations represent the time periodic components $\tilde{u}(t)$ and $\tilde{i}(t)$, i.e., the *AC* components of $u(t)$ and $i(t)$.

The instantaneous power that is being transferred by the voltage and current is given by

$$p(t) = u(t)i(t). \tag{29}$$

By definition the active power P is the mean value of the instantaneous power over the time period of the fundamental frequency. Substituting the expressions (27) and (28) on (29), after some development one arrives at

$$P = \bar{p}(t) = U_0 I_0 + \frac{1}{2}\sum_{\mu=1}^{\infty}\hat{U}_\mu \hat{I}_\mu \cos(\alpha_\mu - \beta_\mu)\cdot \tag{30}$$

From the equation above one observes that only the *DC* and the components with same frequency contribute to the average active power. If only the *DC* components are available, the active power is given solely by the first equation term, whereas if there are no such components the summation of the voltage and current products with the same harmonic order define the power. Considering no DC components, the fundamental components of sinusoidal voltage and current are given by

$$u_1(t) = u(t) - \sum_{\mu=2}^{\infty}\hat{U}_\mu \cos(\mu\omega t + \alpha_\mu) = \hat{U}_1 \cos(\omega t + \alpha_1) \tag{31}$$

$$i_1(t) = i(t) - \sum_{\nu=2}^{\infty}\hat{I}_\nu \cos(\nu\omega t + \beta_\nu) = \hat{I}_1 \cos(\omega t + \beta_1)\cdot \tag{32}$$

The harmonic components $\tilde{u}(t)$ and $\tilde{i}(t)$ denoted by the summations on both equations above are considered distortions of the fundamental sinusoidal wave forms

$$u(t) = u_1(t) + \tilde{u}(t), \quad \tilde{u}(t) = \sum_{\mu=2}^{\infty}\hat{U}_\mu \cos(\mu\omega t + \alpha_\mu) \tag{33}$$

$$i(t) = i_1(t) + \tilde{i}(t), \quad \tilde{i}(t) = \sum_{\nu=2}^{\infty}\hat{I}_\nu \cos(\nu\omega t + \beta_\nu)\cdot \tag{34}$$

2.2.2 Nature of reactive power.

The standard *Wechselstromgrössen - Zweileiter Stromkreise* (1994) defines the reactive power as the mean value of the product of the voltage delayed by a quarter of its fundamental period times the current

$$Q = \bar{q}(t) = \frac{1}{T}\int_t^{T+t} u(t - \frac{T}{4}) i(t) dt, \quad T = \frac{2\pi}{\omega}\cdot \tag{35}$$

It represents the amount of energy that cannot be converted being exchanged between electric and magnetic fields, i.e. source and load, over the time.

More general discussion about the nature of the electrical powers will be avoided at this point. For the purposes of this work the definitions presented above are considered sufficient. Furthermore, this is a very extensive matter that is once and again treated in the literature Späth (2000). Unsymmetrical systems requiring determination of positive, negative and zero sequences will also not be dealt here, since faulty conditions are not going to be treated.

2.2.3 Complex power.

Taking equations (15) to (18) for the currents one can derive the following expressions for the voltage and current effective complex vectors

$$\underline{U} = U(\cos\varphi_u + j\sin\varphi_u) = Ue^{j\varphi_u} \quad (36)$$

$$\underline{I} = I(\cos\varphi_i + j\sin\varphi_i) = Ie^{j\varphi_i}, \quad (37)$$

By multiplying (36) by the complex conjugate of (37) yields

$$\underline{U}\cdot\underline{I}^* = UIe^{j(\varphi_u-\varphi_i)} = UI\cos\varphi + jUI\sin\varphi \cdot \quad (38)$$

Thus, the complex power $\underline{S}$ can be defined as

$$\underline{S} = \underline{U}\cdot\underline{I}^* = P + jQ \cdot \quad (39)$$

The same can be deduced for the 3-phase power. Furthermore, due to the fact that in steady-state, the instantaneous values equal the average values, and the powers can be computed using the instantaneous values of the complex vectors as follows

$$\underline{S} = P + jQ = \frac{3}{2}\{\underline{u}\cdot\underline{i}^*\}, \quad (40)$$

where the average active and reactive powers are the respective real and imaginary parts of (40). The instantaneous active and reactive power values can be written as functions of the scalar components of voltages and currents on the *dq*-axis as follows

$$p(t) = \frac{3}{2}\Re\{\underline{u}\cdot\underline{i}^*\} = \frac{3}{2}\{u_d i_d + u_q i_q\} \quad (41)$$

$$q(t) = \frac{3}{2}\Im\{\underline{u}\cdot\underline{i}^*\} = \frac{3}{2}\{u_q i_d - u_d i_q\} \cdot \quad (42)$$

3. Electric generators

The next step in energy conversion in a wind turbine is the transformation of the mechanical energy of the rotating masses into electrical energy by means of an electrical machine. To produce the required electromagnetic torque as a reaction of the electrical subsystem to the mechanical driving torque in order to absorb the input mechanical energy, the generator must be excited, i.e., magnetic energy must be available in the air-gap between stator and rotor. Therefore, the magnetic field is responsible for coupling the electrical and mechanical subsystems and consequently reactive power is required in order to magnetise the generator. The exact way in which the electromechanical energy conversion process takes place in a given electrical machine depends on its construction. The distribution of the magnetic induction on the air-gap depends on its geometry, on the distribution of the winding conductors and on the iron's magnetic characteristics. Although the detailed theoretical deduction of the rotating field electrical machines is usually found in the literature Leonhard (1980); Lipo (1995); Müller (1977) the basics of this in the case of an induction machine is repeated here for a completion of the text.

3.1 Induction machine.

Wound rotor induction and synchronous machines are the mostly used generator types in wind power plants. Induction machines are very well known, and one of the most employed types in several areas, due mainly to its robustness, power-to-volume ratio and relative low cost. However, induction machines do not possess separated field circuits as do synchronous or DC machines. These are lumped together with the armature circuit in the stator windings so that a decoupled control of torque and flux, or active and reactive powers, requires some mathematical treatment of the machine model. However, this is no longer a drawback thanks to the development of the power electronics and microprocessors of the last 20 years. Today, high dynamics and reliable machine drives can be assembled with relative low cost.

There are basically two types of induction machines, the squirrel cage and the wound rotor. In squirrel cage induction machines the rotor circuit is made up of aluminum or copper bars cast in the rotor iron and short-circuited by non-accessible end-rings. The required magnetization is normally drawn from outside through the stator windings. This can be accomplished using a capacitor bank parallel to the machine terminals in a self-excited scheme or feeding the stator with an inverter or by simply connecting the machine terminals to the grid supply. The squirrel cage induction machine is currently the industry's horse power.

The construction of the wound rotor induction machine is likely the squirrel cage one. Instead of short-circuited bars, the windings in the rotor slots end up on copper slip-rings mounted to the shaft. The connection to the static part is accomplished through fixed carbon brushes that make pressure contact by means of adjustable springs to the turning rings. In this way the rotor circuit is also available to external connections like additional rotor resistors for startup, speed control, slip power recovery schemes or controlled voltage and frequency sources. The wound rotor induction machine was often used as a rotating frequency converter in the past but became obsolete upon the development of semiconductors and static converters. Its renaissance took place first on pump storage hydroelectric power plants, where high power variable speed drives that operate as motor and generator were required. The reduced power converters needed in the speed control within a slip power recovery scheme makes the wound rotor machine very attractive. Later on, it established a place on the market as a doubly-fed generator for wind power plants.

3.1.1 Rotating field electrical machines

Assuming a 2-pole induction machine having 3-phase symmetrical windings $a - a'$, $b - b'$ and c-c' with the respective magnetic axis a, b and c 120° spatially separated from each other, mounted on the stator and on the rotor. The stator windings are fed by symmetrical 3-phase currents like (5) to (7). The rotor is assumed cylindrical and the air-gap radial dimension g constant, i.e., eccentricity and slot harmonics as well as end-winding effects are not taken into account. Permeability of the iron is much greater than that of the air, so that the flux density is concentrated on the air-gap crossings and saturation effects in the iron can be neglected.

3.1.2 Steady-state equivalent circuit.

Based on the self inductances and on the flux linkages, the voltage equations under steady-state conditions can be written in complex form as follows

$$\underline{u}_S = R_S \underline{i}_S + j\omega_S \underline{\Psi}_S \tag{43}$$

$$\underline{u}'_R = R'_R \underline{i}'_R + j\omega_R \underline{\Psi}_R, \tag{44}$$

where ω_R is the slip frequency and the flux linkages are given by

$$\underline{\Psi}_S = L_S \underline{i}_S + L_m \underline{i}'_R \tag{45}$$

$$\underline{\Psi}_R = L'_R \underline{i}'_R + L_m \underline{i}_S. \tag{46}$$

From the direct proportionality of the flux density to the current vectors the magnetizing current vector i_μ can be computed as the vector addition of the stator and rotor current vectors related to the stator side as follows

$$\underline{i}_\mu = \underline{i}_S + \left(\frac{N_R}{N_S}\right)\underline{i}_R = \underline{i}_S + \underline{i}'_R, \tag{47}$$

where N_S and N_R are the stator and rotor number of turns. By substituting the imaginary coefficients with the respective mutual and leakage inductive reactances

$$X_m = \omega_S L_m \tag{48}$$

$$X_\sigma = \omega_S L_\sigma \tag{49}$$

and the magnetising current (47) in the voltage and flux linkage equations yields

$$\underline{u}_S = R_S \underline{i}_S + jX_{\sigma_S} \underline{i}_S + jX_m \underline{i}_\mu \tag{50}$$

$$\frac{\underline{u}'_R}{s} = \frac{R'_R}{s}\underline{i}'_R + jX'_{\sigma_R}\underline{i}'_R + jX_m \underline{i}_\mu \tag{51}$$

where the slip is defined as $s = \frac{\omega_R}{\omega_S}$.

Departing from the magnetic induction resulting from the superposition of the stator and rotor magnetic induction waves and from the rotation induced voltage, the main flux linkage vector $\underline{\Psi}_m$ induces the stator and rotor internal voltages, respectively $\underline{e}_S$ and $\underline{e}'_R$

$$\underline{e}_S = -j\omega_S \underline{\Psi}_m = -j\omega_S L_m \underline{i}_\mu, \tag{52}$$

$$\underline{e}'_R = -js\omega_S \underline{\Psi}_m = -js\omega_S L_m \underline{i}_\mu. \tag{53}$$

These expressions show that both stator and rotor induced voltages are related by the slip s as follows

$$\underline{e}'_R = \left(\frac{N_R}{N_S}\right)\underline{e}_R = s\underline{e}_S. \tag{54}$$

The complex voltage drop over the stator winding can be determined by substituting (52) in the voltage equation (50).

$$\underline{u}_S = R_S \underline{i}_S + jX_{\sigma_S} \underline{i}_S - \underline{e}_S \cdot \tag{55}$$

Similarly, by substituting (53) in the rotor voltage equation (51) and multiplying both sides of the equation by s gives

$$\underline{u}'_R = R'_R \underline{i}'_R + jsX'_{\sigma_R} \underline{i}'_R - \underline{e}'_R \cdot \tag{56}$$

The single phase equivalent circuit of the induction machine based on equations (50) and (51) as well as on equations (55) and (56) is depicted in figure 2.

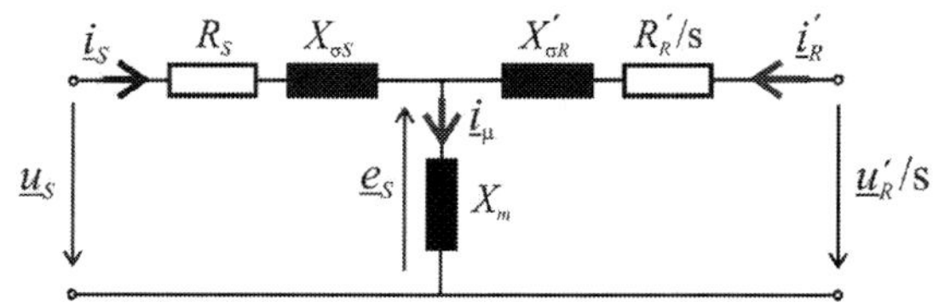

Fig. 2. Equivalent circuit of the induction machine

The constant mains voltage drives the stator current and the machine magnetisation. The resulting rotor induced voltage has to be counter-balanced by means of the voltage drop over the rotor impedance and by the terminal voltage. The interaction of the stator and rotor currents, due to the superposition of the magnetic induction waves, results in the magnetizing current as shown in expression (47).

Taking the equation (56) and considering short-circuited rotor bars as in the cage induction machine, i.e., $\underline{u}'_R = 0$ yields

$$\underline{0} = R'_R \underline{i}'_R + jsX'_{\sigma_R} \underline{i}'_R - \underline{e}'_R \cdot \tag{57}$$

In this particular case the complex voltage drop over the rotor impedance must equal the rotor induced voltage. Figure 3 shows the space vector diagram of the stator and rotor values for a squirrel cage induction machine.

The stator power factor angle φ_S is defined as the angle between the terminal voltage and the current vectors. The rotor power factor angle φ_R between the rotor induced voltage and rotor current vectors is a function of the rotor parameters and of the slip given by

$$\varphi_R = \arctan\left(s \frac{X_{\sigma_R}}{R_R}\right) \cdot \tag{58}$$

From this expression and from figure 3, one finds that by increasing the slip, i.e., loading the machine, results in an increase in the rotor power factor angle, theoretically up to $\varphi_R = \pi / 2$ for $s = \infty$, and a decrease on the stator power factor angle φ_S. Towards no-load condition the stator power factor decreases to nearly zero while the rotor power factor is nearly unity.

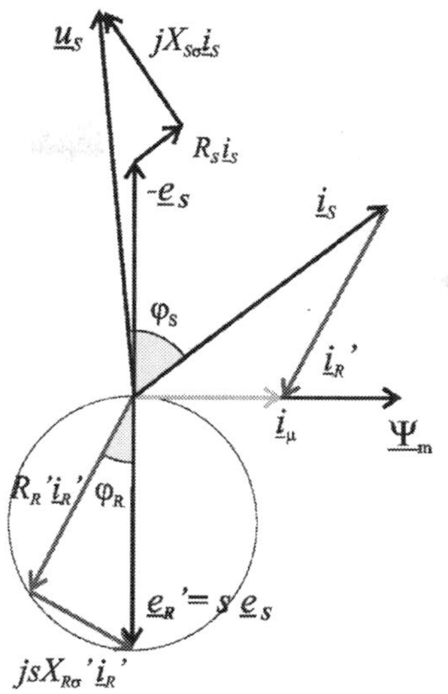

Fig. 3. Current space vectors in an induction machine

3.2 Energy and power balance.

In a physical system the variation of stored magnetic energy plus losses accounts for the difference between the input and the output energy variations

$$dW_{mag} + dW_{loss} = dW_{ele} - dW_{mech}, \tag{59}$$

where W is the energy, the subscripts *mag*, *ele* and *mech* denote its magnetic, electric and mechanical forms, respectively, and *loss* stands for inherent process losses, i.e., the energy converted into heat. The electromechanical energy conversion process is based on the energy conservation principle pointed out by this expression.

To keep a constant excitation level, i.e., maintaining the amount of magnetic energy constant, any variation in the input energy to the system must be followed by a variation in the system output thereby accounting for the process losses as shown in the following equation

$$dW_{loss} = dW_{ele} - dW_{mech}, \tag{60}$$

A detailed description of the electromechanical energy conversion process can be found in the literature Müller (1977).

As power is defined as the rate of change of energy

$$p(t) = \frac{dW}{dt} \Rightarrow dW = p(t)dt, \tag{61}$$

equation (60) can be rewritten more explicitly as follows

$$p_{loss}dt = uidt - m_e\omega_m dt, \tag{62}$$

where p_{loss} represents the instantaneous power losses, u and i stand for terminal voltage and current and m_e and ω_m are the electromagnetic torque and the mechanical angular speed of the shaft, respectively.
Defining the average power as the average energy exchange over the time interval Δt, the equation (62) takes the form of a difference equation

$$P_{loss}\Delta t = P_{ele}\Delta t - P_{mech}\Delta t, \tag{63}$$

where P_{ele} and P_{mech} are the average electrical an mechanical powers during Δt, respectively. Eliminating the time in the expression (63) leads to the power balance equation

$$P_{loss} = P_{ele} - P_{mech} \, . \tag{64}$$

3.2.1 Active power.

Considering the motor operation of the induction machine as the positive direction for the active power flow, the power balance equation (64) for steady-state can be assumed. The power losses will be defined and situated according to their cause and where they take place. Therefore, stator and rotor copper losses as well as the mechanical losses due to friction and windage on the shaft will be presented herein. In order to do so, intermediary power values in different locations are defined like the air-gap power, the rotor input power on the rotor terminals and the internal mechanical power. Iron losses will be neglected here for the sake of simplicity and explained later, when they will be taken into account in the optimization procedure.
The stator input instantaneous electrical power p_S is computed as shown in (41) using equation (55)

$$p_S = \frac{3}{2}\Re\{\underline{u}_S\underline{i}_S^*\} = \frac{3}{2}\Re\{R_S\underline{i}_S\underline{i}_S^* + jX_{\sigma_S}\underline{i}_S\underline{i}_S^* - \underline{e}_S\underline{i}_S^*\} \, . \tag{65}$$

The first term in (65) accounts for the energy dissipated as heat over the stator windings resistance resulting in the stator copper losses

$$P_{Cu_S} = \frac{3}{2}R_S\hat{I}_S^2 \, . \tag{66}$$

By subtracting (66) from (65) results in the air-gap power p_g given by the third term in (65), since the second term is a pure imaginary number

$$p_g = p_S - P_{Cu_S} = -\frac{3}{2}\Re\{\underline{e}_S\underline{i}_S^*\} \, . \tag{67}$$

The power balance for the rotor circuit can be found in the following relation

$$p_g + p_R = p_{mech} + P_{Cu_R} \, , \tag{68}$$

where p_R is the input rotor power on the slip-ring terminals, p_{mech} is the internal mechanical power and P_{CuR} stands for the rotor copper losses. The input rotor power is computed in much the same was as the stator input power. From the voltage equation (56)

$$p_R = \frac{3}{2}\Re\{\underline{u}'_R \underline{i}'^*_R\} = \frac{3}{2}\Re\{R'_R \underline{i}'_R \underline{i}'^*_R + jsX'_{\sigma_R} \underline{i}'_R \underline{i}'^*_R - \underline{e}'_R \underline{i}'^*_R\} \cdot \tag{69}$$

By analogy to the stator, the rotor copper losses are given by

$$P_{Cu_R} = \frac{3}{2} R'_R \hat{I}'^2_R \cdot \tag{70}$$

On substituting (69) and (70) in (68) yields

$$p_{mech} - p_g = -\frac{3}{2}\Re\{\underline{e}'_R \underline{i}'^*_R\} \cdot \tag{71}$$

Application of the current relation (47) and the induced voltage relation (54) in (71) allows for the following development

$$p_{mech} - p_g = -\frac{3}{2}\Re\{s\underline{e}_S(\underline{i}_\mu - \underline{i}_S)^*\} = \frac{3}{2}s\Re\{\underline{e}_S \underline{i}^*_S\} \cdot \tag{72}$$

The internal mechanical power can then be expressed as a function of the air-gap power and of the slip

$$p_{mech} - p_g = -sp_g \Rightarrow p_{mech} = (1-s)p_g \cdot \tag{73}$$

Taking the equation (73) and substituting it in (68) leads to the following result

$$p_R = -sp_g + P_{Cu_R} \cdot \tag{74}$$

From (73) and (74) one may conclude that the distribution of the power transferred through the air-gap depends on the slip. In relation to the wound rotor machine, it translates into the ratio between the rotor and stator active powers

$$s = -\frac{p_R - P_{Cu_R}}{p_S - P_{Cu_S}} \cdot \tag{75}$$

Equation (75) is a basic relation that fosters an understanding on how the wound rotor induction machine operates. The negative sign comes from the motor convention used.

For motoring condition ($P_S > 0$) power is fed from the rotor ($P_R < 0$) under sub-synchronous operation ($s > 0$) and from the mains to the rotor ($P_R > 0$) under super-synchronous operation ($s < 0$). For generating condition ($P_S < 0$) power is fed to the rotor under sub-synchronous operation and from the rotor to the mains under super-synchronous operation. These results are summarised in table 1.

In the case of a squirrel cage induction machine, where the power on the rotor terminals equals zero, the slip provides the relation between the rotor copper losses and the air-gap power. In other words, the amount of air-gap power given by the slip must cover the rotor copper losses

$$s = \frac{P_{Cu_R}}{p_g} \cdot \tag{76}$$

Operating Condition	Motor ($P_S > 0$)	Generator ($P_S < 0$)
Sub-synchronous ($s > 0$)	$P_R < 0$	$P_R > 0$
Super-synchronous ($s < 0$)	$P_R > 0$	$P_R < 0$

Table 1. Power flow under different operating conditions

The mechanical losses account for the friction and windage losses during machine operation. The first is a function of the squared angular speed while the second is a function of the cubic angular speed. The separation of these losses is a complex task as well as the determination of the respective proportionality factors. Therefore, a single friction coefficient proportional to the power 2.5 of the angular speed will be considered

$$p_f = B_G \omega_m^{2.5} \cdot \tag{77}$$

And on being subtracted from the internal mechanical power, this results the output mechanical power delivered to the load by the rotating shaft p_{load}

$$p_{load} = p_{mech} - p_f \cdot \tag{78}$$

3.2.2 Reactive power.

The reactive power flow is responsible for the maintenance of the magnetic energy stored in the air-gap as well as for the building up of the flux linkages required for the energy conversion process.

The reactive power flowing through the stator terminals is computed using (42) on (55) and results in the amount of reactive power required for building up the stator flux linkage $\underline{\Psi}_S$

$$q_S = \frac{3}{2}\Im\{\underline{u}_S \underline{i}_S^*\} = \frac{3}{2}\Im\{R_S \underline{i}_S \underline{i}_S^* + jX_{\sigma_S} \underline{i}_S \underline{i}_S^* - \underline{e}_S \underline{i}_S^*\} \cdot \tag{79}$$

From equation (79) one sees that only the second and third terms are reactive powers; the second term being responsible for creating the stator leakage flux $\underline{\Psi}_{\sigma S}$

$$q_{\sigma_S} = \frac{3}{2}\Im\{jX_{\sigma_S} \underline{i}_S \underline{i}_S^*\} \tag{80}$$

and its average value is then

$$Q_{\sigma_S} = \frac{3}{2} X_{\sigma_S} \hat{I}_S^2 \cdot \tag{81}$$

The remaining term in (79) represents the one which actually contributes to the magnetization of the machine.

$$q_S - q_{\sigma_S} = -\frac{3}{2}\Im\{\underline{e}_S\underline{i}_S^*\} = -\frac{3}{2}\Im\{\underline{e}_S(\underline{i}_\mu - \underline{i}'_R)^*\} \cdot \tag{82}$$

According to (47), (82) goes into

$$q_S - q_{\sigma_S} = -\frac{3}{2}\Im\{\underline{e}_S\underline{i}_\mu^*\} + \frac{3}{2}\Im\{\underline{e}_S\underline{i}'^*_R\} \cdot \tag{83}$$

The first part of the equation (83) stands for the actual amount of reactive power involved in building up the mutual flux linkage $\underline{\Psi}_m$ and according to (52), the magnetising reactive power can be written as follows

$$q_m = -\frac{3}{2}\Im\{\underline{e}_S\underline{i}_\mu^*\} = \frac{3}{2}\Im\{j\omega_S L_m \underline{i}_\mu \underline{i}_\mu^*\} \cdot \tag{84}$$

The average magnetising power can then be written as

$$Q_m = \frac{3}{2}X_m \hat{I}_\mu^2 \cdot \tag{85}$$

The second part stands for the contribution of the rotor current vector towards magnetisation, and if one defines it as in Vicatos & Tegopoulos (1989) as the amount of reactive power being delivered to the air-gap

$$q_g = \frac{3}{2}\Im\{\underline{e}_S\underline{i}'^*_R\}, \tag{86}$$

the stator reactive power can be expressed as follows

$$q_S = q_{\sigma_S} + q_m + q_g \cdot \tag{87}$$

Analogously to the stator side the rotor reactive power is the imaginary part of the apparent power in the rotor terminals that contributes to building up the rotor flux linkage $\underline{\Psi}_R$

$$q_R = \frac{3}{2}\Im\{\underline{u}_R\underline{i}'^*_R\} = \frac{3}{2}\Im\{R'_R\underline{i}'_R\underline{i}'^*_R + jsX'_{\sigma_R}\underline{i}'_R\underline{i}'^*_R - \underline{e}'_R\underline{i}'^*_R\} \cdot \tag{88}$$

On analysing equation (88) one notices that the first term is a real number whereas the second stands for the reactive power required for building up the rotor leakage flux $\underline{\Psi}_{\sigma R}$

$$q_{\sigma_R} = \frac{3}{2}\Im\{jsX'_{\sigma_R}\underline{i}'_R\underline{i}'^*_R\} \cdot \tag{89}$$

The rotor average reactive leakage power is given by

$$Q_{\sigma_R} = \frac{3}{2}sX'_{\sigma_R}\hat{I}'^2_R \cdot \tag{90}$$

The third term is, similarly to the stator side in (83), a contribution of the rotor side to the machine magnetisation. Applying the induced voltage relation (54) and substituting the expression (86) yields

$$q_R - q_{\sigma_R} = -\frac{3}{2}\Im\{\underline{e}'_R \underline{i}'^*_R\} = -s\frac{3}{2}\Im\{\underline{e}_S \underline{i}'^*_R\} = -sq_g \,. \tag{91}$$

From equations (87) and (91) the following relation between stator and rotor reactive powers can be found

$$q_R - q_{\sigma_R} = -s(q_S - q_{\sigma_S} - q_m) \Rightarrow s = -\frac{q_R - q_{\sigma_R}}{q_S - q_{\sigma_S} - q_m} \,. \tag{92}$$

It can be concluded that the reactive powers on the stator and rotor sides are related by the slip, similarly to the active powers except by the reactive power required by the magnetisation.

3.3 The doubly fed induction generator

One of the preferable solutions employed as wind turbine generator is the wound rotor induction machine with the stator windings directly connected to the network and rotor windings connected to controllable voltage source through the slip-rings, also known as the doublyfed induction generator (DFIG), the object of this work. As already mentioned, one of the most attractive features of this drive is the required power electronics converter rated part of the generators nominal power, reducing the acquisition costs, inherent losses and harmonic pollution as well as volume.

The other most employed type of drive that shares the market with the DFIG is the gear-less high pole-numbered synchronous generator (SG). It is also an elegant and successful solution for the wind energy branch. The stator windings present high number of poles enabling the generator to turn with mechanical speeds of the same order of the turbine rotor. Thus there is no need for a gear-box and the generator shaft is directly connected to the turbine axis.

However, this drive requires a full rated power electronics converter between the stator and the grid in order to convert the generated variable voltage and frequency to the net constant values. This latter assumption, allied to the fact that the machine requires a very large diameter to accommodate the high number of poles, makes the manufacturing and assembling process costly. Furthermore, the drive train and generator must be dimensioned in order to experience the turbine torque peaks at this speed range. On the other hand, the fast turning DFIG, due to its small number of poles, experiences reduced torque values compared to the gear-less SG and is produced in series by several manufacturers, since it is not a particularly costly process.

In addition, the DFIG can be operated as a synchronous machine, in that it is magnetized through the rotor, but has the advantage of not having the stiff torque versus speed characteristics. In this way it is possible to "slip" over the synchronous speed, thus avoiding mechanical and electrical stresses to the drive train and network. In the synchronous operation of the DFIG the resulting magnetic induction vector direction is not coupled to the rotor position as it is in the synchronous machine (due to the construction of a DC exciting circuit or permanent magnet on the rotor). Neglecting the nut harmonic effects and considering concentrated windings, feeding 3-phase currents with slip frequency to the rotor windings, not only the amplitude also the position of the rotor field vector can be varied synchronously with the stator field vector, independently of the rotor position or speed. In other words, active and reactive power, i.e. electromagnetic torque and excitation, can be controlled decoupled from each other and from rotor angle position Leonhard (1980). For the reasons pointed out above, the DFIG is one of the most used generator types in MW class wind power plants according to statistical figures disclosed in Germany recently Rabelo & Hofmann (2002). There is also an increasing tendency to employ the DFIG in upcoming higher powered turbines. Other promising variants that do not require the power electronics converter using the synchronous generator combined with a hydrodynamic controlled planetary gear in order to keep the synchronous speed left recently the prototype phase and are being already manufactured in series Rabelo et al. (2004).

3.3.1 Simplified analysis.

In order to explain the operation of a wound rotor induction machine, some simplifications on the steady-state circuit will be provided. This is merely for better comprehension of the machine operation and do not invalidate the theory presented to this point. Deviations from the complete original model will be pointed out.

Neglecting the voltage drop over the stator winding resistance and the stator leakage reactance in the equation (55), which is a reasonable assumption in the case of high powered machines, yields

$$\underline{u}_S = -\underline{e}_S = j\omega_S \underline{\Psi}_m = j\omega_S L_m \underline{i}_\mu \cdot \tag{93}$$

As a result, the induced voltage vector $\underline{e}_S$ has the same amplitude and opposite direction of the terminal voltage vector $\underline{u}_S$. And the magnetising flux vector $\underline{\Psi}_m$, as well as the magnetizing current, $\underline{i}_\mu$ lags the terminal voltage by 90°.

These first assumptions lead to a reduction of the machine's equivalent circuit as presented in figure 4.

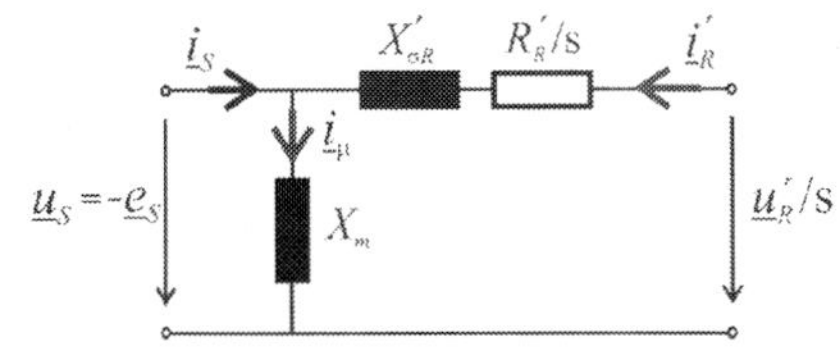

Fig. 4. Simplified equivalent circuit of the induction machine.

The new simplified voltage equation for this circuit is easily deduced

$$\underline{u}'_R = R'_R \underline{i}'_R + jsX'_{\sigma_R} \underline{i}'_R + s\underline{u}_S, \tag{94}$$

as well as the rotor current

$$\underline{i}'_R = \frac{\underline{u}'_R - s\underline{u}_S}{R'_R + jsX'_{\sigma_R}}. \tag{95}$$

With regard to what happens with the additional rotor resistance, a controllable voltage source applies similar voltage drop over the resistance in the rotor terminals. The basic idea is to counter balance the induced voltage by different slip values applying suitable values of the voltage to the rotor terminals, i.e. the slip-rings, in order to control speed and/or torque so as to keep the rotor current under acceptable values. Hence, the voltage source ceiling value depends on the desired operating range. For different rotor voltage values, different base or synchronous speeds are also given. The base slip s_0 can be found by setting the rotor current vector to zero for a respective rotor voltage on the equation (95)

$$\underline{u}'_{R0} = s_0 \underline{u}_S \Rightarrow s_0 = \frac{\underline{u}'_{R0}}{\underline{u}_S}. \tag{96}$$

The admissible values for the rotor current normally take place over the speed range for a short-circuited rotor. Therefore, the voltage difference on the numerator of equation (95) must lie within the same range of the voltage drop in the rotor windings for the machine with short-circuited rotor, as shown in the figure 5.

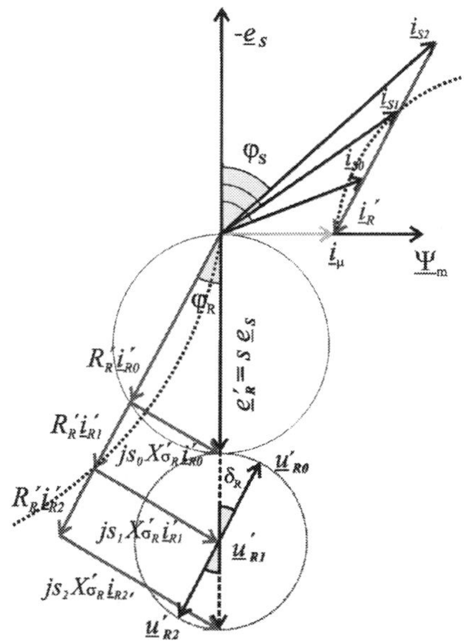

Fig. 5. Variation of the rotor voltage

The diagram of the figure 5 presents the variation of the rotor terminal voltage along three values $\underline{u}'_{R0}$, $\underline{u}'_{R1}$ and $\underline{u}'_{R2}$ in phase with the rotor current emulating the voltage drop over an external resistance. The rotor induced voltage $\underline{e}'_R$ in a squirrel cage machine is represented by the continuous line while the rotor induced voltage in a doubly-fed machine is increased by the higher slip values and represented by the dashed line. The voltage drops over the rotor complex impedance is depicted for each of these values.

One may assume the operating point number 1 as the original value for the machine with short-circuited rotor, i.e. $\underline{u}_{R1} = 0$, by slip $\underline{s}_1$ and stator and rotor currents $\underline{i}_{S1}$ and $\underline{i}'_{R1}$, respectively. A positive increase in the rotor voltage to $\underline{u}'_{R0}$ forces the rotor current to a new value $\underline{i}'_{R0}$, according to (95) and the stator current to $\underline{i}_{S0}$, according to (47), as well as the slip to s_0. It means that the speed is increased and a reduced voltage $s_0\underline{e}_S$ is induced in the rotor circuit. Similarly, a negative increase in the rotor terminal voltage to $\underline{u}'_{R2}$ forces the rotor and stator currents to $\underline{i}'_{R2}$ and $\underline{i}_{S2}$, respectively. The speed is reduced and the induced voltage in the rotor side increases to $s_2\underline{e}_S$. The fact that the rotor voltage and currents are in phase means that only active power is flowing between the controllable voltage source and the rotor circuit.

The dotted arcs point out the loci of the stator (Heyland circle) and rotor currents, taking the stator parameters into consideration. Furthermore, the internal stator's induced voltage and the magnetising current also deviate slightly from the assumed constant values due to the voltage drop over the stator winding. In comparison to the cage machine, the doublyfed induction machine possesses a family of Heyland circles, depending on the imposed rotor voltage. This degree of freedom allows for determining the current loci for desired slip values. The rotor voltage vector can be chosen in such a way as to ensure that only the imaginary components of the rotor and stator currents vary, as can be deduced from (95). In this case, machine magnetisation may be influenced independently of speed. The machine can be overand under-excited through the rotor circuit in the same way as a synchronous machine. Depending on the available ceiling voltage and on the operating point the machine may be fully magnetised through the rotor or even assume capacitive characteristics.

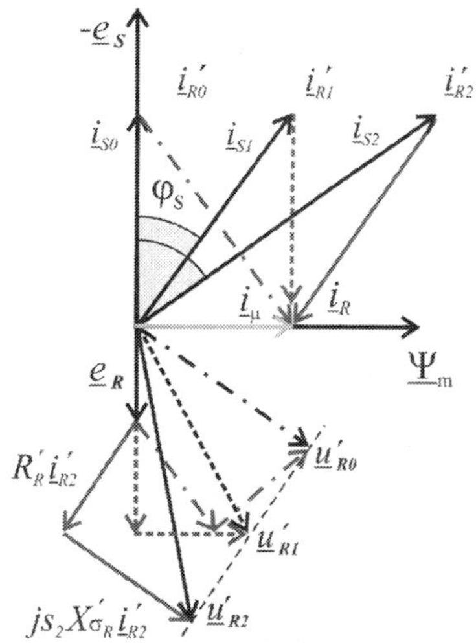

Fig. 6. Magnetisation between stator and rotor

The figure 6 shows this kind of operation for 3 distinct operating points, where the rotor voltages $\underline{u}'_{R0}$, $\underline{u}'_{R1}$ and $\underline{u}'_{R2}$ are applied to the rotor terminals in order to determine the way the machine is being magnetised. The real component of the currents is kept constant in order to maintain a constant torque or active power. The resulting stator and rotor current vectors for all situations according to (47) are also depicted. For the operating point 2, one notices that the rotor current is demagnetising so that the machine is under-excited. The imaginary component of the resulting stator current compensates the demagnetisation referring the required reactive power from the network. In situation 1 magnetisation is carried out through the stator circuit and the rotor current vector is aligned with the internal induced voltage. In the case 0, the rotor current assumes the magnetising current. The resulting stator current possesses only a real component and the stator power factor equals one. If the rotor voltage angle is further increased in this direction the machine will be over-excited and the imaginary component of the stator current will be capacitive.

4. LC-Filter and mains supply

The basic electrical circuits theory is used in modeling the LC-filter and the mains supply at the output of the mains side inverter. Initially, the inverter and the mains are considered ideal symmetrical 3-phase voltage sources, $\underline{u}_n$ and $\underline{u}_N$, respectively. The LC-filter composed of the filter inductance and capacitance, L_f and C_f, together with the filter resistance R_f, build the first mesh. The network impedance $\underline{Z}_N = R_N + j\omega_N L_N$ between the capacitor filter and the mains voltage source builds the second mesh. Lastly, one gets a T-circuit that is similar to the induction machine equivalent circuit, but instead of a magnetising inductance the filter capacitance as shown in figure 7.

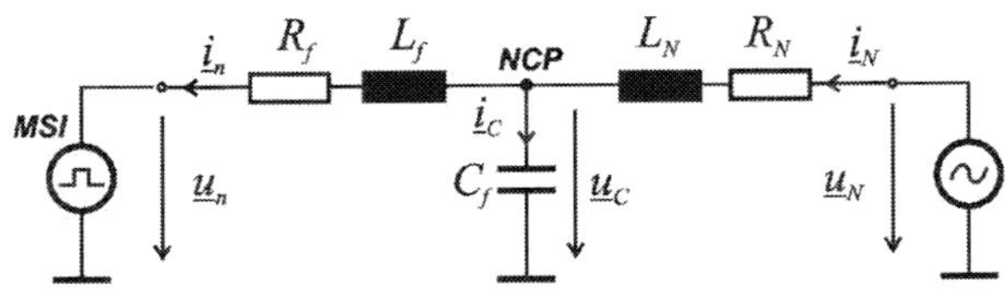

Fig. 7. LC-filter and mains supply equivalent circuit

4.1 Steady state analysis

Considering the equivalent circuit 7, the steady state voltage and current equations of the LC-filter and of the mains supply can be written as

$$\underline{u}_C = R_f \underline{i}_n + jX_f \underline{i}_n + \underline{u}_n \tag{97}$$

$$\underline{u}_N = R_N \underline{i}_N + jX_N \underline{i}_N + \underline{u}_C \tag{98}$$

$$\underline{i}_C = j\frac{\underline{u}_C}{X_C} = \underline{i}_N - \underline{i}_n, \tag{99}$$

where $X_f = \omega_N L_f$, $X_N = \omega_N L_N$ and $X_C = \dfrac{1}{\omega_N C_f}$ are the respective inductive and capacitive reactances of the filter inductor, network and filter capacitor.

If one neglects the voltage drop over the mains impedance, the voltage over the capacitor becomes equal to the net voltage. Under this assumption voltage equations above can be simplified to

$$\underline{u}_N = R_f \underline{i}_n + jX_f \underline{i}_n + \underline{u}_n \tag{100}$$

$$\underline{u}_N = \underline{u}_C \tag{101}$$

4.1.1 Active power flow.

The power flowing from the network to the MSI output can be computed as it was for the generator side using expressions (41) and (42) in equation (100). The active power is given by

$$p_n = \frac{3}{2}\Re\{\underline{u}_N \underline{i}_n^*\} = \frac{3}{2}\Re\{R_f \underline{i}_n \underline{i}_n^* + jX_f \underline{i}_n \underline{i}_n^* + \underline{u}_n \underline{i}_n^*\} \cdot \tag{102}$$

The active power flowing at the net connecting point (NCP) is composed of the power losses in the filter resistance

$$P_{Cu_f} = \frac{3}{2} R_f \hat{I}_n^2, \tag{103}$$

and the contribution of the mains-side inverter (MSI) measured at its output

$$p_n' = \frac{3}{2}\Re\{\underline{u}_n \underline{i}_n^*\} \cdot \tag{104}$$

Based on expression (99) and the fact that the capacitor filter current presents no active component, one may conclude that the active current components of the inverter and network must be the same. Hence, the active power flowing to or from the network is equal the active power being delivered at the NCP in equation (102) and is given as

$$p_N = \frac{3}{2}\Re\{\underline{u}_N \underline{i}_N^*\} = \frac{3}{2}\Re\{\underline{u}_N \underline{i}_n^*\} = p_n \cdot \tag{105}$$

4.1.2 Reactive power flow.

The reactive power at NCP is the remaining imaginary part of (102)

$$q_n = \frac{3}{2}\Im\{\underline{u}_N \underline{i}_n^*\}, \tag{106}$$

consisting of the contribution of the inductor filter

$$Q_L = \frac{3}{2} X_f \hat{I}_n^2, \tag{107}$$

and the inverter's reactive power contribution

$$q_n' = \frac{3}{2}\Im\{\underline{u}_n \underline{i}_n^*\} \cdot \tag{108}$$

The capacitor contribution is purely reactive and can be easily computed as

$$q_C = \frac{3}{2}\Im\{\underline{u}_N \underline{i}_C^*\}, \tag{109}$$

where the current flowing through the capacitor is given by (99). Developing (109) yields

$$Q_C = -\frac{3}{2}\frac{\hat{U}_N^2}{X_C}, \tag{110}$$

The *LC*-filter reactive power share due to its passive components, namely the inductor and capacitor, can be summarised by the following expression

$$Q_F = Q_L + Q_C = \frac{3}{2} X_f \hat{I}_n^2 - \frac{3}{2} \frac{\hat{U}_N^2}{X_C} \cdot \tag{111}$$

The total reactive power flowing to or from the network can be computed as

$$q_N = \frac{3}{2} \Im\{\underline{u}_N \underline{i}_N^*\} \cdot \tag{112}$$

Once again taking equation (99) and remembering that the capacitor current possesses only a reactive component, the total reactive current is composed by the MSI and the filter capacitor's reactive current components.

Substituting the current relation in (112) and using the relations pointed out in the reactive power expressions above, one may see that the total reactive power is composed by MSI and *LC*-filter contributions

$$q_N = \frac{3}{2} \Im\{\underline{u}_N (\underline{i}_n + \underline{i}_C)^*\} \Rightarrow q_N = q_n + q_C \cdot \tag{113}$$

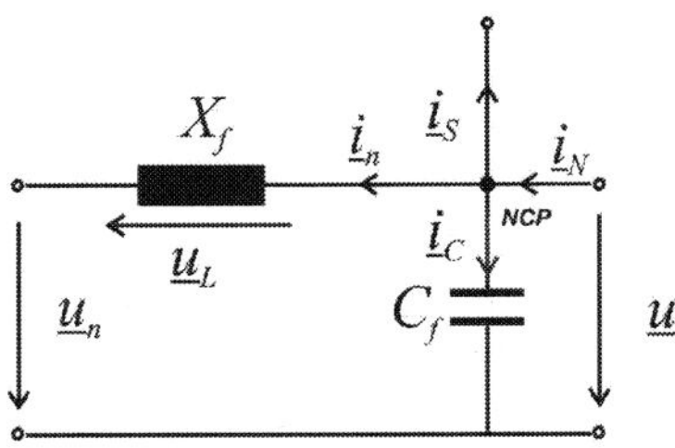

Fig. 8. LC-filter and mains supply equivalent circuit

4.1.3 Generator contribution.

If the DFIG stator terminals are connected to the NCP as shown in the equivalent circuit in figure 8, one has to newly compute the power flow balance. Let us now consider the stator current flowing from the NCP node. According to Kirchoff current's law the node equation is then

$$\underline{i}_N = \underline{i}_C + \underline{i}_n + \underline{i}_S \cdot \tag{114}$$

The active and reactive powers being delivered to the network in this situation can be easily computed substituting expression (114) in the equations (105) and (112), respectively. If we develop this equation further, based on the considerations made on the capacitor current and the power expressions derived in this section, we have

$$p_N = \frac{3}{2} \Re\{\underline{u}_N (\underline{i}_n + \underline{i}_S)^*\} = p_n + p_S \cdot \tag{115}$$

$$q_N = \frac{3}{2}\Im\{\underline{u}_N(\underline{i}_C + \underline{i}_n + \underline{i}_S)^*\} = q_C + q_n + q_S \cdot \quad (116)$$

Both these equations are very important in fostering further development of the optimization procedures in the next chapters.

4.1.4 Simplified analysis.

For a better understanding of the MSI operation together with the output LC-filter, some simplifications are featured. The first is the above-mentioned consideration that the capacitor is constant and equal the net voltage. The second is to neglect the filter resistance as shown in the simplified equivalent circuit in figure 8. Under these assumptions and according to the equivalent circuit, the voltage difference between the converter output and the net voltage is the voltage drop over the filter inductance

$$\underline{u}_n = \underline{u}_N - \underline{u}_L = \underline{u}_N - jX_f \underline{i}_n \cdot \quad (117)$$

If one orients the reference coordinate system oriented to the net voltage space vector and vary the active current component, active power can be delivered to or consumed from the network by MSI. According to (117) the voltage drop over the filter inductance is always in quadrature with the mains voltage, if the inverter output current is in phase with it, i.e., $\underline{i}_n$ possesses only a real part

$$\underline{u}_L = jX_f \underline{i}_n \cdot \quad (118)$$

The MSI output voltage $\underline{u}_n$ required to impose the output current can thus be determined. This situation is depicted in left figure 9.

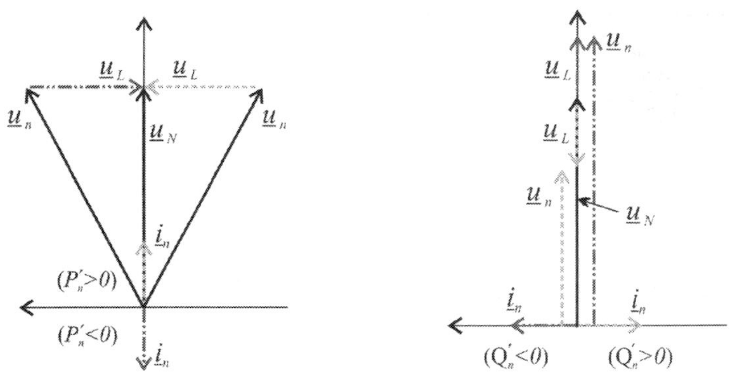

(a) Active Current (b) Reactive Current

Fig. 9. Phasor diagram for active (a) and reactive (b) currents

According to equation (104) the active power production or consumption, i.e., whether negative or positive, depends on the active current component's sign.

For the disconnected generator stator, where $\underline{i}_S$ =0, substituting (117) in (104) and considering that no active power can be produced or consumed by the inductor, the active power at the MSI is equal to the active power in the network .

$$p'_n = \frac{3}{2}\Re\{(\underline{u}_N - \underline{u}_L)\underline{i}_n^*\} = \frac{3}{2}\Re\{\underline{u}_N\underline{i}_n^* - \underline{u}_L\underline{i}_n^*)\} = p_N, \quad (119)$$

Allowing the reactive current to vary in the quadrature axis results in voltage drops over the filter inductor parallel to the net voltage, whose sense depends on that of the current. The required inverter output voltages in order to impose these currents are in phase with the mains voltage vector and can be computed based on (117). Since the active current component is zero, the reactive power production or consumption, i.e., negative or positive, depends on the sign of the reactive current component, as per (108). The phasor diagram for this situation is found in right figure 9.
Again considering the disconnected generator, if one substitutes (117) in (108), we have the expression

$$q'_n = \frac{3}{2}\Im\{(\underline{u}_N - \underline{u}_L)\underline{i}_n^*\} = q_n - q_L \,. \quad (120)$$

Hence, besides the capability to deliver and consume active power, the MSI is able to work as a static synchronous compensator or a phase shifter, influencing the net voltage and the power factor in order to produce or consume reactive power. Equation (119) and (120) can be used for the design of the MSI.

5. System topology and steady state power flow

The common DFIG drive topology depicted in figure 10 shows the stator directly connected to the mains supply while the rotor is connected to the rotor-side inverter. A voltage DC-link between the rotor and mains-side inverter performs the short-term energy storage between the generator rotor and the network. The LC-filter at the MSI output damps the harmonic content of the output voltage and current. The bi-directional converters, i.e., inverter/rectifier operation, enable the active and reactive power flow in both directions. Within the sub-synchronous speed range, the active power flows from the grid to the rotor circuit whereas within the super-synchronous speed range, it flows from the rotor to the grid. The sub-syncrhonous operation mode is illustrated in figure 11.

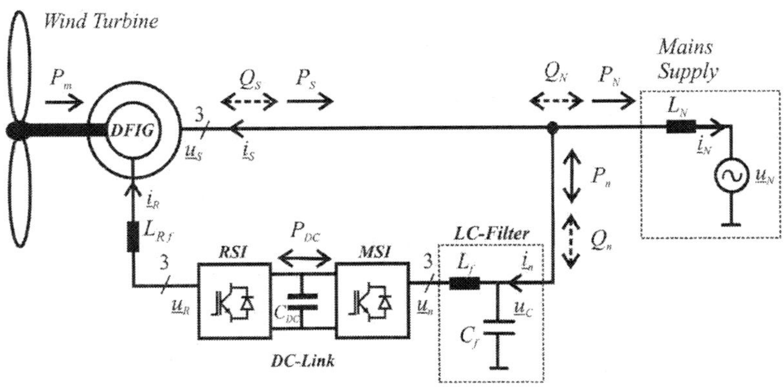

Fig. 10. 3-phase schematic

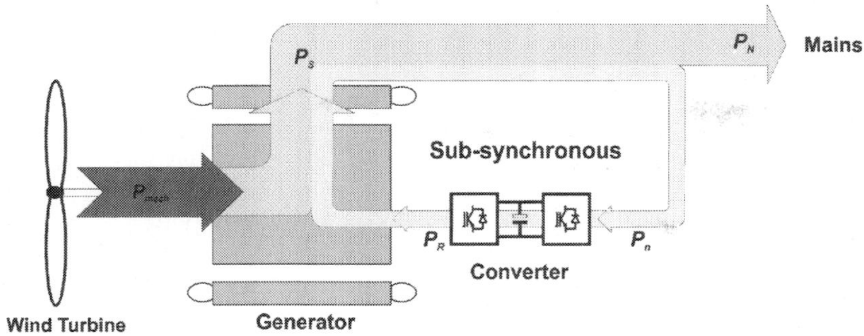

Fig. 11. Active power flow in sub-synchronous operation

5.1 Active power flow

The active power P_R flowing through the rotor circuit, also named slip power, is proportional to stator active power P_S and to the slip s, as denoted by expression (75). Neglecting the copper losses, the slip relation becomes

$$P_R = -sP_S \cdot \tag{121}$$

For a constant DC-link voltage U_{DC}, neglecting the losses in the converter, the output power in the MSI P_n must be equal the rotor power, thereby guaranteeing the energy balance

$$P_n = P_R \cdot \tag{122}$$

The total active power delivered to the network P_N is the sum of the stator power plus the MSI output power at the net connecting point.

$$P_N = P_S + P_n \cdot \tag{123}$$

Now, substituting (121) and (122) in (123) yields

$$P_N = (1-s)P_S \cdot \tag{124}$$

5.2 Reactive power flow

From equation (84) and the simplifications assumed for the equivalent circuit depicted in figure 4, the required reactive power Q_m to be delivered to the generator for the rated magnetization can be found as

$$Q_m = \frac{3}{2}\frac{\hat{U}_S^2}{X_m} \cdot \tag{125}$$

Besides controlling the active power flow, the bi-directional switches on the inverters enable the phase displacement between converter output current and voltage allowing for the generation or consumption of reactive power, as shown in figure 12. Therefore, the machine excitation can be also carried out through the rotor circuit so that Q_m can be delivered by the stator, rotor, mains supply and the RSI. According to equation (92), neglecting the leakage reactive powers, the following expression can be written for the magnetising reactive power

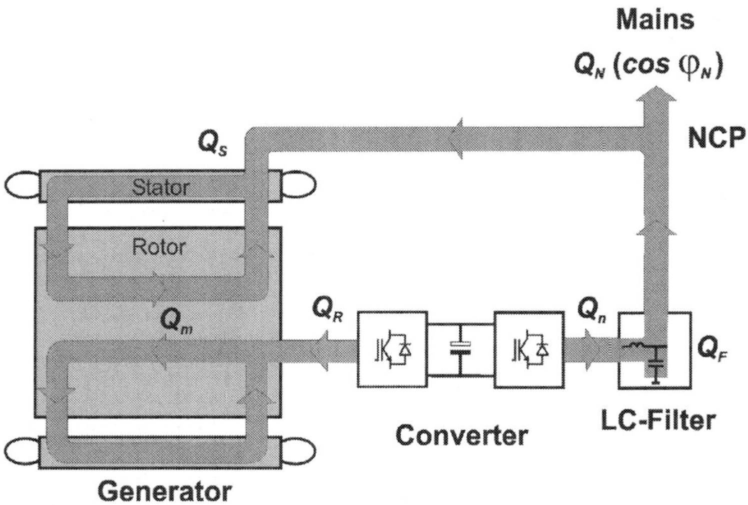

Fig. 12. Reactive power flow

$$Q_m = Q_S - \frac{Q_R}{s} \Rightarrow Q_S = Q_m + \frac{Q_R}{s}. \tag{126}$$

The equation (126) above shows how the reactive power in the stator side can be determined by the reactive power fed to the rotor side and influences the stator power factor. It also points out the required values for the rotor reactive power in order to impose unity power factor to the stator side

$$Q_S = 0 \Rightarrow Q_R = -sQ_m \cdot \tag{127}$$

In this way the full machine magnetisation is accomplished by the rotor side. Also capacitive (leading) power factors can be imposed to the stator terminals provided that

$$Q_R < -sQ_m \cdot \tag{128}$$

Hence, the dimensioning of the inverter depends also on the desired power factor control range and has to be extended in order to accommodate the additional reactive power that flows through the rotor circuit. This is also true for the MSI if power factor correction or voltage regulation is required.

The reactive power contributions to the MSI Q_n and to the LC-filter capacitance Q_C can be also taken into account and influence the power factor in the NCP.

The total reactive power delivered to or consumed by the network Q_N at the NCP is given by the addition of the DFIG stator, MSI and LC-filter contributions, as pointed out in expression (116).

Substituting the stator reactive power given by equation (126) yields

$$Q_N = Q_C + Q_n + Q_m + \frac{Q_R}{s}, \tag{129}$$

where Q_R and Q_n are available controllable reactive powers. Choosing the filter capacitance in order to compensate the generator power factor, i.e. $Q_C = -Q_m$ expression (129) is reduced

$$Q_N = Q_n + \frac{Q_R}{s}. \tag{130}$$

On the other hand, using a simple L-filter gives

$$Q_N = Q_n + Q_m + \frac{Q_R}{s}, \tag{131}$$

With the derived equations for the active and reactive power flow capability diagrams for the DFIG drive can be drawn as it was carried out in Santos-Martin et al. (2008) including the contribution of the mains-side inverter and LC-filter.

The possibility of distributing the reactive power between stator and rotor allows for the splitting of the magnetising current in stator and rotor reactive current components. Depending on the operating point, i.e., on the active currents and on the winding resistances, the reactive current may be smartly distributed so as to reduce system losses.

The optimisation of the power flow aims to reduce inherent losses with a view to improving drive efficiency. The efficiency can be determined by computing the power losses corresponding to the amount of energy that is transformed into heat during the conversion process. In an electrical drive, losses are present as electrical losses due to the current flowing through the involved circuits and to semiconductors switching in the inverter, as iron losses in the magnetic circuit of the electrical machine, transformer and filter cores, and as mechanical losses due to friction and windage. Hence, the problem consists in minimising the losses by manipulating one or more optimisation variables. The required control structure and controllers design in order to perform the reactive power splitting is described in Rabelo et al. (2009).

6. Conclusion

The basics of electrical drives in rotating *dq*-coordinate system was introduced as well as the classical definition of active and reactive powers leading to the complex power. Electrical power generation, specifically the induction generator and later the doubly-fed induction machine and the power flow were discussed. The simplified analysis showed the possible operating ranges of the doubly-fed induction generator drive for production of active and reactive powers.

The drive system topology allows for an independent control of the reactive power in the generator and mains side as well as the decoupling from the active power due to the mains voltage or flux orientation. The suggested optimisation is based on the controlled distribution of the reactive power flow in the drive components. The reactive power can be defined the rate of magnetic and electrical energy exchange between the generator, the mains supply and the inverters. It is a function of the reactive current and of the voltage amplitude and, therefore, closely related to process losses. For this reason, it proves to be a very suitable optimization variable.

7. References

Leonhard, W. (1980). *Regelung in der elektrischen Energieversorgung*, 6th edn, B.G. Teubner, Stuttgart.

Lipo, T. (1995). *Vector Control of Electrical Machines*, 1st edn, Clarendon Press, Clarendon.

Müller, G. (1977). *Elektrische Maschinen - Theorie rotierender elektrischer Maschinen*, 4th edn, VEB Verlag Technik, Berlin.

Rabelo, B. & Hofmann, W. (2002). Optimal reactive power splitting with the doubly-fed induction generator for wind turbines, *DEWEK Conference Proceedings*, Wilhemshaven.

Rabelo, B., Hofmann, W., Silva, J., Gaiba, R. & Silva, S. (2009). Reactive power control design in doubly-fed induction generators for wind turbines, *IEEE Transactions on Industrial Electronics* . Vol.56, No.10, pp.4154-4162, October 2009.

Rabelo, B., Hofmann,W., Tilscher, M. & Basteck, A. (2004). A new topology for high powered wind energy converters, *EPE PEMC Conference Proceedings*, Riga.

Santos-Martin, D., Arnaltes, S. & Amenedo, J. R. (2008). Reactive power capability of doubly-fed asynchronous generators, *ELSEVIER Electric Power Systems Research*. Vol.78, Issue 11, pp.1837-1840, November 2008.

Späth, H. (2000). *Leistungsbegriffe für Ein- und Mehrphasensysteme*, 1st edn, VDE Verlag, Berlin, Offenbach.

Vicatos, M. & Tegopoulos, J. (1989). Steady state analysis of a doubly-fed induction generator under synchronous operation, *IEEE Transactions on Energy Conversion* . Vol.4, No.3, pp.495-501, 1989.

Wechselstromgrössen - Zweileiter Stromkreise (1994). Din norm 40110 teil 1, Deutsches Institut für Normung.

6

Control Methods for Variable Speed Wind Energy Converters

Sourkounis, Constantinos and Ni, Bingchang
Ruhr-University Bochum
Germany

1. Introduction

By the utilization of wind energy, wind energy converters are preferably designed to provide electrical energy. The electrical energy is, in most cases, fed into the electrical grid. The kinetic energy of flowing air masses is converted into rotational mechanical energy with the help of uplift or air resistance effects at wind rotors. This energy is transmitted through a drive shaft and, in most cases, additionally through a gear box, to feed the generator. The generator transforms the mechanical energy into electrical energy. Depending on the generator system, the energy is then, either directly or through a power electronic device, fed into the electrical grid.

For the efficiency of the wind energy utilization and the durability of the energy conversion chain of the wind energy converter, it is of essential significance that the stationary and dynamical operation behavior of each component are adjusted to each other and to optimize the operational behavior of the whole energy conversion chain. In the first instance, the stationary and dynamical operation behavior of the wind energy converter must, on the one hand, meet the demand of the wind energy conversion process (aerodynamic process) and, on the other hand, the demand of the electrical supply network. Above that, from the system side of view, basic conditions concerning the operational behavior of the energy conversion chain have to be considered.

The basic requirement to meet the above mentioned demands on the operational behavior can be realized by the basic set up, e. g. the variable speed operation, of the energy conversion chain. The optimal operational behavior can be finally set with the help of the control and of the operation management. Depending on the structure and the operating characteristic of the energy conversion chain of wind energy converters, the type and structure of the control unit and of the operation management varies.

In the frame of this chapter, the theoretical fundamentals of wind energy converter controls will be presented and explained. Among others the approach to lay out a mathematical model of the energy conversion chain of wind energy converters, which reproduces the stationary and dynamical behavior, will be described. Furthermore, the basic structure of the control and the operation management will be presented and explained, whereat a differentiation between the task of the control and operation management will take place.

For the control, different conventional methods will be described. For this purpose the controller design and layout, based on control technique methods, will be presented. Additionally, newly studied control methods will be presented. They will be compared to the conventional control methods. For comparing the efficiency of the wind energy conversion, the fluctuation of the delivered electrical power and therefore the system perturbation, and the load spectrum in the energy conversion chain will serve as criteria.

2. Energy conversion chain of wind energy converters

The operating behaviour of the energy conversion chain of wind energy converters is significantly influenced by the wind rotor and the mechanical-electrical energy converter system, which generally comprises of a generator and the connection system to the electrical grid. Two basic categories can be distinguished; "Constant speed systems," where an asynchronous machine is coupled to the electrical grid, can be compared with "variable speed systems," which feature a decoupling of the generator frequency resp. speed from the grid frequency by means of a dc-link inverter. Nowadays, variable speed systems are realized preferably by a synchronous generator machine (see Figs. 1c and d) which is coupled to the electrical grid by a dc-link inverter, or by means of a doubly fed asynchronous machine (see Fig. 1b).

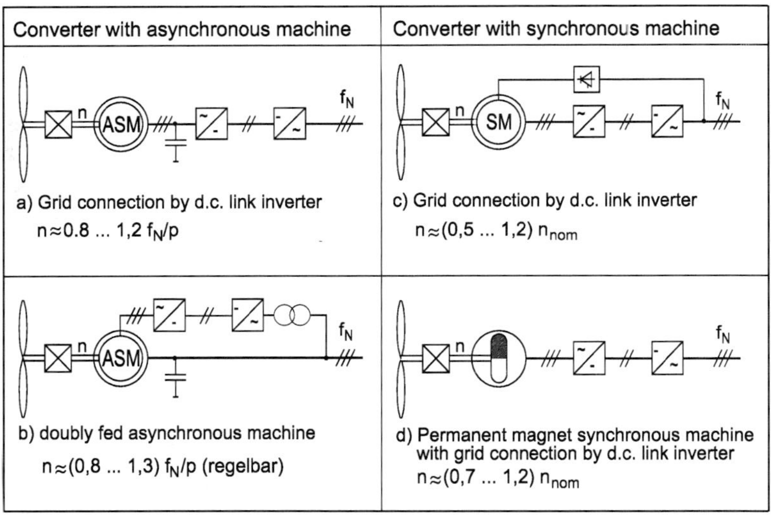

Fig. 1. Applied Generator Systems for variable speed wind energy converters

The necessity of variable speed operation can be attributed directly to the general requirements of the process of wind energy utilization. They can be formulated as follows:

- operation at maximum possible power coefficient c_p of the wind rotor,
- reduction of wind caused power fluctuations in the drive train of the wind energy converter and
- reduction of the resulting mains pollution.

By decoupling the generator speed and thus the wind rotor speed from the grid frequency, the speed of the wind rotor can be adjusted dynamically to the prevailing wind speed, so that the wind rotor is able to operate at the maximum power point (MPP) (see Fig. 2).

At the same time, variable speed wind energy converters offer the possibility to smoothen short-time wind caused power fluctuations by utilizing the rotating masses (e.g. the wind

rotor) as kinetic energy storage. When the wind speed increases, the rotor speed must be increased too so that the wind rotor operates at the MPP (see Fig. 2).
Part of the wind power is then stored in the rotor masses. When the speed is reduced at wind slacks, the kinetic energy stored in the rotor masses is transformed into electrical energy, so that alterations (fluctuations) of the supplied electrical power and thus of the torque in the drive train are reduced. This requires a suitably designed speed/power control and operation management.

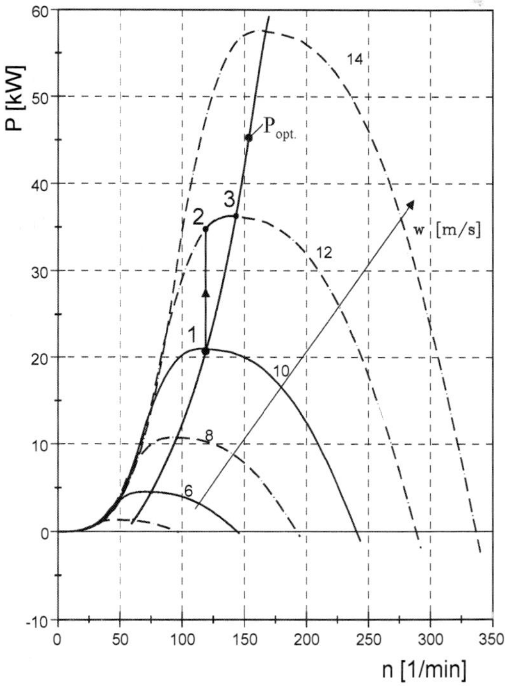

Fig. 2. Power map example of a wind rotor

Based on the requirement to achieve maximum energy yields, the primary control-engineering task is to provide a dynamical control of the respective optimal operating speed. At the same time, the wind caused power fluctuations at the wind rotor shall be smoothed by utilizing the storage effect of the rotating mass and thus avoids mains pollution in form of flicker effects. Furthermore, the power or torque fluctuations in the drive train are reduced. To minimize cumulated loads and thus enable an increase of the life of drive train components, the torsional vibrations, possibly caused by load peaks, must be damped as well.

3. Control path and basic control structure

The structure of a mechanical-electrical drive train of variable speed wind energy converters is generally suitable for a two-level type basic control structure. The inner control defines

the operating behavior of the electrical subsystem and provides a dynamical impression of the electrical counter-torque in accordance with the demand. The internal control has the task to dampen torsion vibrations in the drive train induced by load peaks. The process control (e.g. speed control) is realised in the second level. The process control shall improve the quality of the energy conversion process at the wind rotor by adjusting the optimal speed for the prevailing wind speed dynamically. At the same time, the set point torque value shall be regulated dynamically in such a way that wind caused power fluctuations in the drive train of the wind energy converter are smoothed.
Following considerations are exemplarily based on the drive train of a wind energy converter with synchronous generator. The generator is connected to the electrical grid by a dc-link inverter (Fig. 3).

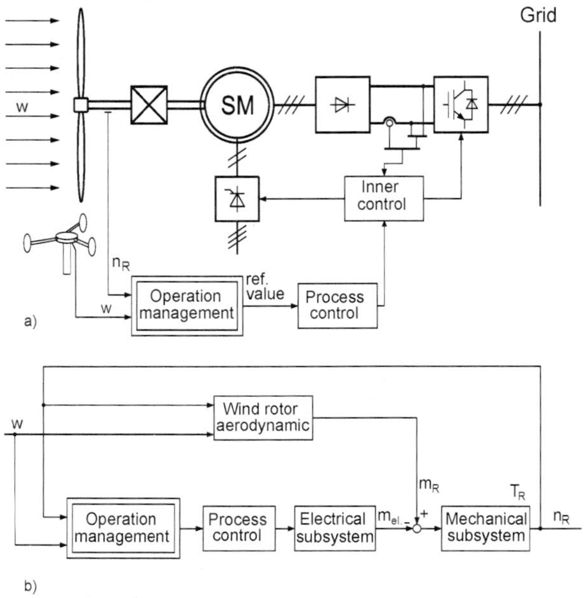

Fig. 3. Drive train structure of variable speed wind energy converter: a) function diagram; b) block diagram

The basic structure of the drive train can be split into a mechanical and an electrical subsystem as shown in Figs. 3 and 4. Based on the general differential equations of motion, the dynamic behavior of the mechanical subsystem is presented by the mathematical model of a two-mass spring system (see Fig. 4) simplified by the assumption that the mass of the shaft is small enough to be neglected. Besides others this simplification means that the mechanical subsystem of the drive train or its operating behavior respectively can be identified by a single dominant natural frequency.
For control-engineering purposes, this structure simplification of the mechanical subsystem, which is in real a multi-mass oscillator, has to be carried out according to Klöckner [4]. Contrary to other methods this method makes sure, that the reduction is carried out solely physically. The physical interpretability is retained and the remaining natural oscillation

frequencies are also exactly retained. The procedure according to Klöckner can be applied successively, until the subsystem is reduced to a two-mass oscillator. In the following, the first mass is identified by the rotor mass of the drive motor and the second mass by the rotating mass of the process machine for simplification purposes.

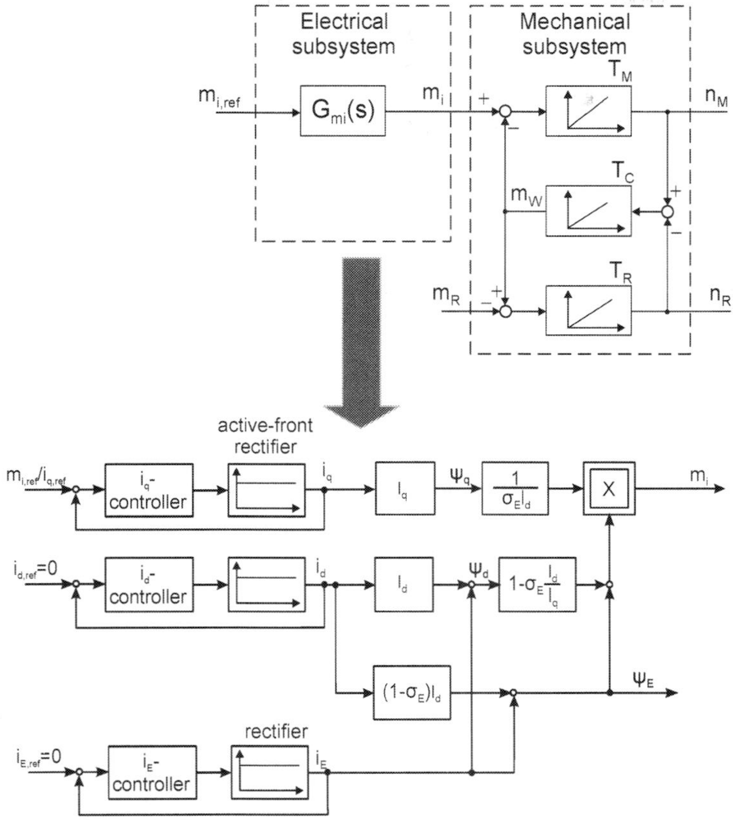

Fig. 4. Simplified mathematical model of the electro-mechanical drive train

The following simplified equation system provides the modelling basis:

$$
\begin{aligned}
M_i - M_W &= J_M \frac{d\omega_M}{dt}, \\
M_W &= C(\varphi_M - \varphi_R) = C\int(\omega_M - \omega_R)\,dt, \\
M_W - M_R &= J_R \frac{d\omega_R}{dt}.
\end{aligned}
\tag{1}
$$

After normalization of the motor speed n_M, the load speed resp. rotor speed n_R, and the shaft torque m_W, the system is transformed into state space representation. Air gap torque m_i and rotor torque m_R are the input variables.

$$\begin{bmatrix} \dot{n}_M \\ \dot{m}_W \\ \dot{n}_R \end{bmatrix} = \begin{bmatrix} 0 & -\frac{M_{nom}}{2\pi n_{nom}} \cdot \frac{1}{J_M} & 0 \\ \frac{2\pi n_{nom}}{M_{nom}} \cdot C & 0 & -\frac{2\pi n_{nom}}{M_{nom}} \cdot C \\ 0 & \frac{M_{nom}}{2\pi n_{nom}} \cdot \frac{1}{J_R} & 0 \end{bmatrix} \cdot \begin{bmatrix} n_M \\ m_W \\ n_R \end{bmatrix} + \begin{bmatrix} \frac{M_{nom}}{2\pi n_{nom}} \cdot \frac{1}{J_M} & 0 \\ 0 & 0 \\ 0 & -\frac{M_{nom}}{2\pi n_{nom}} \cdot \frac{1}{J_R} \end{bmatrix} \cdot \begin{bmatrix} m_i \\ m_R \end{bmatrix} \tag{2}$$

By introduction of the relevant time constants, the mechanical quantities (mass inertia and spring stiffness) can be replaced. With the acceleration time T_M of the motor

$$T_M = \frac{2\pi n_{nom}}{M_{nom}} J_M , \tag{3}$$

the acceleration time T_R of the load mass resp. of the rotor (e.g. wind rotor)

$$T_R = \frac{2\pi n_{nom}}{M_{nom}} J_R , \tag{4}$$

the mass inertia ratio v

$$v = \frac{J_M}{J_R} = \frac{T_M}{T_R} \tag{5}$$

and the time constant T_C for spring stiffness C,

$$T_C = \frac{M_{nom}}{2\pi n_{nom}} \cdot \frac{1}{C} . \tag{6}$$

The following is applicable for the eigen angular frequency ω_{ef} of the mechanical subsystem resp. its relevant time constant T_{ef} [5], [6]:

$$\omega_{ef} = \frac{1}{T_{ef}} = \sqrt{C\left(\frac{1}{J_M} + \frac{1}{J_R}\right)} = \sqrt{(1+v)\frac{C}{J_M}} \tag{7}$$

The state space representation adopts the following form by introducing the time constant:

$$T_{ef} = \frac{1}{\omega_{ef}} = \sqrt{\frac{T_C T_M}{(1+v)}} . \tag{8}$$

From state space representation, the transfer function between air gap torque m_i resp. shaft torque m_W can be derived. After execution of the Laplace transformation

$$\begin{bmatrix} \dot{n}_M \\ \dot{m}_W \\ \dot{n}_R \end{bmatrix} = \begin{bmatrix} 0 & -\frac{1}{T_M} & 0 \\ \frac{T_M}{(1+v)T_{ef}^2} & 0 & -\frac{T_M}{(1+v)T_{ef}^2} \\ 0 & \frac{1}{T_R} & 0 \end{bmatrix} \cdot \begin{bmatrix} n_M \\ m_W \\ n_R \end{bmatrix} + \begin{bmatrix} \frac{1}{T_M} & 0 \\ 0 & 0 \\ 0 & -\frac{1}{T_R} \end{bmatrix} \cdot \begin{bmatrix} m_i \\ m_R \end{bmatrix}, \tag{9}$$

following transfer function is obtained:

$$m_w(s) = \frac{1}{v+1} \cdot \frac{1}{s^2 T_{ef}^2 + 1} \cdot m_i(s) + \underbrace{\frac{v}{v+1} \cdot \frac{1}{s^2 T_{ef}^2 + 1} \cdot m_R(s)}_{m_R^*}$$
$$m_w(s) = \frac{1}{v+1} \cdot \frac{1}{s^2 T_{ef}^2 + 1} \cdot (m_i(s) + v m_R(s)). \tag{10}$$

The transfer function with a drive shaft torque m_W as output variable is the basis for examinations on the operating behaviour of the mechanical subsystem (see Fig. 4) with respect to the propagation of load peaks from the process and excitation of torsion vibrations. From the transfer function a direct correlation arises between load peak propagation in the drive train and the mass inertia ratio v. With increasing mass inertia ratio v at constant total mass inertia, higher torque peaks can be expected in the time response of the shaft torque. The same behaviour of the mechanical subsystem was visible in connection with the excitation affinity of torsion vibrations in the drive train.
Considerations about the operating behaviour of the mechanical subsystem are based on the assumption, that the air gap torque impression by the electrical subsystem is free of feedbacks to the mechanical subsystem.
The electrical subsystem which is part of the basic structure of the drive train consists of the electrical part of the electromechanical energy converter (motor resp. generator), the electrical grid and the frequency inverter with internal current control. In modern converters this is generally realized in compliance with the basic concept of field-oriented control, which features a decoupled control of the two orthogonal components of three-phase systems (Fig. 4). Flux- and torque building components are part of the dq-coordinate system constantly rotating with the rotary field. In the nominal speed range (constant flux), the operating behavior of the electrical subsystem is described by the transfer behavior (control behavior) of the control circuit for the torque-building current components i_q resp. the air gap moment M_i. Since overshoot is generally not allowed at reference jumps, the reference reaction of the control loop and thus the operating behavior of the electrical subsystem can be approached by a 1st order lag function (PT_1 element) (see Fig. 5). The substitution time constant T_{Str} of the approached transfer function for the electrical subsystem corresponds to the control rise time of the torque-generating current component's control loop.

4. Process control and operation management for variable speed wind energy converters

As mentioned before, the wind energy converter must be operated within a wide speed range in the course of the wind energy transformation process. This means that the often

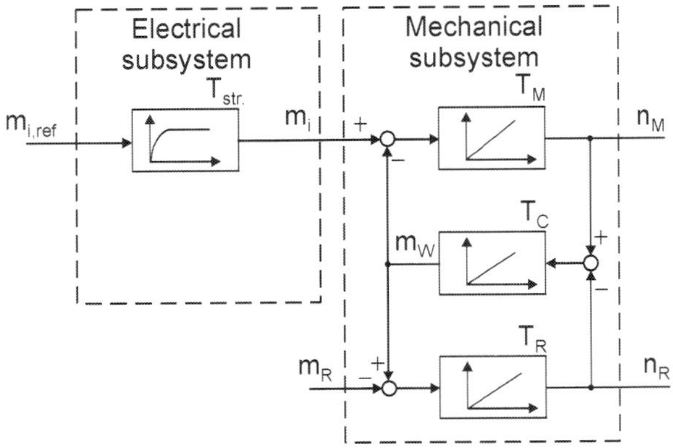

Fig. 5. Simplified mathematical model of the electromechanical drive train

applied linearization of the operating point is not permitted in this case. Another special characteristic of the process is, from a control-engineering point of view, that every change of the wind speed affects the disturbance effect of the non-linear transfer function of the wind rotor on the one hand, and requires an adjustment of the nominal value by displacing the "optimal working point" on the other.

Variable speed operation makes it possible to store part of the energy generated by the wind in the rotating mass for a short time period. At the same time, the speed shall be adjusted to the prevailing wind speed, so that the wind rotor is operated at maximum power coefficient. Task of the process control/operation management is, to provide a reference value for the dc-link current $I_{z,soll}$ resp. for the shaft torque at any operation state $x(n_R, w)$, given by the rotor speed n_R and the wind speed w, on the basis of which the internal regulation controls the electrical subsystem.

Besides the wind rotor and the mechanical-electrical energy transformation system, the operating behavior of the wind energy converter is determined by the power control and in the broadest sense by the operation management too. Power conversion at the wind rotor is described by the coefficient c_P, which reaches its maximum only at a certain tip-speed ratio. A maximum energy yield can therefore only be achieved by adjusting the rotor speed to the prevailing wind speed. At the same time variable speed operation allows to utilize the rotating masses as energy storage to smoothen short-time, wind caused power fluctuations.

To ensure an as high as possible energy yield, a quick adjustment to maximum power conversion at the wind rotor must be aimed for. On the other hand, this has negative effects on the frequency spectrum of the power output. From the aspect of operation at maximum power coefficient and low ripple of the yielded power P_{el}, different power control methods are compared in the following.

4.1 Optimal rotor speed control

The correlation between wind speed and optimal speed n_{opt} can be derived from the characteristic operation field resp. c_P-λ-characteristic of the wind rotor (see Fig. 6). For this purpose, the tip-speed ratio at which the power coefficient c_P becomes maximal must be determined from the c_P-λ-characteristic. On this basis, the optimal speed can be calculated on the assumption that the instantaneous wind speed can be measured at good accuracy.

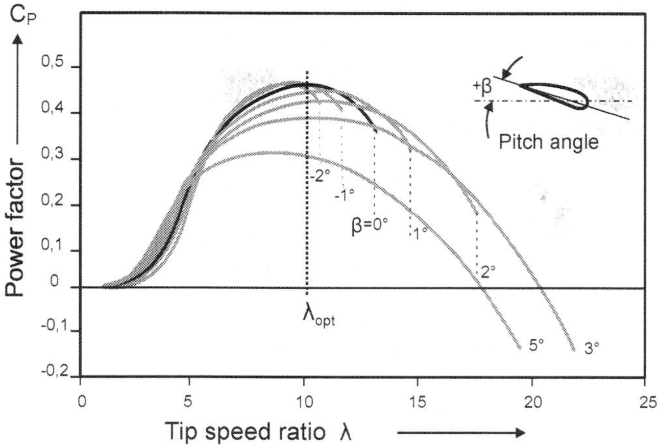

Fig. 6. Characteristic operation field of wind rotor

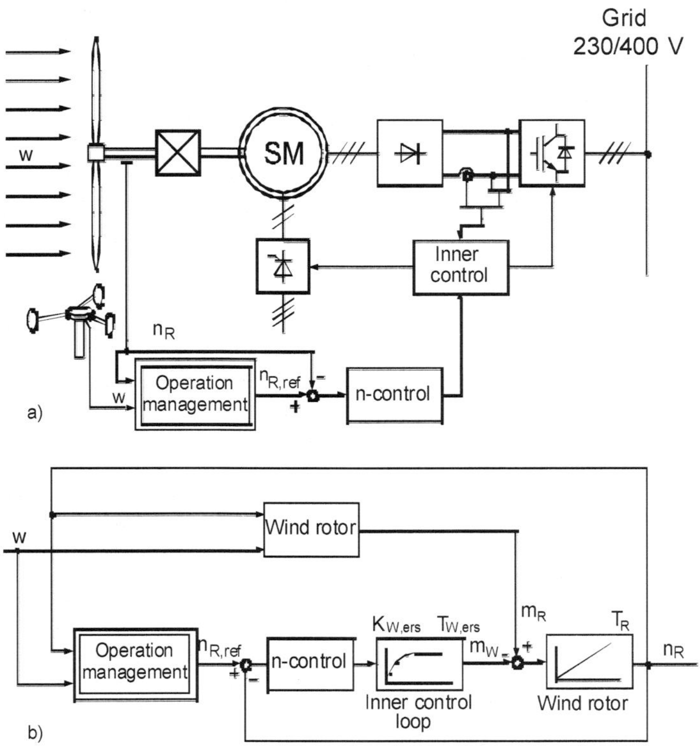

Fig. 7. Control of the optimal speed: a) principle of operation; b) block diagram of the controlled system

$$n_{R,opt} = \frac{1}{2\pi R_R} \lambda_{opt} \, w \tag{11}$$

with

R_R : the radius of the wind rotor,

λ_{opt} : the optimal tip-speed ratio,

$n_{R,opt}$: the optimal operating speed of the wind rotor, and

w : the prevailing wind speed.

Fig. 7 shows the function of the control procedure examined in this paper. When the wind speed w accelerates quickly, the actual value of the speed is not able to follow the nominal value n_{opt} due to the large mass inertia. The control difference at the input of the speed controller (n-controller) causes a step-type change of the power output since the speed controller attempts to accelerate the control procedure by relieving the load from the generator. At decelerating wind speed the reciprocal control procedure takes place.

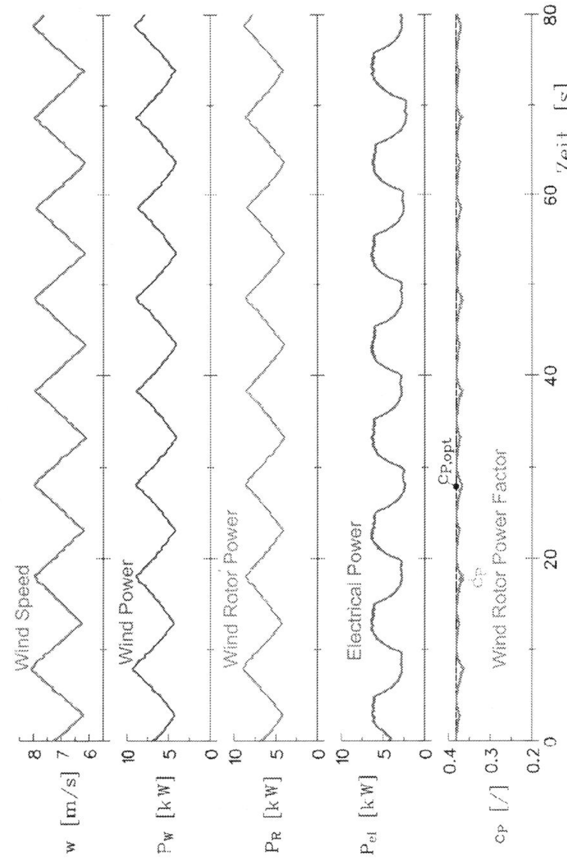

Fig. 8. Time response of parameters of the wind energy converter with speed control

In order to avoid fluctuations in the yielded power P_{el} caused by these control procedures, alterations in the course of the electrical generator counter-torque M_{el} during transition must be kept low. This can be achieved by admitting a limited speed difference between reference and actual values. A ramp function (T_{rp}), fulfilling the equation of movement of the simplified mechanical system, shall reflect the necessary time response of the reference value.

$$m_R - m_{el} = p\, T_R\, n_G \tag{12}$$

Examinations with the ramp-type time behaviour of the wind speed show relatively strong power fluctuations at low dynamical deviations of the c_P-value from the maximum value $c_{P,opt}$ = 0,388. This becomes obvious by the time response of the power coefficient (see Fig. 8). The control of the optimal speed features a good stationary behavior without continuous deviations. This is based on the condition that the wind speed can be recorded exactly and no alterations of the rotor characteristics occur.
Ice formation at the blades, deformation of the blades or alteration of the surface roughness influence the characteristics, so that the optimal tip-speed ratio is deferred to higher or lower values respectively.

4.2 Direct power control

Another possibility to adjust the rotor speed to the actually prevailing wind speed can be realized by presetting the reference value of the yielded power $P_{el,soll}$. This procedure determines the desired time response of the electrical power directly (see Fig. 9).
Based on wind and rotor speed, the operation management determines the tip-speed ratio for the actual operating point. Hereby, it can determine by means of the c_P-λ-characteristic, whether the plant is operating at maximum power, and, if applicable, adjust the power set point value $P_{el,soll}$ by a defined and steady transition. The calculation of the reference power value is carried out on the basis of the wind speed averaged over the time period Δt_v. The time period Δt_v and the kind of average determination have to be adapted to the desired power spectrum and the size of the wind rotor. Since the storage capacity of the rotating masses increases with their size, a wider time frame for averaging the yielded electrical power can then be selected.
A random extension of the time frame would only consider part of the demands on the operating behavior of the wind energy converter. The longer the considered time frame is, the farther away is the objective to adjust the optimal speed (resp. λ_{opt}, $c_{p,opt}$) quickly. On the other hand, the electrical power features an almost smoothed time behavior, which prevents the mechanical components from being damaged. Furthermore, in comparison to the speed control, this method enables a reduction of power fluctuations caused by speed control procedures (see Figs. 6 and 7), depending on the length of the time period.
For stationary operation, exact knowledge of the degree of efficiency over the whole speed range of the plant must be available besides exact recording of the wind speed, so that a stationary deviation of the power coefficient from the maximum value is avoided.

4.3 Self-adjusting maximum

Due to the fact that the air flow is affected by the wind rotor, an exact measuring of the wind speed by anemometers fitted to the wind energy converter is not possible and because of the inhomogeneous two-dimensional wind speed profile, measurements outside of the area influenced by the rotor are useless for the power control of a wind energy converter. A power control method which is able to adjust the operation point at a maximum c_P-value without the need to measure the wind speed is therefore of utmost importance for practical applications. Such a method is described in the following.

For this power control method, the power gradient is the decisive criterion whether an operation at maximum power coefficient is present or whether a speed adjustment is necessary. The operation management changes the reference speed value n_{opt} by Δn. After the speed reference value has been adjusted, the yielded power is measured and compared with the one prior to the control procedure. If the power alteration is positive, the speed is

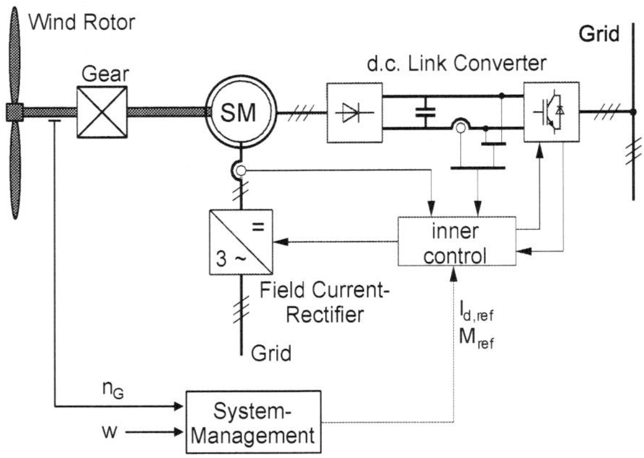

Fig. 9. Operating principle of average value power control

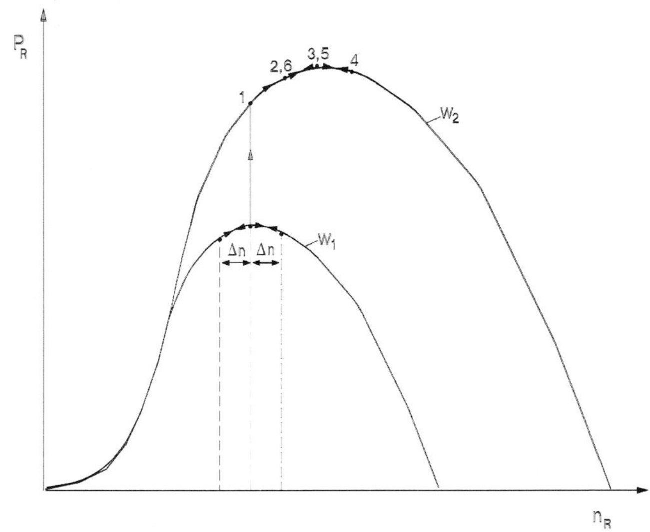

Fig. 10. Principle function of the search algorithm in the operation characteristic field of the wind rotor

changed by Δn in the same direction (Fig. 10). As long as the gradient of the power is positive, the same procedure is repeated. If a negative power gradient occurs, the prefix of the speed change is altered for the next step. The search procedure maintains the prefix of the speed change as long as the achieved power changes are positive.
In stationary operation, the speed oscillates around the optimal speed value of n_{opt}, in the most favourable case with an amplitude of $\Delta n/2$ resp. Δn. In order to minimize stationary offsets as well as power oscillations due to control procedures, the step width Δn of the search algorithm must be kept as small as possible. This has, however, the consequence of longer convergence periods, so that short-time wind fluctuations cannot be utilized optimally. The wind rotor is not able to yield 100 p.c. of the maximum possible power P_W due to the slow speed adjustment.

4.4 Speed-dependent torque control

A control procedure for power resp. speed control in the partial load operation range, which does not entail any unnecessary control procedures, is the "speed-controlled torque reference value setting" [8]. Based on the requirement for a constant tip-speed ratio λ - the ratio of the rotary speed of the rotor tip to the wind speed w - the torque M_R developed by the wind rotor at the then prevailing wind speed can be described as a function of the speed n_R.

$$M_R = \frac{P_R}{\omega_R} = \frac{\frac{1}{2} c_P \rho \pi R^2 w^3}{2\pi n_R} = 2\frac{c_P \rho \pi^3 R^5}{\lambda^3} \cdot n_R^2 , \quad \text{mit } \lambda = \frac{2\pi R n_R}{w} \tag{13}$$

$$M_R = K \cdot n_R^2$$

The procedure entails inaccuracies and alignment problems associated with the control:

- For a one-time adjustment of the control by

$$M_{el} = \eta_G \eta_{Getr} 2 \frac{c_P \rho \pi^3 R^5}{\lambda^3} \cdot n_R^2 , \tag{14}$$

  exact knowledge of the parameters is necessary, which are assumed to be constant for the whole speed and power range. This requires measuring investigations which are combined with high technical efforts.
- Alterations of the parameters resp. displacement of the optimal operation point due to increased roughness of the rotor blade surface, bending or twisting of the rotor blades cannot be taken into account.

4.5 Stochastic Dynamic Optimisation (StoDO)

Another approach to WEC control is the stochastic dynamic optimisation [2] with iterative adaptive wind speed probability distribution, given the general name of "iterative self-adapting system management" (ISSM) [3].
This procedure assesses informative values of the actual wind speed data on the short-term or medium-term wind power capability at a given location on the basis of the hitherto characteristics of the wind speed (resp. frequency distribution). Non-representative instantaneous values of the wind speed of minor importance can thereby be taken into

account for the calculation of the actual power reference value. Compared with known procedures, the new procedure shall avoid unnecessary control actions and thus power fluctuations due to short-time, for the location not representative alterations of the wind speed. At a short-term wind gust the control does not change the operation point of the plant, a readjustment of the system is omitted and the optimal operation point can be regained much quicker as soon as the wind gust abated. This promises a higher energy yield at optimally adjusted control. The extent of the increased energy yield is directly connected with the gust frequency at the plant location. The more gusts are to be expected, the higher is the additional yield by using the ISSM.

The steady operation at maximum possible power coefficient entails a dynamic adjustment of the rotor speed to the prevailing wind speed resp. after every wind speed change, which presumes a good reference-variable response of the control circuit (speed control circuit). In order to smooth the power fluctuations due to alterations in the wind speed, a control procedure optimized for the disturbance reaction is essential.

The mentioned two demands on the technical system (drive train of the wind energy converter) and the resulting control-engineering requirements disagree with respect to the dynamic behavior of the wind energy converter. It is therefore obvious to formulate the problem as an optimisation task.

As a criterion for the compliance with these aims, two indices are implemented [1]:

A deviation from the power average is indicated by

$$P-\mathrm{Index} := \frac{1}{m}\sum_{i=1}^{m}\frac{\left|\bar{P}-P_i\right|}{\bar{P}} = \frac{1}{m}\sum_{i=1}^{m}\frac{\left|\bar{I}_d-I_{d,i}\right|}{\bar{I}_d}, \tag{15}$$

with an electrical power $P_i=U_d I_{d,i}$ at a measuring point of time i as well as the average electrical power:

$$\bar{P} = \frac{1}{m}\sum_{i=1}^{m}P_i \tag{16}$$

Here, I_d is the dc-link current and U_d the assumed constant dc-link voltage. The deviation from $c_{p,opt}$ is described by

$$c_p-\mathrm{Index} := \frac{1}{m}\sum_{i=1}^{m}\frac{\left|c_{p,opt.}-c_{p,i}\right|}{c_{p,opt}} \tag{17}$$

with $c_{p,i}=c_p(\lambda_i)$. The $c_{p,opt}$-value describes the maximum degree of efficiency of the transformation of wind energy into mechanical energy for a specific rotor profile.

These indices reach smaller values, the more the respective aim is maintained. The weighted sum

$$M = g(P-\mathrm{Index}) + (1-g)(c_p-\mathrm{Index}) \tag{18}$$

must therefore be minimized, whereby g with $0<g<1$, is the weighting factor for both indices simultaneously.

The algorithm of the "stochastic optimization" [8] is taken as a basis to solve the herewith formulated optimisation task. This algorithm is used to model uncertain problem-relevant incidents in form of random variables, the exact value of which is indeed not yet known at the time of optimal determination, but their probability allocation is presumed as being real.

The optimisation task then comprises the determination of decision variables in such a way, that the expected value of a quality functional is minimized, under consideration of the usual problem conditions and probability allocations of the random variables. If the wind energy converter is taken as stochastic process, the stochastic optimization method is a very good approach to solve this problem.
A suitable mathematical representation resp. modelling has been carried out for this purpose. Fig. 11 shows a scheme of the wind energy converter as stochastic process together with the operation management which determines the reference value. State quantities of the process are wind velocity w and speed n_R, whereby only the speed is influenced by the plant itself. The wind speed is exposed to a stochastic alteration (disturbance) z. The dc-link current $I_{d,soll}$ serves as control variable (resp. reference variable), which is determined by the operation management on the basis of a control law evaluation. The control law is hereby not static but continuously updated, although considerably slower than the power set point value. Hereby statistical data about the measuring values of the wind velocity and the dc-link current (active power) are used which are continuously updated during operation. This has the effect that the control adjusts itself automatically to long-term and medium-term alterations of the wind conditions. The plant therefore adapts itself to the different wind conditions independently which arise in the cause of a day or even a year, e.g. by the effect of a sea-land breeze.

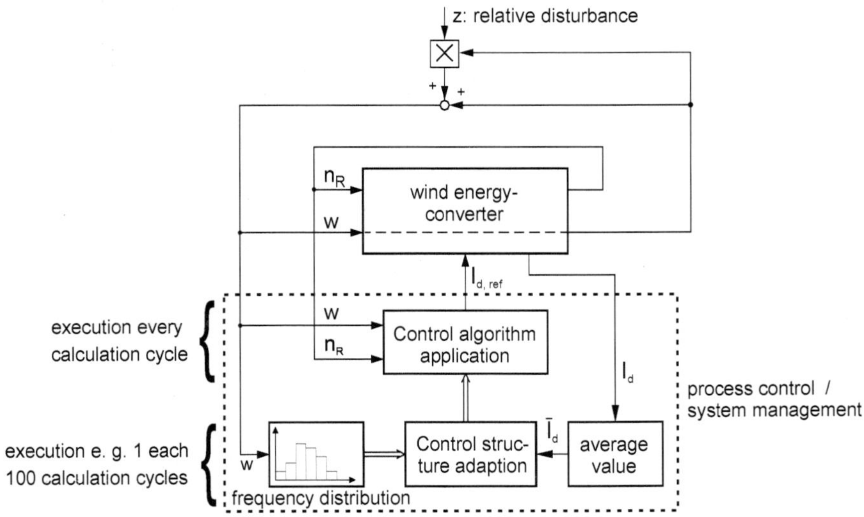

Fig. 11. Block diagram of an iterative self-adapting system management

To make sure that an implementation on a digital computer hardware is basically possible, a time-discrete modelling with a constant step width of $h := t_{i+1} - t_i$ has been used as a basis.
The step width is defined by the sampling rate, which means by the time between two available measuring values for w and n_R. At the same time, the measuring values w, n_R and I_d are graded in amplitude-discrete manner and linearly interpolated, if applicable, so that a control law $I_{d,soll}(n_R, w)$ can be established in look-up table form. Optimal $I_{d,soll}$-values can be found by finite search. For simplification purposes, I_d is determined for (18) only on the basis of former current values, according to

$$\overline{I}_{d,i+1} := \alpha I_{d,mess} + (1-\alpha)\overline{I}_{d,i} \tag{19}$$

with an actual measuring value $I_{d,mess}$ and $0<\alpha<1$ which indicates how quickly I_d adapts itself to new measuring values.

$n_{R,i}$ and w_i are the state variables of the process. The transfer $(n_{R,i},w_i)$ -> $(n_{R,i+1},w_{i+1})$ is influenced by the stochastic disturbance of the wind z, the nominal value of the dc-link current $I_{d,soll}$ as control variable of the operation management as well as by resulting speed alterations of the wind energy converters. The disturbance variable z defines the relative change of the wind speed (see Fig. 11):

$$z_i = \frac{w_{i+1} - w_i}{w_i} \tag{20}$$

The transition probabilities for w_i -> w_{i+1} result from conditional probabilities $P(z \mid w)$, that means the probability of a disturbance z under the condition, that the last measured wind speed is w. If a measurement (z, w) is available, i.e. the last measured wind speed is w and the alteration of the actual wind speed z, the following is applicable:

$$P_{neu}(\tilde{z} \mid \tilde{w}) := \begin{cases} (1-\alpha)P_{alt}(\tilde{z} \mid \tilde{w}) + \alpha & \text{für} \quad (\tilde{z},\tilde{w}) = (z,w) \\ (1-\alpha)P_{alt}(\tilde{z} \mid \tilde{w}) & \text{für} \quad (\tilde{z},\tilde{w}) \neq (z,w) \end{cases}, \tag{21}$$

with α from (19).

By the terms of "Operations Research" [8] of which this algorithm is taken, deviations from the average current value and from the optimal power coefficient are defined as costs. From this the cost function can be determined, which indicates the contribution of a state (n_R,w) at a dc-link current I_d and a disturbance z in (18):

$$\varphi(n_R, w, I_d, z) = g\left|\frac{I_d - \overline{I}_d}{\overline{I}_d}\right| + (1-g)\left|\frac{c_p\left(2\pi R\frac{f(n_R,w,I_d)}{w+wz}\right) - c_{p,opt}}{c_{p,opt}}\right|. \tag{22}$$

Due to the stochastic disturbance, the cost function cannot be minimized at any time. The objective is indeed to keep the costs minimal within a time average, which means to minimize the expected value of the sum of the costs

$$\underset{z}{E}\left\{\sum_{\kappa=0}^{K-1}\varphi(n_R(\kappa), w(\kappa), I_{d,\kappa}, z(\kappa))\right\}, \tag{23}$$

whereby the expected value is calculated over all z, and K is the end time of the process. Here the system shall indeed not be transferred into an end state. Instead it shall operate arbitrarily long without any time limitation or a certain state to be reached. That means that the time horizon K is infinite [2].

In this case, the determination of a time-invariant control law is possible. It is a requirement for the convergence of the algorithm that the probability distribution of the disturbance variable is time-invariant. This can be assumed for the case under examination since convergence, that means a constant control law, is only aimed at in case of continuous frequency distribution. If the frequency distribution is changed, the algorithm converges to a new, adjusted control law. On this basis, the algorithm of the stochastic optimization has been specifically expanded by a new optimization algorithm for the application described here, which is called *stochatic dynamic optimization*.

The resulting infinite sum of the expected remaining costs in (23), e.g. the expected value of the costs incurred until the end state K= ∞ is reached at given nominal value $I_{d,soll}(\kappa)$, κ=k,...,K-1, can be dissolved by applying the principle optimality. The expected minimal remaining costs, that means the remaining costs to be expected at optimal control law are determined by

$$V(n_R(k),w(k),k) = \min_{I_{soll}(k)} \left\{ \mathop{E}_{z(k)} \left[\begin{array}{l} \varphi(n_R(k),w(k),I_{d,soll}(k),z(k)) + \\ V(f(n_R(k),w(k),I_{d,soll}(k)),w(k),k+1) \end{array} \right] \right\}, \tag{24}$$

that means minimization is reduced to one single step.
This procedure is based on the principle to always reuse the stored values $V(n_R,w)$ independent of k, instead of $V(n_R,w,k)$, so that $V(n_R,w)$ is iteratively improved for all (n_R, w). This entails, however, an unlimited increase of $V(n_R, w)$, since always positive values are added to $V(n_R, w)$ in (24) and φ does not approach 0. Therefore, a weighted sum is introduced in (24):

$$V(n_R(k),w(k),k) = \min_{I_{soll}(k)} \left\{ \mathop{E}_{z(k)} \left[\begin{array}{l} \zeta\varphi(n_R(k),w(k),I_{d,soll}(k),z(k)) + \\ (1-\zeta)V(f(n_R(k),w(k),I_{d,soll}(k)),w(k)) \end{array} \right] \right\} \tag{25}$$

with $0 < \zeta < 1$.
The discount factor ζ indicates to what an extent the costs of the actual state $f(n_R(k),w(k), I_{d,soll}(k),z(k))$ are decisive for the newly calculated expected remaining costs $V(n_R(k),w(k))$. This way it has influence on how quickly the remaining costs and thus the control law are changed. For practical applications this means that the parameter z determines how quickly the operation management of the plant is able to adapt itself to other wind conditions, e.g. to a changed frequency distribution $P(z \mid w)$.
To determine the minimum, only those dc-link currents $I_{d,soll}(k)$ are considered, which maintain a given limitation for the approximated speed in the next operation state $f(x(k), I_{d,soll}(k), k)$. At the same time, a speed limitation is realized by the electrical counter-torque.
All $V(x(k), k)$ and the control law $Rg(x(k), k)$ are determined as follows [3]:

For all discrete $x(k)$, e.g. all $n_R(k)$, $w(k)$,
  For all $I_{d,soll}(k)$:
    Calculate $\tilde{n}_R(k+1) := f(n_R(k), w(k), I_{d,soll}(k))$,
    If $n_{min} \le \tilde{n}_R(k+1) \le n_{max}$:
      Calculate:

$$\begin{aligned} J_k(I_{d,soll}(k)) &= \mathop{E}_{z(k)}\Big[\zeta\varphi(n_R(k),w(k),I_{d,soll}(k)) + \\ &\quad (1-\zeta)V(f(n_R(k),w(k),I_{d,soll}(k)),k+1)\Big] \\ &= \sum_{z(k)}\Big[\zeta\varphi(n_R(k),w(k),I_{d,soll}(k)) + \\ &\quad (1-\zeta)V(f(n_R(k),w(k),I_{d,soll}(k)),k+1)\Big]P(z(k)|w(k)) \end{aligned} \tag{26}$$

Chose smallest $J_k(I_{d,soll}(k))$, apply

$$V(\boldsymbol{x}(k),k) := \min_{I_{d,soll}(k)} J_k\left(I_{d,soll}(k)\right) =: J_k\left(\hat{I}_{d,soll}(k)\right)$$
$$\underline{Rg(\boldsymbol{x}(k),k) := \hat{I}_{d,soll}(k)} \tag{27}$$

Repeat!

Here, P(z(k) | w(k)) is the probability of a disturbance z(k) of the wind speed on the condition that the actual wind speed is w(k). This is defined by measurement in form of a table of relative frequencies. At a constant frequency distribution the calculation must only be repeated until the remaining costs V and the control law Rg have reached the optimal value. This becomes obvious when the control law is no more subjected to a change. In case of the wind energy converter, however, the frequency distribution changes. Therefore the calculation must be repeated successively to realize constant updating of the control law.

The value $I_{d,soll}$, for which the expected value in (25) becomes minimal, is recorded in the control law $Rg(n_R,w)$. The minimum in (25) can be found by discretization in form of finite search. Since the instantly calculated expected remaining costs contain information from the past, the stored old values $V(n_R,w)$ can be overwritten by the newly calculated ones, so that the time parameter k is eliminated and memory space saved. $V(n_R,w)$ is recorded in table form for the time and amplitude-discrete values of n_R and w. The reference current value for the given measuring values n_R und w is determined by linear interpolation of the control law $Rg(n_R,w)$.

5. Comparison of the control methods

The examined control concepts are compared in this chapter on the basis of simulation research work. For this purpose, the wind energy converter has been subjected to different wind speeds. For better reproducibility and thus better comparability of the results for a certain time range, a periodic behavior of the wind speed realized by a delta function has been used for the comparing examinations. The wind power gradient (WLG) which is relevant for the control of the wind energy converter has been adjusted for a wide range of values [1] by variation of the frequency. The WLG specifies the alteration of the covered-area related wind power per time unit and is a degree for the wind turbulence.

Furthermore, results of comparing examinations are also presented for realistic time curves of the wind speed. For this purpose, random time responses of the wind speed have been generated in accordance with an algorithm [1]. This algorithm allows for a location-specific parameterisation on the ground roughness and the so-called dynamic factor, so that stochastic characteristics of the wind speed could be used for locations in the mountains which is featured by wind speed alterations of high dynamics, as well as for a flat country location with even wind speeds. The test bench examinations have been carried out with parameters of a variable speed wind energy converter of 22 kVA nominal power. The wind energy converter model features the following characteristic data:

- Generator:
  Nominal power: 22 kVA
  Conductor voltage: 325 V
  Stator frequency: variable
  Nominal speed: 1500 rpm
- Windrotor:
  Power: 22 kW
  Rotor radius: 4,9 m
  Optimal tip-speed ratio: 4,3
  Optimal power coefficient: 0,3818
  Mass inertia of the rotors: 4,8 kgm^2 (related to the generator speed)

Gear ratio: 23

The methods of speed control and direct power control have been compared with the control algorithm of the discrete stochastic dynamic optimization.

In case of the speed control, the operation management calculates the speed n_R according to equation (1) at which the wind rotor generates the maximum possible power which is then given as reference value. Assuming that the optimal tip-speed ratio is always constantly λ_{opt} and the instantaneous value of the wind speed can be measured at sufficient exactness, the speed can be adjusted proportionally to the prevailing wind speed w. Herewith the wind rotor is always operated at maximum possible power coefficient $c_{P,opt}$. This becomes evident by the time response of the power coefficient in Fig. 12. The time response of the electrical power output shows, however, changes beyond wind-induced power fluctuations. The control procedures actuated to adjust the optimal speed cause additional dynamic alterations of the electrical power via the control element (dc-link converter) - depending on the layout of the speed controller - which superpose the wind caused power fluctuations. Changes in the electrical power output cause mains pollutions (for example, flicker effects) and impose high loads on the mechanical drive train components.

Particularly short-time fluctuations of wind speed initiate control procedures. But even in case of a very quick adaptation to a new optimal speed for this short time interval, they do not contribute to a considerable increase in energy yield (see Fig. 15). Improvements in this respect are shown by the methods of direct power control (methods "2" and "3"). Compared with the speed control, they avoid to a large extent oscillations of the electrical power during the control procedure (see Fig. 13). At the same time efforts are followed up when calculating the nominal value to reduce the influence of short-time peaks in the wind speed response. In case of a linear averaging, the wind speed is averaged via a defined time interval. The average is used to calculate the nominal power value. When the averaging is carried out according to the Weibull-distribution, a weighted average value is calculated from the last taken measuring values of w in line with their probability density. Hereby the time response of the power coefficient c_P shows remarkable deviations from the maximum possible value with these control procedures.

Compared with the hitherto examined control procedures, the iterative self-adapting system management (ISSM) on the basis of the algorithm of discrete stochastic dynamic optimization features a high process quality. This is supported by a constant power output (electrical power) and a high power coefficient (see Fig. 14). The time responses of the system variables prove, that the control law adapts itself iteratively to the wind speed distribution and fulfils the given process requirements.

In order to provide a better comparability of the research results for stochastic wind speed characteristic, the results have been quantified and evaluated by means of the above introduced indices.

Fig. 15 shows c_P-Index und P-Index of the compared control methods for two different location-specific wind characteristics.

The speed control features a very low c_P index and a high P-index, the same as in case of the deterministic response of the wind speed. The P-index can be reduced by the direct power control methods (methods "2" and "3"). On the other hand, a higher c_P-index must be accepted. By a suitable selection of the weighting factor g, a reduction of the P-index or c_P-index may be granted higher priority in case of the iterative adapted process control methods. The sum of the introduced indices implemented as quality criterion is hereby lower for all possible values of weighting factor g, than in case of the other examined procedures which entails a higher process quality. The minimization of the indices sum is the aim of the optimization task by formulating the cost function.

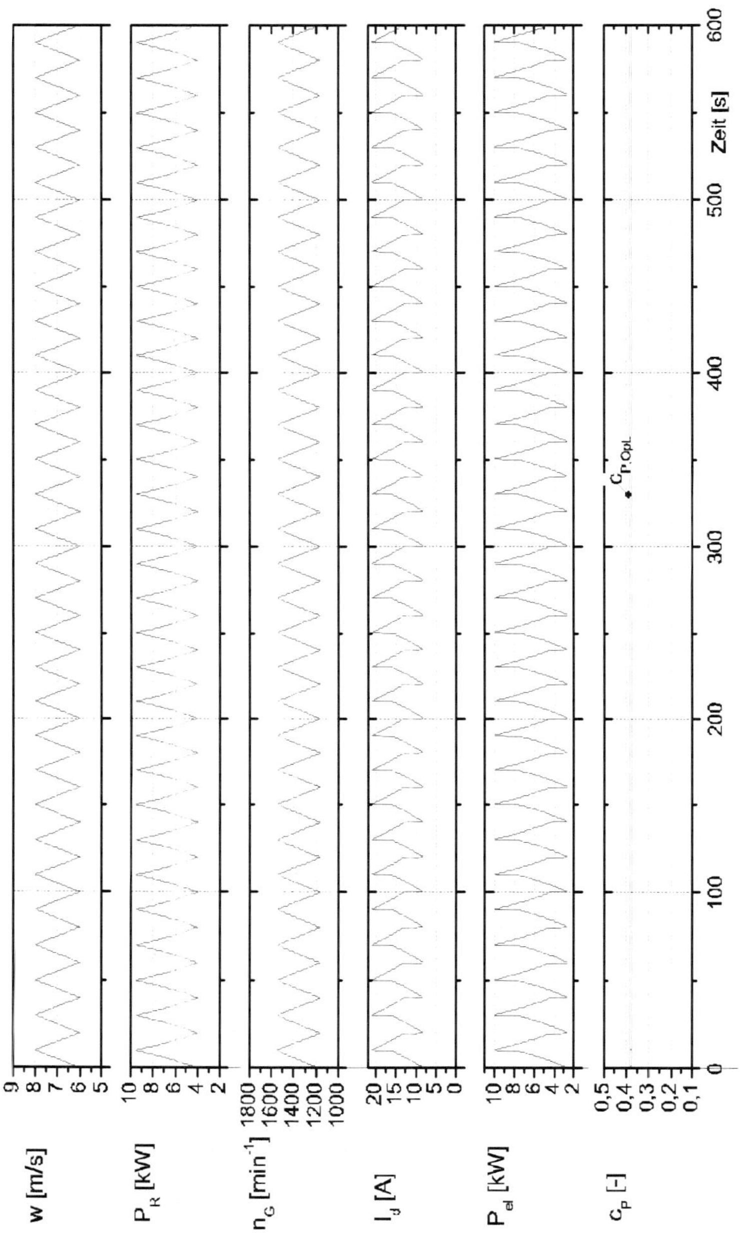

Fig. 12. Time response of system variables at speed control (method "1"); deterministic response of the wind speed

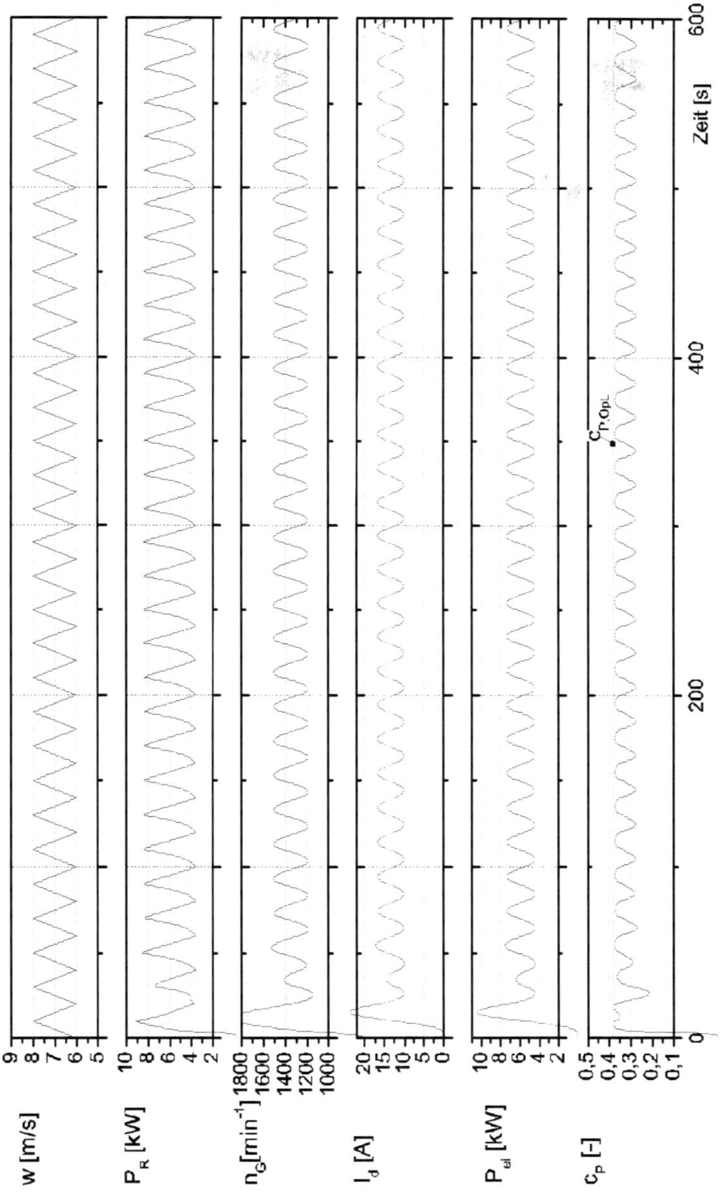

Fig. 13. Time response of system variables at direct power control (method "2) deterministic response of the wind speed

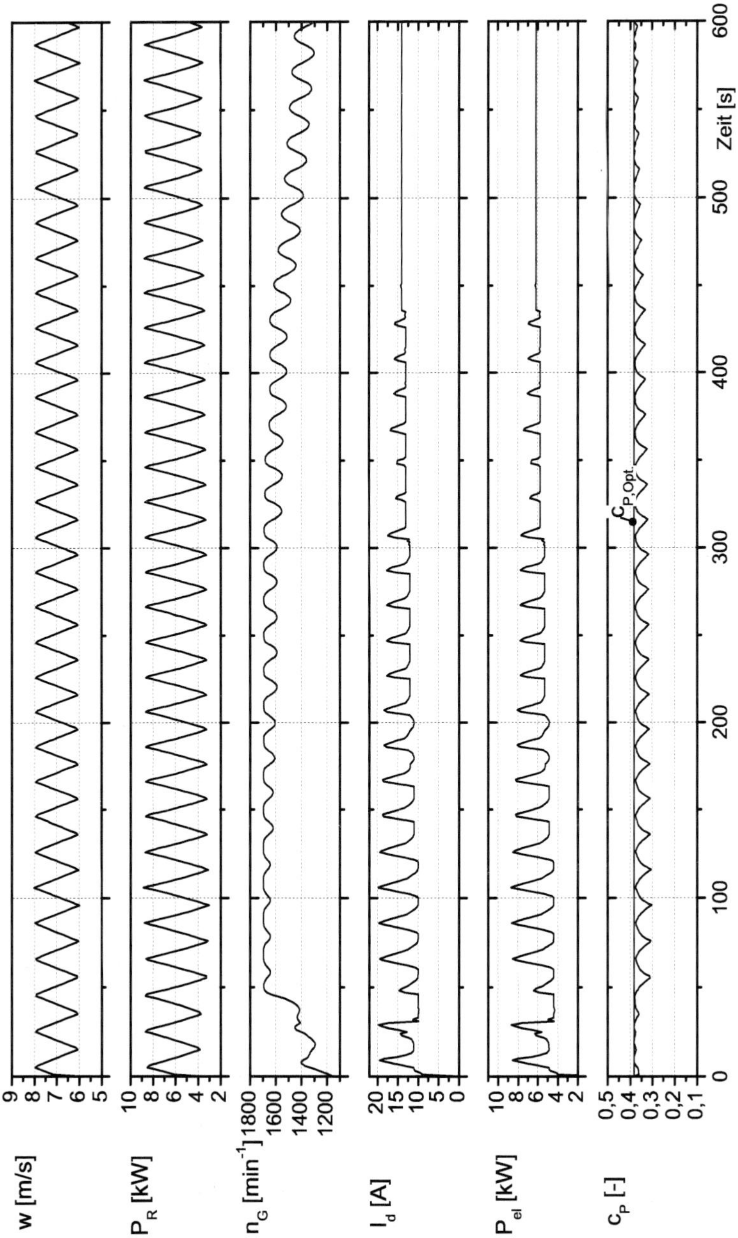

Fig. 14. Time response of system variables at iterative self-adapting system management (stochastic optimization); deterministic response of the wind speed

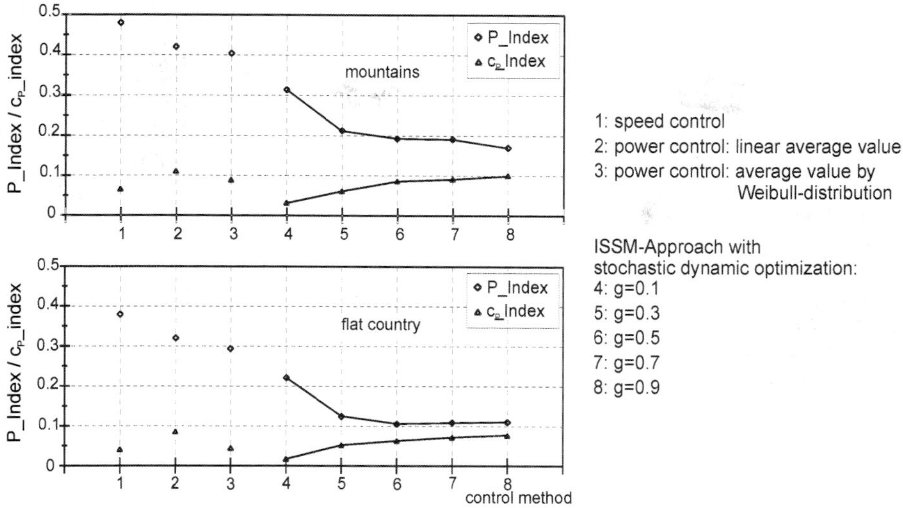

Fig. 15. Quantitative comparison of research results by means of P-index and c_P-index for two characteristic locations

The frequency distribution of the load collective in the mechanical drive train has been used for further evaluation of the results of the comparing examinations. It has been prepared by classification of the time response of the shaft torque. The shaft torque collective of the ISSM method shows a remarkable reduction of the load cycles with high amplitude as compared with collectives of the other control methods, which implies an increase of usability and life (see Fig. 16).

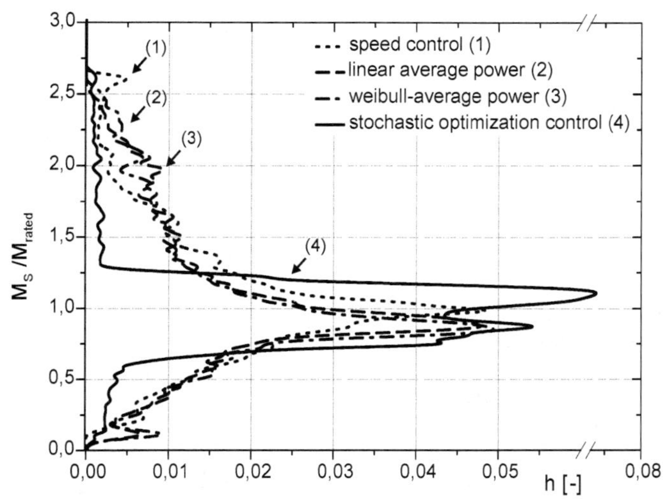

Fig. 16. Relative frequency distribution of the shaft torque at the examined control methods

6. References

[1] Sourkounis, C. (1994/1995) Windenergiekonverter mit maximaler Energieausbeute am leistungsschwachen Netz, *Dissertation*, TU Clausthal; Papierflieger, Clausthal-Zellerfeld

[2] Sourkounis, C.; Bekehermes, M. (2000). Iterativ adaptierende Betriebsführung für drehzahlvariable Windenergiekonverter, *Atp - Automatisierungstechnische Praxis 3*, Oldenbourg Verlag

[3] Sourkounis, C. (2004). Drehzahlelastische Antriebe unter stochastischen Belastungen, Habilitationsschrift, TU Clausthal; Papierflieger, ISBN 3-89720-737-0, Clausthal-Zellerfeld

[4] Klöckner, J. (1979). Berechnung des Schwingungsverhaltens von gedämpften, schwach nichtlinea-ren Schwingerdaten unter Berücksichtigung der Reduktion der Freiheitsgrade, *Dissertation*, TU Berlin

[5] Krüger, M. (1994). Eine anwendungsorientierte Methode zum Entwurf von Zustandsregelungen für elektro-mechanische Hochleistungs-Antriebssysteme. *Dissertation*, TU Clausthal 1994, Fortschritt-Bericht VDI, Reihe 20, Nr. 157, VDI-Verlag, 1994

[6] Goslar, M. (1998). Ein Beitrag zur anwendungsorientierten Zustandsregelung elektrischer Hoch-leistungsantriebe, *Dissertation*, TU Clausthal

[7] Sourkounis, C.; Beck, H.-P. (2003). Electromecanical Power Converter Systems for 5 MW-off-Shore Wind Energy Converters, *Proceedings of EPE 2003*, ISBN 90-75815-07-7 Toulouse

[8] Ernst-Cathor, J. (1987). Drehzahlvariable Windenergieanlage mit Gleichstromzwischen-kreis-Umrichter und Optimum-suchendem Regler, *Dissertation*, TU Braunschweig

[9] Neumann, K. (1977). *Operations Research, Band II*, Hanser, München, Wien

[10] Sourkounis, C.; Ni, B. (2005). Optimal Control Structure to Reduce the Comulative Load in the Drive Train of Wind Energy Converters, *European Conference on Power Electronics and Applications (EPE)* Dresden

[11] Sourkounis C.; Ni, B. (2006). Drive Train Control for Wind Energy Converter Based on Stochastic Dynamic Optimization, *IEEE Industrial Electronics Conference (IECON)*, Special Session: Self-optimizing systems and advanced control, Paris, France

[12] Wang, Q., Chang, L. (2004). An Intelligent Maximum Power Extraction Algorithm for Inverter-Based Variable Speed Wind Turbine Systems, *IEEE Transactions on Power Electronics*. Vol. 19, NO. 5

[13] Toedorescu, R., Blaabjerg, F. (2004). Flexible Control of Small Wind Tubines With Grid Failure Detection Operating in Stand-Alone and Grid-Connected Mode, *IEEE Transactions on Power Electronics*. Vol. 19, NO. 5

[14] Carlson, O., Ulén, E. (1996). Torque Control of Synchronous and Induction Generators for Variable Speed Operation of Wind Turbines, *European Union Wind Energy Conference*, Göteborg Sweden

7

Small Wind Turbine Power Controllers

Dr. Horizon Gitano-Briggs
University Science Malaysia
Penang, Malaysia

1. Introduction

One of the greatest challenges associated with wind power is the unpredictable character of the wind. Even at the best wind sites, those with steady reasonably high speed wind, there are variations in speed and direction of the wind which affect the ability of the wind turbine to deliver power. Larger wind turbine systems have complex control systems which automatically track changes in wind direction and speed, and adjust turbine orientation, blade pitch, and generator gearing to maintain the desired electrical output. Small turbine systems are typically much less sophisticated, however they generally still have some form of control to improve their longevity and power production. The main purposes of a controller in a wind energy system are (in order of priority):

1. Prevent damage to the wind turbine
2. Prevent damage to the load
3. Maximize power production

We will concern ourselves here with smaller wind turbine systems, figure 1, which we will define somewhat arbitrarily as those systems rated at 10kW or less. Such small systems in the past have been predominantly designed for sturdiness, with robust mechanical controls, and relatively modest overall performance. With advances in microcontrollers, and electronic power switching components, however, the level of sophistication of theses small wind systems has been steadily improving. In this chapter we will investigate the controls associated with small wind turbine systems, culminating in a detailed description of the Peak Power Tracking controller utilizing a microcontroller running an impedance matching DC-DC converter between the turbine and the load.

Although the main focus of our discussion is the control of wind turbines, the techniques and components described here also find use in other alternative power systems, such as solar photovoltaic systems, where they can improve the overall system power extraction. Many alternative power systems incorporate several different power sources, such as wind and solar. Such "hybrid" systems have the advantage of greater power availability (periods of low wind are often characterized by clear skies and good solar power, while strong winds are often associated with cloud activity). We will therefore use examples from both solar and wind power systems in the discussion of the relevant techniques.

1.1 Wind turbine systems

A wind turbine converts wind power into shaft power to drive an electrical generator or other load (Van Baars & Bongers 1992). In smaller electrical wind power systems the most

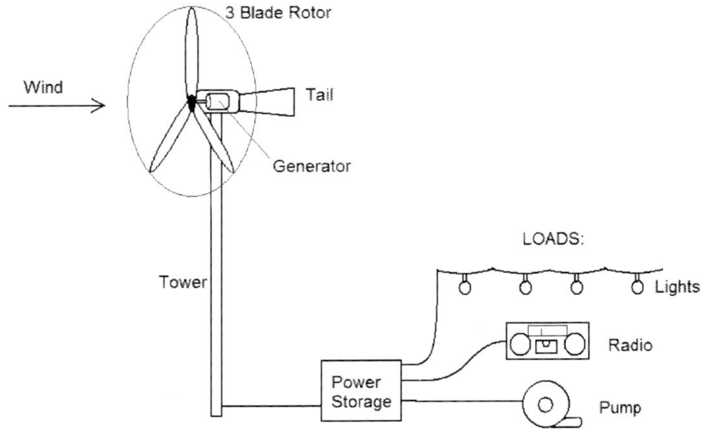

Fig. 1. Schematic diagram of a typical small wind turbine power system.

common configuration is a 3-blade, horizontal axis wind turbine directly driving a 3-phase permanent magnet generator.
The output power of a wind turbine is function of wind velocity cubed. It can be described mathematically by (1).

$$P = 0.5 C_P \rho A V_{wind}^3 \quad (1)$$

where P is the power in Watts, C_P is the dimensionless coefficient of performance, ρ is air density (kg/m³), A is rotor swept area (m²), and V_{wind} is the wind speed (m/s) (Jonson 1985). Generally both swept area and air density are taken as constants while the control of the wind is beyond the scope of this text. Therefore, to maximize the output power of a

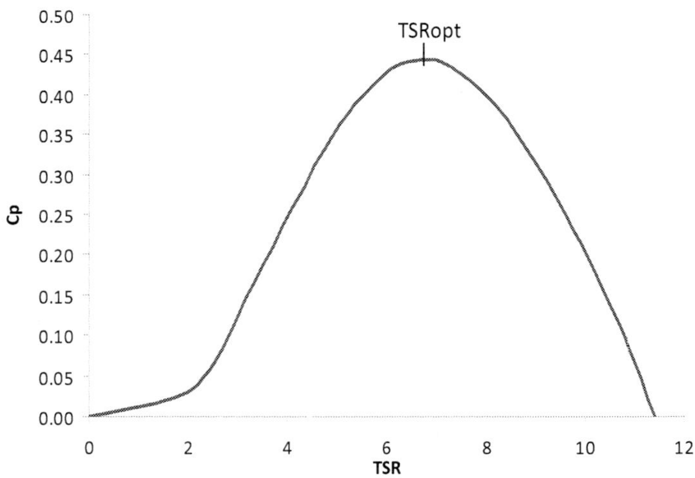

Fig. 2. Coefficient of performance as a function of tip-speed-ratio for a wind turbine.

wind turbine we have to optimize the value of C_P. The coefficient of performance is a strong, non- linear function of the "Tip Speed Ratio" as seen in Figure 2. It depends on such factors as the number of blades, and the pitch and shape of the blades. The maximum value C_P can attain is theoretically around 57%, however the highest performing turbines generally only attain values in the 40% range, while values in the 30% range are more common.

The tip-speed-ratio is the linear speed of the outer extremity of the turbine blade divided by the wind speed. For a turbine of radius R, and rotor speed ω_{rotor} (in radians/sec) the tip-speed-ratio, TSR, is thus:

$$TSR = \frac{\omega_{rotor} R}{V_{wind}} \tag{2}$$

Where V_{wind} is the speed of the wind at the turbine hub height. In order to maximize C_p we can vary the load on the turbine to achieve a rotor speed ω_{opt} of:

$$\omega_{opt} = \frac{TSR_{opt} V_{wind}}{R} \tag{3}$$

The optimum tip-speed ratio TSR_{opt} is a constant dependent on turbine blade design for fixed blade turbines. Thus for every wind speed there will be a different rotor speed for maximum power production. The optimum TSR value is around 5 to 7 for most small turbines, meaning that the tip of the turbine blade will have a linear velocity around 6 times the wind speed.

Most small wind turbines have 2 or 3 fixed blades rotating about a horizontal axis directly driving a permanent magnet generator. The control system generally consists of a mechanical system for furling the turbine in high winds, and with perhaps a power controller and electrical breaking system. Bergey Windpower's 10kW Excell turbine is a notable exception, having a mechanical blade pitch control system which adjusts the pitch of the blades as a function of rotor speed (Manwell et al., 2002). Larger turbines commonly have 2 or 3 adjustable pitch blades, and 1 or 2 gear stages between the rotor and the generator. A larger number of blades gives a lower operating speed, and greater starting torque facilitating turbine starting in low speed winds.

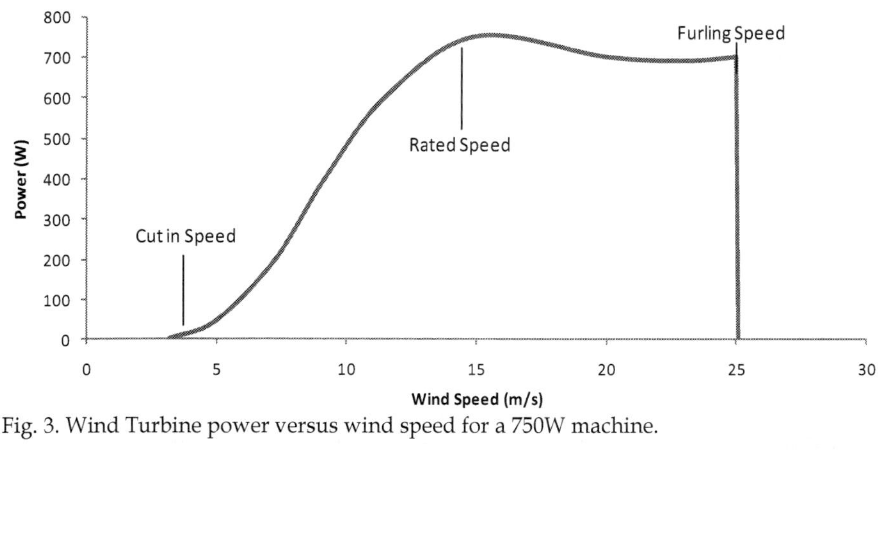

Fig. 3. Wind Turbine power versus wind speed for a 750W machine.

Figure 3 shows the power produced from a typical small wind turbine as a function of wind speed. Below the "cut in speed" there is not enough wind power to overcome friction, thus no power is produced. Above the cut-in speed the power increases rapidly to the "rated speed". Generally the turbine produces its rated power at the rated wind speed, though some manufacturers quote numbers differently. This turbine is rated at 750W in a 15 m/s wind. The system has been optimized for operation around 15 m/s, and though the wind power continues to increase above this point, the turbine power drops off slightly as the blades begin stalling. The turbine has a maximum operating speed (furling speed) of 25 m/s, above which it is stopped to avoid damage to the machine.
Several mechanisms are used to stop turbine rotation including rotation of the turbine out of the wind, either vertically (tilt) or horizontally (yaw), frictional breaks, electrical breaking and blade pitch adjustments to cause the blades to stall, i.e. furl. Most small machines use a mechanical system to rotate the turbine out of the wind and may use electrical breaking, while larger machines furl the blades and have frictional breaks as well. A comparison of large and small wind turbine configurations is given in figure 4. While there is a wide range of variation among both groups, and overlap between them, these illustrate the general trends.

	LARGE TURBINES	SMALL TURBINES
Power Rating (kW)	>100kW	10kW or less
Power Utilization	Grid	Batteries or small load (ie. Pump)
Gearing	2-stage or CVT	Direct drive or 1 stage fixed
Generator	Synchronous AC	Permanent magnet AC
Design Point	High Power production	Low Cost/Maintenance
Blades	Variable Pitch	Typically fixed pitch
Yaw Control	Motorized Control	Free - Tail vane
Breaking	Pitch, Frictional	Tilt, Electrical

Fig. 4. Typical characteristics of large and small wind turbine systems

2. Load – source matching

In power systems a power source drives a load. A common concept in power transmission is that of "impedance matching" or load - source matching. The overall system will be most efficient when the power source matches the characteristics of the load. In the case of grossly divergent load - source characteristics an impedance matching intermediary, such as a gear box for example, may be used. A classic example of this is the mechanical power drive system of internal combustion engine vehicles. Internal combustion engines tend to operate at a few thousand revolutions per minute (rpm), while the wheels of a typical vehicle are spinning several hundred rpm. To improve the overall vehicle performance a transmission with an overall gearing ratio of around 10 couples the engine to the wheels, allowing the engine to operate at higher speeds, and lower torques while delivering the lower speeds and higher torques required at the wheels. Gearing systems are often used in larger turbines to match the low speeds and high torques of the turbine to the higher speeds, and lower torques of the generator. In small wind turbine systems we will be making use of another electrical impedance matching device: the DC-DC converter.

2.1 DC-DC converters

DC-DC converters are essentially the direct current equivalent of transformers in alternating current circuits. They can convert electrical power from one voltage and current level to another, while of course obeying the conservation of energy: as voltage is increased current must necessarily be decreased. Transformers take in alternating currents, and convert them to changing magnetic fields in a primary winding. A magnetically linked secondary winding converts the magnetic flux back into an alternating current. The secondary winding voltage can be either greater or less than the primary winding, figure 5, depending on the relative number of turns of wire in the coils. The output voltage is proportional to the input voltage times the ratio of the number of turns in the input (N_1) to the output (N_2), thus $V_{out} = V_{in} \times (N_2/N_1)$. The currents follow the inverse relationship, i.e. $I_{out} = I_{in} \times (N_1/N_2)$. Smaller transformers tend to be less efficient, around 85%, while well designed high power transformers may have efficiencies as high as 98% (Calwell & Reeder, 2002).

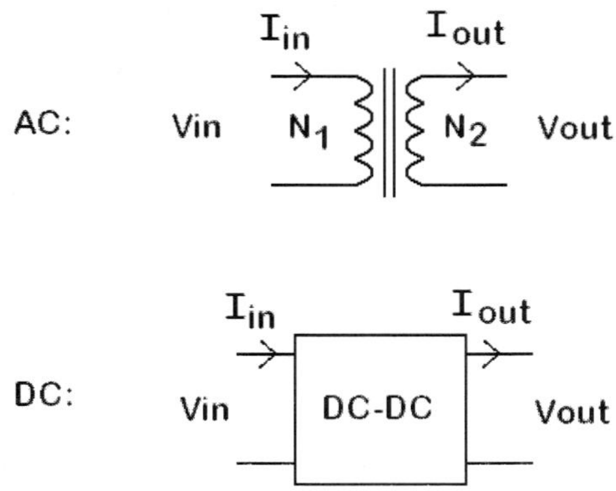

Fig. 5. AC and DC voltage/current converters

A DC-DC converter in which the output voltage is greater than the input voltage is usually referred to as a "boost" converter. In a boost converter the output current will always be less than the input current. If the output current is greater than the input current (and thus the output voltage is lower than the input) the converter is referred to as a "buck" converter. A boost converter is the DC equivalent of a step-up transformer in AC circuits, and the buck is analogous to the step down transformer. In the past DC-DC conversion was performed using a motor-generator device, or by first converting to AC, using a transformer to step the voltage up/down, then rectifying. With advances in the power and speed of power switching devices such as power FETs and IGBTs, DC-DC converters have become smaller, less expensive and more efficient. Higher frequency operation of the switching device allows the use of smaller inductors, reducing the size, weight and cost of these components. Today peak efficiencies of DC-DC converters are around 85 to 95%. These improvements have allowed DC-DC converters greater application in a variety of fields, including alternative energies where both boost and buck converters are common components.

2.2 DC-DC internal structure

A DC-DC buck, or step down, converter is essentially a Pulse Width Modulated (PWM) connection between the source and load (Agrawal 2001). A transistor or similar switching element is placed between the supply voltage and the load and cycles on/off at a high frequency. When the transistor is "on" current will flow from the source to the load. When the transistor is turned off current stops flowing. Various inductive and capacitive components can be placed in the circuit to smooth out the current flow. If the transistor is in the on state most of the time, the voltage output voltage is close to the source voltage. As the duty cycle, or on time divided by cycle period, decreases, so to does the output voltage. Thus as the relative width of the "on" pulse feeding the transistor is increased, the output increases. For a maximum output the transistor will be left on (ie. 100% duty cycle).

In a boost DC-DC converter a transistor is intermittently connecting the input voltage to an inductor, grounded at the opposite end. When the transistor is on current will flow into the inductor towards ground. Once the transistor is switched off the current decrease in the inductor will induce a high voltage on the open end of the inductor. This voltage can be fed through a diode into a capacitor, charging it to a voltage higher than the original input voltage. This boosted voltage can then be used to drive a higher voltage load. Both the frequency, duty cycle can be varied to adjust the output voltage of a boost converter.

3. DC-DC boost converter in alternative energy systems

The use of a simple boost converter can greatly enhance the power capture from alternative energy systems in several ways: 1) they can boost low voltages up to levels capable of charging batteries or running higher voltage loads, 2) they can impedance match sources and loads, and 3) they can be adaptively adjusted to provide peak-power extraction from variable power sources such as wind turbines.

One of the simplest applications of a DC-DC converter in an alternate energy system is a boost converter used to power loads when the supply voltage is below the required load voltage. For example, if we have a small photovoltaic power system with a maximum panel voltage of 16V charging batteries at 12V, the system will only extract power when the panel voltage is above the battery charging voltage Vb. The solar panel will only produce the full 16V in full, nearly normal sunlight. This will limit power production to only a few hours a day when the panel voltage exceeds Vb, as shown in figure 6. Even in partial sun or when the sun is not exactly normal to the panel, it can still produce power, but at a lower voltage.

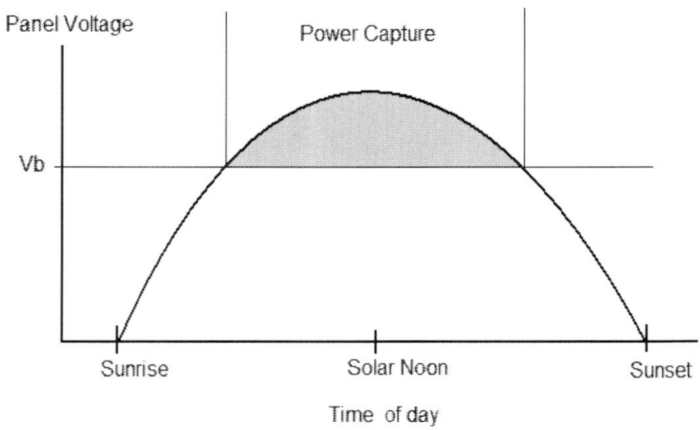

Fig. 6. Power production from a solar panel connected to a battery of voltage Vb.

If a boost DC-DC converter is placed between the solar panel and the batteries, the panel's voltage can be pumped up to charge the batteries even when the panel voltage is less than

Vb, as shown in figure 7. The batteries will now be charged any time the panel voltage exceeds the DC-DC converter's "turn on" voltage, Von, typically a few volts. This greatly increases the power production from the overall system.
In many photovoltaic power systems where a DC-DC converter is not used more solar cells are added in series to boost the voltage. A 12V battery charging panel will typically operate at 18V open circuit. As solar cells are quite expensive, this represents an excellent area of application for the DC-DC boost converter.

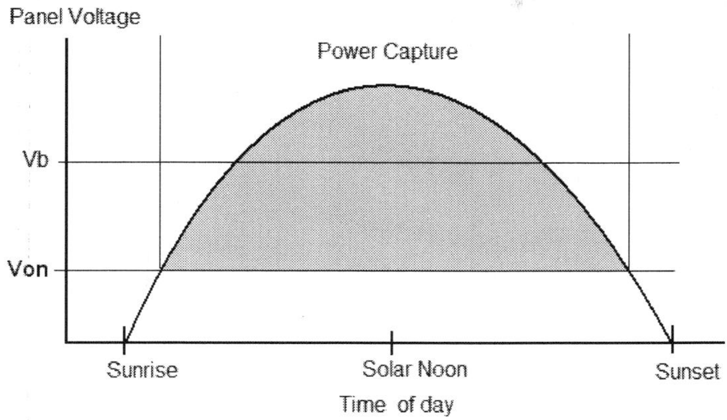

Fig. 7. Power production from solar power system with a DC-DC boost converter.

Depending on the voltage supplied to the DC-DC boost converter there are 3 major modes of operation. Figure 8 shows the output voltage (Vout) of an actual DC-DC boost converter connected to a resistive load as a function of input voltage (Vin). When the supplied voltage is lower than the converters "turn on" voltage Von, the output voltage of the circuit is below the input voltage, as in region 1 in the figure. This will not drain batteries connected as a load as there is a diode in series with the output of the converter. Once the supplied voltage exceeds Von, approximately 2.5V for this unit, the boost converter maintains 14V on its output, as in region 2. Finally when the input voltage exceeds the output voltage the circuit is bypassed, and the input voltage is transmitted to the output through a diode with a small voltage drop, as in region 3.
When applied to a 12V battery load, operation in region 1 will give no power extraction. In region 2 it depends on the power supplied and the efficiency of the converter. Converter efficiencies are typically above 75%, with well designed units in the 90-95% range. When the converter is bypassed, i.e. in region 3, there is typically just one diode drop between the input and output.

3.1 DC-DC buck converter in alternative energy systems

In an AC circuit transformers are often used to impedance match sources to loads. For example if you have a 220VAC low current supply, but require high current and low voltage to operate several 36W 12V halogen bulbs, you would likely use a 220 to 12V step down transformer between the supply voltage and the bulbs. Similarly both buck and boost DC-DC converters can be used to match up disparate voltage loads and sources.

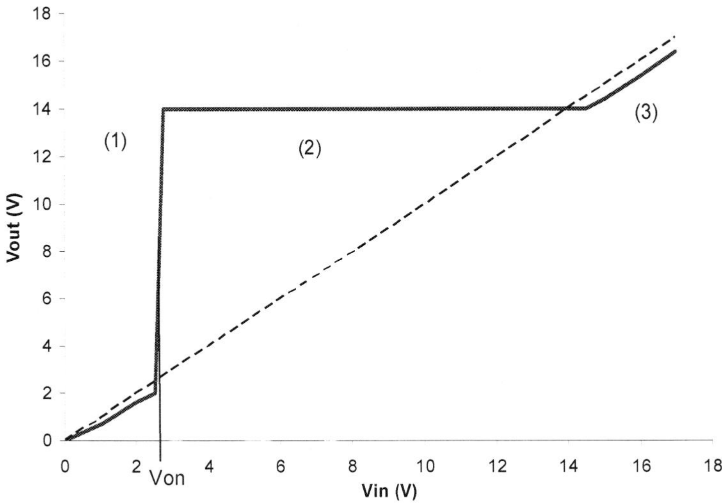

Fig. 8. DC-DC boost converter output voltage versus input voltage

One example of this is a wind turbine system in India. The wind turbine generates 48VDC and the villagers use it to charge 12V batteries for domestic use powering lights, radios and televisions. The batteries are carried to the charging station at the base of the wind turbine, and left there to charge. Ideally the batteries could be placed in series of 4 batteries for direct charging from the turbine, however this proved impractical in practice due to random charging and retrieval schedules. Instead a DC-DC converter was placed between the 12V batteries and the turbine, as in figure 9. The 500W turbine can supply 10A at its rated speed, and an 80% efficient DC-DC converter will deliver 32A at 12V to the batteries, providing 400W for charging.

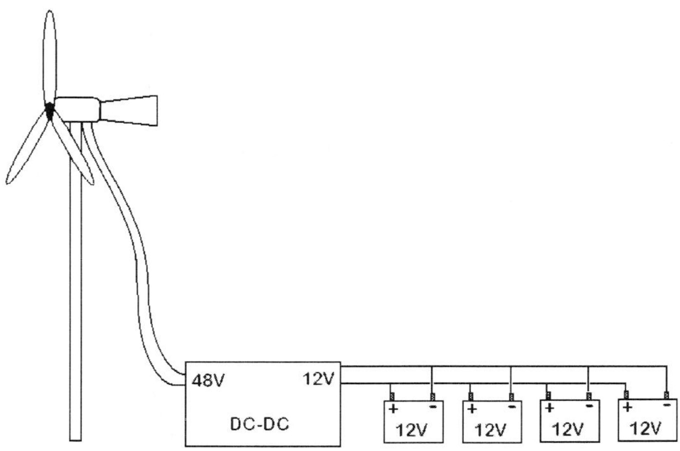

Fig. 9. Wind Turbine with DC-DC Step Down Converter charging batteries

If, alternatively, the 12V batteries were directly connected to the turbine in parallel, the current draw from the batteries would be more than the turbine could supply, slowing the turbine, reducing its power output, and likely stalling the turbine, giving essentially no power production.

The DC-DC system has another advantage over the "4 batteries in series" system in that it can extract power and charge the batteries even if the turbine voltage is below 48V, as in the previous example. The operation of this system is illustrated in figures 10 and 11. In Figure 10 we see the current and Power of the turbine as a function of turbine speed for its rated wind speed. The current and power go to zero when the turbine produces an open circuit voltage of about 100V at its no load speed of approximately 475 rpm. The ideal operating point is indicated at the maximum power point occurring at a rotor speed of ½ of the no-load speed, or 238 rpm. The voltage is proportional to rotor speed, and has reduced to 50V at this maximum power point, and the current has risen to 10A.

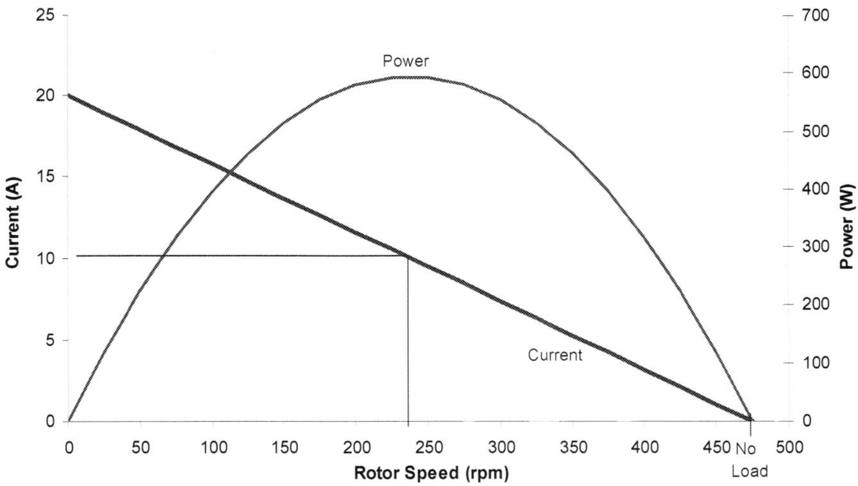

Fig. 10. Wind Turbine Power and Current versus rotational speed at rated wind speed.

The load line of a single battery and 20 batteries in parallel is shown in figure 11. The batteries draw about 1.5A each at a terminal voltage of 14V which corresponds to 56 rpm rotor speed. If the batteries are placed in parallel they will draw a total current of 30A at 14V. This current is well in excess of what the turbine could provide at this voltage, bogging the turbine down at about 50 rpm.

At this point the tip-speed-ratio is so small that the turbine torque is very low, and the turbine will likely stall completely. The single battery appears to be a better match, however the charging current of approximately 15A is excessive for the batteries used. With the DC-DC converter inserted between the turbine and the 20 batteries the load line of the batteries can now be matched to draw 10A at 50V from the turbine. The output voltage of the DC-DC converter will be near 14V, supplying the batteries with about 28A total, or just over 1.4A per battery.

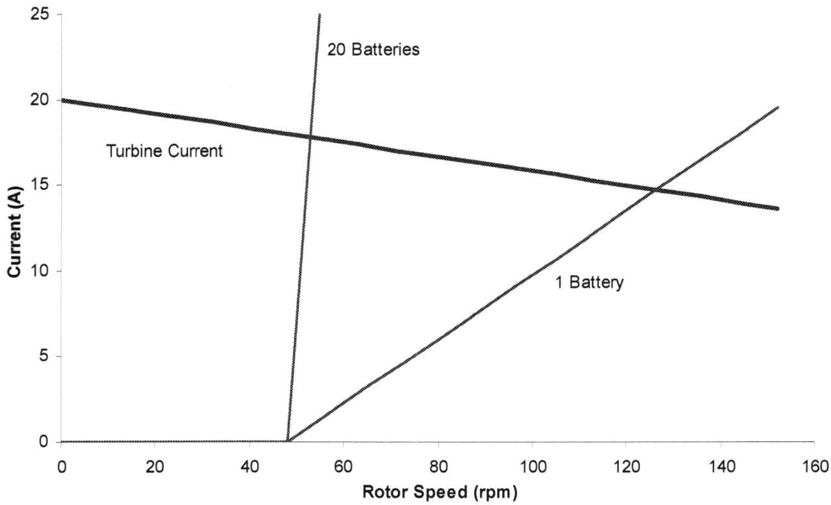

Fig. 11. Load current (1 and 20 batteries in parallel) and turbine current versus rotor speed

The above is an idealized case where the wind speed was taken to be a constant. As the wind speed varies, so too will the maximum power extractable, and the appropriate turbine speed. An intelligent controller can be applied to the DC-DC converter to adjust the impedance between the turbine and the batteries to maintain the maximum power extraction over a wide range of speeds. This is the so-called peak power tracking controller.

4. Detailed description of a peak power tracking controller

A 1.2kVA peak-power tracking controller was developed for a 48V turbine to charge varying loads, anticipating a typical output voltage of 14V for the charging of 12V batteries. The controller used a buck circuit as shown in figure 12 between the turbine and the load. It is composed of single switching element, and an L-C filter.

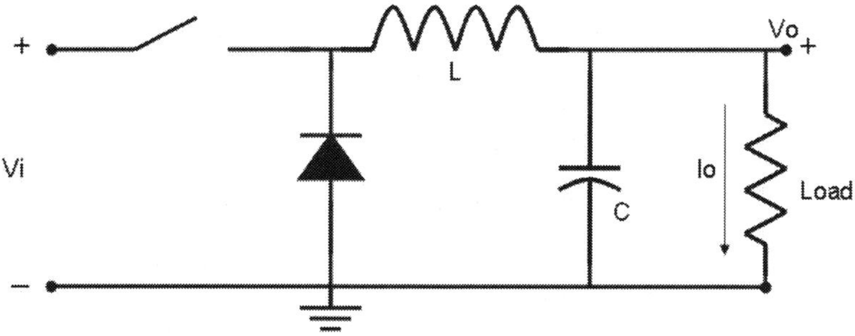

Fig. 12. The basic elements of the buck circuit used in the peak-power tracking controller

The buck circuit provides an average output voltage (Vo) that is less than the average input voltage (Vi) by the duty cycle of the circuit (D), which is the ratio of the switching element conduction time to the full period time. The function of the inductor and the capacitor in the circuit is to smooth the output current and voltage, respectively (Rashid 1993).

$$V_o = DV_i \tag{4}$$

Switching time plays a significant role in the efficiency of buck circuit, as well as the size and cost of the overall circuit, it must therefore be considered in the design. Higher frequencies allow for the use of smaller inductive components, reducing cost, size and weight.

An Insulated Gate Bipolar Transistor (IGBT) was used as the switching element of the circuit. It was chosen because of its ability to handle a high currents as well as its fast switching speed. To switch at the fastest rate possible the IGBT is driven by an IGBT driver chip. The PWM duty cycle is set by the microcontroller based on the operating conditions of the turbine and load. To enhance the switching time of the circuit a schottky freewheeling diode was used (Schelle & Castorena, 2006). All power components were sized appropriately for 1.2 kVA operation.

A voltage divider is placed across the output of the controller for sensing of the output voltage. The voltage must be divided down to a range appropriate for the input of the controller, in this case 0 to 5V. A hall-effect current sensor in series with the load is used to measure the output current of the circuit. The final buck circuit is shown in Fig. 13.

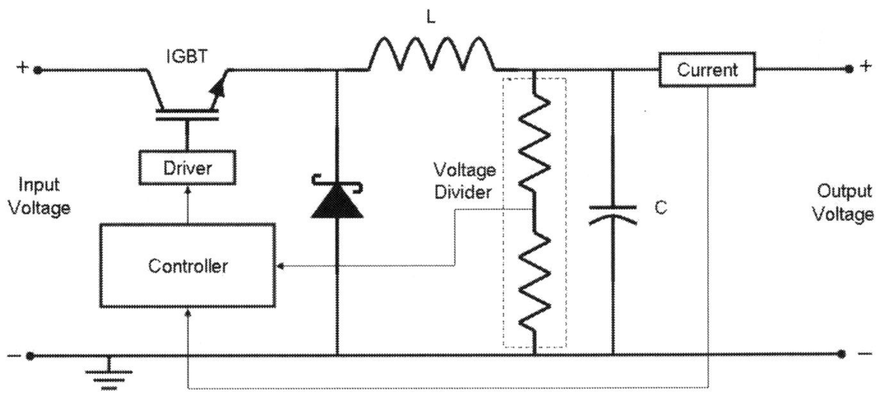

Fig. 13. The final schematic of the buck circuit

Varying the duty cycle of the switching element changes the wind turbine loading level. Zero duty cycle means that the switching element is always OFF and no load is connected to the wind turbine. This will cause the turbine to rotate at its maximum rotational speed and give no power output. Increasing the duty cycle decreases the output impedance of the system, which adds more load to the wind turbine; thereby slowing the rotor. The rotor reaches the minimum speed when duty cycle is one.

A simple model suffices to determine the relationship between the duty cycle of the DC-DC converter and the expected shaft speed. Assuming a constant power input and efficiency the output electrical power I^2R should also be constant. The resistance is essentially an average between the connected load resistance R weighed by the switch 'on' time and an infinite resistance weighed by the switch 'off' time, thus the effective average resistance is R/D where D is the duty cycle. For a Permanent Magnet (PM) DC generator such as ours, the relationship between the generator current and rotor speed is linear decreasing from the stall current I_{stall} at rotor speed $\omega = 0$, to zero current at the unloaded speed ω_{NL}. Mathematically we would write:

$$I = I_{stall} - I_{stall}\left(\frac{\omega}{\omega_{NL}}\right) \tag{5}$$

Knowing that I^2 times the effective resistance R/D is constant allows us to write:

$$I_{stall}^2\left(1-\frac{\omega}{\omega_{NL}}\right)^2\frac{R}{D} = C \tag{6}$$

Where C is a constant. Rearranging to solve for the rotor speed ω we have:

$$\omega = \omega_{NL} - \omega_{NL}\left(\frac{CD}{I_{stall}^2 R}\right)^{0.5} \tag{7}$$

This shows that the rotor speed will decrease as the square root of the duty cycle. The factor $\omega_{NL}\,(C\,/\,I_{stall}^2\,R)^{0.5}$ represents the speed decrease from ω_{NL} at a duty cycle of unity.

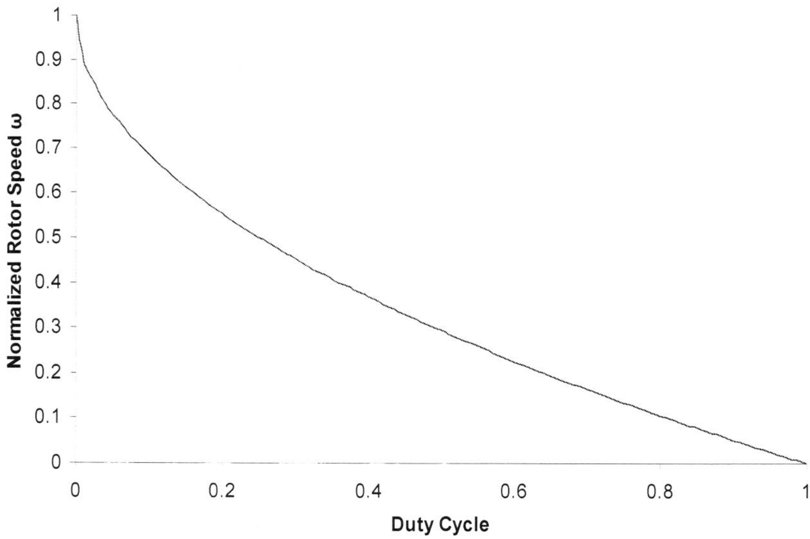

Fig. 14. Rotor speed versus duty cycle for an ideal DC-DC converter and load

If we assume that the load can completely stall the rotor, i.e. $\omega = 0$ at $D = 1$, then we can rewrite the relationship as:

$$\omega = \omega_{NL}\left(1 - D^{0.5}\right) \tag{8}$$

This relationship is shown graphically in Fig. 14 where rotor speed has been normalized, i.e. $\omega_{NL}=1$.

4.1 Breaking system

Some environmental conditions, for instance hurricanes, include very high wind speeds. This may result in turbine runaway and destruction of the wind turbine. Additionally, it may be necessary to periodically stop the wind turbine spinning, if there is a mechanical or electrical fault or even during periodic maintenance. Thus some kind of braking system for the wind turbine is required.

Many turbines include a mechanical braking design (i.e. furling or tilting the axis of the turbine away from incoming wind). Alternatively, frictional beaks may be incorporated into the wind turbine. The disadvantage of such system is the short life expectancy and additional system cost and complexity of these systems (Rajambal et al., 2005).

Another option is to use electrical braking. Under high wind conditions heavy dump loads may be connected directly to the generator to slow the turbine enough causing the blades to "stall" aerodynamically. Once stalled, the torque produced by the turbine is greatly reduced, and it will spin at a very slow speed or stop.

The braking system design is shown in Fig. 15. It consists of a relay that connects the generator to a dump load when activated. In sufficiently strong winds the generator output will exceed a pre-set trigger level, i.e. breaking threshold. During operation the controller is continuously measuring the generator voltage (and thus speed) via the voltage divider. When the generator voltage exceeds the breaking threshold the breaking relay is activated, connecting a high power, low resistance dump load directly to the generator output. An anti-shock inductor has been placed in series with the dump load to avoid arcing at the contacts.

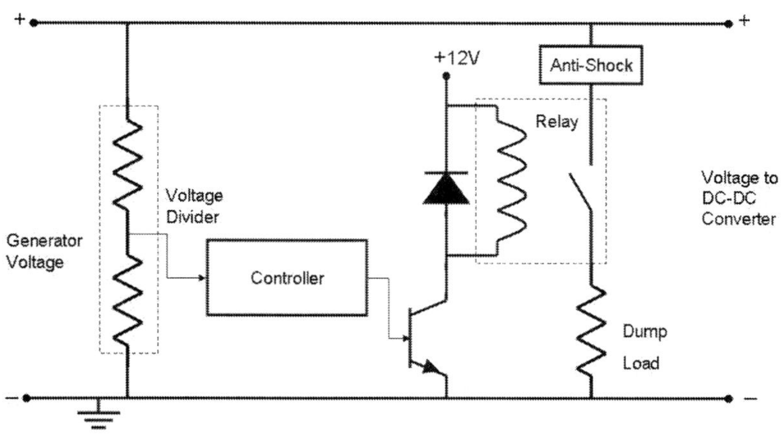

Fig. 15. The schematic of the braking circuit

Stalling the wind turbine to avoid run-away is the primary purpose of the breaking system. However it can also be used to avoid the over charging the load (i.e. batteries) during high winds. If the controller senses that the batteries are being over charged, it will reduce the duty cycle, reducing the charging current. This will also unload the wind turbine and cause it to rotate faster. If the turbine is near the maximum speed a dump load can be added instead, slowing the turbine and reducing its power production. The presence of different dump loads is helpful in the reducing the turbine power by converting it to heat in the dump loads. Alternatively some useful intermittent loads, such as pumps, can be used instead of dump loads, only being activated at high wind speeds.
The response of the braking circuit is shown in Fig. 16. Breaking is initially overridden manually, and the generator is allowed to run over the breaking threshold limit. At approximately 7 seconds breaking is re-enabled, and the controller immediately goes into breaking mode, reducing the rotor speed to less than 30 rpm. At intervals of approximately 10 seconds the controller releases the breaks allowing the turbine to spin up until the threshold is again exceeded, and break is then reapplied. This continues until the wind speed is reduced below the breaking threshold, and normal operation ensues.

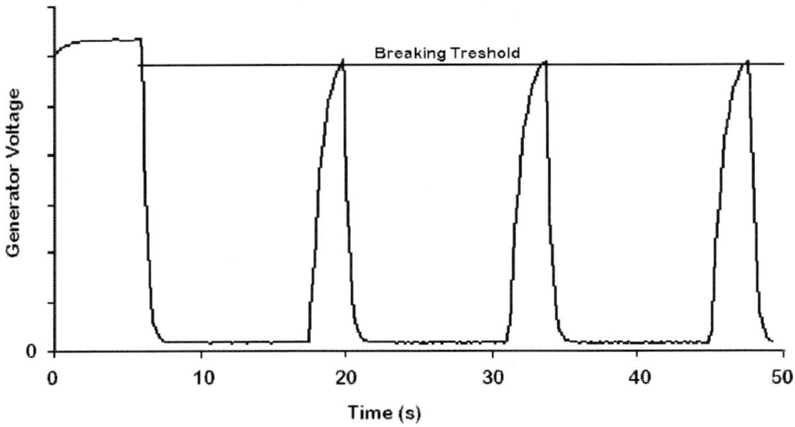

Fig. 16. Breaking system response. Breaking system was overridden before t = 7 sec.

4.2 Controller algorithm

A micro-controller is responsible for measuring the generator voltage and the output power of the DC-DC converter, and then making decisions based on these measurements. The controller used for this project is the Microchip PIC16F818, however a wide variety of microcontrollers are available on the market which would suffice for such a controller. This microcontroller was chosen for its low cost, hardware PWM generator and five analog-to-digital converter channels. In operation the PWM frequency is held constant, and the duty cycle is controlled either automatically or manually depending on a mode select switch. In manual mode, the duty cycle is varied manually allowing flexibility for system testing. In automatic mode the controller governs the duty cycle.
At this point we have a controller capable of adjusting the PWM duty cycle, thereby changing the load on the rotor and the rotor speed. At each wind speed we should be able to

pick the optimum duty cycle in order to operate the turbine at the optimum tip-speed-ratio, thereby extracting the maximum power possible with the given turbine. While this could involve measurements of wind speed, rotor speed and a detailed map of turbine performance, there is no guarantee that this would lead to peak-power extraction. For one as the turbine ages the performance will vary due to a number of factors including blade warpage and contamination. A number of different control approaches have been tired (Simones et al., 1997). One of the simpler and more robust algorithms for maintaining peak-power extraction is the "gradient method".

In the gradient method of power control, the duty cycle is periodically changed by a small increment. If the resulting power is greater then the previous measured power, then the subsequent change in duty cycle will be in the same direction. If instead the resulting power is less than the previous measured power, the direction of duty cycle change will be reversed. What we are accomplishing is a measurement of the gradient of the power with duty cycle. Any change that is in the direction of increasing power is continued. If the change decreased the power then we reverse directions.

This "gradient" method algorithm, shown in figure 17, will result in operation around the peak of the power curve, and will also allow quick adaptation to varying wind speeds without measuring either wind speed or turbine rpm directly.

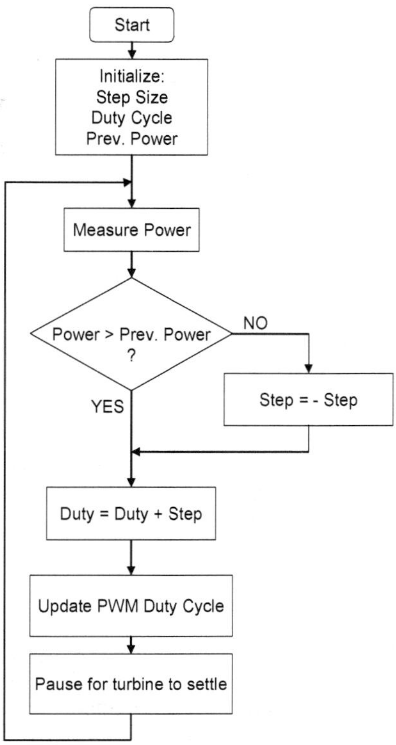

Fig. 17. Gradient method of peak power tracking

4.3 System testing

In order to confirm the controller's ability to vary the load on the generator and change the system speed, the controller was tested on a generator test bed consisting of a DC motor driving a permanent magnet AC generator which is then rectified and sent to the controller. Changes in wind speed are simulated by varying the voltage to the DC motor. Test bed characteristics are shown in figure 18 for low and high simulated wind speeds. In each case the current varies linearly from the stall current at zero volts to zero current at the no-load voltage. The power output follows a parabolic curve with a maximum at ½ of the no-load voltage.

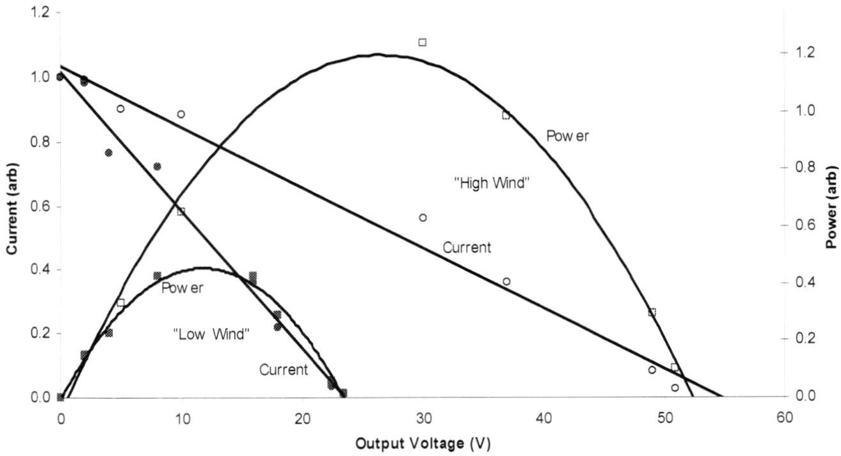

Fig. 18. Current (circles) and power (squares) versus generator voltage for various loads high and low simulated wind speeds

Two tests were performed to check the ability of the controller to track the power starting from both high and low duty cycles. The output of the DC-DC converter was connected to a constant load and the controller was manually set to the appropriate initial condition, then switched into peak-power tracking mode. In the first test, shown in Fig. 19, the controller starts from a high duty cycle. Initially the rotor is overloaded and turning at less than 500 rpm. The circuit decreases the duty cycle, allowing rotor to speed up, increasing the output voltage and thus power.

Fig. 20 shows another run beginning from a very low duty cycle. As the generator is lightly loaded it operates at high speed, around 2300 rpm. Because the duty cycle is low the output voltage is also low, around 7V. As the duty cycle is increased the output voltage increases and the generator is loaded, slowing down. The controller continues to increase the duty cycle, overshooting the peak power until the system speed drops to below 1000 rpm. At this point the power decrease is sufficient to cause the controller to reverse directions, and unload the generator, increasing the system speed. This continues until the system is operating near the peak power output at approximately 1250 rpm. At a constant input "wind speed" the system will oscillate back and fourth about this maximum power point.

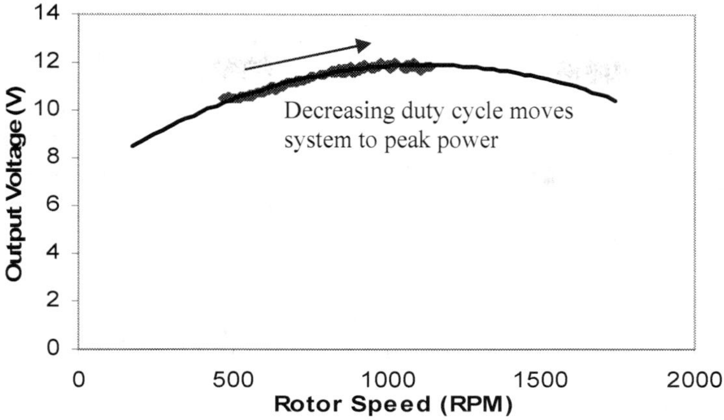

Fig. 19. Peak Power tracking starting from high duty cycle at constant load

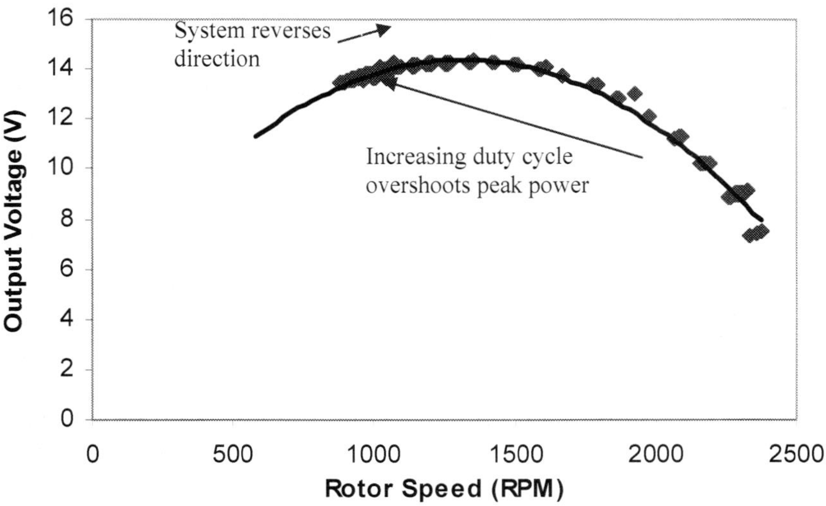

Fig. 20. Peak Power tracking starting from low duty cycle at constant load

Some scatter in the output data can be seen, more noticeably in figure 20, due to the mechanical system dynamics. When the duty cycle is increased the electrical output increases nearly instantly. The increased load does not immediately change the rotor speed, however, due to rotor inertia. Thus each step up in duty cycle results in an 'arc' up as the voltage output increasing, and to the left as the system gradually slows to a new steady state operating point. Similarly decreases in duty cycle will result in arcs down and to the right. As data was taken at a higher frequency than the controller loop cycle time these 'arcs' then deviate from the actual steady-state operating curve. It is important to give the system time to settle into a new steady state operating point before another measurement is taken, least the controller confuse these temporary deviations with actual power increases or decreases. The actual minimum settling time will depend on rotor speed and inertia and is on the order of several seconds for a 1.2kVA system.

4.4 System performance

For verification of the earlier derived relationship between rotor speed and duty cycle (equation 8) the speed of the rotor was measured as a function of duty cycle while connected to a constant load. The load was not sufficient to completely stall the generator; however the resulting data, figure 21, follows the shape of equation 8.

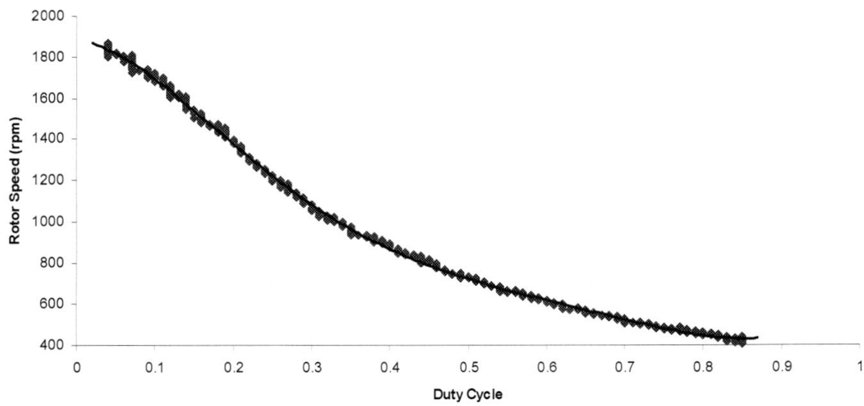

Fig. 21. The effect of the duty cycle on the rotor speed at a constant load resistance

Fig. 22 shows the system response to varying simulated wind speeds. Initially the system is started at a low duty cycle and high wind speed, V_1.
It quickly stabilizes at the optimum power point near 1200 rpm and 13.5V output. The DC motor is then given less power simulating a second, lower wind speed. At this lower wind speed, V_2, the controller increases the duty cycle further and the system again stabilizes as the new peak power, now around 900 rpm and 12V. Finally, a lower wind speed is tested and the system once again finds a peak power at 500 rpm and 10.2 V.
Fig. 23 shows the efficiency of the DC-DC converter as a function of duty cycle. The efficiency varies non-linearly with duty cycle at constant load, and constant PWM frequency. The inductor has a significant effect on the efficiency, because of its influence on the output current, and due to its resistance it will suffer I^2R losses as well. The efficiency of the controller can be divided into three regions. At very low duty cycles, less than 20%, the discharge time of the inductor is less than the OFF time of the PWM signal. This results in a discontinuity of the output current and lost efficiency. The discontinuity time decreases duty cycle is increased, because the OFF time of the PWM signal decreases. Above 20% duty cycle the inductor current never reaches zero because the discharge time of the inductor becomes greater than the IGBT OFF time. From about 20% to 80% the buck circuit is fully functioning as step-down-converter with an average efficiency of approximately 76%.

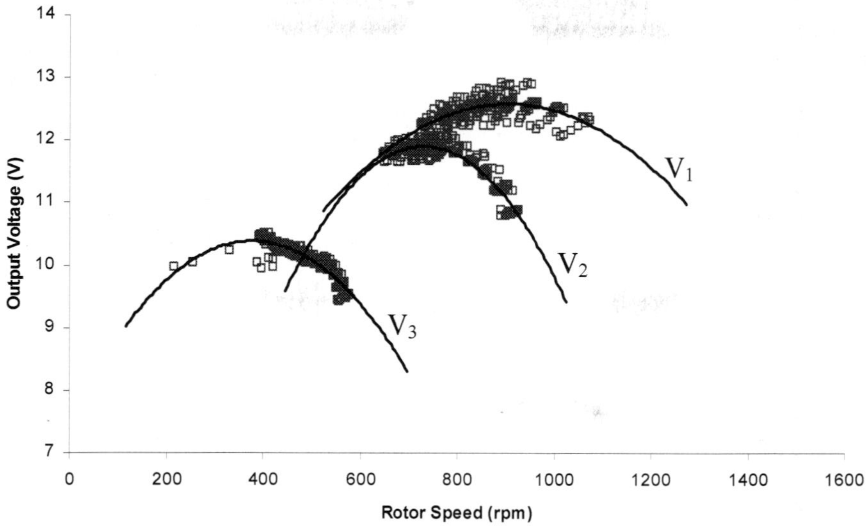

Fig. 22. Peak power tracking at three different simulated wind speeds and constant load

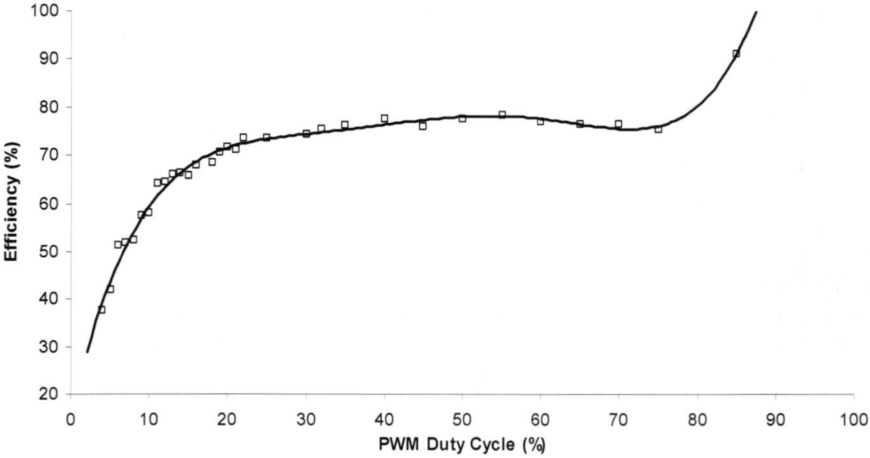

Fig. 23. Efficiency of the DC-DC converter versus duty cycle at constant load and PWM frequency

Above 80% duty cycle the efficiency jumps to 90% because the OFF time of PWM is insufficient to drive the IGBT fully OFF. In this case the DC-DC converter begins to act as low impedance resistive link between the generator and the load.

5. Peak power tracking controller modelling and economics

To investigate the gains possible with a peak-power tracking controller, we can model the turbine-controller-load system at a number of wind speeds. First, we'll start by defining the wind turbine as a 3-blade, 4 meter diameter, nominal 1kW machine which generates 57Vat 17A in an 8 m/s wind. The Cp of the machine is given in figure 24, and has a maximum of 0.3 at a TSR of 6.

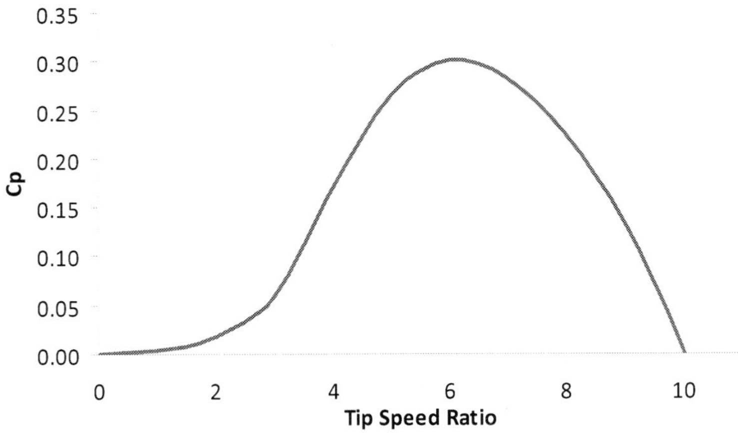

Fig. 24. Coefficient of performance versus Tip Speed Ratio

We will assume a binned wind speed probability of 40% at 4 m/s, 20% at 8 m/s and 10% at 12 m/s. Taking the turbine power to be as given previously in (1) we get the curves of figure 25 as a function of rotor speed.

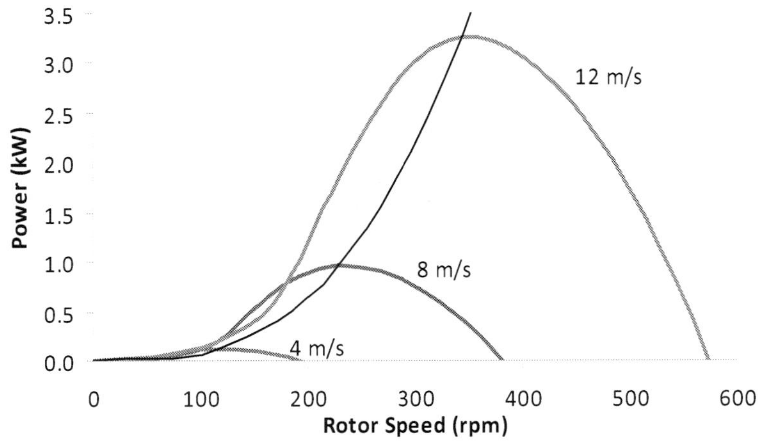

Fig. 25. Power as a function of turbine speed at 3 different wind speeds.

A fine line connects the peak power at each wind speed, giving us the peak power curve.
For the loads we have selected 4 different scenarios:

1. A resistive load matching the turbine power at 8m/s
2. A battery load matching the turbine power at 8 m/s
3. A poorly matched battery load
4. A DC-DC peak power tracking controller and any of the above loads.

In an 8m/s wind both the resistive and the well matched battery load pull 17A from the turbine at a rotational speed of 229 rpm, extracting 965W. The poorly matched battery load draws higher current, slowing the turbine down to about 180 rpm, and extracting only 890W. Figure 26 shows the current deliverable by the turbine as a function of turbine speed for the 3 wind speeds, and the currents drawn by the loads. The system will operate where the load line crosses the source current. In the figure we can see that the well matched loads cross the 8m/s current source line at 229 rpm.

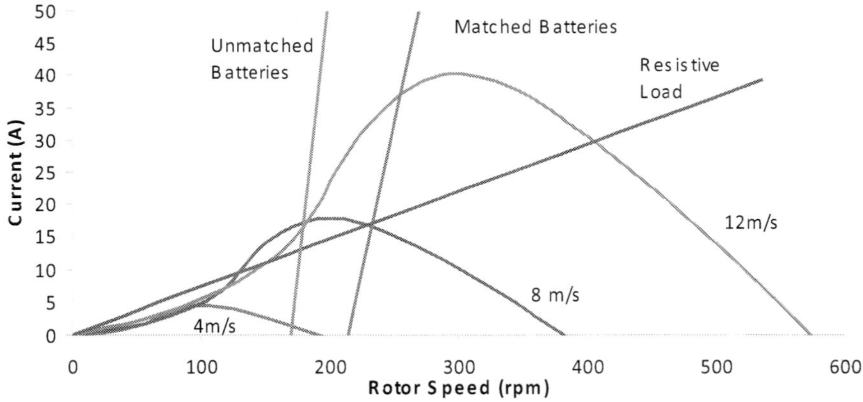

Fig. 26. Current of the turbine at selected wind speeds with 3 different loads.

The resistive load has a current which is directly proportional to turbine speed (and thus voltage). Both of the battery loads only begin drawing current once the turbine voltage exceeds the battery's terminal voltage. As a load is applied to the turbine it will slow down, and produce more current. As the turbine slows the tip to speed ratio will decrease, initially providing more torque. However, if turbine slows to below ½ of the no-load speed the torque begins to drop off, and so too will the current.
If we recast the data of figure 26 in terms of power, we get the figure 27.
Clearly the matched resistive load and battery load are operating the turbine at its peak power output in an 8m/s wind. At 12m/s, however, the matched battery load bogs the turbine down extracting about 2.2kW, far less than the maximum 3260W possible. The resistive load can't load the turbine enough, causing it to operate at a higher than optimal speed, extracting just over 3kW. The poorly matched battery load bogs the turbine down even worse than the matched battery load at 12m/s, extracting less than 800W.
Although the power produced from a 4 m/s wind is much less than 8 m/s, the lower wind speed occurs much more often. Both battery loads extract too little current to operate the turbine efficiently, and the resistive load in this case is extracting more current than optimum, completely stalling the turbine.

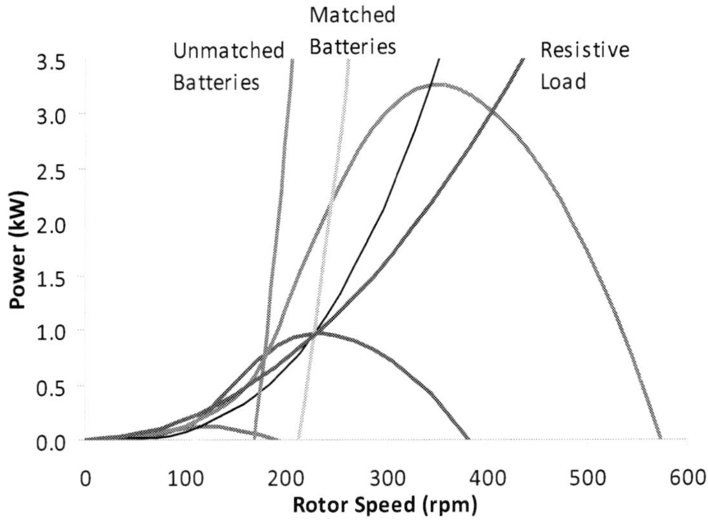

Fig. 27. Power of the turbine and loads versus turbine speed

Wind Speed (m/s)	Probability %	Power Extraction (W) Unmatched Battery	Resistive	Matched Battery	Peak-Power Tracker
4	40	60	10	0	109
8	20	790	965	965	869
12	10	770	3020	2180	2931
	Total	**259**	**499**	**411**	**510**

Fig. 28. Comparison of the power extractions of 4 loads from the turbine.

The peak power tracking controller will always maintain operation at the turbines peak power point, operating along the thin line linking the peak powers of the turbine's curves. If we take the Peak Power tracking controllers efficiency to be 90%, we will always extract 90% of the maximum power in any wind speed in the operational range. Power extraction results for the 4 loads are compared in figure 28.

The perfectly matched resistive load gives the highest power extraction of the directly connected loads with 499W. The matched battery bank extracts 411W while the poorly matched battery bank extracts only 259W. The peak power tracker will adjust the power extraction allowing the turbine to operate at it's maximum power at all 3 wind speeds. This gives the peak power tracker a power yield of 510W. While the wind energy system designer would strive to carefully match the power source and load, any changes over time, even something as subtle as the blades becoming dirty or battery ageing, can cause a miss match between the turbine and load. Thus the mismatched battery case is a more common scenario than one might initially suspect.

One of the important aspects this highlights is that most loads even if carefully matched to a turbine's output at a given wind speed are likely to be poorly matched for other wind speeds. Additionally the Peak-power tracking controller can allow the turbine to operate at its maximum efficiency over a wide range of wind speeds and loads. Despite less than unity efficiency peak power trackers can extract more power than "straight through" systems.

The turbine described in the above section will likely cost several thousand dollars to buy and erect. Taking 10,000$ as the assembled turbines cost, and assuming a lifetime of 10 years if we couple it to the well matched battery load we would produce a total of 10 x 356 x 24 x 411/1000 or about 31,100 kWh of energy, for a unit energy cost of 28.5 cents per kWh. Using the peak power tracker we would have produced 43,600 kWh. Taking the cost of the controller to be 250$, using it would give a unit energy cost of 23.5 cents per kWh making it the most economical system overall. Perhaps a more realistic comparison would be with the poorly matched system which only generated 22,100 kWh for a unit energy cost of 45.2 cents per kWh. In this case use of the peak-power controller would almost double the energy production.

6. Conclusions

Almost all wind energy systems require a controller to prevent damage to the turbine and the load. With advances in power electronics and microcontrollers, inexpensive yet sophisticated power controllers can be produced which are also capable of enhancing the power extraction of the overall system. These controllers may use DC-DC converters to boost the voltage of a given power source up to a higher voltage to match the requirement of a load, such as in the case of the solar panel or wind turbine operating below the charging voltage of a battery pack. A DC-DC converter, either buck or boost, may also be used in impedance matching a power source to a given load with an improvement in the overall system efficiency. With modern microcontrollers and high-frequency power switching devices a peak-power tracking controller can be developed which includes many functions such as load over charging protection, load prioritization, and turbine breaking. Such a peak-power tracking controller can greatly enhance the overall power production of a wind turbine system, and may cost a small fraction of the price of the turbine.

With reasonable care the wind power system can be modelled and the cost/benefit ratio of a controller can be determined. In many applications the additional cost of a sophisticated controller is justified by the increased power yield of the system.

7. Ongoing work

With increasing costs of primary fuel sources and the recent explosion of inexpensive small wind turbines from China, small wind power systems are bound to become more common. As the price of electronic power components comes down, and the sophistication of small microcontrollers increases, powerful and inexpensive "intelligent" controllers can help protect investments in alternative energy systems and extract more power from existing equipment. We are continuing long-term field testing of controllers to verify their overall performance and power extraction improvements.

Additional work in this area will consist of overall system optimization and balancing, including battery voltage and storage capacities, power transmission and conversion, not just from the power sources, but also at the end-use points.

8. References

Agrawal J. P. (2001). *Power Electronic Systems Theory and Design*, Prentice Hall, Upper Saddle River, New Jersey 2001.

Calwell C. & Reeder T. (2002). *Power Supplies: A Hidden Opportunity for Energy Savings* NRDC

Gitano H.; Taib S. & Khdeir M. (2008). Design and Testing of a Low Cost Peak-Power Tracking controller for a Fixed Blade 1.2 kVA Wind turbine. *Electrical Power Quality and Utilisation*, Vol.14, No. 1, July 2008, pg. 95-101, ISSN 1234-6799

Johnson G. (1985). *Wind Energy Systems*, ISBN 0-13-957754-8

Manwell J. F.; McGowan, J. G. & Rogers, A. L. (2002). *Wind Energy Explained: Theory Design and Application*, pg. 321-367, ISBN, J. Wiley and Sons

Patel M. R. (2006). *Wind and Solar Power Systems Design and Optimization*, pg. 68-108, ISBN, Taylor and Frances Group

Rajambal K. ; Umamaheswari B. & Chellamuthu C. (2005). *Electrical Braking of Large Wind Turbines*. [Online]. Available: http://www.sciencedirect.com

Rashid M. H. (1993). *Power Electronics Circuits, Devices, and Applications*, Prentice hall, Upper Saddle river, New Jersey 1993

Schelle D. & Castorena J. (2006). *Buck-Converter Design Demystified*. Maximum Integrated Products, Sunnyvale, Cliff. [Online]. Available: http://www.powerelectronics.com

Simones M. ; Bose B. K. ; & Spiegel R. J. (1997). Fuzzy Logic Based Intelligent Control of a Variable Speed Cage Machine Wind Generation System, *IEEE Trans. on Power Electronics*, vol. 3, pp. 87-95, Jan. 1997

Van Baars G. E. & Bongers P. M. M. (1992). "Wind turbine control design and implementation based on experimental models" in *Proc. 1992 Decision and Control Conf.*, pp. 2454-2459.

8

Analysis and Investigation of the Inverter for Energy Transfer from Small Wind Power Plant to Common Grid

Romuald Luczkowski and Roman Muszynski
Poznan University of Technology
Poland

1. Introduction

There are many different solutions of wind power systems sized for different power values (El-Tamaly et al.). The examples of the low power systems are shown in Fig.1. In the A structure (Chikaraishi et al., 1990) the induction generator along with the fully controlled rectifier is used. In the B system (Higuchi et al., 2000) a synchronous generator, diode rectifier and pulse converter for DC voltage control are applied. In the C system (Tyc, 2003; Muszynski & Pilacinski, 2006) the energy is obtained using the large vehicle alternator.
The last solution can be considered in the plant of the lowest power range as the competitive one regarding two previous systems. It is foreseen as a non-industrial application for coupling the amateur-made plants with one-phase common grid.

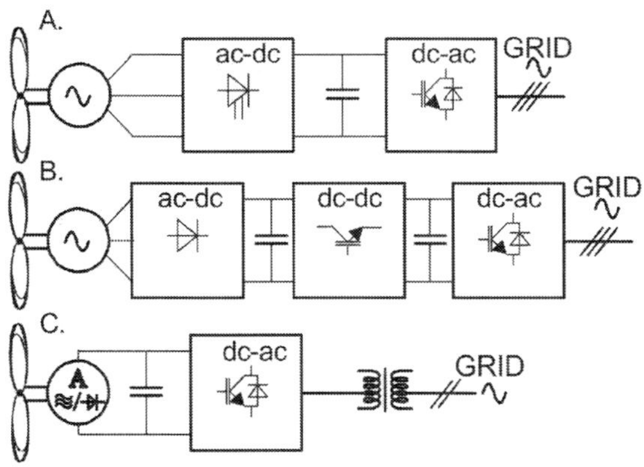

Fig. 1. The small wind energy systems

Need of "home-built" small wind power plants in many countries (Piggott, 2005) is motivation for the problem investigation. Such the plants of power one-to-several kW are built on the base of low speed permanent magnet synchronous machines with axial flux or on the base of car-, bus- or truck-alternators. Operation of these plants with accumulator bank only does not satisfy enthusiasts of own renewable energy. People would like to have higher financial effect that will be reached if the produced energy can be transferred to the public network. There are not industrial inverters of such small power on the market. A trial of solution of this problem on the base of voltage source inverter with hysteresis current controller was undertaken.

In the chapter the problems of transfer of energy from small wind plant to the electrical network are studied. Firstly, the investigated system is described (Section2). Then, the theoretical and simulation analysis of operation of the system is given (Section 3), where the physical phenomena as well as problems related to its operation and design are widely described. Next the first laboratory plant is presented (Section 4). At the end, the original, cost and reliability oriented design of the converter is provided (Section 5).

2. The investigated system

The scheme of the considered wind power system is shown in Fig.2 (solution C in Fig.1). The system (Tyc, 2003; Muszynski & Pilacinski, 2006) consists of the wind turbine, alternator, RC filter, voltage source inverter, VSI, and matching transformer, TR. The transformer gives not only the voltage adjustment of the source to the grid but is necessary also from point of view of safety. As persons without suitable conscience of electric shock can approach the equipment of small wind power plant by, it must be separated from the grid voltage.

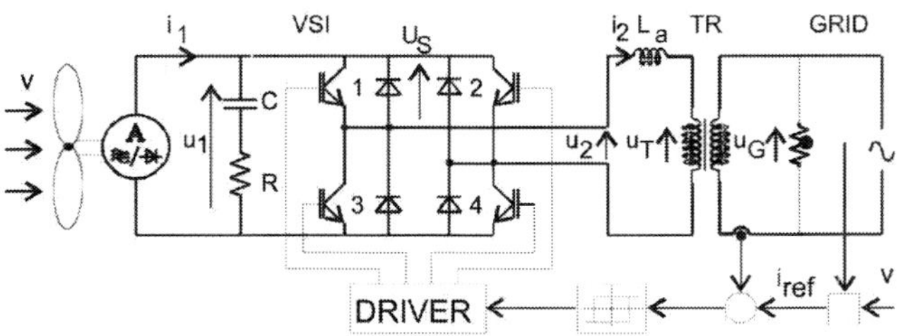

Fig. 2. Block diagram of the investigated system

The system from Fig.2 can not operate without filter between the source and load circuits. Direct switching both the circuits generates over-voltage of infinite value due to different initial values of currents flowing in the branches with inductances. Filter softens the switching phenomena.

It turns out (Section 3.4) that, for decreasing of the current slope rate and switching frequency of the inverter, the additional inductance into the output circuit should be inserted.

In the system the hysteresis-based output current control has been implemented. To obtain the unity power factor, the phase of current reference signal i_{ref} should be opposite regarding to the phase of the network voltage. To provide the operation at the maximum power, the amplitude of signal should be adequate to wind velocity v.

The structure of the system (Fig.2) is the same as the structure used in the variable speed induction motor drive with inverter and hysteresis current controller (Mohan et al., 2003). However, there is fundamental difference in regard of parameter proportion between both the systems. It changes cardinal the conditions of design and operation of the system for transfer energy from small power plant to the common grid, in comparison with drive system where the energy flows from source of high power to receiver of smaller power.

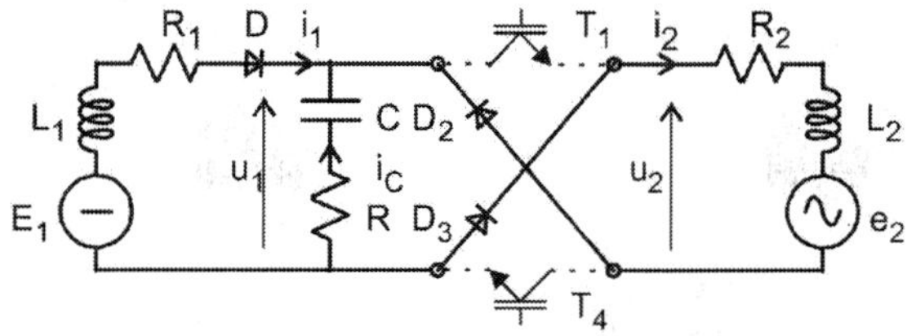

Fig. 3. Circuit configuration during D2 and D3 conducting (solid) as well as during transistor T1 and T2 conducting (dashed line)

The system: *Small power source - Inverter with hysteresis current controller - Common grid* from the Fig.2 and 3 can be characterized as follows:

- The voltage (EMF) E_1 of the source (alternator) is small.
- The short circuit inductance L_1 of the source is great.
- The requested value of filter capacitance C is great, in order to decrease voltage fluctuation.
- The frequency of the resonance in the circuit L_1C is low.
- The natural inductance L_2 of the output circuit is small.
- The natural current slope rate and frequency of the hysteresis controller operation are high.

Because of above properties and values of the parameters the conditions of design and operation of the examined system are significantly more difficult than for the converter drive supplied from the grid. Damping of the resonance oscillation in the input circuit requires additional filter resistance. Decreasing of the current slope rate and frequency of the inverter operation requires additional choke in the output circuit. All the additional elements influence the power loss as well as the reliability and cost of the system.

The properties of the system as well as its operation and specific phenomena were studied by means of theoretical analysis and simulation investigation.

The following basic data and parameters of the system were taken into account:

- alternator: $E_1 = 28$V, $L_1 = 0.175$mH,
- amplitude of the grid voltage $U_{Gm} = 311$V,
- grid with transformer (voltage ratio $n_T = 18.3$) reduced to the low voltage side:
- $E_{2m} = 17$V, $L_T = 0.058$mH, $R_2 = 2.7\text{m}\Omega$,
- additional inductance in output circuit: $L_a = 0.08$mH,
- resistance of the additional choke $R_a = 0$,
- total inductance of the output circuit $L_2 = L_T + L_a = 0{,}138$mH
- amplitude of sinusoidal reference current: $I_{2m} = 50$A,
- hysteresis band: $\Delta i = 6$A,
- filter: $C = 1$mF, $R = 0.4\Omega$.

In the chapter are given values of only those parameters, which differ from the above basic data. Transistors and diodes of the converter are assumed to be ideal (zero voltage drop and zero turn-off as well as turn-on time).

3. Theoretical analysis and simulation of operation of the system

3.1 Operation of the system with basic parameters

The figures 4, 5, 6 and 7 present one set of the quantities registered during simulating the system for basic parameters that give stable but not optimal operation of the converter.

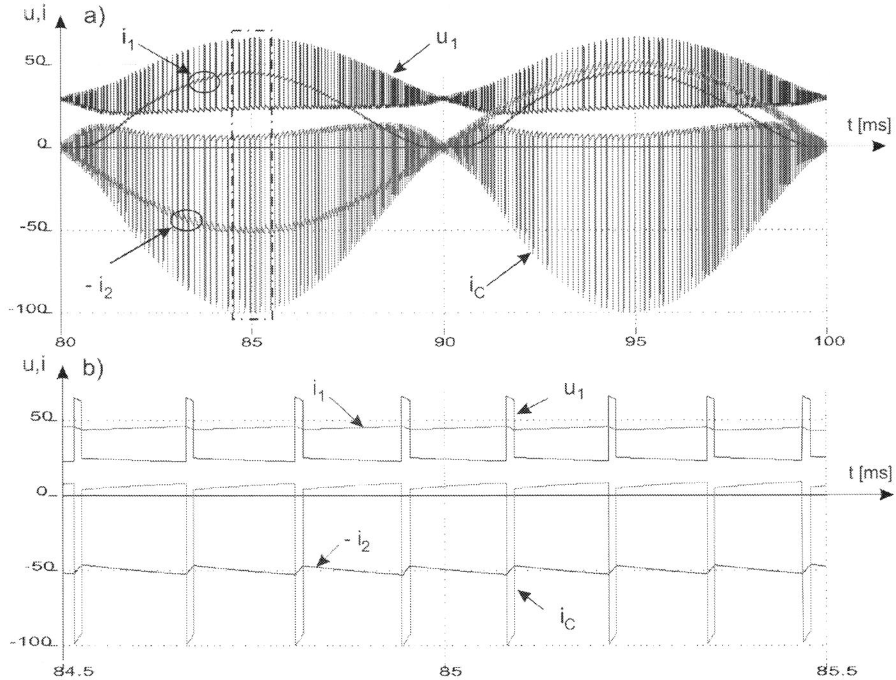

Fig. 4. The course of filter voltage u_1 and current i_C as well as current of source i_1 and receiver i_2 (Fig.a) and time extension of the course in region of current maximum (Fig.b) for basic data

It is seen from Fig.4a that the receiver current i_2 is sinusoidal. It follows the reference current $i_{ref}\ n_T$ with the error not greater than hysteresis band. The shape of the source (alternator) current i_1 (Fig.4a) complies in approximation with the rectified receiver current i_2. It can be also recognized as the near *sin* function of twofold frequency with DC component equal to half of its amplitude. The high frequency ripple of the current i_1 due to hysteresis controller operation is smaller than the receiver current one. Precise relation between the currents shown in Fig.4 is

$$i_2 = i_1 + i_C \tag{1}$$

during transistor conducting and

$$-i_2 = i_1 + i_C \tag{2}$$

during diode conducting, according to Fig.3. The change of sign in the equations is in accordance with the switching the circuit configuration.
The course of i_C in Fig.4 consists of the long positive-going pulses of small value (transistor on state) and the very short negative-going ones of very high value (diode conducting). The height of the negative pulses equals in approximation to the twofold absolute instantaneous

value of the reference current. It can be substantiated as follow. At the end of transistor conducting $i_C \approx 0$ and according to (1) is

$$i_2 = i_1 = i_{ref} n_T \tag{3}$$

In the transistor switch off state (diode switch on state) two currents are flowing into the filter: source current i_1 forced by inductance L_1 and current i_2 forced by inductance L_2. Then according to (2) is

$$i_C = -(i_1 + i_2) \approx -2i_{ref} n_T. \tag{4}$$

The filter voltage u_1 is an important quantity in the system, as it appears on terminals of alternator, transistors, diodes and output circuit in accordance with states of the switches. It can be obtained using equations for the input circuit of the converter. For the alternator branch the following equation is valid:

$$u_1 = E_1 - L_1 \frac{di_1}{dt} - i_1 R_1 - u_D, \tag{5}$$

which for the period of transistor conducting, can be simplified to the form:

$$u_1 \approx E_1, \tag{6}$$

as $u_D \approx 0$ was assumed and, moreover, voltages on resistance R_1 and inductance L_1 are negligible due to small values of the resistance and current derivative. For the filter branch is:

$$u_1 = u_C - i_C R. \tag{7}$$

Due to small value of i_C during transistor conducting, the formula (7) takes form:

$$u_1 \approx u_C. \tag{8}$$

Therefore, according to (6) and (8), during transistor conducting the filter voltage u_1 and capacitor voltage u_C have approximately the same values equal to EMF E_1 of the source. During diode conducting the equation (7) takes the form

$$u_1 \approx E_1 - i_C R = E_1 + 2i_2 R = E + 2i_{ref} n_T R. \tag{9}$$

The commutation overvoltage can be obtained in the form:

$$r = \frac{u_1}{E_1} = 1 + 2\frac{i_2 R}{E_1}. \tag{10}$$

The highest value of u_1 is reached in the region of current maximum. Taking into consideration, $I_{Cm} = 2I_{refm} n_T = 2 \cdot 50$A, result is $u_m \approx 28 + 100 \cdot 0{,}4 = 68$V for the data of the simulated plant. This value is visible in Fig.4.
Figure 5 shows the voltages of the system on the output side. The output voltage u_2 of the inverter equals to the filter voltage u_1 during transistor on state and $-u_1$ for transistor off state. The grid side voltage u_G of the transformer is sinusoidal. Its amplitude equals to 311V

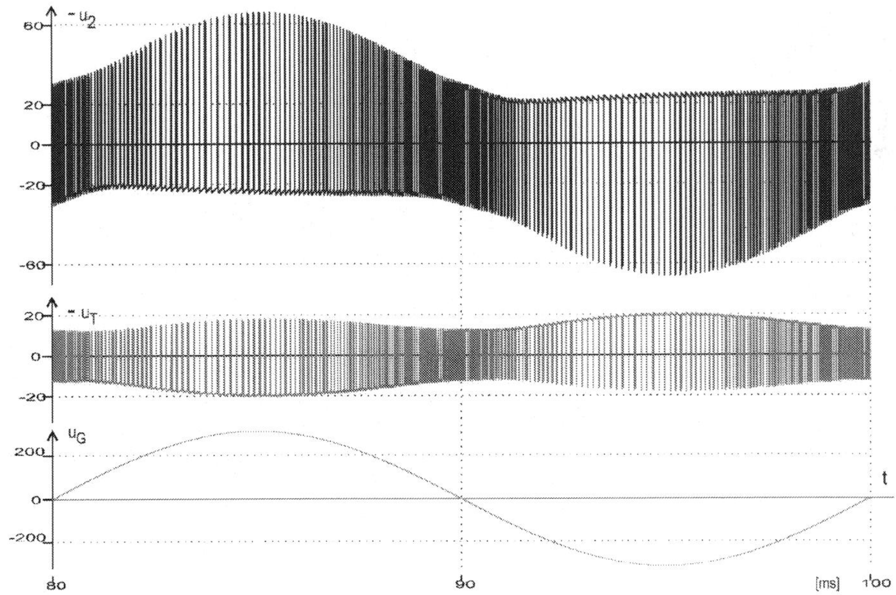

Fig. 5. The course of the output, u_2, transformer, u_T, and grid, u_G, voltages for basic data

and phase is opposite regarding to the phase of the reference current i_{ref}. The shape of the inverter side voltage u_T of the transformer is intermediate between the shape of u_2 and sinusoidal one, according to the equations:

$$u_T = u_2 - L_a \frac{di_2}{dt} \tag{11}$$

and

$$u_T = -\frac{u_G}{n_T} + i_2 R_2 + L_T \frac{di_2}{dt}, \tag{12}$$

where the negative sign at the voltage u_G denotes the phase change between both the sides of the transformer. If the small value of the transformer resistance R_2 can be neglected then on the base of (11) and (12) is:

$$u_T \approx \frac{L_T}{L_a + L_T} u_2 - \frac{L_a}{L_a + L_T} \frac{u_G}{n_T}. \tag{13}$$

For the simulated system is L_T = 0.058mH, L_a = 0.08mH and according to these data the shape of u_T in Fig.5 is influenced by u_2 with the weight of 42% and by u_G/n_T with the weight of 58%. It must be taken into account that in formulas (12) and (13) the phase of voltage u_T is opposite in relation to the phase of the voltage u_G. The additional inductance L_a is not profitable from the voltage point of view as it produces voltage drop

$$\Delta u_a = u_2 - u_T \approx \frac{L_a}{L_a + L_T}\left(u_2 + \frac{u_G}{n_T} \right). \qquad (14)$$

It deteriorates the voltage matching of source (alternator) to the receiver (grid). The additional inductance L_a is introduced to the system in order to decrease current slope rate and operation frequency of the inverter. This problem is separately analysed in Section 3.4.
In efficiency analysis the power loss in the choke must be also taken into consideration, as there is not choke without resistance in the real system.
The high frequency voltage impulses u_T presented in Fig.5 induce also additional loss in the magnetic circuit of the transformer, which is not taken into consideration in the presented model.
The course of the transistor (and diode) voltage is shown in Fig.6. Impulses of u_s are the fragments of the voltage u_1 from Fig.4a, referring to the transistor off state. Hence, the maximum value of switch voltage can be obtained using formula (9) and maximum value of current I_{2m} or I_{refm}.
The power transferred between the separate elements of the system is registered in Fig.7. It is calculated as a simply product of voltage and current without any farther processing of the signals. In result of it the power registration in Fig.7 contains the impulse according to the course of current and voltage in Fig.4 and 5. Especially great and fast changes of the power take place on the inverter side terminals of the transformer (p_T). The differences between the power in the separate points of the system, visible in the Fig.7, arise due to loss in the filter resistance (difference p_1 - p_T) and - in the transformer resistance (difference p_T - p_G). The inverter as well as the additional choke is assumed for the simulation as the loss-less element.

3.2 The case of filter with reduced resistance

In order to reduce the converter switching overvoltage given by formula (10) and loss in the filter, decrease of the filter resistance is analysed.
Lack of the resistance in the filter is a case of its extremely reduction. The filter voltage during converter switching can be obtained on the base of circuit equations, as the formula (11) is not valid for the case without resistance.

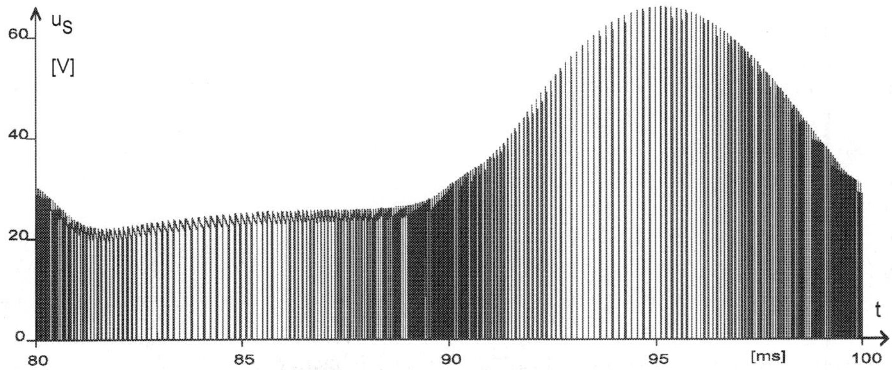

Fig. 6. The course of switch module voltage u_S for basic data

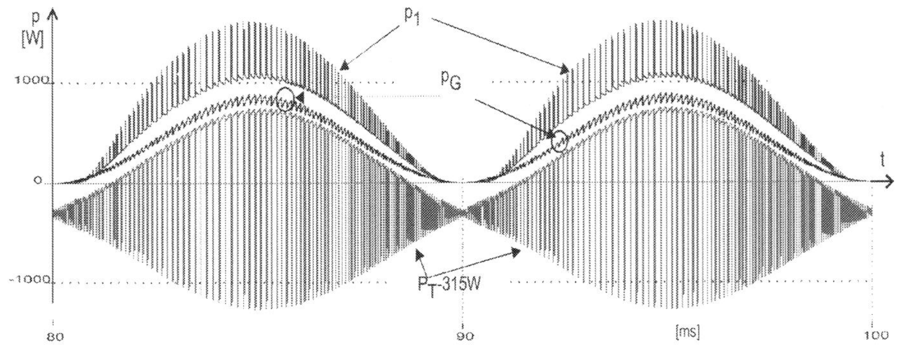

Fig. 7. The course of power on terminal of alternator, p_1, transformer, p_T, and grid, p_G, for basic data

If the transistors are switching-off then the diodes are starting with conducting and the current i_2 flowing in the branch with output inductance decreases rapidly. In result of it the energy accumulated in the inductance decreases as well. Figure 3 shows configuration of the circuits at this stage. The following equations describe dynamical state:

$$L_1 \frac{di_1}{dt} = E_1 - u_1 \tag{15}$$

$$L_2 \frac{di_2}{dt} = -(u_1 + e_2). \tag{16}$$

The rate of change of current i_2 on the base of (15) and (16) is:

$$\frac{di_2}{dt} = -\frac{E_1 + e_2}{L_2} + \frac{L_1}{L_2} \cdot \frac{di_1}{dt} \tag{17}$$

It is seen that the derivative of i_2 is great, as the EMFs E_1 and e_2 act in the same direction. At this e_2 has a great value equal to the amplitude E_{2m} (i_2 and e_2 are in the phase opposition). On the contrary, the rate of change of current i_1 is small, as on the right side of equation (15) is negative difference of two values, which are approached one to other.

In the state of system shown in Fig.3 the source 1 supplies the circuits with energy $\int E_1 i_1 dt$ and the grid receives energy $\int e_2 i_2 dt$. The initial conditions of the state and the small changes of currents and voltages (due to short duration of the state) allow assuming $E_1 \approx e_2$ and $i_1 \approx i_2$ in the above integrals. Therefore, it can be stated that these small energies of source and receiver balance one with other. In result of it, inner balance of energy in the converter circuits consists in its flowing from choke L_2 to the capacitor C. Decrease of choke current i_2 results in increase of filter voltage u_1. If the current i_2 decreases by hysteresis band Δi then the energy accumulated in inductance L_2 decreases by the value:

$$\Delta E = \frac{L_2 I_{2m}^2}{2} - \frac{L_2 (I_{2m} - \Delta i)^2}{2} = L_2 \left(I_{2m} \cdot \Delta i - \frac{\Delta^2 i}{2} \right), \tag{18}$$

where I_{2m} is the value of current before switching-off the transistors. The amplitude of output current of the converter is taken in the formula in order to obtain the maximum voltage increase. The energy flows to the capacitor. The increase Δu of capacitor voltage u_1 is bounded with the energy according to the equation:

$$\Delta E = C\frac{(u_1 - \Delta u)^2}{2} - C\frac{u_1^2}{2} = C\left(u_1 \cdot \Delta u + \frac{\Delta^2 u}{2}\right). \tag{19}$$

Comparison of (18) and (19) gives the equation:

$$L_2\left(I_{2m} \cdot \Delta i - \frac{\Delta^2 i}{2}\right) = C\left(u_1 \cdot \Delta u + \frac{\Delta^2 u}{2}\right). \tag{20}$$

The increase of capacitor voltage calculated from (20) is:

$$\Delta u = \sqrt{u_1^2 + 2\frac{L_2}{C}\left(I_{2m} \cdot \Delta i - \frac{\Delta^2 i}{2}\right)} - u_1. \tag{21}$$

At the and of the transistor off state the filter, transistor, source and receiver voltage is:

$$U_m = u_1 + \Delta u = \sqrt{u_1^2 + 2\frac{L_2}{C}\left(I_{2m} \cdot \Delta i - \frac{\Delta^2 i}{2}\right)}. \tag{22}$$

For the initial value of $u_1 \approx E_1$ the overvoltage written like (10) is

$$r = \frac{U_{1m}}{E_1} = \sqrt{1 + 2\frac{L_2}{E_1^2 C}\left(I_{2m} \cdot \Delta i - \frac{\Delta^2 i}{2}\right)}. \tag{23}$$

In formulas (18) and (19) the terms $\Delta^2 i/2$ and $\Delta^2 u/2$ can be neglected. In this case instead of (20) is

$$L_2 I_{2m} \cdot \Delta i = C u_1 \cdot \Delta u \tag{24}$$

and instead of (21) is

$$\Delta u = \frac{L_2 I_{2m} \cdot \Delta i}{C u_1}. \tag{25}$$

Filter containing capacitor without serial resistance is a good solution from point of view of softening switching phenomena. In this case the voltage jump Δu during diode conducting is small. For $R = 0$ and for the basic data of remaining parameters the jump according to (21) is Δu=1,38V,only, and according to (25) is Δu=1,5V.

Unfortunately, the operation without filter resistance can be inadmissible. Decrease of the resistance results in decrease of the overvoltage according to the formula (10) and of the power loss. The decreased voltage drop on filter resistance is visible in Fig.8 in comparison with Fig.4a. But the resistance can not be too small due to resonance phenomenon.

The resonance arises in the circuit containing filter capacitance C and source inductance L_1. It produces oscillations of voltage supplying the converter. Hysteresis shaping of the current separates efficaciously output circuit from the capacitor. In these conditions, inductance L_2 of the circuit does not influence resonance. Frequency of the oscillation is calculated now according to formula:

$$f = \frac{1}{2\pi\sqrt{L_1 C}} \tag{26}$$

and can be only a few times greater than the network voltage frequency. Amplitude of oscillation can be so great, that the voltage u_1 becomes in some periods of time much less than the receiver EMF e_2. It disturbs significantly the process of current forming and is unacceptable.

Such the situation is shown in Fig.9b. For the data of the simulated system the frequency is f =381Hz according to (26). It can be confirmed in Fig.8b. According to the oscillating value of u_1 the switching frequency changes in a wide range.

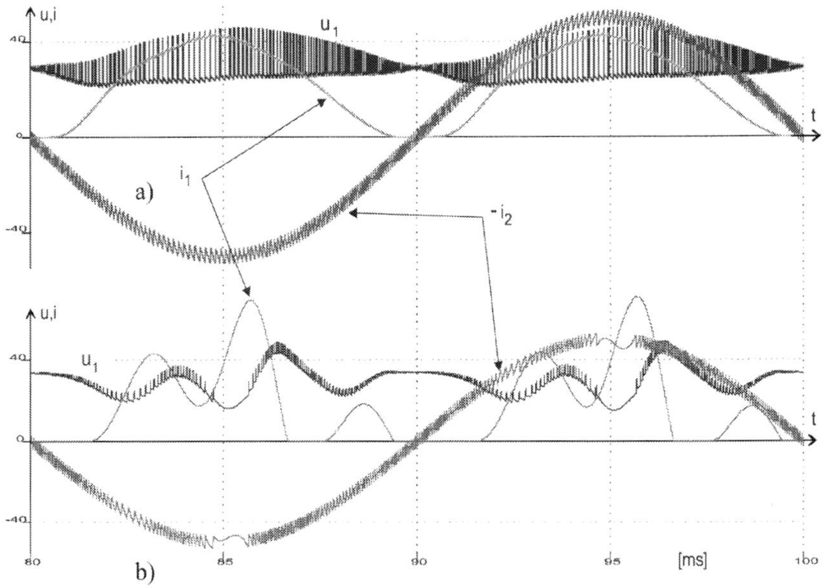

Fig. 8. The course of filter voltage, u_1, source, i_1, and receiver i_2 currents at filter resistance decreased to the value: a) R = 0,2Ω and b) R = 0,005Ω

The oscillation can be suppressed by resistor connected into the resonating circuit. In order to minimise power loss, the resistor should be in the branch with capacitor where the current is smaller than in alternator branch.

In order to have full no-oscillating transient in the input circuit with parameters L_1, C and R_1+R, the sum of resistances must fulfil the condition

$$R_1 + R > 2\sqrt{\frac{L_1}{C}} \tag{27}$$

which for basic data gives

$$R_1 + R > 2\sqrt{\frac{0,175 \cdot 10^{-3}}{10^{-3}}} = 0,84\Omega.$$

Figure 8a shows, that satisfying operation is yet at the resistance four times smaller.

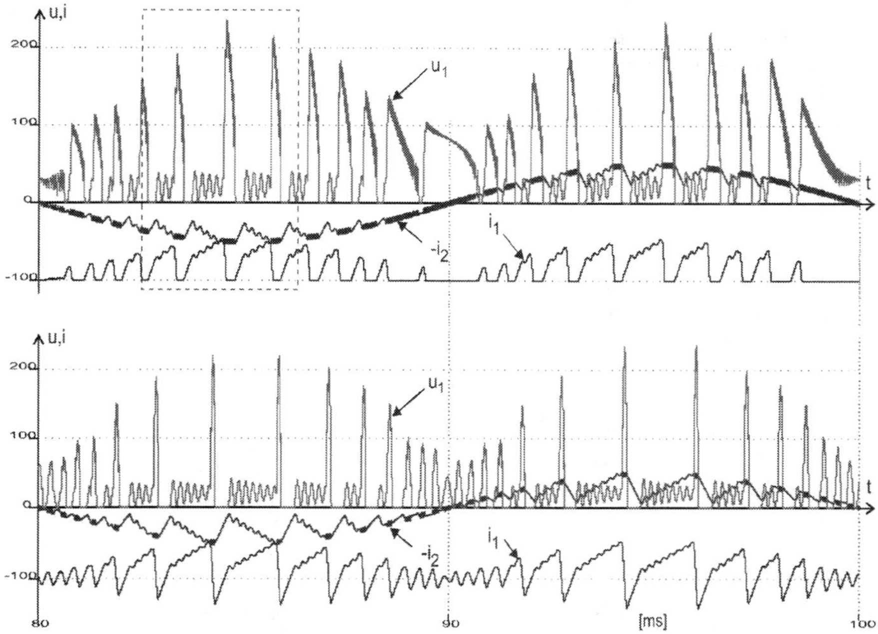

Fig. 9. The course of filter voltage, u_1; i_1 (with offset equal to - 100A) and i_2 - currents in the case of non-damped resonance, taking place at zero filter resistance R and decreased filter capacity to the value C =10μF, with (upper figure) and without (lower figure) the diode in the circuits of the source

3.3 Phenomena at small filter capacitance

In order to avoid unprofitable low frequency oscillation in the input circuit giving unacceptable disturbance of current forming, the filter capacitance was decreased significantly, at zero filter resistance. It was expected that oscillation of high frequency would not disturb of the current forming in spite of lack of damping the oscillation.

Figure 9 shows the phenomena in the system in the case of zero filter resistance and of distinctly decreased value of the filter capacitance in relation to its basic value (Fig.4a). Due to small value of capacitance, there are changing in turn two states in the system: long duration transistor on state and very high frequency switching state. They can be good analyzed using the time extension of fragment from upper Fig.9 shown in Fig.10.

The long duration on state begins in the time point 1 in Fig.10. In this moment, at zero value of source current i_1 and at the filter voltage u_1 equal to or smaller than the actual value of

EMF e_2 (Fig.3), the hysteresis comparator switches on the transistor, because the current i_2 becomes smaller than admissible one (outside the hysteresis band).
In the interval 1-2 the source current i_1 increases with small slope rate dependent on source inductance L_1 and charges the filter capacitor. The current i_2 decreases with the small slope rate dependent on inductance $L_2=L_T+L_a$ as the filter voltage is still smaller than the EMF e_2. In the moment 2 the filter voltage exceeds e_2.
In the interval 2-3 the increase of both the currents i_1 and i_2 occurs as a result of positive difference between E_1 and e_2, with the small slope rate dependent on the sum of inductance L_1+L_2. At the same time the resonance rises up between capacitance and parallel connection of input L_1 and output L_2 inductance. There is visible resonance oscillation of the filter voltage u_1 as well as of the currents i_1 and i_2 in Fig. 9 and 10. The parallel resonance has the frequency

$$f = \frac{1}{2\pi\sqrt{\frac{L_1 L_2}{L_1 + L_2} C}} \tag{28}$$

which for C = 10μF and for basic value of inductance L_1 and L_2 of the system equals to 5.73kHz. The resonance is insignificant damped due to transformer resistance R_2. Duration time of the on state is the longer the higher is the actual value of the reference current. In the time point 3 the current i_2 reaches the upper value of the hysteresis band and in this moment begins the very high frequency switching state.

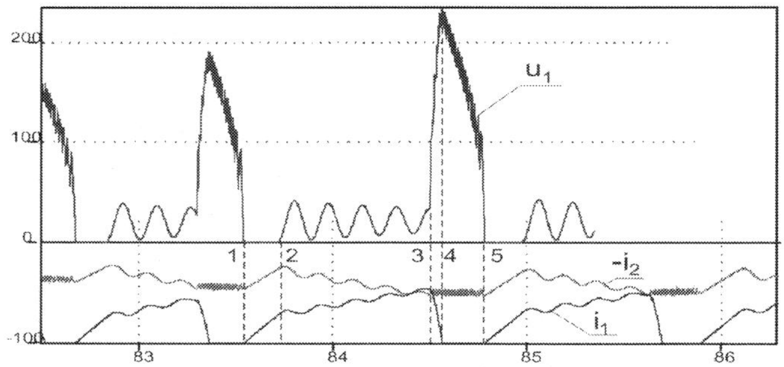

Fig. 10. The case from the upper Fig.9 with time extension in the region of the current maximum

The process of fast increase of the filter voltage in the period 3-4 is initialized when the transistors switch off first one and the diodes start with conducting. Into the filter two currents are flowing: source current i_1, forced by inductance L_1, and current i_2, forced by inductance L_2. Both the currents at the beginning of this time period equal approximately to the actual value of the reference current $i_{ref}\,n_T$. As the capacitance of filter is relatively small its voltage increases rapidly. It leads to the fast decay of the source current i_1 due to great value of difference $u_1 - E_1$. In result of it the current i_1 totally disappears. The energy accumulated in inductance L_1 supplies the filter and causes a high overvoltage on the filter, in spite of discharging the capacitor during next transistor on states. In the time point 4 the voltage u_1 reaches its maximum value.

During the interval 4-5 the filter capacitor discharges gradually i.e. with each period of switching the voltage u_1 becomes lower as the energy flows from the filter to the output circuit. At the end the capacitor is completely discharged. The process repeats, as the end point 5 is a new start point 1. In the intervals 3-4-5 the filter voltage u_1 can be many times greater than the receiver EMF e_2. In the case from Fig.9 and 10 the voltage exceeds 200V at the maximum value of the current. Due to great value of difference u_1-e_2 the very high slope ratio during increasing (transistor on state) and decreasing (transistor off state) of the current i_2 is in the interval 3-4-5. This is the very high frequency switching state, which is visible as "bold" fragments of the current and voltage shapes in the Fig.9 and 10. Thickness of the current line equals to the width of the hysteresis band. The maximum frequency in the figures exceeds 110kHz.

Neither the long duration on state nor the very high frequency switching one is permissible in the system. The first state gives the long duration error of the current, the second one generates the high switching loss in the transistors and diodes. Then the capacitance of the filter should be sufficiently great in order to eliminate the unprofitable phenomena described above. For the investigated plant the capacitance should be at least several hundred μF.

The course of the same phenomena without the diode in the input circuit is shown in the Fig. 9. It relates to the DC generator instead of alternator with diode bridge. The lower figure 9 is similar to upper one. The difference consists in the negative value of the source current i_1 that is reached in the time when the filter voltage u_1 is greater than the source EMF E_1. The increase of the current in the contrary direction takes place at the cost of energy accumulated in the filter capacitor. Therefore, the phase with the very great value of the filter voltage as well as the state of high switching frequency is shorter than in the case with diode. However, the long duration on state is longer as the increase of the current starts from the negative value. It results in the very great error of the current i_2, which is visible in the lower Fig.9.

It can be stated that the system with source containing diode operates a little better than the system with diode-less source.

3.4 Current slope rate and switching frequency

Changing the operation frequency in some range is a disadvantage of converter with direct forming of the current wave. From point of view of loss in the power electronics elements the maximum switching frequency must be limited. The below analysis aims to express the switching frequency as a function of system parameters.

The maximum frequency can be find among two cases of the operation of the system from Fig.2 and 3:

- case 1: the reference current crossing zero is from the negative to positive value (or inversely),
- case 2: the reference current reaches maximum (or minimum) value.

The both cases are illustrated in Fig.11. Duration of increase as well as decrease of the current can be obtained from the geometrical relations. In the Figure 11 the letter S denotes slope of the reference current curve, S_1 and S_3 – slopes of the output current i_2 during its increasing, but S_2 and S_4 – during decreasing, for first and second case, respectively.

Duration of the separate phases of the current change, shown in Fig.11, can be expressed in the following way:

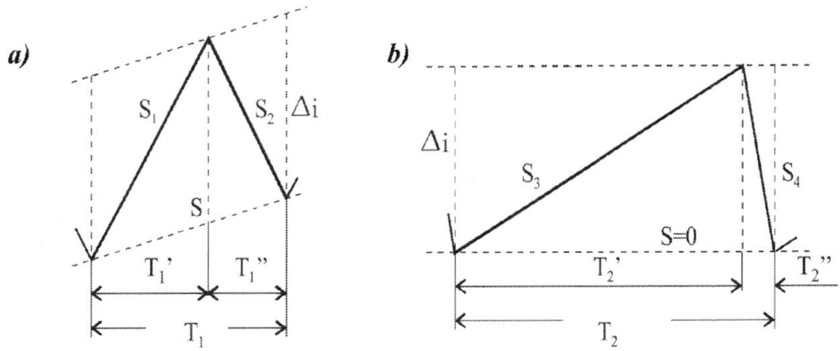

Fig. 11. Fragment of the current course: zero (a–case1) and maximum (b–case 2) neighbourhood

$$T_1^{'} = \frac{\Delta i}{S_1 - S}, \quad T_1^{''} = \frac{\Delta i}{|S_2| + S}, \quad T_2^{'} = \frac{\Delta i}{S_3}, \quad T_2^{''} = \frac{\Delta i}{|S_4|}. \tag{29)-(32}$$

Then the periods for case 1 and 2 are:

$$T_1 = T_1^{'} + T_1^{''} = \frac{S_1 + |S_2|}{(S_1 - S)(|S_2| + S)}\Delta i \tag{33}$$

and

$$T_2 = T_2^{'} + T_2^{''} = \frac{S_3 + |S_4|}{S_3 \cdot |S_4|}\Delta i, \tag{34}$$

respectively. The slope S of the reference current equals to maximum value of its derivative:

$$S = \omega I_{2m} \tag{35}$$

for the case 1 and equals to zero for the case 2. At this $\omega = 2\pi f$ is the pulsation of the network voltage and reference current.
The slopes of the current i_2 for the separate phase of its changing equal to the resultant voltage, acting in the circuit, divided by its inductance and are:

$$S_1 = \frac{u_1}{L_1}, \quad S_2 = \frac{u_1}{L_2}, \quad S_3 = \frac{u_1 - E_{2m}}{L_2}, \quad S_4 = \frac{u_1 + E_{2m}}{L_2}. \tag{36)-(39}$$

Inserting the formulas (35) - (39) into (33) - (34) ones, after simply mathematical transformation, the following periods can be obtained:

$$T_1 = \frac{2u_1 L_2 \cdot \Delta i}{u_1^2 - (\omega L_2 I_{2m})^2}, \quad T_2 = \frac{2u_1 L_2 \cdot \Delta i}{u_1^2 - E_{2m}^2}. \tag{40)-(41}$$

The switching frequencies are:

$$f_1 = \frac{u_1^2 - (\omega L_2 I_{2m})^2}{2u_1 L_2 \cdot \Delta i}, \qquad f_2 = \frac{u_1^2 - E_{2m}^2}{2u_1 L_2 \cdot \Delta i}. \tag{42)-(43}$$

The term $\omega L_2 I_{2m}$ represents voltage drop on the leakage inductance of the receiver and E_{2m} represents amplitude of the receiver EMF. As $\omega L_2 I_{2m} \ll E_{2m}$ the frequency $f_2 \ll f_1$. It means that the highest frequency is for the case a) from Fig.11 i.e. when the output current and voltage are crossing zero line. It can be also noticed in Fig. 4, 5, 6, 7, 8, 9 and 10. The relation $u_1 > E_{2m}$ is a condition of operation of the system. Then the second term in the numerator of (42) can be neglected and the maximum frequency of inverter operation can be written (with accuracy sufficient for practice) in the form:

$$f_{\max} = \frac{u_1}{2L_2 \cdot \Delta i}. \tag{44}$$

For the $u_1 = E_1$, $L_2 = L_T$ and basic data of the system the maximum frequency is 40,2kHz.
Formula (44) shows the next problem of the small plants with hysteresis forming the current. The grid inductance together with inductance eventual transformer between inverter and grid is small and gives high frequency, unacceptable, even when the hysteresis band is wide. For decrease of frequency the external choke should be added. The value of its inductance must be chosen in compromised way, taking into consideration the loss in the choke and voltage drop (14), which deteriorate the efficiency of the converter.
Formula (44) shows also that high values of filter voltage u_1 (overvoltage) are unprofitable also from point of view of operation frequency, whose maximum value is proportional to u_1. Very high frequency can be noticed during time periods with overvoltage registered in Fig.8b, 9 and 10.

4. Laboratory plant

The laboratory converter was built on the base of IGBT module of SKM 75 GB 124 D type with IR2110 gate driver. In the control system the hysteresis comparator LM339 with integrated circuit CD4041 was used. There was network transformer with diode rectifier instead of alternator on the source side.
The capacity of filter was 4,4mF. Filter was without resistance as the source resistance damped sufficiently oscillations in the input circuit. The output transformer had leakage inductance equal to 0,07mH. In order to decrease switching frequency the inductance of 1mH was serial added. The additional inductor decreased efficiency of the energy transfer to about 50% at output power about 100W.
Figure 12 shows operation of the system. The great rate of current slope during diode conducting in the region of maximum i_2 is visible. Switching frequency changes from about 970 Hz to about 7500 Hz when the reference current changes from maximum to zero. Due to great value of inner impedance of source, the filter voltage changes by a few volts according to current pulsation, in spite of great value of filter capacitance.

5. Cost and reliability oriented design of the converter

5.1 The need of compromised optimization of the system

Preliminary theoretical analysis as well as simulation and laboratory investigations of the inverter (Muszynski & Pilacinski, 2006; Muszynski & Pilacinski, 2007; Luczkowski &

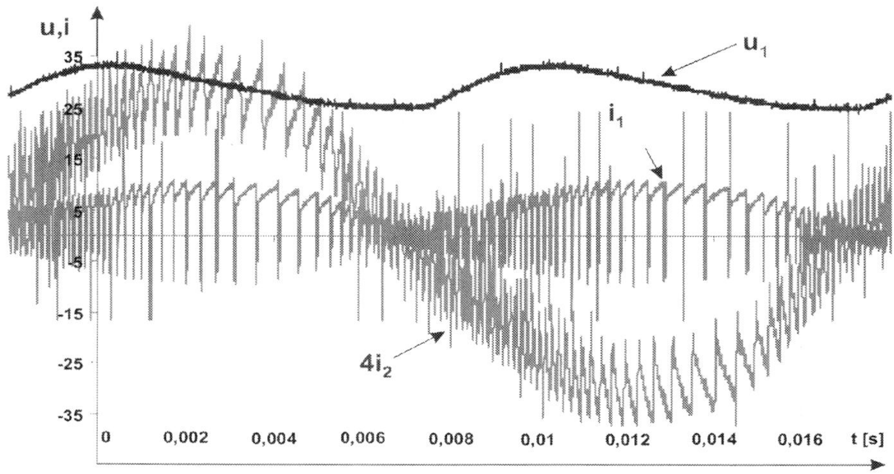

Fig. 12. The course of filter voltage u_1, current i_1 of source and current i_2 of receiver in the laboratory plant at I_{2m}= 8A and Δi=5A

Muszynski, 2007) allowed identifying the problems. In the system there is very closed correlation of the circuit and control parameters with the reliability and efficiency of its operation. Every choice of the design parameters has influence on capital cost, on power loss (exploitation effects) and on the level of the reliability. The problem is composed as the system has many design parameters, the partial criteria have different physical nature and their values can be found only for the separate combination of the parameters by means of simulation.

This section presents methodology of designing the converter with consideration of the above mentioned problems.

If the values of the source parameters: EMF E_1 and inductance L_1 as well as of the grid and transformer parameters: their short circuit inductance and resistance are given, then at least four other parameters should be chosen during designing. They are the following parameters: filter capacitance C and resistance R in the input circuit, hysteresis band Δi of the controller and additional inductance L_a in the output circuit. The filter resistance R is needed for damping the resonance oscillation in the input circuit while additional choke L_a is necessary in the output circuit for decreasing the current slope rate and frequency of the inverter operation.

These four parameters influence many quantities and indexes of the system. Among them are the over-voltage and voltage class of all elements, frequency and damping decrement of the oscillation in the input circuit, current slope rate and switching frequency of the inverter. Therefore, they influence capital cost, power loss (exploitation effect) as well as reliability of the system.

Some of the requirements are opposite. For instance, introduction of the additional inductance L_a is profitable from point of view of the current slope rate and switching frequency. But due to voltage drop on the choke (its inductance and resistance) the voltage adjustment of source (alternator) to the receiver (grid) becomes worse and the additional power is dissipated.

Due to above feature of the designed system the special compromised its design is proposed.

5.2 Optimization methodology

In order to consider during designing the above mentioned requirements of different nature a special generalized optimization criterion

$$G = pY = \min \tag{45}$$

was used where p is the penalty function and Y is the one year cost

$$Y = C_C + C_L \tag{46}$$

having two components: C_C equal to the capital cost of the system divided by number of years of the plant operation live and C_L equal to the cost of energy lost in the converter per one year of its operation.

The penalty function considers unreliability of the converter and has the form

$$p = 2 - H \tag{47}$$

where H is the two-exponential desirability function (Harrington, 1965) in the form

$$H = \exp\left[-\exp(-h)\right]. \tag{48}$$

The function H (Fig. 13) has many good properties (Harrington, 1965) and is suitable for reliability evaluation. It equals to 1 (practically for $h \geq 5$) if the operation of the converter with given combination of parameters is totally reliable (acceptable) and equals to 0 (practically for $h \leq -2$) if the operation is totally unreliable (unacceptable).

As a measure of quality of inverter operation (reliability) can be used index

$$Q = \frac{\Delta i_r}{\Delta i_d} \tag{49}$$

where Δi_d is the desired hysteresis band of the current and Δi_r is the really reached band. The operation is fully reliable if the controller is able to keep the current in the hysteresis band. For this case $Q \leq Q_a$ where Q_a is the totally acceptable value of the quality index. Above fully acceptable value Q_a begins operation with deteriorated forming the current. User of the plant decides about the value Q_a as well as about the value Q_u at which the operation is treated as fully acceptable or unacceptable (unreliable).

The quality index Q is transformed into the dimensionless variable h used in (48). The transformation can be linear according to formula

$$h = a + bQ \tag{50}$$

where constants a and b are calculated from the conditions: if $Q = Q_a$ then $h = 5$ and if $Q = Q_u$ then $h = -2$, which allow to obtain the characteristic shown in Fig. 13 (Harrington, 1965). As result of it the coefficients are:

$$a = -\frac{2Q_a + 5Q_u}{Q_a - Q_u}, \qquad b = \frac{7}{Q_a - Q_u}. \tag{51)-(52}$$

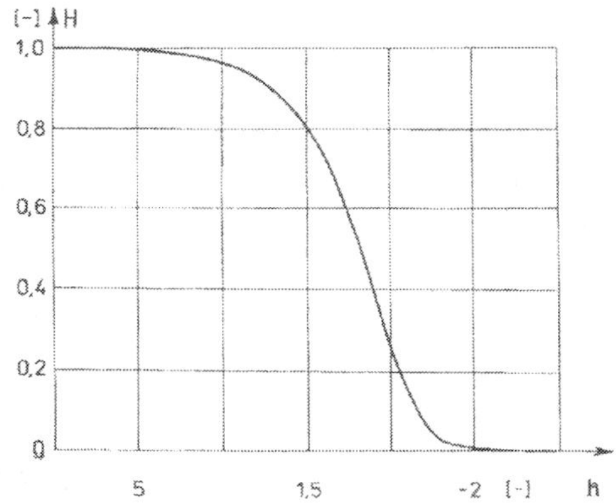

Fig. 13. The desirability function

If for instance the fully reliable value Q_a = 0.5 and the fully unreliable one Q_u = 2 then for this choice the coefficients are: a = 7.333 and b = -4.667.

5.3 Course of design

During optimizing the data are exchanged between three blocks in Fig.14.

In the SIMULATION MODEL of the converter for each set of parameters R, C, Δi and L_a the values of dissipated energy e, maximum filter voltage U_m, maximum frequency f_m of transistor switching and effectively reached band Δi_e of current forming are obtained. These values together with the model parameters C and L_a are base for calculating all quantities used in formulas (46) to (50) and finally the generalized optimization criterion G. For

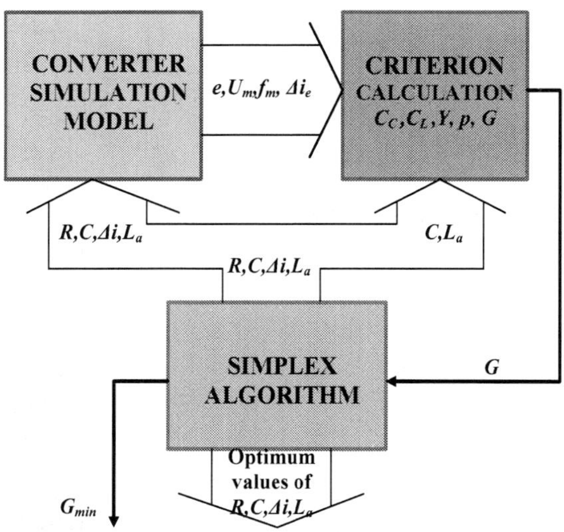

Fig. 14. Block diagram of the optimization

simulation of the system the TCad Power Electronic Simulation Software (Nieznanski et el., 1996) was used.
The formulas as well as dependencies for obtaining the capital cost C_C and the cost C_L of lost energy are programmed into the block CRITERION CALCULATION shown in Fig. 14.
For searching after the optimum solution in sense of (45) the SIMPLEX ALGORITHM of multidimensional optimization (Spendley et al., 1962) was used. It generates the successive sets of the inverter model parameters until the minimum value of the generalized index G_{min} is reached. The simplex method is proper for optimization using modeling as it gives possibility to reduce notably the needed number of simulations.
The course of optimisation is shown in Fig. 15. For the incoming parameters R = 0.48 Ω, C = 1 mF, Δi = 6 A and L_a = 0.08 mH the algorithm has found after 16 steps the outgoing optimum set: R = 0.112 Ω, C = 0.94 mF, Δi = 3.49 A and L_a = 0.0882 mH. At this the generalized index (45) has been reduced from 75 € to 51.8 €.

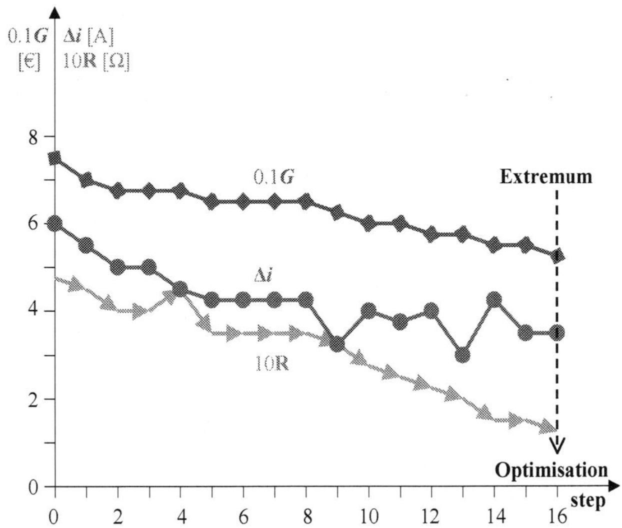

Fig. 15. Course of the values of generalized criterion, current band of the controller and filter resistance during optimization

6. Conclusions

The specific proportion of the parameters in the system: *Small power source - Inverter with hysteresis controller - Grid* causes unprofitable phenomena and difficulties with choice of the filter parameters, hysteresis band and other parameters assuring its good operation.
The theoretical analysis of the phenomena in the circuits of the converter as well as its simulation and laboratory investigations were carried over.
The system needs the great value of filter capacitance and the resistance for damping the low frequency resonance oscillation in the input circuit and the additional inductance for decreasing the current slope rate and switching frequency in the output circuit.

The design of the system is more difficult than the design of the same structure drive system where the energy flows through the inverter from high power source to the motor.
The task of design of the inverter for energy transfer from small renewable source to common grid should be solved with consideration of power loss, operation reliability and cost of the system. The requirements are bounded here very close one with another and partially are opposed.
The methodology of designing the converter with consideration of the above-mentioned problems was presented.
The solution consists in suitable aggregating the separate requirements in one generalized optimization criterion as well as in using two-exponential desirability function for reliability evaluation and the effective optimization method in order to decrease the simulation number.
The proposed designing gave notable improvement of the inverter properties at small number of optimization steps.

7. References

Chikaraishi, H., Hayashi, Y., Sato N. (1990). *A variable speed control of the induction generator without speed sensor for wind generation.* Trans. IEE Japan. Vol. 110-D. No.6., pp. 664-672.

El-Tamaly, A.M., Enjeti, P.N., El-Tamaly, H.H. *An improved approach to reduce harmonics in the utility interface of wind, photovoltaic and fuel cell power systems.* APEC, Vol. 2.

Harrington, E.C. (1965). The desirability function, Industry quality control, Vol. 21, No 10, pp. 494 - 498.

Higuchi, Y., Yamamura, N., Ishida, M., Hori, T. (2000). An improvement of performance for small-scaled wind power generating system with permanent magnet type synchronous generator. Industrial Electronics Society, 2000. 26th Annual Conference of the IEEE, IECON 2000, Vol. 2., pp. 1037-1043.

Luczkowski, R., Muszynski, R. (2007). Cost and Reliability Oriented Design of the Converter for Small Wind Power Plant, *Proceedings of the Colloquium on Reliability in Electromagnetic Systems*, May 2007, Paris, France, paper on Conference CD.

Mohan, N., Undeland, T.M., Robbins, W.P. (2003). *Power electronics.* John Wiley & Sons, inc.

Muszyński, R., Piłaciński, J.(2006). *Small wind power plant with alternator and voltage source inverter.* Proc. of the Int. Workshop on Renewable Energy Based Units and Systems, St.Petersburg, June Russia 2006.

Muszynski, R.., Pilacinski, J. (2007). Investigation of the inverter for energy transfer from small renewable source to common grid, *Proceedings of the 16th Int. Conference on Electrical Drives and Power Electronics,* The High Tatras, September 2007, Slovakia, paper on Conference CD.

Nieznanski, J., Iwan, K., Szczęsny, R., Ronkowski, M. (1996). Tcad for Windows. High Performance Power Electronic Simulation Software, SOFTECH, Gdansk.

Piggott H. (2005). *How to build wind turbine. The axial flux windmill plans,* Scoraig Wind Electric, UK.

Spendley W., Hext, G.R., Himsworth, F.R. (1962). The sequential application of simplex designs in optimisations - an evolutionary operations, Technometrics, Vol.4, pp. 441–448.

Tyc, M. (2003). The physical model of co-operation of small wind plant with the 1-phase network. The MSc work (consultant R. Muszyński), Poznań University of Technology, Institute of Electrical Engineering and Electronics, Poznan 2003 (in polish).

9

Control Strategies for Variable-speed Fixed-pitch Wind Turbines

Bunlung Neammanee, Somporn Sirisumrannukul and Somchai Chatratana[1]
Department of Electrical Engineering, Faculty of Engineering
King Mongkut's University of Technology North Bangkok
[1]National Science and Technology Development Agency
Thailand

1. Introduction

This chapter deals with a design of controllers for variable-speed fixed-pitch wind turbines. The chapter begins with an introduction to wind turbine control and the detail of nonlinear models of the wind energy conversion system (WECS). Linearization around a set of equilibrium points is presented to obtain a linear parameter variable model from a nonlinear system. Open-loop characteristics for small signals are then described to analyze dynamic behaviors under different operating conditions. The control objectives for variable-speed fixed-pitch wind turbines detailed in this chapter consists of three operating conditions:

1. Maximization of extracted energy: The wind turbine should extract wind energy at the highest efficiency to obtain the highest energy conversion ratio. Three alternative maximum peak power tracking (MPPT)-based algorithms for fixed pitch wind turbines are introduced. The first algorithm is guided by a torque reference (Mirecki et at., 2004). The second method searches an optimal operating point from the slope of the power-rotational speed curve. The last method is based on the control objective derived from a fuzzy rule base.
2. Limitation of extracted energy with active stall with rotational speed control: The aim of this control is to limit stresses on the turbine while minimizing the power fluctuations around a constant value, normally around the nominal power.
3. Control of MPPT and stall regulation at the overlapping region: This operating condition is effective with constant rotational speed control.

The controller objectives, controller schemes and controller designs are discussed in detail. The developed controllers for fixed-pitch wind turbines are based on a speed and torque-feedback control scheme. The proper design of the reference signal allows accurate tracking of each control strategy along the entire operating range. Hardware and software implementation for the control algorithms are explained. The case studies were carried out with two laboratory experiments with a developed wind turbine simulator: 1) three MPPT algorithms and active stall regulation with rotational speed control, and 2) an MPPT algorithm with a grid-connected converter. The chapter is concluded in the last section. The references are also provided for further research and studies.

2. Model of variable speed fixed-pitch wind energy conversion system

2.1 Modeling of rotor blade characteristics

A wind speed generally varies with elevation of the blades (i.e., every single spot on the turbines may not have the same wind speed). Modeling wind speed taking into account all different positions on the blades could be, therefore, very difficult. For this reason, a single value of wind speed is normally applied to the whole wind turbines. Modeling the rotor blade characteristic requires the tip speed ratio (TSR) and the relationship of torque and power coefficient versus TSR. The TSR is obtained from

$$\lambda = \lambda(v_t, \omega_t) = \frac{\omega_t R}{v_t} \tag{1}$$

where λ = tip speed ratio [rad.s]
v_t = wind speed [m/s]
ω_t = rotational speed [rad/s]
R = blade radius [m]

The power captured by the blades, P_{turb}, can be calculated by

$$P_{turb} = \frac{\rho}{2}\pi R^2 v_t^3 C_P(v_t, \omega_t, \beta) \tag{2}$$

The aerodynamic torque acting on the blades, T_a, is obtained by

$$T_a = \frac{\rho}{2}\pi R^3 v_t^2 C_T(v_t, \omega_t, \beta) \tag{3}$$

where ρ = air density [kg/m^3]
C_P = power coefficient [-]
C_T = torque coefficient [-]
β = pitch angle [degree]

If C_P is known, the aerodynamic torque can also be calculated from

$$T_a = \frac{\rho}{2}\pi R^2 v_t^3 C_P(v_t, \omega_t, \beta) / \omega_t \tag{4}$$

It can be seen from the above two equations that C_T and C_P are a function of λ and β. In this chapter, β is kept constant; namely, the pitch angle is fixed and this is generally true for small and medium sized wind turbines. Therefore, C_T and C_P depend only upon λ.

Figure 1 shows a relationship of C_T versus TSR of a typical 3 kW, three blade horizontal axis wind turbine with a rotor diameter of 4.5 m (Kojabadi et at., 2004). This curve represents an important characteristic in determining the starting torque of the wind turbine. In general, this curve is available from the manufacture or obtained from a field test. With this curve, C_P, indicating the efficiency of power conversion of the rotor blades, can be calculated by multiplying C_T with λ. Figure 1 also shows the C_T-λ profile corresponding to the C_P-λ curve. It is important to note that the power and torque coefficient of a wind turbine depends on aerodynamic design of the blades.

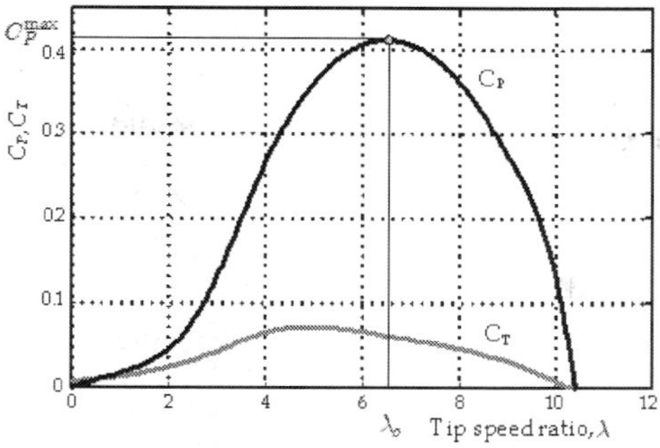

Fig. 1. C_T-λ and C_P- λ characteristics of a typical, 3 kW three blade horizontal wind turbine

2.2 Mathematical model of drive train

The most important part of the drive train is the modeling of the turbine itself as the other parts can be analyzed with common methods. Aerodynamic torque of the turbine is a nonlinear function with respect to the tip speed ratio and the pitch angle. This relation, $C_T(v_t, \omega_t, \beta)$, may be modeled with splines or with look-up tables. The C_T-λ curve is linearized with respect to wind speed, rotational speed, and pitch angle (in the pitch controlled turbine) at a given operating point. To analyze the turbine with linear methods, the non-linear torque function in (5) can be linearized by taking into account only the first terms of the Taylor series (H. Vihriälä, 2002).

$$\Delta T_a = \theta \Delta v_t + \gamma \Delta \omega_t + \kappa \Delta \beta \tag{5}$$

where

$$\theta = \left.\frac{\partial T_a}{\partial v_t}\right|_{O.P.} = C_o \omega_t (2C_{T,o} - \lambda_o \frac{\partial C_T}{\partial v_t})\Big|_{O.P.} \tag{6}$$

$$\gamma = \left.\frac{\partial T_a}{\partial \omega_t}\right|_{O.P.} = C_o R \omega_o \left.\frac{\partial C_T}{\partial \omega_t}\right|_{O.P.} \tag{7}$$

$$\kappa = \left.\frac{\partial T_a}{\partial \beta}\right|_{O.P.} = C_o \left.\frac{\partial C_T}{\partial \beta}\right|_{O.P.} \tag{8}$$

$$C_o = \frac{\rho}{2}\pi R^3 \tag{9}$$

$$\lambda_o = \frac{\omega_{t,o} R}{v_{t,o}} \tag{10}$$

With linearization around an operating point, the product of the derivative of aerodynamic torque and wind speed may be regarded as an external disturbance. If the pitch angle control is adopted as a means of control, the linearized coefficient κ must be calculated. The dynamics of the pitch actuator may be modeled with the first order dynamics. The

derivatives term $\partial T_a / \partial \beta$ can be obtained from blade design calculations or by identification from tests. Since this chapter deals with the fixed pitch configuration, no further discussion will be made on pitch control modeling.
Aerodynamic torque on a wind turbine is a nonlinear function of λ and β. But for the fixed pitch wind turbines, its pitch angle is constant and therefore, the aerodynamic torque depends only on tip speed ratio. Areodynamic torque in (5) can be rewritten as

$$\Delta T_a = \theta \Delta v_t + \gamma \Delta \omega_t \tag{11}$$

The wind power drive train can be modeled as one mass with the assumptions that there is no interaction between the drive train and tower dynamics, and no gravitational force that acts on the blade which causes periodic excitation. The mathematical model of the drive train consists of aerodynamic torque (12), auxiliary torque (13) and rotational acceleration (14).

$$T_a = J_t \dot{\omega}_t + B_t \omega_t + T_g \tag{12}$$

where J_t = inertia of turbine [kg.m²]
B_t = frictional coefficient of turbine [N.s/m²]
T_g = generator torque [N.m]

$$T_{aux} = T_a - B_t \omega_t \tag{13}$$

$$\dot{\omega}_t = \frac{1}{J_t}(T_{aux} - T_g) \tag{14}$$

If T_g is an input and ω_t is an output of the system, (12) and (13) are rearranged to give

$$T_a - T_g = (\dot{\omega}_t J_t + B_t \omega_t) \tag{15}$$

$$T_a = \theta v_t + \gamma \omega_t \tag{16}$$

From (15) and (16) the block diagram can be drawn as in Fig. 2.

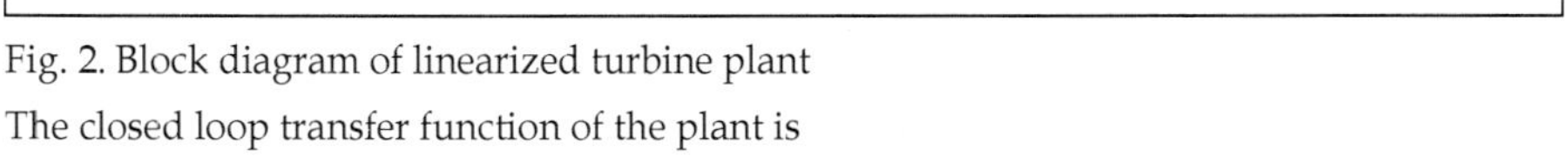

Fig. 2. Block diagram of linearized turbine plant

The closed loop transfer function of the plant is

$$\frac{\omega_t(s)}{T_g(s)} = \frac{1}{(-J_t s - B_t + \gamma)} \tag{17}$$

As the pole of the system is $s = (\gamma - B_t)/J_t$ with $|\gamma| \gg |B_t|$, the sign of the pole depends on the value of γ, which indicates the slope of the C_T-λ curve. This means the system will be unstable if γ is positive (the operating point lies in the left hand side of maximum torque of Fig. 1). Hence, if a linear controller is employed, the controller must be designed to compensate the effect of a positive γ.

2.3 Aerodynamic torque observer

A standard state-space model of a plant can be expressed by (18) and (19).

$$\dot{x} = Ax + Bu \tag{18}$$

$$y = Cx \tag{19}$$

The system can be transformed from the continuous domain to the discrete domain by the z transformation with zero order hold (ZOH), as given in (20) and (21).

$$y[k+1] = \Phi x[k] + \Gamma u[k] \tag{20}$$

$$y[k] = cx[k] \tag{21}$$

The closed-loop observer structure of (22) and (23) is

$$\hat{x}[k+1] = \Phi \hat{x}[k] + \Gamma u[k] + Kc(x[k] - \hat{x}[k]) \tag{22}$$

$$\hat{x}[k+1] = (\Phi - Kc)\hat{x}[k] + \Gamma u[k] + Ky[k] \tag{23}$$

The block diagram with a torque observer in the discrete form is shown in Fig.3. The problem is to calculate the observer gain K of the plant. The mathematical model of the drive train consists of aerodynamic torque (12), auxiliary torque (24) and rotational acceleration (25).

$$T_{aux} = T_a - B_t \omega_t \tag{24}$$

$$\dot{\omega}_t = \frac{1}{J_t}(T_{aux} - T_g) \tag{25}$$

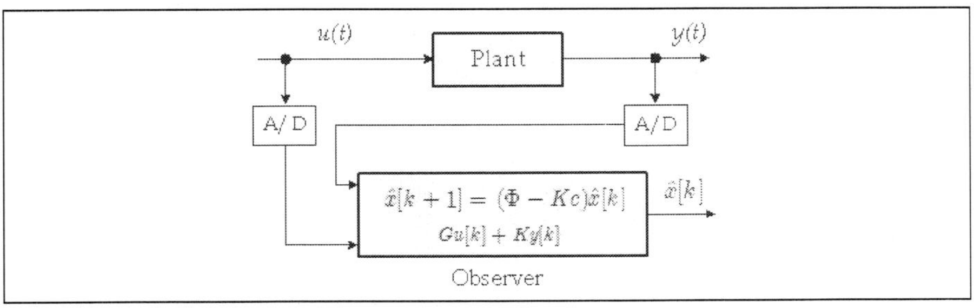

Fig. 3. Block diagram of observer and plant in discrete form

To derive a dynamic torque equation, it is assumed that the aerodynamic torque has much slower variation than a sampling rate, T_s, (Cardenas-Dobson et al., 1996). That is

$$\dot{T}_a = 0 \tag{26}$$

Rearranging (25) and (26) gives

$$\begin{bmatrix} \dot{\omega}_t \\ \dot{T}_{aux} \end{bmatrix} = \begin{bmatrix} 0 & \frac{1}{J_t} \\ 0 & 0 \end{bmatrix} \begin{bmatrix} \omega_t \\ T_{aux} \end{bmatrix} - \begin{bmatrix} \frac{1}{J_t} \\ 0 \end{bmatrix} T_g \tag{27}$$

$$y(t) = [1 \quad 0] \begin{bmatrix} \omega_t \\ T_{aux} \end{bmatrix} \tag{28}$$

From (27) and (28), the observe gain K can be calculated. A mechanical torque observer can then be implemented on a digital signal controller (DSC) board.

3. Control objectives

The wind turbine is an energy converter device that captures energy from the wind and converts it into useful work. Almost all of the wind energy conversion systems are connected to the grid of electric power networks. Although the main objective of wind turbine operation is to optimize energy capture, other technical and environmental objectives should also be satisfied such as mechanical loads, power quality standards, acoustic emission, obstruction, *etc.* (Fernando et al., 2007). These objectives are actually a tradeoff among each other. Thus, the WECS should find a well balanced compromise among them. The control objective of this chapter is emphasized on optimal energy capture.

Due to the requirement in speed control, four wind velocities separate the operation into three operating regions as shown in Fig. 4, which represents a typical power curve of a wind turbine. The cut-in velocity ($v_{cut\text{-}in}$) is defined as the wind speed at which the turbine starts to generate the power. Below this wind speed, it is not efficient to turn on the turbine. The transition speed $v_{\omega N}$ is the wind speed that the operating point starts to move from maximum power curve of region I to the rated power in region III The rated velocity (v_{rated}) is the wind speed at which the turbine reaches its rated turbine power. The cut-out velocity ($v_{cut\text{-}out}$) is the maximum wind speed at which the wind turbine can still operate. Beyond this wind speed, the rotor has to be locked to keep the blades, the electrical generator and other components from reaching damage.

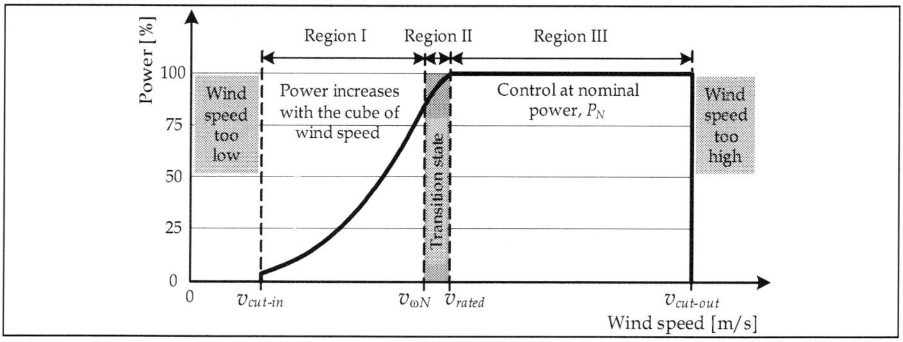

Fig. 4. Operation regions of wind turbine

Region I covers a wind speed range between v_{cut-in} and $v_{\omega N}$ and is referred as the below rated wind speed region. The available power is defined as the power in the wind passing through the rotor area multiplied by the maximum power coefficient $C_P^{\max}$, that is

$$P_{av} = C_P^{\max} P_V = \frac{1}{2}\rho\pi R^3 C_P^{\max} v^3 \tag{29}$$

The control objective in region I is to extract the maximum power from the wind. Therefore, the ideal power curve in this region is a function of wind speed cube defined by (12b). Region II covers a wind speed range between $v_{\omega N}$ and v_{rated} and is referred as a transition between maximum power curve of region I and rated power in region III. In this region, if the rotor speed is so high that it reaches the speed limit, it will generate acoustic noise. Therefore, this region may not exist if the rotor speed is below the limit. Region III covers a wind speed range between v_{rated} and $v_{cut-out}$ and is referred as the above-rated wind speed region. As the mechanical power generated by the rotor is limited, the main task of a controller in this region is to keep output at its rated power (Fernando et al., 2007). Operation regions of interest for wind turbine control of this chapter are those of I and III.

4. Control schemes with variable-speed fixed-pitch wind turbine

As maximum energy conversion is achieved at the fixed optimum angle, keeping the tip speed ratio at λ_o becomes a necessary condition. The constraint $\lambda = \lambda_o$ indicates that the rotor speed must change proportionally to wind speed; to be specific, $\omega_o = \lambda_o v_t / R$. A wind turbine controller uses electronic converters to transfer the output energy from the generator, which produces the counter torque to arrive at ω_o and λ_o, respectively. Thus, the turbine can be controlled to operate at different points, for instance, to track the optimum speed as wind speed fluctuates. The maximum power locus on the torque-speed plane can be obtained from

$$T_o = \frac{1}{2\lambda_o^3}\rho\pi R^5 C_P^{\max}\omega_o^2 = c\omega_o^2 \tag{30}$$

where T_o= optimum torque [N.m]

$$c = \frac{1}{2\lambda_o^3}\rho\pi R^5 C_P^{\max}$$

ω_o= optimum rotational speed [rad/s]

Equation (30) presents a parabola equation between the torque and rotational speed. In below-rated wind speed region, the turbine is controlled to track this $C_P^{\max}$ locus. Thus, variable-speed control strategies essentially differ in the way power is limited in the above-rated wind speed (Munteanu et al., 2008).

This section describes control trajectories to optimize energy capture from the fixed pitch wind turbine in the below-rated and above-rated wind speed regions. Methods to maximize the power output of wind turbines in the below-rated wind speed region are 1) torque reference-based MPPT algorithm, 2) searching-based MPPT algorithm and 3) fuzzy-based MPPT algorithm. These three methods are explained in more detail in next section. In the above-rated wind speed region, there are two methods to limit the output power from the

wind turbine at a specified output: 1) passive stall control and 2) active stall with rotational speed control.
In the below-rated wind speed region, especially, between $v_{cut\text{-}in}$ and v_{rated} (see Fig. 5), the wind turbine is programmed to operate along the quadratic curve AB with the C_P^{max} path. The turbine is operated at variable speed in this region. In the above-rated wind speed, when the wind speed exceeds the nominal level, the angle of attack will increase and thus approaches stalling. The air stream separates into larminar flow and eddy current flow at some parts of blade profile. Depending upon the angle of attack, the lift force is reduced but the drag force increases. This phenomenon is called passive stall.
Without a mechanism to turn the blades around the axis (i.e., pitch angle is constant), rotational speed control is another method to control the wind turbine to regulate output power in the stall region. For instant, when the controller reduces the rotational speed, the power coefficient will decrease to limit the output power at the nominal value as shown in the path BC of Fig. 5. The wind turbine is operated at variable speed throughout its operational range. If the wind speed approaches to the value at which the nominal power is produced, further torque development at the rotor must be limited. In general, the active stall is more efficient than the passive stall for energy regulation (Erich Hau, 2002).

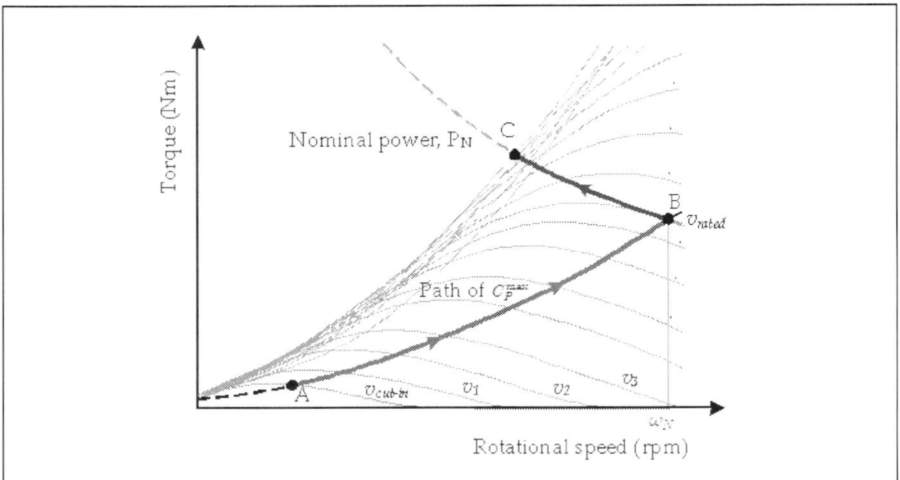

Fig. 5. Active stall with rotational speed control (path BC)

5. Maximum peak power tracking algorithms

The main purpose of MPPT algorithms is to maintain operating points on C_P^{max} for any wind speeds in the below-rated wind speed region (see Fig. 6). The operating points can be on the positive slope (the left side of C_P^{max}), zero slope (at C_P^{max}), and negative slope (the right side of C_P^{max}). If an operating point is in the positive slope region, the controller will move it to the right towards the optimum tip speed ratio, λ_o. This can be achieved by decreasing load current which results in an increase in the rotational speed. Conversely, if the operating point lies on the right hand side of the peak, the load current has to be increased, resulting in a decrease in the rotational speed. With this principle, the operating point can be maintained at C_P^{max} for various wind speeds.

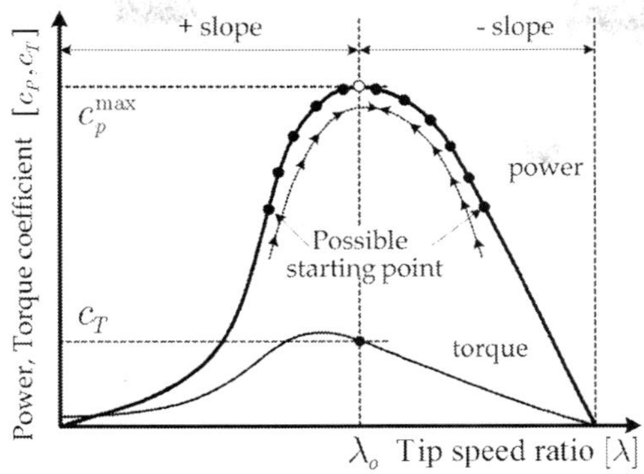

Fig. 6. Power and torque with MPPT tracking process

5.1 Torque reference-based MPPT algorithm

This method requires the wind turbine characteristics (e.g., R, C_P^{max} and λ_o). The below-rated wind speed reference torque, T_{be}^{ref}, can be calculated by substituting (1) into (4).

$$T_{be}^{ref} = k_t \omega_t^2, \quad P_a < P_{rated} \tag{31}$$

where $k_t = \frac{1}{2}\rho\pi R^3 C_P^{max}$

λ_o= tip speed ratio at C_P^{max} [rad.s]

The reference torque for the above-rated wind speed, T_{ab}^{ref}, is calculated from the rated power, P_{rated}.

$$T_{ab}^{ref} = P_{rated} / \omega_t, \quad P_a \geq P_{rated} \tag{32}$$

The controller uses (31) and (32) as the reference torque to control the plant. The torque reference-based MPPT algorithm can be alternatively achieved by testing the wind turbine to find the optimal torque-rotation speed curve with various wind speeds as shown in Fig. 7 a). The reference torque trajectory can be mathematically written as a function of torque and rotational speed or it can be stored as a look-up table which is easy to be programmed in a microcontroller or a DSC board. Figures 7 a) and b) show a reference torque trajectory and the associated output power in the below- and above-rated wind speed regions, respectively (Morimoto et al., 2005).

Control block diagram of torque reference-based MPPT

The torque reference-based MPPT block diagram is shown in Fig. 8. This system receives the rotational speed from the plant. The rotational speed is sent to the torque reference look-up table to interpolate the aerodynamic torque reference, T_{ref}. This reference will be sent to a load regulation loop (lower right corner of the figure), where a proportional-integral (PI) compensator is used to control the load current as desired and improve system stability.

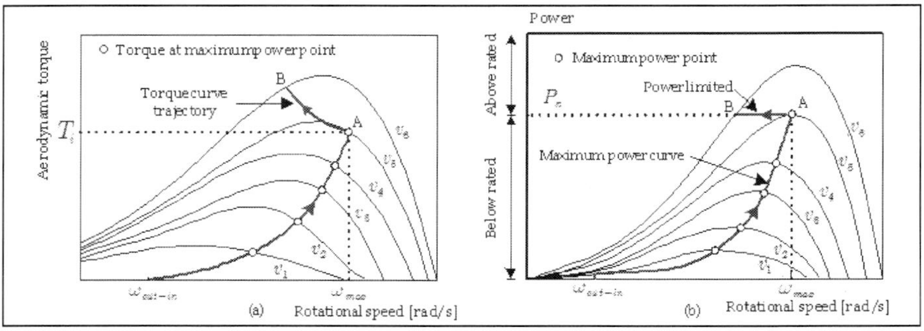

Fig. 7. a) Optimum torque trajectory and b) optimal power trajectory

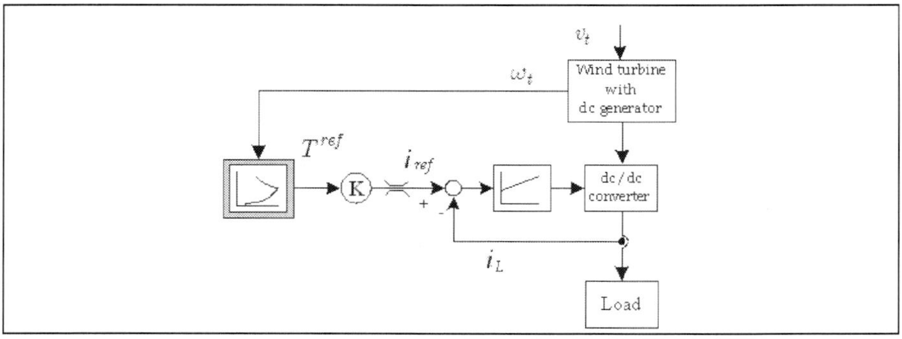

Fig. 8. Control block diagram of torque reference-based MPPT

5.2 Searching-based MPPT algorithm

This algorithm brings the operating point toward C_P^{max} by increasing or decreasing the rotational speed step by step. This tracking methodology is called the perturbation and observation method. To limit the output power at a specified power limit by stall regulation, the controller will reduce the rotational speed until the power coefficient reduces to the power limit. If the output is lower than the power limit, the controller will increase the rotational speed until the power matches the power limit. The control flowchart of the maximum power tracking system in Fig. 9 illustrates the details of decision processes based on the tracking procedure in Fig. 7. If the rotational speed is higher than the cut-in speed, $\omega_{cut\text{-}in}$, the MPPT controller will start the procedure. If a given perturbation leads to a positive or negative slope, the next perturbation increases or decreases the rotational speed until the slope becomes zero (i.e., maximum power point is reached). An updated load current reference, i_{ref}, for each sampling period, T_s, and an instantaneous power slope are calculated by (33) and (34), respectively (Morimoto et al., 2005) and (A.M De Broe et al., 1999).

$$i_{ref}[(k+1)T_s] = i_{ref}[kT_s] + M\frac{\Delta p[(k)T_s]}{\Delta \omega_t[(k)T_s]} \tag{33}$$

$$slope = \frac{\Delta p[(k)T_s]}{\Delta \omega_t[(k)T_s]} \tag{34}$$

where M = updated factor [-]
$i_{ref}[kT_s]$ = current at kT_s [A]

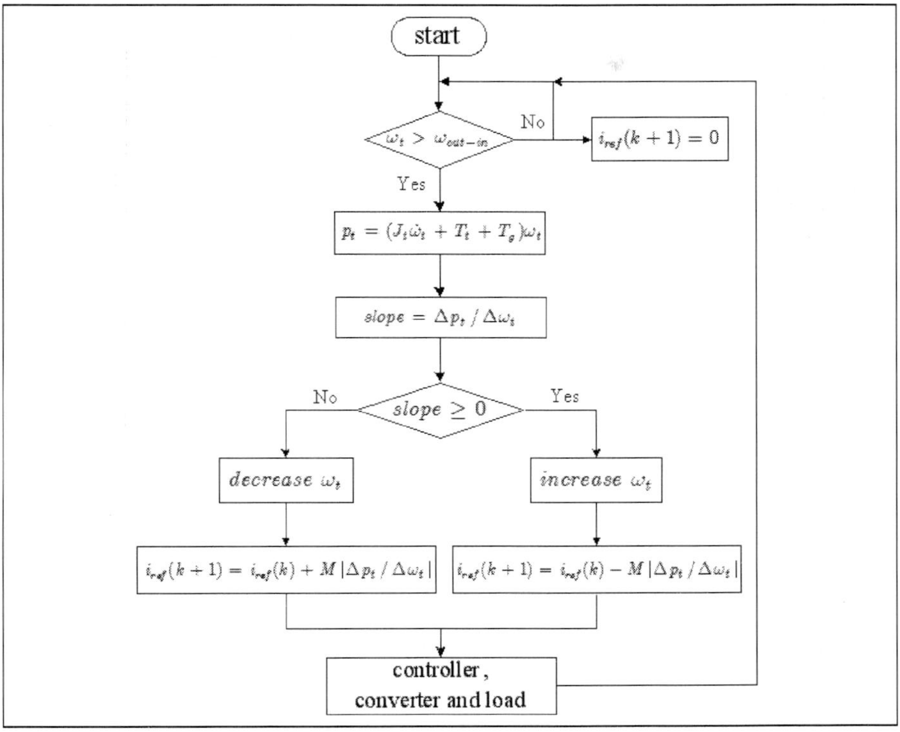

Fig. 9. Searching-base MPPT algorithm flowchart

Control block diagram of searching-based MPPT

To implement the MPPT and stall regulation control in a DSC controller unit, the total power Pt should be written in more detail as given in (35).

$$P_t = J_\omega \dot{\omega}_t \omega_t + P_e + T_f \omega_t + V_b i_g + i_g^2 R_g \tag{35}$$

The overall block diagram of MPPT and stall regulation control system is shown in Fig. 10. The top part of the diagram which is enclosed in the dotted line represents the MPPT controller built from (35) for the below-rated wind speed region. The below-rated wind speed control receives the generator current, voltage and rotational speed from the plant as inputs and use them to calculate the slope of the power-speed curve. The rate of change of power is compared with the reference (zero rate of change of power). The error is multiplied by a dc gain to generate the current reference for the PI controller of the plant control system. The lower part of the block diagram in Fig. 10, which is enclosed by the dotted line, represents the stall control for power limit. This controller receives the magnitude of the total power value from the below-rated control block diagram. If the instantaneous power is greater than the rated power, the selecting switch will be changed to the lower pole. This PI compensator in the load regulation loop (lower right hand corner of the figure) keep the output power within the rated power value.

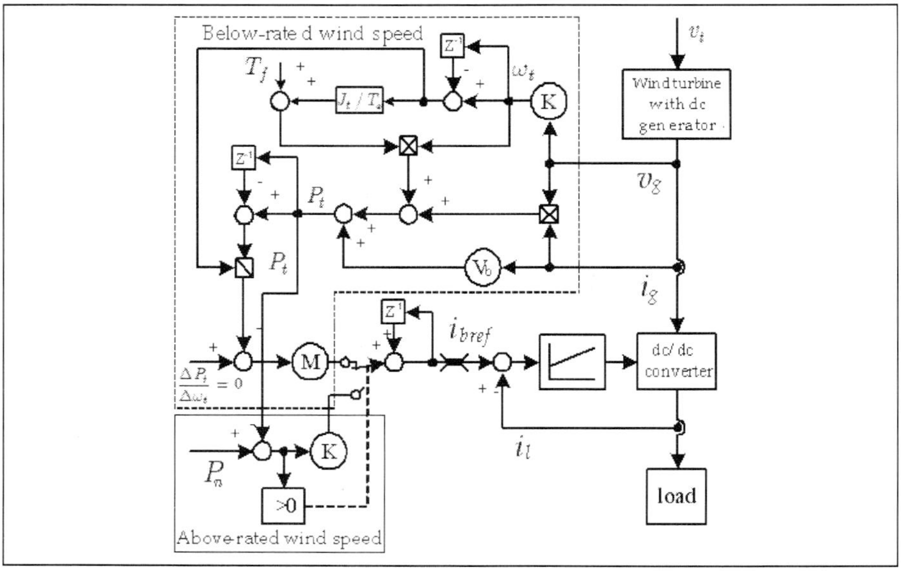

Fig. 10. Block diagram of searching-based MPPT

5.3 Fuzzy-based MPPT algorithm

A fuzzy logic control (FLC) algorithm is characterized by "IF-THEN" rules. The algorithm is suitable for wind turbine control with complex nonlinear models and parameters variation. Like the second algorithm, the fuzzy-based MPPT uses the perturbation and observation to track the maximum output power in the below-rated wind speed without knowledge of wind turbine characteristic. The input variables of the fuzzy-based MPPT are the rotational speed and the aerodynamic torque observer $\hat{T}_a$. In the above-rated wind speed, the FLC uses a torque reference calculated from (6) to limit the output power at the power limit. Another two input parameters, $\Delta\hat{T}_a$ and $\Delta\omega_t$, are used to limit torque and speed fluctuation. The MPPT with FLC uses $\hat{T}_a$ to classify the operating regions.

5.3.1 Fuzzy logic control design for fixed pitch wind turbine

In the below-rated wind speed region, the main control objective is to maximize energy capture from the wind. The wind turbine characteristics has a single maximum power coefficient, C_P^{max}, at the tip speed ratio λ_o. When the rotor operates at constant speed, the power coefficient will be at maximum at only one wind speed and therefore to achieve the highest annual energy capture, the value of the power coefficient must be maintained at the maximum level all the time (Yaoqin et al., 2002). In order to meet this objective, the controller calculates the reference torque using (36) and uses the error between this reference torque and feedback aerodynamic torque as the main input feedback.

$$T_{be}^{ref} = \frac{\rho}{2}\pi R^5 C_P^{\max}\frac{1}{\lambda_o^3}\omega_t^2 = k_T\omega_t^2 \tag{36}$$

where $k_T = \frac{\rho}{2}\pi R^5 C_P^{\max}\frac{1}{\lambda_o^3}$

The FLC based operation in this region can be explained by the below-rated torque control path in Fig. 11 a). Initially, it is assumed that the operating point is at point a, which corresponds to $C_P^{\max}$ operating point for a wind speed of 6 m/s. If the wind speed reduces to 4 m/s, the operating point will move to b with the same rotational speed, ω_1. For that wind speed, the FLC moves the operating point from b to c, which is the optimal position for the 4 m/s wind speed, by increasing the generator torque to reduce the rotational speed to ω_2. If the wind speed immediately increases from 4 m/s to 6 m/s, the operating point will move up to d, where the power is not maximum. Hence, the FLC will have to move the operating point to point a by reducing the generator torque and therefore increases the rotational speed. The control law for this region is summarized below:

$$\text{If } T_a > T_{be}^{ref}\text{, reduce } \Delta T_g \text{ to increase } \omega \tag{37}$$

$$\text{If } T_a < T_{be}^{ref}\text{, increase } \Delta T_g \text{ to reduce } \omega \tag{38}$$

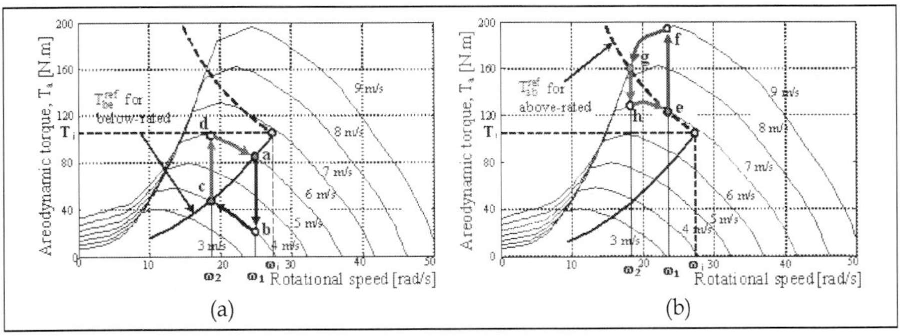

Fig. 11. a) Torque control path in below-rated wind speed region b) Torque control path in above-rated wind speed region

In the above-rated wind speed region, the main control objective is to limit the energy capture at a specified value. Typical methods are pitch control, active stall regulation, and passive stall (Muljadi et al., 1998). For a fixed pitched wind turbine, the active stall with rotational speed control is of interest. The output power can be limited by controlling its tip speed ratio until it is stalled. The controller can limit the output power by reducing the power coefficient, which can be done by moving the operating point to the left or right of ($\lambda_o, C_P^{\max}$). The controller will keep the operating point in the left side of (λ_o, $C_P^{\max}$) to avoid the system operating in higher rotational speeds. The controller calculates the reference torque based on (39) and can be shown by the dash line in Fig. 11 b). The controller uses the error between the reference and aerodynamic torques as the main input feedback.

$$T_{ab}^{ref} = \frac{P_{rated}}{\omega_t} \tag{39}$$

The intersection of the two regions gives the coordinate ($\omega_{\max}$, T_i), where

$$\omega_{\max} = \sqrt[3]{P_{rated} / k_T} \quad \text{and} \quad T_i = k_T^{1/3} P_{\max}^{2/3} \tag{40}$$

In Fig.11 b), the operating point is assumed to be at point e when the wind speed is 7 m/s. If the wind speed jumps to 9 m/s, the aerodynamic torque will increase accordingly, forcing the operating point to move to f. The FLC will increase the generator torque to reduce the rotational speed. The operating point will then come to a new optimum position, g. Conversely, if the wind speed reduces to 7 m/s, a transition from g to h can be expected. Thus the FLC will reduce the generator torque to increase the rotational speed to move the operating point to e, based on the reference torque path. The control law for this region is

$$\text{If } T_a > T_{ab}^{ref}, \text{ increase } \Delta T_g \text{ to reduce } \omega \tag{41}$$

$$\text{If } T_a < T_{ab}^{ref}, \text{ reduce } \Delta T_g \text{ to increase } \omega \tag{42}$$

Note that other constrains to be taken into account for the fuzzy logic control are, for example, changes in the rotational speed and torque.

5.3.2 Fuzzy logic control

In the fuzzification process, the relevant numerical parameters are linguistically converted into equal-base symmetric triangles (for the output) and trapezoidal membership functions (Bimal, 2000). The membership functions for six input parameters consisting of E_{be} and E_{ab}, defined in (43), Figure 12 shows the membership functions of 6 variables, namely E_{be}, E_{ab}, T_a, ΔT_a, ω_t, and $\Delta\omega_t$. The aerodynamic torque observer described in section 3.2 is used to observe aerodynamic torque, $\hat{T}_a$, in order to construct four inputs E_{be}, E_{ab}, T_a, and ΔT_a. The rotational speed ω_t is measured by 1000-pulse rotary encoder to obtain the rotational speed and the change of rotational speed $\Delta\omega_t$ as the other inputs of FLC. The membership function for one output parameter, ΔT_g, is shown in Fig. 13. All the membership functions are normalized in a range of [-1,1].

$$E_{be} = T_{be}^{ref} - T_a \quad \text{and} \quad E_{ab} = T_{ab}^{ref} - T_a \tag{43}$$

The fuzzy rule base that associates the fuzzy output to the fuzzy inputs is derived from the system behavior. It basically contains the knowledge acquired by designers as fuzzy rules and is expressed in forms of IF-THEN rules. The sole objective of the fuzzy rules designed here is to keep the wind turbine operating at the optimal point by using torque control for the two regions (Simoes et al., 1997).

In the below-rated wind speed region, the FLC knows how to operate in this region by observing whether the aerodynamic torque is lower than T_i (40). In this region, only three

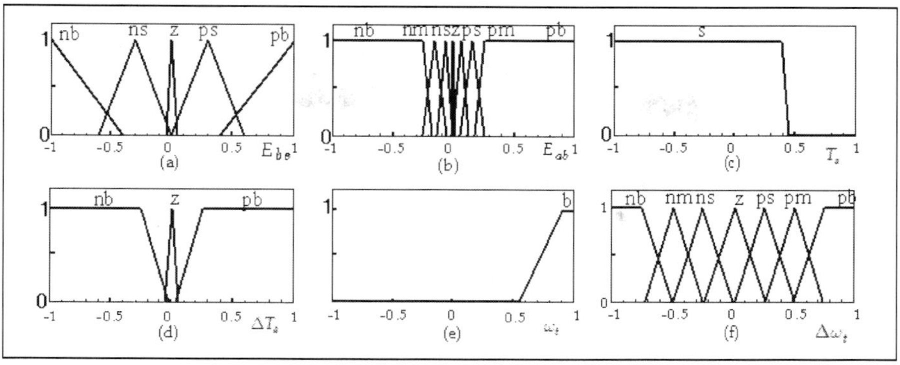

Fig. 12. Membership function input variables of a) E_{be}, b) E_{ab}, c) T_a, d) ΔT_a, e) ω_t, f) $\Delta\omega_t$

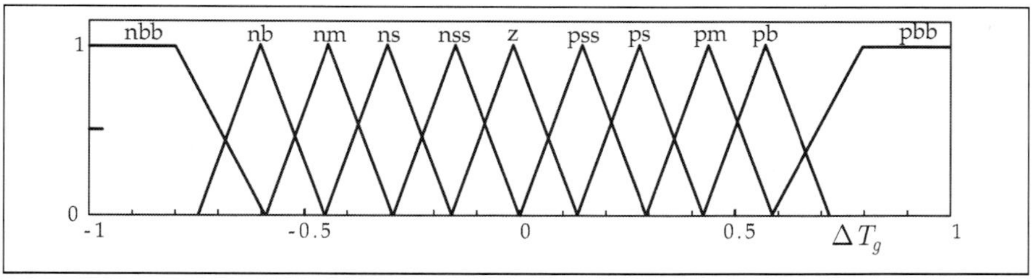

Fig. 13. Membership function output variables ΔT_g

where

nbb =	negative big big	nb =	negative big	nm =	negative medium
ns =	negative small	nss =	negative small small	z =	zero
pss =	positive small small	ps =	positive small	pm =	positive medium
pb =	positive big	pbb =	positive big	b =	big
s =	small				

input parameters are needed to determine the output: T_a, $\Delta\omega_t$ and E_{be}. The FLC tracks the torque reference given in (36) to obtain the maximum peak power. The relationship between $\Delta\omega_t$ and E_{be} generates ΔT_g with 35 rule base when T_a is s (see Fig. 12 c).

In the above-rated wind speed region, when the aerodynamic torque is grater than T_i (i.e., T_a is ns), the FLC tracks the torque reference by (39) and uses five input parameters, E_{ab}, $\Delta\omega_t$, ΔT_a, T_a and ω_t , to generate ΔT_g. The first three parameters are used to create the 105 rule base whereas T_a decides in which region the controller will operate. The final parameter is used for over-speed protection; to be specific, if ω_t is b, then ΔT_g is pbb.

Defuzzification is a process that converts the fuzzy set representing the overall conclusion into a real number after the evaluation of the rule base module. There are various types of defuzzification but a widely used one is the center of gravity (or centroid) defuzzification method, which determines the center of gravity of output membership function. If the discrete fuzzy set is applied in Fig. 14, the center of gravity of output membership function can be obtained from (Simoes et al., 1997)

$$U_0 = \sum_{i=1}^{n} U_i \mu(U_i) / \sum_{i=1}^{n} \mu(U_i) \tag{44}$$

where U_i = area of final fuzzy value at i

$\mu(U_i)$ = fuzzy output value i

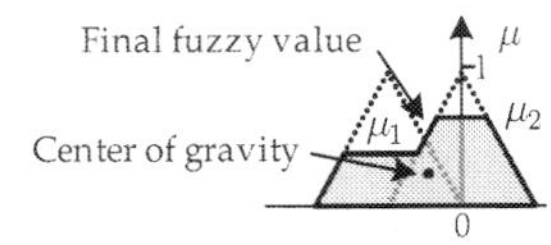

Fig. 14. Discrete fuzzy defuzzification for center of gravity method

5.3.3 Fuzzy logic control block diagram

The FLC block diagram of the system is shown in Fig. 15 (Neammanee et al., 2006). This system receives the current, voltage and rotational speed from the plant and then sends them to the torque observer to estimate the aerodynamic torque, $\hat{T}_a$. The FLC calculates ΔT_g using the predefined membership functions, fuzzy rule base and defuzzification to update the output torque and output command (Neammanee et al., 2007).

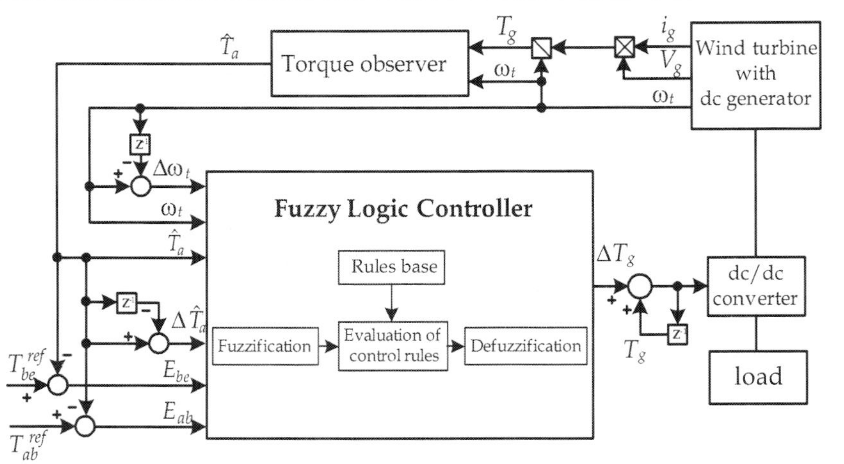

Fig. 15. Block diagram of fuzzy logic controller wind turbine

6. Case studies

The test system is composed of two main parts: 1) a developed wind turbine simulator on the left side of Fig. 16. and 2) a purposed wind turbine controller on the right side. The wind turbine simulator consists of a torque control inverter connected to a 7.5 kW induction motor and voltage and current sensors, data acquisition and a DSC controller board. The

DSC controller board uses a high performance 16 bits dsPIC30f6010, which combines the advantage of a high performance microcontroller and high computation speed digital signal processors (Huynh, P. & Cho, B.H., 1996). The software used to control the simulator was implemented in this DSC linked with a personal computer via two RS232 ports: one for transferring wind speed data to the DSC board and the other for sending parameters (e.g., P_e , i_g, v_g, ω_g) to the computer. The proposed wind turbine controller on which the MPPT algorithms are implemented consists of a dc/dc converter connected between a generator and a load, voltage and current sensors, a data acquisition unit and a DSC controller with the same performance as the one used in the simulator (Neammanee et al., 2007).

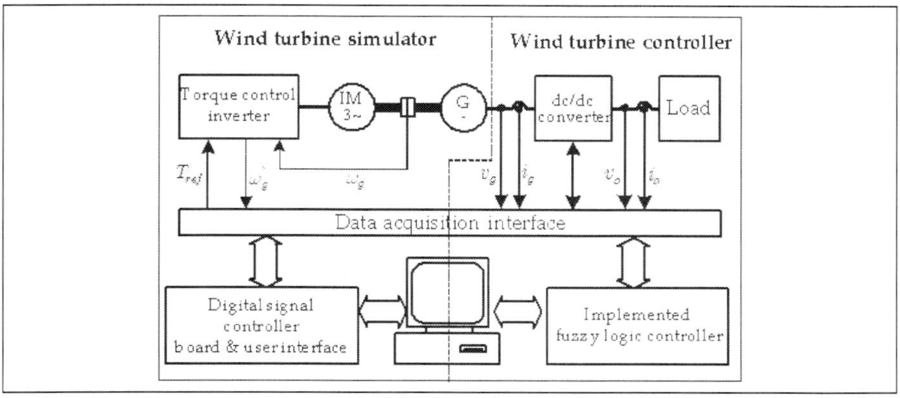

Fig. 16. Test system

6.1 MPPT algorithm and active stall regulation with rotational speed control

Figures 17-19 show control trajectories of the three MPPT-based algorithms with five different wind speeds: 3, 3.5, 4, 4.5 and 5 m/s. As can be seen from the figures, in the below-rated wind speed (less than 700W), the MPPT with torque reference controller succeeds in tracking the maximum power for each wind speed with the lowest rotational speed variation. When the wind speed is stepped up from 4.5 to 5 m/s, the system starts to limit the output power at 700W. If the wind speed increases beyond 5 m/s, the operating point will move to the left hand side of the previous one with a constant output power of 700W but with higher aerodynamic torques. During the move, the controller decreases the rotational speed until the wind turbine has been stalled. It can be seen that there is a high fluctuation in the output power in the above-rated wind speed.

The searching-based MPPT controller tracks the maximum peak power at each wind speed with highest rotational speed fluctuation in the below-rated wind speed. Around the output power limit (700 W), it is observed that this algorithm has a slightly lower power fluctuation than that of the first algorithm. Note that the second algorithm uses the lower part of the control diagram in Fig. 8 to calculate the reference torque in the above-rated wind speed. Referring to Fig. 15, the fuzzy based MPPT controller tracks the maximum peak power with the lowest power fluctuation as well as fairly low rotational speed fluctuation in the below-rated wind speed. The tracking time for the three control algorithms is compared in Fig. 20. It is clearly seen that the torque reference-based MPPT and the fuzzy-based MPPT are fastest in the below-rated wind speed and the above-rated wind speed, respectively (Neammanee et al., 2008).

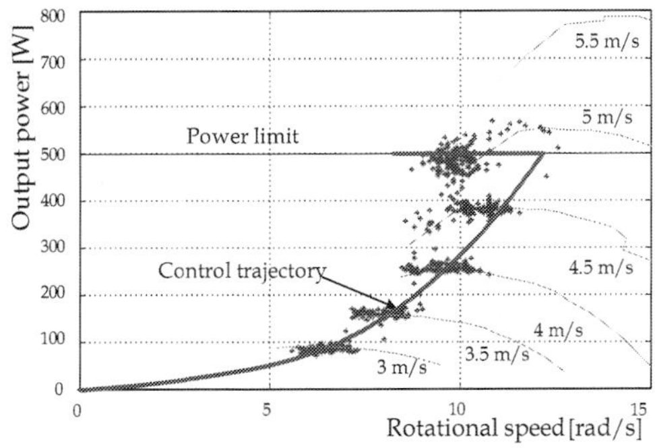

Fig. 17. Control trajectory of torque reference-based MPPT algorithm

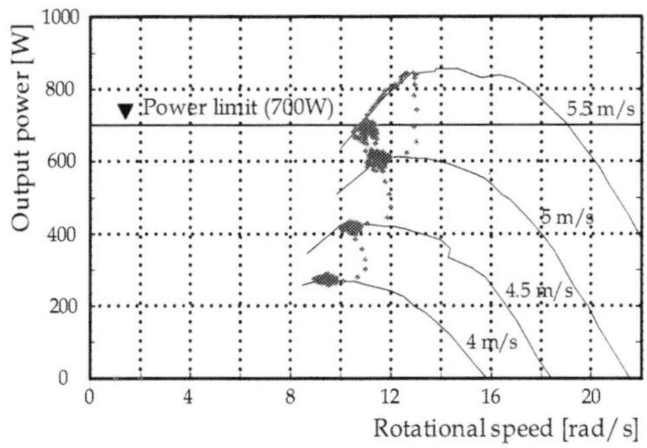

Fig. 18. Control trajectory of searching-based MPPT searching algorithm

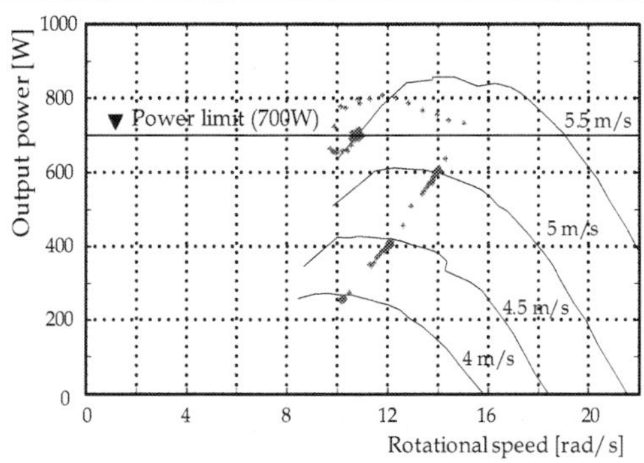

Fig. 19. Control trajectory of fuzzy-based MPPT algorithm

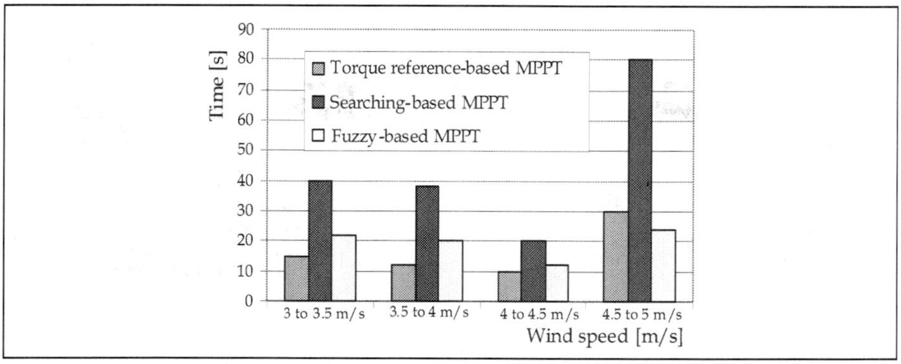

Fig. 20. Tracking time of three algorithms

6.2 MPPT algorithm with grid connected converter

There are three main parts of the test system shown in Fig. 21: 1) a 7.5 kW wind turbine simulator source on the left hand side of the figure, 2) a purposed MPPT controller with a high performance double-interleaved dual boost converter for the wind turbine in the middle, and 3) a 3 kW single phase grid connected converter on the right side. The wind turbine simulator is used to test the searching-based MPPT with the double-interleaved dual boost converter.

The purpose of this experiment is to confirm that the MPPT controller can be used with the DIDB converter to maximize power. This experiment was tested with various input wind speeds to the wind turbine simulator, coupled to a dc generator connected to the grid via the DIDB converter. The wind simulator was started at a wind speed of 4m/s, stepped up to 4.5, 5, 5.5 and 6 m/s respectively and run until steady state. The MPPT controller would capture a maximum power of 260, 380, 580, 700 and 900 W respectively. Figure 23 a) shows the relationship between the output power and time with various wind speeds. It can be seen from Fig. 23 b) that in this case, the MPPT controller can manage to keep C_P at the optimum

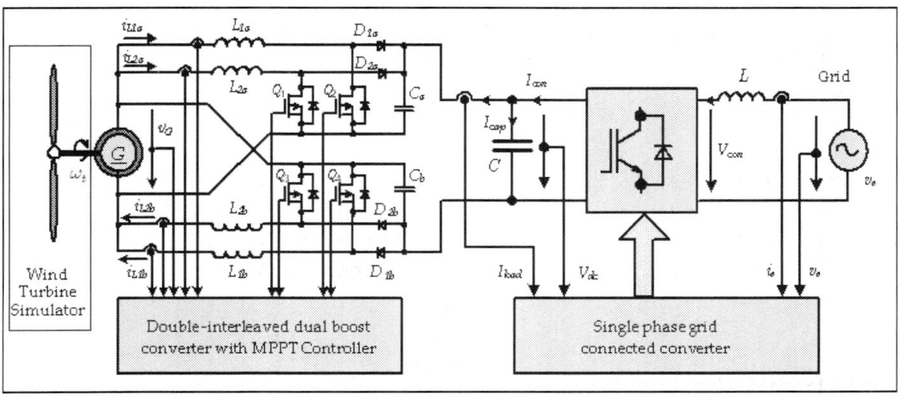

Fig. 21. Schematic diagram of test system

Fig. 22. Hardware of test system

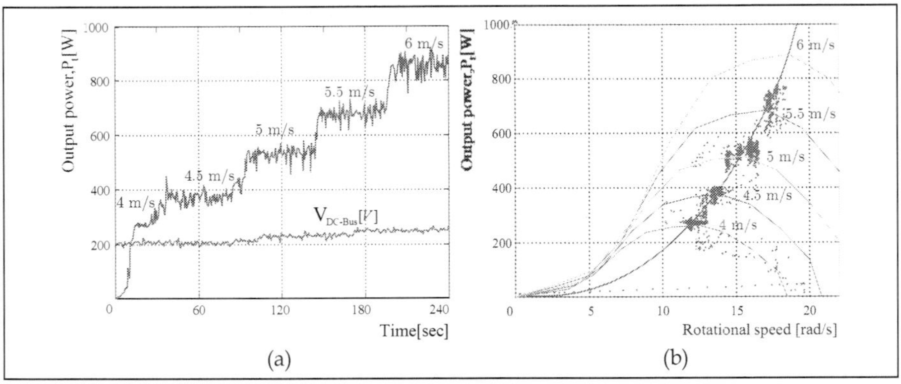

Fig. 23. a) 2-kW DIDB converter with MPPT controller b) Control trajectory of MPPT controller with various wind speeds

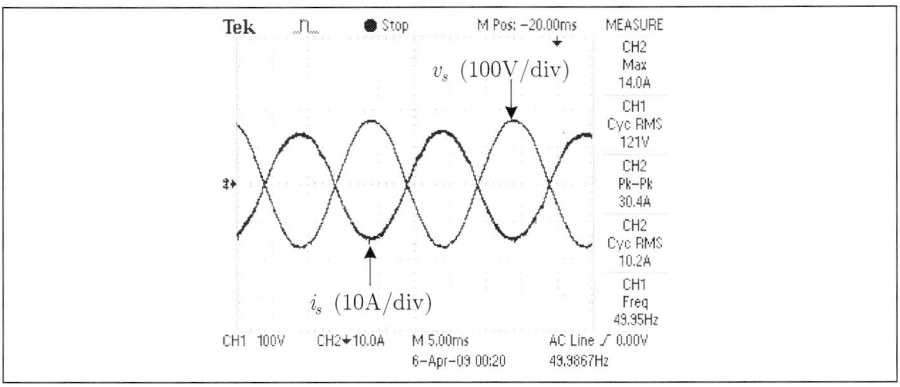

Fig. 24. Line voltage, v_s and current, i_s, measured by oscilloscope

value (λ_o) for the 5 step wind speeds. This experiment also reveals that the DIDB converter and the MPPT controller can be combined to achieve maximum power tracking. Figure 24 shows the phase voltage v_s and current i_s measured by oscilloscope on the grid side in inverting mode under steady state operation. It is clearly seen that phase current has a low total harmonic distortion. The phase voltage and current are 180° out of phase, indicating this converter generates only active power to the grid. The output power and dc bus voltage for real wind speed input data, v_t, to the wind turbine simulator are shown in Fig. 25. It is clear that the MPPT controller can track the maximum (Kajangpan et al., 2009).

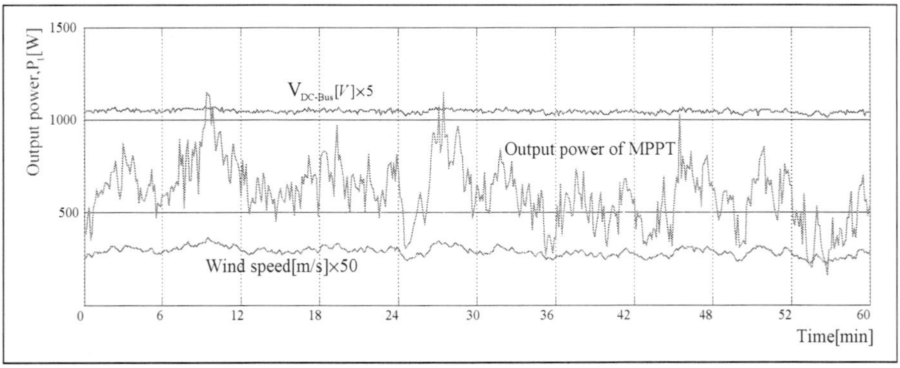

Fig. 25. Wind speed, output power of MPPT and dc-link voltage

7. Conclusion

This chapter is emphasized on control strategies for a variable speed fixed-pitch wind turbine with the main objective to optimize energy capture in below- and above-rated wind speed regions. Two main laboratory experiments were conducted: 1) three maximum peak power tracking (MPPT) algorithms and active stall regulation with rotational speed control, and 2) an MPPT algorithm with a grid-connected converter. All the MPPT algorithms were implemented on a low cost DSC board and tested with a developed wind turbine simulator. The first algorithm tracks the maximum power using a torque reference obtained from the wind turbine characteristic. The second method is based on the observation that the power versus rotational speed curve has a single well defined peak. Therefore, a necessary condition for the speed being at the maximum power point is that the first derivative of the power respect to the rotational speed is zero. The third algorithm employs a fuzzy logic as the key controller.

It can be concluded from the first experimental results that the MPPT with torque reference offers fastest tracking time in the below-rated wind speed region and the MPPT with fuzzy logic is favored in terms of power fluctuation and tracking time in the above-rated wind speed region. Although the second MPPT algorithm has the slowest tracking time and the highest rotational speed fluctuation, it is attractive for a small amount of computational resource and therefore low cost for implementation. Although the FLC possesses many advantages as described above, the implementation of the FLC is quite complicated, especially in the software part. For example, the implementation of the FLC with 140 rule base requires 1) plant knowledge to construct the rule base, 2) simulation process before implementation and 3) large memory storage for control program.

The second experimental results confirm that a developed grid connected converter with an MPPT algorithm can deliver the output power with low input current ripple and high gain. The system can track the maximum output power with step and various wind speeds and can regulate the dc bus voltage with nearly sinusoidal line side current, near-unity power factor and low harmonic distortion.

The proposed methodology can be extended to develop an adaptive FLC algorithm by the knowledge-base system for data recording in intelligent updated memory (IUM) to reduce the generator current and recording time. By applying the MPPT controller in the initial mode of tracking, the maximum power operating points can be located. These data will be stored in an IUM. When the data in the IUM cover a specified range in the below-rated wind speed region, the adaptive FLC will use these data as an input to maximize the output power within the range. The adaptive FLC, which consists of a MPPT controller, a torque observer, an IUM recorder and a FLC, can be implemented on a low cost, high performance digital signal controller board with a microcomputer for data acquisition and control verification.

8. Acknowledgement

The authors would like to express sincere thank to Energy Policy and Planning Office, Ministry of Energy, Thailand, and Faculty of Engineering, King Mongkut's University of Technology North Bangkok (KMUTNB) for financial supports.

9. References

A. Mirecki, X. Roboam, F. Richardeau (2004). Comparative Study of Maximum Power Strategy in Wind Turbines, *IEEE Trans. Ind. Electronics*. Vol. 2, 4-7 May., pp. 993-998.

A.M De Broe, S.Drouilhet, V. Gevorgian (1999). A Peak Power Tracker for Small Wind Turbines in Battery Charging Applications, *IEEE Transactions on Energy Conversion*, Vol. 14, No. 4, Dec., pp. 1630-1635.

B. Neammanee, K. Krajangpan S. Sirisumrannukul and S.Chatratana (2006). Fuzzy Logic Based Optimal Energy Capture for Wind Energy Conversion Systems, *The 10 Conference of the Electrical Power Supply Industry (CEPSI),* Mumbai, India.

B. Neammanee, S. Sirisumranukul, S. Chatratana (2007). Development of a Wind Turbine Simulator for Wind Generator Testing. *International Energy Journal.* pp. 21-28.

B. Neammanee, K. Krajangpan, S. Sirisumrannukul and S. Chatratana (2008), Maximum Peak Power Tracking-Based Control Algorithms with Stall Regulation for Optimal Wind Energy Capture, the industry applications transaction, *The Institute of Electrical Engineering of Japan (IEEJ),* Vol. 128, No.4, Japan, 1 Apr., pp.: 411-417.

Bimal K. Bose (2000). Fuzzy Logic and Neural Networks, *IEE Inds. App. Magazine*, May/June. pp.57-63.

E. Muljadi, K. Pierce and P. Migliore (1998). Control Strategy for Variable-Speed, Stall-Regulated Wind Turbines, American Control con. Pennsylvania, USA, June, pp. 1710-1714.

Erich Hau (2002). *Wind turbines*. Spring-Verlag, ISBN-10 3-540-24240-6, Berlin Heidelberg.

Fernando, D. Bianchi; Hernan De Battista & Ricardo, J. Mantz. (2007). *Wind Turbine Control Systems*, Springer-Verlag, ISBN-13:9781846284922, London.

H. Vihriälä (2002). *Control of Variable Speed Wind Turbine*. Ph.D. Thesis. Tampere University of Technology.

H.M. Kojabadi, Liuchen Chang, T. Boutot (2004). Development of a Novel Wind Turbine Simulator for Wind Energy Conversion Systems Using an Inverter-Controlled Induction Motor, *Energy Conversion IEEE Tran.* on Vol. 19, Issue 3, Sep., pp. 547-552.

Huynh, P. and Cho, B.H. (1996). Design and analysis of a Microprocessor-Controlled Peak-power-Tracking System, *Aerospace and Electronic Systems, IEEE Tran.* Vol. 32, Issue 1, Jan., pp.182 -190.

Iulian Munteanu; Antoneta Iuliana Bratcu, Nicolaos-Antonio Cutululis ; Emil Ceanga (2008). *Optimal Control of Wind Energy Systems*, Springer-Verlag, ISBN-978-1-84800-080-3, London.

Jia Yaoqin, Yang Zhongqing, Cao Binggang (2002). A New Maximum Power Point Tracking Control Scheme for Wind Generation, *IEEE Con. on Power System Technology*, Vol. 1, 13-17 Oct. pp.: 144-148.

Korawit Kajangpan, Bunlung Neammanee and Somporn Sirisumranukul (2009). High Performance Grid Connected Converter with Double-Interleaved Dual Boost Technique and MPPT Control for Wind Turbine," *World Renewable Energy Congress 2009 - Asia, The 3rd International Conference on "Sustainable Energy and Environment (SEE 2009)"* Bangkok, Thailand, 18-23 May, pp.: 718-722.

Morimoto, S.; Nakayama, H. Sanada, M.; Takeda, Y.(2005). Sensorless Output Maximization Control for Variable-Speed Wind Generation System Using IPMSG, *Ind. Appl., IEEE Trans*, Vol. 41, Issue 1, Jan.-Feb. pp.:60- 67.

Morimoto, S.; Nakayama, H. Sanada, M.; Takeda, Y.; (2005). Sensorless Output Maximization Control for Variable-Speed Wind Generation System Using IPMSG. Ind. Appl., IEEE Trans Vol. 41, Issue 1, Jan.-Feb. pp.:60- 67.

Marcelo Godoy Simoes, Bimal K. Bose, Ronald J. Spiegel, (1997). "Design and Performance Evaluation of a Fuzzy-Logic-Based Variable-Speed Wind Generation System." *Industrial Application IEEE Trans.* on Vol. 33 No. 4 (Jul./Aug.,) pp. 956-965.

Marcelo Godoy Simoes, Bimal K. Bose and Ronald J. Spiegel, (1997). Design and Performance Evaluation of a Fuzzy-Logic-Based Variable-Speed Wind Generation System. *Industrial Application IEEE Trans.* on Vol. 33 No. 4 (Jul./Aug.,) pp. 956-965.

R. Cardenas-Dobson, G.M. Asher and G. Asher (1996). Torque Observer for the Control of Variable Speed Wind Turbines Operating Below Rated Wind Speed, *Wind Engineering*, Vol. 20, No.4, pp. 259-285.

10

Real-time Physical Simulation of Wind Energy Conversion Systems

Iulian Munteanu[1,2], Antoneta Iuliana Bratcu[1,2],
Seddik Bacha[1] and Daniel Roye[1]
[1]*Grenoble Electrical Engineering Laboratory (G2Elab),*
BP 46, 38402 Saint-Martin d'Hères
[2]*"Dunărea de Jos" University of Galați,*
47 Domnească, 800008- Galați
[1]*France*
[2]*Romania*

1. Introduction

The real-time wind turbine simulators are enabling the testing of wind energy conversion systems' (WECS) control units and of the associated control algorithms in a controlled environment. It is known that a simulator can predict the behaviour of an industrial system subjected to certain operating conditions before its real-world implementation. But, the main motivation of using physical simulators for wind energy applications, resides within another argument. It is about the fact that a dedicated laboratory setup can provide something that does not exist in real-world applications: controllable wind velocity. So, a wind turbine physical simulator offers a prime mover which behaves as a "wind-turbine-powered-like shaft" (Nichita et al., 2002), allowing the static and dynamic characteristics replication of a mathematically-modelled wind turbine. This is the great advantage of such simulator as it replaces very expensive parts - such as the turbine rotor and the drive train - operating in a stochastic environment by conveniently-controlled electrical motors, operating in a controlled environment. This allows the repetitive experiments being carried out independently from wind and meteorological conditions in a most safe laboratory environment. Besides its prime mover, all of the physical WECS elements are present in the physical simulator as they are in the real-world application. Therefore, the electromechanical part of the generation chain exhibits the real phenomena presented in a wind power system. In this way the above-listed advantages are not faded by replicating a simplified electrical behaviour and a WECS-dedicated control unit will interact with a genuine wind power system.

When analyzing the concerned literature, one can note that the preliminary experimental validation of WECS control laws is always performed on wind turbine simulators. This is a reason for quite rich literature being dedicated to this subject. One can find two types of papers dealing with small-scale WECS simulators for different generation configurations. The first category is composed of works focusing on test rig building aspects (Leithead et al., 1994; Battaioto et al., 1996; Rodriguez-Amenedo et al., 1998; Diop et al., 1999a; Akhmatov et al., 2000; Cardenas et al., 2001; Rabelo & Hofmann, 2002; Teodorescu & Blaabjerg, 2004; Steurer et al., 2004). Some works underlining the role of test rigs for preliminary validation of WECS control laws compose a second category (Enslin & van Wyk, 1992; Cárdenas et al., 1996; Kana et al., 2001; Munteanu et al., 2005; Camblong et al., 2006; Munteanu et al., 2008b). One should note that the list is far from being exhausted.

This chapter aims at providing the reader with possible answers to the questions related to the manner in which a WECS physical simulator can be build, including implementation details about how its different elements must be chosen and how its effectiveness can be assessed. All these issues are dealt with in the next section. The third section contains a comprehensive example in the form of a case study that applies the theoretical guidelines introduced previously. The last section, the fourth, is dedicated to conclusion. The chapter is completed by an appendix section. Even if this chapter concerns mainly the physical simulation of the horizontal-axis wind turbines (HAWT), the presented principles can be used without significant changes for the vertical-axis wind turbines.

2. How to build a WECS simulator?

2.1 Concepts

This section mainly refers to the prime mover rebuilding (in the sense of its behaviour replication), whereas the other parts are assembled by using the same methods and equipment as in the real-world application. As already stated, the turbine rotor should be replaced by an electrical servomotor which behaves as the former. To fulfil that purpose, the electrical motor is somehow driven by the wind turbine mathematical model, provided that an adequate model is available.

Of course that the entire simulator can be built using only hardware elements (*i.e.*, by using analogic integrated circuits for implementing the turbine model), but taking into account the computing power of the digital systems, the turbine model is implemented as software in the quasi-totality of the cases. Therefore, the WECS simulator has a physical part - which develops power - and a software part - the controlling model - connected in closed loop.

Conceptually speaking, a system containing a software model interacting directly with a hardware control unit has emerged in the last years as hardware-in-the-loop (HIL) simulator (Hanselmann, 1996). It is clear that the two closed-loop subsystems exchange only information one with each other. This kind of system has been extensively used for developing (fast prototyping) and testing control structures for mechanical equipments.

Concerning the WECS simulator, the interaction between the simulated plant (turbine aerodynamics) and the physical part (servomotor) suppose not only the information transfer but also the existence of the associated physical variables, as the servomotor develops power. This version of the concept is often called power hardware-in-the-loop (PHIL) simulation (Wu et al., 1996). The main difference with respect to the original concept is the existence of a power interface between the simulated plant and the so-called hardware under test. Even if the WECS physical simulator finds itself in the PHIL category, for sake of simplicity the term HIL will be used in the following.

2.2 Methodology

The simulator building approach presented in this work relies upon the general concepts, terminology and methodological aspects introduced by Nichita et al. (1998) and reused in Munteanu et al. (2008a). Even if it is dedicated entirely to the WECS simulators, the present

text uses some more general concepts and variables, which are valid for an entire class of industrial systems (Andreica et al., 2009).
Beside the WECS to be simulated, one must define an associated class of the operating conditions to be analyzed or of the control problems to be solved. Correspondingly, one can generally consider suitable input and output vectors (*e.g.*, the wind velocity and the output power), as Figure 1 depicts. Irrespectively of the actual WECS configuration, the intended analysis is focused on a precise subsystem, denoted in the HIL-related literature as *hardware under test*. So, the basic idea used in HIL structures generally supposes that the original plant can be naturally divided into two subsystems which interact one with the other. Generally speaking, the two subsystems are chosen in order to fulfil some simulation efficiency criterion. In most of the cases, the first subsystem is such that the closed loop experiments are very expensive and deterministic experiments are almost impossible and it represents the prime mover. Therefore, it will be this subsystem whose behaviour must be replaced by a physical simulator; consequently, it will be called an *emulated physical system* (EPS). This implies that EPS is the only part of the plant that is mandatory to model. In the WECS case, this can be the turbine rotor and can include the drive train. The second subsystem will exist in the HIL simulator exactly as it is in the original plant, thus allowing laboratory experiments under realistic conditions. Being the object of research undertaking the control action, it will be further called in this text an *investigated physical system* (IPS).

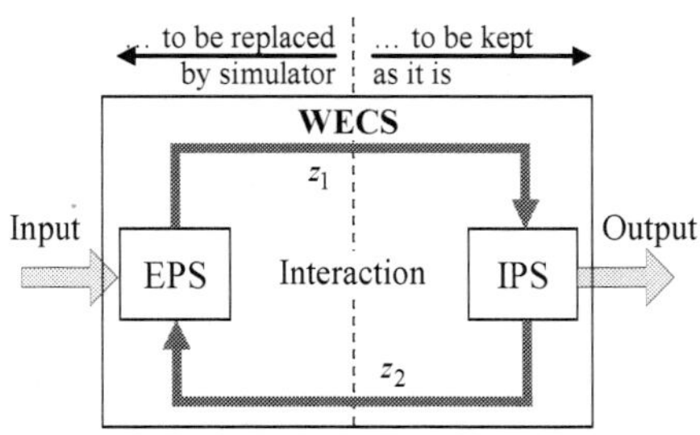

Fig. 1. Original plant to simulate illustrating the interaction between the IPS and the EPS

The interaction between EPS and IPS, corresponding to a power transfer between them, is characterized by a pair of so-called *interaction variables*, further denoted as z_1 and z_2.
Supposing that the EPS is the prime mover, the energy flows towards IPS. Having made this assumption, the interaction from the EPS point of view is depicted in Figure 1. The physical nature of the interaction variables depends on the original system in a biunique manner. z_1 is the cause variable, whose variation initiates the energy imbalance, and z_2 is the response variable, common to the EPS and IPS. By virtue of their coupling, their product has always power dimension.
Now, concerning the WECS physical simulation, in the quasi-totality of cases available in the literature, the building of an electro-mechanical simulator is intended. Hence, the WECS is split between EPS and IPS at one of its rotating shafts. So, the two interaction variables are the shaft torque and the rotational speed, and EPS will contain at least the aerodynamics subsystem. This is not, of course, the sole simulator configuration that can be chosen.
The solution employed to build a physical simulator is to replace the EPS by the so-called *real-time physical simulator* (RTPS). The IPS remains the same as in the WECS, as its

behaviour study represents one of the main purposes of the HIL simulator building. The RTPS must offer the "natural" environment for IPS and must replicate the EPS behaviour and the interaction EPS-IPS. In this way the resulted HIL simulator will approximate the original WECS dynamics. In short, the RTPS must physically provide one of the interaction variables based on the measure of the other one and, of course, on the EPS model. This goal is achieved by means of a tracking loop at the output of the RTPS, which in some works (Munteanu, 2006) is called the *effector* (EFTs in Figure 2); the controlled variable is called driving variable and the measured one - response variable.
The effector reference is established by a model of the EPS; its input is established by an algorithm dedicated to the resource synthesis (*e.g.*, wind speed). This model is embedded as software subsystem in the so-called *real-time software simulator* (RTSS). In conclusion, the RTPS includes the real-time software simulator (EPS modelling) and the tracking loop for physical replication of the controlled interaction variable. This structure is given in Figure 2.

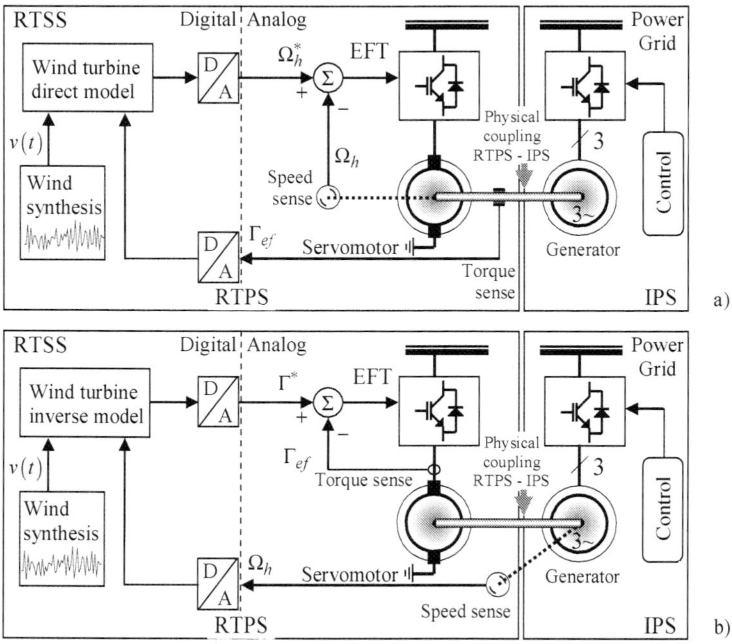

Fig. 2. Example of WECS HIL electromechanical simulator structures: a) the driving variable is an effect; b) the driving variable is a cause

When choosing the driving and response variables, two situations may happen, as follows. Let us consider a first case, when the driving variable is an output or a state of the WECS. So, it is about controlling an effect variable (of z_2 type), and the model implemented in the RTSS is strictly causal and is obtained directly from the EPS model, fed by a measure of the cause variable (z_1). For example, this effect variable can be a rotational speed if it is about an electromechanical simulator or a voltage if it is about an electrical simulator. An example is given in Figure 2a). There is also a second case when the driving variable is a cause variable (of z_1 type). In this event, the model implemented in the RTSS is non-causal (the EPS inverse model) and is fed by a measure of the effect, z_2. This effect variable can be a mechanical torque in an electromechanical HIL simulator and the associated software implementation implies the EPS model being rewritten. An example is given in Figure 2b).

Both of the two above-described cases have disadvantages, which can affect the simulator reliability. In the first case the effector dynamic is quite slow, whereas getting the second case into practice is difficult because temporal derivatives must be computed, increasing the measurement noise. Also, in Figure 2 one can note that the response variable is affected by the transducer dynamic and the driving variable by the effector dynamic. Therefore, these variables have slightly modified instantaneous values, affecting the accuracy with which the HIL simulator emulates the WECS. Of course that for ensuring good simulator performance, these dynamics, together with the computation inside the RTSS, must be sufficiently faster than the dynamic of the EPS.

As stated before, there is not just a single way of building HIL simulators for WECS. Usually - and it is also the case chosen in this chapter - it is considered that IPS and EPS interact by means of the rotating high-speed shaft. Thus, the RTPS physical part is based on a rotating electrical machine (servomotor), either DC motors (Battaioto et al., 1996; Rabelo et al., 2004), or AC machines offering similar performances (Steurer et al., 2004; Munteanu et al., 2005).

The IPS is typically based on synchronous or induction machine and may include power electronics converters and control systems in order to implement the variable-speed operation. The interaction variables in this case are the rotational speed, $\Omega_h \equiv z_2$, and the mechanical "effective" torque of the high-speed shaft, $\Gamma_{ef} \equiv z_1$, with the high-speed shaft dynamical characteristic being the RTPS output. The EPS consists of aerodynamics and drive train. The algorithm within RTSS will thus implement the associated models and also the wind velocity as a stochastic sequence with statistical parameters depending on a certain wind site. Models of various deterministic test signals can also be implemented.

In the speed-driven case, a measure of the high-speed shaft torque is required. A computed value of the generator electromagnetic torque is often preferred in this case (Munteanu et al., 2008b). In the torque-driven case, the effector needs a torque feedback. In most of cases a measure of the servomotor electromagnetic torque is available starting from currents measure (*e.g.*, the armature current in the case of a DC motor).

To conclude, between the two above-described cases one can remark some differences. Concerning the software, the aerodynamics model is inversed in the torque-driven case, while in the speed-driven case it is written "as is" in the RTSS. The associated hardware (effector) is configured as follows. The torque-driven case has a single control level - the very fast servomotor current (torque) loop. In the other case, there is a supplementary outer speed control loop; the speed controller should impose sufficiently fast dynamics to the coupling servomotor-generator in order to ensure sufficiently small simulation errors. In the following some more in-depth simulator building aspects will be presented.

2.3 Rigid drive train case

As stated before, a preliminary EPS short modelling stage is necessary. The aerodynamic subsystem of a fixed-pitch HAWT can be modelled in average by means of the interaction between air masses and the turbine rotor (Burton et al., 2001). The turbine model outputs the wind torque based on the wind velocity, v, and the low-speed shaft rotational speed, Ω_l. The rotor aerodynamic performance is generally described by means of the power coefficient, C_p, which is a unimodal function of the tip speed ratio (Figure 3a), if assuming constant parameters of the air stream (air density, Reynolds numbers, *etc.*). If R denotes the blade length, the tip speed ratio is defined as:

$$\lambda = \Omega_l \cdot R/v \tag{1}$$

The C_p curve is constant for fixed-pitch turbines. When the wind speed varies, the power curves shapes reproduce the C_p shape (Wilkie et al., 1990), as shown in Figure 3b).

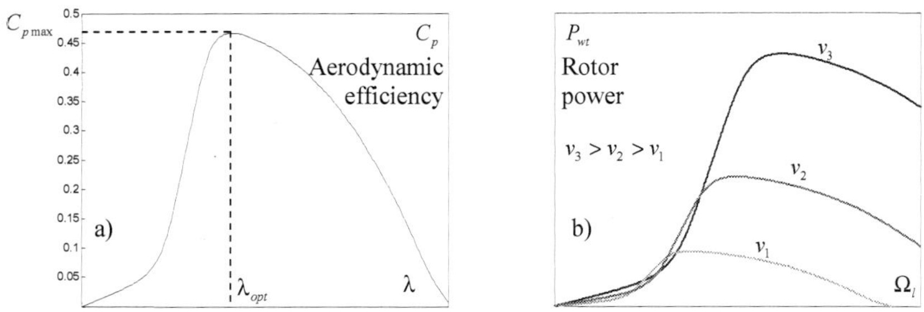

Fig. 3. a) Efficiency and b) corresponding power characteristics for a HAWT-based WECS

The corresponding wind torque is given by the relation (Wilkie et al., 1990):

$$\Gamma_{wt}(\Omega_l, v) = 0.5 \cdot \pi \cdot \rho \cdot v^2 \cdot R^3 \cdot C_\Gamma(\lambda) + \Gamma_{ts} - \Gamma_{fv} - \Gamma_s, \tag{2}$$

where $C_\Gamma(\lambda) = C_p(\lambda)/\lambda$ denotes the torque coefficient, Γ_{ts} represents the torque generated by the tower shadow effect, Γ_{fv} is the viscous friction torque and Γ_s is the static friction torque. Other elements can be added into relation (2), for example dynamical effects such as induction lag or spatial filter, in order to obtain a better approximation whenever needed (Rodriguez-Amenedo et al., 1998). If the turbine blades are pitchable, the wind torque is computed based on a supplementary input variable - the blades pitch, β - $\Gamma_{wt}(\Omega_l, v, \beta)$.

The drive train is the interaction device between the turbine rotor and the electrical generator. A rigid drive train generally consists of a multistage helical or spur gear-based speed multiplier (together with the associated shafts), modelled in average by a multiplication ratio i and efficiency η. For modelling purposes, the dynamics of the rigid drive train are rendered either at the low-speed or at the high-speed shaft, thus obtaining the so-called one-mass model (Wilkie et al., 1990). The motion equation for the latter case is:

$$J_h \cdot \dot{\Omega}_h = \Gamma_R(v, \Omega_h) - \Gamma_G(\Omega_h), \tag{3}$$

where $\Omega_h = i \cdot \Omega_l$ is the high-speed shaft rotational speed, $\Gamma_R = \eta \cdot \Gamma_{wt}/i$ is the high-speed shaft torque and Γ_G is the electromagnetic torque provided by the electrical generator. The turbine inertia rendered at the high-speed shaft is $J_h \approx J_{wt} \cdot \eta/i^2 + J_G$, with J_{wt} and J_G being the turbine rotor and electrical generator inertias respectively. Relations (2) and (3) compose a model of the EPS.

Now, being given a test rig composed of a rigid coupling servomotor-generator with an inertia $J_{sim} = J_G + J_{SM}$, its associated motion equation can be written:

$$J_{sim} \cdot \dot{\Omega}_{sim} = \Gamma_{SM} - \Gamma_G(\Omega_h), \tag{4}$$

where Γ_{SM} and J_{SM} are the servomotor torque and inertia and Ω_{sim} is its rotational speed. One wants that this mechanical assembly (simulator) rotates exactly as the WECS described by (3) when subjected to the same generator torque, Γ_G. Therefore, it is imposed that $\Omega_{sim} \equiv \Omega_h$ and $\dot{\Omega}_{sim} \equiv \dot{\Omega}_h$. By subtracting equations (4) and (3), one obtains the necessary value of the servomotor torque for fulfilling the above conditions:

$$\Gamma^*_{SM} = \Gamma_R(v, \Omega_{sim}) - (J_h - J_{sim}) \cdot \dot{\Omega}_{sim} \tag{5}$$

Equation (5) shows that the torque value imposed to the servomotor is computed by subtracting the dynamical torque from the wind torque (at the high-speed shaft). The former variable is computed by estimating the simulator rotational speed gradient. The latter variable is calculated using a synthesized value of the wind speed, a measure of the simulator rotational speed and the model from (2). So, equation (5), together with a wind speed model, is to be implemented into RTSS for the torque-driven case (see Figure 4a).

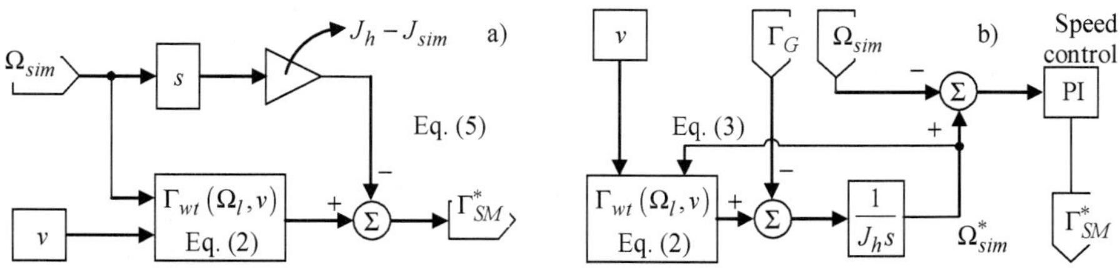

Fig. 4. RTSS configuration for WECS having rigid drive train: a) torque-driven case, b) speed-driven case

If a speed-driven scheme is required, one must impose to the servomotor-generator assembly the rotational speed value computed by integrating equation (3). Of course that a measure of the generator torque, Γ_G, should be available. The structure to implement, when the simulator speed controller is also embedded into the RTSS, is sketched in Figure 4b).

From a systemic viewpoint, the electromechanical part of WECS can be regarded as having two inputs, the wind speed and the electromagnetic (generator) torque, and one output, the rotational speed. Therefore, both wind speed and electromagnetic torque influence the rotational speed through two different channels (with different dynamics). The two above-described simulator structures should replicate the WECS behaviour for both influence channels. The simulation performances can be qualitatively assessed in the frequency domain if considering the linearized model of WECS around a typical operating point.

Figure 5 shows the relative position of the simulator characteristics with respect to the original WECS, for the two cases (torque- and speed-driven) and for both influence channels: the wind speed to rotational speed channel (Figure 5a) and the generator torque to rotational speed channel (Figure 5b). The influence of each channel has been studied independently of the other channel. These characteristics have been obtained by numerical simulation for a linearized low-power WECS, and do not contain the additional lags induced by transducers, neither the simulation time step itself. Only the servomotor torque loop dynamics have been considered. Globally, one remarks that the simulation is valid until certain frequency. This value depends on the actual parameters of the rotational speed gradient estimator (see Figure 4a) and on those of the rotational speed controller (Figure 4b).

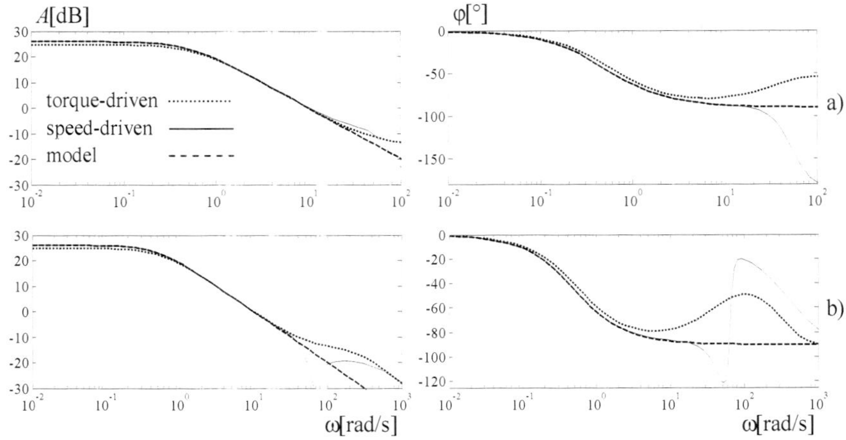

Fig. 5. The simulator frequency characteristics versus the original WECS model (dashed - model, solid - speed-driven case, dotted - torque-driven case): a) wind speed to rotational speed transfer, b) electromagnetic torque to rotational speed transfer

When analyzing Figure 5a), concerning the torque-driven case, one can remark a steady-state error in the gain, due to the nonzero dynamic friction of the simulator shaft. This means that a slight difference from the WECS rotational speed may appear, and may change with the operating point. For the same case, one can also note the leading effect in the phase characteristic, meaning that the high-frequency wind variations (turbulence) will not reproduce correctly the genuine rotational speed variations. However, the inherent lags present when a physical implementation is achieved can alleviate this aspect.

As regards Figure 5b), the characteristics have been traced for a lager frequency domain as the input torque variations can be significantly faster than the wind speed turbulences. For the torque-driven case one can restate the remarks above. For the speed-driven case one may expect bandwidth reduction when physical implementation is achieved. This figure lays out some limitations, particularly if the simulator is intended to be used as a WECS control laws benchmark. As the generator torque is the control input, one should not test WECS controllers designed with too large bandwidths. Otherwise, the designed controllers cannot be directly transferred to the real-world applications.

2.4 Flexible drive train case

The same aerodynamic subsystem as in the rigid drive train case is considered; therefore the same model can be used. The flexible drive train dynamics are described by the following equations (Akhmatov, 2003):

$$\begin{cases} \dot{\Omega}_l = 1/J_{wt} \cdot \Gamma_{wt}(\Omega_l, v) - i/(\eta \cdot J_{wt}) \cdot \Gamma \\ \dot{\Omega}_h = 1/J_G \cdot \Gamma - 1/J_G \cdot \Gamma_G \\ \dot{\Gamma} = K_s(i\Omega_l - \Omega_h) + B_s\left(i\dot{\Omega}_l - \dot{\Omega}_h\right) \end{cases}, \tag{6}$$

where K_s and B_s are respectively the stiffness and the damping coefficients of the spring, i is the speed multiplier ratio and η is the drive train efficiency.

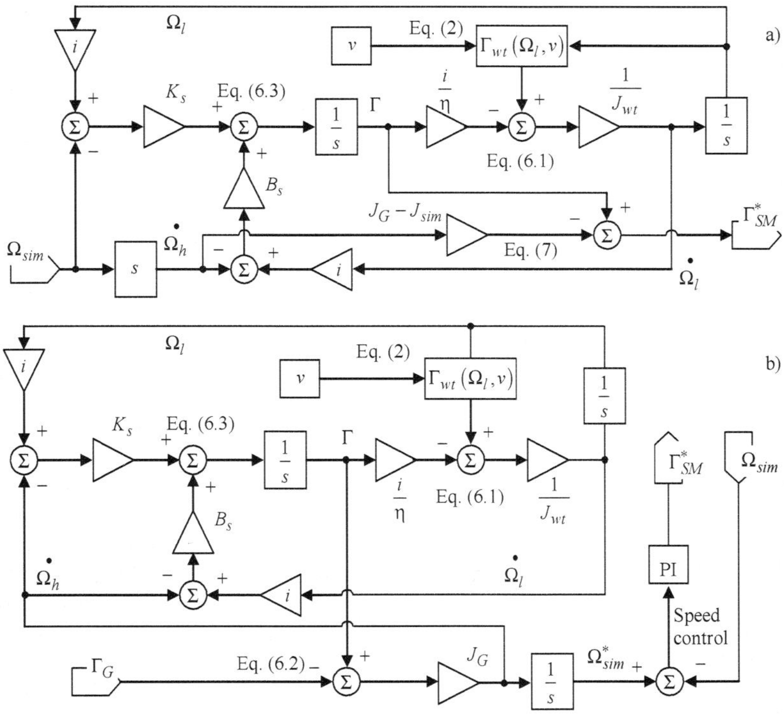

Fig. 6. RTSS configuration for WECS having flexible drive train: a) torque-driven case, b) speed-driven case

In the torque-driven case, the servomotor's torque reference is obtained based on measuring both the servomotor's rotational speed and its gradient, *i.e.*, Ω_h and $\dot{\Omega}_h$, supposing that the servomotor-generator assembly emulates perfectly the high-speed shaft, *i.e.*, $\Omega_{sim} \equiv \Omega_h$ and $\dot{\Omega}_{sim} \equiv \dot{\Omega}_h$. By subtracting the second equation of (6) from equation (4) one obtains:

$$\Gamma^*_{SM} = \Gamma - (J_G - J_{sim}) \cdot \dot{\Omega}_h, \tag{7}$$

where Γ is obtained by integrating the first and the third equation of system (6). Like in the rigid drive train case, equation (7) shows that the torque value imposed to the servomotor is the difference between the internal torque and the dynamical torque, computed by estimating the simulator's rotational speed gradient. The corresponding block diagram to implement in the RTSS is given in Figure 6a).

In the speed-driven case, the servomotor's rotational speed reference is obtained by integrating the system of equations (6). A measure of the generator torque, Γ_G, should be available. The block diagram to implement, in the case when the simulator speed controller is also embedded into the RTSS, is shown in Figure 6b).

2.5 Which elements to employ?

The simulator is centred on the electromechanical assembly. This is made up by rigidly coupling similarly-sized servomotor and electrical generator (in power and speed) - element 1 in Figure 7. The generator has the same type as in the real WECS and determines the configuration of the power circuit ensuring the electrical power transfer to the grid/load - elements 2 and 3 in Figure 7. Corresponding to the power circuit structure one may employ a control circuit or digital system dedicated to the power flow control. Various measuring devices such as encoders, current and voltage transducers, and some other power elements, such as insulation transformers or circuit breakers are likely to be used (*e.g.*, item 4 in Figure 7). All these elements, composing and controlling the power generation system, form together what is called in the general HIL simulation methodology as the IPS. For example, if the WECS is squirrel-cage induction generator-based, a back-to-back power electronics converter must be used, if it is about a permanent-magnet synchronous generator (PMSG), the power circuit may include a diode rectifier, DC-DC and DC/AC converters, and so on.

In order to control the servomotor torque, a power electronics device - depending on the servomotor type - must be employed (*e.g.*, item 5 in Figure 7). The servomotor voltages, currents and rotational speed must be supervised; therefore the associated transducers must be present in the so-called effector. For example, if the servomotor is a DC machine, a full-bridge chopper should be used for driving purposes. Using this hardware, a current loop can be built using classical control algorithms implemented on a digital signal processor.

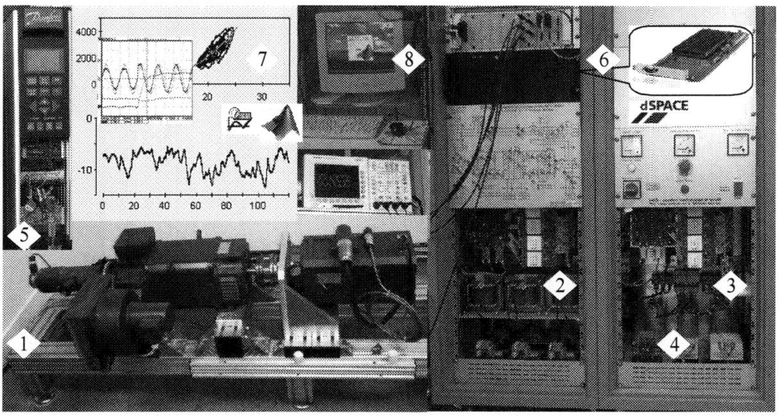

Fig. 7. Components of the WECS simulator: 1 - electromechanical assembly; 2, 3 - back-to-back power electronics converter; 4 - generic power elements; 5 - servomotor drive; 6 - digital system; 7 - user interface; 8 - host computer

According to Section 2.2, the plant to be simulated, EPS, must be translated into an algorithm, RTSS. Additionally, this latter contains wind speed models, electrical generator torque estimation, measures (inputs) filtering and output conditioning modules and should be implemented into a sufficiently fast digital system (element 6 in Figure 7). One must note that its computing step time (hundreds of microseconds in this case) is critical to the physical simulator performances. Beside this, the software-hardware loop overall lag contains some delays due to transducers, anti-aliasing filters and I/O system. Most of the applications encountered in the literature are supported by advanced, fast and flexible digital systems allowing rapid prototyping and changing of the software simulator. The dSPACE (*e.g.*, DS 1103 PPC) (Teodorescu & Blaabjerg, 2004), RTDS (RTDS, 2009) or RT-LAB (RT-LAB, 2009) systems are among these. Frequently, one uses the same systems for implementing the IPS-related control algorithms, also. Multiple interfaces (generically represented by element 7 in Figure 7) allowing the building of the software applications and the supervision of the WECS simulator may be used in conjunction with these digital systems. Element 8 in Figure 7 represents the host computer that supports these interfaces.

2.6 Is the WECS simulator well-performing enough? Errors analysis

A WECS real-time simulator is a laboratory tool very useful for applicative research envisaging control subsystem design or grid interfacing. In this context, the assessment of the simulator performance and the analysis of the simulation errors must be performed before the simulator is effectively used (Diop et al., 2000). Accumulation of errors begins in the modelling phase. Ideally, a physical simulator implements the adopted model, this is why the physical system modelling must correspond to the goal of simulator-based experiments. As regards the simulation errors, they can be minimized if properly configuring the real-time computing system. Unlike these two kinds of errors, it is not obvious how to minimize the tracking errors due to real-world implementation, as they depend on the way of choosing the driving variable. In the following, these latter errors in WECS simulators are analysed in the frequency domain, based on the linearized model of wind turbine (Munteanu et al., 2008a).

The linearized models of a wind turbine around a conveniently chosen steady-state operating point can be seen in Figure 8, where the influence of the two exogenous signals – the wind speed as a perturbation and the electromagnetic torque as a control input – on the plant's output have been represented. The plant's output is the high-speed rotational speed, provided that, for sake of simplicity, the case of a rigid drive train is considered. Notation $\Delta\bullet$ denotes variations of the variable around the point of linearization. The transfer from the wind speed to the high-speed shaft rotational speed can be identified in Figure 8a), whereas the transfer from the electromagnetic torque to the same rotational speed is shown in Figure 8b). The two transfer functions will be denoted by H_{wt}^{v} and $H_{wt}^{\Gamma_G}$ respectively.

In Figure 8 the following coefficients have been introduced:

$$K_{\Gamma v} = \overline{\frac{\partial \Gamma_{wt}}{\partial v}}, \; K_{\Gamma \Omega} = \overline{\frac{\partial \Gamma_{wt}}{\partial \Omega_h}}, \tag{8}$$

where notation $\overline{\bullet}$ denotes the value in the steady-state point of linearization. In Figure 8a) is considered that the rotational speed variations should be reflected in a change of the control input, Γ_G. This influence is modelled by means of a transfer function denoted by

$H_l(s)$, therefore the only input of the system is the perturbation, Δv. Simple calculations allow obtaining the transfer function of the channel wind speed to rotational speed as:

$$H_{wt}^{v}(s) = \frac{\Delta\Omega_h(s)}{\Delta v(s)} = \frac{K_{\Gamma v}}{H_l(s) + J_h s - K_{\Gamma\Omega}} \tag{9}$$

The same applies for the block diagram in Figure 8b), where the influence of the perturbation has been cancelled in order to allow the electromagnetic torque to rotational speed transfer being mathematically described as:

$$H_{wt}^{\Gamma_G}(s) = \frac{\Delta\Omega_h(s)}{\Delta\Gamma_G(s)} = -\frac{1}{J_h s - K_{\Gamma\Omega}} \tag{10}$$

The accuracy of the real-time simulation is judged in relation to the capacity of the simulator to replicate the original WECS's behaviour on both transfer channels from the exogenous inputs to the system's output. In the following, the linearized models of the real-time simulation diagrams for both the torque-driven case and the speed-driven case are deduced, in order to be compared to relations (9) and (10). In this context, one can develop an errors analysis and emphasize the conditions in which these errors are minimized.

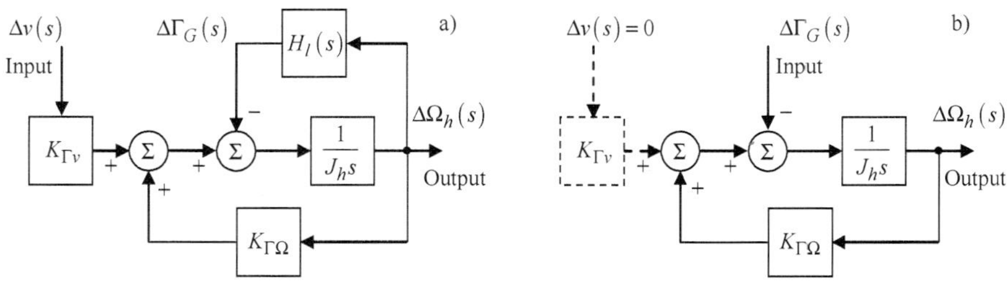

Fig. 8. Linearized model of the wind turbine: a) wind speed to rotational speed transfer; b) electromagnetic torque to rotational speed transfer

Figure 9 presents the linearized models of the real-time simulation diagram in the torque-driven case for both input-output channels, *i.e.*, from the wind speed to the simulator's rotational speed (Figure 9a) and from the electromagnetic torque to the simulator's rotational speed (Figure 9b). The two transfer functions will be denoted by H_{td}^{v} and $H_{td}^{\Gamma_G}$ respectively. H_0^{Γ} is the transfer function of the servomotor torque realization.
Some simple algebra allows obtaining the transfer functions of the two influence channels in the torque-driven case as follows:

$$H_{td}^{v}(s) = \frac{\Delta\Omega_{sim}(s)}{\Delta v(s)} = \frac{K_{\Gamma v} \cdot H_0^{\Gamma}(s)}{J_{sim}s + H_l(s) + H_0^{\Gamma}(s)\left((J_h - J_{sim})s - K_{\Gamma\Omega}\right)} \tag{11}$$

$$H_{td}^{\Gamma_G}(s) = \frac{\Delta\Omega_{sim}(s)}{\Delta\Gamma_G(s)} = \frac{1}{J_{sim}s + H_0^{\Gamma}(s)\left((J_h - J_{sim})s - K_{\Gamma\Omega}\right)} \tag{12}$$

A good quality of the real-time replication is obtained when the simulator's transfer functions H_{td}^{v} and $H_{td}^{\Gamma_G}$ approach the analogous genuine WECS's transfer functions, H_{wt}^{v}

and $H_{wt}^{\Gamma_G}$ respectively. This reflects the replication of the WECS rotational speed, Ω_h, by the simulator's rotational speed, Ω_{sim}. By comparing relations (9) and (11), respectively (10) and (12), one can remark that it is sufficient that the transfer function H_0^{Γ} to approach 1 in order to have both $H_{td}^{v} \to H_{wt}^{v}$ and $H_{td}^{\Gamma_G} \to H_{wt}^{\Gamma_G}$, *i.e.*, good real-time replication of the WECS's input-output behaviour. The requirement $H_0^{\Gamma} \to 1$ means to ensure a very fast dynamic response of the torque loop, which is realistically achievable.

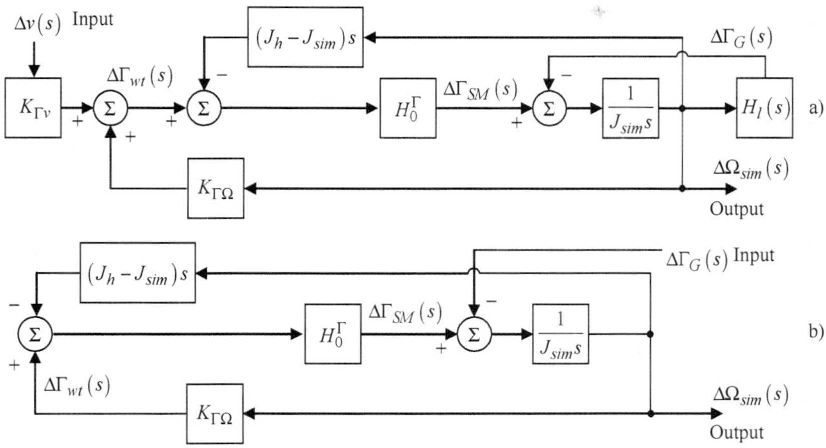

Fig. 9. Linearized model of the real-time simulation diagram in the torque-driven case: a) wind speed to simulator's rotational speed transfer; b) electromagnetic torque to simulator's rotational speed transfer

As regards the speed-driven case, one can see in Figure 10 the linearized models of the real-time simulation diagram for both input-output channels, *i.e.*, from the wind speed to the simulator's rotational speed (Figure 10a) and from the electromagnetic torque to the simulator's rotational speed (Figure 10b). The associated transfer functions will be denoted by H_{sd}^{v} and $H_{sd}^{\Gamma_G}$ respectively. Notation R_{Ω} corresponds to the transfer function of the rotational speed controller, whereas H_0^{Γ} keeps the meaning of transfer function of the servomotor torque realization.

Supposing that the requirement $H_0^{\Gamma} \to 1$ is fulfilled, after some calculations, the transfer functions of the two influence channels in the speed-driven case result as follows:

$$H_{sd}^{v}(s) = \frac{\Delta\Omega_{sim}(s)}{\Delta v(s)} = \frac{K_{\Gamma v}}{\left(\dfrac{J_{sim}s + H_l(s)}{R_{\Omega}} + 1\right)(J_h s - K_{\Gamma\Omega}) + H_l(s)} \tag{13}$$

$$H_{sd}^{\Gamma_G}(s) = \frac{\Delta\Omega_{sim}(s)}{\Delta\Gamma_G(s)} = -\frac{1}{\dfrac{J_{sim}}{R_{\Omega}}s + 1}\left(\frac{1}{J_h s - K_{\Gamma\Omega}} + \frac{1}{R_{\Omega}}\right) \tag{14}$$

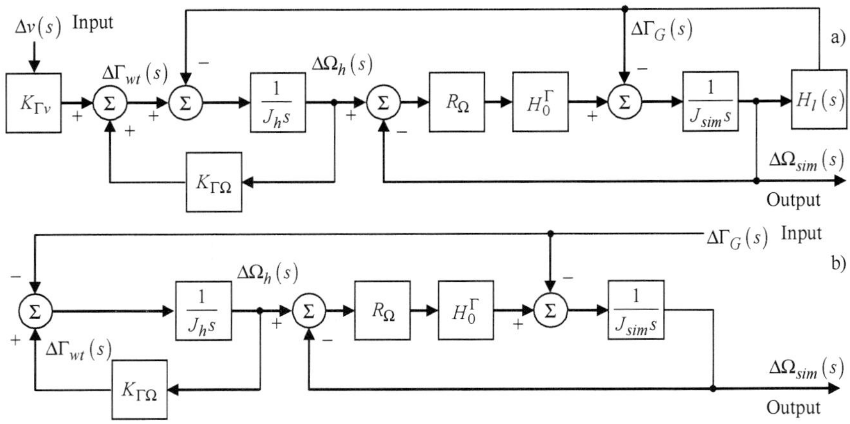

Fig. 10. Linearized model of the real-time simulation diagram in the speed-driven case: a) wind speed to simulator's rotational speed transfer; b) electromagnetic torque to simulator's rotational speed transfer

Following an inference analogous to the torque-driven case, one can remark that it is sufficient for the transfer function of the rotational speed controller to have a high gain in order to obtain a satisfactory input-output behaviour replication. Indeed, according to relations (13) and (14) compared with (9) and (10) respectively, if $R_\Omega \to \infty$, then both $H^v_{sd} \to H^v_{wt}$ and $H^{\Gamma_G}_{sd} \to H^{\Gamma_G}_{wt}$.
Note that a similar approach can be developed for the case when the mechanical transmission consists of a flexible drive train.
As a conclusion, the simulation error is generally expressed by the difference of the rotational speeds between the WECS and the real-time simulator under the same wind velocity sequence and the same control input sequence respectively. Thus, the errors can be frequency-domain characterized by the following relation:

$$H^{v,\Gamma_G}_{\varepsilon}(s) = H^{v,\Gamma_G}_{wt}(s) - H^{v,\Gamma_G}_{td,sd}(s) , \tag{15}$$

where, in the case of a rigid drive train, $H^{v,\Gamma_G}_{wt}(s)$ is given by relations (9) or (10) and $H^{v,\Gamma_G}_{td,sd}(s)$ stands for one of the transfer functions in relations (11), (12), (13) or (14). In general, the error $A^{v,\Gamma_G}_{\varepsilon}[\text{dB}] = 20 \cdot \log\left(\left|H^{v,\Gamma_G}_{\varepsilon}(j\omega)\right|\right)$ increases with the frequency in the EFT bandwidth, but for higher frequencies, the error is continuously decreasing due to the strictly causal nature of the system $H^{v,\Gamma_G}_{\varepsilon}(s)$ (Diop et al., 2000). The error in speed-driven simulators grows faster with the frequency than in the torque-driven simulator case. An improvement can be brought by the use of fast-dynamic servomotors.
A general expression of minimizing the simulation dynamic errors is by means of an integral criterion defined on a large time horizon. In this case, the minimization of the error power spectrum appears as well suited. The power spectra of the exogenous signals - wind velocity and electromagnetic torque - strongly influence the error. Figures 11 a) and b) respectively show the relative position of these spectra to the corresponding error frequency characteristic. The EFT bandwidth must be chosen such that it includes the wind velocity

spectrum and the WECS bandwidth, meanwhile ensuring small enough values of errors $A_{\varepsilon}^{v,\Gamma_G}(\omega_{EFT})$. Consequently, two integral criteria can be defined, each of which contains the error expressed by the absolute value of the quantity in relation (15) weighted by the power spectrum of the corresponding exogenous signal (Diop et al., 2000):

$$I^{v,\Gamma_G} = \frac{1}{\pi} \cdot \int_0^{\infty} \left| H_{wt}^{v,\Gamma_G}(j\omega) - H_{td,sd}^{v,\Gamma_G}(j\omega) \right| \cdot S_{vv,\Gamma_G\Gamma_G}(\omega) d\omega \tag{17}$$

The problem of simultaneously minimizing the index associated to the wind velocity and that of the electromagnetic torque is an issue under investigation.

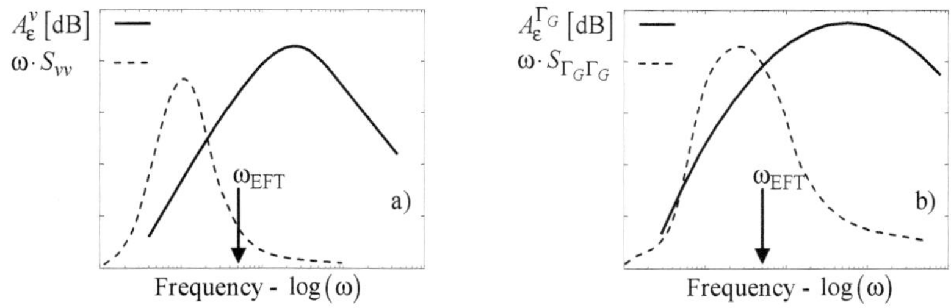

Fig. 11. Performance assessment of the simulation accuracy by frequency characterization of the model errors: a) in relation to the perturbation input (wind speed); b) in relation to the control input (electromagnetic torque)

3. Case study: PMSG-based WECS real-time physical simulation

The wind energy conversion topologies are various and one can note significant differences from a case to another (*e.g.*, the one using doubly-fed induction generator *vs.* the one with a PMSG). The control system structure depends in general on the particularities of the delivery point (if it is about a strong or weak grid, isolated load, *etc.*). The final scope of a WECS simulator being the testing of the control algorithms, it usually contains the necessary actuators exactly as they are in the real-world system. In this way, the IPS is fixed and the experimental rig can simulate a precisely identified class of WECS based on a given topology, whereas its software simulation capabilities are those which confer flexibility.

3.1 Description of the WECS to simulate

In this case study, the approached WECS has a horizontal-axis fixed-pitch 3-bladed rotor and is connected to an infinite power grid. The rotor is placed upwind and is oriented normally on the wind by means of a vane. The mechanical transmission consists of a rigid single-step speed multiplier. The power generation structure is based on a PMSG whose stator is grid-connected by means of a back-to-back power electronics converter. The decoupling from the grid is thus achieved and the variable-speed regime is made possible. For details concerning both the aerodynamic and the electrical features of the WECS, the reader is sent to the Appendix. As Figure 12 shows, two kinds of controllers ensure the transfer of the converted power to the electrical grid. The first one implements the PMSG

torque/speed control by acting on the generator-side converter, whereas the second one acts on the grid-side converter aiming at maintaining the DC-link voltage at an imposed level.
In the context of the methodology presented in the previous sections, the above described system is the original one, for which the control problem formulation is generally complex. The reason is that not only the output power must be controlled, *i.e.*, by the grid-side converter, but also the behaviour of the coupling turbine-generator. Therefore, the control objective depends on the WECS regime: in partial load one must ensure the maximisation of the power captured from the wind by using the variable-speed capability, while in full load the power limitation is mandatory (Burton et al., 2001).
The WECS modelling assumptions are related to its rated power. This case study is focused on a low-power WECS (less than ten kilowatts). Thus, as the blades length is relatively small, many effects due to rotor interaction with the wind stream or with the tower are negligible. Also, if the rotor's rotational speed is large enough *vs.* the lateral wind speed component dynamics, a scalar model can be used for the wind speed (Burton et al., 2001).

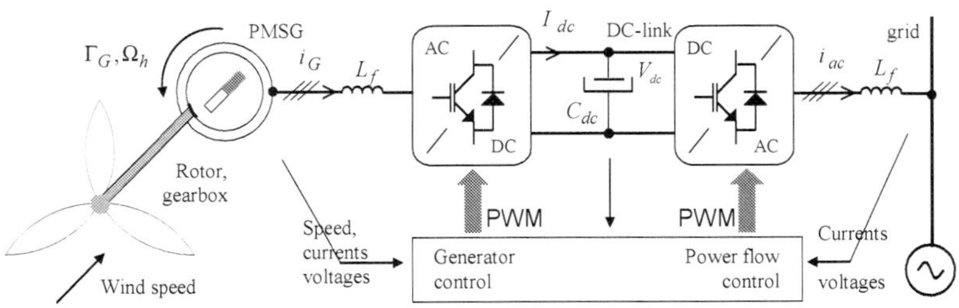

Fig. 12. Topology of the WECS to simulate

As general requirement, the simulator has to replicate the dynamic behaviour of various wind turbines belonging to the described WECS class under variable wind conditions and in different operating regimes. The rated power of the generator within the test rig is limited, therefore the simulator must offer the possibility of changing the scale factors in order to emulate wind turbines of various power sizes and also the possibility of switching between various control laws. Flexibility is also reflected by the facility of changing the parameters of the wind turbine and those of the wind site. Design of a friendly user interface enables the manipulation of the simulation parameters and experimental results.

3.2 Building of the RTPS

3.2.1 Splitting of the original system and identification of the interaction variables

Taking account of considerations stated in §2 the interaction EPS-IPS takes place at the high-speed (electrical generator) shaft, by means of mechanical interaction variables, namely the mechanical torque and the high-speed shaft rotational speed. Consequently, the subsystem under test (IPS) contains the electrical part of the WECS, which is taken as it is from the original system, *i.e.*, the PMSG, the AC/DC/AC converter and the grid. The aerodynamic interactions and mechanical transmission behaviours must be simulated, so be included into the subsystem to be simulated (EPS). Synthetic wind velocity will be used to excite the EPS model (already detailed in §2.3), thus ensuring controllable test conditions.

3.2.2 Effector configuration

The information exchange between the software simulator and the IPS is possible by means of the physical part of the RTPS. In this case study the chosen driving variable is the high-speed shaft mechanical torque, Γ_R, therefore the response variable will be the high-speed shaft rotational speed, Ω_h, which is measured by means of an encoder. The effector is implemented by a current-controlled permanent-magnet DC motor, rigidly coupled to the PMSG shaft, which is required to replicate the mechanical characteristics of the WECS high-speed shaft. The DC motor torque control ensures a very fast EFT dynamic. As the driving variable is a cause variable, the model implemented in the RTSS is a non-causal one. The DC motor torque, Γ_{SM}, is proportional with the armature current and must emulate the evolution of Γ_R, such that the rotational speed of the simulator, Ω_{sim}, to replicate the real dynamical evolution of the response variable, Ω_h. By using equations (3) and (5) one obtains the DC machine (EFT) reference torque:

$$\Gamma_{SM}^* = \eta \cdot \Gamma_{wt}/i - \left(J_{wt} \cdot \eta/i^2 - J_{sim}\right) \cdot \dot{\Omega}_{sim} , \tag{17}$$

where the wind torque, Γ_{wt}, results from relation (2) in which a synthesized wind speed, v, is used. Finally, the model to implement in the RTSS is composed of relations (2) and (17).

3.2.3 RTSS design

In Figure 13 one can remark the structure of the WECS simulator, where the IPS and the RTPS have been represented. One can note the two main components of the RTPS: the hardware part (the effector and the speed transducer) and the software part (the real-time software simulator, RTSS). Being implemented based on the MathWorks xPC Target® (MathWorks, 2008), the RTSS must perform the following tasks:

- to compute a wind speed sequence by using a pseudo-random model or a deterministic one (Nichita et al., 2002), and to read the value of the response variable, Ω_h,
- to implement the EPS model - relations (2) and (17) - and
- to compute the reference value of the driving variable, Γ_{SM}^*, which is sent to the EFT.

These are coded within a MATLAB®/Simulink® diagram; this is compiled using Real-Time Workshop and C compiler on the host PC and downloaded on the target PC via TCP/IP.

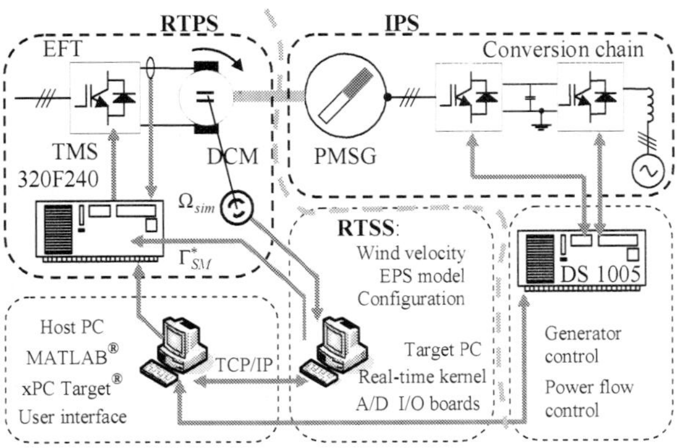

Fig. 13. RTPS and IPS identified on the schematic of the WECS simulator

Further, the target PC runs the real-time application and exchanges information with the hardware part of the RTPS by means of a PCI I/O interface (PCI–DAS1200/JR and PCIM–DDA 06/16 boards - Measurement Computing, 2008). The driving variable, Γ_{SM}, is implemented physically by using the DC machine, namely based upon the proportionality between the torque and the armature current. Thus, a current tracking loop is designed, whose components can be viewed in Figure 13: a current transducer, a numerical PID controller supported by a fixed-point DSP, TMS320F240, and an actuator composed of a 4-quadrant PWM IGBT chopper connected with a diode rectifier. The settling time of the DC-machine current is 4 ms, corresponding to a sufficiently fast EFT dynamic. For more details about the RTPS specification the reader is sent to the Appendix.

3.3 Building of the IPS

3.3.1 PMSG control

In the case of grid-connected WECS, the wind turbine generator control objective is to maximize the output power in partial load and to limit it at the rated value in full load. This regime is obtained by variable-speed operation, *i.e.*, by imposing a suitable generator rotational speed reference, depending on the turbine operating regime: in partial load this reference is outputted by a MPPT algorithm, while in full load it is generally given by a power regulation controller. Thus, the power control is entirely achieved by means of the generator control. A control structure called self-controlled synchronous machine is built for the PMSG vector control (Bose, 2001). It employs an AC-DC converter and an absolute position encoder. By imposing the PMSG stator voltages, the output current and hence the machine torque is controlled by separately controlling the Park current components, i_{sd} and i_{sq}. On each d and q channels anti-windup PI controllers are used, which are tuned by the well-known modulus criterion, thus ensuring an optimal trade-off between the settling time and the overshoot of the closed-loop step response (Åström & Hägglund, 1995). The PMSG speed control relies on employing an outer loop over the q channel control loop, which is based on a PI controller - designed according to the symmetrical optimum criterion - outputting the electromagnetic torque reference, hence the i_{sq} reference value.

3.3.2 Control of the power transfer to the grid

The DC-to-AC power transfer control is ensured by the grid-side converter (see Figure 12). The plant is composed of a DC-link bus, a DC-AC converter and the electrical grid. The 3-phased currents are fed into the mains while controlling the DC-link voltage at a constant value. If the considered WECS is part of a wind farm, of high impedance, it must also be able to supply reactive power to the grid in order to compensate voltage variations at the point of common coupling. Two current control loops are built for the q and d Park components respectively; an outer voltage control loop is in charge with regulating the DC-link voltage, V_{dc}, at an imposed value (Gombert et al., 2006). The DC-link voltage controller provides the i^*_{acq} reference, while a nonzero value of i^*_{acd} corresponds to reactive power to be supplied to the grid. The inner loops output the grid voltage references in the (d,q) frame, which are further transformed into a 3-phased system and applied to the inverter. These loops are all based upon anti-windup PI controllers which are tuned according to the modulus optimum criterion. A software 3-phased phase-locked loop is used in order to provide the grid electrical angle (Kaura & Blasko, 1997; Rabelo & Hofmann, 2002).

3.3.3 Hardware and software support of the control loops within the IPS

The control loops described above are supported by some widely-used hardware: three-leg IGBT power converters and associated gate drivers, current and voltage sensors, LC filters, incremental encoder and electrical connection equipment (Figure 12). As regards the software component of the control loops, this is supported by a dSPACE modular hardware around the DS 1005 PPC board (dSPACE, 2008), whose I/O system allows information being exchanged in both logic and analogic form. The program on DS 1005 runs two tasks:

- a *main (interruption) task*, with the highest priority, in charge with real-time operations;
- a *background task*, which updates the DSP memory by means of an user interface.

Simulation diagrams implementing the two tasks have been built in MATLAB®/Simulink® on the dSPACE host PC computer, then transferred to the target processor by using the Real Time Workshop and C compiler. The diagrams allow the main parameters being modified on-line. A ControlDesk® panel and an oscilloscope allow visualizing the interest variables.

3.4 Errors analysis

Figure 14 displays information about the accuracy of the simulation results provided by the real-time simulator and allows the stability of the simulation being assessed. Thus, the simulation errors are evaluated over a sufficiently large frequency domain (three decades), which must include the spectra of the exogenous signals, according to the method presented in §2.6. The measured frequency response in rotational speed of the real-time simulator is compared with its "should-be" value, *i.e.*, that of the model, with respect to both exogenous signals, the wind speed (frequency-bounded stochastic disturbance) and the electromagnetic torque (control input). Variable-frequency sinusoidal variations of the exogenous signals around a typical steady-state point were applied to the simulator in order to obtain its frequency response. An operating point in the vicinity of the maximum power point was chosen, which corresponds to an average wind speed of 8 m/s, electromagnetic torque –5.4 N·m and rotational speed 245 rad/s. For sake of correctness, the model of the rotational speed transducer effectively used on the test rig has been included in the WECS model.

By analyzing the simulator's frequency response with respect to the wind speed, along with a typical spectral density of the wind turbulence - exhibiting most of energy between 0.1 and 1 rad/s (Burton et al., 2001) - one can remark that the simulation errors are reasonably small for almost the whole frequency range considered, which totally includes the wind speed spectrum (Figure 14a). Therefore, the simulator replicates almost exactly the original system subject to the wind speed variations.

The same type of results has been obtained with respect to the electromagnetic torque - see Figure 14b). In this case the amplitude errors are larger, but however almost constant (3 dB) over the whole frequency range of interest, therefore easy to compensate by control action if needed. The electromagnetic torque has a controllable dynamic, essentially following the dynamic of the main disturbance, *i.e.*, of the wind turbulence, except the high-frequency variations for reasons of minimizing the fatigue loads (Burton et al., 2001). The error evaluation is therefore relevant in almost the same frequency range as in the case of wind turbulence. Both sets of results exhibit increases of the error at high frequency, but this is not critical as the disturbances have low content of high-frequency spectra.

Concerning the stability of the real-time simulation schemes implemented, it can be verified if the emulation dynamic is sufficiently fast with respect to the dynamics to be emulated. The emulation dynamic is expressed by the settling time of the effector. In this case study, the effector responds in about 4 ms, in other words, it has a bandwidth of about 1000 rad/s, which largely includes the dynamics to be emulated (Figure 14). In conclusion, the delays introduced by the emulation in the frequency range of interest can reasonably be neglected.

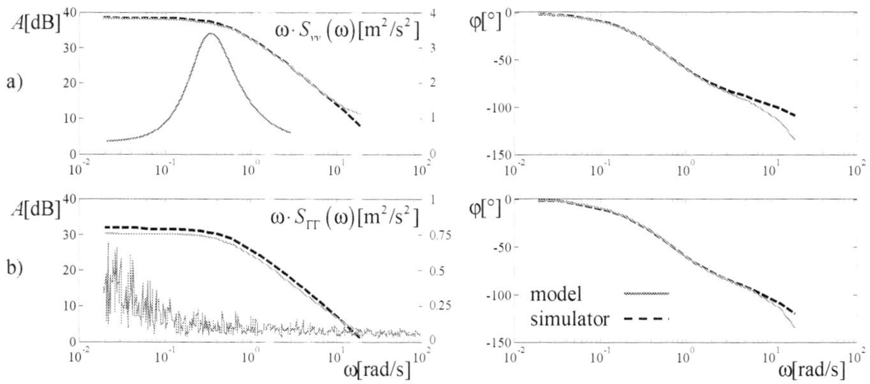

Fig. 14. Measured frequency responses - amplitude and phase lag - of the simulator in comparison with those of the WECS model: a) wind velocity to rotational speed transfer, b) electromagnetic torque to rotational speed transfer

3.5 Performance evaluation

3.5.1 Replication of the main operating regimes: partial load and full load

In the partial-load regime - which is in this case study delimited by the rated wind speed 11 m/s - the output power depends proportionally on the wind velocity cubed and the rotational speed varies proportionally with the wind velocity, whereas the aerodynamic efficiency is at its maximum value. The full-load regime is characterized by the wind velocity going beyond the rated value. In this regime, the rotational speed is adjusted such that to limit the output power at rated; meanwhile, the power coefficient significantly decreases with the wind speed. Figure 15 presents the replication of the steady-state characteristics of the studied WECS, where the two main operating regimes are identified.

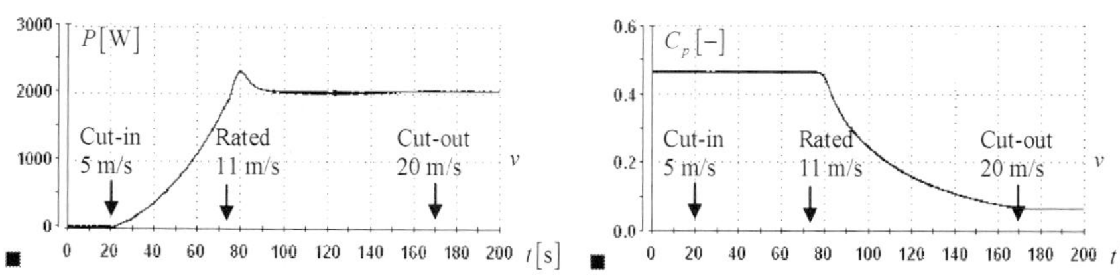

Fig. 15. WECS steady-state characteristics, as replicated by the real-time simulator: a) provided power *vs.* wind speed; b) power coefficient *vs.* wind speed

The two regimes are respectively managed by two PI controllers providing the rotational speed references. The aim of the partial-load controller is to maintain the power coefficient at its maximum value based on the wind velocity and rotational speed measures. At the

rated wind velocity, the control is switched to the full-load controller, which aims at regulating the WECS output power at its rated value by changing the rotational speed in response to the measured active power fed to the grid. The reader should note that these are not the only control designs that can be envisaged for a WECS (Munteanu et al., 2008a).
Figure 16 shows how the real-time simulator replicates the WECS partial-load operation when the average wind speed is around 8 m/s. The oscilloscope captures show: a) variable-speed operation under variable wind velocity with medium turbulence intensity of I=0.15, obtained by using the von Karman spectrum in the IEC standard (Nichita et al., 2002); b) behaviour at trapeze-shaped wind gusts under variable-speed operation. Figure 16a) shows that the active power supplied to the grid - negative by convention - depends obviously on the wind velocity. In variable-speed operation the wind torque contains dynamic elements due to the rotational speed variations (Figure 16b), whereas the wind velocity variations are compensated such that λ and C_p are maintained around their optimal values.

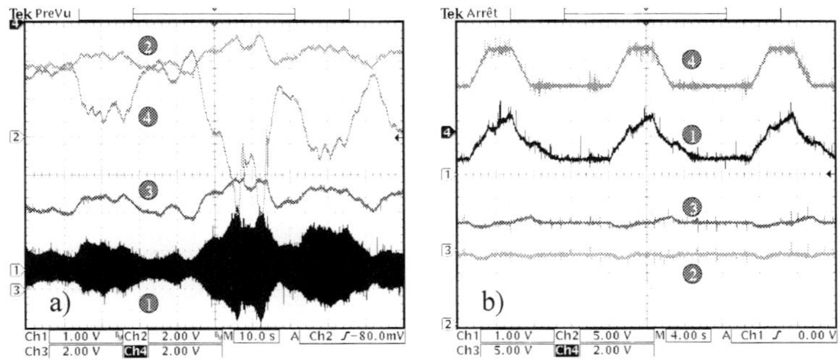

Ch1, i_{ac1}, 5 A/V Ch2, v, 2 m/s/V Ch1, Γ_{SM}, 10 Nm/V Ch2, C_p, 0.05 /V
Ch3, Ω_h, 50 rad/s/V Ch4, P, 200 W/V Ch3, λ, 2 /V Ch4, v, 2 m/s/V

Fig. 16. Real-time replication of the WECS partial-load operation: a) under variable wind speed sequence; b) under trapeze-shaped wind gusts

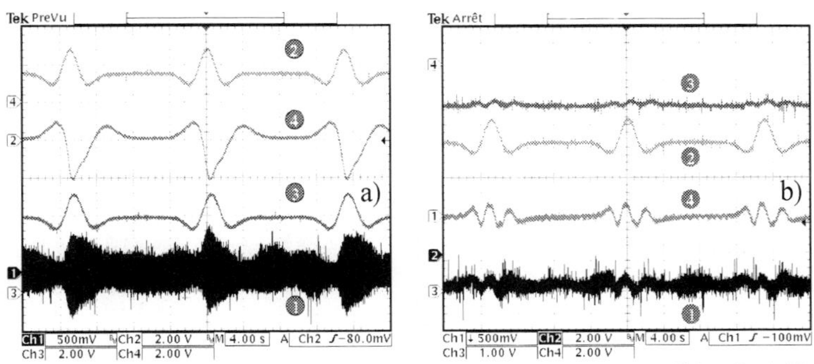

Ch1, i_{ac1}, 5 A/V Ch2, v, 2 m/s/V Ch1, I_{dc}, 5 A/V Ch2, v, 2 m/s/V
Ch3, Ω_h, 50 rad/s/V Ch4, P, 200 W/V Ch3, Ω_h, 50 rad/s/V Ch4, P, 250 W/V

Fig. 17. Real-time replication of the WECS operation under extreme-operating wind gusts: a) in partial load; b) in full load

Globally speaking, the real-time simulations show realistic performance under both turbulent wind and extreme-operating gusts.
Figure 17 contains results concerning the real-time simulation of the WECS behaviour at bell-shaped, *i.e.*, extreme-operating, wind gusts (Burton et al., 2001), under variable-speed operation in partial-load (Figure 17a) and in full-load regime respectively (Figure 17b). The full-load controller acts on the rotational speed, Ω_h, aiming at limiting the active power at its rated value. Thus, when the wind velocity increases, Ω_h is decreased in order to reduce the power coefficient as the turbine rotor enters in deeper stall. The output power, the DC-link power and the mean value of I_{dc} are kept constant irrespective of the wind variations.

3.5.2 Grid power transfer

The results presented above have been obtained when reactive power generation is not required, *i.e.*, the reactive power reference is zero. The evolution of the main variables involved in the real-time simulation of the WECS-grid interaction when imposing a nonzero level of reactive power can be seen in Figures 18 and 19.

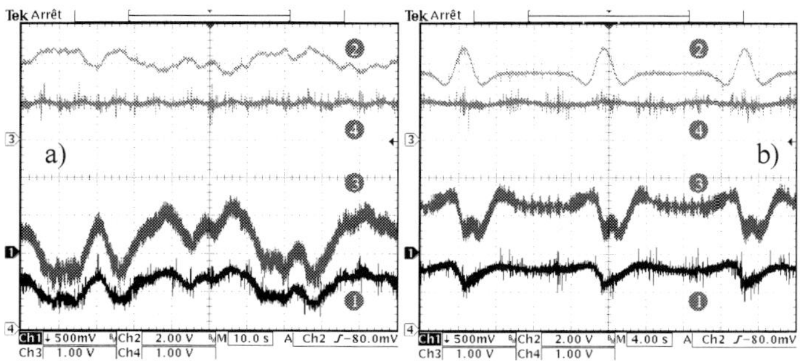

Ch1, I_{dc}, 5 A/V Ch2, v, 2 m/s/V
Ch3, i_{sq}, 5 A/V Ch4, V_{dc}, 75 V/V

Ch2, Ch3 same zero

Fig. 18. RTPS - power grid interaction: a) under variable wind speed; b) under wind gusts

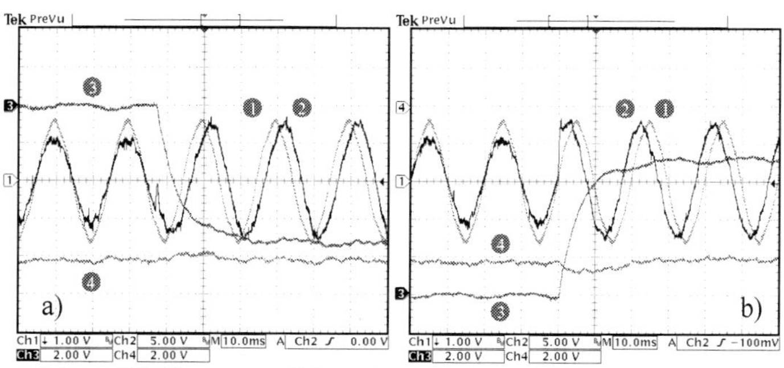

Ch1, v_{ac1}, 25 V/V Ch2, i_{ac1}, 5 A/V
Ch3, Q, 200 VAR/V Ch4, P, 200 W/V

a, b: Ch1, Ch2 same zero
a: Ch3, Ch4 same zero

Fig. 19. RTPS - power grid interaction: a) reactive power injection into the grid; b) reactive power absorption from the grid

Results presented in Figures 18 a) and b) correspond to conditions of variable wind speed and gusts of wind speed respectively. The DC-link current, I_{dc}, varies with the wind, as the active power variation follows the wind speed variation and the DC-link voltage is maintained almost constant. The variations of the q-component of the PMSG stator current are proportional with the electromagnetic torque. Figures 19 a) and b) display the behaviour of the grid electrical variables at the point of coupling, v_{ac1} and i_{ac1}, in both cases, of reactive power injection into and, respectively, absorption from the grid. The phase lag between voltage and current is modified accordingly. The reactive power amount corresponds to imposing a nonzero reference of i_{acd}, yielding a larger value of i_{ac1}.

4. Conclusion

This chapter has presented the essentials concerning the building of WECS physical simulators. It has tried to give answer to some basic practical questions. First, why using and how to build WECS simulators. Then, which elements to employ in their construction and how close can behave a physical simulator with respect to the original WECS. Last but not least, an example embedding a complete generation chain has been approached, for a better orientation of the reader throughout the complex challenge which is the WECS physical simulator building. The replication of wind turbines behaviour in a laboratory-controlled environment is done by closed-loop connecting hardware and software components in a hardware-in-the loop structure. The management of the digital system and the control of the electromechanical components are key issues in obtaining good simulator performances. In short, this complex system is built in order to replicate the turbine rotor behaviour, while subjected to controllable wind velocities, whereas the electrical conversion chain remains exactly as it is (at a certain scale) in the real wind power system.

Even if the subject has been extensively approached in the literature, it has not been depleted yet of scientific interest. In this context, an in-depth simulation errors analysis may uncover interesting theoretical results. As concerns the practical aspects, the connection of such device with a physically-simulated microgrid (Gaztañaga et al., 2007) may boost its capabilities and enlarge its range of applications.

5. Appendix

List of acronyms

EFT	effector
EPS	emulated physical system
HAWT	horizontal-axis wind turbine
IPS	investigated physical system
(P)HIL	(power) hardware-in-the-loop
PMSG	permanent-magnet synchronous generator
RTPS	real-time physical simulator
RTSS	real-time software simulator
WECS	wind energy conversion system

WECS characteristics in the case study

- *aerodynamic features*: fixed-pitch 3-bladed HAWT turbine, rated output power 2 kW at v=11 m/s, cut-in wind speed 5 m/s, R=1.7 m, J_{wt}=1.5 kg·m^2, maximal power coefficient $C_{p_{max}} = 0.47$ at optimal tip speed ratio λ_{opt}=7.4, 6.5x rigid drive train;

- *electrical components*: PMSG Siemens 1FT6084-8SK71-1TGO, rated values of: power - 6 kW, speed - 3000 rpm, torque - 22 Nm, voltage - 135 V; inertia 0.0048 kg·m², 4 poles, $R_s = 173.77\ \text{m}\Omega$, $L_{sd} = 0.8524\ \text{mH}$, $L_{sq} = 0.9515\ \text{mH}$, $\Phi_f = 0.1112\ \text{Wb}$; IGBT Semikron SKM 50 GB 123D-based three-leg voltage source inverters, 5 kW, 10 kHz; Semikron SKHII22A gate drivers; DC-link V_{dc}=450 V, C_{dc}=2200 µF/800V; V_{ac}=127/230 V - 50Hz, L_f=350 $\mu H / 5A$.
- *RTPS features*: 1:1 simulator, DC machine Siemens 1G 5 106-0EH-3UV1, rated values of: power - 6.4 kW, speed - 3470 rpm, torque - 20.2 Nm, voltage - 310 V; inertia 0.02 kg·m², torque constant 0.87 Nm/A, electromechanical time constant 4.6 ms; 10 kW 3-phased bridge diode rectifier, 4-quadrant 5 kW chopper based on IGBT Semikron SKM 50 GB 123D; incremental optical encoder 1024 lines; sampling time T_s=100 μs.

6. References

Akhmatov, V. (2003). *Analysis of dynamic behaviour of electric power systems with large amount of wind power*, Ph.D. Thesis, Technical University of Denmark, ISBN 87-91184-18-5, Denmark

Akhmatov, V.; Knudsen, H. & Nielsen, A.H. (2000). Advanced simulation of windmills in the electric power supply. *International Journal of Electrical Power and Energy Systems*. 22, 6, 421-434, ISSN 0142-0615

Andreica, M.; Bacha, S., Roye, D., Munteanu, I., Bratcu, A.I. & Guiraud, J. (2009). Stand-alone Operation of Cross-flow Water Turbines. *Proceedings of IEEE International Conference on Industrial Technology - ICIT 2009*, pp. 463-468, ISBN 978-1-4244-3507-4, Churchill, Victoria, Australia, February 10-13 2009, IEEE Press

Åström, K.J. & Hägglund, T. (1995). *PID controllers: theory, design and tuning, 2nd edn.*, Instrument Society of America, ISBN 978-1556175169, U.S.A.

Battaioto, P.E.; Mantz, R.J. & Puleston, P.F. (1996). A wind turbine emulator based on a dual DSP processor system. *Control Engineering Practice*, 4, 9, 1261-1266, ISSN 0967-0661

Bose, B.K. (2001). *Modern power electronics and AC drives*, Prentice-Hall, ISBN 978-0130167439, Englewood Cliffs, NJ, U.S.A.

Burton, T.; Sharpe, D., Jenkins, N. & Bossanyi, E. (2001). *Wind energy handbook*, John Wiley & Sons, ISBN 978-0471489979, New-York

Camblong, H.; Martinez de Alegria, I., Rodriguez, M. & Abad, G. (2006). Experimental evaluation of wind turbines maximum power point tracking controllers. *Energy Conversion and Management*, 47, 18-19, 2846-2858, ISSN 0196-8904

Cárdenas, R.; Asher, G.M., Ray, W.F. & Peña, R. (1996). Power limitation in variable speed wind turbines with fixed pitch angle. *Proceedings of International Conference on Opportunities and Advances in International Electric Power Generation*, pp. 44-48, ISBN 0-85296-655-5, Durham, U.K., March 18-20, IEE, London

Cárdenas, R.; Peña, R., Asher, G.M. & Clare, J.C. (2001). Experimental emulation of wind turbines and flywheels for wind energy applications. *Proceedings of 9th European Conference on Power Electronics and Applications - EPE 2001*, (CD-ROM), ISBN 90-75815-06-9, Graz, Austria

Diop, A.D.; Nichita, C., Belhache, J.J., Dakyo, B. & Ceangă, E. (1999). Modelling variable pitch HAWT characteristics for a real time wind turbine simulator. *Wind Engineering*, 23, 4, 225-243, ISSN 1991-8763

Diop, A.D.; Nichita, C., Belhache, J.J., Dakyo, B. & Ceangă, E. (2000). Error evaluation for models of real time wind turbine simulators. *Wind Engineering*, 24, 3, 203-221, ISSN 1991-8763

dSPACE (2008). *DS 1005 PPC Board Manual*. Available: http://www.dspace.de/ww/en/inc/home/products/hw/modular_hardware_introduction/processor_boards/ds1005.cfm, October 2009

Enslin, J.H.R. & van Wyk, D. (1992). A study of a wind power converter with micro-computer based maximal power control utilizing an over-synchronous electronic Scherbius cascade. *Renewable Energy*, 2, 6, 551-562, ISSN 0960-1481

Gaztañaga, H.; Etxeberria-Otadui, I., Ocnaşu, D. & Bacha, S. (2007). Real-Time Analysis of the Transient Response Improvement of Fixed-Speed Wind Farms by Using a Reduced-Scale STATCOM Prototype. *IEEE Transactions on Power Systems*, 22, 2, 658-666, ISSN 0885-8950

Gombert, C.; Ocnasu, D., Bacha, S., Roye, D. & Besanger, Y. (2006). Test of a PWM controller using a Real-Time Digital Simulator. *International Review of Electrical Engineering*, CD-ROM ISSN 1827- 6679

Hanselmann, H. (1996). Hardware-in-the-loop simulation testing and its integration into a CACSD toolset. *Proceedings of the 1996 IEEE International Symposium on Computer-Aided Control System Design*, pp. 152-156, ISBN 0-7803-3032-3, Dearborn, MI, U.S.A., September 15-18 1996, IEEE Press

Kaura, V. & Blasko, V. (1997). Operation of a phase locked loop system under distorted utility conditions. *IEEE Transactions on Industry Applications*, 33, 1, 58-63, ISSN 0093-9994

Leithead, W.E.; Rogers, M.C.M., Connor, B., Pierik, J.T.E., Van Engelen, T.G. & O'Reilly, J. (1994). Design of a controller for a test-rig for a variable speed wind turbine. *Proceedings of the 3rd IEEE Conference on Control Applications, vol. 1*, pp. 239-244, ISBN 0-7803-1872-2, Glasgow, U.K., August 24-26 1994, IEEE Press

Mathworks (2008). *xPC Target User's Guide*. Available: www.mathworks.com/support/product/XP/productnews/xpc_target_ug_Nov_07_2003.pdf, October 2009

Measurement Computing (2008). *I/O Boards Manual*. Available: http://www.measurementcomputing.com, October 2009

Munteanu, I. (2006). *Contributions to the optimal control of wind energy conversion systems*, Ph.D. Thesis, "Dunărea de Jos" University of Galaţi, Romania

Munteanu, I.; Bratcu, A. I., Cutululis, N. A. & Ceangă, E. (2008a). *Optimal Control of Wind Energy Systems – Towards a Global Approach*. Springer-Verlag, ISBN 978-1-84800-079-7, London

Munteanu, I.; Cutululis, N.A., Bratcu, A.I. & Ceangă, E. (2005). Optimization of variable speed wind power systems based on a LQG approach. *Control Engineering Practice*, 13, 7, 903-912, ISSN 0967-0661

Munteanu, I.; Seddik, B., Bratcu, I.A., Guiraud, J. & Roye, D. (2008b). Energy-Reliability Optimization of Wind Energy Conversion Systems by Sliding Mode Control. *IEEE Transactions on Energy Conversion*, 23, 3, 975 -985, ISSN 0885-8969

Nichita, C.; Diop, A.D., Belhache, J.J., Dakyo, B. & Protin, L. (1998a). Control structures analysis for a real time wind system simulator. *Wind Engineering*, 22, 6, 275-286, ISSN 1991-8763

Nichita, C.; Luca, D., Dakyo, B. & Ceangă, E. (2002). Large band simulation of the wind speed for real time wind turbine simulators. *IEEE Transactions on Energy Conversion*, 17, 4, 523-529, ISSN 0885-8969

Rabelo, B. & Hofmann, W. (2002). DSP-based experimental rig with the doubly-fed induction generator for wind-turbines. *Proceedings of the 10th International Power Electronics and Motion Control Conference - EPE-PEMC 2002*, (CD-ROM), ISBN 953-184-046-6, Cavtat & Dubrovnik, Croatia, September 9-11 2002

Rabelo, B.; Hofmann, W. & Gluck, M. (2004). Emulation of the static and dynamic behaviour of a wind turbine with a DC-machine drive. *Proceedings of the 35th Power Electronics Specialists Conference - PESC '04*, vol. 3, pp. 2107-2112, ISBN 0-7803-8399-0, Aachen, Germany, June 20-25 2004, IEEE Press

Rodriguez-Amenedo, J.L.; Rodriguez-Garcia, F., Burgos, J.C., Chincilla, M., Arnalte, S. & Veganzones, C. (1998). Experimental rig to emulate wind turbines. *Proceedings of the ICEM Conference*, vol. 3, pp. 2033-2038, Istanbul, Turkey, September 2-4 1998

RTDS (2009). *RTDS 2009 Hardware Overview*. RTDS Technologies Inc. Available: http://www.rtds.com/hardware.htm, October 2009

RT-LAB (2009). Available: http://www.opal-rt.com/product/rt-lab-professional, October 2009

Steurer, M.; Li, H., Woodruff, S., Shi, K. & Zhang, D. (2004). Development of a unified design, test, and research platform for wind energy systems based on hardware-in-the-loop real time simulation. *Proceedings of the 35th Annual IEEE Power Electronics Specialists Conference*, pp. 3604-3608, ISBN 0-7803-8399-0, Aachen, Germany, June 20-25 2004, IEEE Press

Teodorescu, R. & Blaabjerg, F. (2004). Flexible control of small wind turbines with grid failure detection operating in stand alone and grid connected mode. *IEEE Transactions on Power Electronics*, 19, 5, 1323-1332, ISSN 0885-8993

Wilkie, J.; Leithead, W.E. & Anderson, C. (1990). Modelling of wind turbines by simple models. *Wind Engineering*, 14, 4, 247-274, ISSN 1991-8763

Wu, X.; Lentijo, S. & Monti, A. (2004). A novel interface for power-hardware-in-the-loop simulation. *Proceedings of the IEEE Workshop on Computers in Power Electronics*, pp. 178–182, ISBN 0-7803-8502-0, Urbana, Illinois, 15-18 August 2004, IEEE Press

Part B
The Variability of Wind Power

11

Variability and Predictability of Large-Scale Wind Energy in the Netherlands

A.J. Brand, M. Gibescu and W.W. de Boer
Energy research Centre of the Netherlands, Delft University of Technology & KEMA Netherlands

1. Introduction

This chapter presents in a national context energy balancing requirements due to the variability and the limited predictability of wind energy in the thermal energy system of the Netherlands. In addition options to reduce these requirements are discussed. To this end 7.8 GW of wind power capacity in a system with 35 GW of total capacity is considered. The balancing requirements due to the cross-border flow of wind energy (export of domestic wind energy or import of foreign wind energy) however are not covered as these require an international context (ETSO, 2008). In addition the potential benefits of an intra-day market are not explored.

This chapter is organized as follows. First, section 2 presents various scenarios for wind and other energy capacity in the Netherlands, and introduces the structure of the Dutch electricity market. Next, section 3 gives a short overview of studies which addressed balancing energy reduction options in the contexts of the electricity markets in Denmark, Germany and Spain. Section 4 continues with the modeling of wind variability and wind predictability and its relevance to wind energy integration. Sections 5 and 6 then present the balancing energy requirements due to wind variability and limited wind predictability. Subsequently, section 7 discusses options to reduce the extra balancing energy requirements, which options include short-term forecast updates, aggregation, pumped storage, compressed air energy storage, fast start-up units, inverse offshore pump accumulation system, and wind farm shut-down strategies. Finally, section 8 summarizes the results.

2. Energy scenarios and market structure

2.1 Synopsis

In order to study balancing energy requirements in the future in the Netherlands, various energy scenarios were developed. These are presented in section 2.2, with attention for wind energy production capacity (paragraph 2.2.1), total electricity production capacity (paragraph 2.2.2), and flexibility of production (paragraph 2.2.3). The future structure of the Dutch electricity market is presented in section 2.3. The material in this section has been published in greater detail in de Boer et al., 2007; Gibescu et al., 2008b; and Gibescu et al., 2009.

2.2 Energy scenarios

2.2.1 Wind energy capacity

Offshore wind energy growth scenarios were developed that are consistent with the renewable policy goals in the Netherlands over the period up to the year 2020. Based on these rough estimates, on the onshore wind farm placement in the year 2006, and on the pending applications for environmental permits for offshore wind farms, the most likely locations and installed capacities were chosen for the years 2010, 2015 and 2020. In addition, three offshore wind energy scenarios were created: Low, Basic and Advanced. Only one scenario was created for onshore wind installed capacity. The scenarios are summarized in table 1.

		Year	
	2010	2015	2020
Low Offshore	720	2010	3800
Basic Offshore	1180	3110	6030
Advanced Offshore	1520	4110	8000
Onshore	1750	1800	1800

Table 1. Scenarios for offshore and onshore wind capacity in MW in the Netherlands

The aim of the Dutch government (from the 2004 policy) was to have 20% of demand served with help of renewable energy in the year 2020. The scenario Advanced will cover this completely with wind energy (given capacity factors of 25% and 37% respectively for onshore and offshore). Since this is an optimistic view of wind energy growth, the Basic scenario is employed in this study.

The offshore locations of wind farms for the scenario Basic Offshore 2020 were derived from the requests for permits for wind farms in the North Sea as filed by early 2006.

2.2.2 Total electricity capacity

Scenarios for the total electricity capacity in the Netherlands were developed by considering the total production plant in the year 2005, and estimating the retirement and addition of plant by the years 2010-2015-2020. The resulting total capacity break-up for the year 2020 is shown in table 2.

	Production in the Netherlands for several scenarios					
	2005	**2020**				
Type of power production	KEMA database	basic scenario	gas scenario	coal scenario	high growth scenario	low growth scenario
	MW	MW	MW	MW	MW	MW
Gas motor	1.450	1.950	1.950	1.950	2.260	1.680
Gas turbine	890	1.200	1.200	1.200	1.390	1.040
STAG of Combi	11.690	17.470	18.920	15.570	19.950	15.310
Conventional: boiler + ST (gas)	2.100	360	360	360	360	360
Conventional: boiler + ST (coal)	4.180	5.630	4.180	7.530	6.510	4.850
Nuclear	450	450	450	450	450	450
Waste and biomass	390	520	520	520	610	450
Wind	390	7.800	7.800	7.800	10.400	4.800
Total production	**21.540**	**35.380**	**35.380**	**35.380**	**41.930**	**28.940**

Table 2. Installed power in the Netherlands for several growth scenarios in 2020

As to the conventional production, on basis of the current practice, it is assumed that power plants can operate at 150% in respect to the original design. Their capacity is expected to decrease from 21 GW in the year 2005 to 9 GW in the year 2020. In addition, it is expected that most of the coal fired power plants and gas-fired combined cycle plants are still operating in the year 2020.

As to new production capacity five scenarios - each covering the years 2010–2015–2020 - were set up: basic, gas, coal, high growth, and low growth. The following parameters were considered: economic growth (respectively 1, 2 and 3% per year), fuel mix (basic scenario with current gas-to-coal ratio 1.0:3.5, a gas-and-coal reign scenario), and intensity of wind energy (see section 2.2.1).

In the basic scenario the control capabilities will be dominated the Combined Heat and Power (CHP) plants because the major growth of the capacity will most probably come from these plant. Power plants build after the year 2000 have better control capabilities: ~ 8% of nominal power per minute for gas, and ~3%/min for coal. The range of power change capability for CHP plants is 50% or more.

In the other scenarios the control capabilities differ slightly. For the coal scenario the rate of power change capabilities will be somewhat lower and for the gas scenario it will slightly higher.

2.2.3 Flexibility of production

Flexibility of production is required in order to follow the expected wind power variations, and to compensate unexpected wind power variations. This warrants a certain margin and rate of change capability, primary for the Programme Responsible Parties (PRPs) and secondly for the Transmission System Operator (TSO). The flexibility of production is defined in terms of: rate of change of the total capacity, amount of regulating (i.e. spinning) power and reserve power, rate of change of the spinning reserve units, and start time of the remaining units that are not delivering power during the load following cycle. Most of these terms depend on the operating point in the load following cycle and on the types of power units operating in that operating point.

A maximal ramping capability of $8\%P_{nom}$/min is expected for gas-fired units and 3% for coal-fired units. In the year 2020 the morning shoulder (i.e. the difference between off-peak and peak load) is expected to cover approximately 10 GW with a maximal required ramp rate of 60 MW/min. The gas fired power units are expected to carry this ramping load. This implies that a minimum of 10 GW of gas-fired units have to be spinning. If they have an average rate of change of 4%/min, then 400 MW/min can become available. This is enough to handle the expected variability due to load.

2.3 Structure of the electricity market

In the Netherlands wind power has been fully integrated in the day-ahead and imbalance market structures since the year 2001, and this situation is not expected to change in the future. Market participants known as Programme Responsible Parties (PRPs), governing a portfolio consisting of both renewable and conventional energy resources, submit to the Transmission System Operator (TSO) balanced schedules for energy delivered to and absorbed from the system during a 15-minute interval known as Programme Time Unit (PTU). This arrangement provides some insulation from the full exposure to imbalance charges for the wind producer, as conventional units in the PRP's portfolio may act to correct energy programme deviations due to wind variability and limited predictability.

3. International experience

3.1 Overview

This section presents a short overview of studies on balancing energy reduction options in the contexts of the electricity markets in Denmark, Germany and Spain. Section 3.2 starts with a short survey of international experiences with instruments for balancing the variability and forecasting errors introduced by large-scale wind energy in a power system. The focus is on wind power forecast updates (paragraph 3.2.1), aggregation of wind power (paragraph 3.2.2), energy storage (paragraph 3.2.3), and wind farm control (paragraph 3.2.4). In addition, the design of balancing markets is addressed in subsection 3.3.

3.2 Technology

3.2.1 Wind power forecast updates

The quality of wind power forecasts significantly improves as the forecast horizon decreases (Lange and Focken, 2005). The state-of-the-art indicates that the capacity normalized root mean square error (cRMSE) may reach a minimum value of 2...3% for a lead time of 2 hours before delivery (Krauss et al., 2006). For example in Germany this significant improvement in the accuracy of wind power forecasts consequently allowed for a better commitment and dispatch of the other generation units (Krauss et al., 2006). By doing so, the reserves held for wind power were decreased and the resulting surplus power could be offered by the conventional units in for example the intra-day market. Also a more efficient use was made of the available ramping capabilities of different units.

3.2.2 Aggregation of wind power

Aggregation of wind power over a larger geographical area, apart from smoothing out variability, improves the quality of the forecast because of the partly uncorrelated character of the forecast errors (Lange & Focken, 2005; von Bremen et al., 2006). As a result, both the reserves held and the reserves actually applied in a control area are decreased. Balancing wind power across control areas is even more efficient (Krauss et al., 2006).

3.2.3 Energy storage

Due to the relatively high investment costs of large-scale energy storage technologies, storage has to be multi-functional and market-driven, rather than employed only in order to reduce imbalances resulting from wind energy.

In the Netherlands, several studies were devoted to cost-benefit analysis for large scale energy storage systems (Ummels et al., 2008; de Boer et al., 2007). In particular an energy storage system has been proposed that would provide the following functions (de Boer et al, 2007):

- Download capacity for wind power at night during high wind and light load periods;
- Download capacity at night for base-load units that cannot be switched off, coupled with additional production capacity during peak load;
- Extra production capacity during periods with cooling water discharge restrictions for conventional plants; and
- Primary action.

Section 7.4 describes the benefits of such a system when it is used to perform the first function.

3.2.4 Wind farm control

Although in a technical sense clustering of wind farms into a virtual power plant may provide benefits for active power management and reactive power control, it is not economically attractive to operate such a plant for power balancing if the market design penalizes curtailment, as shown in Germany (Wolff et al., 2006). However, occasional use of wind farms to provide downward regulating power may be attractive during certain periods, e.g. when the surplus price is negative.

3.3 Balancing market design

As to the market design for balancing services, there are major differences between various countries (Verhaegen et al., 2006), where each market design has an unique impact on how balancing is actually provided. For example, there are differences in the institutional environment where the responsibility for taking care of imbalances arising from wind power either is assigned to a system operator (Germany, Spain, and Denmark for onshore wind power) or to a market party (the Netherlands, United Kingdom and Denmark for offshore wind power). Also, differences exist in the rules of use and provision of balancing services. In the following a number of developments are listed.

In the past years progress has been made to increase the liquidity of intra-day markets. Gate closure times of about one hour ahead of delivery (such as in the Netherlands) are sufficient to increase the accuracy of wind energy predictions to an acceptable level. This is in addition to the single-buyer balancing market, which is operated by the Transmission System Operator (TSO).

Power systems with dual imbalance pricing are problematic for wind energy due to the high penalties imposed, e.g. in the United Kingdom. To minimize imbalance costs, market parties should aggregate their production portfolios (Gibescu et al., 2008a).

If market parties employ wind power forecasts without being made responsible for balancing, their aim would be to optimize financial gains rather than to minimize their imbalance. This is why in such cases aggregated wind power forecasts have to be managed by the TSO.

There is a clear trend in Europe towards more cross-border balancing, which certainly promises advantages for wind power (Verhaegen et al., 2006). Balancing geographically larger control areas will provide benefits for wind power, not only because of overall decreased variability and increased predictability, but also because of larger market volumes and larger balancing resources.

Finally it is noted that in all European countries the present organization of support schemes - which to date remains the major source of revenues for wind power producers - discourages the use of curtailment as a balancing instrument. Controlling the power output of wind farms must therefore be considered as an option from a power system operations perspective, since the opportunity loss by curtailment is significant.

4. Wind modeling aspects of wind energy integration

4.1 Outline

This section presents the modeling of wind variability and wind predictability and its relevance to wind energy integration. First, section 4.2 critically reviews existing methods to generate wind power time series for integration studies. Next, the sections 4.3 and 4.4 present a new method to create measured respectively forecasted wind speed time series. And finally in section 4.5 the method to create wind power time series is explained. The methods described in the sections 4.3-4.5 were developed for this purpose by the authors (Brand, 2006; Gibescu et al., 2006; Gibescu et al., 2009).

4.2 Existing methods

A wind power integration study requires wind power time series originating from wind speed time series, where wind speed comprises measured and forecasted data. In addition the spatial correlation of wind speeds between sites must be taken into account because, as wind farms will be concentrated in areas with favorable wind conditions, their outputs will be strongly correlated. The resulting cross-correlations are essential when assessing the system-wide variability and predictability in large-scale wind production, and in turn affect the system requirements for reserve and regulation energy.

Three different methods to generate wind time series can be identified, namely by using actually measured wind speed time series, by using synthesized wind time series data (Doherty & O'Malley, 2005), or by using a combination of measured and synthesized wind speed time series (Giebel, 2000; Holttinen, 2005; Norgard et al., 2004). Valued against the requirements for integration studies these methods fall short for the creation of *both* realistic measured and forecasted wind power time series.

In order to correctly account for the spatial and temporal correlations of wind in an area, the method in section 4.3 derives the relevant statistical properties of the interpolated series from measured wind speeds. To this end assumptions are made only regarding the Markov property and the exponential decay of covariance with distance. In addition, this method uses 15-minute averaged wind speed in order to accurately model the balancing market in the Netherlands.

Two methods to generate wind forecasts can be identified, namely by using real wind forecasts (Lange & Focken, 2005) or by using synthesized wind forecasts (Norgard et al., 2004; Söder, 2004).

In order to correctly account for the limitations in a forecasting method and for the degree of uncertainty, in section 4.4 real wind forecasts are used. Unlike the alternative, this approach does not require assumptions on the distribution, correlation and increase of wind speed forecasting errors.

4.3 Measured wind speed

4.3.1 Historical wind data

Wind speed was modeled using historical wind data. To this end wind speed data sets were obtained from the Royal Dutch Meteorological Institute (KNMI). The data comprise 10-minute wind speed averages with a resolution of 0.1 m/s for 16 locations in the Netherlands and its coastal waters (six onshore, four coastal and six offshore; see figure 1) measured between 31 May 2004 and 1 June 2005. In addition, 10-minute wind speed standard deviations are available for the onshore locations and are estimated for the offshore locations. (The standard deviations are used in the height transformation in section 4.3.2.) The chosen time series reflects the spatial distribution of present and future installed wind power in the Netherlands.

4.3.2 Height transformation

Sensor height where wind speed was measured may differ between locations. The standard method to transform to hub height is to employ the logarithmic vertical wind speed profile

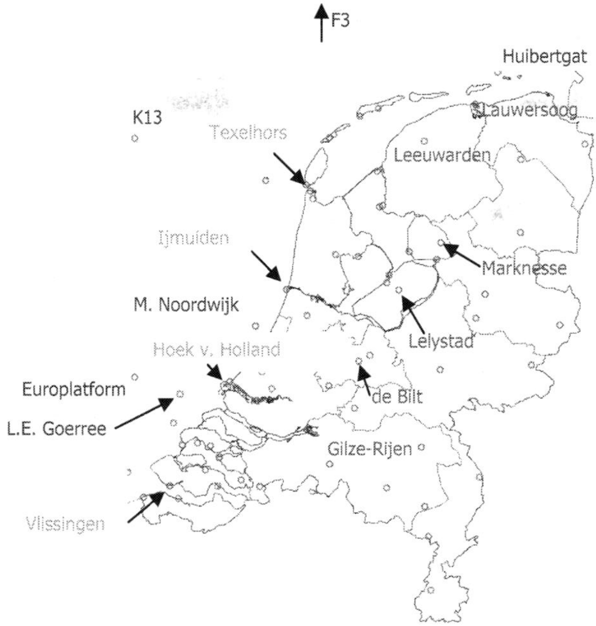

Fig. 1. Onshore, coastal and offshore wind speed measurement sites in this study

in combination with the surface roughness length (e.g. Walker & Jenkins, 1997). The local surface roughness length however is difficult to estimate. For this reason Brand, 2006, has eliminated this need. Instead, two location-dependent parameters are used: the friction velocity u_* and the average Monin-Obukhov length L_{esti}. The friction velocity is estimated from the 10-minute wind speed standard deviation which for most locations is available. If not, for an offshore location the friction velocity is estimated from the vertical wind speed profile. The Monin-Obukhov length is estimated by the average value that follows from the positive average heat flux that has been found over the North Sea and over the Netherlands, implying that the average vertical wind speed profile is stable (Brand & Hegberg, 2004).
Given the 10-minute average wind speed $\mu(z_s)$ and standard deviation $\sigma(z_s)$ at sensor height z_s, the estimates of the wind speed average and standard deviation at hub height z_h are:

$$\mu_{u,esti}\left(z_h\right)=\mu_u\left(z_s\right)+\sigma_u\left(z_s\right)\left(\ln\left(\frac{z_h}{z_s}\right)+5\frac{z_h-z_s}{L_{esti}}\right) \tag{1}$$

and

$$\sigma_{u,esti}\left(z_h\right)=\sigma_u\left(z_s\right), \tag{2}$$

where L_{esti} is the location-dependent average Monin-Obukhov length.
If only $\mu(z_s)$ is available, and provided that the location is offshore, the estimates of the wind speed average and standard deviation at hub height are

$$\mu_{u,esti}\left(z_h\right) = \mu_u\left(z_s\right) + 2.5u_*\left(\ln\left(\frac{z_h}{z_s}\right) + 5\frac{z_h - z_s}{L_{esti}}\right) \quad (3)$$

and

$$\sigma_{u,esti}\left(z_h\right) = 2.5u_*; \quad (4)$$

where u_* is determined from

$$\mu_u\left(z_s\right) - 2.5u_*\left(\ln\left(\frac{z_s g}{A u_*^2}\right) + 5\frac{z_s}{L_{esti}}\right) = 0, \quad (5)$$

and $g = 9.81\ m/s^2$ is the gravitational acceleration and $A = 0.011$ is Charnock's constant.

4.3.3 Averaging-time transformation

A transformation from 10 to 15-minute averages is required by the design of the Dutch balancing market and is accomplished as follows: If μ_k, μ_{k+1}, μ_{k+2} etc are the consecutive 10-minute wind speed averages, then m_k m_{k+1} etc. are the consecutive 15-minute wind speed averages:

$$m_{k,esti} = \frac{2\mu_{3(k-1)/2+1} + \mu_{3(k-1)/2+2}}{3} \text{ and } m_{k+1,esti} = \frac{\mu_{3(k-1)/2+2} + 2\mu_{3(k-1)/2+3}}{3}.$$

4.3.4 Interpolation

4.3.4a Introduction

This section describes how wind speed at given locations is sampled conditionally on the wind speed at measurement locations. To this end a multivariate Gaussian model is used, in combination with assumptions on the spatial and the temporal covariance structure. In addition, a variance-stabilizing transformation is used.

4.3.4b Approach and assumptions

Consider the natural logarithm W(x, t) of the wind speed at a location x and time t, where t = (d, k) is defined by the day of the year d and the time of day k. There are two reasons for taking the logarithm. First, there is a pronounced heteroscedasticity (i.e. increasing variance with the mean) in the wind speeds, which is stabilized by the log transformation (section 9.2 in Brockwell and Davis, 1991). Second, upon taking logarithms the (multivariate) normal case is reached, which allows one to make extensive use of conditioning.
Following Brockwell and Davis, 1991, a random vector $\underline{X}$ is considered which is distributed according to a multivariate normal distribution with mean vector $\underline{\mu}$ and covariance matrix $\underline{\underline{\Sigma}}$. Supposing that $\underline{X}$ is partitioned into two sub-vectors, where one corresponds to the sampled data and the other to the observed data, and, correspondingly, the mean vector and covariance matrix, then the following may be written:

$$\underline{X} = \begin{pmatrix} \underline{X}^{(1)} \\ \underline{X}^{(2)} \end{pmatrix} \text{ and } \underline{\mu} = \begin{pmatrix} \underline{\mu}^{(1)} \\ \underline{\mu}^{(2)} \end{pmatrix} \text{ with } \underline{\underline{\Sigma}} = \begin{pmatrix} \underline{\underline{\Sigma}}_{11} & \underline{\underline{\Sigma}}_{12} \\ \underline{\underline{\Sigma}}_{21} & \underline{\underline{\Sigma}}_{22} \end{pmatrix}. \quad (6)$$

If $\det(\Sigma_{22}) > 0$, then the conditional distribution of $\underline{X}^{(1)}$ given $\underline{X}^{(2)}$ is again multivariate normal, and the conditional mean and the conditional covariance matrix are:

$$\underline{\mu}^{(1)} + \underline{\Sigma}_{12}\underline{\Sigma}_{22}^{-1}\left(\underline{X}^{(2)} - \underline{\mu}^{(2)}\right) \quad \text{and} \quad \underline{\Sigma}_{11} - \underline{\Sigma}_{12}\underline{\Sigma}_{22}^{-1}\underline{\Sigma}_{21}. \tag{7}$$

As to the log wind speeds W(x, t) at location x and time t = (d, k), the following model is proposed:

$$W(x,t) = \mu(x,k) + \varepsilon(x,t), \tag{8}$$

where μ is a deterministic function representing the daily wind pattern by location and ε is a zero-mean random process representing the variations around the mean. Note that it has been assumed that μ depends on time only through the time of day k. In other words, the model does not include seasonal effects. (This assumption was checked and found to be reasonable in an analysis aimed at finding any other trend or periodic component, in particular a seasonal, in the 1-year data set.)

Figure 2 shows the average daily wind pattern for the 16 measurement locations. Since the lower curves correspond to onshore and the higher curves to offshore sites, the figure suggests that a daily effect is modeled which varies smoothly with geographical location. An onshore site is found to have a typical pattern with a maximum around midday, whereas an offshore site has a much flatter daily pattern, with a higher overall average. A coastal site falls in between.

The mean log wind speed μ(x, k) is estimated at all measurement locations by the daily averages shown in figure 2. Estimates for the locations of interest within the convex hull formed by the measurement sites were obtained by using linear spatial interpolation. On the other hand, for locations outside that hull, nearest neighbor interpolation was used. The results are shown as dotted lines in figure 2.

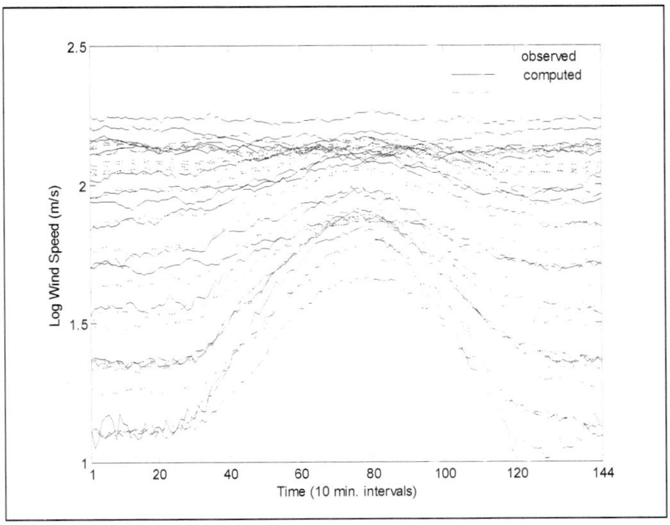

Fig. 2. Daily wind speed pattern for measured and interpolated sites

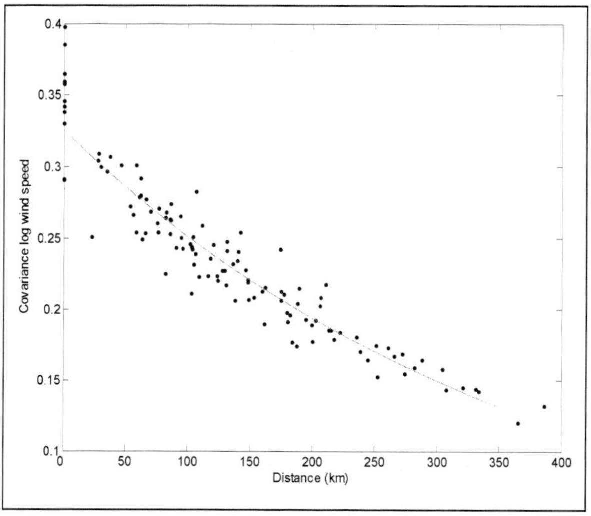

Fig. 3. Wind speed covariance versus site distance for 16 measurement sites

As to the model for the random part $\varepsilon(x, t)$, as explained above, a zero-mean, multivariate normal distribution is assumed for the log wind speeds minus the daily pattern. Figure 3 shows the sample covariance between the log wind speeds at all pairs of (measurement) locations versus the distance between them. From the displayed decay and the assumption that covariance vanishes at very large distances, it is reasonable to propose an exponential decay with distance:

$$\mathrm{Cov}\left(\varepsilon\left(x_i, t\right), \varepsilon\left(x_j, t\right)\right) = \alpha_0 \exp\left(-\beta\left\|x_i - x_j\right\|\right) \tag{9}$$

where $\|.\|$ denotes the Euclidean distance. To be able to sample wind speed time series, temporal dependence must be taken into account. Similar to equation (9), the following covariance is proposed:

$$\mathrm{Cov}\left(\varepsilon\left(x_i, t\right), \varepsilon\left(x_j, t-1\right)\right) = \alpha_1 \exp\left(-\beta\left\|x_i - x_j\right\|\right) \tag{10}$$

The parameters α_0, α_1 and β are jointly estimated by a least squares fit. The fit for α_0 and β is shown in figure 3, where α= 0.32 and $1/\beta$= 392.36 km. The latter term is known as the characteristic distance. By transforming the parameters of this decay fit from logarithmic to pure wind speeds, and by inspecting the correlation coefficients (i.e. covariance normalized by the product of the two standard deviations) between location pairs, a value of 610 km is obtained for the characteristic distance. This value is in line with the 723 km reported in Chapter 6 of Giebel, 2000, which is based on measurements from 60 locations spread throughout the European Union, and the 500 km reported in Landberg et al., 1997, and Holttinen, 2005, using Danish only and Scandinavian data, respectively. This suggests that these values are generic.

A final assumption is the Markov property for the sampled time series: it is assumed that conditionally on W(x,t-1), W(x,t) does not depend on W(x,t-2), W(x,t-3), etc. Consequently, it is not needed to specify the covariance between $W(x_i,t)$ and $W(x_j,s)$ when $s-t > 1$.
It should be noted that since the equations 9 and 10 do not depend on time, any daily or seasonal changes in the covariance structure are ignored. Such effects have been tried to identify, but it was found that they were not very large, and not particularly systematic; hence, they would not have a substantial effect on the time series that the method ultimately generates.

4.3.4c Interpolation scheme

The interpolation scheme is as follows. At each stage, a collection of normal random variables is conditionally sampled on some other normal random variables. The mean and the covariance structure of all random variables is fully described, and therefore the general theory from equations 6 can be used, where subset (1) denotes the unobserved wind speeds at time t, and subset (2) denotes both observed wind speeds at times t and t-1, and unobserved, but already interpolated values at time t-1.
Once the log wind speeds for the locations of interest are sampled, these are exponentiated to obtain the wind speeds. Of course, the time series produced in this way will reflect the assumptions that were made, but this does not mean that they will look like samples from the multivariate log-normal distribution. The method provides nothing more than linear interpolations of the measured time series, and so their Weibull character will be preserved to a great extent.
The effectiveness of the method is evaluated by using cross-validation: leaving one measurement location out of the data set and using the remaining n-1 locations to "re-create" it. First, it is verified that the method preserves the marginal Weibull parameters. As an illustration, figure 4 shows the histogram of the original data for the coastal location IJmuiden together with a Weibull fit of the original and the interpolated data. As expected, some smoothing has occurred in the interpolated data due to the weighted averaging, but not much. Second, it is verified whether or not the method reproduces the (auto-)covariance structure of the original data. Figure 5 shows the lag-one auto-correlations for the original and cross-validated data, with the straight line indicating a perfect match. Even though some over- and underestimation of the auto-covariances can be observed from figure 5, there does not seem to be any structural bias.

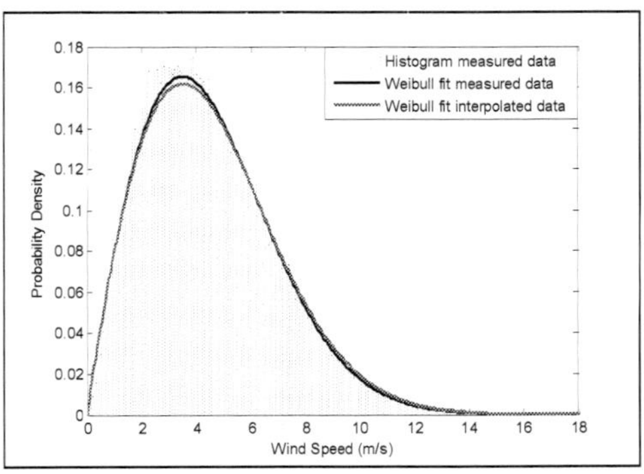

Fig. 4. Wind speed histogram and fit to Weibull distribution at the location IJmuiden

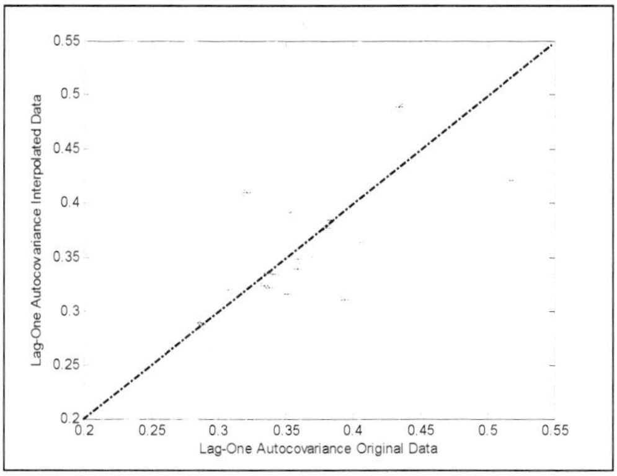

Fig. 5. Lag-one auto-covariance, original versus interpolated wind speeds

As to limitations of this method, it should be kept in mind that the interpolation weights are determined by the assumption of the exponential decay of the covariance with distance. As a consequence, if this decay does not hold, the covariance structure of the generated series will not be correct. In addition, the estimated time series are only as good as the input data allows. For instance, under more complex terrain, measured data at closer distances would be required to correctly track local changes in wind behavior.

4.4 Forecasted wind speed

The 15-minute average wind speed forecast time series are generated for locations where measurements are available. These forecasts originate from the wind power forecasting method AVDE (Brand and Kok, 2003); a physical forecasting method with an output statistics module. In an operational sense, AVDE is a post-processor to the high-resolution atmospheric model HiRLAM or any weather prediction model that delivers the required input data (two horizontal wind speed components, temperature and pressure in two vertical levels on a horizontal grid covering the sites to be considered) in the required format (GRIB). If wind speed and/or wind power realizations are available, the output statistics module of the AVDE can be used in order to compensate for systematic errors in the forecasts. The forecasts are meant to guide wind producers in a day-ahead market, and are completed at 12:00 the previous day, thus carrying an increasing delay of 12 to 36 hours. By employing a method similar to the one used for the spatial interpolation of wind speed measurements, appropriately correlated forecast error time series are generated for the wind farm locations. Since the variability of wind forecast errors over successive time intervals is not analyzed, it is assumed that, conditional on the forecast errors at the observed locations, the forecast errors at the computed locations at time t are independent of the errors experienced at time t-1.

Figure 6 presents the geographical locations of the seven wind speed forecast sites together with the projected offshore wind farm locations for the year 2020 and the current density of onshore wind energy capacity by province in the Netherlands.

Similar to the wind speeds, the forecast errors are modeled as the sum between a deterministic term, derived from the average daily pattern (figure 7), and a random term, which obeys a covariance matrix derived from the exponential fit presented in figure 8. Note

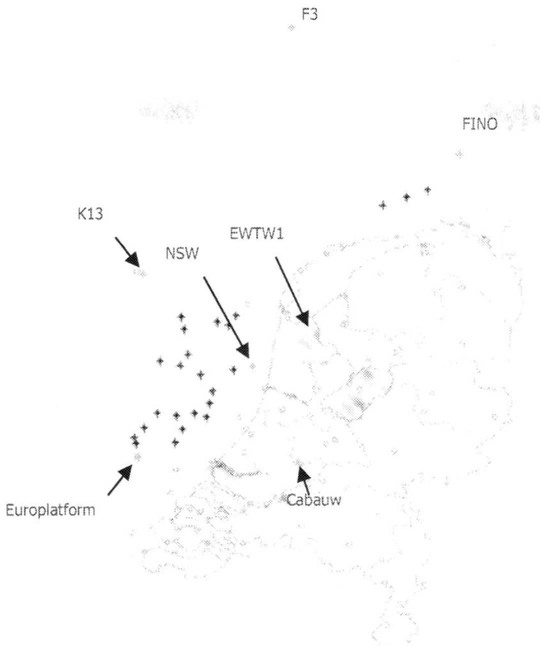

Fig. 6. Wind speed forecast sites (labeled), onshore (shaded grey) and offshore (black stars) wind farm sites for the Basic 2020 scenario

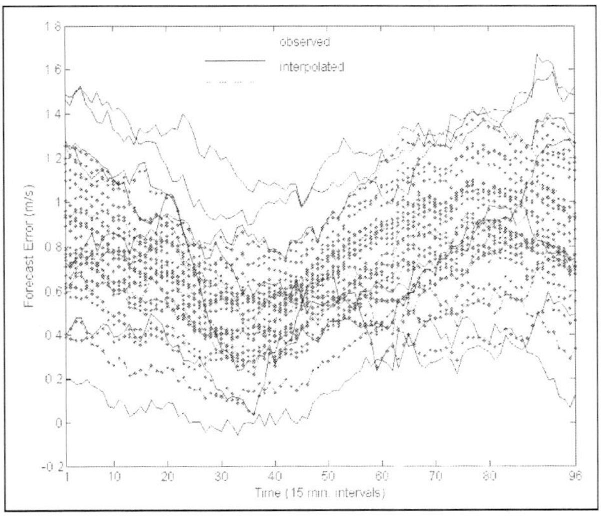

Fig. 7. Daily wind speed forecast error pattern for measured and interpolated sites

that the logarithmic transformation was not necessary here because the variance of the forecasting error does not significantly increase with its mean. In order to correctly take into account the changes in the covariance structure due to the look-ahead time, 24 × 4 = 96 separate exponential decay curves were fitted as shown in figure 8.

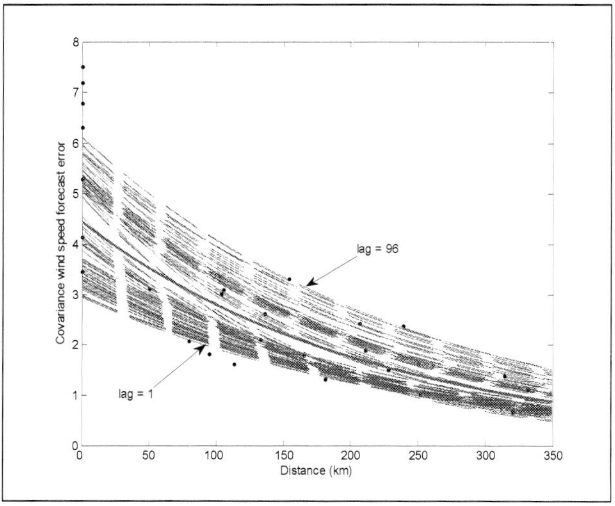

Fig. 8. Wind speed forecast error covariance versus distance for various forecast horizons

4.5 Wind power

4.5.1 Multi-turbine power curve

For each location wind power has been created using regionally averaged power curves, which depend on the area covered with wind turbines and the standard deviation of the wind speed distribution at the location. As the name suggests, regional averaging provides the average power of a set of wind turbines placed in an area where the wind climate is known, assuming the turbines do not affect each other. The multi-turbine curve is created by applying a Gaussian filter to a single-turbine power curve, and is not to be confused with a wind farm power curve, which brings the wind shadow of turbines into account.

Although inspired by and having the same effect as the Gaussian filter in the multi-turbine approach of Norgard and Holttinen, 2004, the standard deviation in the new filter correctly originates from the local wind climate alone. Unlike the Norgard–Holttinen method, the filter does not require estimating the turbulence intensity, which incidentally is a measure of variation in a 10-minute period in a given location rather than a measure of variation in the same 10-minute period at different locations. Nor does the method apply a moving block average to the wind speed time series with the time slot arbitrarily based on the local average wind speed.

Figure 9 shows an example of a multi-turbine power curve as constructed for an offshore wind farm of installed power 405 MW at a location where the standard deviation of the wind speed is 4.6 m/s. The width σ_F of the Gaussian filter is given by an estimate for the standard deviation that describes the regional variation of wind speeds at different locations in the same wind climate (appendix A in Gibescu et al., 2009)

$$\sigma_F = \sigma\sqrt{\frac{1}{2}\left(1-\exp\left(-\frac{d_{ave}}{D_{decay}}\right)\right)}, \quad (12)$$

where σ is the standard deviation of the wind speed distribution, d_{ave} is the average distance between the locations and D_{decay} is the characteristic distance of the decay of correlation (as estimated in section 3). If the individual locations are not known, as is the case in this study, an estimate for d_{ave} is (appendix B in Gibescu et al., 2009):

$$d_{ave} = \frac{2}{3}\sqrt{\frac{A}{\pi}\left(1+\frac{2}{\sqrt{M}}\right)}, \quad (13)$$

where A is the area of the region and M is the number of locations in that area. In this study, the area relates to a province for the onshore wind power and to an individual wind farm for the offshore wind power. The area of an individual farm is approximated by the area of a rectangle whose sides depend on the number of turbines, the rotor diameter and the spacing between turbines.

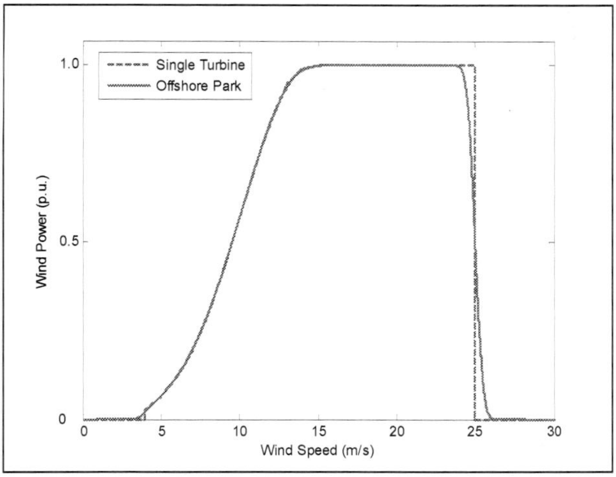

Fig. 9. Example of an aggregated power curve

The method to determine the regional variation of wind speeds at different locations in the same wind climate was verified by using the measured data introduced in section 4.3. The method to determine the multi-turbine power curve for a given area is still in need of verification data.

4.5.2 Aggregation levels

Aggregating the power of the individual wind farms at the system level gives a good initial estimate for the degree of variability and predictability that come with large-scale wind energy. It however ignores the real situation where wind power is integrated by several sub-levels, as owned and operated by the individual market parties. To that effect, seven PRPs are defined, each owning a unique combination of installed power and geographical spread of onshore and offshore wind farms, as described in table 3. For reasons of confidentiality, these parties have fictitious names; however, the installed power are consistent with the current and planned developments in the Netherlands.

PRP	Offshore (MW)	Onshore (MW)	Total (MW)
Anton	881	840	1721
Berta	1792	593	2385
Cesar	800	0	800
Dora	2520	140	2660
Emil	40	0	40
Friedrich	0	92	92
Gustav	0	135	135
System	6033	1800	7833

Table 3. Programme Responsible Parties (PRP) in the Basic 2020 scenario

5. Impact of extra variability due to wind

In this section the balancing energy requirements due to wind variability are presented for the scenario with 7.8 GW of installed wind power in the Netherlands in the year 2020.

Given the locations and installed power for future wind farms, the estimation method of the sections 4.3 and 4.4 is used in combination with the aggregated power curve of section 4.5 to compute the average wind power generated per 15-minute time interval for the duration of a year. By differentiating the wind power time series an estimate is obtained of the variability of aggregated power across 15-minute time intervals and above. This quantity and its sign are of interest because simultaneous load and wind variations are to be balanced by the remaining conventional generation units via the up- or down-ramping of their outputs.

Table 4 presents the 99.7% confidence intervals and the extreme values (smallest and largest) of the 15-minute, 30-minute, 1-hour and 6-hour variations at the system level. The sorted positive and negative variations in wind power over various time ranges are shown in figure 10. Based on the 99.7% confidence interval, the system-wide variations across 15-minute intervals are in the range of ±14% of the installed power for this scenario.

Table 5 shows the statistics of the 15-minute variations for each of the seven PRPs individually. These variations are in the range of ±12–26% of the power installed by the PRP, depending on the geographical spread of its locations. The collective requirement for balancing 15-minute variations becomes approximately ±16% of the system's installed capacity, which is 2% more than the requirement at the system level.

Time range	Minimum (MW)	Maximum (MW)	99.7%Conf.Int. (MW)
15 min	−2411	2883	−1090.8 to 1054.2
30 min	−2411	2883	−1252.9 to 1309.6
1 hour	−3133	3634	−1968.0 to 1846.0
6 hour	−7211	6790	−5157.8 to 5105.4

Table 4. Statistics of wind variability in the Basic 2020 scenario

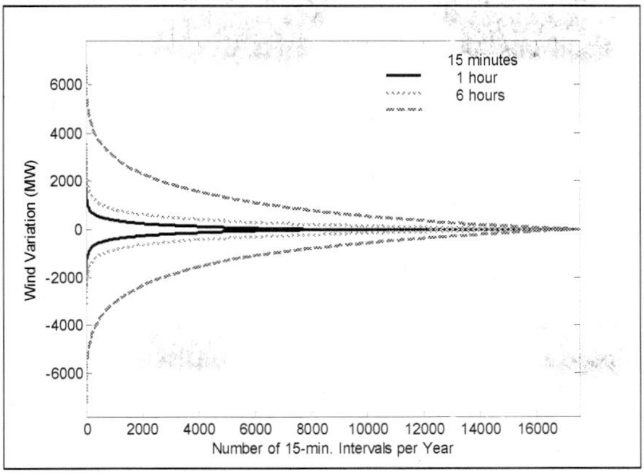

Fig. 10. Variations in 7830 MW aggregated wind power

PRP	Minimum (MW)	Maximum (MW)	99.7%Conf.Int. (MW)
Anton	-367	481	-201 to 197
Berta	-755	825	-323 to 327
Cesar	-482	448	-191 to 188
Dora	-1264	1420	-461 to 484
Emil	-36	38	-9 to 9
Friedrich	-48	42	-22 to 24
Gustav	-74	78	-26 to 27
Total PRPs	-3026	3332	-1233 to 1256
System	-2411	2883	-1091 to 1054

Table 5. Statistics of 15-minute variability at the PRP level

6. Impact of limited wind predictability

In this section balancing energy requirements due to the limited predictability of wind are presented for the scenario with 7.8 GW of installed wind power in the Netherlands in the year 2020.

To this end a statistical analysis is performed of the forecasting error as aggregated over the wind production of the Netherlands. The time series of forecasted 15-minute average wind power include different day-ahead forecasts issued at 24, 18, 12 and 6 hours before delivery. System reserve is allocated among online generators to account for equipment outages and uncertainties in load and wind forecast errors. Obviously the higher the forecast uncertainty, the larger the amount of reserve needed to achieve the same reliability level. Figure 11 shows the normalized histogram for the system-aggregated forecast error, together with the fit to a double-exponential probability density function, which was found to be a more accurate analytical representation of the data than the normal distribution.

Table 6 shows the predictability at the system level in terms of the 99.7% confidence interval plus the average, standard deviation, minimum and maximum of the imbalance. (Imbalance is equal to wind power forecast error.) Statistics for positive and negative imbalance are presented in the second and third rows of table 6, respectively. Based on the 99.7% confidence interval, the positive (up-regulation or reserve) balancing energy requirement is about 56%, and the negative (down-regulation) requirement is about 53% of the installed capacity.

Table 7 shows the predictability at the PRP level. Balancing energy requirements for an individual PRP are in the range 45–82% of its installed capacity for up-regulation or reserve, and 46–72% for down-regulation.

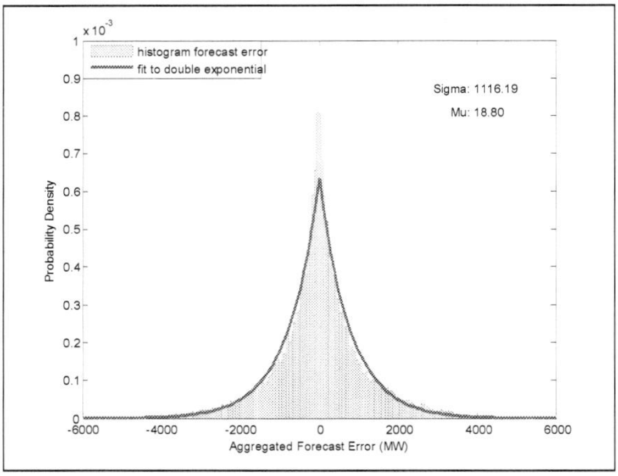

Fig. 11. Aggregated forecast error histogram and probability density function for 7830 MW installed wind power

Imbalance (MW)	Minimum	Maximum	99.7% Conf.Int.	Mean	St.Dev.
Total	−5366	5692	−4112.9 to 4370.6	18.8	1116.2
Positive	0	5692	1.2 to 4765.2	789.8	821.1
Negative	−5366	0	−4471.8 to −1.0	−754.7	790.5

Table 6. Statistics of wind predictability at the system level

PRP	Minimum (MW)	Maximum (MW)	99.7% Conf.Int. (MW)
Anton	−1057	1156	−787 to 773
Berta	−1604	1621	−1239 to 1242
Cesar	−696	798	−575 to 660
Dora	−1941	2169	−1577 to 1726
Emil	−35	39	−26 to 26
Friedrich	−78	67	−56 to 45
Gustav	−116	109	−85 to 74
Total PRPs	−5527	5959	−4345 to 4546
System	−5366	5692	−4113 to 4371

Table 7. Statistics of predictability at the PRP level

The collective requirement is about 58% of the total capacity for up-regulation or reserve, and about 56% for down-regulation. These collective requirements are larger by up to 3% than if the balancing actions were taken at the system level.

In addition, a statistical analysis is performed of the power forecast error by look-ahead time, which in this case varies between 48 and 144 PTUs (i.e. 15-minute intervals), or 12–36 hour. The results are shown in figure 12, expressed in percentage with respect to the installed power. As expected, the performance of the forecast degrades slightly with look-ahead time. The best values obtained are for the 12-hour-ahead forecast, where the standard deviation is 12.7%, and the 99.7% confidence interval is [-50% to +48%]. The standard deviation for the 36-hour-ahead forecast goes up to 17.2%.

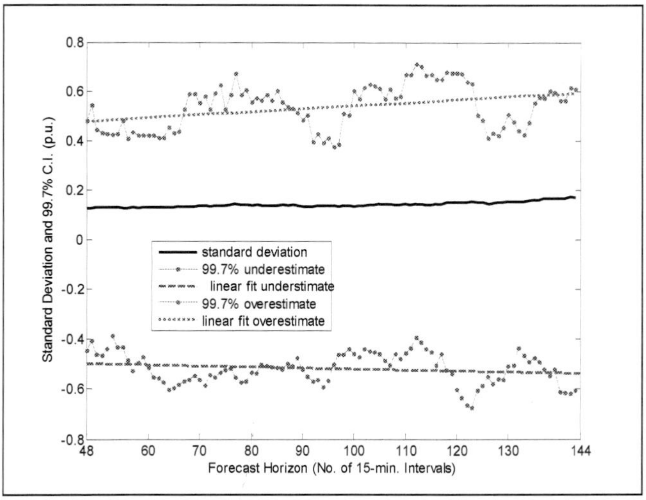

Fig. 12. Standard deviation and 99.7% confidence interval of the aggregated forecast error by forecast horizon

	MW	$P_{average}$ (%)	$P_{installed}$ (%)
RMSE	1116.2	32.8	14.2
MAE	772.3	22.7	9.8

Table 8. Overall predictability statistics for 7830 MW of wind power

Table 8 presents the overall forecast error measures: the root mean square error (RMSE) and the mean absolute error (MAE); see Madsen et al., 2005. The values are presented both in absolute and in percentage with respect to the average power and with respect to the nominal power. The percent RMSE value of 14.2% for the system level is smaller compared with the 17–19% for the single wind farm level, and the percent MAE (9.8%) is also smaller compared with the 12–14%, both reported in Madsen et al., 2005, in percent of installed power for lead times between 12 and 36 hour. The values calculated in percentage with respect to the average power (equal to 3434.5 MW for the 7800 MW installed capacity scenario) are understandably higher. Note that in the presence of an intra-day market, the aggregated forecast errors could drop to about half of the day-ahead values, as simulated in Ummels et al., 2007.

7. Instruments for balancing wind energy

7.1 Outline

This section contains a critical discussion on options to reduce the extra balancing energy requirements for the scenario with 7.8 GW of installed wind power in the Netherlands in the year 2020. The following instruments for balancing wind power forecasting errors are analyzed: short-term forecast updates and aggregation (section 7.2), pumped storage, compressed air storage and fast start-up units (section 7.3), and inverse pumped accumulation (section 7.4). In addition, a wind farm shut-down strategy is discussed in section 7.5.

7.2 Short-term forecast updates and aggregation

7.2.1 Influence of forecast lag on system imbalance

The accuracy of wind power forecasts is evaluated by comparing the forecasted values to the produced amounts. The key indicator is the capacity normalised mean of the absolute forecast error (cNMAE) (Madsen et al., 2005). As table 9 shows, the impact of bad day-ahead forecasts can be alleviated by making use of forecast updates. This clearly shows the importance of continuous wind power forecast updates, which will also allow for a better allocation of the forecast errors within the operation of other generation units in the system.

Forecast lag before delivery	cNMAE [%]	
	Min	Max
24 hours	10.5	13.5
18 hours	10.0	13.0
12 hours	9.0	11.5
6 hours	8.5	11.5

Table 9. Capacity normalized mean absolute forecast error (cNMAE) for different day-ahead forecasts (Duguet and Coelingh, 2006)

Another indicator for the forecast accuracy is the capacity normalized standard deviation of the wind power forecast error (cNRMSE). As shown in figure 13, the cNRMSE is found to drop to half between the forecasts performed at 36 hours and 3 hours before delivery.
It should however be noted that neither the NMAE nor the NRMSE of forecasts based on numerical weather prediction models reduce to zero if the forecast lag approaches present time because of the intrinsic uncertainty in these models. Such a reduction however can be achieved if online production data is included in the forecasts, as is done in figure 13, also showing the cNRMSE for the 0 to 6 hours before delivery.

7.2.2 Aggregation of forecast errors at different levels

The impact of aggregation of wind power on imbalance due to wind power forecast errors is investigated on the basis of forecasts issued 24 hours before the day of delivery. Two aggregation levels are considered: the system level and the Programme Responsible Party (PRP) level. The PRP level consists of seven individual market parties, each with some wind power as part of their portfolio; see table 3. The hypothesis is that a central aggregation would allow internal cancelling out of forecast errors. It is found that this indeed is the case: aggregation at the system level requires about 6% less overall reserves for the compensation of forecast errors (this is the percent reduction in the length of the confidence interval, as computed from table 10).

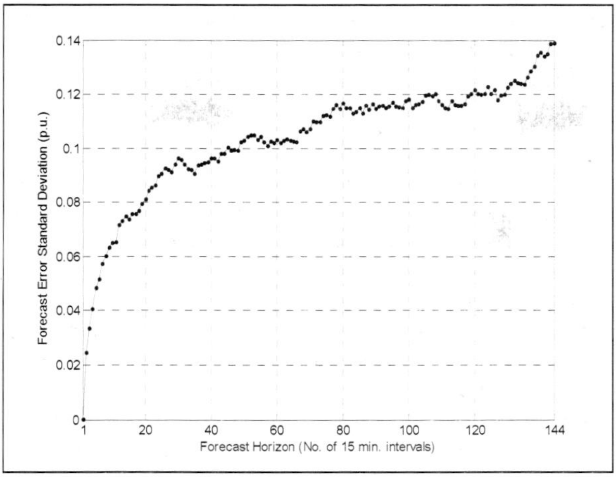

Fig. 13. Capacity normalized standard deviation (cNRMSE) of the wind power forecast error for 7800 MW of wind power

	Wind power forecast error		
	Max. [MW]	Min. [MW]	99.7% Conf.Int. [MW]
Sum PRPs	+5257	-5450	[-3754 ... 4071]
System	+5148	-5326	[-3482 ... 3907]
Difference	109	-124	[-272 ... 164]

Table 10. Range and confidence interval of wind power forecast error for the sum of individual PRPs and the system

7.3 Pumped storage, compressed air energy storage and fast start-up units

7.3.1 Methodology

In this section, pumped storage and compressed air technologies of similar energy content are compared. It is assumed that the storage system does not participate in any market trading, in order to focus on the effectiveness of various technologies in reducing imbalances arising from wind speed forecasting errors. As a further simplifying assumption, the storage system is allowed to reverse operation between consecutive Programme Time Units (PTU), i.e. from charging to discharging and vice versa, depending on the sign of the forecast error.

The following comparably-sized storage systems are analyzed for the scenario with 7800 MW installed wind power, and a 24-36 hour lead time for the wind forecasts:

- A pumped storage (PS) system of 10.08 GWh, charging time 8 hours, hence 1260 MW installed power, with a 0.81 round-trip efficiency, i.e. equal 0.9 pumping and generating efficiencies, with efficiencies independent of charging levels.
- A compressed air energy storage (CAES) system of 7.2 GWh, charging time 8 hours, with a 0.8 compression efficiency and a 1.4 charge efficiency factor, which means that the amount of energy that can be generated at full discharge is 7200×1.4 = 10.08 GWh, thus equal to the pumped storage.

In addition, the effect of 852 MW of installed fast start-up units on the reduction of negative imbalances (less wind power than predicted) is analyzed. Negative imbalances are considered more dangerous to system reliability than positive ones, which can ultimately be taken care of by curtailing excess wind production. The fast start-up units are supposed to complement the pumped storage and so the value of 852 MW was chosen as equal to the standard deviation of the imbalance remaining in the system after the implementation of the 10.08 GWh pumped storage system. It is assumed that the fast start-up power can be switched on or off in increments of 2 MW, and reacts to correct imbalances whose absolute value is bigger than 200 MW. This prevents an unnecessarily large number of start-ups and shut-downs in cases when the imbalance is less than 200 MW and can thus be covered from the spinning reserve carried by conventional units on-line. It is assumed that the fast start-up units are open-cycle gas turbines (OCGT), and hence are capable of starting and ramping up to their installed capacity within one PTU, i.e. 15-minute time interval.

7.3.2 System level aggregation

As an illustration, figure 14 shows a 52-day (5000 PTUs) sample from the yearly time series for the original and the reduced imbalance after the application of a 10.08 GWh pumped storage system in combination with 852 MW installed capacity from fast start-up units.
Results from the comparison of the various technologies are summarized in table 11, which shows the reduced standard deviation and the average positive and negative imbalances, all in terms of per unit with respect to their original values.

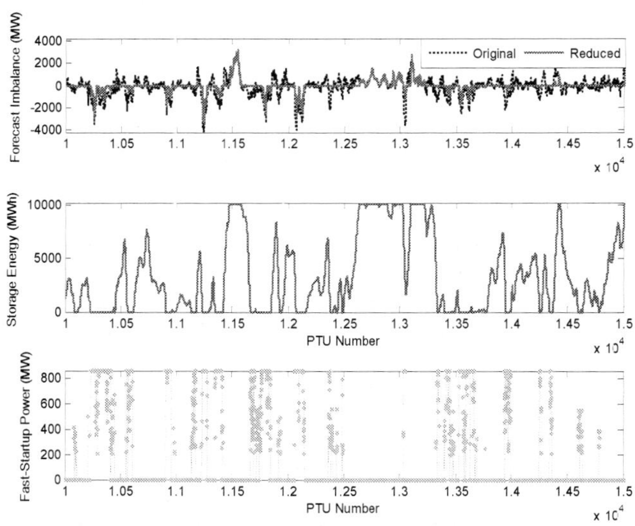

Fig. 14. Time series for 52 days of forecast imbalance, state-of-charge and fast start-up power with a 10 GWh pumped storage system and 852 MW of open-cycle gas turbines

In addition, the reduced 99.7% confidence intervals for the aggregated forecast error are shown in the last row. As a reference point, the original confidence interval before applying any storage was [-3948 ... 3441] MW.

	10080 MWh Pumped Storage	7200 MWh CAES	10080 MWh PS 852 MW Fast Start-Up
σ [p.u.]	0.84	0.84	0.66
μ^+ [p.u.]	0.61	0.68	0.61
μ^- [p.u.]	0.68	0.64	0.25
99.7% C.I. [MW]	[-3530..3310]	[-3421..3326]	[-2678..3310]

Table 11. Statistical properties for the system imbalance reduction

From table 11 it can be seen that it is easier for the pumped storage system to take care of positive (excess wind) imbalances. This is because the 0.9 pumping and generating efficiencies lead to consuming 111% more energy than stored from the positive (excess wind) forecasting errors, whereas only 90% of the stored energy can effectively be used when discharging to cover for negative (deficit wind) errors. The overall standard deviation has been reduced by the CAES system to the same value as in the case of the PS system, i.e. 84% (from 1013 to 852 MW). By contrast, the CAES system, thanks to its charge efficiency factor of 1.4, is slightly better at taking care of negative imbalances than a PS system of comparable installed capacity. However, unlike PS, a CAES "discharge" implies burning of fuel (gas) and hence extra emissions and higher operating costs.

The technology for diabatic CAES systems is available and already has been applied successfully, e.g. the Huntorf plant in Germany, already in operation for about 20 years. In the Netherlands there are a small number of caverns (unused salt domes) which can be used for CAES. However these caverns are more favorable for storing gas or CO_2. For this reason it is concluded that CAES development in the Netherlands will be hard and will have to compete with other technologies.

The last column of table 11 shows the results for the pumped storage and fast start-up units combination. The resulting reduction in average negative imbalance is to 25% of its original value, which is achieved with an average of 6.5 start-ups per day. The reduction in positive imbalance is naturally the same as that without the fast start-up units, whereas the overall standard deviation is now reduced to 66% (667 MW).

7.3.3 PRP level aggregation

The installed 7800 MW wind power is now distributed over seven market parties at the Programme Responsible Party (PRP) level; see table 3. In order to facilitate comparison with the results for the system level aggregation, the installed storage and fast start-up capacities are allocated proportionally to the installed wind power of each PRP. These installations are now controlled to correct the individual imbalances due to forecasting errors as experienced by each PRP. Figure 15 shows the reductions in negative imbalance for system versus PRP level aggregation, for various technologies, and increasing values of storage capacities, up to 30 GWh. From this figure it can be noted that installing storage and/or fast start-up units to be controlled for reducing the imbalance at system level is slightly more advantageous than at the PRP level in terms of reducing the average negative imbalance. The advantage stays approximately constant regardless of storage capacity, with the largest difference experienced for the PS and fast start-up combination, at about 0.12 p.u., which translates to 86 MW. By contrast, installing storage and fast start-up units to be controlled for reducing the imbalance at PRP level is slightly more advantageous than at the system level in terms of reducing the total spread -- or standard deviation -- of the imbalance. The advantage increases with higher storage capacities, with the largest difference being about 0.025 p.u., which translates to about 25 MW.

It is concluded that differences between system and market participant level aggregation seem insignificant from the standpoint of being more or less effective in reducing the forecast imbalance. It follows that decisions about where storage units are to be installed and how they are to be controlled will probably be governed more by geographical constraints, considerations about economies of scale, ease of accounting, technical accessibility and confidentiality of control signals.

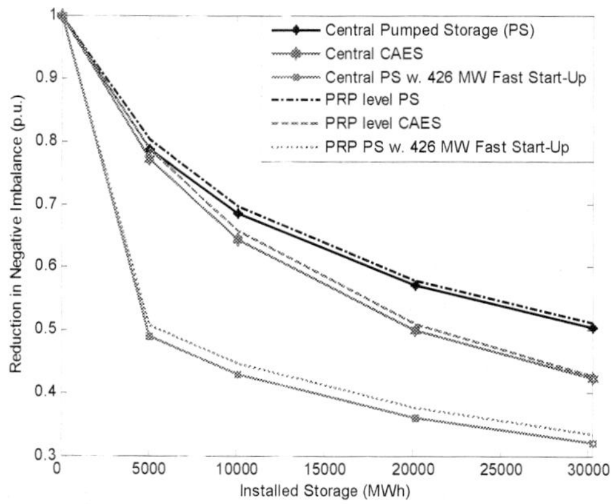

Fig. 15. Reductions in average negative imbalance for PS, CAES and PS with 426 MW fast start-up units for PRP versus system level aggregation

7.4 Inverse offshore pump accumulation system

Next an inverse offshore pump accumulation system (IOPAC) is considered (de Boer et al., 2007). The IOPAC has been proposed to be stationed on an "Energy Island", that is an artificial island at sea consisting of a ring of dikes (6x10 km^2) enclosing a deep dredged reservoir. The intended depth of the reservoir is 50 m below sea level, and the water level of the reservoir will typically vary between -32 and -40 m. In figure 16 an artist impression of the "Energy Island" is shown.

The IOPAC is equipped with a control system that aims to prevent the storage from being saturated, and thus makes it more capable of reducing the imbalance due to wind energy. The strategy of the control system involves selling power during peak hours (8 am to 11 pm) if the water level is too low, and buying power during off-peak hours if the water level is too high. In this way the control system keeps the water level around the half-full operating point (around -36 m).

Energy is bought from or sold to the intra-day market, assuming any needed energy volume is available within the limits of the storage device. This is feasible; whether a surplus or a shortage of energy in the reservoir can be noticed by simply monitoring the water level. The lower part of figure 17 shows this correction signal (where positive power means bought from the market and used to charge the storage system).

Fig. 16. Artist impression of the Energy Island with an Inverse Offshore Pump Accumulation (IOPAC)

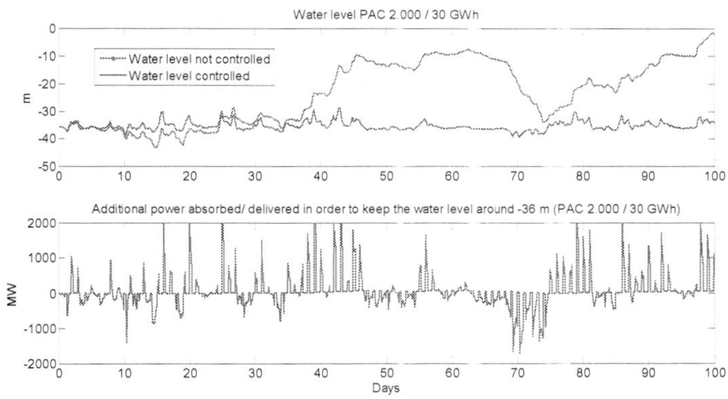

Fig. 17. Impression of average power absorbed/delivered in order to maneuver the IOPAC around the half-full operating point

The effectiveness of the IOPAC in alleviating imbalance is analyzed by considering a pump/turbine power rating of 2000 MW, with a storage energy capacity of 30 GWh. Initial imbalances result from wind power forecasts issued at noon before the day of delivery (with a lag of 12-36 hours). The upper part of figure 17, showing the impact of the control system on the variation of the water level as a response to these imbalances, reveals that the control system prevents the water level from drifting away. Figure 18 shows the original imbalance, the imbalance after employing the IOPAC alone, and the imbalance for the IOPAC with intelligent control strategy. The imbalance reduction, measured in terms of per unit standard deviation with respect to the base-case, ranges from 0.714 (IOPAC alone) to 0.697 (IOPAC with intelligent control). Some imbalance remains after the application of IOPAC, but this could be handled by fast responding conventional units and/or combining these measures with short-term wind power forecasts.

The proposed intelligent IOPAC is therefore shown to alleviate imbalances due to wind power forecast errors.

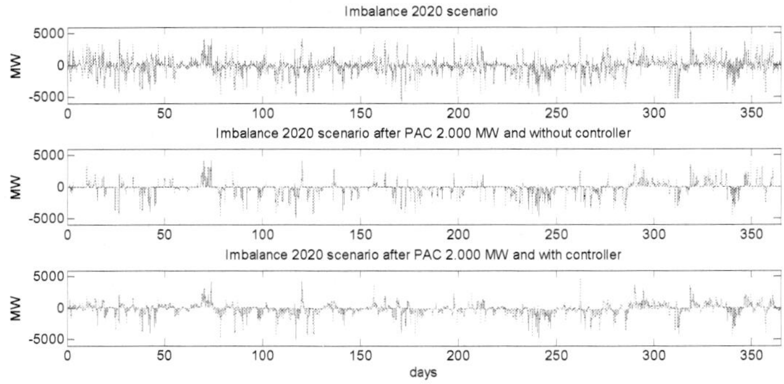

Fig. 18. Imbalance before and after the application of IOPAC storage, 2000 MW installed power, with 30 GWh energy capacity

7.5 Wind farm shut-down strategies

Finally, the impact of wind farm shut-down strategy on the imbalance is analyzed. Two shut-down strategies are considered. In the abrupt shut-down strategy, a wind farm is shut down within 10 seconds if the 15-minute averaged wind speed exceeds 25 m/s, and is started up again within 10 seconds if the 15-minute average wind speed is less then 22 m/s. In the gradual shutting-down strategy, on the other hand, the power of a wind farm varies linearly with the wind speed between full load and zero if the 15-minute averaged wind speed is between 23 and 32 m/s.

In practice the abrupt shut-down strategy or a close variant is applied, where the turbines in a wind farm may either shut down individually based on their individual wind speed measurements or collectively based on one central wind speed measurement system in the wind farm. In this analysis a whole wind farm shuts down based on its 15-minute averaged wind speed. Also, in reality the shutting-down and starting-up times may differ from the value of 10 s employed in this analysis. These choices however do not strongly affect the conclusions.

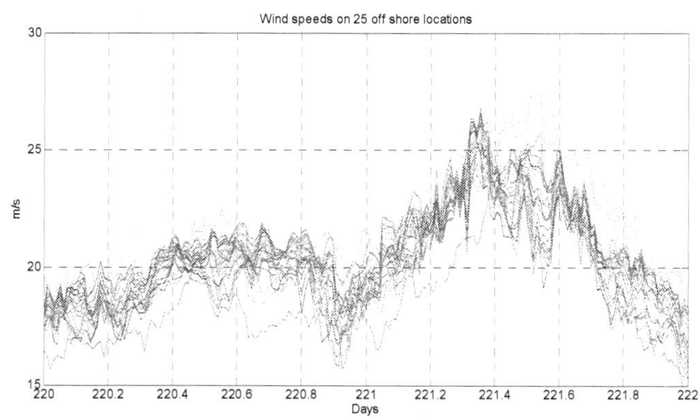

Fig. 19. The wind speed at the 25 offshore locations during one day with wind speeds exceeding 25 m/s

In addition the 15-minute average wind speeds at the wind farm locations foreseen in the Netherlands for the period of one year were considered. It is found that on three days the wind speed exceeds 25 m/s in at least one location. Day 222 was selected for further analysis; the wind speeds are presented in figure 19. It should be noted that the wind speed does not exceed 25 m/s at all wind farms, so that not all wind farms shut down.
As shown in figures 20 and 21, with the abrupt shut-down strategy, there are large differences between the forecasted and the produced wind power, and, as a consequence, larger imbalances. With the gradual shut-down strategy, the differences are significantly less. It was found that with gradual shut-down the imbalance due to forecasting errors is reduced by more than 50% as compared to abrupt shut-down. In addition, benefits in terms of decreasing the variability in the wind power output can also be observed in figure 20.

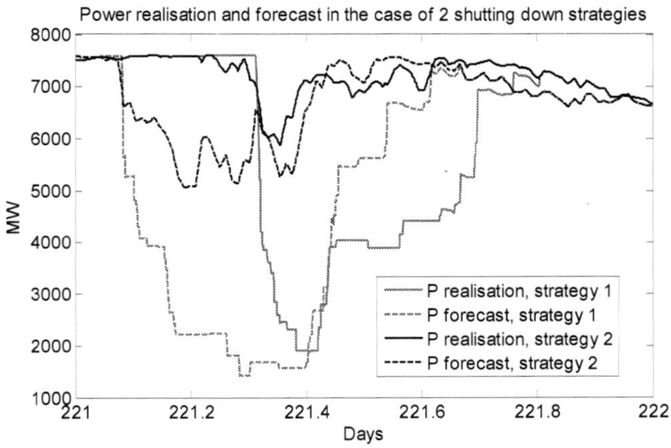

Fig. 20. Wind power production and forecast during the day with wind speeds over 25 m/s and after applying the abrupt and gradual shut-down strategies

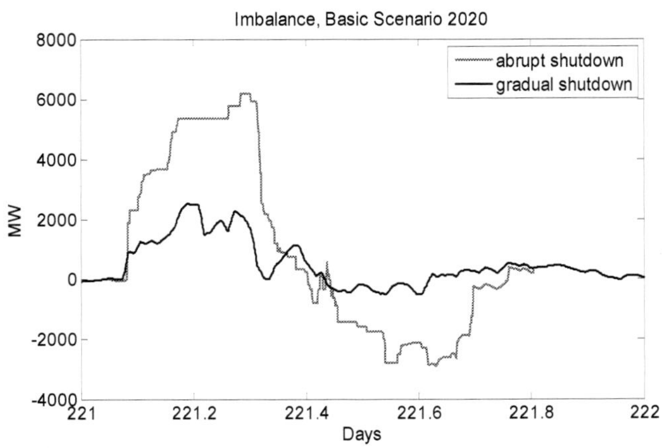

Fig. 21. Remaining forecast imbalance during the day with wind speeds over 25 m/s and after applying the abrupt and gradual shut-down strategies

The concept of shutting down the wind farm gradually as a linear function of the (high) wind speed is therefore found to be significantly better than abrupt shut-down.

8. Summary

A statistical interpolation method to generate time series of system- and participant-aggregated wind power production and forecast values has been presented. The method takes into account the spatial and temporal correlations among multiple sites, as derived from the measurement and forecast data. In addition, a method for deriving park-aggregated power curves with smooth cut-in and cut-out, that takes into account the local wind climate, was introduced.

In the scenario with 7.8 GW of wind power in the Netherlands in the year 2020, at the system level the imbalance energy requirements due to wind variations across 15-minute intervals are ±14%, while the imbalance energy requirements due to forecast errors vary between 53% down-regulation and 56% up-regulation or reserve. If 15-minute variations and forecast errors are compensated individually by market participants, the collective balancing energy requirements are larger: ±16% for the 15-minute wind variations, and between 56% down-regulation to 58% up-regulation or reserve for the forecast errors.

Various alternatives for balance control as required by the addition of wind energy in the Netherlands have been investigated. From this analysis it is concluded that the following instruments for balance control are most applicable: continuously updated short-term forecasts, pumped-accumulation storage systems, fast start-up units, and gradual shut-down strategies for wind farms.

9. Acknowledgements

This work was funded by the Dutch Ministry of Economic Affairs partly under the EOS-LT project "RegelDuurzaam" and partly under the BSIK programme We@Sea, project "Systeemintegratie en balanshandhaving". The measured wind data originate from KNMI, ECN and DEWI. The HiRLAM data originate from KNMI. The authors would like to thank the project partners from Ecofys, TenneT and KEMA for their valuable comments. The authors like to thank Jan Coelingh from Ecofys for providing the data in table 9.

10. References

de Boer W.W., Brand A.J. & Gibescu M. (2007), System balancing with 6 GW offshore wind energy in the Netherlands - Analysis of current methods for balancing, *Ecofys*, Un-numbered report.

de Boer W.W., Verheij F.J., Zwemmer D. & Das R. (2007), The energy island - An inverse pump accumulation station, *EWEC 2007*, Milan, May 2007.

Brand A.J. (2006), Observed and predicted wind speed time series in the Netherlands and the North Sea, *Energieonderzoek Centrum Nederland (ECN)*, Report ECN-C--06-007.

Brand A.J. & Hegberg T. (2004), Offshore wind atlas, *Energieonderzoek Centrum Nederland*, Report ECN-CX--04-136.

Brand A.J. & Kok J.K. (2003), Aanbodvoorspeller Duurzame Energie Deel 2: Korte-termijn prognose van windvermogen, *Energieonderzoek Centrum Nederland*, Report ECN-C-03-049.

von Bremen L., Tambke J., Saleck N. & Heinemann D. (2006), Confidence in large-scale offshore wind farming: wind power predictability and stable grid integration of 25 GW German wind power, *Proc. Sixth International Workshop on Large-Scale Integration of Wind Power and Transmission Networks for Offshore Wind Farms*, Delft, October 2006, pp. 277-284.

Brockwell P.J. & Davis R.A. (1991), *Time Series: Theory and Methods (2nd ed.)*, Springer: New York , ISBN 0387974296.

Doherty R. & O'Malley M. (2005), A new approach to quantify reserve demand in systems with significant installed wind capacity, *IEEE Transactions on Power Systems* 2005; 20, pp. 587–595.

Duguet B. & Coelingh J. (2006), *Simulated imbalance of 8000 MW wind power*, Ecofys, Rapport Wind04071.

ETSO (2008), European Wind Integration Study (EWIS) - Towards a successful integration of wind power into European electricity grids, *European Transmission System Operators, Final Report*, www.etso-net.org (September 2009).

Gibescu M., Ummels B.C. & Kling W.L. (2006), Statistical wind speed interpolation for simulating aggregated wind energy production under system studies, *Proceedings of the 9th International Conference on Probabilistic Methods Applied to Power Systems*, Stockholm, June 2006, 7 pp.

Gibescu M., Brand A.J. & de Boer W.W. (2008a), System balancing with 6 GW offshore wind energy in the Netherlands - Instruments for balance control, *Proc. Seventh International Workshop on Large-Scale Integration of Wind Power and Transmission Networks for Offshore Wind Farms*, Madrid, April 2008, 7 pp.

Gibescu M., van Zwet E.W., Kling W.L. & Christie R.D. (2008b), Optimal bidding strategy for mixed-portfolio producers in a dual imbalance pricing system, *Proc. 16th Power Systems Computation Conference*, Glasgow, July 2008.

Gibescu M., Brand A.J. & Kling W.L. (2009), Estimation of variability and predictability of large-scale wind energy in the Netherlands, *Wind Energy, Volume 12 Issue 3*, pp. 241-260.

Giebel G. (2000), *On the benefits of distributed generation of wind energy in Europe*, PhD Dissertation, Carl von Ossietzky University, Oldenburg.

Holttinen H. (2005), Hourly wind power variations in the Nordic countries, *Wind Energy, Volume 8*, pp. 173–195.

Krauss C., Gräber B., Lange M. & Focken U. (2006), Integration of 18 GW wind energy into the energy market - Practical experiences in Germany, *Proc. Sixth International Workshop on Large-Scale Integration of Wind Power and Transmission Networks for Offshore Wind Farms*, Delft, October 2006, pp. 55-59.

Landberg L., Hansen M.A., Vesterager K. & Bergstrom W. (1997), Implementing wind forecasting at a utility, *Risø National Laboratory*, Report Risø-R-929(EN).

Lange M. & Focken U. (2005), *Physical Approach to Short-Term Wind Power Production*, Springer Verlag, Berlin, ISBN 3-540-25662-8.

Madsen H., Pinson P., Kariniotakis G., Nielsen H.A. & Nielsen T.S. (2005), Standardizing the performance evaluation of short-term wind power prediction models, *Wind Engineering*, Vol. 29, No. 6, 2005, pp. 475–489.

Norgard P., Giebel G., Holttinen H., Söder L. & Petterteig A. (2004), Fluctuations and predictability of wind and hydro power, *Risø National Laboratory,* Report Risø-R-1443(EN).

Norgard P. & Holttinen H. (2004), A multi-turbine power curve approach, *Nordic Wind Power Conference,* Göteborg, 2004; 5 pp.

Söder L. (2004), Simulation of wind speed forecast errors for operation planning of multi-area power systems, *International Conference on Probabilistic Methods Applied to Power Systems,* Ames, 2004, pp. 723–728.

Ummels B.C., Gibescu M., Pelgrum E., Kling W.L. & Brand A.J. (2007), Impacts of wind power on thermal generation unit commitment and dispatch, *IEEE Transactions on Energy Conversion,* Vol. 22, Issue 1, March 2007, pp. 44-51.

Ummels B.C., Pelgrum E. & Kling W.L. (2008), Integration of large-scale wind power and use of energy storage in the Netherlands' electricity supply, *IET Renewable Power Generation,* 2(1), March 2008, pp. 34-46.

Verhaegen K., Ummels B.C., Belmans R.J.M. & Kling W.L. (2006), Comparison of support schemes and market designs for wind power, *Proc. Sixth International Workshop on Large-Scale Integration of Wind Power and Transmission Networks for Offshore Wind Farms,* Delft, October 2006, pp. 39-47.

Walker J.F. & Jenkins N. (1997), *Wind Energy Technology.* Wiley: New York, ISBN 0471960446.

Wolff M., Mackensen R., Füller G., Lange B., Rohrig K., Fischer F., Hofmann L., Heier S. & Valov B. (2006), Advanced operating control for wind farm clusters, *Proc. Sixth International Workshop on Large-Scale Integration of Wind Power and Transmission Networks for Offshore Wind Farms,* Delft, October 2006, pp. 188-195.

12

Variability of Wind and Wind Power

Joaquin Mur-Amada and Ángel Bayod-Rújula
Zaragoza University
Spain

1. Introduction

The installed capacity for wind power is increasing substantially in response to the worldwide interest in low-emissions power sources and a desire to decrease the dependence on petroleum.

The European Union directive 2009/28/EC enforces the mandatory target of a 20 % share of energy from renewable sources in overall Community energy consumption by 2020 and a mandatory 10 % minimum target to be achieved by all Member States for the share of biofuels in transport petrol and diesel consumption by 2020. These targets may require between 30 and 40 % of the electricity in the European Union to come from renewable energy sources by 2020.

In the U.S., the world's top wind producer (The Guardian, 2008), wind currently makes up just one percent of the energy supply. Wind power generation share is expected to grow up to 20% in the USA by 2030 (EnerNex, 2006; U.S. D.O.E., 2008). Moreover, many U.S. states have legislated similarly ambitious renewable energy portfolio standards.

These goals were set without regard for the fact that many in the scientific community have concluded a theoretical wind penetration limit of only 20% due to the degradation of system reliability (DeCarolis et al., 2005). The Department of Energy states that there is no fundamental technical reason why 20 percent of wind energy cannot be assimilated into the grid by 2030. To help make its point, the agency debunks the reliability myth in its fact sheet on Wind Energy Myths (U.S. D.O.E. 2005).

According to (Feldman, 2009), the renewable output could leap to 40 percent of the Irish electricity share; in Denmark to 33 percent; in Portugal to 28 percent; and in Germany and Greece to 25 percent. After 2020, a higher proportion may be needed. A significant amount of this renewable electricity is likely to come from wind, and the variability of this power needs to be managed.

With this amount of wind generation, the future electricity markets could be very different to those of today: instead of thermal power stations dominating the system, the market could be dominated by large amounts of price-insensitive nuclear and wind power, combined with highly intermittent output from the wind farms (Pöyry PLC, 2009).

The extent of uncertainty and variability in wind generation makes this resource different from the traditional, dispatchable generation resources, with the result that wind power generation cannot be readily integrated into standard system operating procedures (IEA, 2009). At relatively low levels of installed capacity, wind turbines and the output from large wind farms can essentially be absorbed into traditional system operations without

degrading system reliability. At the current higher projected levels of penetration, wind power requires more sophisticated mechanisms to maximize its participation in the power system without penalizing it for the unavoidably intermittent nature of its resource (Sovacool, 2009).
In some real time markets, and at low levels of penetration, wind is treated as negative load (J. Apt, 2007). But the perception of wind power is changing from being considered a negative load to a capacity resource. Wind power can be integrated into system and market operations as a generating resource that could provide not only energy but also capacity and ancillary services (Piwko et al., 2004).

1.1 Relationship between wind power variability and its forecast

Wind varies in space and in time. The forecasts try to predict this variation from climate dynamics and from the systematic behavior of the weather.
The inherent uncertainty and availability of the meteorological data decreases the accuracy of the next day's wind generation forecast (Anderson et al., 2008). The issue of uncertainty in wind generation can be addressed by improving the accuracy of forecasting the wind resource. In addition, advances in wind forecasting and turbine controls suggest that wind power can participate in ancillary service markets.
The simplest forecast of a stochastic process is the persistence principle: the expected value during next period $[t, t+\Delta t]$ is the average value of the previous period $[t-\Delta t, t]$. The mean square error of the persistence prediction is just the variance of the average process during the interval $[t-\Delta t, t]$.
Thus, the ratio of the forecast root squared error to the standard deviation of the variable predicted is just the performance of the forecast relative to the persistence model.
Usually, numerical weather prediction significantly outperforms persistence for horizons longer than 6 hours. For shorter horizons, statistical methods can be more adequate due to the lack of a dense network of weather sensors near the wind farms. In such cases, the characterization of wind power variability is essential.

1.2 Influence of the wind variability on the grid

Wind power presents the most economically viable renewable solution, apart from hydro power (DeCarolis et al., 2005). The utility system is designed to accommodate load fluctuations, which occur continuously. This feature also facilitates accommodation of wind plant output fluctuations when wind penetration is low.
In order to make the long-term growth of wind generation possible, the variability and the intermittency of wind power must be managed (Boyle, 2007). In Denmark, Northern Germany, and parts of Spain, wind supplies 20% to 40% of electric loads without sacrificing reliability.
Generally, wind power forecast are targeted to optimize the hourly power dispatch. However, the electric system has to cope with instantaneous variation of load, generation and equipment trips. Such variations are usually unpredictable and they are usually considered deviations from the expected power tendency.
The variability of wind power has several negative effects on the reliability and system operation of the electric grid as well as wind project economics (Constantinescu et al., 2009). The stability of the electric grid depends upon reliable and consistent power generation that is balanced to the load through unit commitment (interhour), economic dispatch (intrahour),

and regulation (intraminute), and wind power is counterproductive towards that effort. Due to its stochastic nature, wind generation is not dispatchable and therefore cannot be called upon to serve load. As a result, the capacity credit of a wind farm is very small—even as low as 8% of nameplate capacity according to a probabilistic loss-of-load analysis (Milligan, 2000).

Greater reliance on wind power requires more ancillary services, especially responsive reserves, to match the lost generation and ramp rate of wind turbines when the wind dies down. One study of a typical wind farm indicated ramp rates up to 4.4% of capacity per second (Parson et al., 2001), implying a large burden on ancillary services. In fact, an empirical analysis of increasing wind penetration in the Pacific Northwest U.S. demonstrated that reserve requirements increased with the square of installed wind capacity, and the need for total reserve capacity doubled after just 2500 MW of installed wind capacity (Parsons et al., 2004).

In some applications, the estimation of wind power variability can be as important as its prediction. The forecasted values are usually the hourly wind or the hourly power generated by one turbine or a cluster of them. Most forecast models predict only the average hourly wind or wind power and they are intended for the requirements of the electricity markets. Only a few models characterize the uncertainty of the forecast and quantify the instantaneous fluctuations inside the time period.

The main applications which benefit from variability assessment are:

- The control of wind turbines and wind farms, where the short timescales involved makes forecasting unpractical.
- Automatic generation control, automatic voltage control or changes in automatic taps on transformers, more concerned with interhourly variations, that are quite difficult to predict.
- Improving the power quality, more related to voltage variations and hence, instantaneous power variations.
- Optimum sizing of storage devices or of running reserves in isolated or weak systems, where electricity must be supplied even in the worst case scenario.
- In risk assessment and the safe operation of the grid. The system must cope with unexpected outcomes when the forecast has big uncertainty.

1.3 Geographic diversity on wind power

Both the generated power and the forecast error decrease as more wind power producers are aggregated. Due to the geographic dispersion of wind generators, some power variations and prediction errors can be partially cancelled by other errors in other locations.

On the one hand, the forecast errors can be very low in wide geographic areas. The power balance can be met provided the electricity transmission networks are strong enough to carry the undispatchable generated power from remote areas (Matevosyan, 2006; Holttinen et al., 2007).

On the other hand, many quality parameters of the grid must be met locally and the reinforcement of the electricity networks is costly and, sometimes, it is not feasible. In islands, the power balance cannot rely on geographical diversity and other measures are required to counteract the wind power variability.

The weather conditions may remain stable for relative long periods among shift weather changes. Quick local turbulent fluctuations are tougher to predict in time and place than

some smooth weather evolution. In fact, a good parameter of the accuracy of the prediction is the error relative to the variance of the random variable to be predicted.
However, the wind power forecast accuracy is usually referred to the total installed wind power considered in the prediction. These figures must be considered with caution:

- A 15% prediction error of the hourly power one day ahead of a single wind farm can be an accurate forecast (Martí et al., 2006, Ramirez-Rosado et al., 2009).
- A 15% prediction error of the hourly power one day ahead in a big system is a poor forecast (Juban et al., 2008).

The variance of the wind power decreases when increasing the time period of the measure or the spatial diversity of the wind generators. Therefore, increasing the time or space horizon of the predictions lowers the absolute prediction error. The standard deviation to mean ratio, called coefficient of variation (CV) (Parsons et al., 2001; Wan et al. 2002), is also sensitive to the geographic, the time averaging and the prediction horizon, as can be seen in Table 1.
The table 1 compares output at the start and end of the indicated time period in terms of the percentage of total generation from each turbine group. Std. Dev. is the abbreviation for standard deviation. CV stands for coefficient of variation, the ratio of standard deviation respect the mean of wind power.
The power spectral density identifies which frequencies of variation are contributing to the variance (Apt, 2007). The coherence indicates the degree of partial cancellation of the oscillation among the turbines at different frequencies (Sørensen et al. 2008). These two magnitudes can explain the effect of the geographic and time averaging. In the spatial domain, high frequencies smooth out a wind farm's aggregate power output since the coherence of the turbines' outputs is low, while low spatial frequencies cause a coherent variation in the farm's turbines.

	14 Turbines (%)	61 Turbines (%)	138 Turbines (%)	250+ Turbines (%)
1-Second Interval				
Average CV	0.4	0.2	0.1	0.1
Std. Dev.	0.5	0.3	0.2	0.1
1-Minute Interval				
Average CV	1.2	0.8	0.5	0.3
Std. Dev.	2.1	1.3	0.8	0.6
10-Minute Interval				
Average CV	3.1	2.1	2.2	1.5
Std. Dev.	5.2	3.5	3.7	2.7
1-Hour Interval				
Average CV	7.0	4.7	6.4	5.3
Std. Dev.	10.7	7.5	9.7	7.9

Table 1. Wind generation variability as a function of the number of generators and time interval (from "20% wind energy by 2030: Increasing Wind Energy" (U.S. D.O.E, 2008)

1.4 Demand response and wind variability

To a large extent, load exhibits similar characteristics -uncertainty and variability- to wind power (Apt, 2007). Load patterns though, have been more extensively studied for many years and so are better understood and more accurately forecasted than the wind resource. The purpose of this effort in load modelling is to understand load patterns well enough to operate the power system through the control of individual generation and transmission facilities, in order to serve load and maintain system reliability.

Thus, load is extensively modelled and *other* facilities are controlled to serve load, with relatively little effort made to control load itself. This trend is not absolute, as there are traditional utility mechanisms, such as interruptible contracts and direct load control, to reduce load at times when system reliability would otherwise be threatened. There is also persistent interest in developing mechanisms for more dynamic load response for both reliability and economic purposes.

Recent efforts to allow load to be more responsive to system conditions and a more active participant in electricity markets arise for multiple reasons (Cappers et al., 2009). In addition to giving customers incentives to decrease their demand in the short run to improve system reliability during times of system peak, demand response can be used in the long term to decrease required capacity expansion and lower total costs. Demand response is also an important and essentially absent element in electricity markets. If it were to be more widely implemented, market efficiency would be likely to improve. Many efforts are being made to demonstrate the feasibility and the convenience of smart grids.

1.5 Estimated cost of wind power variability

The estimated cost of the uncontrollability of the wind resource has been widely researched from governmental policy makers to Independent System Operators (Porter et al., 2007). An understanding of the impacts of the variable sources of renewable energy must take into account the wider issues associated with managing electricity systems (Lamont, 2008).

Modern integrated networks are designed to cope with 'shocks' such as the sudden loss of large thermal power stations and with uncertainties in consumer demand, such as those caused by televised sports events. As the tools to deal with these are already available, the key question is the extent to which the introduction of large amounts of wind energy will increase the overall uncertainty in matching supply and demand. This extra uncertainty means that additional short-term reserves are needed to guarantee the security of the system (Matos, 2009).

The variability of wind power also adversely affects wind project economics. A stochastic power source like wind is inherently less valuable than a deterministic source. Net payments to wind generators are reduced by balancing-costs to compensate for unfulfilled obligations to generate power (Holttinen, 2008).

(Hirst et al, 2004) determined that net payments in the USA to wind farms vary from $32/MWh for very small wind farms to less than $10/MWh for farms larger than 2000 MW.

(Holttinen et al., 2007) states that at wind penetrations of up to 20% of gross demand (energy), system operating cost increases arising from wind variability and uncertainty amounted to about 1~4 €/MWh. This is 10% or less of the wholesale value of the wind energy.

The cost of grid reinforcements due to wind power is very dependent on where the wind power plants are located relative to load and grid infrastructure. The grid reinforcement costs from studies vary from 50 €/kW to 160 €/kW in (Holttinen et al., 2007). The costs are

not continuous; there can be single very high cost reinforcements, and there can also be differences in how the costs are allocated to wind power.
According to (Milborrow, 2009), the extra cost in the UK of these reserves -with wind providing 20% of electricity consumption- is unlikely to be more than £1.20/MWh on electricity bills (a little over 1% on domestic bills). With 40% of electricity provided by wind, the corresponding figure would be £2.80/MWh.
A second costs of wind variability is the backup cost for periods of very low wind resource along extensive geographical areas. (Recharge, 2009) reported that the calculations made by Oxford University Environmental Change Institute showed that between 1970-2003, low wind speeds all across the whole UK —those too slow to generate energy— occurred simultaneously in the country only one hour per year on average. In other words, the total wind power in the UK varies, but a 0% generation is quite unlikely event. According to (Milborrow, 2009), wind energy does not require the introduction of special back-up provisions in the UK and its back-up costs are modest. Though the study was focused in the UK, the results are relevant worldwide, as characteristics of wind are broadly similar.
All generating plants make use of a common pool of backup plant that is typically around 20% of the peak demand on the electricity network. When wind is introduced, system operators do not rely on the rated power of all the installed wind farms being available at the times of peak demand, but a lower amount - roughly 30% of the rated capacity at low penetration levels, falling to about 15% at high penetration levels. This lower 'capacity credit' gives rise to a modest 'backup cost'. 'Constraint costs' arise when the output from the wind turbines exceeds the demand on the electricity network. They are unlikely to arise until wind energy is contributing around 25% of electricity requirements.
Overall, it is concluded that the additional costs associated with variability - with wind power providing up to about 40% of all electricity, are quite small. For example, if wind provides 22% of electricity in the UK by 2020, variability costs would increase the domestic electricity price by about 2%, according to (Milborrow, 2009). Further increases in the level of wind penetration beyond that point are feasible and do not rely on the introduction of new technology.
There are numerous technical innovations at various stages of development that can mitigate the costs associated with variability. Improved methods of wind prediction are under development worldwide and could potentially reduce the costs of additional reserve by around 30%. Most other mitigation measures reduce the costs of managing the electricity network as a whole. 'Smart grids', for example, cover a range of technologies that may reduce the costs of short-term reserves; additional interconnections with Europe, including 'Supergrids' also deliver system-wide benefits and aid the assimilation of variability.
Electric cars hold out the prospect of reduced emissions for the transport network as a whole and could act as a form of storage for the electricity network -for which the electricity generator would not have to pay.
With current technology, wind power plants can be designed to meet industry expectations such as riding through voltage dips, supplying reactive power to the system, controlling terminal voltage, and participating in system operation with output and ramp rate control.
Although some aspects of the management of wind variability can be controversial and costly, many utilities agree that there is no insuperable technical reason why high proportions of wind energy cannot be assimilated into the system (UK National Grid, 2009). There is a large body of literature on the topic (IEA, 2006) and the steady growth of wind power, worldwide, indicates that it is seen as a robust choice for reducing greenhouse gas emissions.

2. Classification of wind and wind power oscillations

Oscillations due to wind speed variations can be classified according to their rate and their spatial extent. Slow fluctuations (in the range of tens of minutes and hours) are mainly due to meteorological dynamics and they are highly correlated among near wind farms.

Fast fluctuations have lower spatial correlation (Nichita et al., 2002; Petru & Thiringer, 2002) and wind gust and turbulence effects are smoothed in the output of the wind farm. A swift gust arriving simultaneously to all turbines dispersed in an area of kilometres is also very uncommon. The correlated component of wind is estimated in (Rosas, 2003) from the Davenport type and Schlez and Infield's decay factors (Schlez & Infield, 1998), showing that coherence for distances greater than 100 m is bellow 10^{-20} at tower shadow frequencies (between 0.5 Hz and 2 Hz). In fact, the coherence for the usual dimension of a wind farm is low for oscillations quicker than 0,001 Hz (Sørensen et al., 2007). Thus, the fluctuations quicker than 10 minutes are low correlated among the turbines a wind farm.

Fast fluctuations of power output can be divided into cyclic components (tower shadow, wind shear, modal vibrations, etc.), weather dynamics and events (connection or disconnection of the turbine, change in generator configuration, etc.). Oscillations from a few minutes to power supply frequency are mainly linearly uncorrelated and their sum across a wind farm can be estimated using stochastic analysis commonly applied in other areas such as multipath fading in communication channels, clutter and target cross section in radars, interference in communication systems, etc. However, this approach is novel in wind energy. In (Welfonder, 1997) a transfer function of the wind farm power output respect the output of a single turbine is estimated with gain N at low frequencies ($f < 0.03$ Hz) and gain $\sqrt{N}$ at high frequencies ($f > 0.09$ Hz). Micro-meteorological and dynamic fluid models can predict the correlated fluctuations which predominate at the very low frequencies. Fully correlated fluctuations scales proportional to the number of turbines N whereas linearly uncorrelated fluctuations cancel partially among turbines and they scale up typically in a factor $\sqrt{N}$, according to the central limit theorem.

The influence of blade position in a single turbine power output has been widely analyzed in the literature (Gordon-Leishman, 2002; Dolan & Lehn, 2005; Dolan & Lehn, 2006). According to (Cidrás et al. 2002), a very steady and very uniform wind in conjunction with a weak electrical network is necessary for synchronization to happen driven by voltage drops. Moreover, simultaneous tower shadow at all turbines in a wind farm is a very uncommon event. If the turbine oscillations are not synchronized, the rotor positions have random phases and the crossing of a blade in front of the turbine tower is a random event with constant probability rate. As wind characteristics are similar inside the farm, the magnitude of the cyclic components would be similar in all turbines.

Experimental measures (Mur-Amada et al., 2003) have shown that the cuasi-periodic aerodynamic torque oscillations are transformed into stochastic oscillations of electric power due to the complex wind turbine dynamics. The power oscillations of frequency bellow 0,03 Hz are greatly reduced at the farm output and the main source of variability in the wind power injected in the grid is the meteorological dynamics.

To sum up, this chapter focuses on the spatial and temporal variability of the wind for short horizons. Fast fluctuations of wind power during continuous operation of wind turbine are not an issue for utilities. Only in isolated or weak networks, or in the simultaneous tripping event of big amounts of wind power they are an issue. These results can be extended to several wind farms since the uncorrelated components increases when the distances among turbines become bigger.

3. Turbulence overview

3.1 Turbulence characterization

At a very basic level, a turbulence flow can be interpreted as a population of many eddies (vortices), of different sizes and strengths, embedded in another and forever changing, giving a random appearance to the flow (Fig. 1). Two variables then play a fundamental role: d, the characteristic diameter of the eddies, and $\hat{u}$, their characteristic orbital velocity. Since the turbulent flow consist in many eddies, of varying sizes and speeds, $\hat{u}$ and d do not assume each of a single value but vary within a certain range. In stationary, homogeneous and isotropic turbulence, that is, a turbulent flow that statistically appears unchanging in time, uniform in space and without preferential direction, all eddies of a given size (same d) behave more or less in the same way and can be thought of sharing the same characteristic velocity $\hat{u}$. In other words, we make the assumption that $\hat{u}$ is a function of d (Fig. 2).

Fig. 1. Drawing of a turbulent flow by Leonardo da Vinci (1452-1519), who recognized that turbulence involves a multitude of eddies at various scales.

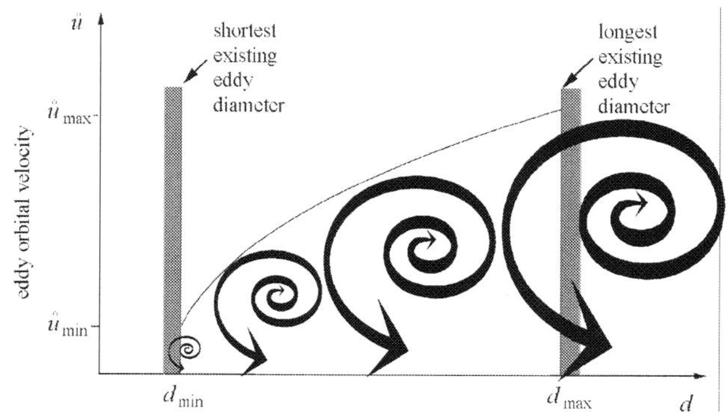

Fig. 2. Eddy orbital velocity versus eddy length scale in homogeneous turbulence. The largest eddies spin the fastest. Modified from (Cushman-Roisin, 2007; Watson, 2003).

3.2 Wind spectra

The fluctuations in the wind can be thought of as resulting from a composite of sinusoidally varying winds imposed on the mean steady wind. These sinusoidal variations will have a variety of frequencies and amplitudes. The term 'spectrum' is used to describe functions of frequency. Thus the function that characterizes turbulence as a function of frequency is

Fig. 3. Schematic of eddies as a function of height at atmospheric boundary layer. Modified from (Watson, 2003).

known as a 'spectral density' function. Since the average value of any sinusoid is zero, the amplitudes are characterized in terms of their mean square values. This type of analysis originated in electric power applications, where the square of the voltage or current is proportional to the power. The complete name for the function describing the relation between frequency and amplitudes of sine waves making up the fluctuating wind speed is therefore 'power spectral density' or PSD for short.
There are three points of particular importance to note regarding PSD's.

- The wind variance due to fluctuations within a frequency range may be found by integrating the $PSD^{+}_{Uwind}(f)$ in that range.
- The integral of $PSD^{+}_{Uwind}(f)$ over all frequencies is equal to the total wind variance.
- If two-sided $PSD(f)$ are used instead of $PSD+(f)$, the variance is twice the $PSD(f)$ if only positive frequencies are used in the integration domain.

Power spectral densities are often used in dynamic analyses. The variations of wind in the stream direction of the flow are usually characterized through the PSD of the longitudinal component of the instantaneous speed. A number of power spectral density functions are used as models in wind energy engineering when representative turbulence power spectral densities are unavailable for a given site. The mathematical forms for along-wind velocity spectra which are currently used in major current, or recent, wind codes and standards (Tamura et al., 2003) are due to von Karman, Kaimal and Davenport.

3.3 Coherence models

IEC 61400-1, 2005 specifies the coherence function (in absolute value) $\gamma_{IEC}(f)$ for two points r and c separated a distance d in the rotor plane according to:

$$\gamma_{IEC}(f) = \exp\left[-A\sqrt{\left(\frac{d_{rc}\cdot f}{\langle U_{wind}\rangle}\right)^2 + \left(0{,}12\frac{d_{rc}}{\ell_{Uwind}}\right)^2}\right] \quad (1)$$

IEC 61400-1 recommends $A \approx 12$; (Frandsen et al., 2007) recommends $A \approx 5$ and (Saranyasoontorn et al., 2004) recommends $A \approx 9{,}7$. Note that this expression is intended for points in the area swept by the blades. Therefore, it is not intended for estimating the wind coherence at different turbines.

(Schlez & Infield, 1998) derived an empirical model, based mainly on measurements with 18 m high masts with distances up to 102 m in the Rutherford Appleton Laboratory, UK. According to them, the coherence decreases exponentially at a site-specific rate respect wind travel time. The decay constants for lateral and longitudinal directions are, A_{ong} and A_{lat}, respectively. A_{ong} is the decay factor when the flow is longitudinal (α_{rc}= 0). A_{lat} is the decay factor when the flow is lateral, i.e. when the wind direction is perpendicular to the line between points r and c (α_{rc}= π/2 rad).

Thus, a compound decay constant A_{rc} can be estimated according to (2) for any arbitrary disposition of points r and c (see Fig. 4).

$$A_{rc} = \sqrt{\left[A_{long}\cos(\alpha_{rc})\right]^2 + \left[A_{lat}\sin(\alpha_{rc})\right]^2} \tag{2}$$

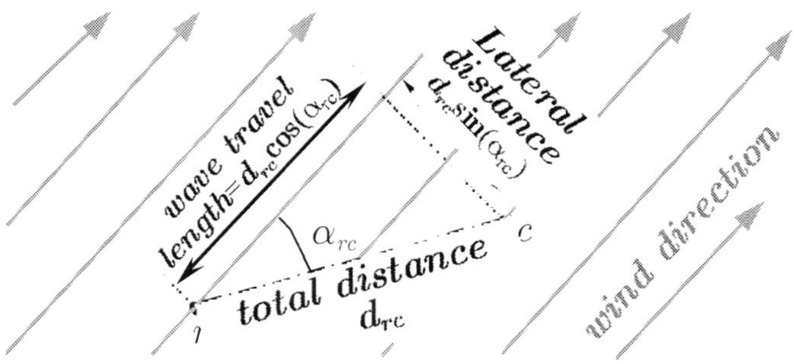

Fig. 4. Definition of distance d_{rc} and angle α_{rc} between the points r and c.

(Schlez and Infield, 1998) recommended for the Rutherford Appleton Laboratory $A_{long} \approx$ (15±5) $\sigma_{Uwind} / \langle U_{wind} \rangle$ and $A_{at} \approx$ (17,5±5)(m/s)$^{-1}\sigma_{Uwind}$, where σ_{Uwind} is the standard deviation of the wind speed in m/s.

(Saranyasoontorn et al., 2004) adjusted a coherence model from experimental data in LIST Test site at Bushland, Texas.

(Sørensen et al., 2008) fitted the lateral and longitudinal decay factors, A_{long} = 4 and A_{lat}= $\langle U_{wind} \rangle$ /(2 m/s), respectively, from measurements at 80 m height with up to 1.2 km distances in Høvsøre, Denmark. With these parameters, Sørensen et al. used a complex rooth coherence $\vec{\gamma}_{rc}(f)$ (adding an average phase delay to the absolute squared coherence $\left|\vec{\gamma}_{rc}^2(f)\right|$ proposed by Schlez and Infield, 1998). Since complex coherence is used, the phase of the coherence indicates the average delay between wind fluctuations at different points. Sørensen et al., 2002, used the Taylor's "frozen turbulence" model to compute the average time delay τ_{rc} as the time difference between the arrival to the points r and c of a flat wind wave front travelling at average wind speed.

$$\tau_{rc} = \frac{d_{rc}\cos(\alpha_{rc})}{\langle U_{wind} \rangle} \quad (3)$$

where d_{rc} is the distance between points r and c (see Fig. 4).
Finally, the expression of the complex root coherence $\vec{\gamma}_{rc}(f)$ for Høvsøre is:

$$\vec{\gamma}_{rc}(f) = \exp\left[-f\left(A_{rc}\frac{d_{rc}}{\langle U_{wind} \rangle} + j2\pi\tau_{rc}\right)\right] \quad (4)$$

It can be derived that at low frequencies $\vec{\gamma}_{rc}(f)$ tends to unity with zero phase (fully positive correlated fluctuations) and at high frequencies $\vec{\gamma}_{rc}(f)$ tends to zero with random phase (uncorrelated fluctuations).

4. Wind turbine torque

Power oscillations are the ultimate response of generators to torque fluctuations due to spatial and temporal wind variations.
The turbine torque can be estimated from blade theory. Since either the blade section, neither the relative speed nor the angle of attack is constant along the blade from the root to the tip, torque must be integrated along the blade elements. The lift and drag coefficients for the whole blade can be parameterized for blade tip conditions.
A further simplification is to consider a torque coefficient $C_q(\lambda,\theta)$ depending only on the pitch angle θ and on tip speed ratio λ. In this work, the tip speed ratio is referred to an equivalent wind speed since the wind conditions vary along the swept area:

$$\lambda = R\,\Omega_{\text{rotor}}/\,U_{eq'} \quad (5)$$

where R is the rotor radius, Ω_{rotor} is the rotor angular speed and U_{eq} is the equivalent wind speed. In a first approximation, U_{eq} is the longitudinal wind speed component averaged along the swept area provided the shaft is aligned with the wind (Handsen et al., 2003 Handsen et al., 2007).
Thus, the turbine torque is:

$$T_{rotor} = \tfrac{1}{2}\rho_{air}\pi R^3 U_{eq}^2 C_q(\lambda,\theta) \quad (6)$$

where ρ_{air} is the air density.
In a second approximation, U_{eq} is defined as the wind speed applied to (5) and (6) which produces the same aerodynamic torque T_{rotor} than the real wind field. Alternatively, the torque coefficient can be computed as the power coefficient divided by the tip speed ratio, $C_q(\lambda,\theta) = C_P(\lambda,\theta)/\lambda$.

4.1 Definition of the equivalent wind, equivalent turbulence and effective quadratic turbulence

a)Equivalent wind

The *equivalent wind* U_{eq} is an artifice defined as the uniform wind which would produce the same torque as the real wind field. According to (6), it can be computed from real torque as (7):

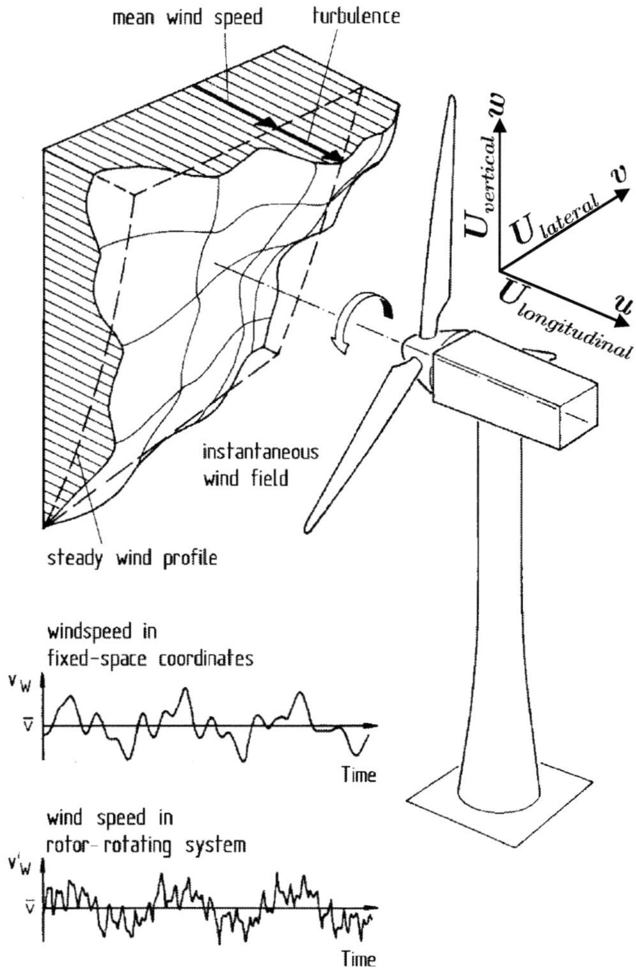

Fig. 5. Effect of an uneven wind-speed distribution over the swept rotor area on the upwind velocity of the rotating rotor blades. The lagrangian motion coordinates are added assuming the turbine is aligned with the wind. Taken from (Handsen et al., 2003).

$$U_{eq} = \sqrt{\frac{2\ T_{rotor}}{\rho_{air}\pi R^3 C_q(\lambda,\theta)}} \tag{7}$$

where $C_q(\lambda,\theta)$ is the turbine torque coefficient, T_{rotor} is the torque in the low speed shaft of the wind turbine, R is the rotor radius, Ω_{rotor} is the rotor angular speed and ρ_{air} is the air density. Since the wind varies along the swept area (wind distribution is irregular), the tip speed ratio λ must be computed also from (5).

The *simplification* of using an *equivalent wind* is huge since the non-stationary three-dimensional wind field is approximated by a signal which *produces the same torque*. Apart form accelerating notably the simulations, U_{eq} describes in only one signal the effect of the turbulent flow in the drive train.

The actual wind speed U_{wind} is measured at a point by an anemometer whereas the equivalent wind speed U_{eq} is referred to the rotor surface (or more precisely, to the turbine torque). Since the Taylor's hypothesis of "frozen turbulence" is usually applicable, the spatial diversity of wind can be approximated to the pointwise time variation of wind times its mean value, $\langle U_{wind} \rangle$, and hence U_{eq} can be considered a low-pass filtered version of U_{wind} (plus the rotational sampling effect due to wind shear and tower shadow effect).

On the one hand, the meteorological science refers to the actual wind speed U_{wind} since the equivalent wind U_{eq} is, in fact, a mathematical artifice. On the other hand, turbine torque or power is customarily referred to the equivalent wind U_{eq} instead of the 3-D wind field for convenience.

A good introduction about the equivalent wind can be found in (Martins et al. 2006). The complete characteristics of the wind that the turbine will face during operation can be found in (Burton et al., 2001).

The equivalent wind speed signal, $U_{eq}(t)$, just describes a smoothed wind speed time series at the swept area. For calculating the influence of wind turbulence into the turbine mechanical torque, it has to be considered the wind distribution along the swept area by a vector field (Veers, 1988). Blade iteration techniques can be applied for a detailed analysis of torques and forces in the rotor (Hier, 2006).

The anemometer dynamic response to fast changes in wind also influences measured wind (Pedersen et al., 2006). Most measures are taken with cup anemometers, which have a response lengths between 1 and 2 m, corresponding to a frequency cut-off between f_c = (10 m/s)/10 m = 1 Hz and f_c = (10 m/s)/20 m = 0,5 Hz for 10 m/s average speed.

Apart from metrological issues, the spatial diversity of turbulent wind field reduces its impact in rotor torque. Complete and proved three dimensional wind models are available for estimating aerodynamic behavior of turbines (Saranyasoontorn et al., 2004; Mann, 1998; Antoniou et al., 2007). Turbulent models are typically used in blade fatigue load.

From the grid point of view, the main effect of spatial diversity is the torque modulation due to wind shear and tower shadow (Gordon-Leishman, 2002). Vertical wind profile also influences energy yield and it is considered in wind power resource assessment (Antoniou et al., 2007).

5. Fundaments of the rotor spatial filtering

The idea in the rotor wind model is to generate an equivalent wind speed which can be applied to a simplified aerodynamic model to simulate the torque on the wind turbine shaft. The rotor wind filter includes the smoothing of the wind speed due to the weighted averaging over the rotor.

The input of this filter is the wind U_{wind} which would be measured at an anemometer installed at the hub height and the output is the estimated equivalent wind, U_{eq1}, which is a smoothed version of the measured wind.

Neglecting the periodic components, the rotor block smoothing of wind turbine can be expressed as a wind turbine admittance function defined as:

$$|H_1(f)|^2 = \frac{PSD_{Uwind}(f)}{PSD_{Ueq_1}(f)} \tag{8}$$

where $PSD_{Uwind}(f)$ is the power spectral density of the wind measured at a point and $PSD_{Ueq_1}(f)$ is the power spectral density of the equivalent wind (without the periodic components due to the cuasi-deterministic variation of torque with rotor angle).

The wind spectrum $PSD_{Uwind}(f)$ is equivalent to low-pass filters with a typical system order $r' = 5/6$ (i.e., the spectrum decays a bit slower than the output of a first-order low pass filter). Power output decreases quicker than the pointwise wind at $f > 0.01$ Hz (Mur-Amada et al. 2003) and this is partially due to the spatial distribution of turbulence, the high inertia and the viscous-elastic coupling of turbine and generator through the gear box (Engelen, 2007). Complex vibration dynamics influence power output and a simple model with two coupled mass (equivalent to a second-order system) is insufficient to represent the resonance modes of blades and tower.

The square modulus of the filter can be computed from the filter Laplace transform $H_1'(s)$:

$$|H_1(f)|^2 = H_1'(j2\pi f)[H_1'(j2\pi f)]^* \tag{9}$$

The phase of the filter indicates the lag between the wind at the anemometer and at the turbine hub. The phase of the filter does not affect $PSD_{Ueq_1}(f)$ since wind process is stationary and, accordingly, the phase is arbitrary. The lag difference of equivalent wind among turbines at points r and c will be considered through complex coherence $\vec{\gamma}_{rc}(f)$, irrespective of the argument of $H_1(f)$.

The frequencies of interest for flicker and blade fatigue are in the range of tenths of hertz to 35 Hz. These frequencies correspond to sub-sound and sound (inertial subrange) and they have wavelengths comparable to the rotor diameter. The assumption that such fluctuations correspond to plane waves travelling in the longitudinal direction and arriving simultaneously at the rotor plane is not realistic. Therefore, quick fluctuations do not reach the rotor disk simultaneously and fluctuations are partially attenuated by spatial diversity. In brief, $H_1'(s)$ is a low-pass filter with meaningless phase.

The smoothing due to the spatial diversity in the rotor area is usually accounted as an aerodynamic filter, basically as a first or second order low-pass filter of cut-off frequency $\sim 0{,}1224\langle U_{wind}\rangle / R$ respect an ideal and unperturbed anemometer measure (Rosas, 2003). For multimegawatt turbines, the rotor filters significantly fluctuations shorter than one minute with a second order decay (cut-off frequency in the order of 0,017 Hz). The turbine vibrations are much more noteworthy than the turbulence at frequencies higher than 0,1 Hz.

The presence of the ground surface hinders vertical development in larger eddies. The lateral turbulence component is responsible for turbulence driven wind direction changes, but it is a secondary factor in turbine torque fluctuations. Moreover, according IEC 61400-1, 2005, vertical and transversal turbulence has a significantly smaller length scale and lower magnitude. Thus, the vertical and lateral component of turbulence averaged along the turbine rotor can be neglected in turbine torque in the first instance.

6. Equivalent wind of turbine clusters

6.1 Average farm behavior

Sometimes, a reduced model of the whole wind farm is very useful for simulating a wind farm in the grid. The behavior of a network with wind generation can be studied supplying the farm equivalent wind as input to a conventional turbine model connected to the equivalent grid.

The foundations of these models, their usual conventions and their limitations can be seen in (Akhmatov & Knudsen, 2002; Kazachkov & Stapleton, 2004; Fernandez et al, 2006). The average power and torque in the turbines and in the farm are the same on per unit values. This can be a significant advantage for the simulation since most parameters do not have to be scalled. Notice that if electrical values are not expressed *per unit*, currents and network parameters have to be properly scalled.

For convenience, all the N turbines of a wind farm are represented with a single turbine of radius R_{farm} spinning at angular speed Ω_{farm}. The equivalent power, torque, wind, rotor speed, pitch and voltage are their average among the turbines of the farm. Thus, the equivalent turbine represents the average operation among the farm turbines.

If the turbines are different or their operational conditions are dissimilar, the averages are weighted by the turbine power (because the aim of this work is to reproduce the power output of farms). Elsewhere, the farm averaged parameters can by approximated by a conventional arithmetic mean.

6.2 Model based in equivalent squared wind

Assuming that the equivalent wind at the different wind turbines behaves as a multivariate Gausian process with spectral covariance matrix:

$$\Xi_{U_{eq}}(f)=\left[\vec{\gamma}_{ij}(f)\sqrt{PSD_{U_{eq,i}}(f)PSD_{U_{eq,j}}(f)}\right] \tag{10}$$

Thus, the $PSD_{U_{eq,farm}}(f)$ of the equivalent squared wind for the farm can be computed as:

$$PSD_{U_{eq,farm}}(f) = \vec{b}_{farm}^{T}\Xi_{U_{eq}}(f)\vec{b}_{farm} = \sum_{i=1}^{N}\sum_{j=1}^{N} b_i b_j \vec{\gamma}_{ij}(f)\sqrt{PSD_{U_{eq,i}}(f)PSD_{U_{eq,j}}(f)} \tag{11}$$

where $\vec{\gamma}_{ij}(f)$ is the complex coherence of the equivalent wind of turbines i and j at frequency f, and the contribution of the turbine i to the farm wind is b_i.

If all the turbines experience similar equivalent wind spectra - $PSD_{U_{eq,i}}(f) \approx PSD_{U_{eq}}(f)$ - and their contribution to the farm is similar - $b_i \approx 1/N$ - then the following approximate formula is valid:

$$PSD_{U_{eq,farm}}(f) \approx \frac{PSD_{U_{eq}}(f)}{N^2}\sum_{i=1}^{N}\sum_{j=1}^{N}\vec{\gamma}_{ij}(f) \tag{12}$$

Notice that $\vec{\gamma}_{ii}(f) = 1$ and $0 \le \left|\vec{\gamma}_{ij}(f)\right| \le 1$. Since the real part of $\vec{\gamma}_{ij}(f)$ is usually positive or close to zero (i.e., non-negative correlation of fluctuations), $PSD_{U_{eq,farm}}(f)$ is generally between the behavior of perfectly correlated and independent fluctuations at the turbines.

$$\frac{PSD_{U_{eq}}(f)}{N} \lesssim PSD_{U_{eq,farm}}(f) \leq \frac{PSD_{U_{eq}}(f)}{N^2} \tag{13}$$

$$\text{since } 0 \lesssim \mathrm{Re}[\vec{\gamma}'_{ij}(f)]$$

6.3 Equivalent wind of turbines distributed along a geographical area

In (4), a model of complex root coherence $\vec{\gamma}_{rc}(f)$ was introduced based on the works of (Schlez & Infield, 1998) in the Rutherford Appleton Laboratory and (Sørensen et al., 2008) in the Høvsøre offshore wind farm. In (12), a formula was derived assuming all the turbines experience a similar wind and they have similar characteristics.

In this section, the decrease of variability of the equivalent wind of a geographical area due to its spatial diversity is computed in (14) from the variability at a single turbine or a single farm and from the complex root coherence $\vec{\gamma}_{rc}(f)$.

Formula (14) assumes that wind turbines are approximately evenly spread over the area corresponding to the integrating limits. Even though the former assumptions are oversimplifications of the complex meteorological behavior neither it considers wakes, (14) indicates the general trend in the decrease of wind power variability due to spatial diversity in bigger areas. Notice that $PSD_{U_{eq,turbine}}(f)$ is assumed to be representative of the average turbulence experienced by turbines in the region and hence, it must account average wake effects. Even though the model is not accurate enough for many calculations, it leads to expression (19) that links the smoothing effect of the spatial diversity of wind generators in an area and its dimensions.

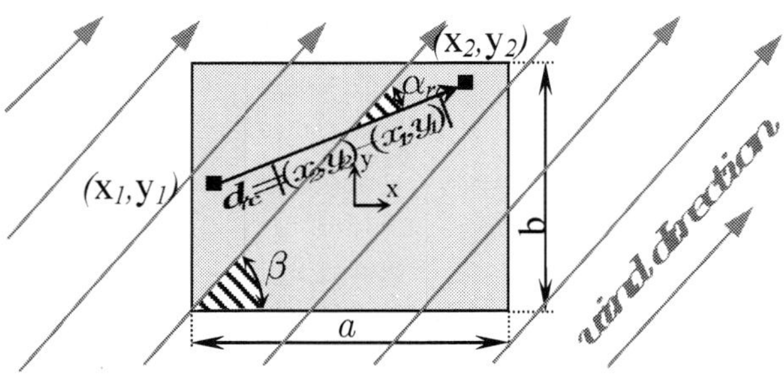

Fig. 6. Wind farm dimensions, angles and distances among wind farm points for the general case.

The coherence $\vec{\gamma}_{rc}(f)$ between points $r = (x_1, y_1)$ and $c = (x_2, y_2)$ inside the wind farm can be derived from Fig. 6 and formulas (2), (3) and (4). The geometric distance between them is $d_{rc} = |(x_2, y_2) - (x_1, y_1)| = [(y_2 - y_1)^2 + (x_2 - x_1)^2]^{1/2}$ and the angle between the line that links the two points and the wind direction is $\alpha_{rc} = \beta - \mathrm{ArcTan}[(y_2 - y_1)/(x_2 - x_1)]$. In the general case, the equivalent wind taking into account the spatial diversity can be computed extending formula (12) to the continuous case:

$$\frac{PSD_{Ueq,area}(f)}{PSD_{Ueq,turb}(f)} \approx \frac{\int_{-b/2}^{b/2}\int_{y_1}^{b/2}\int_{-a/2}^{a/2}\int_{x_1}^{a/2} \vec{\gamma}_{rc}(f,d_{rc},\alpha_{rc})\,dx_2dx_1dy_2dy_1}{\int_{-b/2}^{b/2}\int_{y_1}^{b/2}\int_{-a/2}^{a/2}\int_{x_1}^{a/2} dx_2dx_1dy_2dy_1} \tag{14}$$

where the quadruple integral in the denominator is a forth of the squared area, i.e., $a^2\, b^2/4$.

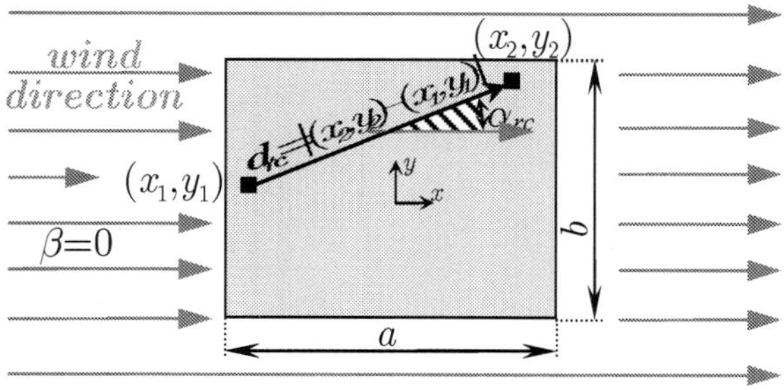

Fig. 7. Wind farm parameters when wind has the x direction (β=0).

Due to the complexity of d_{rc} and α_{rc} and the estimation of $\vec{\gamma}_{rc}(f,d_{rc},\alpha_{rc})$ in formula (4), no analytical closed form of (14) have been found for the general case.
In case wind has x direction as in Fig. 7, then the coherence has a simpler expression:

$$\vec{\gamma}_{rc}(f,d_{rc},\alpha_{rc}) = = \exp\left[\frac{-f}{\langle U_{wind}\rangle}\left(\sqrt{\left[A_{long}(x_2-x_1)\right]^2+\left[A_{lat}(y_2-y_1)\right]^2}+j2\pi(x_2-x_1)\right)\right] \tag{15}$$

The presence of the squared root in (15) prevents from obtaining an analytical $PSD_{Ueq,area}(f)$. In case aA_{long} bA_{lat}, the region can be considered a thin column of turbines transversally aligned to the wind. This is the case of many wind farms where turbine layout has been designed to minimize wake loss (see Fig. 9) and areas where wind farms or turbines are sited in mountain ridges, in seashores and in cliff tops perpendicular to the wind. Since $A_{long}(x_2\text{-}x_1)$ $A_{lat}(y_2\text{-}y_1)$, then $PSD_{Ueq,area}(f)$ can be computed analytically as:

$$\frac{PSD_{Ueq,lat\ area}(f)}{PSD_{Ueq,turb}(f)} \approx f_1\left(\frac{A_{lat}b\ f}{\langle U_{wind}\rangle}\right) \tag{16}$$

where $f_1(x) = 2\left(-1+e^{-x}+x\right)/\,x^2$

In case aA_{long} bA_{lat}, the region can be considered a thin row of wind farms longitudinally aligned to the wind. This is the case of many areas where wind farms are disposed in a gorge, canyon, valley or similar where wind is directed in the feature direction (see Fig. 9). Since $A_{long}(x_2\text{-}x_1)$ $A_{lat}(y_2\text{-}y_1)$, then $PSD_{Ueq,area}(f)$ can be computed analytically as:

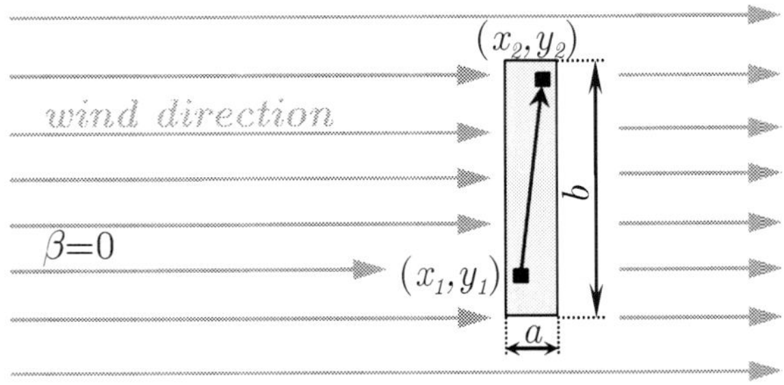

Fig. 8. Wind farm with turbines aligned transversally to the wind.

$$\frac{PSD_{U_{eq,long\ area}}(f)}{PSD_{U_{eq,turb}}(f)} \approx f_2\left(\frac{A_{long} a\ f}{\langle U_{wind}\rangle},\ A_{long}\right) \tag{17}$$

where $f_2\left(\frac{A_{long} a\ f}{\langle U_{wind}\rangle},\ A_{long}\right) = \mathrm{Re}\left\{f_1\left(\frac{(A_{long}+j2\pi)\ a\ f}{\langle U_{wind}\rangle}\right)\right\}$

which can be expressed with real functions as:

$$f_2(\nu, A_{long}) = \frac{(\nu-1)+(\nu+1)\left(\frac{2\pi}{A_{long}}\right)^2 + \left\{\left(1-\left(\frac{2\pi}{A_{long}}\right)^2\right)\mathrm{Cos}\left[\frac{2\pi\nu}{A_{long}}\right] - \frac{4\pi}{A_{long}}\mathrm{Sin}\left[\frac{2\pi\nu}{A_{long}}\right]\right\}e^{-\nu}}{\frac{1}{2}\nu^2\left(1+\left(2\pi/A_{long}\right)^2\right)^2} \tag{18}$$

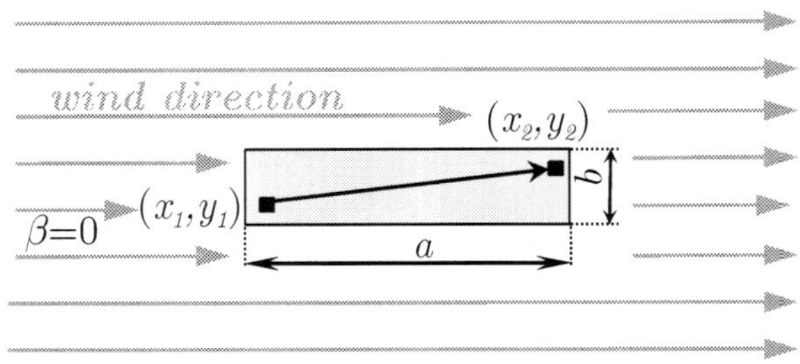

Fig. 9. Wind farm with turbines aligned longitudinally to the wind.

Notice that (17) includes an imaginary part that is due to the frozen turbulence model in formula (4). A wind wave travels at wind speed, producing an spatially average PSD that depends on the longitudinal length a relative to the wavelength. For long wavelengths compared to the longitudinal dimension of the area ($A_{long}\ \ 2\pi$), the imaginary part in (17) can be neglected and (17) simplifies to (16). This is the case of the Rutherford Appleton

Laboratory, where (Schlez & Infield, 1998) fitted the longitudinal decay factor to $A_{long} \approx (15\pm5)\langle U_{wind}\rangle / \sigma_{Uwind}$ for distances up to 102 m.
But when the wavelengths are similar or smaller than the longitudinal dimension, ($A_{long} \lesssim 2\pi$), then the fluctuations are notably smoothed. This is the case of the Høvsøre offshore wind farm, where (Sørensen et Al., 2008) fitted the longitudinal decay factor to A_{long} = 4 for distances up to 2 km. In plain words, the disturbances travels at wind speed in the longitudinal direction, not arriving at all the points of the area simultaneously and thus, producing an average wind smoother in longitudinal areas than in transversal regions.
In the normalized longitudinal and transversal distances have the same order, then (14) can be estimated as the compound of many stacked longitudinal or transversal areas (see Fig. 10):

$$|H_3(f)|^2 = \frac{PSD_{Ueq,rect\ area}(f)}{PSD_{Ueq,turb}(f)} = \frac{PSD_{Ueq,long\ area}(f)}{PSD_{Ueq,lat\ area}(f)} \frac{PSD_{Ueq,lat\ area}(f)}{PSD_{Ueq,turb}(f)} \approx \\ \approx f_1\left(\frac{A_{lat} b\ f}{\langle U_{wind}\rangle}\right) f_2\left(\frac{A_{long} a\ f}{\langle U_{wind}\rangle}, A_{long}\right) \tag{19}$$

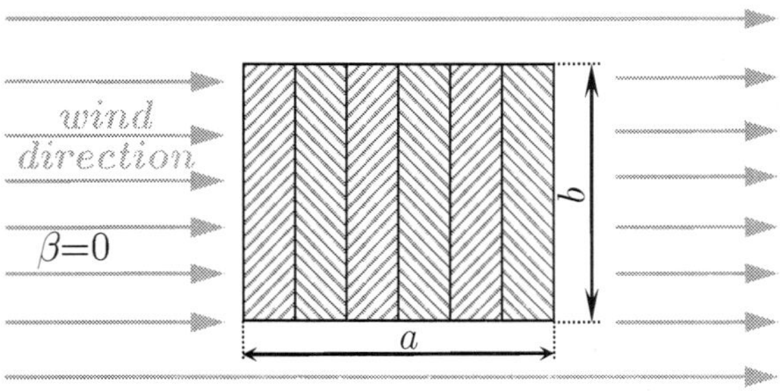

Fig. 10. Rectangular area divided in smaller transversal areas.

The approximation (19) is equivalent to consider the Manhattan distance (L_1 or city-block metric) instead of the Euclidean distance (L_2 metric) in the coherence $\vec{\gamma}_{rc}$ (15):

$$\sqrt{\left[A_{long}(x_2-x_1)\right]^2 + \left[A_{lat}(y_2-y_1)\right]^2} \sim A_{long}(x_2-x_1) + A_{lat}(y_2-y_1) \tag{20}$$

6.4 Equivalent wind smoothing due to turbine spatial layout

Expression (19) is the squared modulus of the transfer function of the spatial diversity smoothing in the area. $H_3(f)$ corresponds to the low-pass filters in Fig. 11 with cut-off frequencies inversely proportional to the region dimensions.
The overall cut-off frequency of the spatially averaged wind is obtained solving $|H_3(f)|^2 = 1/4$. Thus, the cut-off frequency of transversal wind farms (solid black line in Fig. 11) is:

$$f_{cut,lat} = 6.83\frac{\langle U_{wind}\rangle}{bA_{lat}} \tag{21}$$

In the Rutherford Appleton Laboratory (RAL), $A_{lat} \approx (17{,}5\pm5)(\mathrm{m/s})^{-1}\sigma_{Uwind}$ and hence $f_{cut,lat} \approx (0{,}42\pm0{,}12)\langle U_{wind}\rangle \;/\; (\sigma_{Uwind}\, b)$. A typical value of the turbulence intensity $\sigma_{Uwind}/\ \langle U_{wind}\rangle$ is around 0,12 and for such value $f_{cut,lat} \sim (3.5\pm1)/\ b$, where b is the lateral dimension of the area in meters. For a lateral dimension of a wind farm of b = 3 km, the cut-off frequency is in the order of 1,16 mHz.
In the Høvsøre wind farm, $A_{lat}= \langle U_{wind}\rangle / (2\ \mathrm{m/s})$ and hence $f_{cut,lat} \approx 13{,}66/b$, where b is a constant expressed in meters. For a wind farm of b = 3 km, the cut-off frequency is in the order of 4,5 mHz (about four times the estimation from RAL).
In RAL, $A_{long} \approx (15\pm5)\ \sigma_{Uwind} / \langle U_{wind}\rangle$. A typical value of the turbulence intensity $\sigma_{Uwind} / \langle U_{wind}\rangle$ is around 0,12 and for such value $A_{long} \approx (1{,}8\pm0{,}6)$.

$$f_{cut,long} \underset{A_{long}\sim1{,}8}{=} 1{,}1839\frac{\langle U_{wind}\rangle}{a\,A_{long}} \underset{A_{long}=1.8}{=} 0.6577\frac{\langle U_{wind}\rangle}{a} \tag{22}$$

For a wind speed of $\langle U_{wind}\rangle \sim 10$ m/s and a wind farm of a = 3 km longitudinal dimension, the cut-off frequency is in the order of 2,19 mHz.
In the Høvsøre wind farm, $A_{long} = 4$ (about twice the value from RAL). The cut-off frequency of a longitudinal area with A_{long} around 4 (dashed gray line in Fig. 11) is:

$$f_{cut,long} \underset{A_{long}\sim4}{=} 2.7217\frac{\langle U_{wind}\rangle}{a\,A_{long}} \underset{A_{long}=4}{=} 0.6804\frac{\langle U_{wind}\rangle}{a} \tag{23}$$

For a wind speed of $\langle U_{wind}\rangle \sim 10$ m/s and a wind farm of a = 3 km longitudinal dimension, the cut-off frequency is in the order of 2,26 mHz.
In accordance with experimental measures, turbulence fluctuations quicker than a few minutes are notably smoothed in the wind farm output. This relation is proportional to the dimensions of the area where the wind turbines are sited. That is, if the dimensions of the zone are doubled, the area is four times the original region and the cut-off frequencies are halved. In other words, *the smoothing of the aggregated wind is proportional to the longitudinal and lateral distances of the zone* (and thus, related to the square root of the area if zone shape is maintained).
In sum, the lateral cut-off frequency is inversely proportional to the site parameters A_{lat} and the longitudinal cut-off frequency is only slightly dependent on A_{long}. Note that the longitudinal cut-off frequency show closer agreement for Høvsøre and RAL since it is dominated by frozen turbulence hypothesis.
However, if transversal or longitudinal smoothing dominates, then the cut-off frequency is approximately the minimum of $f_{cut,lat}$ and $f_{cut,long}$. The system behaves as a first order system at frequencies above both cut-off frequencies, and similar to ½ order system in between $f_{cut,lat}$ and $f_{cut,long}$.

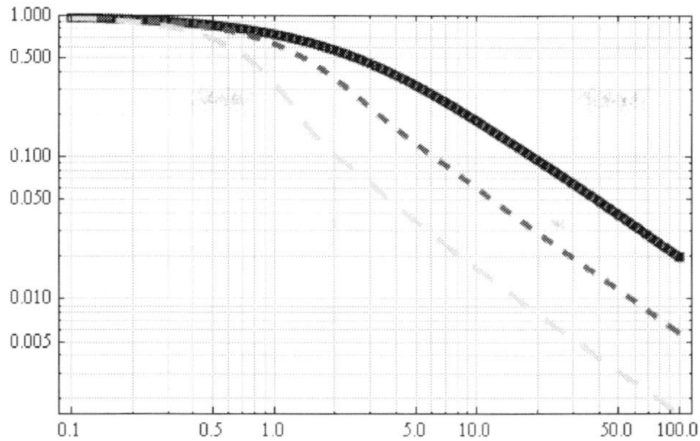

Fig. 11. Normalized ratio $PSD_{Ueq,area}(f)$ / $PSD_{Ueq,turbine}(f)$ for transversal (solid thick black line) and longitudinal areas (dashed dark gray line for A_{long} = 4, long dashed light gray line for A_{long} = 1,8). Horizontal axis is expressed in either longitudinal and lateral adimensional frequency $a\ A_{long} f / \langle U_{wind} \rangle$ or $b\ A_{lat} f / \langle U_{wind} \rangle$.

7. Spectrum and coherence estimated from the weather station network

The network of weather stations provides a wide coverage of slow variations of wind. Many stations provide hourly or half-hourly data. These data is used in the program WINDFREDOM (Mur-Amada, 2009) to compute the wind spectra and the coherences between nearby locations.

Quick fluctuations of wind are more related to the turbine integrity, structural forces and control issues. But they are quite local, and they cancel partially among clusters of wind farm. The slower fluctuations are more cumbersome from the grid point of view, since they have bigger coherences with small phase delays.

The coherence and the spectrum of wind speed oscillations up to 12 days are analyzed, as an illustrative example, at the airports of the Spanish cities of Logroño and Zaragoza. Both cities are located in the Ebro River and share a similar wind regime. The weather stations are 140,5 km apart (see Fig. 12) and the analysis is based on one year data, from October 2008 to October 2009.

The spectrograms in Fig. 13 and Fig. 14 show the evolution of the power spectrum of the signal, computed from consecutive signal portions of 12 days. The details of the estimation procedure can be found in the annexes of this thesis.

Wind spectra and coherence has been computed from the periodogram, and the spectrograms of the signals are also shown to inform of the variability of the frequency content. The quartiles and the 5% and 95% quantiles of the wind speed are also shown in the lower portions of in Fig. 13 and Fig. 14. The unavailable data have been interpolated between the nearest available points. Some measurements are outliers, as it can be noticed from the 5% quantiles in Fig. 13 and Fig. 14, but they have not been corrected due to the lack of further information.

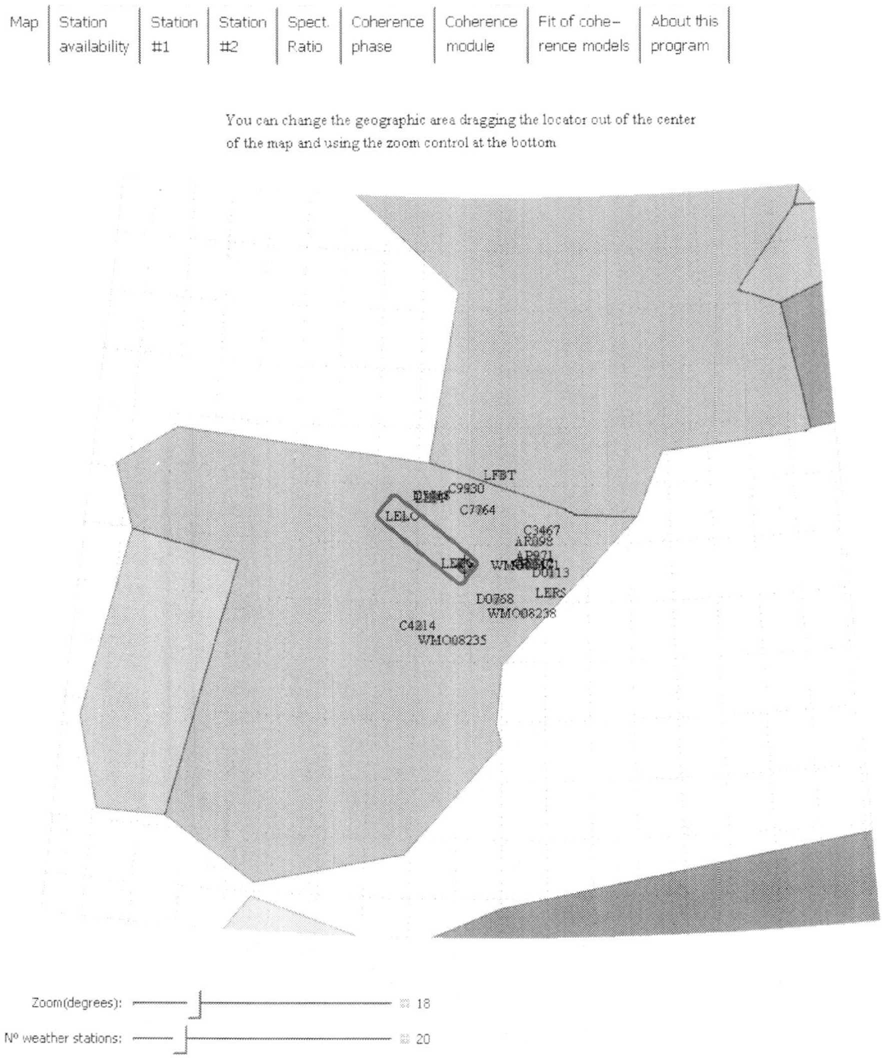

Fig. 12. Map from WINDFREDOM program with the location of Zaragoza and Logroño in the Iberian Peninsula.

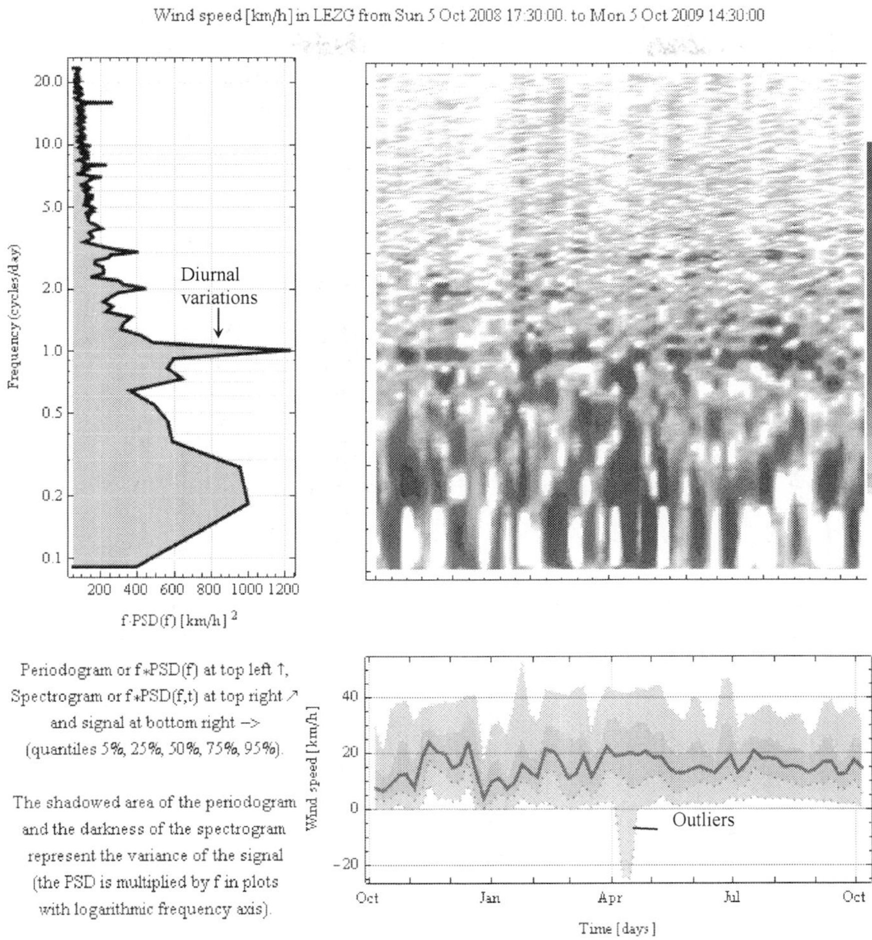

DESCRIPTIVE STATISTICS OF THE TIME SERIES

Number of samples: 17128. sampled at frequency f = 47. samples/day.

Mean μ = 16.292 km/h. Standard deviation σ = 11.481 km/h.

Median = 14.76 km/h. Mode(s) or commonest value(s) = {3.6} km/h.

Weibull fit: shape parameter k = 1.4408 and scale parameter λ = 17.952 km/h.

Kurtosis = 3.303

Fig. 13. Periodogram and spectrogram of Zaragoza airport (Spain) estimated with WINDFREDOM program.

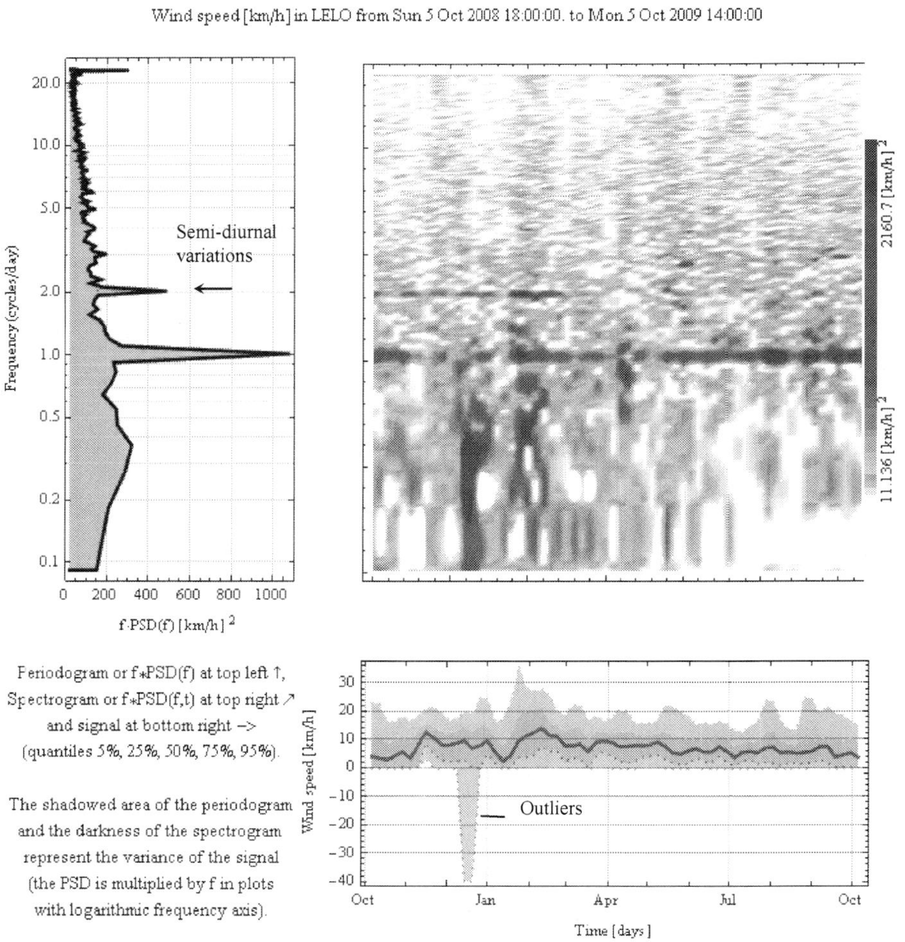

DESCRIPTIVE STATISTICS OF THE TIME SERIES

Number of samples: 13863. sampled at frequency f = 39. samples/day.

Mean μ = 7.8883 km/h. Standard deviation σ = 6.9413 km/h.

Median = 7.56 km/h. Mode(s) or commonest value(s) = {0} km/h.

Weibull fit: shape parameter k = 1.139 and scale parameter λ = 8.2651 km/h.

Kurtosis = 4.4311

Fig. 14. Periodogram and spectrogram of Logroño airport (Spain) estimated with WINDFREDOM program.

The diurnal and semi-diurnal variation peaks can be recognized in clearly in the periodograms of Fig. 13 and Fig. 14 (gray graph on the left) or as dark-bluish horizontal lines in the spectrogram (colour image on the right). The oscillation magnitude is not constant along one year because the horizontal lines get lighter or darker along the time.

The ratio between the periodograms and spectrograms of Fig. 13 and Fig. 14 is shown in Fig. 15. The wind in Zaragoza airport meteorological station (LEZG for short) is the double in average than in the weather station of Logroño airport (LELO for sort). The average ratio is about 0,4~0,6, indicating that the ratio of oscillation amplitudes are around √0,4~√0,6. The coefficient of variation (standard deviation divided by the mean) is 87% in Logroño and 70% in Zaragoza.

The quartiles of the time series at Logroño and Zaragoza (lower graph in Fig. 15) show significant differences. The red shadow indicates the interquartile range of Zaragoza and the thick red line is its median (the blue colours correspond to Zaragoza). The wind in Logroño (in blue) is about half the wind in Zaragoza in average.

The wind variations in each station show different features eventually. Some variations are replicated on the other station but with some non-systematic delay and with different magnitude. These features are the reason of the relatively small coherence of the two stations.

In practice, the oscillations observed in one station are seen, in some extent, in other station with some delay or in advance. The coherence $\vec{\gamma}_{\#1,\#2}$ is a complex magnitude with modulus between 0 and 1 and a phase, which represent the delay (positive angles) or the advance (negative angles) of the oscillations in the second weather station respect the first one (considered the reference). Since the spectrum of a signal is complex, the argument of the coherence $\vec{\gamma}_{rc}(f)$ is the average phase difference of the fluctuations.

The coherence $\vec{\gamma}_{rc}(f)$ indicates the correlation degree and the time pattern of the fluctuations. The modulus is analogous to the correlation coefficient of the spectrum lines from both locations. If the ratio among complex power spectrums shown in Fig. 15 is constant (in modulo and in phase), then the coherence is the unity and its argument is the average phase difference. If the complex ratio is random (in modulo or in phase, then the coherence is null.

However, the wind direction is not considered in this estimation, but it has a great impact on the coherence estimate. The time delay between oscillations τ depends greatly on the wind direction. Thus, the phase difference of the fluctuations, $\varphi = 2\pi f \tau$, can change notably and this would lead to very low coherences. If there are several preferential wind directions, the phase difference can experience great variability. In such cases, a more detailed model – maybe using Markov states indicating prevailing wind directions– is needed.

The red/purple colours in Fig. 15 indicate that phase difference is near 0 up to 0,5 cycles/day (small delay of fluctuations). However, the phase difference at frequencies above 2 cycles/day is quite big, indicating that the timing sequence of the fluctuations has varied along the study period (one year).

Data from Sun 5 Oct 2008 17:30:00. to Mon 5 Oct 2009 14:30:00

The distante between LEZG (station #1) and LELO (station #2) is 140.49 km.

	Latitute	Longitude	Elevation [m]	Series mean	Samples per day
LEZG	41.667	−1.	258. m	16.292 km/h	48
LELO	42.45	−2.333	363. m	7.8883 km/h	36

The periodogram and spectrogram of LELO have been divided by the ones of LEZG:

* A ratio bigger than unity indicate that the fluctuations of frequency f observed in LELO are bigger than in LEZG.

* A ratio smaller than unity indicate that the fluctuations of frequency f observed in LELO are smaller than in LEZG.

In the lower right plot, the quartiles of Wind speed [km/h] in LEZG are shown in red.
The quantiles of Wind speed [km/h] in LELO are shown in blue.

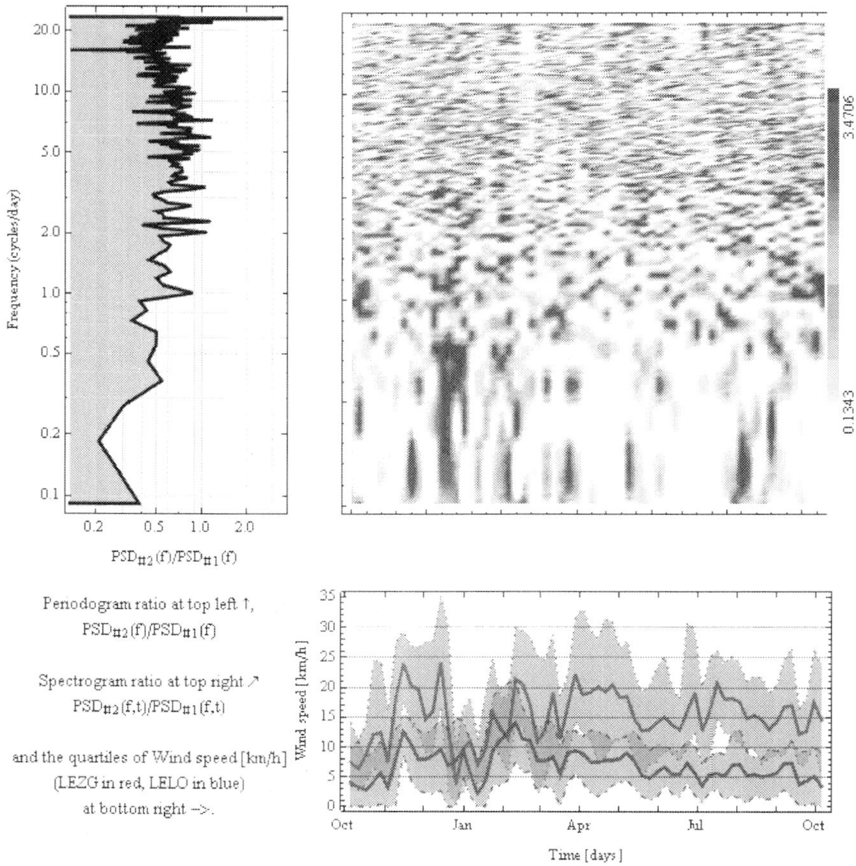

Fig. 15. Periodogram and spectrogram of Logroño airport divided by the ones of Zaragoza airport (estimated by WINDFREDOM program).

The phase difference between the spectrum of LEZG and LELO is computed to estimate the lag between the oscilations in both stations.
* A phase difference positive indicate that the fluctuations of f frequency are observed (in average) first in LEZGand then in LELO and viceversa.
*The average time lag can be computed as lag≈phase/(2π·f) when the phase is expressed in radians.
=> In general, slow fluctuations show a phase proportional to frequency indicating that the slow fluctuations are observed with a constant time delay between stations.
*A random phase indicates that the fluctuations happen first indiferently in any station without a simple temporal link.
=> In general, fast fluctuations show non–systematic phases indicating that they are local.

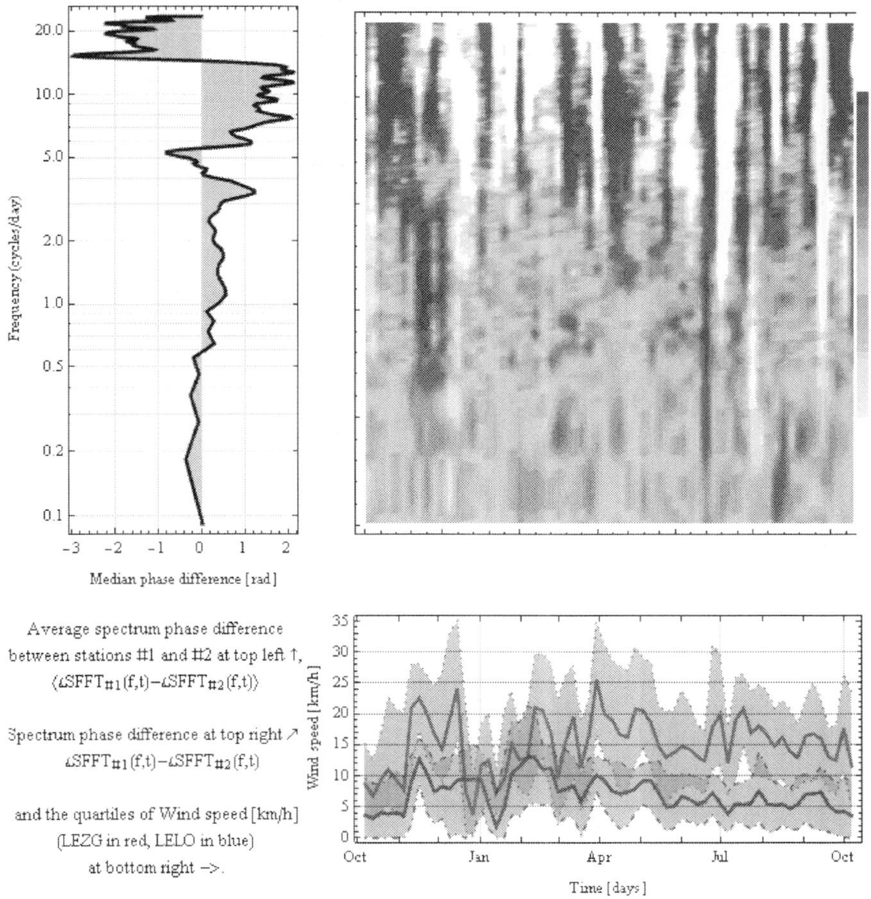

Fig. 16. Phase difference between the periodogram and spectrogram of Zaragoza airport respect the ones of Logroño airport (estimated by WINDFREDOM program).

8. References

2009/28/EC of the European Parliament and of the Council of 23 April 2009 on the promotion of the use of energy from renewable sources and amending and subsequently repealing Directives 2001/77/EC and 2003/30/EC.

Akhmatov, V. & Knudsen, H. (2002). An aggregate model of a grid-connected, large scale, offshore wind farm for power stability investigations-importance of windmill mechanical system. *Electrical Power Systems* 24 (2002) 709-717.

Anderson, C. L.; Cardell, J. B. (2008), *Reducing the Variability of Wind Power Generation for Participation in Day Ahead Electricity Markets*, Proceedings of the 41st Hawaii International Conference on System Sciences - 2008.

Antoniou, I.; Wagner, R.; Pedersen, S.M.; Paulsen, U.S.; Aagaard Madsen, H.; Jørgensen, H.E.; Thomsen, K.; Enevoldsen, P.; Thesbjerg, L., (2007), Influence of wind characteristics on turbine performance, *EWEC 2007*.

Apt, J. (2007) "The spectrum of power from wind turbines", *Journal of Power Sources*, 169, pp. 369–374

Boyle G. (2007). *Renewable Electricity and the Grid. The Challenge of Variability*, September 2007, Earthscan.

Burton T, Sharpe D, Jenkins N, Bossanyi E., *Wind Energy Handbook*. John Wiley & Sons Ltd, England, 2001.

Cappers, P.; Goldman, C.; Kathan, D (2009), Demand response in U.S. electricity markets: Empirical evidence, *Energy* (article in Press).

Cidrás, J.; Feijóo, A.E.; Carrillo González, C. (2002). Synchronization of Asynchronous Wind Turbines, *IEEE Trans, on Energy Conv.*, Vol. 17, No 4, Nov. 2002, pp 1162-1169

Constantinescu, E.M.; Zavala, V.M.; Rocklin, M; Lee, S. & Anitescu, M (2009). *Unit Commitment with Wind Power Generation: Integrating Wind Forecast Uncertainty and Stochastic Programming*, Argonne National Laboratory, September 2009

Cushman-Roisin, B. (2007) *Environmental Fluid Mechanics*, John Wiley & Sons, 2007.

DeCarolis, J. F.; Keith, D.W. (2005). The Costs of Wind's Variability: Is There a Threshold?, *The Electricity Journal*, Volume 18, Issue 1, pp. 69-77.

Dolan, D. S. L. & Lehn, P. W. (2005), Real-Time Wind Turbine Emulator Suitable for Power Quality and Dynamic Control Studies, *International Conference on Power Systems Transients* (IPST'05) in Montreal, Canada on June 19-23, 2005.

Dolan, D. S. L. & Lehn, P. W. (2006), Simulation Model of Wind Turbine 3p Torque Oscillations due to Wind Shear and Tower Shadow, *IEEE Trans. Energy Conversion*, Sept. 2006, Vol. 21, N. 3, pp. 717-724.

EnerNex Corporation, "Final Report - 2006 Minnesota Wind Integration Study Volume 1," Nov. 30, 2006, Available: http://www.puc.state.mn.us/docs/windrpt_vol%201.pdf

Feldman, S. (2009) The Wind Power Variability Myth Gets Debunked, Again, *SolveClimate*, Jul 17th, 2009.

Fernandez, L.M.; Saenz, J.R.; Jurado, F. (2006), Dynamic models of wind farms with fixed speed wind turbines, *Renewable Energy*, 31 (8), pp.1203-1230, Jul 2006.

Frandsen, S.; Jørgensen, H.E. & Sørensen, J.D. (2007), Relevant criteria for testing the quality of turbulence models, *2007 EWEC*, Milan (IT), 7-10 May 2007. (2007) p. 128-132.

Gordon-Leishman, J. (2002). Challenges in Modeling the Unsteady Aerodynamics of Wind Turbines, 21st ASME Wind Energy Symposium, Reno, NV, Jan. 14–17, 2002.

Hansen, A. D.; Iov, F.; Sørensen, P.; Cutululis, N.; Jauch, C.; Blaabjerg, F. (2007). *Dynamic wind turbine models in power system simulation tool DIgSILENT*, Risø National Laboratory, Roskilde, Denmark August 2007, Risø-R-1400(ed.2)(EN).

Hansen, A. D.; Jauch, C.; Sørensen, P.; Iov, F. & Blaabjerg, F. (2003). *Dynamic wind turbine models in power system simulation tool DIgSILENT*, Risø National Laboratory, Roskilde, December 2003, Risø-R-1400(EN).

Heier, S. (2006), *Grid Integration of Wind Energy Conversion Systems*. John Wiley & Sons, New York 2006 (2nd edition).

Hirst, E. & Hild, J. (2004), The Value of Wind Energy as a Function of Wind Capacity, *The Electricity Journal*, Volume 17, Issue 6, pp. 11-20, 2004.

Holttinen, H. (2008), Estimating the impacts of wind power on power systems—summary of IEA Wind collaboration, Environ. Res. Lett., 3, 2, 2008.

Holttinen, H.; Lemström, B.; Meibom, P.; Bindner, H.; Orths, A.; van Hulle, F.; Ensslin, C.; Tiedemann, A.; Hofmann, L.; Winter, W.; Tuohy, A.; O'Malley, M.; Smith, P.; Pierik, J.; Olav Tande, J.; Estanqueiro, A.; Ricardo, J.; Gomez, E.; Söder, L.; Strbac, G.; Shakoor, A.; Smith, J. C.; Parsons, B.; Milligan, M. & Wan, Y. H. (2007). *Design and Operation of Power Systems with Large Amounts of Wind Power. State-of-the-art report*. VTT Technical Research Centre of Finland, 2007.

IEA (2006), IEA Wind Energy Annual Report 2006, July 2006

IEC 61400-1 (2005), *Wind Turbines—Part 1: Design Requirements*, Int. Std., 3rd Ed. 2005-08.

International Energy Agency, (2005). *Variability Of Wind Power And Other Renewables. Management options and strategies*, 2005. Available: http://www.iea.org /textbase/papers/2005/variability.pdf

Juban, J.; Fugon, L. & Kariniotakis G. (2008) Uncertainty Estimation of Wind Power Forecasts, *EWEC* 2008, Brussels, Belgium.

Kazachkov, S. Stapleton, (2004), Modeling wind farms for power system stability studies, *Power Technology Newsletter*, Issue 95, April 2004.

Lamont, A D. (2008). Assessing the long-term system value of intermittent electric generation technologies, *Energy Economics*, 30 (2008), pp. 1208–1231

Mann J. (1998), Wind Field Simulation. *Probabilistic Engineering Mechanics*, vol. 13, nº 4, 1998, 269-282.

Martí, I.; Kariniotakis, G.; Pinson, P.; Sanchez, I.; Nielsen, T. S.; Madsen, H.; Giebel, G.; Usaola, J.; Palomares, A.M.; Brownsword, R.; Tambke, J.; Focken, U.; Lange, M.; Sideratos, G. & Descombes, G. (2006), Evaluation of Advanced Wind Power Forecasting Models - Results of the Anemos Project, *EWEC* 2006, Athens.

Martins, A; Costa, P. C. & Carvalho A. S., (2006) Coherence And Wakes In Wind Models For Electromechanical And Power Systems Standard Simulations, *European Wind Energy Conferences*, EWEC 2006.

Matevosyan, J. (2006), *Wind power integration in power systems with transmission bottlenecks*, KTH, Stockholm, 2006.

Matos, M. A. & Bessa, R. (2009). Operating Reserve Adequacy Evaluation using Uncertainties of Wind Power Forecast, *IEEE Bucharest PowerTech Conference*, June 28th 2009.

Milborrow, D. (2009), *Managing Variability*, A report to WWF-UK, RSPB, Greenpeace UK and Friends of the Earth EWNI, 24 June 2009.

Milligan, M. (2000) Modelling Utility-Scale Wind Power Plants. Part 2: Capacity Credit. *Wind Energy*, 3, pp. 167-206, 2000.

Mur-Amada, J. (2009). WINDFREDOM: WIND variability in the FREquency DOMain. Program under the GNU General Program License, available freely at http://www.windygrid.org/software.

Mur-Amada, J.; Bayod-Rújula, A.A.; Ortiz, S.; Zapata, R. (2003). Power Quality Analysis of Wind Turbines. Part II - Dynamic Analysis, *ICREP* 2003, Vigo.

Nichita, C.; Luca, D.; Dakyo, B. & Ceanga, E. (2002). Large Band Simulation of the Wind Speed for Real Time Wind Turbine Simulators, *IEEE Transs On Energy Conversion*, Vol. 17, No. 4, December 2002.

Parsons, B.; Milligan, M.; Zavadil, B.; Brooks, D.; Kirby, B.; Dragoon, K. & Caldwell, J. (2004), Grid Impacts of Wind Power: A summary of recent studies in the United States, *Wind Energy*, vol. 7, Apr./Jun. 2004, pp 87-108.

Parsons, B.; Wan, Y. & Kirby, B. (2001), Wind Farm Power Fluctuations, Ancillary Services, and System Operating Impact Analysis Activities in the United States, *National Renewable Energy Laboratory*, 2001

Pedersen, T. F.; Dahlberg, J.-Å.; Cuerva, A.; Mouzakis, F.; Busche, P.; Eecen P.; Sanz-Andres, A.; Franchini, S. & Petersen, S. M. (2006), *ACCUWIND -Accurate Wind Speed Measurements in Wind Energy*, Risø National Laboratory, Roskilde, Denmark, July 2006.

Petru, T. & Thiringer T. (2002), Modeling of Wind Turbines for Power System Studies, *IEEE Trans. On Power Systems*, Vol. 17, No. 4, Nov. 2002, pp. 1132 - 1139

Piwko, R.; Boukarikm, G.; Clark, K.; Haringa, G.; Jordan, G.; Miller, N.; Zhou, Y.; Zimberlin, J.; (2004). *The Effects of Integrating Wind Power on Transmission System Planning, Reliability, and Operations*. GE Power Systems Energy Consulting. Schenectady, NY. Available: http://www.nyserda.org /publications/wind_integration_report.pdf

Porter, K.; Yen-Nakafuji D. & Morgenstern, B. (2007), A Review of the International Experience with Integrating Wind Energy Generation, *The Electricity Journal*, Volume 20, Issue 8, October 2007, pp. 48-59.

Pöyry PLC, *Impact of intermittency. How wind intermittency could shape the electricity markets?*, 01 July 2009. Available at http://www.ilexenergy.com/

Ramirez-Rosado, I. J.; Fernandez-Jimenez, L. A.; Monteiro, C; Sousa, J.; Bessa, R. (2009), Comparison of two new short-term wind-power forecasting systems, *Renewable Energy*, 2009 (article in Press).

Rosas, P. (2003). Dynamic influences of wind power on the power system, Ph. D Thesis, Ørsted•DTU, March 2003, ISBN: 87-91184-16-9.

Saranyasoontorn, K.; Manuel, L. & Veers, P. S. (2004), On estimation of coherence in inflow turbulence based on field measurements, *Proceedings of the 2004 ASME Wind Energy Symposium*, pp. 424-434, Reno, Nevada, January 2004.

Saranyasoontorn, K.; Manuel, L. & Veers, P. S. (2004). A Comparison of Standard Coherence Models form Inflow Turbulence With Estimates from Field Measurements, *Journal of Solar Energy Engineering*, Vol. 126 (2004), Issue 4, pp. 1069-1082.

Schlez, W. & Infield, D. (1998), Horizontal, two point coherence for separations greater than the measurement height, *Boundary-Layer Meteor.* 87 (1998), 459-480.

Snieckus, D. (2009). Myth of wind variability problems debunked, *Recharge News*, July 9 2009.

Sørensen, P.; Cutululis, N. A.; Vigueras-Rodríguez, A.; Madsen, H.; Pinson, P.; Jensen, L. E.; Hjerrild, J. and Donovan, M. (2008). Modelling of Power Fluctuations from Large Offshore Wind Farms, *Wind Energy*, 2008.

Sørensen, P.; Cutululis, N. A.; Vigueras-Rodríguez, A.; Madsen, H.; Pinson, P.; Jensen, L. E.; Hjerrild, J. & Donovan, M. (2007), Modelling of Power Fluctuations from Large Offshore Wind Farms, Wind Energy, Oct 2007.

Sørensen, P; Hansen, A. D. & Carvalho Rosas, P. A. (2002). Wind models for simulation of power fluctuations from wind farms, *Journal of Wind Engineering and Industrial Aerodynamics*, Vol. 90, No. 12-15, Dec. 2002, pp. 1381-1402.

Sovacool, B. K. (2009), The intermittency of wind, solar, and renewable electricity generators: Technical barrier or rhetorical excuse?, *Utilities Policy*, 17, (2009), 288–296

Tamura, Y.; Karrem, A.; Solari, G.; Kwok, K. C. S. & Holmes, J. D. (2003). *Report from IAWE Working Group WGE-Dynamic Response*. 2003.

The Guardian, 2008-07-30. *Report finds US is world's top wind producer.* http://www.guardian.co.uk/environment/2008/jul/30/windpower.energy.

U.S. Department of Energy (2008) *20% Wind Energy by 2030. Increasing Wind Energy's Contribution to U.S. Electricity Supply*, July 2008.

U.S. Department of Energy, (2005), Wind Energy Myths, *Wind Powering America Fact Sheet Series*, May 2005.

UK's National Grid, *Operating the Electricity Transmission Networks in 2020.* Initial Consultation. June 2009.

van Engelen, T.G. (2007). Control design based on aero-hydro-servo-elastic linear models from TURBU (ECN), *EWEC 2007.*

Veers PS. *Three-Dimensional Wind Simulation.* Sandia Report, SAND88–0152 l UC–261, 1988.

Wan, Y; Bucaneg, D. (2002). Short-term power fluctuations of large wind power plants, *Journal of Solar Energy Engineering*, 124, pp. 427-431, 2002

Watson, S. (2003), *Introduction to the Atmospheric Boundary Layer*, CREST MSc Flexible & Distance Learning Series. Wind Power Fundamentals. Crest, Lougborough University, 2003.

Welfonder, E.; Neifer, R.; and Spaimer, M.; (1997), Development And Experimental Identification Of Dynamic Models For Wind Turbines, *Control Eng. Practice*, Vol. 5, No. 1, pp. 63-73, 1997.

Impact of Real Case Transmission Systems Constraints on Wind Power Operation

François Vallée - Olivier Deblecker - Jacques Lobry
Electrical Engineering Department, Faculté Polytechnique de Mons
Belgium

1. Introduction

Each investment scenario on a given electrical transmission system must ensure a quality service at the lowest cost (services continuity, system exploitation). In order to answer this major issue of modern networks, it is therefore necessary to compute a faithful representation of the transmission system. In that way, statistical analysis by means of a Monte Carlo simulation (Billinton et al., 1996) - (Papaefthymiou et al., 2006) permits electrical system modeling via the simulation of a large set of representative states. Consequently, Monte Carlo studies permit to obtain coherent exploitation cost and reliability indices for each studied network.

In a near future, stochastic electrical production, and more specially wind generation, is expected to play an important role in power systems. It is therefore imperative to study the impact of this decentralized production source on transmission systems operation constraints. Actually, adequacy studies taking into account wind generation have been extensively developed for the HLI level (load covering with always available transmission system) (Billinton & Bai, 2004) - (Wangdee & Billinton, 2006). From the transmission system point of view, a first reliability study taking into account transmission constraints has been introduced in order to evaluate transmission reinforcement planning associated to large scale wind farms integration (Billinton & Wangdee, 2007). However, that approach was not feasible using normal personal computers and was requiring the use of multiprocessors. Moreover, that proposed study was not considering eventual operation constraints (fatal production, nuclear or high powered thermal units that the producer does not want to stop during the nights...) on classical generation parks.

In the present paper, as reliability and reinforcement analysis are long term studies, stochastic wind generation models are proposed and introduced into an HLII (bulk power system) (Billinton & Wangdee, 2007) - (Allan & Billinton, 2000) non sequential Monte Carlo simulation tool. Thanks to the utilization of that non sequential approach, computing requirements are reduced without worsening the precision of the obtained global indices. Moreover, classical machines and transmission constraints can also be simultaneously considered when facing an increased penetration of wind generation. Therefore, it is believed that the proposed tool will assist system planners and transmission system operators to qualitatively assess the system impact of wind production and to provide adequate input for the managerial decision process in presence of increased wind penetration.

This chapter is organized as follows. In a first part, the methodology used to efficiently introduce wind generation in the HLII simulation tool is explained. Then, hypotheses based on real observations are made in order to introduce wind generation into an economic dispatch with classical parks and transmission constraints. In a third section, wind impact on reliability and reinforcement analysis for transmission systems is computed for an academic test system: the *Ray Billinton Test System (RBTS)* (Billinton et al., 1989). In the following paragraph, the developed simulation tool is applied to the real case Belgian transmission system and points out several reinforcement scenarii in order to safely integrate offshore wind generation. Finally, a conclusion is drawn and points out the major results collected thanks to the introduction of wind generation into HLII analysis taking into account transmission systems constraints.

2. Presentation of existing HLII Monte Carlo simulation tool: Scanner©

2.1 System states generation

The simulation tool Scanner© is the property of *Tractebel Engineering (Gaz de France - Suez)*. Its main objective is to provide technical and economical analysis of development alternatives on a given electrical system. In that way, acceptance (or rejection) criterion is generally based on the following assessment: **"Each investment scenario must ensure a quality service (system exploitation, healthy behavior when facing unexpected outages, continuity of services...) at the lowest cost"**. To answer this issue, a complete analysis of the given transmission system (HLII) is required. Consequently, Monte Carlo simulation tool Scanner© analyses the system evolution as a set of static representative states.

To generate the different system states, the model loops on the 52 weeks of the year (see Fig. 1). During each week, a given number (defined by the user according to the required accuracy on the calculated indices) of system states are generated by mean of the following procedure:

- *Definition of the system state hour during the considered week:* random generation by use of uniformly distributed numbers on the following fixed interval [0, 168] (168 hours during a week);
- *For each generated hour:* uniformly distributed random numbers (V) on the interval [0, 1] are sampled for each element (classical generation units, transformers, lines...) in order to decide its operation state, using the following procedure (Billinton et al., 1996):

  If $V \leq$ Forced Outage Rate (FOR) (Billinton et al., 1996), the element is considered as unavailable;

  If $V > FOR$, the element is considered as fully available;

Concerning the hourly load at each node of the system, its determination is based on the use of the annual peak load value at the considered node; this last one being combined with two modulation diagrams:

- *Diagram of weekly modulation of the annual peak load:* this last one permits to calculate the peak load of the actual week on the basis of the annual peak load value for the considered node. This diagram contains thus 52 modulation rates of the annual peak load value;
- *Diagram of the hourly modulation of the weekly peak load:* it permits to calculate the hourly load for each hour of the week. This diagram contains thus 24 modulation rates of the weekly peak load value.

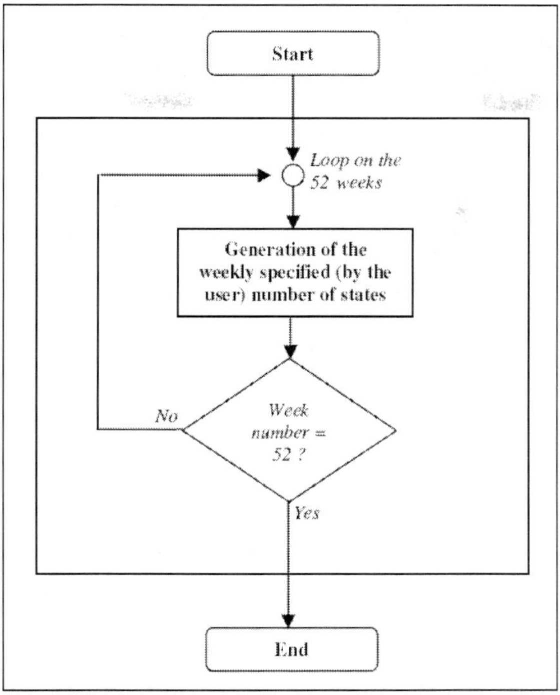

Fig. 1. Algorithm of the system states generation procedure.

The introduction of the load in the investigated simulation tool can thus be considered as 'sequential' (whereas the generation of element states is non sequential). Indeed, thanks to the applied methodology, no random sampling is used to generate the hourly load at each node of the system. More simply, the program just considers, in the weekly modulation diagram, the rate corresponding to the actual week during the simulation process (Fig.1); then, it associates to the generated weekly peak load the rate of the hourly modulation diagram corresponding to the investigated hour of the day.

Finally, as the consumption during one week can change from one day to the other (days of the week, Saturday or Sunday), several diagrams of hourly modulation can be associated to each node during one week. Moreover, seasonal aspects can also be taken into account by defining periods during the year and by changing the set of hourly modulation diagrams associated to each node from one period to the other.

Fig. 2.a illustrates the load behavior during one year (based on a Belgian real case for the year 2000 (Buyse, 2004)) for an annual peak load of 185MW at the considered node. In Fig. 2.b, a zoom is made on one week of the year for the same node and illustrates the consideration of possible change of consumption from one day to the other inside the week.

2.2 System states analysis

Each generated system state must then be analyzed. To proceed to this stage of the process, three steps are consecutively realized for each system state:

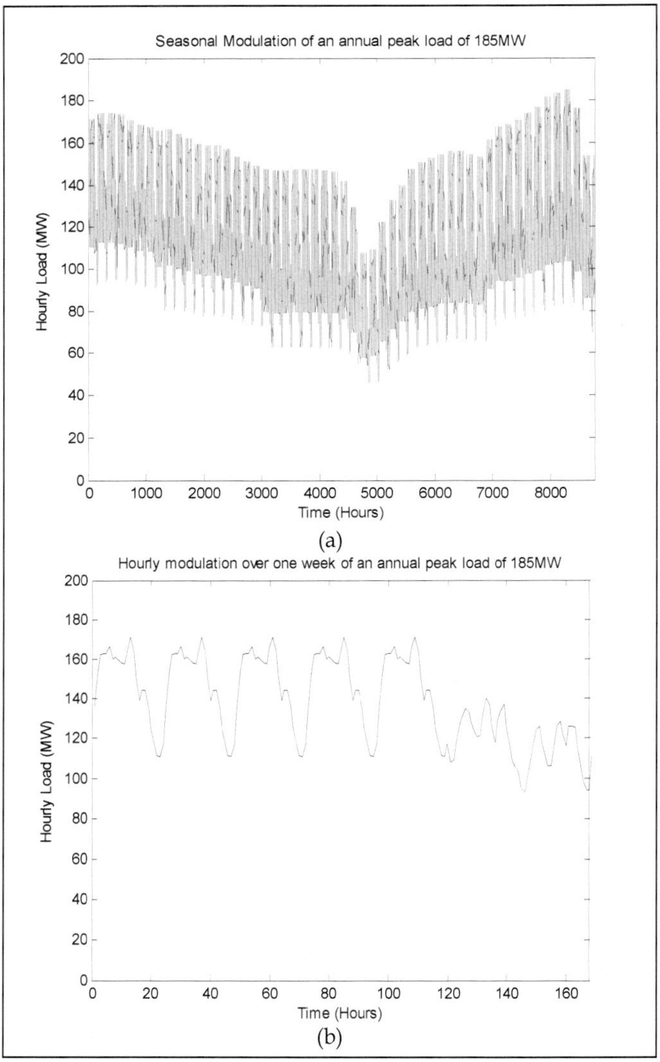

Fig. 2. Seasonal modulation for 185MW peak load value (a) and hourly modulation over one week of the year for the same peak value (b) based on a Belgian real case (Buyse, 2004)

1. *Economic dispatch:* this last one is based on the available production units and is done **without considering transmission facilities availability**. The objective is thus to ensure, at the lowest cost, the hourly load with the available production. Note that **the economic dispatch is taking into account possible constraints on classical units operation**. Consequently, several types of production parks are considered in Scanner© among which:

- Hydraulic production and pumping stations: they are considered as zero cost production in the algorithm and are managed at a weekly time scale;
- Thermal production: three types of constraints are considered for this kind of production. Firstly, **technical minima** (threshold under which the producer does not want to run, for technical reasons, its unit) can be considered. Secondly, **forced units** can be defined by the user. Those entities represent units (such as cogeneration) that have a threshold over which they must always operate when they are available. Finally, in order to take into account high powered thermal or nuclear units that the producer does not want to stop during the week, those machines are considered as **long term units**. They are managed at the weekly time scale and must always run at their technical minimum value during the actual week when they are needed to cover the reference peak consumptions of the week (in the other case, they are supposed entirely unavailable during the entire week).

Finally, the algorithm conducted during the economic dispatch proceeds as follows. In a first step, hydraulic production is used to cover the load (following the orders of the weekly management). Then, technical minimum values of forced and long term (if they are required to cover the reference peak loads during the week) thermal units are considered to satisfy the load (minus hydraulic production). Finally, an economic dispatch of the thermal production (minus the technical minimum values of already considered forced and long term units) is realized to cover the remaining load.

2. *DC Load Flow:* this step realizes the computation of active power flows in transmission lines without considering reactive power.
   In order to solve the DC load flow problem, generated active powers calculated during the economic dispatch are introduced at the connection nodes of the concerned machines. Moreover, the generated hourly consumption for the current state is also taken into account at the required nodes. DC load flow then computes active power flows over the transmission system and permits to take into account transmission constraints. In case of line overflow, step 3 is started. On the opposite, if the optimal solution does not involve overloaded lines, this next step is avoided.
3. *Production rescheduling or load shedding:* this step is only started if the optimal solution of the economic dispatch leads to overloaded lines during step 2. In that case, the solution is "disoptimized" by modifying the production plan (production rescheduling). If this first stage is not sufficient to relieve the overflows, a load shedding procedure is then started in order to limit active power flows.

Finally, note that those three presented steps are repeated to analyze each generated system state.

2.3 Calculated indices and productions

Thanks to the Scanner© tool, it is possible to compute several reliability indices for the studied system.

Among them, the most significant ones are certainly the *Loss of Load Expectation* (*LOLE* in hours/year) (Billinton et al., 1996) (generation system index calculated without considering transmission facilities) and the number of hours (per year) of load shedding (due to transmission overflows). Moreover, hours of overflows are also computed for each transmission line in order to point out the weakest points of the system.

Next to those indices, the annual cost of production (with and without considering elements unavailability) is computed. Moreover, mean production and annual energy generated by each classical unit can also be calculated. Finally, histograms of production can be printed out in order to analyze the utilization of each classical unit.

3. Wind generation modelling for HLII non sequential Monte Carlo simulation

Scanner© does not currently take into account wind production in its algorithm (as well as for the system states generation as for the analysis of those states). However, given the major increase of wind penetration in some countries (like Germany) (Ernst, 2005), this variable kind of production can no more be neglected in technical (and economical) transmission system analysis. Therefore, in the present work, wind production has been implemented in the simulation tool Scanner©. In order to achieve that step, modifications related to the introduction of wind have to impact both major stages of the simulation process: **system states generation and the analysis of these states**.

3.1 Introduction of wind power in system states generation

Before taking into account wind power in the system states generation process, the user has to define three entities related to wind production:

- **Entity 1 (wind parks):** each wind park is practically characterized by its installed capacity, production cost, *FOR* of one turbine, associated wind speed regime and P-W conversion characteristic (Vallée et al., 2008);
- **Entity 2 (wind speed regimes):** they are characterized by *Cumulative Distribution Functions (CDF)* representing different statistical behaviors for wind speed in the studied territory. Those *CDF* can be classical *Weibull* distributions (Vallée et al., 2008) or arbitrary ones. In the latter case, distributions are linearly interpolated in the program on the joint basis of the wind speed step and probability intervals defined by the user. Finally, a name is associated to each different wind speed regime and the user can freely associate a wind speed regime to a wind park by defining the name of this wind speed regime in the characteristics of the wind park (cf. entity 1);
- **Entity 3 (P-W conversion characteristics):** they transform wind speed into production. Practically, the conversion characteristics are linearly interpolated on the joint basis of the wind speed step and the power intervals defined by the user. An example of the linear interpolation related to a classical doubly-fed asynchronous generation structure (Al Aimani, 2004) is given in Fig. 3. Note that a single conversion characteristic is practically applied to an entire park. Finally, the link between wind parks and associated P-W conversion characteristics is made identically as for the wind speed regime case (cf. entity 2).

Once the three basic kinds of entities related to wind production have been defined by the user, the generation process can be started. The algorithm, executed when wind generation is concerned, is presented in Fig. 4. The applied methodology is the following one. During each generated system state, a first loop is started over the defined wind speed regimes and a wind speed per defined characteristic is generated by use of the classical inverse transform method (Vallée et al., 2008). By applying this methodology, it is supposed that wind parks subject to the same wind speed regime are entirely correlated. Based on (Vallée et al., 2008) and (Papaefthymiou, 2006), this approach will lead to the most fluctuating wind power and,

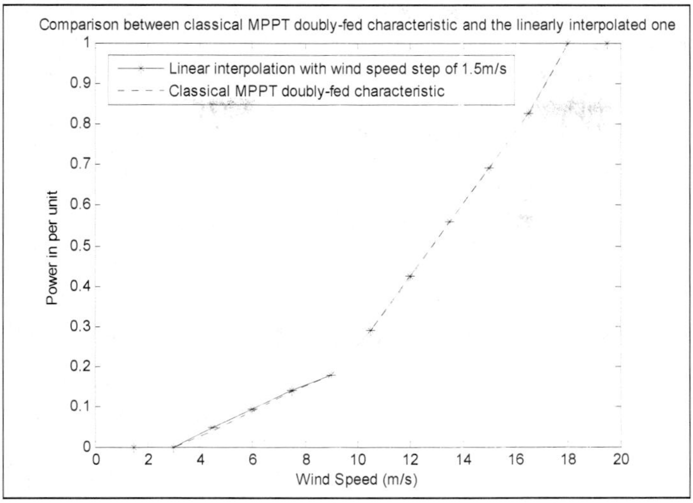

Fig. 3. Comparison between classical *Maximum Power Point Tracking (MPPT)* conversion characteristic (Al Aimani, 2004) and the interpolated one (1.5m/s wind speed step) in p. u.

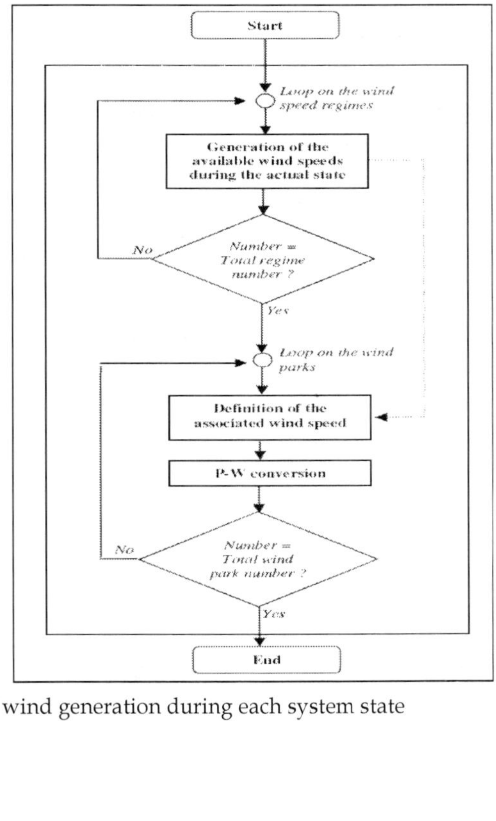

Fig. 4. Algorithm of wind generation during each system state

thus, to the worst case for adequacy studies. Finally, a second loop is made over the defined wind parks. For each wind park, the associated wind speed is decided by taking the one sampled for the wind regime associated to the considered wind park. Then, each wind park production is calculated by introducing the sampled wind speeds in the associated P-W characteristics. Practically, P-W characteristics will be introduced in per unit and the real wind productions will be obtained by multiplying per unit quantities by the *Maximal Available Wind Park Capacity* (*MAWPC*). The *MAWPC* is related to the *Installed Wind Park Capacity* (*IWPC*) by:

$$MAWPC = (1 - FOR) \cdot IWPC \tag{1}$$

By applying (1), possible outages of wind turbines inside a park are considered. Equation (1) supposes that a wind park is made of a sufficiently large number of turbines to consider that the FOR related to one turbine is the same as the one existing for the entire park. This hypothesis is well funded as high powered wind parks connected to transmission systems will be practically composed of a large number of turbines.

Finally, note that the wind production generation process is started back for each system state. Consequently, a *Generated Wind Production Distribution* can be plotted for each wind park *'i'*.

Fig. 5 illustrates the obtained wind production distribution for one 8 MW wind park subject to a *Weibull* wind speed regime (with scale parameter A = 5.25 and shape parameter B = 3.55) and using the P-W conversion characteristic of Fig. 3. This production distribution is logically concentrated over limited wind production as the considered wind speed parameters were quite low in the present case.

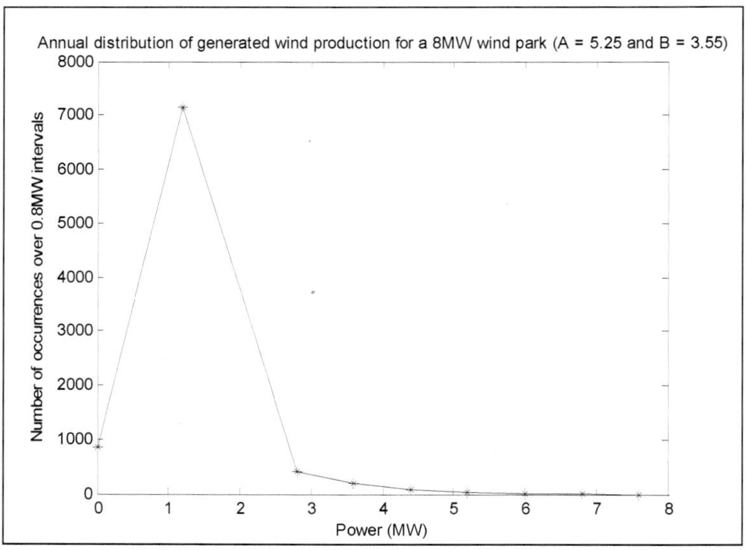

Fig. 5. Simulated annual distribution of generated wind production for a 8 MW wind park (A = 5.25 and B = 3.55)

3.2 Introduction of wind power in system states analysis

The *Generated Wind Production (GWP_i)* represents thus, for each defined park *'i'*, the sampled wind power during the simulated system state. This production must then be taken into account in the system state analysis.

The introduction of wind production into the economic dispatch of Scanner© has so been based on several hypotheses:

- **Hypothesis 1**: it has been considered that wind power was not accurately predictable at the weekly time scale (Ernst, 2005) and could therefore not impact the management of hydraulic and long term thermal (nuclear) units. Those classical units are thus still processed at the weekly time scale without wind impact;
- **Hypothesis 2**: wind power is considered as a must run production with zero cost. This hypothesis is based on the multiple encouraging policies that generally support wind production (Mackensen et al., 2007) - (Maupas, 2006). Consequently, in the economic dispatch, wind production will be directly considered **after** the technical constraints related to **forced and 'having to run' long term thermal units**;
- **Hypothesis 3**: in case of increased wind penetration, the *Transmission System Operator (TSO)* can be forced (like it has already been the case in some German places (Sacharowitz, 2004)) to cut some wind production when facing classical machines constraints (Sacharowitz, 2004). In the proposed algorithm, when encountering such situations, wind production is decreased, for each wind park, proportionally to its available generated power.

The existence of some transmission system operation constraints can thus lead to a reduction of the real produced wind power. Therefore, in the developed algorithm, two quantities related to wind production have been defined **for each wind park**:

- *Real Wind Production (RWP):* it represents the real produced wind power after having taken into account the economic dispatch. A single RWP_i value is associated to each wind park *'i'*;
- *Lost Wind Production (LWP):* it defines the difference between the generated wind production and the real produced one for each considered wind park. A single LWP_i value is thus calculated for each wind park *'i'*;

The algorithm, implemented in order to take into account wind production in the economic dispatch associated to each generated system state, is described in Fig. 6.

Based on hypotheses 1 and 2, wind production is thus used to cover the remaining load after that hydraulic production and technical minima of forced and 'must run' long term thermal parks have been taken into account. If this remaining load L_1 is equal to zero, all the hourly load has already been covered before considering wind generation. In that case, real transmitted wind production RWP_i is set to zero for each wind park *'i'* and their associated lost wind production LWP_i equals their initially generated wind power GWP_i during the considered system state.

On the other hand, if the remaining load L_1 is greater than zero, *Global Generated Wind Production (GGWP)* is taken into account before the remaining classical thermal production (without constraints) and is entirely taken off from load L_1. If the obtained load value L_2 (after consideration of hydraulic production, of forced and 'must run' long term thermal parks technical minima and of wind production) is greater or equal to zero, RWP_i equals the generated wind production GWP_i for each park *'i'* and LWP_i is set to zero (and the classical economic dispatch is pursued). In the case of remaining load value L_2 is negative, it is then

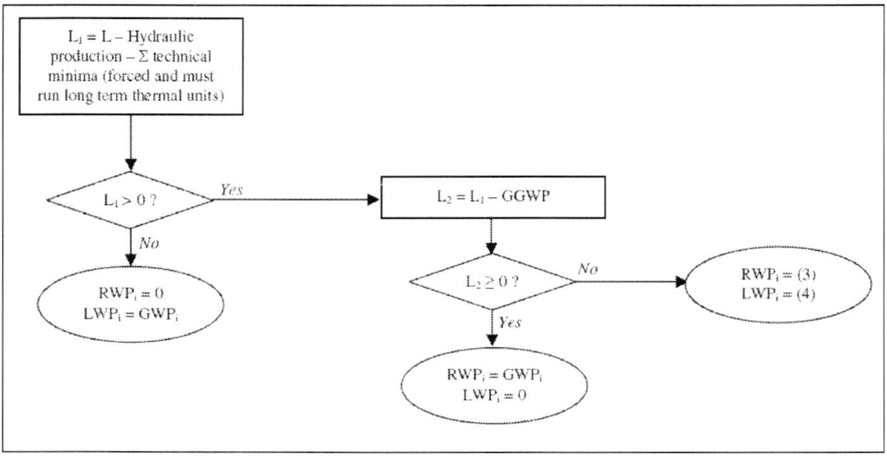

Fig. 6. Algorithm implemented in order to take into account wind production for the economic dispatch of each generated system state

necessary to reduce real transmitted wind production by following hypothesis 3. In order to apply this reduction of wind power, the remaining load L_1 (before introduction of the generated wind production) is taken back. RWP_i and LWP_i associated to each defined wind park *'i'* are obtained via equations (2), (3) and (4):

$$GGWP = \sum_{i=1}^{N} GWP_i \tag{2}$$

$$RWP_i = \frac{GWP_i \cdot L_1}{GGWP} \tag{3}$$

$$LWP_i = GWP_i - RWP_i \tag{4}$$

At the end of the actual system state economic dispatch, real and lost wind productions (RWP_i and LWP_i) are thus defined for each wind park. In order to take into account wind production impact over transmission constraints, calculated RWP_i are then injected at the adequate nodes and the DC load Flow (section 2.2) is launched. If no line overflows are recorded for the computed system state, RWP_i and LWP_i stay unchanged for each wind park *'i'*. On the opposite, in case of overloaded lines, the algorithm of rescheduling/load shedding (section 2.2) is used and RWP_i et LWP_i can have to be modified in order to ensure a safe behavior of the transmission system.

The system state analysis leads thus now to the computation of RWP_i and LWP_i for each wind park *'i'*. This process is set back for each generated system state. Consequently, at the end of the Monte Carlo simulation, histograms of GWP_i, RWP_i and LWP_i can be drawn for each defined wind park *'i'*. Moreover, mean values of generated (GWP_i) and real (RWP_i) exchanged wind powers are calculated, for each wind park, in order to point out the impact of transmission system constraints on wind production. Also note that reliability indices defined in section 2.3 are now taking into account wind production.

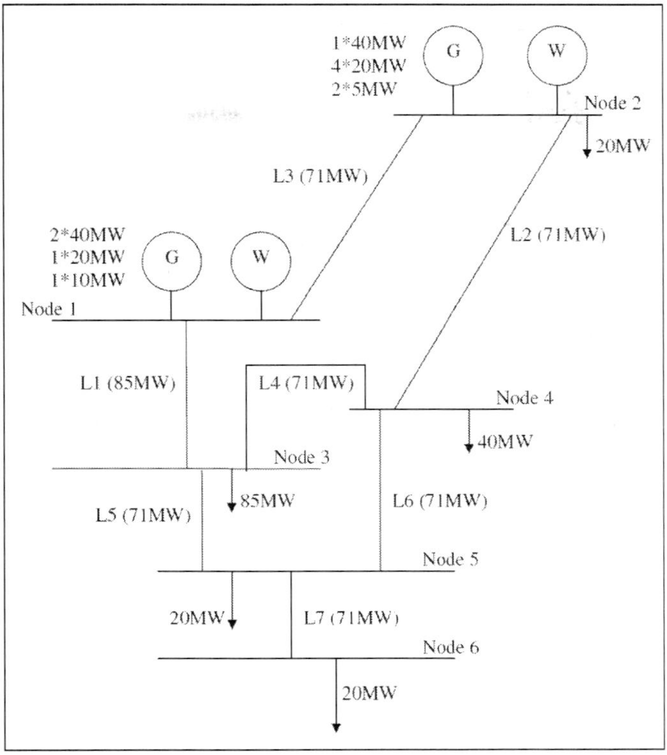

Fig. 7. Implemented version of the *RBTS*

4. Simulations results on a modified RBTS test system

4.1 Wind parks introduced in RBTS generation nodes (Billinton et al., 1989)

In order to point out wind generation behavior in modern transmission systems management, a slightly modified version of the academic *RBTS* test system (Billinton et al., 1989) has been considered. The implemented version of *RBTS* is described in Fig. 7 and differs from the original one for some aspects:

- Peak load is always 185MW but the weekly and hourly modulation diagrams have been based on the Belgian real case (Buyse, 2004) given in Fig. 2;
- Wind generation (48MW) is firstly introduced in generation nodes 1 (24MW) and 2 (24MW) by following the data given in tables 2 and 3. Wind speed regimes are supposed to be *Weibull* ones but differ from node 1 to 2. P-W conversion characteristic is the one depicted in Fig. 3;
- The *RBTS* number of lines has been reduced from 9 (Billinton et al., 1989) to 7 in the present simulation. Indeed, both parallel lines between nodes 1-3 and nodes 2-4 (Billinton et al., 1989) are suppressed in order to point out the interest of Scanner© tool in showing the weak points of the studied system;

- Compared to the initial version of the RBTS (Billinton et al., 1989), operation constraints can be added, here, for classical parks (section 2.2). Consequently, more realistic system states management can be studied in the present case;
- No *FOR* is considered for lines and transformers as their availability is not under the scope of the present study. On the other hand, transmission constraints are taken into account;
- *FOR* is considered for classical production parks and for wind generation.

	Wind regime	A	B
Node 1	*Weibull*	7.10	2.85
Node 2	*Weibull*	9.95	2.75

Table 1. Wind speed regimes considered at nodes 1 and 2 of the *RBTS*.

	Installed capacity (MW)	Connection node
Wind park 1	8	Node 1
Wind park 2	6	Node 1
Wind park 3	12	Node 2
Wind park 4	1	Node 1
Wind park 5	3	Node 2
Wind park 6	4	Node 1
Wind park 7	5	Node 2
Wind park 8	4	Node 2
Wind park 9	5	Node 1

Table 2. Wind generation considered for the modified *RBTS*

In order to face wind generation and transmission system operation constraints, two cases have been simulated:

- *Case 1*: no operation constraints are considered for the classical park: all parks are supposed to be thermal ones without technical minimum value;
- *Case 2*: operation constraints are associated to the defined classical parks. Here, the sum of technical minima is supposed to be 75MW. This represents almost 30% of the installed classical capacity (240MW) in reference to the Belgian real case (Buyse, 2004). Those 75MW are divided in nuclear (30MW), long term thermal (30MW) and forced parks (15MW).

Both cases have been firstly investigated for a 48MW installed wind capacity (table 2). With this first wind penetration, table 3 summarizes real transmitted annual wind energy for each park. It can clearly be observed that, with this reduced penetration, classical park operation constraints have a limited impact on wind as collected energies are similar in both investigated cases.

Moreover, Fig. 8 compares *RWP* for wind park 1 in both investigated cases and clearly confirms an identical behavior of this wind park with or without classical operating constraints (this result can be extended to all the other considered wind parks). For information, *GWP* (identical in both cases as wind data are the same from one simulation scheme to the other) is also plotted and points out that all available wind power is transmitted in the system when wind penetration is low. Finally, note that, due to the limited transfer capacity of line *L1*, 340 annual hours of load shedding (node 3) are computed here but have no impact on wind power. In fact, as computed load shedding situations are quite seldom and not severe in the present case, wind generation is not modified by transmission constraints. Consequently, with the version of *RBTS* presented in Fig. 7, only classical units operation constraints can have an impact on wind generation.

	Case 1	Case 2
Annual energy wind park 1 (GWh/y)	7.5	7.5
Annual energy wind park 2 (GWh/y)	5.5	5.5
Annual energy wind park 3 (GWh/y)	25.0	25.0
Annual energy wind park 4 (GWh/y)	0.9	0.9
Annual energy wind park 5 (GWh/y)	6.3	6.3
Annual energy wind park 6 (GWh/y)	3.7	3.7
Annual energy wind park 7 (GWh/y)	10.4	10.4
Annual energy wind park 8 (GWh/y)	8.2	8.2
Annual energy wind park 9 (GWh/y)	4.8	4.8

Table 3. Annual wind energies with (case 2) and without (case 1) classical park operation constraints for 48MW installed wind capacity

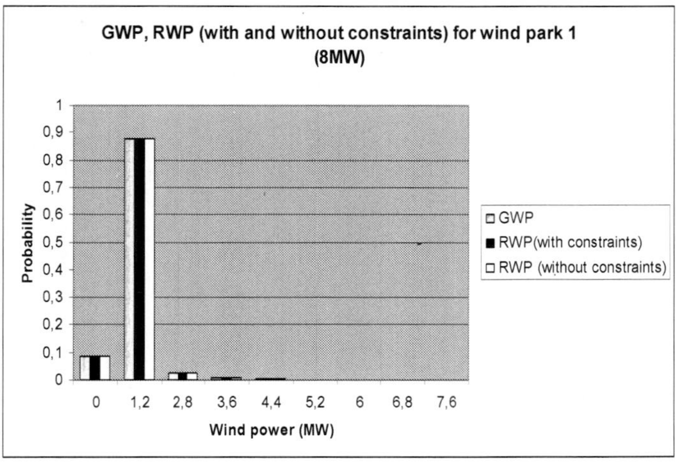

Fig. 8. *GWP* and *RWP* for wind park 1 (8MW) with and without classical operating constraints

A second simulation has then been investigated by doubling wind production at node 2 to 48MW and by setting to zero the one of node 1. Again, simulations have been realized with and without operation constraints on classical parks (cases 1 and 2). In table 4, it can be concluded that, if all wind production is installed in node 2 (greater wind speeds; see Table 1), some situations of wind power excess can be computed. By comparing the results of tables 3 and 4, it can thus be concluded that, with the same global wind penetration, smoothing effects due to wind generation dispersion can have a positive effect on the electrical transmission system management.

	Case 1	Case 2
Annual energy wind park 1 (GWh/y)	0	0
Annual energy wind park 2 (GWh/y)	0	0
Annual energy wind park 3 (GWh/y)	49.2	48.8
Annual energy wind park 4 (GWh/y)	0	0
Annual energy wind park 5 (GWh/y)	12.4	12.3
Annual energy wind park 6 (GWh/y)	0	0
Annual energy wind park 7 (GWh/y)	20.5	20.3
Annual energy wind park 8 (GWh/y)	16.2	16.1
Annual energy wind park 9 (GWh/y)	0.0	0.0

Table 4. Annual wind energies with (case 2) and without (case 1) classical park operation constraints for 48MW installed wind capacity at node 2 (only)

4.2 Adequate repartition of wind parks in the transmission system

In the previous paragraph, as transmission line capacities were sufficiently high, only classical parks operation constraints had an impact on wind generation. In order to also take into account transmission lines constraints, the *RBTS* test system has been voluntarily weakened by reducing the capacity of line *L1* (Fig. 7) to 40 MW.

Fig. 9 presents the simulation results collected in that case with the wind parks defined in table 2 (48 MW installed in nodes 1 and 2). If compared with table 3 (case 2), it can be observed that the power of all wind parks connected to node 1 must be reduced due to the limited capacity of *L1*.

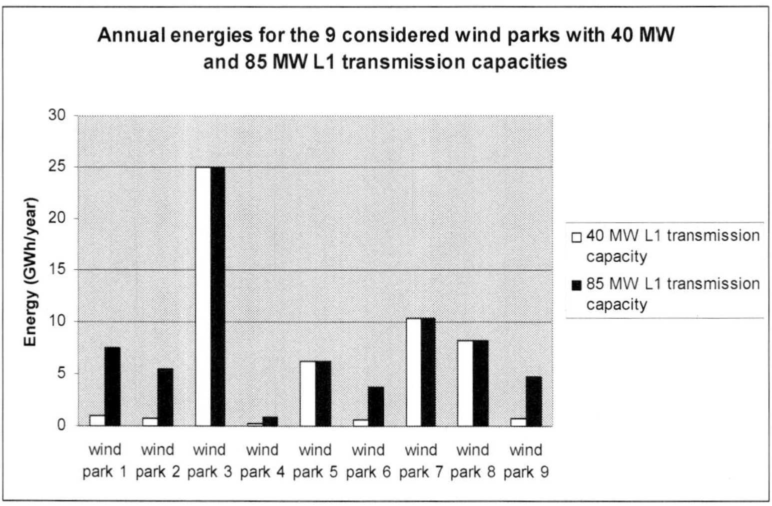

Fig. 9. Annual energy (GWh/year) for the 9 considered wind parks with 40 MW (white) and 85 MW (black) *L1* transmission capacities

In the case of a weakened transmission system, the connection nodes of wind parks take thus a major importance. Indeed, if the considered 48MW of wind generation are now distributed between nodes 2 and 4 (see table 5), the limited transmission capacity of *L1* does no more impact wind power and this last one can be entirely transferred in the network (see table 6). This complete use of wind production was not feasible when some of the defined wind parks (24MW) were directly connected at *L1* (via node 1; see Table 2 and Fig. 9).

	Installed capacity (MW)	Connection node
Wind park 1	8	Node 4
Wind park 2	6	Node 4
Wind park 3	12	Node 2
Wind park 4	1	Node 4
Wind park 5	3	Node 2
Wind park 6	4	Node 4
Wind park 7	5	Node 2
Wind park 8	4	Node 2
Wind park 9	5	Node 4

Table 5. Wind generation considered for the modified *RBTS* test system

Annual energy wind park 1 (GWh/y)	7.5
Annual energy wind park 2 (GWh/y)	5.5
Annual energy wind park 3 (GWh/y)	25.0
Annual energy wind park 4 (GWh/y)	0.9
Annual energy wind park 5 (GWh/y)	6.3
Annual energy wind park 6 (GWh/y)	3.7
Annual energy wind park 7 (GWh/y)	10.4
Annual energy wind park 8 (GWh/y)	8.2
Annual energy wind park 9 (GWh/y)	4.8

Table 6. Annual wind energy for wind parks located in nodes 2 and 4 with limited transmission capacity of *L1* (40MW)

This result points out the utility of the developed tool in order to improve the management of wind generation. Indeed, thanks to the proposed software, the transmission system operator will now be able, not only, to quantify the maximal wind penetration in a given network, but also, to propose an adequate distribution of wind parks connection nodes. However, for this last point, note that environmental concerns for the establishment of wind parks must still be taken into account.

5. Wind generation management in a real case transmission system

In order to point the utility of the developed tool for investments studies in modern networks, we have applied the proposed program to the real case Belgian transmission system. The major issue for this network concerns the large scale integration of offshore wind power. In that way, two projects (for an installed capacity of 630 MW) are actually built in the North Sea and are going to lead to the connection of respectively 300 MW at the 150 kV Slijkens connection node and of 330 MW at the 150 kV Zeebrugge node. Initially, the transmission capacity from Slijkens and Zeebrugge towards Brugge is highly sufficient as it reaches 800 MW. However, as illustrated in Fig. 10 (Van Roy et al., 2003), the integration of offshore wind power associated with the importation of electricity from France towards the Netherlands can lead to the apparition of congestions between Rodenhuize (Gent) and Heimolen (Antwerpen). Such a result is confirmed with our developed simulation tool as an increase of congestion hours over the line between Rodenhuize and Heimolen can be observed in Fig. 11 when 200 MW of wind power are installed in the North Sea and that 1 GW is imported form France towards the Netherlands. Simultaneously, the increase of installed offshore wind power does not change the amount of critical hours over the Slijkens - Brugge and Zeebrugge - Brugge lines. This last result confirms thus that the major issue of Belgian wind integration is mainly related to possible congestion hours inside the country (between Gent and Antwerpen).

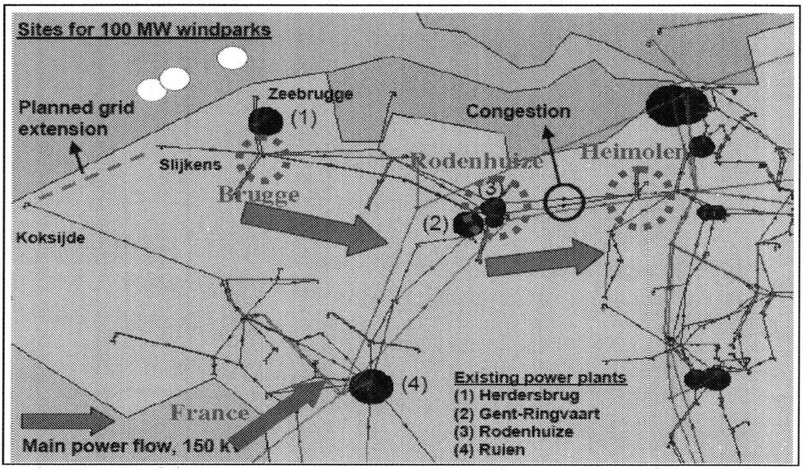

Fig. 10. Major active power flows over the Belgian transmission system after the large scale integration of offshore wind power

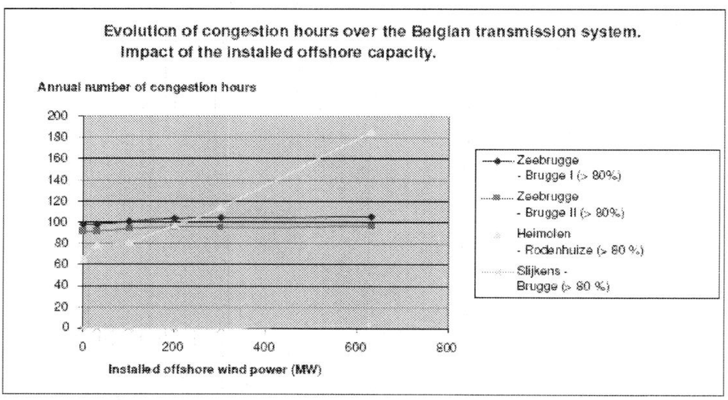

Fig. 11. Evolution of congestion hours over major transmission lines in the Belgian high voltage system. Impact of the installed offshore capacity

In order to improve the offshore wind power integration and to consequently reduce the number of congestion hours over the Rodenhuize-Heimolen line, a grid extension of 150 MW between Koksijde and Slijkens was proposed (dashed curve in Fig. 10). With this new 150 kV line, simulation results (Fig. 12) clearly confirm a reduction of congestion hours between Gent and Antwerpen when the importation level is limited (and that the installed offshore wind power reaches 630 MW). However, after an increase to 2 GW of the electricity

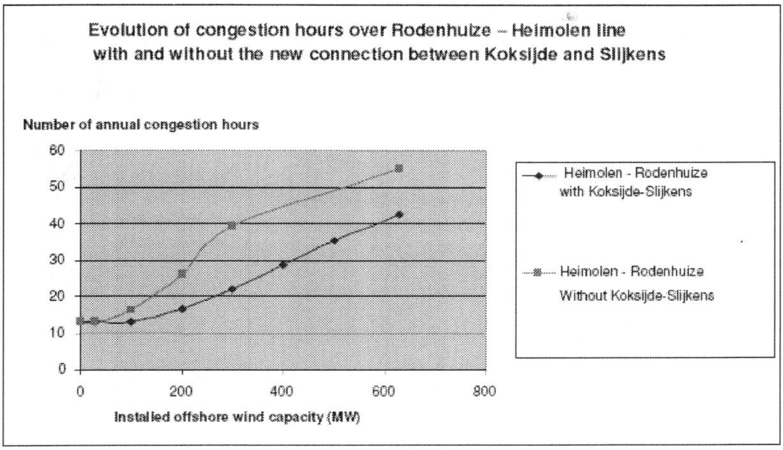

Fig. 12. Evolution of congestion hours between Rodenhuize and Heimolen with and without the added connection Koksijde-Slijkens (importation level of 1 GW and 630 MW installed offshore wind power)

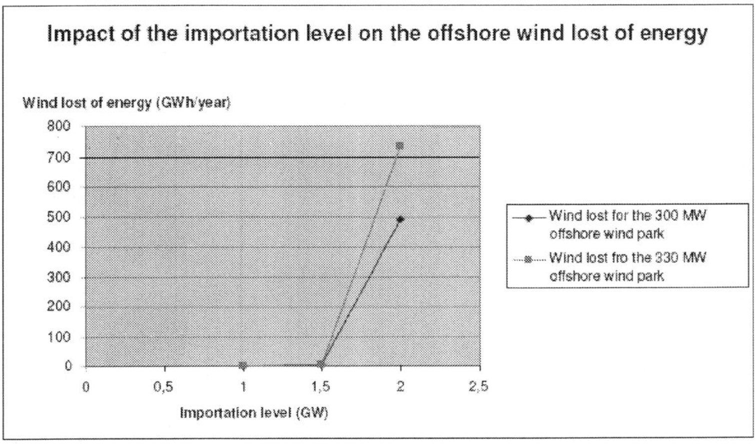

Fig. 13. Impact of the importation level on the offshore lost of energy (installed capacity set to 630 MW)

exchange between France and the Netherlands, not only a reduction of the transmitted wind power can be computed (Fig. 13) but it can also be observed that the number of congestion hours dramatically increases over the Rodenhuize-Heimolen line (Fig. 14). Therefore, in the context of large scale interconnected European networks, it will obviously be necessary to imagine new reinforcements over the Belgian transmission system (connection of Zeebrugge node to the 380 kV network or reinforcement of the Heimolen-Rodenhuize line).

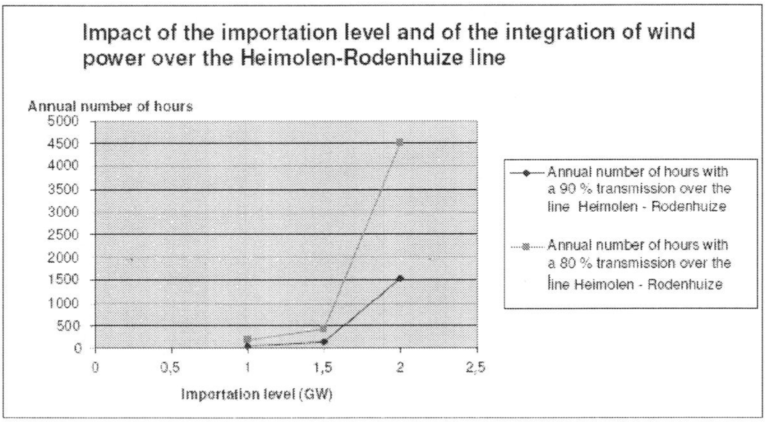

Fig. 14. Impact of the importation level and of the offshore wind power (installed capacity set to 630 MW) over the Heimolen-Rodenhuize line

Finally, it can thus be concluded that the proposed simulation tool permits to study reinforcement scenarii taking into account large scale integration of wind power. In that way, the developed program is thus perfectly suitable for the recent and future developments to be made over modern transmission systems.

6. Conclusion

In this chapter, wind generation has been introduced into a transmission system analysis tool. This last one was composed of two parts: system states generation (non sequential Monte Carlo simulation) and analysis (economic dispatch, DC load flow and eventual load shedding). In order to take into account wind generation in this simulation tool, each part had thus to be modified. Finally, a useful bulk power system analysis software taking into account wind generation has been developed and has permitted to study the impact of wind generation not only on reliability indices but also on the management of the classical production park. In that way, situations of forced wind stopping were pointed out due to increased wind penetration and transmission system operation constraints. Moreover, the interest of the proposed software was demonstrated by adequately determining reinforcements to be made in order to optimize large scale wind penetration in modern real case electrical systems.

7. References

Al Aimani S. (2004). Modélisation de différentes technologies d'éoliennes intégrées à un réseau de distribution moyenne tension, *Ph.D. Thesis*, Ecole Centrale de Lille, chap.2, pp.24-25, Dec. 2004.

Allan R.N., Billinton R. (2000). Probabilistic assessment of power systems, *Proceedings of the IEEE*, Vol. 22, No.1, Feb. 2000.

Billinton R., Kumar S., Chowdbury N., Chu K., Debnath K., Goel L., Kahn E., Kos P., Nourbakhsh, Oteng-Adjei J. (1989). A reliability test system for educational purposes - Basic data. *IEEE Trans. On Power Systems,* Vol. 4, No. 3, Aug. 1989, pp. 1238-1244.

Billinton R., Chen H., Ghajar R. (1996). A sequential simulation technique for adequacy evaluation of generating systems including wind energy. *IEEE Trans. On Energy Conversion*, Vol. 11, No. 4, Dec. 1996, pp.728-734.

Billinton R., Bai G. (2004). Generating capacity adequacy associated with wind energy. *IEEE Trans. On Energy Conversion*, Vol. 19, No. 3,Sept. 2004, pp. 641-646.

Billinton R., Wangdee W. (2007). Reliability-based transmission reinforcement planning associated with large-scale wind farms. *IEEE Trans. On Power Systems*, Vol. 22, No. 1, Feb. 2007, pp. 34-41.

Buyse H. (2004). Electrical energy production. *Electrabel documentaion*, available web site: www.lei.ucl.ac.be/~matagne/ELEC2753/SEM12/S12TRAN.PPT, 2004.

Ernst B. (2005). Wind power forecast for the German and Danish networks. *Wind Power in Power Systems*, edited by Thomas Ackerman, John Wiley & Sons, chap.17, pp.365-381, 2005.

Mackensen R., Lange B., Schlögl F. (2006). Integrating wind energy into public power supply systems - German state of the art. *International Journal of Distributed Energy Sources*, Vol. 3, No.4, Dec. 2007.

Maupas F. (2006). Analyse des règles de gestion de la production éolienne : inter-comparaison de trois cas d'étude au Danemark, en Espagne et en Allemagne. Working paper, *GRJM Conference*, Feb. 2006.

Papaefthymiou G. (2006). Integration of stochastic generation in power systems. *PhD. Thesis*, Delft University, chap. 5 & 6, June 2006.

Papaefthymiou G., Schavemaker P.H., Van der Sluis L., Kling W.L., Kurowicka D., Cooke R.M. (2006). Integration of stochastic generation in power systems. *International Journal of Electrical Power & Energy Systems*, Vol. 18, N°9, Nov. 2006, pp. 655-667.

Sacharowitz S. (2004). Managing large amounts of wind generated power feed in - Every day challenges for a German TSO and approaches for improvements. *International Association for Energy Economics (IAEE)*, 2004 North American Conference, Washington DC, USA, 2004.

Vallee F., Lobry J., Deblecker O., (2008). System reliability assessment method for wind power integration. *IEEE Trans. On Power Systems*, Vol. 23, No. 3, Aug. 2008, pp. 1288-1297.

Van Roy P., Soens J., Driesen Y., Belmans R. (2003), Impact of offshore wind generation on the Belgian high voltage grid, *European Wind Energy Conference (EWEC)*, Madrid, Spain, June 2003.

Wangdee W., Billinton R. (2006). Considering load-carrying capability and wind speed correlation of WECS in generation adequacy assessment. *IEEE Trans. On Energy Conversion*, Vol. 21, No. 3, Sept. 2006, pp. 734-741.

14

Wind Power at Sea as Observed from Space

W. Timothy Liu, Wenqing Tang, and Xiaosu Xie
Jet Propulsion Laboratory, California Institute of Technology, USA

1. Introduction

With the increasing demand of electric power and the need of reducing greenhouse gas emission, the importance of turning wind energy at sea into electric power has never been more evident. For example, China is vigorously studying and pursuing the potential of wind energy to lessen dependence of coal consumption (McElroy et al., 2009). The White Paper on Energy (DTI, 2007) lays out an ambitious plan to the British Parliament in meeting the Renewables Obligation with offshore wind energy. The paper posted a challenge not only to Denmark, the leader of European offshore wind energy, but also to the world. New technology has also enabled floating wind-farms in the open seas to capture the higher wind energy and reduce the environmental impact on the coastal regions. Detailed distribution of wind power density (E), as defined in Section 4, at sea is needed to optimize the deployment of such wind farms. The distribution is discussed in Section 5.

Just a few decades ago, almost all ocean wind measurements came from merchant ships. However, the quality and geographical distribution of these wind reports were uneven. Today, operational numerical weather prediction (NWP) also gives us wind information (Capps & Zender, 2008), but NWP depends on numerical models, which are limited by our knowledge of the physical processes and the availability of data. Recently, spacebased microwave sensors are giving us wind information with sufficient temporal and spatial sampling, night and day, under clear and cloudy conditions. Results from the most advanced passive sensor, which measures only wind speed, and active sensor, which measures both speed and direction, will be discussed. The principles of wind retrievals by active and passive microwave sensors are described in Section 2 and 3 respectively. The dependence of wind speed on height above sea level and on atmospheric stability is discussed in Section 6 and 7.

2. Scatterometer

The capability of the spacebased scatterometer in measuring wind vector at high spatial resolution is discussed by Liu (2002) and Liu and Xie (2006). The scatterometer sends microwave pulses to the Earth's surface and measures the backscatter power. Over the ocean, the backscatter power is largely caused by small centimeter-scale waves on the surface, which are believed to be in equilibrium with stress (τ). Stress is the turbulent momentum transfer generated by vertical wind shear and buoyancy. Liu and Large (1981) demonstrated, for the first time, the relation between measurements by a spacebased scatterometer and surface stress measured on research ships. Although the scatterometer has been known to measure τ, it has also been promoted as a wind-measuring instrument. The geophysical data product of the scatterometer is the equivalent neutral wind, U_N, at 10 m height (Liu and Tang 1996), which, by definition, is uniquely related to τ, while the relation between τ and the actual winds at the reference level depends on atmosphere stability and ocean's surface current. U_N has been used as the actual wind, particularly in operational weather applications. The difference between the variability of stress and wind is assumed to be negligible because the marine atmosphere has near neutral stratification, and that the magnitude of ocean current is small relative to wind speed over most ocean areas. Because stress is small-scale turbulence generated by buoyancy and wind shear, its magnitude should have strong spatial coherence with sea surface temperature and its direction should show influence by current. These features that are driven by ocean processes may not be fully represented in winds that are subjected to larger-scale atmospheric factors, as discussed by Liu and Xie (2008) and Liu et al. (2010).

NASA launched a Ku-band scatterometer, QuikSCAT, in June 1999. Level-2 data at 12.5 km resolution are obtained from the Physical Oceanography Distributed Active Archive Center. Seven years of the data, from June 2002 to May 2009 (coincide with radiometer data as discussed in Section 3), organized in wind vector cells along satellite swath, are binned into uniform 1/8 degree grids over global oceans and fitted to the Weibull distribution for the 7 year periods. There is hardly any in situ stress measurement. Even for winds, there is no in situ measurement that could represent the range of scatterometer data, particularly at the high and low ends, to evaluate the probability density function (PDF) from which E is derived.

3. Microwave radiometer

Ocean surface wind speed can also be derived from the radiance observed by a microwave radiometer. It is generally believed that wind speed affects the surface emissivity indirectly through the generation of ocean waves and foam (Hollinger, 1971; Wilheit, 1979). Radiometers designed to observe the ocean surface operate primarily at window frequencies, where atmospheric absorption is low. To correct for the slight interference by tropospheric water vapor, clouds, and rainfall and, to some extent, the effect of sea surface temperature, radiances at frequencies sensitive to sea surface temperature, atmospheric water vapor, and liquid water are also measured (Wentz, 1983). The Advanced Microwave Scanning Radiometer-Earth Observing System (AMSR-E), on board of NASA's Aqua satellite, was launched in May 2002 and has been measuring ocean parameters including wind speed and sea surface temperature. These parameters averaged to 0.25° grids for ascending and descending paths were obtained from Remote Sensing System.

4. Power density

The Weibull distribution (Gaussian and Rayleigh distributions are special cases of it) has been often used to characterize the PDF of wind power (e.g., Pavia & O'Brien 1986). A two parameters Weibull distribution has the PDF (p) as a function of wind speed U,

$$p(U) = (k/c)(U/c)^{k-1} \exp[-(U/c)^k] \quad (1)$$

where k is the dimensionless shape parameter, and c is the scale parameter. A number of methods to estimate Weibull parameters exist, with negligible difference in the results (Monahan, 2006). We used the simplest formula:

$$c = \overline{U}/\Gamma(1+1/k) \tag{2a}$$

$$k = (\overline{U}/\sigma)^{1.086} \tag{2b}$$

where $\overline{U}$ is the mean, σ is the standard deviation of wind speed, and Γ is the gamma function. The available wind power density E (which is proportional to U^3) may be calculated from the Weibull distribution parameters as

$$E = ½ \rho c^3 \Gamma(1+3/k) \tag{3}$$

where ρ is the air density. E is essentially the kinetic energy of the wind.
We will analyze PDF and E, which will provide the characteristics of not only the means and the frequencies of strong wind, but also the variation and higher moments critical in relating the non-linear effects of wind on electric power generation capability.

5. Geographic distribution

Scatterometer climatology in forms of mean wind (e.g., Risien & Chelton, 2006), frequency of strong wind (Sampe & Xie, 2007), and power density (Liu et al., 2008a) have been produced before. The PDF of 7 year of wind speed at 10 m height above oceans between 75° latitudes (Fig. 1) shows the slight difference between QuikSCAT and AMSR-E. AMSR-E, which peaks at 7.5 m/s, has more high wind than QuikSCAT, which peaks at 7 m/s. The global distributions of E (Fig. 2 and 3) are very similar, with AMSR-E data giving a slightly larger dynamic range.
The distributions of E, as shown in Fig. 2 and 3, confirm the conventional knowledge: strongest E is found over the mid-latitude storm tracks of the winter hemisphere, the relatively steady trade winds over the tropical oceans, and the seasonal monsoons. At mid-

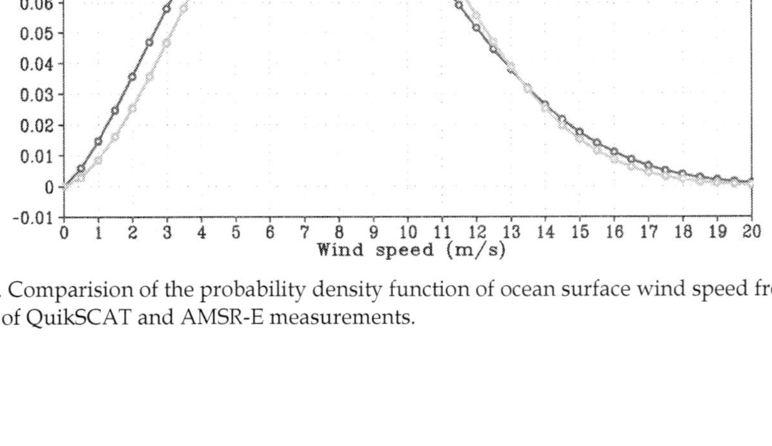

Fig. 1. Comparision of the probability density function of ocean surface wind speed from 7 years of QuikSCAT and AMSR-E measurements.

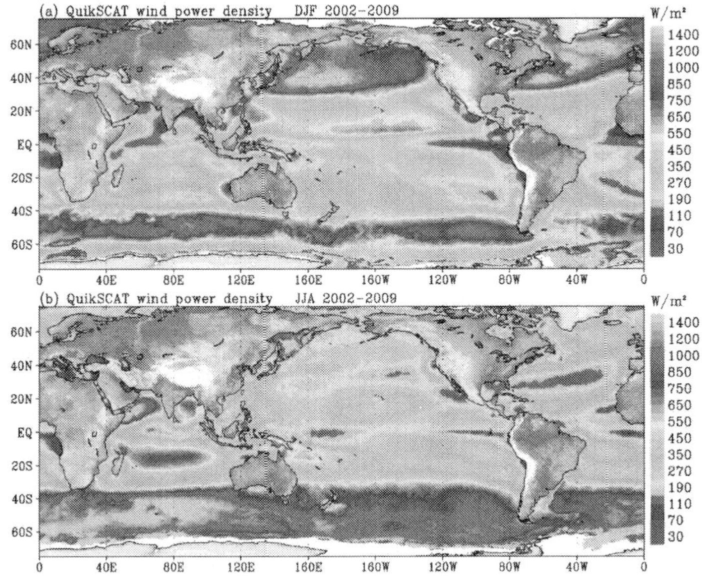

Fig. 2. Distribution of power density of ocean surface wind (10 m) from QuikSCAT for (a) boreal winter (December, January, and February) and (b) boreal summer (June, July, and August).

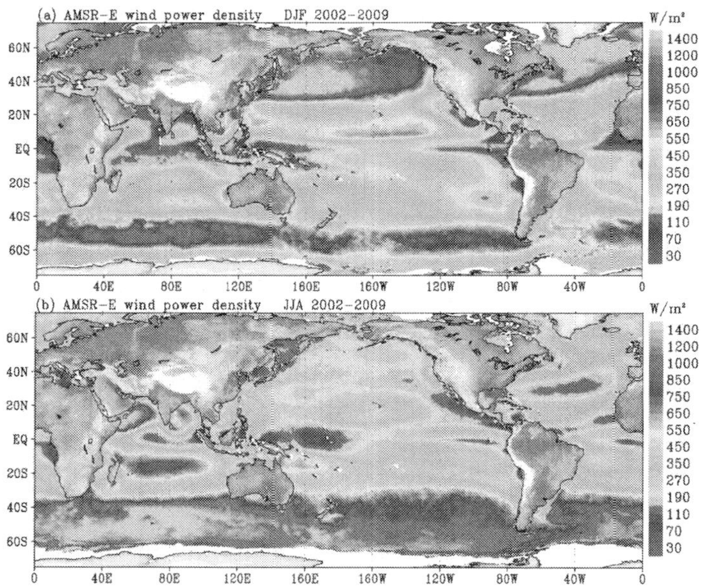

Fig. 3 Same as Fig. 2, but from AMSR-E.

latitude in the winter hemisphere, E is much larger than those in the tropics, making the display of the major features with the same color scale extremely difficult. The trade winds, particularly in the western Pacific and Southern Indian oceans are stronger in winter than summer, but the seasonal contrast is much less than those of the mid-latitude storm track. In the East China Sea, particularly through the Taiwan and Luzon Strait, the strong E is caused by the winter monsoon. In the Arabian Sea and Bay of Bengal, it is caused by the summer monsoon. In the South China Sea, the wind has two peaks, both in summer and winter. QuikSCAT data also reveal detailed wind structures not sufficiently identified before. The strong winds of transient tropical cyclones are not evident in E derived from the seven-year ensemble.

Because space sensors measure stress, the distribution reflects both atmospheric and oceanic characteristics. Regions of high E associated with the acceleration of strong prevailing winds when defected by protruding landmasses are ubiquitous. Less well-know examples, such as the strong E found downwind of Cape Blanco and Cape Mendocino in the United States and Penisula de La Guajira in Columbia, stand out even on the global map. Strongest E is observed when the along-shore flow coming down from the Labrador Sea along the west Greenland coast as it passes over Cape Farewell meeting wind flowing south along the Atlantic coast of Greenland. Strong E is also found when strong wind blows offshore, channeled by topography. The well-known wind jets through the mountain gap of Tehuantepec in Mexico and the Mistral between Spain and France could be discerned in the figures. Alternate areas of high and low E caused by the turbulent production of stress by buoyancy could also be found over mid-latitude ocean fronts, with strong sea surface temperature gradient (e.g., Liu & Xie, 2008), particularly obvious over the semi-stationary cold eddy southeast of the Newfoundland.

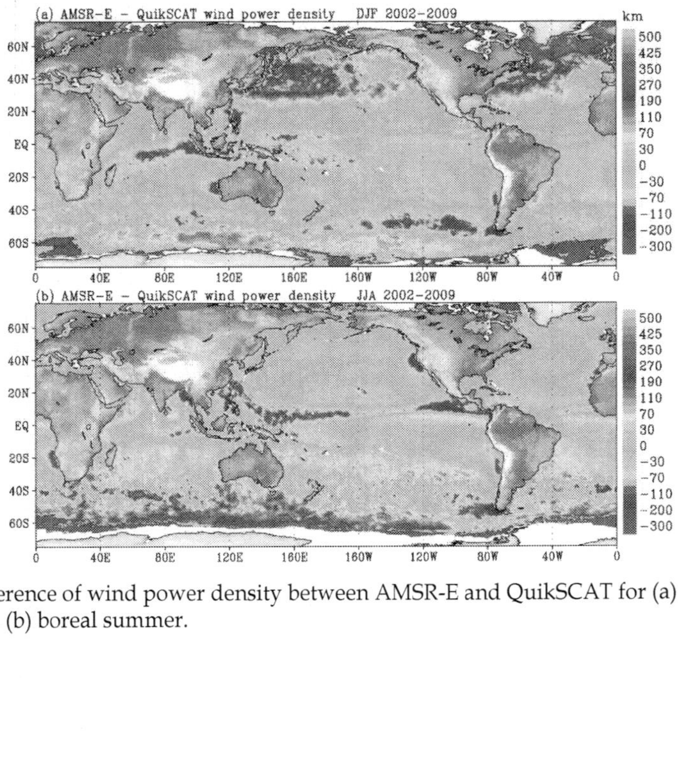

Fig. 4. Difference of wind power density between AMSR-E and QuikSCAT for (a) boreal winter and (b) boreal summer.

Fig. 4 shows that E from AMSR-E is higher than that from QuikSCAT in the winter hemisphere at mid to high latitudes of both Pacific and Atlantic, and slightly lower in the tropics. The large differences around Antarctica may be due to contamination of scatterometer winds by ice.

6. Height dependence

The analysis, so far, is based on the equivalent neutral wind at 10 m, the standard height of scientific studies. The effective heights of various designs of the wind turbines, from the lower floating turbine that spins around a vertical axis to the anchored ones that spin around a horizontal axis, are likely to be different. The turbine height dependence has been well recognized (e.g. Barhelmie, 2001). There is a long history of studying the wind profile in the atmospheric surface (constant flux) layer in term of turbulent transfer. The flux-profile relation (also called similarity functions) of wind, as described by Liu et al. (1979), is

$$\frac{U-U_s}{U_*}=2.5(\ln\frac{Z}{Z_0}-\psi)=\frac{1}{\sqrt{C_D}} \tag{4}$$

where U_s is the surface current, $U_*=(\tau/\rho)^{1/2}$ is the frictional velocity, ρ is the air density, Z_0 is the roughness length, Ψ is the function of the stability parameter, and C_D is the drag coefficient. The stability parameter is the ratio of buoyancy to shear production of turbulence. The effect of sea state and surface waves (e.g., Donelan et al. 1997) are not included explicitly in the relation. U_* and Z_0 are estimated from the slope and zero intercept respectively of the logarithmic wind profile. The drag coefficient is an empirical coefficient in relating τ to ρU^2 (Kondo 1975, Smith 1980, Large & Pond, 1981) and is expressed as a function of wind speed. An alternative to using the drag coefficient is to express Z_0 as a function of U_*. For example, Liu and Tang (1996) incorporated such a relation in solving the similarity function. They combined a smooth flow relation with Charnock.s relation in rough flow to give

$$Z_o=0.11\frac{\upsilon}{U_*}+0.011\frac{U_*^2}{g} \tag{5}$$

where ν is the kinematic viscosity and g is the acceleration due to gravity.
In general oceanographic applications, the surface current is assumed to be small compared with wind and the atmosphere is assumed to be nearly neutral. With the neglect of U_s and Ψ in (1), U becomes U_N by definition. The wind speed at a certain height z (U_z) relative to U_N at 10 m, U_{10}, is given by

$$\frac{U_z}{U_{10}}=1+2.5\sqrt{C_D}\ln(\frac{z}{10}) \tag{6}$$

and z is in meter. Fig 5 shows the variation of wind speed at 80 m as a function of wind speed at 10 m, under neutral conditions for three formulations of the drag coefficient. For example, the 80 m wind exceeds 10 m wind by 5% and 20% at wind speed of 10 m/s and 30 m/s respectively, according to the drag coefficient given by Kondo (1975).

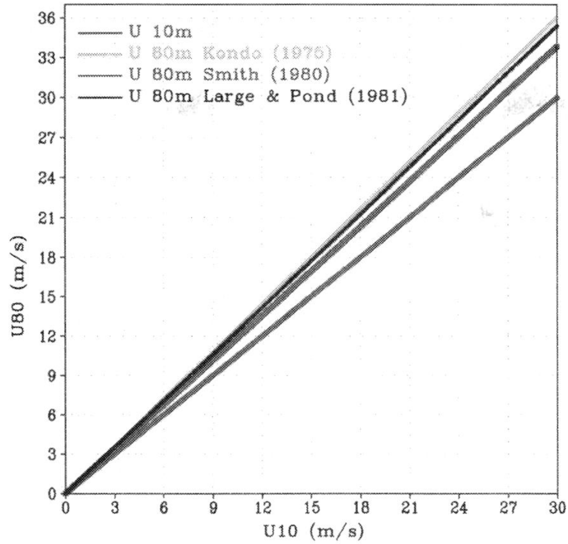

Fig. 5. Wind speed at 80 m height as a function of wind speed at 10 m under neutral stability for three formulations of drag coefficient.

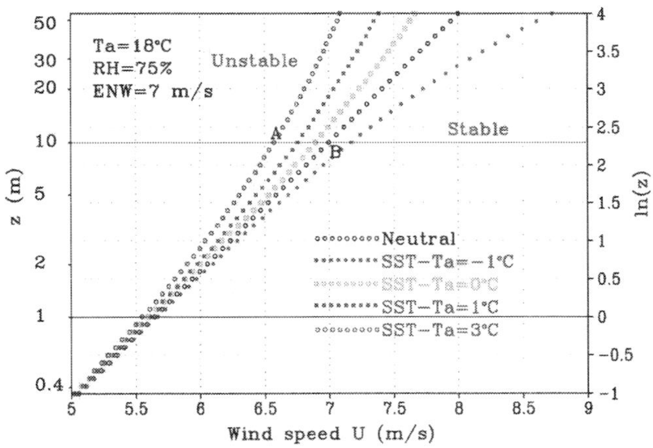

Fig. 6. Comparison of wind profiles under various stability conditions.

7. Stability dependence

Typical wind profiles at various stabilities are shown in Fig. 6. At a given level, U_N is larger than the actual wind under unstable condition but lower under stable condition.
From (4) the difference Between U_N and the actual wind U is

$$\delta U = U_N - U = 2.5U * \psi \tag{7}$$

As described by Liu et al. (1979) and the computer program in Liu and Tang (1996), the flux profile relations for wind, temperature, and humidity could be solved simultaneously for inputs of wind speed, temperature, and humidity at a certain level and the sea surface temperature to yield the fluxes of momentum (stress), heat, and water vapor. The value of Ψ is a by-product. Using U_N provided by QuikSCAT, sea surface temperature from AMSR-E, air temperature, and humidity from the reanalysis of the European Center for Medium-range Weather Forecast, δU at 10 m averaged over a three years period, for January and July, are computed and shown in Fig. 7. The distribution of stability effect on wind speed closely follows the distribution of sea-air temperature difference shown in Fig. 8.

U_N is higher than U in the unstable regions and lower in stable regions. U_N is higher than U by as much as 0.7 m/s in January over the western boundary currents. It is also higher than U over the intertropical convergence zone, the south Pacific convergence zone, and the South Atlantic convergence zone. U_N is lower than U in stable regions, such as over the circumpolar current and in northeast parts of both Pacific and Atlantic.

8. Future potential and conclusion

One polar orbiter could sample the earth, at most, two times a day and may introduce error in E because of sampling bias, as discussed by Liu et al. (2008b) in constructing the diurnal cycle with data from tandem missions. There are three scatterometers in operation now. QuikSCAT or the similar scatterometer on Oceansat-2 launched recently by India, will covered 90% of the ocean daily, and the Advanced Scatterometer (ASCAT) on the European Meteorology Operational Satellite (METOP) will covered similar area in two days, as showed in Fig. 9.

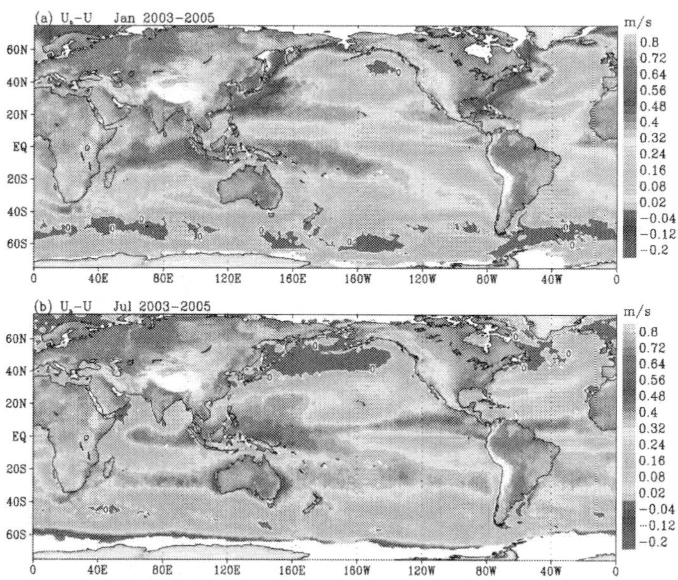

Fig. 7. Difference between equivalent neutral wind and actual wind at 10 m for (a) Januray and (b) July.

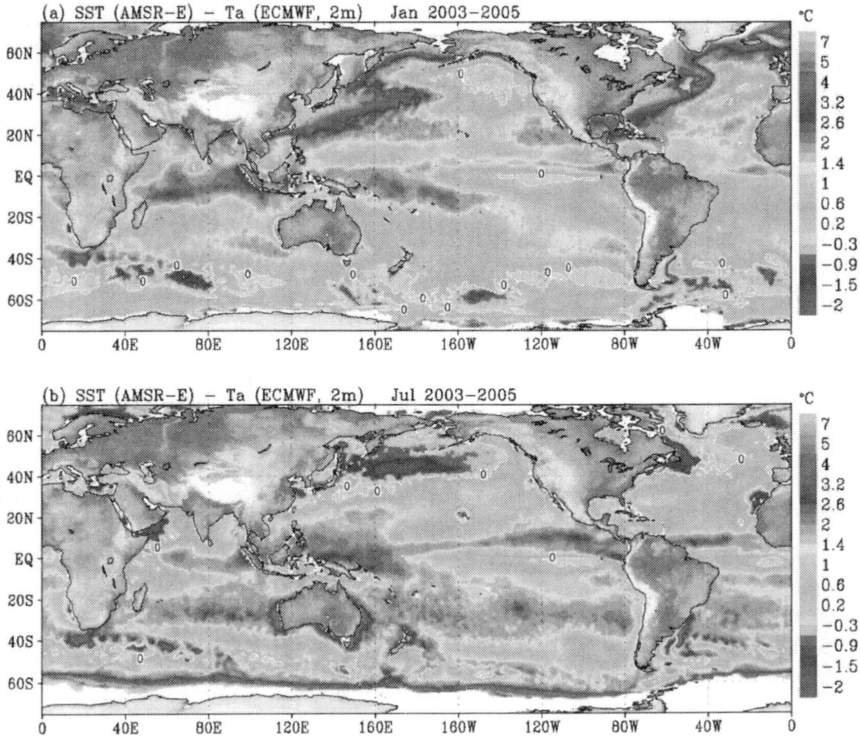

Fig. 8. Difference between sea surface temperature and air temperature (2 m) for (a) January and (b) July.

QuikSCAT alone could resolve the inertial period required by the oceanographers only in the tropical Oceans, but the combination of QuikSCAT and ASCAT will cover the inertial period at all latitudes, as shown in Fig. 10. Even the combination of QuikSCAT and ASCAT would not provide six hourly revisit period, as required by operational meteorological applications, over most of the oceans. The addition of Oceansat-2 brings the revisit interval close to 6-hour at all latitudes. The scatterometer on Chinese Haiyang-2 satellites, approved for 2011 launch, will shorten the revisit time or will make up the sampling loss at the anticipated demise of the aging QuikSCAT. As shown in Fig. 9 and 10, the combination of these missions will meet the 6 hourly operational NWP requirement in addition to the inertial frequency required by the oceanographers.

Deriving a consistent merged product may need international cooperation in calibration, and maintaining them over time may require political will and international support. It remains a technical challenge to generate electricity by wind off shore and transmit the power back for consumption efficiently, but satellite observations could contribute to realize the potential.

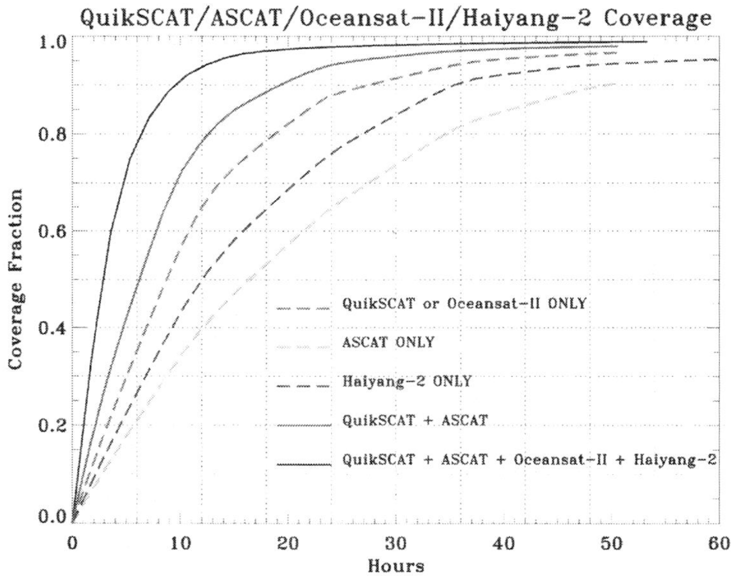

Fig. 9 Fractional coverage, between 70°N and 70°S by various tandem missions as a function of time.

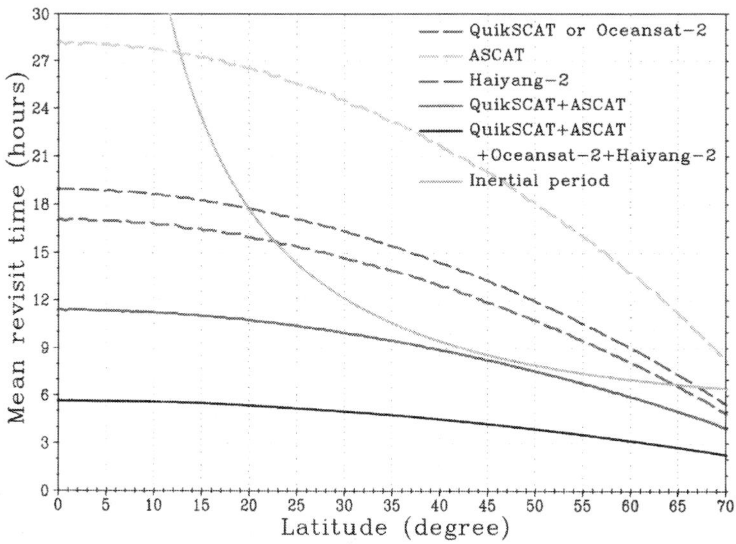

Fig. 10 The latitudinal variation of zonally averaged revisit interval for various tandem missions.

9. Acknowledgment

This study was performed at the Jet Propulsion Laboratory, California Institute of Technology under contract with the National Aeronautics and Space Administration (NASA). It was jointly supported by the Ocean Vector Winds and the Physical Oceanography Programs of NASA.

10. References

DTI, 2007: Meeting the Energy Challenge: A White Paper on Energy, Department of Trade and Industry, 341 pp. The Stationary Office, London, United Kingdom.

Barthelmie, R. J., 2001: Evaluating the impact of wind induced roughness change and tidal range on extrapolation of offshore vertical wind speed profiles. Wind Energy, 2001; 4:99-105 (DOI: 10.1002/we.45).

Capps, S.B., and C.S. Zender, 2008: Observed and CAM3 GCM sea surface wind speed distributions: Characterization, comparison, and bias reduction. J. Clim., 21, 6569-6585.

Donelan, M.A., W.M. Drenan, and K.B. Katsaros, 1997: The air-sea momentum flux in conditions of wind sea and swell. J. Phys. Oceanogr., 27, 2087-2099.

Hollinger, J. P. 1971. Passive microwave measurements of sea surface roughness. IEEE Trans. Geosci. Electronics GE-9:165-169.

Kondo, J., 1975: Airsea bulk transfer coefficients in diabatic conditions. Bound-Layer Meteor., 9, 91-112.

Large, W.G., and S. Pond, 1981: Open ocean momentum flux measurements in moderate to strong winds. J. Phys. Oceanogr., 11, 324-336.

Liu, W.T., 2002: Progress in scatterometer application, J. Oceanogr., 58, 121-136.

Liu, W.T., and W.G. Large, 1981: Determination of surface stress by Seasat-SASS: A case study with JASIN Data. J. Phys. Oceanogr., 11, 1603-1611.

Liu, W.T., and W. Tang, 1996: Equivalent Neutral Wind. JPL Publication 96-17, Jet Propulsion Laboratory, Pasadena, 16 pp.

Liu, W.T., and X. Xie 2006: Measuring ocean surface wind from space. Remote Sensing of the Marine Environment, Manual of Remote Sensing, Third Edition, Vol. 6, J. Gower (ed,), Amer. Soc. for Photogrammetry and Remote Sens. Chapter 5, 149-178.

Liu, W.T., and X. Xie, 2008: Ocean-atmosphere momentum coupling in the Kuroshio Extension observed from Space. J. Oceanogr., 64, 631-637.

Liu, W.T., K.B. Katsaros, and J.A. Businger, 1979: Bulk parameterization of air-sea exchanges in heat and water vapor including the molecular constraints at the interface. J. Atmos. Sci., 36, 1722-1735.

Liu, W.T., W. Tang, and X. Xie, 2008a: Wind power distribution over the ocean. Geophys. Res. Lett., 35, L13808, doi:10-1029/2008GL034172.

Liu, W.T., W. Tang, X. Xie, R. Navalgund, and K.Xu, 2008b: Power density of ocean surface wind-stress from international scatterometer tandem missions. Int. J. Remote Sens., 29(21), 6109-6116.

McElroy, M.B., X. Lu, C.P. Nielsen, and Y. Wang, 2009: Potential for wind-generated electricity in China. Science, 325, 1378-1380.

Monahan, 2006: The probability distribution of sea surface wind speeds. Part I: theory and SeaWinds observations. J. Clim., 19, 497-520.

Pavia, E. G., and J. J. O.Brien, 1986: Weibull statistics of wind speed over the ocean, J. Clim. Appl. Meteorol., 25, 1324-1332.

Risien, C. M., and D. B. Chelton, 2006: A satellite-derived climatology of global ocean winds. Remote Sens. Environ., 105, 221-236.

Sampe, T., and S-P. Xie, 2007: Mapping high sea winds from space: a global climatology. Bull. Amer. Meteor. Soc., 88, 1965-1978

Smith, S.D., 1980: Wind stress and heat flux over the ocean in gale force winds. J. Phys. Oceanogr., 10, 709-726.

Wentz, F. J. 1983: A model function for ocean microwave brightness temperatures. J. Geophys. Res., 88, 1892-1908.

Wilheit, T. T. 1979: A model for the microwave emissivity of the ocean.s surface as a function of wind speed. IEEE Trans. Geoscience Electronics GE-17, 244-249.

Part C
The Grid Integration Issues

15

Methods and Models for Computer Aided Design of Wind Power Systems for EMC and Power Quality

Vladimir Belov[1], Peter Leisner[2,3], Nikolay Paldyaev[1],
Alexey Shamaev[1] and Ilja Belov[3]
[1]Mordovian State University, 430000, Saransk,
[2]SP Technical Research Institute of Sweden, Box 857, 501 15 Borås,
[3]School of Engineering, Jönköping University, Box 1026, SE 551 11, Jönköping,
[1]Russia
[2,3]Sweden

1. Introduction

In off-grid wind power systems (WPS) a power source generates the power which is comparable to the consumed power. Solving electromagnetic compatibility (EMC) problems in such a WPS is directly related to power quality issues. High levels of low- and high-frequency conducted emissions in a WPS worsen the quality of consumed electric power, increase power losses, and adversely affect reliability of connected appliances. The indicated problems should be addressed in the WPS design phase. Here, power quality and EMC related criteria have to be given a high rank when choosing the structure and parameters of a WPS.

The mission of this chapter is to provide grounds for practical application of both a mathematical model of WPS and a method for parametric synthesis of a WPS with specified requirements to EMC and electric power quality.

The present chapter is focused on a simulation-based spectral technique for power quality and EMC design of wind power systems including a power source or synchronous generator (G), an AC/DC/AC converter and electronic equipment with power supplies connected to a power distribution network. A block diagram of a typical WPS is shown in Fig. 1 (EMC Filters Data Book, 2001), (Grauers, 1994).

Three-phase filter 1 is connected to the generator side converter in order to suppress current harmonics caused by the rectifier circuit. An output Γ-filter placed after the AC/DC/AC converter comprises inductance *L* and capacitor *C*. It is designed for filtering emissions caused by pulse-width modulation (PWM) in the AC/DC/AC converter.

Single-phase filter 2 (shown with the dash line) is connected to the load side inverter. It protects the load from low frequency current harmonics impressed by the AC/DC/AC converter.

A synchronous generator and an AC/DC/AC converter are the key elements of a WPS. The AC/DC/AC converter is a source of low-frequency conducted emissions. They cause voltage distortions at the synchronous generator output, thereby reducing the quality of the supplied voltage and increasing active losses. The pulse-width modulation (PWM) in the

AC/DC/AC converter is the main source of high-frequency emissions as well as single-phase non-linear loads, such as a switch mode power supply (SMPS). High-frequency emissions create EMC problems in a WPS.

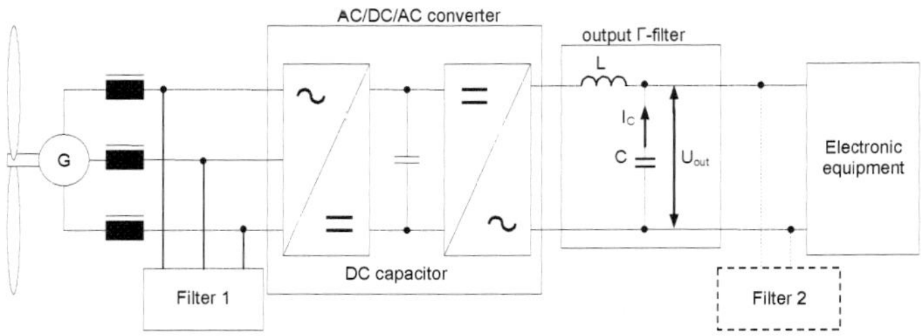

Fig. 1. Block diagram of a wind power system

The described problems of EMC and power quality can be solved on the basis of a complex approach, via designing a filtering system.

Parametric synthesis of the system of harmonic, EMC and active filters constitute an important practical task in variant design of WPS.

The task of computer aided design of the filtering system can be solved through application of the simulation-based spectral technique (Belov et al., 2006). The spectral technique utilizes multiple calculations of current and voltage spectra in the nodes of WPS during the power quality and EMC design procedure. It essentially differs from the filter design methods based on the insertion loss technique (Temes et al, 1973), since it can search for WPS frequency response and for the corresponding filter circuit given the EMC and power quality requirements for WPS. Change in the WPS frequency response during design is reflected in the spectral technique. In the proposed spectral technique, power converters and power supplies are described with complete non-linear models.

A general WPS includes a number of AC/DC/AC converters. Therefore, *a WPS modeling methodology* is developed that computes the WPS frequency response. The modeling methodology developed for a general multi-phase electric power supply system has the following features:

- Operation of all switching elements is implemented in the WPS model, for arbitrary cascade circuits including bridge converters in single-phase, three-phase and, generally, *m*-phase realizations.
- Modelling of a three-phase and, generally, an *m*-phase synchronous generator is performed according to complete equations written in *dq0* co-ordinates.

Mathematical modelling of power quality and EMC in the WPS is performed on the basis of the multi-phase bridge-element concept (B-element concept), (Belov et. al., 2009). This concept corresponds well both to the structure and to the operation principles of an AC/DC/AC converter, being efficiently tied both to the transient phenomena in electrical machines and to the PWM techniques.

Mathematical models of single- and three-phase devices in WPS are obtained as a particular case of multi-phase B-element concept. In the complete model of a WPS, the AC/DC/AC converter is represented in *m*-phase co-ordinate system, whereas electro-mechanical converters are represented in *dq0* co-ordinates, thereby contributing to modelling efficiency and validity of the results; it will be demonstrated by computational experiments, performed for the WPS including an active filter integrated into the voltage inverter of the AC/DC/AC converter

2. Spectral technique for power quality and EMC design of wind power systems

The problem of EMC and power quality design of the WPS shown in Fig. 1 may include calculation of filter 1 and filter 2 which can be either active or harmonic filters, as well as any additional filter installed in the WPS. The steps of the simulation-based spectral technique will thus be formulated on the example of a general filter.

Calculation of the filter includes an optimization procedure. Objective function and constraints are defined based on application reasons. For example, the total reactive power Q of the filter capacitors defines the volumetric dimensions of the filter, which in some applications is an important design criterion. Minimization of the total reactive power of filter capacitors can be performed for a passive harmonic filter (Belov et al., 2006). Active and hybrid filters also include capacitors. In this case, minimization of the total reactive power of filter capacitors can be performed along with solving the optimal control problem.

The filter optimization problem includes constraints regarding EMC and power quality in WPS nodes. Power quality in WPS is presented by electric power quality indices, *THD* and *DPF*. The constraints relate the filter component values to the electric power quality indices. Constraints can be specified e.g. for the capacitors' peak voltage and the WPS frequency response. The latter addresses the EMC requirements.

The spectral technique for power quality and EMC design includes the following steps (see Fig. 2).

Step 1. *Specifying WPS structure and parameters.* WPS elements are defined by component values (resistance, inductance and capacitance), electrical characteristics (*e.g.* SG total power), and control parameters (*e.g.* commutation delay of an AC/DC converter).

Step 2. *Specifying desired power quality.* Desired power quality in WPS is presented by THD_D and DPF_D, specified according to power quality regulations. They are brought to a matrix $EPQ_{desired}$. Each row in *EPQ*-matrix corresponds to a node in WPS, and each column corresponds to a power quality index.

Step 3. *Specifying desired EMC.* In order to identify EMC problem in WPS, a designer uses regulations for conducted emissions, related to the equipment's power supplies connected to WPS.

Step 4. *Calculation of voltage and current spectra.* The calculation procedure utilizes a complete mathematical model of WPS to reflect essential non-linear processes in elements of WPS. A set of ordinary differential equations with discontinuous right-hand sides is numerically solved in time domain. The FFT technique is then used for calculating current and voltage spectra in WPS.

Step 5. *Forming an updated EPQ-matrix.* Calculated voltage and current spectra are used for forming an updated *EPQ*-matrix ($EPQ_{updated}$). *THD* and *DPF* are calculated according to the following well-known equations in the node of WPS where power quality is monitored:

$$THD = \left(\sum_{n=2}^{N} U_n^2\right)^{1/2} / \; U_1 , \qquad (1)$$

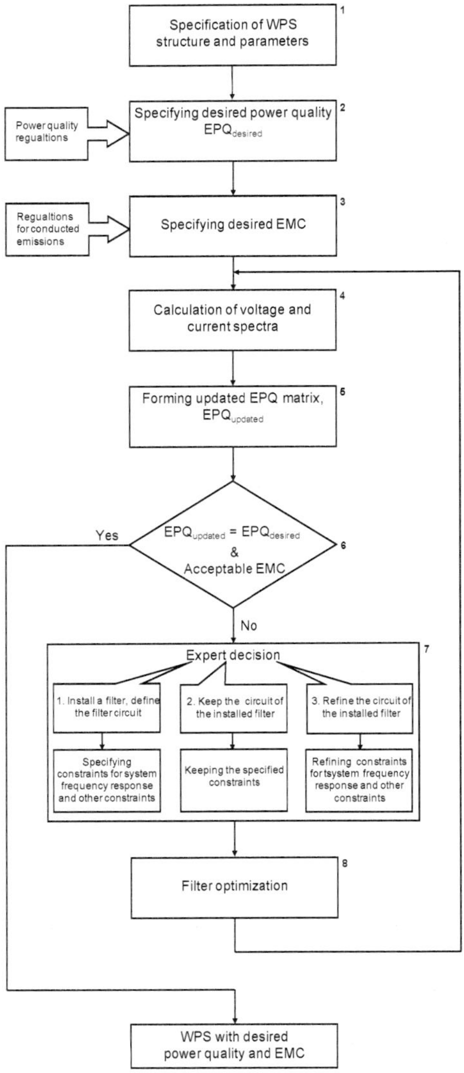

Fig. 2. Block diagram of the simulation-based spectral technique

$$DPF = (\sum_{n=1}^{N} U_n \cdot I_n \cdot \cos\varphi_n) / (\sum_{n=1}^{N} U_n^2 \sum_{n=1}^{N} I_n^2)^{1/2} \tag{2}$$

Step 6. *Comparing $EPQ_{updated}$ with $EPQ_{desired}$ and identifying EMC/power quality problem.* The desired *EPQ*-matrix is subtracted from the updated *EPQ*-matrix. If the matrix difference contains elements with the absolute values smaller than tolerance values specified for each power quality index, then the power quality problem has been solved. Additionally, voltage and/or current spectra at the power supplies' output have to be compared with EMC regulations for conducted emissions. If a power quality and/or an EMC problem are identified, an expert decision has to be taken. Otherwise, the design process is finished.

Step 7. *Expert decision.* At the first pass of the algorithm the expert decision is installing a filter in the node of WPS with a poor power quality or EMC The choice of filter circuit and the filtered frequencies depends on the *EPQ*-matrices, the tolerance values, and the rms-values of current harmonics. Constraints for the WPS frequency response are specified by the designer. Some other constraints can be included, e.g. for the filter capacitors' peak voltage. These constraints will be used in the filter optimization procedure along with the power quality requirements defined in step 2. At the next passes of the algorithm two types of expert decision are possible. One of them is direct passing to step 8 with current and voltage spectra calculated in step 4 as the new input data for filter optimisation. Since the filter circuit has not been refined, the constraints are unchanged. The other expert decision is refinement of the filter circuit. In case of designing a passive filter, a resonant section can be added to the filter circuit. For an active or a hybrid filter, the refinement of the filter circuit would consist *e.g.* in adding passive components. Refinement of the filter circuit might lead to changing the constraints.

Step 8. *Filter optimisation.* The non-linear model is replaced by an algebraic model of WPS including the filter. Filter component values are determined by solving a non-linear programming problem, given the constraints for power quality indices (defined by $EPQ_{desired}$), for EMI (the WPS frequency response), and other constraints. The total reactive power of filter capacitors can be used as the minimization criterion-minimized.

Checking the filter performance is implemented by passing to step 4, where current and voltage spectra are calculated taking into account the power filter designed in step 8. Passing to step 4 can be explained by loss of some properties of WPS due to the simplified algebraic model neglecting non-linear properties of the filter in step 8. $EPQ_{updated}$ is then compared to $EPQ_{desired}$, and emission levels are compared to EMC regulations (step 6). A new expert decision is made (step 7), *etc.*

An example of application of the presented spectral technique, including a harmonic filter optimization is provided in (Belov et al., 2006). The optimization method was chosen from (Himmelblau, 1972).

3. Multi-phase electric power supply system modeling methodology

3.1 Multy-phase system elements and modeling requirements

Multi-phase electric power supply systems with the number of phases $p > 3$ have a number of advantages as compared to conventional three-phase systems. They include lower installed power of ac-machines at fixed dimensions, more compact power transmission line at equal carrying power, lower current loading per phase to result in lower-power semiconductor devices and more compact control equipment, wider range of speed control, and lower level of noise and vibration for electrical machines. Analysis and design methods for multi-phase electric power supply systems have been addressed by a number of authors, e.g. in (Binsaroor et al., 1988), (Toliyat et al., 2000). However, they are still not well developed for independent power supply systems, where generated power and the consumed power are comparable. The tendency to consider multi-phase systems in the variant design (Brown et al., 1989) should be addressed in the modelling software enabling a common CAD environment for electric power supply systems, consisting of elements with different number of phases in feeding lines.

A general multi-phase electric power supply system includes a p-phase synchronous generator SG (p is a prime integer), power transmission line, an AC/DC/AC power converter and an ac-load. Additional multi-phase filters can be included in the power system.

Reducibility to a canonical form is the main requirement the mathematical models of power system's elements have to satisfy. It is related to the properties of a CAD tool designed for integrating element models to the complete mathematical model of the electric power supply system. The canonical form for mathematical models of the p-phase system elements is given by equation (3):

$$\frac{d}{dt} i_{dq0} = Bu_{dq0} + D \tag{3}$$

where i_{dq0} is the vector of phase currents of a multi-phase element in $dq0$ rotating coordinate system obtained by transformation from the multi-phase stationary coordinate system; B is the matrix containing the parameters of the modelled element (SG, AC/DC/AC converter, ac-load, and filters); u_{dq0} is the vector of phase voltages in the node where the multi-phase element is connected to the power transmission line, D is the vector dependent of i_{dq0}. Vector D is recalculated in each step of integration of equation (3).

Integrating of the models of the system elements into the complete mathematical model of the multi-phase electric power supply system is performed by means of equation (4) for multi-phase power transmission line (Belov, 1993):

$$U_{dq0} = -(K_1 + LK_2B_\Sigma)^{-1}(LK_2D_\Sigma + (LK_2' + RK_2)I_{dq0\Sigma}), \tag{4}$$

where $K_2' = dK_2 / dt$, $L = diag(L_j)$, $R = diag(R_j)$, $j = 1, 2, \ldots, N$, and N is the number of nodes in the power transmission line. Block-matrices K_1 and K_2 describe the structure of the power transmission line. Being multiplied by the vector of node voltages $U_{dq}0$, sub-diagonal block-matrix K_1 expresses the potential difference between the neighbouring nodes. The number of columns and the number of rows in this block-matrix are equal to the number of nodes in the power transmission line. An element of block-matrix K_1 is a non-zero matrix when an interconnecting line is connected to the corresponding nodes. Super-diagonal block-matrix K_2 is for reduction of the node currents contained in vector $I_{dq0\Sigma}$ to a base coordinate system. The base coordinate system can be a $dq0$ system of any synchronous generator. This is the case when several generators are connected to the power transmission line. The rows and columns of matrix K_2 are constructed similarly to matrix K_1. The row corresponding to the interconnecting line where the selected generator is connected is filled with only non-zero matrices, since the current of this line is the sum of all node currents. A node current, in its turn, is the sum of currents through the system's elements, connected to a particular node of the power transmission line. In the other rows of block-matrix K_2, non-zero matrices are placed on the crossing with the columns corresponding to the nodes, whose currents are summated in a particular interconnecting line. Block-matrices K_1, K_2 are built of matrices $\begin{bmatrix} \cos\delta_{ij} & \sin\delta_{ij} \\ -\sin\delta_{ij} & \cos\delta_{ij} \end{bmatrix}$ and $\begin{bmatrix} \cos\delta_{ij} & -\sin\delta_{ij} \\ \sin\delta_{ij} & \cos\delta_{ij} \end{bmatrix}$, where δ_{ij} is determined from equation (5):

$$(d\delta_{ij} / dt) = \omega_i - \omega_j, \tag{5}$$

where ω_i and ω_j are the angular speed values for rotating *dq0* coordinate systems in the neighboring nodes of the power transmission line.

3.2 Model of multi-phase synchronous generator

For mathematical modelling of power system elements with arbitrary number of electrical phases, it is necessary to have a general operator C_P. It can be obtained in the process of construction of the mathematical model for the *p*-phase synchronous generator (*p* is a prime integer, which can be represented as $p = 2k+1$, $k = 1, 2, 3, 5, \ldots$) (Belov et al., 2004). Let us consider SG with $p+3$ magnetically coupled circuits: *p* phase windings which are stationary and three rotor circuits which are rotating: one drive circuit and two damping circuits. The phase windings are supposed to be connected to the power transmission line. The drive circuit is connected to an *emf* source. The equilibrium equations for voltages in the rotor and stator circuits can then be written in a matrix form as follows:

$$\frac{d\Psi}{dt} + Ri + e = 0, \tag{6}$$

where $R = diag(r, r, \ldots, r, r_D, r_Q, r_r)$ is a diagonal matrix, r is the resistance of each phase winding, r_D and r_Q are the resistances of the damping circuits, r_r is the resistance of the drive circuit; $i = [i_1, i_2, \ldots, i_p, i_D, i_Q, i_r]^T$ is the unknown vector of currents in the SG circuits, i_j, $j = 1, 2, \ldots, p$, are the phase currents, i_D and i_Q are the currents in the dumping circuits, i_r is the current in the drive circuit; $e = [e_1, e_2, \ldots, e_p, 0, 0, -e_r]^T$ is the known voltage vector, e_1, e_2, ..., e_{2k}, e_p are the voltages in the nodes of connection of the phase windings to the power transmission line, e_r is the voltage of the drive circuit; Ψ is the magnetic linkage vector for SG circuits, such that $\Psi = Li$, and L is the square matrix of the coefficients of self-induction and mutual induction for SG circuits, which is a symmetric matrix for physical reasons. The elements of matrix L are π - or 2π -periodic functions of angle γ. Angle γ is the angle between the magnetic axis of the first phase winding of SG and the longitudinal rotor axis at counter clockwise rotation.

Equations (6) have periodic coefficients. With the help of the 2π -periodic transformation matrix $C_P = C_P(\gamma)$, equation system (6) can be rewritten in a rotating *dq0* coordinate system, which corresponds well to a symmetric nature of the synchronous generator SG with poles. Matrix C_P has a dimension of $p \times p$ and is given by equation (7):

$$C_{P\,ij} = \begin{cases} 1/p, & \text{if } i = 1; \\ (2/p)\cos(\gamma - (j-1)n\rho), & \text{if } i = 2n-1, n \geq 2; \\ -(2/p)\sin(\gamma - (j-1)n\rho), & \text{if } i = 2n, n \geq 1. \end{cases} \tag{7}$$

In equation (7), $n \in \mathrm{N}$, $\rho = 2\pi / p$ is the angle between magnetic axes of adjacent phase windings. Here, $\det(C_P) = \pm\sqrt{\frac{2^{p-1}}{p^p}}$, i.e. determinant of matrix C_P is not equal to zero at all γ. Matrix $C = diag(C_P, 1, 1, 1)$ with a dimension of $(p+3)\times(p+3)$ can be obtained by complementing matrix C_P by an identity matrix. Let us now pre-multiply equation system (6) by matrix C keeping the following designations: $i_{dq0} = Ci$, $e_{dq0} = Ce$, $i_{dq0} = \left[i_0, i_{1,d}, i_{1,q}, i_{2,d}, i_{2,q}, \ldots, i_{k,d}, i_{k,q}, i_D, i_Q, i_r\right]^T$, and $e_{dq0} = \left[e_0, e_{1,d}, e_{1,q}, e_{2,d}, e_{2,q}, \ldots, e_{k,d}, e_{k,q}, e_D, e_Q, e_r\right]^T$. Performing a number of trigonometric calculations results in equation system (8) below, where $\omega = \dot{\gamma} = d\gamma / dt$ is the rotor angular speed.

$$\begin{cases}
(r + L_0 \frac{d}{dt}) i_0 = -e_0 \\
(r + L_d^{(1)} \frac{d}{dt}) i_{1,d} - \omega L_q^{(1)} i_{1,q} + m_{sr} \frac{d}{dt} i_r + m_{sD} \frac{d}{dt} i_D - \omega m_Q i_Q = -e_{1,d} \\
(r + L_q^{(1)} \frac{d}{dt}) i_{1,q} + \omega (L_d^{(1)} i_{1,d} + m_{sr} i_r + m_{sD} i_D) + m_Q \frac{d}{dt} i_Q = -e_{1,q} \\
(r + L_d^{(2)} \frac{d}{dt}) i_{2,d} - \omega L_q^{(2)} i_{2,q} = -e_{2,d} \\
(r + L_q^{(2)} \frac{d}{dt}) i_{2,q} + \omega L_d^{(2)} i_{2,d} = -e_{2,q} \\
\qquad \cdots\cdots\cdots\cdots\cdots\cdots \\
\frac{p}{2} m_{sD} \frac{d}{dt} i_{1,d} + (r_D + L_D \frac{d}{dt}) i_D + m_{rD} \frac{d}{dt} i_r = 0 \\
\frac{p}{2} m_{sQ} \frac{d}{dt} i_{1,q} + (r_Q + L_Q \frac{d}{dt}) i_Q = 0 \\
\frac{p}{2} m_{sr} \frac{d}{dt} i_{1,d} + (r_r + L_r \frac{d}{dt}) i_r + m_{rD} \frac{d}{dt} i_D = e_r
\end{cases} \tag{8}$$

Coefficients L_0, $L_d^{(s)}$, $L_q^{(s)}$, s = 1, 2, ..., *k*, are now independent of angle γ. The coefficients of mutual induction between the phase windings can be sub-divided into *k* different sets (upper index in parentheses) according to the angle between the magnetic axes of phase windings that can take *k* different values: ρ, 2ρ, ..., $k\rho$, $k = \left[\frac{p}{2}\right]$. E.g., the first set includes mutual inductances of the windings, which are adjacent to each another along the stator bore, i.e. the angle between their magnetic axes equates ρ radians.

Symbol *m* with a lower index is used in equation system (8) for designation of the other coefficients of mutual induction. Symbols *s, r, D, Q* represent stator windings, drive circuit and two damping circuits (corresponding to *d*-axis and *q*-axis in the *dq0* coordinate system), respectively.

Extraction from equations (8) of the vector of time derivatives of SG phase currents leads to the canonical form of the SG model, expressed by equation (3).

3.3 Model of multiphase AC-load

Let us now make use of the transformation matrix C_P for derivation of an ac-load model in the form of equation (3). Differential equation expressing for ac-load in phase coordinates is represented by equation (9):

$$u = Ri + L\frac{d}{dt}i \qquad (9)$$

where $R = diag(r_1,...,r_p)$; $L = diag(l_1,...,l_p)$; $u = \lfloor u_1,...,u_p \rfloor^T$; $i = \lfloor i_1,...,i_p \rfloor^T$; r_j, l_j, u_j, i_j are the resistances, inductances, voltages and currents, respectively, for each phase of the ac-load, j = 1, ..., p, $p \geq 3$.

Pre-multiplying equation (9) by matrix C_P and assuming $C_P u = u_{dq0} = \lfloor u_0, u_{1,d}, u_{1,q},..., u_{k,d}, u_{k,q} \rfloor^T$ and $C_P i = i_{dq0} = \lfloor i_0, i_{1,d}, i_{1,q},..., i_{k,d}, i_{k,q} \rfloor^T$ results in equation (10), written in a dq0 coordinate system:

$$u_{dq0} = Si_{dq0} + M\frac{d}{dt}i_{dq0}, \qquad (10)$$

where $S = C_P R C_P^{-1} + C_P L \frac{d}{dt} C_P^{-1}$, $M = C_P L C_P^{-1}$.

Finally, pre-multiplying equation (10) by matrix M^{-1} would give the mathematical model of the ac-load in the canonical form (see equation (3)), where $B = M^{-1}$, and $D = -M^{-1}Si_{dq0}$.

4. Complete mathematical model of wind power system with integrated active filter

4.1 Statement of problem

In this section, a complete mathematical model will be built for the WPS presented in Fig. 1. The canonical form of mathematical models of the power system elements is given by (3). Integration of the models of the system elements into the complete mathematical model of the multi-phase electric power supply system is performed by means of equation (4).

Based on both the first Kirchoff's law and equation (3) for the *n*-th node of the power transmission line, equation (11) can be written under assumption that there are *k* elements connected to this node:

$$\frac{d}{dt}I_{dq0_{\Sigma n}} = \left(\sum_{i=1}^{k} B_{in}\right)U_{dq0_n} + \sum_{i=1}^{k} D_{in} = B_{\Sigma n}U_{dq0_n} + D_{\Sigma n} \qquad (11)$$

It follows from (11) that in (4), B_Σ is the block-matrix constructed from matrices $B_{\Sigma n}$, vector D_Σ is constructed from sub-vectors $D_{\Sigma n}$, and vector I_{dq0_Σ} is constructed from sub-vectors $I_{dq0_{\Sigma n}}$.

The problem to be solved is thus formulated as developing of mathematical models of the AC/DC/AC converter and SMPS in the form (3), with the following integration of the developed element models into the mathematical model of the WPS by using equation (4).

Note that the active filter 1 model in the form (3) is derived in (Belov et al., 2005). It is also proposed to replace harmonic filter 2 with an active filter integrated into the control system of the load side voltage inverter (hereafter referred to as the voltage inverter), which compensates low frequency voltage and current harmonics. Moreover, the proposed active filter enables filtering of high frequency conducted emissions from a single-phase non-linear load, in the studied case represented by a switch mode power supply (SMPS) of electronic equipment. The active filter integrated in the AC/DC/AC converter is based on the PWM technique.

4.2 Multi-phase B-element concept

It was shown in (Belov, 1998), (Belov et al., 2005), (Belov, 1993) that most of electric power conversion circuits can be represented by a combination of bridge elements or B-elements, independently of the number of electrical phases. Thus, an m-phase B-element can be defined both by an equivalent graph and by a state scale for the graph edges. The principle of how to construct the equivalent graph of a B-element is shown in Fig. 3.

In general, the equivalent circuit of B-element contains $2m+2$ switching elements (see Fig. 3). Among them, $2m$ switching elements are the operating elements. Besides, there are a shunt switching element ($2m+1$) and a switching element referenced as $2m+2$, which is connected in series with the resistive-inductive load (R_{load}, L_{load}). Designations R^i and L^i stand for resistance and inductance of the feeding transformer winding, respectively; R_{S_1}, R_{S_2}, ..., $R_{S_{2m+2}}$ and L_{S_1}, L_{S_2}, ..., $L_{S_{2m+2}}$ are the resistances and inductances of typical branches of the B-element (see Fig. 3). The bridge element is fed from the p-phase ac-power transmission line (p is a prime integer). Here, $i = 1, 2, ..., m$ are the nodes where the bridge element is connected to the power transmission line. It is clear, that $m \le p$, and when a single-phase B-element is connected to the phase voltage, it is assumed that $m = 2$.

The equivalent directed graph has three vertices O, D, and E connected with $2m+2$ edges. The graph is obtained by removing some of the nodes from the equivalent circuit in order to combine the B-element typical branches (including the switching elements S_1, ..., S_{2m}) with the branches representing the feeding transformer windings.

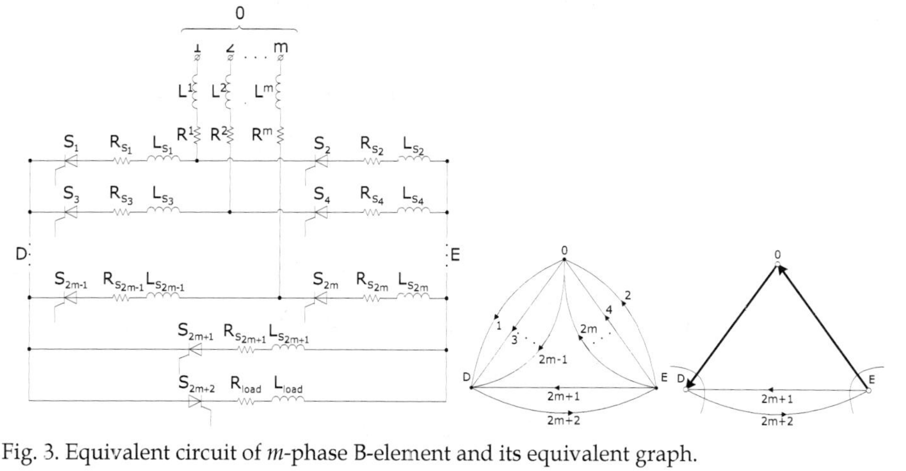

Fig. 3. Equivalent circuit of m-phase B-element and its equivalent graph.

The state scale Λ_S is a row-matrix consisting of $2m$+2 elements. Each element characterizes the state of a single switching element, and equal to either 1 (closed) or 0 (open). The state scale should be related to equivalent graph's edges, with 1 indicating a conducting edge and 0 indicating a non-conducting edge.
The B-element concept is a convenient tool for mathematical modelling of multi-phase semiconductor rectifiers, inverters, active filters and complex combinations thereof. Mathematical models of single- and three-phase devices are obtained as a particular case of multi-phase B-element concept.

4.3 Integrated active filter

A functional block diagram of the control system of the voltage inverter is shown in Fig. 4. It includes two control channels: channel 1 (low frequency) and channel 2 (high frequency). Channel 1 affects switches $S_3^{(1)}$ and $S_4^{(1)}$, and channel 2 affects switches $S_1^{(1)}$ and $S_2^{(1)}$. Such a channel division is implemented in order to simplify the control system and reduce the total cost of the switching elements (transistors). Superscript "(1)" distinguishes the switches in the AC/DC/AC converter from those in the voltage inverter.
Channel 1 forms rectangular pulse train with a duration of 180 electrical degrees, and with the fundamental frequency 60 Hz set by sinusoidal voltage U_{sin}, according to the following algorithm. Signal U_{sin} is supplied to the input of comparator K_1. In case of positive input signal, a high-level control signal is formed at the output of K_1, which is in its turn supplied both to repeater D_3 and to logic inverter D_4 (element NOT). The signal from D_3 is supplied to switch $S_3^{(1)}$ to close it. Logic inverter D_4 inverts the signal from the K_1 output and supplies it to switch $S_4^{(1)}$, thereby to open it. In case of negative input signal, a low-level control signal is formed on the output of comparator K_1, which, in its turn, passing through D_3 and D_4 opens switch $S_3^{(1)}$ and closes switch $S_4^{(1)}$.

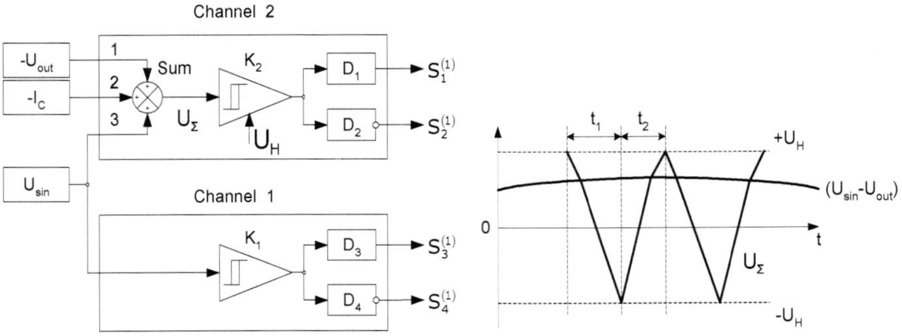

Fig. 4. Functional block diagram of the voltage inverter control system and formation of the high frequency pulse train in channel 2

Channel 2 forms high frequency control pulses with parameters depending on the inverter output voltage deviation from the reference voltage, (U_{sin} - U_{out}). The instant output voltage deviation corresponds to the total conducted interference in the node of the load (SMPS) connection. Channel 2 is used for controlling the level of harmonic voltages on the inverter output. The high frequency control channel, switches $S_1^{(1)}$ and $S_2^{(1)}$ along with the output Г-filter of the voltage inverter are the elements of the integrated active filter.

The active filter operates against conducted emissions in the following manner. Input 1 of summator Sum is supplied with the signal U_{out} from the voltage sensor placed on the inverter output. Input 2 is supplied with current I_C from the current sensor, corresponding to the current through the capacitor of the output Γ-filter (see Fig.1 and Fig. 4). Input 3 of summator Sum is supplied with voltage U_{sin}. The summed signal is supplied to comparator K_2. If the input signal exceeds the threshold level $+U_H$, a high-level control signal is formed on the K_2 output, which is then supplied both to repeater D_1 and to logic inverter D_2. The signal of D_1 is supplied to switch $S_1^{(1)}$, to close it. Logic inverter D_2 inverts the signal and supplies it to switch $S_2^{(1)}$ thereby opening this switch. If the input signal is below the threshold level $-U_H$, a low-level control signal is formed in the K_2 output, which by means of blocks D_1 and D_2, opens switch $S_1^{(1)}$ and closes switch $S_2^{(1)}$.

The process of forming the high frequency pulse train in channel 2 is shown in Fig. 4. Let us consider this process during the time interval where the positive half-wave of the inverter output voltage is formed. Here switch $S_3^{(1)}$ is closed and switch $S_4^{(1)}$ is open, and the high-level control signal is formed in the K_2 output (switch $S_1^{(1)}$ is closed, and switch $S_2^{(1)}$ is open). This state of the switches enables discharging of the capacitor of the output Γ-filter through $S_1^{(1)}$ and $S_3^{(1)}$, to result in lowering the voltage on the load.

This shows us the role of the output Γ-filter as a part of the integrated active filter. The voltage signal corresponding to $-I_C$ is added to the difference (U_{sin} - U_{out}), and the summed signal U_Σ starts to decrease down to the level $-U_H$. When the summed signal in the K_2 input reaches level $-U_H$, a low-level control signal is formed in the K_2 output, so that switch $S_1^{(1)}$ is closed whereas switch $S_2^{(1)}$ is opened. At this moment of time current I_C changes the direction and the summed signal U_Σ in the K_2 input starts to increase up to level $+U_H$. When comparator K_2 changes its state, a high-level control signal is formed in its output, so that switch $S_1^{(1)}$ is opened and switch $S_2^{(1)}$ is closed.

Described above is the single operating cycle of channel 2, where t_1 and t_2 are the time intervals of $S_1^{(1)}$ and $S_2^{(1)}$ operation, respectively. On the time interval where the negative half-wave is formed, switch $S_4^{(1)}$ is closed and switch $S_3^{(1)}$ is opened, so that the whole picture of pulse train formation is just mirrored. The frequency of operating cycles depends on the output Γ-filter parameters and the level of U_H of comparator K_1.

4.4 Mathematical model of the AC/DC/AC converter

An equivalent circuit of the AC/DC/AC converter presented in Fig. 5 is built according to the multi-phase bridge-element concept. The studied AC/DC/AC converter is described by two bridge elements (*B*-elements) with typical branches composed from a switching element, a resistor and an inductor. The equivalent circuit of the voltage inverter is represented by *B*-element B_1, whereas the equivalent circuit of the rectifier is represented by *B*-element B_2. Superscript "1" is assigned to the elements of the single-phase voltage inverter, and superscript "2" is assigned to the elements of the three-phase rectifier.

The following designations are used in Fig. 5: 1, 0 are the points of connection of the AC/DC/AC converter to the single-phase power line; *a*, *b*, *c* are the points of connection of the AC/DC/AC converter to the power generator; $R_1^{(1)}$, $R_0^{(1)}$ are the output resistances of the inverter; $L_1^{(1)}$, $L_0^{(1)}$ are the output inductances of the inverter; $S_i^{(1)}$, $R_{S_i}^{(1)}$, $L_{S_i}^{(1)}$ are the switching element, resistance and inductance, respectively, placed in the *i*-th (i = 1, 2, 3, 4) branch of the inverter bridge; $C_5^{(1)}$ is the capacitor to accumulate electric power, performing as battery (DC capacitor in Fig. 1); $S_5^{(1)}$, $R_{S_5}^{(1)}$, $L_{S_5}^{(1)}$ are the elements found in the circuit of battery $C_5^{(1)}$; $S_1^{(X)}$, $R_{S_1}^{(X)}$, $L_{S_1}^{(X)}$ are the elements corresponding to the output choke of the rectifier; $S_i^{(2)}$, $R_{S_i}^{(2)}$, $L_{S_i}^{(2)}$ are the elements placed in the i-th (i = 1, 2,…, 6) branch of the three-phase rectifier bridge; $R_1^{(2)}$, $R_2^{(2)}$, $R_3^{(2)}$ are the input resistances of the rectifier; $L_1^{(2)}$, $L_2^{(2)}$, $L_3^{(2)}$ are the input inductances of the rectifier; $S_7^{(1)}$, $R_{S_7}^{(1)}$, $L_{S_7}^{(1)}$ are the elements found in the circuit of the voltage sensor in the rectifier output.

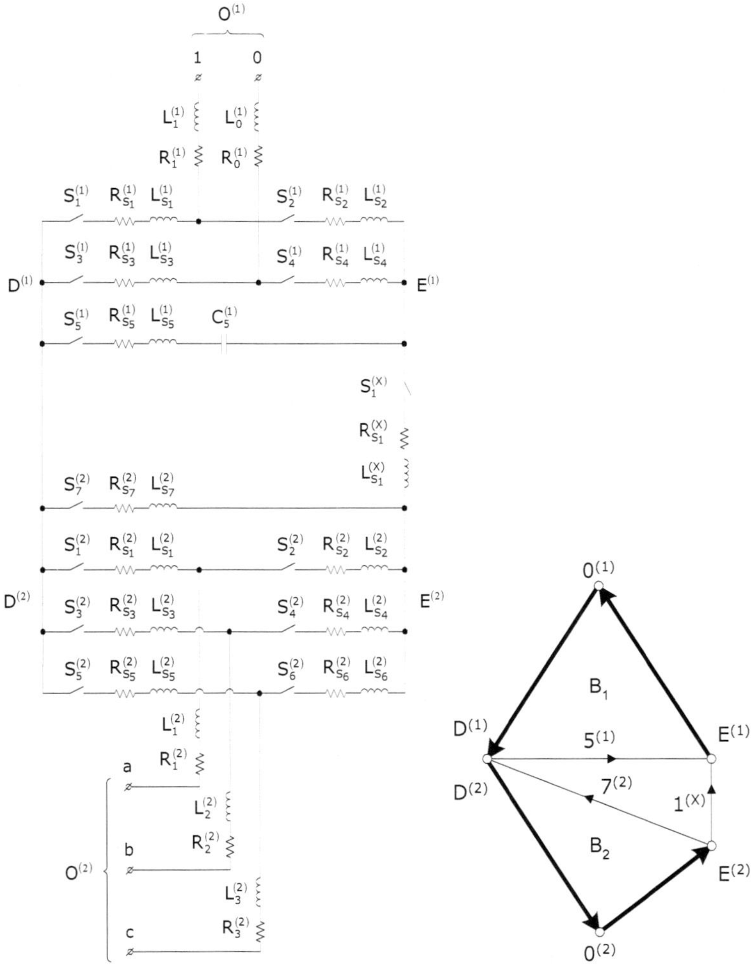

Fig. 5. Equivalent circuit of the AC/DC/AC converter, and its equivalent graph.

In Fig. 5 a graph is presented, that corresponds to the equivalent circuit of the AC/DC/AC converter. The main principles of forming the graph are reported in (Belov et al., 2005). Thickened arcs in the graph include the elements connected in series, in the equivalent circuit. For example, arc $O^{(1)}D^{(1)}$ consists of two adjacent edges. The first edge includes elements $L_1^{(1)}$, $R_1^{(1)}$, $L_{S_1}^{(1)}$, $R_{S_1}^{(1)}$, $S_1^{(1)}$ connected in series, and the second edge includes elements $L_0^{(1)}$, $R_0^{(1)}$, $L_{S_3}^{(1)}$, $R_{S_3}^{(1)}$, $S_3^{(1)}$.

Graphs of *B*-elements B_1 and B_2 are combined in the graph of the AC/DC/AC converter by means of chord $1^{(X)}$. Below is the equation for voltages related to the graph edges:

$$V = U_S + ZI + NU_{abc} + MU_C \tag{12}$$

where $V = \left[\left(V^{(1)}\right)^T, \left(V^{(2)}\right)^T, \left(V^{(X)}\right)^T\right]^T$, $I = \left[\left(I^{(1)}\right)^T, \left(I^{(2)}\right)^T, \left(I^{(X)}\right)^T\right]^T$. Here $V^{(1)}$, $V^{(2)}$ are the vectors of voltages, $I^{(1)}$, $I^{(2)}$ are the vectors of currents related to the graph edges corresponding to B_1 and B_2; $V^{(X)}$, $I^{(X)}$ are the voltage and current, respectively, related to edge $1^{(X)}$. Besides, $U_S = \left[\left(U_S^{(1)}\right)^T, \left(U_S^{(2)}\right)^T, \left(U_S^{(X)}\right)^T\right]^T$, and $U_S^{(1)}$, $U_S^{(2)}$ are the vectors of voltages on the switching elements of B_1 and B_2 respectively; $U_S^{(X)}$ is the voltage on the switching element of chord $1^{(X)}$ (note that $U_S^{(X)} = 0$, since this edge is always conducting); $U_{abc} = \left[\left(U_{abc}^{(1)}\right)^T, \left(U_{abc}^{(2)}\right)^T\right]^T$, $U_{abc}^{(1)}$, $U_{abc}^{(2)}$ are the vectors of phase voltages in the nodes of connection of the first and of the second *B*-elements to the single-phase power line and to the three-phase power generator, respectively (note that both the second and the third elements of vector $U_{abc}^{(1)}$ are equal to zero); $Z = j\omega L + R$, and R, L are the matrices containing resistances and inductances, respectively, related to the graph edges; $R = diag(R^{(1)}, R^{(2)}, R^{(X)})$, and $R^{(1)}$, $R^{(2)}$ are the matrices containing resistances related to the graph edges of the respective *B*-elements, and $R^{(X)} = R_{S_1}^{(X)}$ is the resistance of edge $1^{(X)}$; $L = diag(L^{(1)}, L^{(2)}, L^{(X)})$, and $L^{(1)}$, $L^{(2)}$ are the matrices containing inductances related to the graph edges of the respective *B*-elements, and $L^{(X)} = L_{S_1}^{(X)}$ is the inductance of edge $1^{(X)}$; U_C is the voltage on $C_5^{(1)}$; M is the vector which forms the 5th column in the (13×13) unity matrix; $N = \begin{bmatrix} diag(N^{(1)}, N^{(2)}) \\ N^{(X)} \end{bmatrix}$, $N^{(X)}$ is the (1×6) null row matrix. Matrices $R^{(1)}$ and $R^{(2)}$ have the following structure:

$$R^{(1)} = \begin{bmatrix} R_{S_1}^{(1)} + R_1^{(1)} & -R_1^{(1)} & 0 & 0 & 0 \\ -R_1^{(1)} & R_{S_1}^{(1)} + R_1^{(1)} & 0 & 0 & 0 \\ 0 & 0 & R_{S_3}^{(1)} + R_2^{(1)} & -R_2^{(1)} & 0 \\ 0 & 0 & -R_2^{(1)} & R_{S_4}^{(1)} + R_2^{(1)} & 0 \\ 0 & 0 & 0 & 0 & R_{S_5}^{(1)} \end{bmatrix},$$

$$R^{(2)} = \begin{bmatrix} R_{S_1}^{(2)}+R_1^{(2)} & -R_1^{(2)} & 0 & 0 & 0 & 0 & 0 \\ -R_1^{(2)} & R_{S_2}^{(2)}+R_1^{(2)} & 0 & 0 & 0 & 0 & 0 \\ 0 & 0 & R_{S_3}^{(2)}+R_2^{(2)} & -R_2^{(2)} & 0 & 0 & 0 \\ 0 & 0 & -R_2^{(2)} & R_{S_4}^{(2)}+R_2^{(2)} & 0 & 0 & 0 \\ 0 & 0 & 0 & 0 & R_{S_5}^{(2)}+R_3^{(2)} & -R_3^{(2)} & 0 \\ 0 & 0 & 0 & 0 & -R_3^{(2)} & R_{S_6}^{(2)}+R_3^{(2)} & 0 \\ 0 & 0 & 0 & 0 & 0 & 0 & R_{S_7}^{(2)} \end{bmatrix}.$$

Matrices $L^{(1)}$ and $L^{(2)}$ have a structure similar to matrices $R^{(1)}$ and $R^{(2)}$. Matrices $N^{(1)}$ and $N^{(2)}$ are the following:

$$N^{(1)} = \begin{bmatrix} 1/2 & 0 & 0 \\ -1/2 & 0 & 0 \\ -1/2 & 0 & 0 \\ 1/2 & 0 & 0 \\ 0 & 0 & 0 \end{bmatrix}, \quad N^{(2)} = \begin{bmatrix} -1 & 0 & 0 \\ 1 & 0 & 0 \\ 0 & -1 & 0 \\ 0 & 1 & 0 \\ 0 & 0 & -1 \\ 0 & 0 & 1 \\ 0 & 0 & 0 \end{bmatrix}.$$

In order to determine the currents related to the conducting edges of the graph, and the voltages of the switching elements on the non-conducting graph edges, matrix operator C is constructed as follows:

$$C = \left[\begin{array}{c:c:c} C_B^{(1)} & 0 & 0 \\ \hdashline 0 & C_B^{(2)} & 0 \\ \hdashline C_X^{(1)} & C_X^{(2)} & 1 \end{array}\right],$$

where $C_B^{(1)}$, $C_B^{(2)}$ are the circuit matrices for independent directed cycles, formed by the chords of corresponding B-elements, $\left[C_X^{(1)}, C_X^{(2)}, 1\right]$ is the row-matrix for the independent directed cycle corresponding to chord $1^{(X)}$. All non-zero block-matrices forming matrix C are built on the basis of state scales $\Lambda_D^{(1)}$, $\Lambda_E^{(1)}$, $\Lambda_T^{(1)}$, $\Lambda_D^{(2)}$, $\Lambda_E^{(2)}$, $\Lambda_T^{(2)}$ corresponding to the tree edges of the respective B-elements. Particularly, $C_X^{(1)} = \Lambda_T^{(1)}$ and $C_X^{(2)} = \Lambda_T^{(2)}$. The tree edges are collected one by one from adjacent edges that form the thickened arcs of B-elements.

Let us consider the formation of state scales on the example of B-element B_1. A state scale is a row-matrix, with the number of elements equal to the number of edges in a B-element. First, the main state scale $\Lambda_S^{(1)}$ is formed corresponding to the states of switching elements of all B-element edges. After that, the state scales $\Lambda_D^{(1)}$ and $\Lambda_E^{(1)}$ are formed for the tree edges, which are incident to the fundamental cut-sets $D^{(1)}$ and $E^{(1)}$, respectively, see Fig. 5. State scale $\Lambda_D^{(1)}$ is constructed from state scale $\Lambda_S^{(1)}$ by placing in $\Lambda_D^{(1)}$ the first odd (by counting) non-zero element of $\Lambda_S^{(1)}$, and by assigning zero values to the rest elements.

Similarly, row-matrix $\Lambda_E^{(1)}$ is constructed from $\Lambda_S^{(1)}$, by leaving unchanged the first even (by counting) non-zero element of $\Lambda_S^{(1)}$, and by letting the other elements be zero-valued. Thus, the state scale of the tree becomes the following: $\Lambda_T^{(1)} = \Lambda_D^{(1)} + \Lambda_E^{(1)}$. For example, let us assume the edges 1, 4 and 5 to be conducting. In this case, $\Lambda_S^{(1)} = [1 \;\; 0 \;\; 0 \;\; 1 \;\; 1]$, $\Lambda_D^{(1)} = [1 \;\; 0 \;\; 0 \;\; 0 \;\; 0]$, $\Lambda_E^{(1)} = [0 \;\; 0 \;\; 0 \;\; 1 \;\; 0]$, $\Lambda_T^{(1)} = [1 \;\; 0 \;\; 0 \;\; 1 \;\; 0]$.

In order to construct block $C_B^{(1)}$ in block-matrix *C*, corresponding to *B*-element B_1, it is necessary to perform the following operations. State scale $\Lambda_D^{(1)}$ is subtracted from the odd rows (except the last one) of the unity matrix of the appropriate size, whereas state scale $\Lambda_E^{(1)}$ is subtracted from the even rows of the unity matrix. State scale $\Lambda_T^{(1)}$ should then be added to the last row of the obtained matrix. As result,

$$C_B^{(1)} = \begin{bmatrix} 0 & 0 & 0 & 0 & 0 \\ 0 & 1 & 0 & -1 & 0 \\ -1 & 0 & 1 & 0 & 0 \\ 0 & 0 & 0 & 0 & 0 \\ 1 & 0 & 0 & 1 & 1 \end{bmatrix}.$$

By rearranging the rows matrix *C* can be presented by two blocks, i.e. both by matrix C_L (for independent directed cycles, containing only conducting edges of the graph) and by matrix C_0 (for independent directed cycles, containing at least one non-conducting edge). Matrices C_L and C_0 contain the rows of matrix *C* with the numbers corresponding to the numbers of the elements valued to 1 in state scales Λ_L and Λ_O, respectively. State scales $\Lambda_L^{(1)}$ and $\Lambda_L^{(2)}$ corresponding to the closed circuits in the graphs of B_1 and B_2 are formed according to the following algorithms: $\Lambda_L^{(1)} = \Lambda_S^{(1)} - \Lambda_T^{(1)}$ and $\Lambda_L^{(2)} = \Lambda_S^{(2)} - \Lambda_T^{(2)}$. State scale Λ_L, corresponding to the closed circuits in the graph of the AC/DC/AC converter is formed as $\Lambda_L = [\Lambda_L^{(1)}, \Lambda_L^{(2)}, \Lambda_L^{(X)}]$, where $\Lambda_L^{(1)}$ and $\Lambda_L^{(2)}$ are state scales of *B*-elements B_1 and B_2, respectively, and $\Lambda_L^{(X)}$ is the state scale of the graph edge $1^{(X)}$; here $\Lambda_L^{(X)} = 1$, since this edge is always conducting. State scale Λ_O corresponding to the open circuits in the graph of the AC/DC/AC converter is formed according to logic expression, $\Lambda_{Oi} = \begin{cases} 1, & if \; \Lambda_{Li} = 0 \\ 0, & if \; \Lambda_{Li} = 1 \end{cases}$,

where *i* is the number of a state scale element, $i = 1, 2, ..., 13$.

According to the second Kirchoff's law,

$$\begin{bmatrix} C_L \\ C_O \end{bmatrix} V = 0 \tag{13}$$

The currents related to the edges of the graph of the AC/DC/AC converter are given by (14):

$$I = C^T I_C = [C_L^T, C_O^T] \cdot \begin{bmatrix} I_L \\ I_O \end{bmatrix} = C_L^T I_L \tag{14}$$

where I_C is the vector of currents related to all independent directed cycles in the graph, I_L and I_O are the vectors of currents related to closed and open circuits in the graph, respectively. By definition, elements of vector I_O are always set to zero. Substitution of (12) and (14) into (13) leads to (15):

$$\begin{bmatrix} C_L \\ C_O \end{bmatrix} U_S + \begin{bmatrix} C_L \\ C_O \end{bmatrix} Z C_L^T I_L + \begin{bmatrix} C_L \\ C_O \end{bmatrix} N U_{abc} + \begin{bmatrix} C_L \\ C_O \end{bmatrix} M U_C = 0 \tag{15}$$

Let us now rewrite (15) in time domain, taking into account $Z = Lp + R$ and $C_L U_C = 0$, with p as the differentiation operator:

$$L_L \frac{d}{dt} I_L + R_L I_L + N_L U_{abc} + M_L U_C = 0 \tag{16}$$

$$U_O + L_O \frac{d}{dt} I_L + R_O I_L + N_O U_{abc} + M_O U_C = 0 \tag{17}$$

where $U_O = C_O U_S$, $L_L = C_L L C_L^T$, $R_L = C_L R C_L^T$, $N_L = C_L N$, $M_L = C_L M$, $L_O = C_O L C_L^T$, $R_O = C_O R C_L^T$, $N_O = C_O N$, $M_O = C_O M$. Equation (16) can be modified into (18), which is being subsequently substituted into (17) results in equation (19) for currents $I_L(t)$ related to conducting edges and for voltages $U_O(t)$ across the non-conducting edges of the graph.

$$\frac{d}{dt} I_L = -L_L^{-1}(R_L I_L + N_L U_{abc} + M_L U_C) \tag{18}$$

$$\begin{cases} \frac{d}{dt} I_L = Z_L I_L + L_{N_L} U_{abc} + L_{M_L} U_C \\ U_O = Z_O I_L + L_{N_O} U_{abc} + L_{M_O} U_C \end{cases} \tag{19}$$

where $Z_L = -L_L^{-1} R_L$, $Z_O = L_O L_L^{-1} R_L - R_O$, $L_{N_L} = -L_L^{-1} N_L$, and $L_{N_O} = L_O L_L^{-1} N_L - N_O$, $L_{M_L} = -L_L^{-1} M_L$, $L_{M_O} = L_O L_L^{-1} M_L - M_O$. System of equations (19) should be supplemented with differential equation (20) for voltage on capacitor $C_5^{(1)}$:

$$\frac{d}{dt} U_C = C_{C_L} I_L \tag{20}$$

where $C_{C_L} = \frac{1}{C_5^{(1)}} M^T C_L^T$. Combination of the first equation from system (19) with equation (20) leads to system of equations (21):

$$\begin{cases} \frac{d}{dt} \left(\begin{array}{c} I_L \\ \hline U_C \end{array} \right) = \left(\begin{array}{c} L_{N_L} \\ \hline 0 \end{array} \right) U_{abc} + \left(\begin{array}{c|c} Z_L & L_{M_L} \\ \hline C_{C_L} & 0 \end{array} \right) \left(\begin{array}{c} I_L \\ \hline U_C \end{array} \right) \\ U_O = Z_O I_L + L_{N_O} U_{abc} + L_{M_O} U_C \end{cases} \tag{21}$$

Multiplication of the first equation in system (21) by C_L^T and taking into account (14), result in (22):

$$\frac{d}{dt}I = C_L^T Z_L I_L + C_L^T L_{N_L} U_{abc} + C_L^T L_{M_L} U_C \tag{22}$$

In order to obtain the phase currents, equation (22) is left-sided multiplied by matrix $F = diag\left(F^{(1)}, F^{(2)}, 0\right)$, which consists of blocks, as follows:

$$F^{(1)} = \begin{bmatrix} 1 & -1 & 0 & 0 & 0 \\ 0 & 0 & 0 & 0 & 0 \\ 0 & 0 & 0 & 0 & 0 \end{bmatrix} \text{ and } F^{(2)} = \begin{bmatrix} 1 & -1 & 0 & 0 & 0 & 0 & 0 \\ 0 & 0 & 1 & -1 & 0 & 0 & 0 \\ 0 & 0 & 0 & 0 & 1 & -1 & 0 \end{bmatrix}.$$

Equation (23) is the result of the left-sided multiplication:

$$\frac{d}{dt}I_{abc} = F_{ZL} I_L + F_{LNL} U_{abc} + F_{LML} U_C \tag{23}$$

where $I_{abc} = \left[\left(I_{abc}^{(1)}\right)^T, \left(I_{abc}^{(2)}\right)^T\right]^T$, and $I_{abc}^{(1)}$, $I_{abc}^{(2)}$ are the vectors of phase currents in the nodes where B_2 and B_1 are connected to the corresponding power lines; $F_{ZL} = FC_L^T Z_L$, $F_{LNL} = FC_L^T L_{N_L}$ and $F_{LML} = FC_L^T L_{M_L}$.

Let us now make transformation of the three-phase coordinate system to the rotating $dq0$ coordinate system. This is done by means of matrix operator C_{dq}:

$$\frac{d}{dt}\left(C_{dq}^{-1} I_{dq}\right) = F_{ZL} I_L + F_{LNL} C_{dq}^{-1} U_{dq} + F_{LML} U_C \tag{24}$$

where $C_{dq} = diag\left(C_p, C_p\right)$; C_p is the Park's transformation (Park, 1929); $I_{dq} = \left[\left(I_{dq}^{(1)}\right)^T, \left(I_{dq}^{(2)}\right)^T\right]^T$, and $U_{dq} = \left[\left(U_{dq}^{(1)}\right)^T, \left(U_{dq}^{(2)}\right)^T\right]^T$ are the vectors of images of phase currents and voltages in the $dq0$ coordinate system. A number of simple mathematical manipulations with (24) result in (25):

$$\frac{d}{dt}I_{dq} = C_{dq} F_{ZL} I_L + C_{dq} F_{LNL} C_P^{-1} U_{dq} + C_{dq} F_{LML} U_C - C_{dq} \frac{d}{dt}\left(C_{dq}^{-1}\right) I_{dq} \tag{25}$$

According to (20),

$$I_{dq} = C_{dq} I_{abc} = C_{dq} F I = C_{dq} F C_L^T I_L \tag{26}$$

Substitution of (26) into (25) leads to (27):

$$\frac{d}{dt}I_{dq} = C_{dq} F_{LNL} C_{dq}^{-1} U_{dq} + \left\{ C_{dq} F_{ZL} - C_{dq} \frac{d}{dt}\left(C_{dq}^{-1}\right) C_{dq} F C_L^T \right\} I_L + C_{dq} F_{LML} U_C \tag{27}$$

At last, (27) can be represented in canonical form (3):

$$\frac{d}{dt} I_{dq} = BU_{dq} + D \tag{28}$$

where $B = C_{dq} F_{LNL} C_{dq}^{-1}$, $D = \left\{ C_{dq} F_{ZL} - C_{dq} \frac{d}{dt} \left(C_{dq}^{-1} \right) C_{dq} F C_L^T \right\} I_L + C_{dq} F_{LML} U_C$. Combination of (28) with equations (21) results in system of equations (29):

$$\begin{cases} \frac{d}{dt} I_{dq} = BU_{dq} + D \\ \frac{d}{dt} \begin{pmatrix} I_L \\ \hline U_C \end{pmatrix} = \begin{pmatrix} L_{N_L} \\ \hline 0 \end{pmatrix} C_{dq}^{-1} U_{dq} + \left(\begin{array}{c|c} Z_L & L_{M_L} \\ \hline C_{C_L} & 0 \end{array} \right) \begin{pmatrix} I_L \\ \hline U_C \end{pmatrix} \\ U_O = Z_O I_L + L_{N_O} C_{dq}^{-1} U_{dq} + L_{M_O} U_C \end{cases} \tag{29}$$

Equation system (29) is a system of differential and algebraic equations. It represents the mathematical model of the AC/DC/AC converter, connected to the three-phase power generator and to the single-phase power line.

Let us now consider interaction between the model of the AC/DC/AC converter (29) and the mathematical model of the control system of the voltage inverter, and thereby realization of high frequency channel control functions as the main element of the integrated active filter. According to the functional block diagram of the voltage inverter control system presented in Fig. 4, the main state scale for *B*-element B_1 is formed according to the following algorithm:

$$\begin{cases} \Lambda_{S_1}^{(1)} = \begin{cases} 1, & \begin{cases} if\ U_\Sigma > 0 \\ if\ -U_H \le U_\Sigma \le U_H, \quad if\ \left[\Lambda_{S_1}^{(1)}\right]_{prev} = 1 \end{cases} \\ 0, & \begin{cases} if\ U_\Sigma < 0 \\ if\ -U_H \le U_\Sigma \le U_H, \quad if\ \left[\Lambda_{S_1}^{(1)}\right]_{prev} = 0 \end{cases} \end{cases}, \quad \Lambda_{S_2}^{(1)} = \begin{cases} 1, & if\ \Lambda_{S_1}^{(1)} = 0 \\ 0, & if\ \Lambda_{S_1}^{(1)} = 1 \end{cases} \\ \Lambda_{S_3}^{(1)} = \begin{cases} 1, & if\ U_{\sin} > 0 \\ 0, & if\ U_{\sin} \le 0 \end{cases}, \quad \Lambda_{S_4}^{(1)} = \begin{cases} 1, & if\ \Lambda_{S_3}^{(1)} = 0 \\ 0, & if\ \Lambda_{S_3}^{(1)} = 1 \end{cases}, \quad \Lambda_{S_5}^{(1)} = 1 \end{cases} \tag{30}$$

where $\left[\Lambda_{S_1}^{(1)}\right]_{prev}$ is the previous state of $\Lambda_{S_1}^{(1)}$. Notice that $\Lambda_{S_5}^{(1)} = 1$ since there is element $C_5^{(1)}$ in the 5th edge of the graph. The main state scale of B_2 is formed according to the following algorithm:

$$\Lambda_{S_i}^{(2)} = \begin{cases} 1, & if\ U_{S_i}^{(2)} > 0 \\ 0, & if\ I_{S_i}^{(2)} \le 0 \end{cases} \tag{31}$$

where i is the number of the state scale element, $i = 1, 2, ..., 6$.

4.5 Mathematical model of SMPS

An equivalent circuit of a switch mode power supply (Gottlieb, 1994) is shown in Fig. 6.

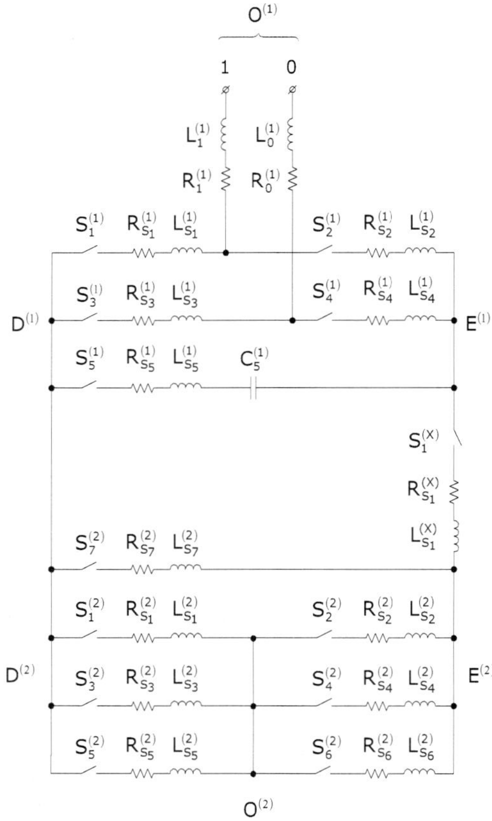

Fig. 6. Equivalent circuit of the SMPS.

The following designations are used in the figure: 0 and 1 are the nodes of connection of the SMPS to the single-phase power line; $R_1^{(1)}$ and $R_0^{(1)}$ are the input resistances; $L_1^{(1)}$ and $L_0^{(1)}$ are the input inductances; $S_i^{(1)}$, $R_{S_i}^{(1)}$, and $L_{S_i}^{(1)}$ are the switching element, resistance and inductance, respectively, in the i-th ($i = 1, 2, 3, 4$) section of the rectifier bridge; $C_5^{(1)}$ is the smoothing capacitor; $S_5^{(1)}$, $R_{S_5}^{(1)}$ and $L_{S_5}^{(1)}$ are the elements present in the smoothing capacitor circuit; $S_1^{(X)}$, $R_{S_1}^{(X)}$ and $L_{S_1}^{(X)}$ are the switching element, resistance and inductance, respectively, of the output choke of the rectifier; elements $S_1^{(2)}$, $R_{S_1}^{(2)}$ and $L_{S_1}^{(2)}$ represent a MOSFET; elements $S_2^{(2)}$, $R_{S_2}^{(2)}$ and $L_{S_2}^{(2)}$ model the load, referred to the primary winding of the SMPS transformer; elements $S_3^{(2)}$, $R_{S_3}^{(2)}$ and $L_{S_3}^{(2)}$ model the damping circuit of the MOSFET; $S_4^{(2)}$, $R_{S_4}^{(2)}$ and $L_{S_4}^{(2)}$ represent the load damper, including a diode; elements $S_i^{(2)}$, $R_{S_i}^{(2)}$ and $L_{S_i}^{(2)}$ are related to the voltage sensors ($i = 5, 6, 7$).

Let us apply the multi-phase bridge-element concept for building of the graph of the SMPS equivalent circuit. It can be shown that the resulting graph would be fully similar to the one presented in Fig. 5. Therefore, the mathematical model represented by (29) can be applied in SMPS analysis with the minor changes shown below. In case of SMPS, matrices $R^{(2)}$, $L^{(2)}$ corresponding to B_2, and matrix N are $R^{(2)} = diag\left(R_{S_1}^{(2)}, R_{S_2}^{(2)}, R_{S_3}^{(2)}, R_{S_4}^{(2)}, R_{S_5}^{(2)}, R_{S_6}^{(2)}, R_{S_7}^{(2)}\right)$, $L^{(2)} = diag\left(L_{S_1}^{(2)}, L_{S_2}^{(2)}, L_{S_3}^{(2)}, L_{S_4}^{(2)}, L_{S_5}^{(2)}, L_{S_6}^{(2)}, L_{S_7}^{(2)}\right)$, $N = \begin{bmatrix} N^{(1)} \\ 0 \end{bmatrix}$, where the structure of matrix $N^{(1)}$ is alike the case of AC/DC/AC converter, and the null-matrix has a size of (8×3). Furthermore, $U_{abc} = U_{abc}^{(1)}$, $I_{abc} = I_{abc}^{(1)}$, $I_{dq} = I_{dq}^{(1)}$, $U_{dq} = U_{dq}^{(1)}$, where $U_{abc}^{(1)}$, $I_{abc}^{(1)}$ are the vectors of phase voltages and currents, respectively, in the node of SMPS connection; $I_{dq}^{(1)}$, $U_{dq}^{(1)}$ are the vectors of images of phase currents and phase voltages, respectively, in the $dq0$ coordinate system. Transition matrix $C_{dq} = C_p$, where C_p is the Park's transformation operator.

The main state scale of B_1 is formed according to the algorithm, given by (31). For the SMPS model, however, the number of the state scale element, $i = 1, 2, 3, 4$. Here, $\Lambda_{S_5}^{(1)} = 1$, since there is element $C_5^{(1)}$ in the 5th edge of the graph of B-element B_1 . According to the operation principles of switching elements included in the edges of the B_2 graph, the main state scale for B_2 is formed with the algorithm provided in (32):

$$\begin{cases} \Lambda_{S_1}^{(2)} = f_S(t),\ \Lambda_{S_2}^{(2)} = 1, \\ \Lambda_{S_3}^{(2)} = \begin{cases} 1, & if\ \Lambda_{S_1}^{(2)} = 0 \\ 0, & if\ \Lambda_{S_1}^{(2)} = 1 \end{cases},\ \Lambda_{S_4}^{(2)} = \begin{cases} 1, & if\ U_{S_4}^{(2)} > 0 \\ 0, & if\ I_{S_4}^{(2)} \le 0 \end{cases}, \\ \Lambda_{S_5}^{(2)} = 0,\ \Lambda_{S_6}^{(2)} = 0,\ \Lambda_{S_7}^{(2)} = 0 \end{cases} \tag{32}$$

where $\Lambda_{S_1}^{(2)}$ is an element of the state scale corresponding to the edge that includes the MOSFET; $f_S(t)$ is the logic function of time supplying the switching function for the MOSFET; element $\Lambda_{S_2}^{(2)}$ corresponds to the edge that includes the SMPS load referred to the primary winding of the SMPS transformer (this edge is always conducting); $\Lambda_{S_3}^{(2)}$ corresponds to the damping circuit of the MOSFET (this edge is conducting, when the edge corresponding to the MOSFET is non-conducting, and vice versa); $\Lambda_{S_4}^{(2)}$ corresponds to the edge including the damping circuit of the load and the diode (the state of this edge depends on the diode state, i.e. conducting or non-conducting), whereas state-scale elements $\Lambda_{S_5}^{(2)}$, $\Lambda_{S_6}^{(2)}$, and $\Lambda_{S_7}^{(2)}$ correspond to the edges designed for voltage measurement respectively on the MOSFET (this edge is always non-conducting), on the load (this edge is always non-conducting) and on the output of the single-phase diode rectifier (this edge is always non-conducting).

4.6 Numerical results

The mathematical model of the integrated active filter is investigated by means of two numerical tests. The task of the first test is to show that the SMPS and the AC/DC/AC converter produce significant conducted emissions in WPS powered from the synchronous generator having a small power.

For demonstration purposes the SMPS is not equipped with an input filter, to simulate the situation when high frequency emissions from a non-linear load are present in the power line. The task of the second test is to check efficiency of the integrated active filter to ensure an appropriate power quality and EMC in the WPS.

Let us first introduce the parameters of the WPS presented in Fig. 1. The power system includes a synchronous generator with nominal power of 1 kW, nominal voltage of 220 V and fundamental frequency of 60 Hz. Parameters of generator are provided in (Belov et. al., 2009).

Harmonic filter 1 is a symmetrical three-phase harmonic filter consisting of three resonant sections tuned to 5th, 7th, 11th harmonics and a parallel capacitor in each electric phase. RLC parameters of the section containing the parallel capacitor are (0.07 Ω, 2.86 µH, 4.5 µF). RLC parameters of the resonant sections are (2.68 Ω, 184.45 mH, 1.26 µF) for resonance frequency 300 Hz, (3.12 Ω, 195.08 mH, 0.74 µF) for resonance frequency 420 Hz, and (16.02 Ω, 24.68 mH, 2.36 µF) for resonance frequency 660 Hz.

A diode rectifier bridge is taken as the rectifier. The diode parameters are the same for all the diodes: $R_{S_1}^{(2)} = ... = R_{S_6}^{(2)} = 1$ mΩ, $L_{S_1}^{(2)} = ... = L_{S_6}^{(2)} = 5.26$ mH. Inductance of the output choke of the rectifier is $L_{S_1}^{(X)} = 44.4$ mH.

The single-phase inverter has the following parameters: the capacitor acting as the battery, $C_5^{(1)} = 1000$ µF; equivalent parameters of the switching elements: $R_{S_1}^{(1)} = ... = R_{S_4}^{(1)} = 10$ µΩ, $L_{S_1}^{(1)} = ... = L_{S_4}^{(1)} = 9.86761$ µH. Parameters of the output Γ-filter are L = 1 mH, and C = 30 µF.

Threshold level U_H in the inverter control system is taken to be 4 mV.

An electronic product with a 0.075 kW SMPS having the switching frequency 25 kHz is considered. Such a circuit is provided in (Gottlieb, 1994) (Verhees, 2003).

Results of numerical testing of the mathematical model are provided in Fig. 7 and Fig. 8, where symbols "1" and "2" correspond to the first test and to the second test, respectively. Analysis of the diagrams reveals high efficiency of the integrated active filter.

The first test resulted in the total harmonic distortion (THD) of voltage on the generator terminals to be THD = 0,069956. In the second test the result was THD = 0,00199. Here a complex effect of the active filter on the conducted emissions in the power network has to be noticed. Conducted emissions both from the SMPS and from the voltage inverter of the AC/DC/AC converter are suppressed. As can be seen in Fig. 8, the higher order harmonics almost disappeared after filtering (white bars). In addition, the weight and the dimensions of the equipment enabling the high frequency channel of the voltage inverter are much less than those of harmonic filter 2 designed for the WPS shown in Fig. 1.

5. Conclusion

The methods and models for computer aided design of wind power systems for electromagnetic compatibility and power quality presented in this chapter can be summarized as follows.

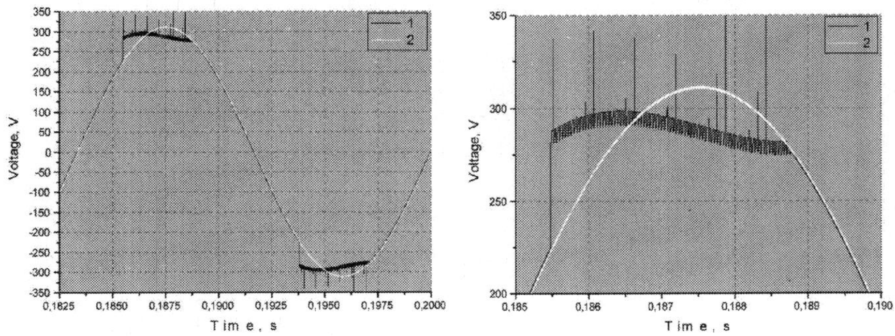

Fig. 7. Voltage diagrams in the SMPS input obtained in tests 1 and 2: the time interval between 0.1825 s and 0.2 s (left) and between 0.185 s and 0.19 s (right).

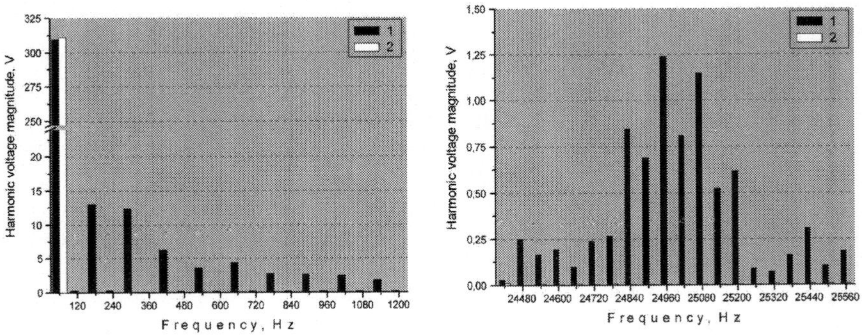

Fig. 8. Voltage spectra in the SMPS input obtained in tests 1 and 2: low frequencies (left) and high frequencies (right).

The iterative spectral technique for power quality and EMC design takes into account changes in the wind power system frequency response during the design procedure. In general, the proposed technique is independent of the circuit of the designed filters. It can be used for designing passive, active and hybrid filters. Further development and generalization of this technique will enable creating universal CAD tools for robust system design.

The multi-phase bridge-element concept for modelling of independent electric power systems with complex power conversion schemes does fully describe the features of wind power systems with AC/DC/AC converter and an SMPS taken as a load. This concept leads to the compact form of the mathematical model of the wind power system represented by the system of ordinary differential equations based on the algebraic calculations and the special algorithm. The algorithm for determining the coefficients of the differential equations enables realization of the PWM technique in the models of control systems for wind power system elements.

The mathematical model of the active power filter integrated into the voltage inverter was derived using the B-element concept. In a wind power system including an AC/DC/AC converter, it is both technically and economically rational to use integrated active filters based on pulse-width modulation. When designing such a filter a very precise choice of parameters is required for the high frequency channel of the voltage inverter control system, which can only be achieved by using the complete mathematical model of the whole power system.

A mathematical model of a multi-phase synchronous generator was introduced as well as a model of a multi-phase ac-load. The system's elements were integrated into the complete model of a multi-phase electric power supply system. This was achieved both by presenting all element models in the canonical form in dq0 coordinate system and by means of the equations for multi-phase power transmission line.

The computational experiments demonstrated the efficiency of the developed mathematical model in designing of the power quality and EMC in the wind power system.

Results of the presented theoretical study and computational experiments supported by the results published earlier by the authors set a framework for computer-aided design of wind power systems with specified power quality and levels of conducted emissions.

The developed methods and models for computer aided design of wind power systems provide a full set of tools for the specialists working with variant design of wind power systems for EMC and power quality, in early design phase. In particular, the developed multi-phase bridge-element concept enables formalization of complex procedures of mathematical modelling multi-cascade circuits of active filters.

The reported iterative spectral technique along with the standard form of models of wind power system elements, based on the multi-phase bridge-element concept and dq0 rotating coordinate system, enable conditions for developing both a user interface and a set of mathematical models for computer aided design tools in the area of wind power system design.

Further theoretical work on developing computer aided design tools for power quality and EMC design could be done in the following areas:

- Realization of fast numerical integration methods for systems of ordinary differential equations.

- Extension of the set of mathematical models via implementation of standard forms of equations for asynchronous motors, DC drives, and various electro-mechanical devices.
- Development of formal criteria for more efficient decision making during the variant analysis within the iterative spectral technique.

6. References

Belov, V. (1993). *Computer-aided EMC Design for Autonomous Power Conversion Systems*, Mordovian State University Press, Saransk (in Russian)

Belov, V. (1998). Mathematical simulation of multiphase electric converting systems, *Journal of Math Modelling*, Vol. 10, No. 10, (1998), 51-63 (in Russian)

Belov, V.; Belov, I.; Nemoykin, V.; Johansson, A. & Leisner, P. (2004) Computer modelling and analysis of EMC in a multi-phase electrical system, *Proceedings of the 3rd national conference on computational electromagnetics - EMB04*, 294-301, Göteborg, Sweden, October 2004

Belov, V.; Paldyaev, N.; Shamaev, A.; Johansson, A.; Leisner, P. & Belov, I. (2005). A complete mathematical model of an independent multi-phase power supply system based on multi-phase bridge-element concept, *WSEAS Transactions on Circuits and Systems*, Vol. 4, No. 9, (September 2005), 1009-1018

Belov, V.; Shamaev, A.; Leisner, P.; Johansson, A.; Magnhagen, B. & Belov, I. (2006). A simulation-based spectral technique for power quality and EMC design of an independent power system, *International Journal of Emerging Electric Power Systems*, Vol. 7, No. 1, (2006), Article 7

Belov, V.; Leisner, P.; Johansson, A.; Paldyaev, N.; Shamaev, A. & Belov, I. (2009) Mathematical modelling of a wind power system with an integrated active filter, *Journal of Electric Power Systems Research*, Vol. 79, No. 1, (2009), 117-125

Binsaroor, A. & Tiwari, S. (1988). Evaluation of twelve phase (multiphase) transmission line parameters, *Journal of Electric Power Systems Research*, Vol. 15, (1988), 63-76.

Brown, D. & Chandrasekaran, B. (1989). *Design Problem Solving: Knowledge Structures and Control Strategies*, Morgan Kaufmann, Los Altos, CA

EMC Filters Data Book, (2001). EPCOS AG Marketing Communications, Munich

Gottlieb, I. (1994). *Power Supplies, Switching Regulators, Invertors, and Converters*, TAB Books, Division of Mc Draw-Hill, New York

Grauers, A. (1994). *Synchronous Generator and Frequency Converter in Wind Turbine Applications: System Design and Efficiency, Lic. thesis, Technical Report No. 175L* Chalmers University of Technology, ISBN 91-7032-968-0, Göteborg, Sweden

Himmelblau, D. (1972). *Applied Nonlinear Programming*, McGraw-Hill, New York.

Ortuzar, M. ; Carmi, R. ; Dixon, J. & Morán, L. (2003) Voltage source active power filter, based on multi-stage converter and ultracapacitor DClink, *Proceedings of IEEE industrial electronics conference - IECON'2003*, 2300 – 2305, Roanoke, Virginia, USA, November 2003

Park, R. (1929). Two-reaction theory of synchronous machines - part 1, *AIEE Transactions*, Vol. 48, (1929), 716-730

Temes, G.; Mitra, S. (1973) *Modern Filter Theory and Design*, John Willey & Sons, 1973

Toliyat, H.; Shi, R. & Xu, H. (2000) A DSP-based vector control of five-phase synchronous reluctance motor, *Proceedings of the IEEE-IAS 2000 annual meeting*, 1759-1765, Rome, Italy, October 2000

Verhees, H. (2003). *TEA1541 SMPS Control IC with Synchronization Function, AN10205-01,* Philips Electronics N.V.

Design of Robust Power System Stabilizer in an interconnected Power System with Wind Power Penetrations

Sarjiya, Cuk Supriyadi A.N and Tumiran
Department of Electrical Engineering, Gadjah Mada University, Indonesia

1. Introduction

In the recent years, renewable electrical energy such as wind power generations, have achieved a significant level of penetration in the power systems due to infinite availability and low impact to environment. However, wind power generation is intermittent in nature. Matching the supply and the demand is often a problem. The power output fluctuations from wind power generations cause a problem of low frequency oscillation, deteriorate the system stability and make the power system operation more difficult. The power frequency and the tie-line power deviations persist for a long duration. In these situations, the governor system may no longer be able to absorb the frequency fluctuations due to its slow response (Elgerd & Fosha, 1970). To stabilize power oscillation, PSS is often used as an effective device to enhance the damping of electromechanical oscillations in power systems. The power system stabilizer is a supplementary control system, which is often applied as part of excitation control system. The basic function of the PSS is to apply a signal to the excitation system, creating electrical torques to the rotor, in phase with speed variation, that damp out power oscillations.

In the past decades, the utilization of supplementary excitation control signals for improving the dynamic stability of power systems has received much attention. Extensive research has been conducted in many fields such as the effect of PSS on power system stability, PSS input signals, PSS optimum locations, and PSS tuning techniques. In (deMello & Concordia, 1969), the concept of synchronous machine stability as affected by excitation control has been examined. This work developed insights into effects of excitation systems and requirement of supplementary stabilizing action for such systems based on the concept of damping and synchronizing torques. These stabilizing requirements included the adjustment of voltage regulator gain parameters as well as the PSS parameters.

Since the primary function of the PSS is to add damping to the power oscillations, basic control theories have been applied to select the most suitable input signal of PSS. Some readily available signals are generator rotor speed, calculated bus frequency, and electrical power. In (Larsen & Swann, 1981), the application of PSS utilizing either of speed, frequency or power input signals has been presented. Guidelines were presented for tuning PSS that enable the user to achieve desired dynamic performance with limited effort. The need for torsional filters in the PSS path for speed input PSS was also discussed. The most PSS

controls today use the generator rotor speed as the feedback input signal. They would provide robust damping over a wide range of operating conditions with minimum interaction (Murdoh et al, 2000).
Simulation studies of PSS effects on inter-area and local modes of oscillations in interconnected power systems have been presented by (Klein et al, 1991) and (Klein et al, 1992). It was shown that the PSS location and the voltage characteristics of the system loads are significant factor in the ability of a PSS to increase the damping of inter-area oscillations. The procedures for the selection of the most effective machines for stabilization have been proposed. In (Abdalla et al, 1984), an eigenvalue-based measurement of relative improvement in the damping of oscillatory modes has been implemented and used as a criterion to find the best candidate machine for stabilizer application. On the other hand, an eigenvector analysis to identify the most effective generating units to be equipped with PSSs in multi-machine systems that exhibit dynamic instability and poor damping of several inter-area modes of oscillations, has been presented in (DeMello et al, 1980).
Nowadays, the conventional lead/lag compensator PSS is widely used by the power system utility (Tse & Tso, 1993). Other types of PSS such as proportional-integral PSS (PI-PSS) and proportional-integral-derivative PSS (PID-PSS) have also been proposed by (Hsu & Hsu, 1986) and (Hsu & Liou, 1987). Several approaches based on modern control theories have been successfully applied to design PSSs. In (Yu & Siggers, 1971), the application of state-feedback optimal PSS has been presented, while an eigenvalue shifting technique for determining the weighing matrix in the performance index has been proposed by (Moussa & Yu, 1972). In (Fleming et al, 1981), a sequential eigenvalue assignment algorithm for selecting the parameters of stabilizers in a multi-machine power system has been presented. In sequential tuning, the stabilizer parameters are computed using repeated application of single-input/single-output (SISO) analysis. In (Zhou et al, 1992), the eigenvalue assignment has been proposed to design the optimal PSS. Besides, the new optimal linear quadratic regulator (LQR) based design has been presented by (Aldeen & Crusca, 1995). It is superior to previously reported LQR approaches. Moreover, PSS designs based on self tuning control (Cheng et al, 1986) and (Lim, 1989), fuzzy-logic system (Hsu & Cheng, 1990) and (Hoang & Tomsovic, 1996), artificial neural network (ANN) (Zhang et al, 1993), (Segal et al, 2000) and (Abido & Abdel_Magid, 1998) have been presented. However, since these techniques do not take the presence of system uncertainties such as system nonlinear characteristics, variations of system configuration due to unpredictable disturbances, loading conditions etc. into consideration in the system modeling, the robustness of these PSSs against uncertainties cannot be guaranteed.
To overcome these problems, H_∞ control has been applied to design of robust PSS configuration by (Chen & Malik, 1995) and (Yan, 1997). In these works, the designed H_∞ PSS via mixed sensitivity approach have confirmed the significant performance and high robustness. In this approach, however, due to the trade-off relation between sensitivity function and complementary sensitivity function, the weighting functions in H_∞ control design cannot be selected easily. Moreover, the order of H_∞controller depends on that of the plant which is different from the conventional lead/lag PSS. Despite the significant potential of control techniques mentioned above, power system utilities still prefer the conventional lead/lag PSS structure. This is due to the ease of implementation, the long-term reliability, etc.
On the other hand, much research on a conventional lead/lag PSS design has paid attentions to tuning of PSS parameters. The parameters of a lead/lag PSS are optimized

under various operating conditions by heuristic methods such as tabu search (Abdel-Magid et al, 2001), genetic algorithm (Abdel-Magid et al, 1999), and simulated annealing (Abido, 2000). Using these approaches, the PSS parameters are obtained so that all of the electro-mechanical mode eigenvalues may be placed at the prescribed locations in the s-plane. In these designs, however, the uncertainty model is not embedded in the mathematical model of the power system. Furthermore, the robust stability against system uncertainties is not taken into consideration in the optimization process. Therefore, the robust stability margin of the system in these works may not be guaranteed in the face of several uncertainties.

To solve this problem, the robust PSS design by a fixed structure with a conventional lead/lag PSS have been proposed [Cuk supriyadi et al, 2008]. In this work, the fixed structure robust PSS design by the H_∞ loop shaping technique is proposed. The normalized coprime factor is used to model system uncertainties. To optimize the control parameters, the performance and robust stability conditions in the H_∞ loop shaping technique are formulated as the objective function. As a result, the proposed PSSs are very robust against various uncertainties. With lower order, the stabilizing effect and robustness of the proposed PSS are almost the same as those of the PSS with high-order designed by H_∞ loop shaping technique. In this works, however, the weighting functions in H_∞ control design cannot be selected easily.

To tackle this problem, a new parameters optimization of robust PSS is proposed. The inverse additive perturbation is applied to represent unstructured system uncertainties. The configuration of PSS is a conventional second-order lead-lag compensator. To tune the PSS parameters, the concept of enhancement of system robust stability margin is formulated as the optimization problem. The genetic algorithm (GA) is applied to solve the problem and achieve the PSS parameters. Simulation studies in the two-area four-machine system with wind farms confirm that the damping effect and robustness of the proposed PSS are superior to those of the compared PSS.

2. System modelling

2.1 Power system model

A two-area four-machine interconnected power system with wind farms in Fig. 1 is used to design PSS. Each generator is represented by a 5th-state transient model. It is equipped with a simplified exciter and PSS with the speed deviation input. L1 and PW1 are load and wind farms in area 1, respectively. L2 and PW2 are load and wind farms in area 2, respectively.

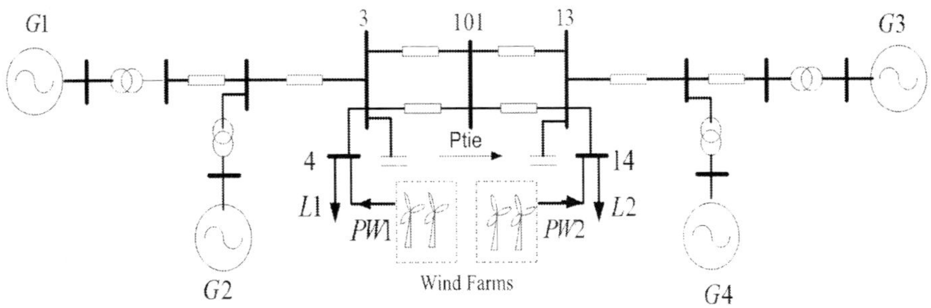

Fig. 1. Two areas four machines power system with wind farms

The linearized state equation of system in Fig. 1 can be expressed as

$$\Delta \dot{X} = A\Delta X + B\Delta u_{pss,i} \tag{1}$$

$$\Delta Y = C\Delta X + D\Delta u_{pss,i} \tag{2}$$

$$\Delta u_{pss,i} = K_{pss,i}(s)\Delta \omega_i \tag{3}$$

Where the state vector $\Delta X = \begin{bmatrix} \Delta\delta & \Delta\omega & \Delta e_d' & \Delta e_q' & \Delta E_{fd} \end{bmatrix}^T$, the output vector $\Delta Y = [\Delta\omega]$, $\Delta u_{pss,i}$ is the control output signal of the PSS no. i ($K_{pss,i}(s)$), which uses only the angular velocity deviation ($\Delta\omega$) as a feedback input signal and i is the number of PSS. Note that the system in (1) is a Multi-input Multi-output (MIMO) system. The proposed method is applied to design a robust PSS *K(s)*. The system of (1) is referred to as the nominal plant *G*.

2.2 Wind power model

2.2.1 Wind velocity model

The output power of wind generator depends on wind velocity. The wind speed model chosen in this study consists of four-component model (Dong-Jiang & Li Wang, 2008), and is defined by

$$V_W = V_{WB} + V_{WG} + V_{WR} + V_{WN} \tag{4}$$

where:

V_{WB} = Base wind velocity
V_{WG} = Gust wind component
V_{WR} = Ramp wind component
V_{WN} = Noise wind component

The base wind velocity component is represented by

$$V_{WB} = K_B \tag{5}$$

Where K_B is a constant, this component is always assumed to be presented in a wind power. The gust wind velocity can be expressed by

$$V_{WG} = \begin{cases} 0 & t < T_{1G} \\ V_{\cos} & T_{1G} < t < T_{1G} + T_G \\ 0 & t > T_{1G} + T_G \end{cases} \tag{6}$$

where:

$V_{\cos}$ = $(MAXG/2)\{1 - \cos 2\pi[(t/T_G) - (T_{1G}/T_G)]\}$
MAXG = the gust peak
T_G = the gust period
T_{1G} = the gust starting time

(1-cosine) gust is an essential component of wind velocity for dynamic studies.

The ramp wind velocity component is described by

$$V_{WR} = \begin{cases} 0 & t < T_{1R} \\ V_{ramp} & T_{1R} < t < T_{2R} \\ 0 & t > T_{2R} \end{cases} \tag{7}$$

where:

V_{ramp} = $MAXR[1-(t-T_{2R})/(T_{1R}-T_{2R})]$
MAXR = the ramp peak
T_{1R} = the ramp start time
T_{2R} = the ramp maximum time

This component may be used to approximate a step change with $T_{2R} > T_{1R}$.
The random noise component can be defined by

$$V_{WN} = 2\sum_{i=1}^{N}[S_V(\omega_i)\Delta\omega]^{1/2}\cos(\omega_i t + \phi_i)t < 0 \tag{8}$$

where:

$\omega_i = (i-1/2)\Delta\omega$
ϕ_i = a random variable with uniform probability density on the interval 0 to 2π

and the spectral density function is defined by

$$S_V(\omega_i) = \frac{2K_N F^2[\omega_i]}{\pi^2[1+(F_{\omega_i}/\mu\pi)^2]^{4/3}} \tag{9}$$

Where K_N (=0.004) is the surface drag coefficient, F(=2000) is turbulence scale, and μ is the mean speed of wind at reference height. Various study have shown that values of N=50, and $\Delta\omega$ = 0.5-2.0 rad/s provide results of excellent accuracy.

2.2.2 Characteristic of wind generator output power

The output power of studied wind generator is expressed by a nonlinear function of the power coefficient Cp as function of blade pitch angle, β, and tip speed ratio, γ.
The tip speed ratio can be described by

$$\lambda = \frac{R_{blade}\omega_{Blade}}{V_W} \tag{10}$$

The power coefficient can be expressed by

$$C_P = (0.44 - 0.0167\beta)\sin\left[\frac{\pi(\lambda-3)}{15-0.3\beta}\right] - 0.0184(\lambda-3)\beta \tag{11}$$

Finally, the output mechanical power of wind generator is

$$P_W = \frac{1}{2}\rho A_r C_P V_W^3 \tag{12}$$

where ρ (=1.25 kg/m³) is the air density and A_r(=1735 m²) is the swept area of blade.

3. Proposed method

3.1 System uncertainties

System nonlinear characteristics, variations of system configuration due to unpredictable disturbances, loading conditions etc., cause various uncertainties in the power system. A controller which is designed without considering system uncertainties in the system modeling, the robustness of the controller against system uncertainties can not be guaranteed. As a result, the controller may fail to operate and lose stabilizing effect under various operating conditions. To enhance the robustness of power system damping controller against system uncertainties, the inverse additive perturbation (Gu et al, 2005) is applied to represent all possible unstructured system uncertainties. The concept of enhancement of robust stability margin is used to formulate the optimization problem of controller parameters.

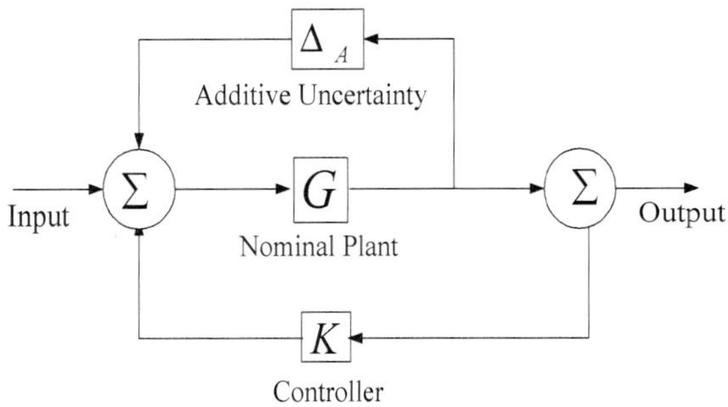

Fig. 2. Feedback system with inverse additive perturbation.

The feedback control system with inverse additive perturbation is shown in Fig.2. G is the nominal plant. K is the designed controller. For unstructured system uncertainties such as various generating and loading conditions, variation of system parameters and nonlinearities etc., they are represented by Δ_A which is the additive uncertainty model. Based on the small gain theorem, for a stable additive uncertainty Δ_A, the system is stable if

$$\|\Delta_A G/(1-GK)\|_\infty < 1 \tag{13}$$

then,

$$\|\Delta_A\|_\infty < 1/\|G/(1-GK)\|_\infty \tag{14}$$

The right hand side of (14) implies the size of system uncertainties or the robust stability margin against system uncertainties. By minimizing $\|G/(1-GK)\|_\infty$, the robust stability margin of the closed-loop system is a maximum.

3.2 Implementation

3.2.1 Objective function

To optimize the stabilizer parameters, an inverse additive perturbation based-objective function is considered. The objective function is formulated to minimize the infinite norm of $\|G/(1-GK)\|_\infty$. Therefore, the robust stability margin of the closed-loop system will increase to achieve near optimum and the robust stability of the power system will be improved. As a result, the objective function can be defined as

$$\text{Minimize} \quad \left\| G/(1-GK) \right\|_{\infty} \tag{15}$$

It is clear that the objective function will identify the minimum value of $\left\| G/(1-GK) \right\|_{\infty}$ for nominal operating conditions considered in the design process.

3.2.2 Optimization problem

In this study, the problem constraints are the controller parameters bounds. In addition to enhance the robust stability, another objective is to increase the damping ratio and place the closed-loop eigenvalues of the electromechanical mode in a D-shape region. The D-shape region can be established to achieve the following objectives.

1. To have some degree of relative stability (Abdel-Magid et al, 1999). The parameters of the controller may be selected to place the electromechanical mode eigenvalue in the left-side of the s-plane by the following function,

$$J_1 = \sigma \le \sigma_{spec} \tag{16}$$

   where σ is the actual real part of eigenvalue and σ_{spec} is desired real part of the dominant inter-area oscillation mode, respectively. The relative stability is determined by the value of σ_{spec}. This will place the closed-loop eigenvalues in a region as shown in Fig. 3.

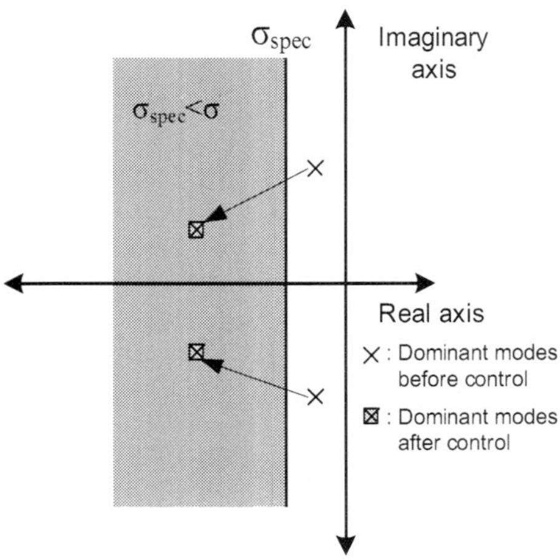

Fig. 3. Region in the left-side of the s-plane where $\sigma \le \sigma_{spec}$

2. To limit the maximum overshoot, the parameters of the controller may be selected by the following function

$$J_2 = \zeta \ge \zeta_{spec} \tag{17}$$

ζ and ζ_{spec} are the actual and desired damping ratio of the dominant inter-area oscillation mode, respectively. This will place the closed-loop eigenvalues in a wedge-shape region in which as shown in Fig. 4.

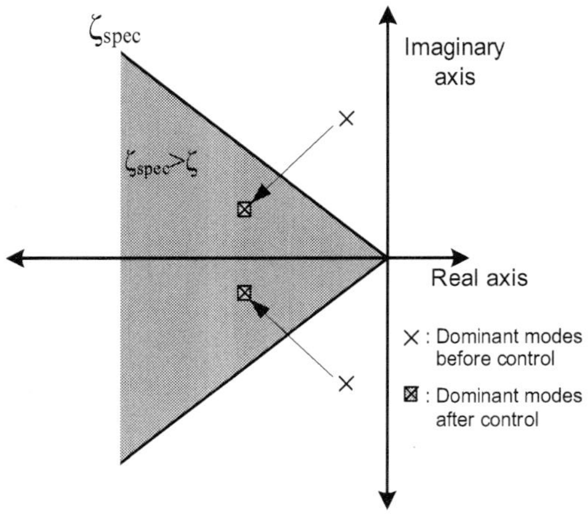

Fig. 4. Wedge-shape region in the s-plane where $\zeta \geq \zeta_{spec}$

Next, the conditions J_1 and J_2 are imposed simultaneously and will place the system closed-loop eigenvalues in the D-shape region characterized by $\zeta \geq \zeta_{spec}$ and $\sigma \leq \sigma_{spec}$ as shown in Fig. 5. It is necessary to mention here that only the unstable or lightly damped electromechanical modes of oscillations are relocated.

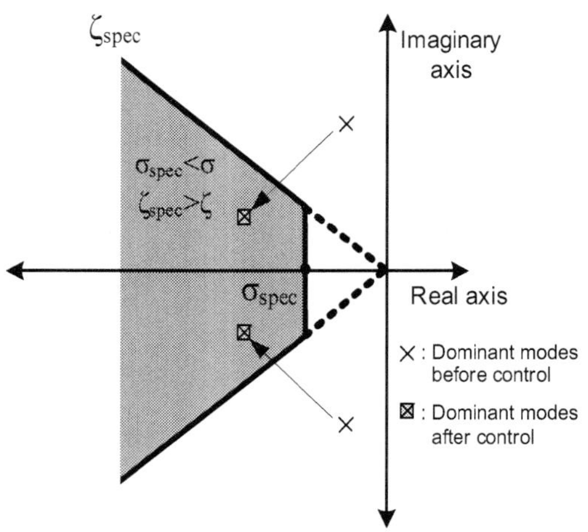

Fig. 5. D-shape region in the s-plane where $\sigma \leq \sigma_{spec}$ and $\zeta \geq \zeta_{spec}$

Therefore, the design problem can be formulated as the following optimization problem.

$$\text{Minimize} \quad \left\| G/(1-GK) \right\|_\infty \tag{18}$$

$$\text{Subject to} \quad \zeta \geq \zeta_{spec}, \sigma \leq \sigma_{spec} \tag{19}$$

$$K_{\min} \leq K \leq K_{\max}$$
$$T_{\min} \leq T \leq T_{\max}$$

where ζ and ζ_{spec} are the actual and desired damping ratio of the dominant inter-area oscillation mode, respectively; σ and σ_{spec} are the actual and desired real part, respectively; $K_{\max}$ and $K_{\min}$ are the maximum and minimum controller gains, respectively; $T_{\max}$ and $T_{\min}$ are the maximum and minimum time constants, respectively. This optimization problem is solved by GA (GAOT, 2005) to search the controller parameters.

3.3 Parameters optimization by GA

In this section, GA is applied to search the controller parameters of PSS with off line tuning. The flow chart of the proposed method is illustrated in Fig. 6. Each step is explained as follows.

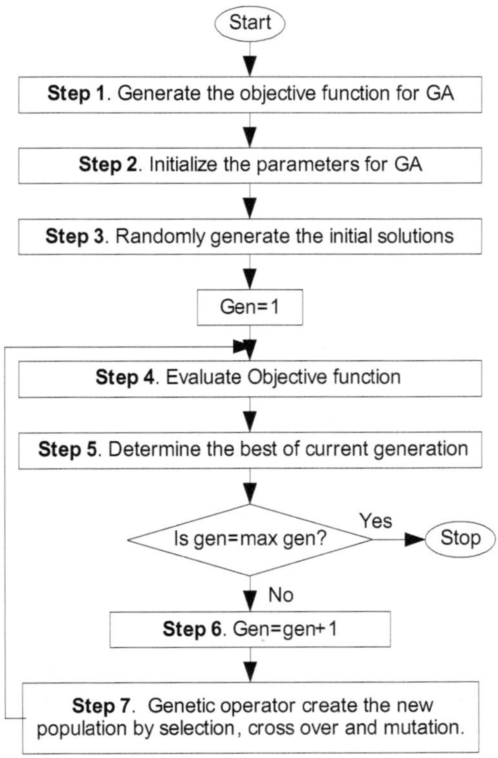

Fig. 6. Flow chart of the proposed design

Step 1. Generate the objective function for GA optimization.
In this study, the performance and robust stability conditions in inverse additive perturbation design approach is adopted to design a robust PSS. The conventional PSS with a 2nd-order lead-lag controller is represented by

$$\Delta u_{pss,i} = K_i \left(\frac{sT_W}{sT_W + 1} \right) \left(\frac{sT_{1,i} + 1}{sT_{2,i} + 1} \right) \left(\frac{sT_{3,i} + 1}{sT_{4,i} + 1} \right) \Delta \omega_i \tag{20}$$

where, $\Delta u_{pss,i}$ and $\Delta \omega_i$ are the control output signal and the rotor speed deviation at the *i-th* machine, respectively; K_i is a controller gain; T_W is a wash-out time constant (s): and $T_{1,i}$, $T_{2,i}$, $T_{3,i}$, and $T_{4,i}$ are time constants (s).

Step 2. Initialize the search parameters for GA. Define genetic parameters such as population size, crossover, mutation rate, and maximum generation.
Step 3. Randomly generate the initial solution.
Step 4. Evaluate objective function of each individual in (18) and (19).
Step 5. Select the best individual in the current generation. Check the maximum generation.
Step 6. Increase the generation.
Step 7. While the current generation is less than the maximum generation, create new population using genetic operators and go to step 4. If the current generation is the maximum generation, then stop.

4. Performance simulation and results

In the optimization, the ranges of search parameters and GA parameters are set as follows: ζ and ζ_{spec} are actual and desired damping ratio is set as 0.1, respectively, σ and σ_{spec} are actual and desired real part of the inter-area oscillation mode is set as -0.1, $K_{i,\min}$ and $K_{i,\max}$ are minimum and maximum gains of PSS are set as 1 and 30, $T_{ji,\min}$ and $T_{ji,\max}$ are minimum and maximum time constants of PSS are set as 0.01 and 1. T_w is set to 10 s. The optimization problem is solved by genetic algorithm. Under the normal operating condition case 1 in Table 1, the robust control parameters (RPSS) are obtained as follows.

$$\begin{aligned}
K_{PSS1} &= 25.05 \left(\frac{0.8638s + 1}{0.7425s + 1} \right) \left(\frac{0.8538s + 1}{0.7227s + 1} \right) \\
K_{PSS2} &= 25.58 \left(\frac{0.5395s + 1}{0.3324s + 1} \right) \left(\frac{0.5124s + 1}{0.3175s + 1} \right) \\
K_{PSS3} &= 12.79 \left(\frac{0.4940s + 1}{0.2545s + 1} \right) \left(\frac{0.4748s + 1}{0.2675s + 1} \right) \\
K_{PSS4} &= 12.50 \left(\frac{0.6235s + 1}{0.1806s + 1} \right) \left(\frac{0.6133s + 1}{0.1350s + 1} \right)
\end{aligned} \tag{21}$$

Table 2 shows the eigenvalue and damping ratio of the dominant inter-area oscillation mode. Clearly, the damping ratio of the oscillation mode of RPSS is improved as designed in comparison with No PSS case.

CASE	1. NOC (Ptie =3.0)	2. HL (Ptie =4.5)	3. HL (Ptie =4.5) with fault	4. HL (Ptie =4.5) and WL with fault
G1	PG=7	PG=9	PG=9	PG=9
G2	PG=6	PG=7.5	PG=7.5	PG=7.5
G3	PG=3.25	PG=5	PG=5	PG=5
G4	PG=3.75	PG=4	PG=4	PG=4
Load	L1=10 , L2=10	L1=12 , L2=13	L1=12 , L2=13	L1=12 , L2=13
Line condition	Two lines of 3-101	Two lines of 3-101	one line of 3-101 is openned at 10 s.	one line of 3-101, 3ø fault at line 3-101

Note: NOC=normal operating condition, HL=heavy load, WL=weak line, G = Generation (pu), L = Load (pu), Base = 900 MVA

Table 1. Operating conditions

Cases	Eigenvalue and damping ratio
No PSS	$-0.0829 \pm j\,3.6759,\ \zeta = 0.023$
With RPSS	$-0.3407 \pm j\,2.4603,\ \zeta = 0.137$

Table 2. Dominant inter-area modes

In the study, the performance and robustness of RPSS are compared with Conventional PSS (CPSS) (Klein et al, 1992). The eigenvalue analysis and nonlinear simulations are carried under four case studies as given in Table 1.

Table 3 shows the eigenvalues and damping ratios of the dominant inter-area oscillation mode. Clearly, the CPSS loses control effect in case 4 (heavy load and weak line). The damping ratio is negative and the system becomes unstable. On the other hand, the damping ratio of the oscillation mode of RPSS is still positive. The damping effect of RPSS is very robust under any operating condition.

Case	NOC	HL & WL
CPSS	-0.1186±j3.6845	0.0452±j2.5707
	0.0322	-0.0176
RPSS	-0.3407±j2.4603	-0.2095±j2.0705
	0.137	0.101

Table 3. Dominant inter-area modes and damping ratio

Fig. 7 depicts wind velocity [Vw1] and [Vw2] of wind farms based on (4). Using (12), wind power generations [PW1] and [PW2] can be shown in Fig. 8. In simulation studies of all cases, wind power generations are injected to buses 4 and 14, respectively. Fig.9 shows tie-line power deviation in case 1 at the normal condition. CPSS and RPSS are able to damp power oscillations due to wind power fluctuations.

Under heavy load condition in case 2, No PSS loses stabilizing effect. It is not able to damp out power fluctuation. The system stability can not be maintained. On the other hand, the PSS is capable of stabilizing power fluctuation. It still retains system stability successfully. Nevertheless, the RPSS provides more damping effects than CPSS as shown in Fig.10. These results signify that the stabilizing effect of RPSS against wind power fluctuations is superior to that of CPSS.

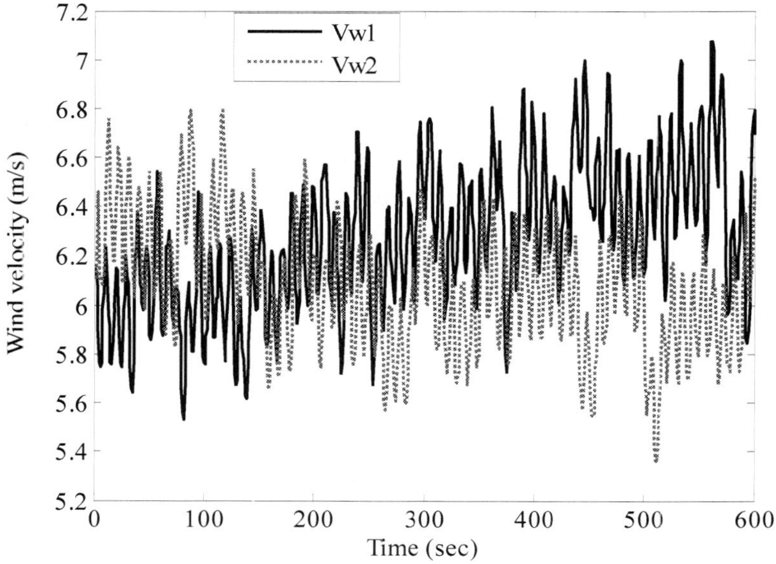

Fig. 7. Wind velocity.

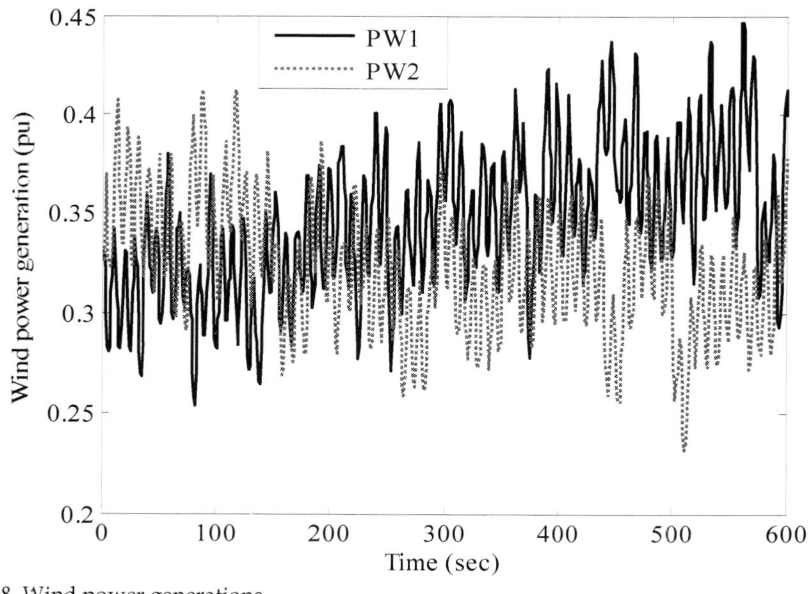

Fig. 8. Wind power generations.

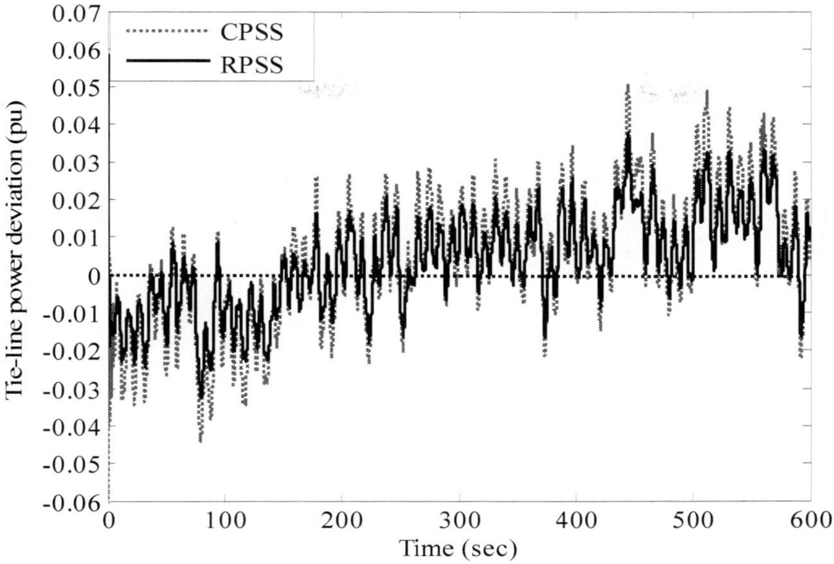

Fig. 9. System responses in case 1 (normal condition).

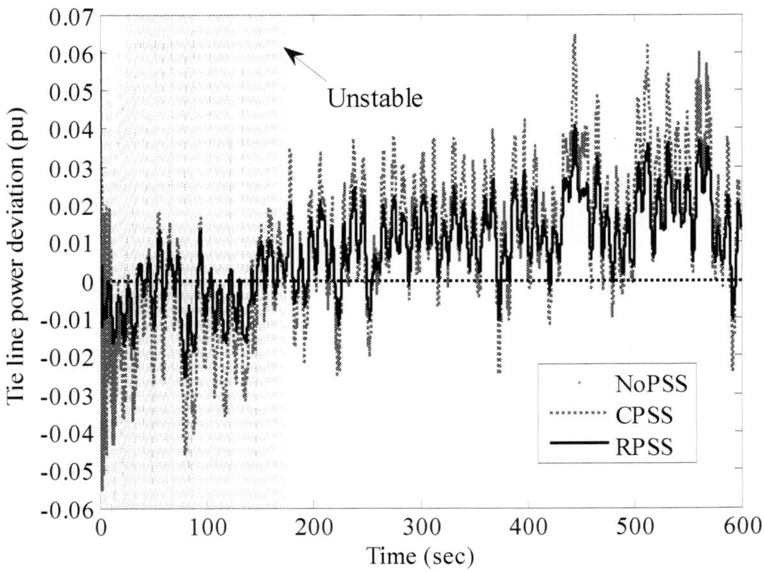

Fig. 10. System responses in case 2 (heavy load).

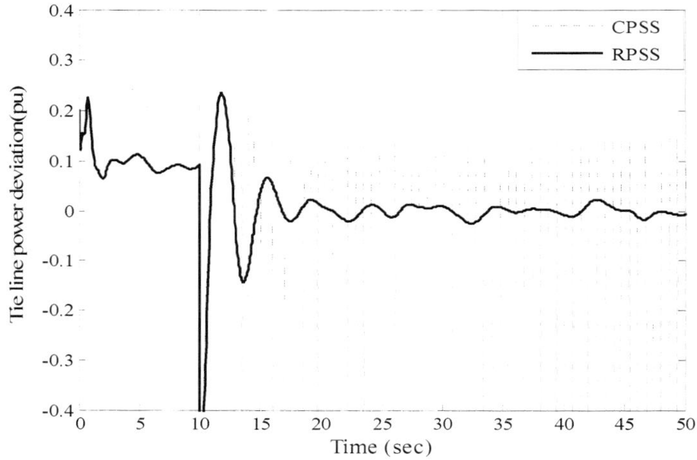

Fig. 11. System responses in case 3 (heavy load).

In case 3, it is assumed that the tie-line power transfers from areas 1 to 2 via two lines of tie-line 3-101, then one line is suddenly opened at 10 s. Simulation result is depicted in Fig. 11. The CPSS is not capable of damping power oscillation and eventually loses stabilizing effect. On the other hand, the RPSS is very robust against this situation. The power oscillation can be stabilized effectively.

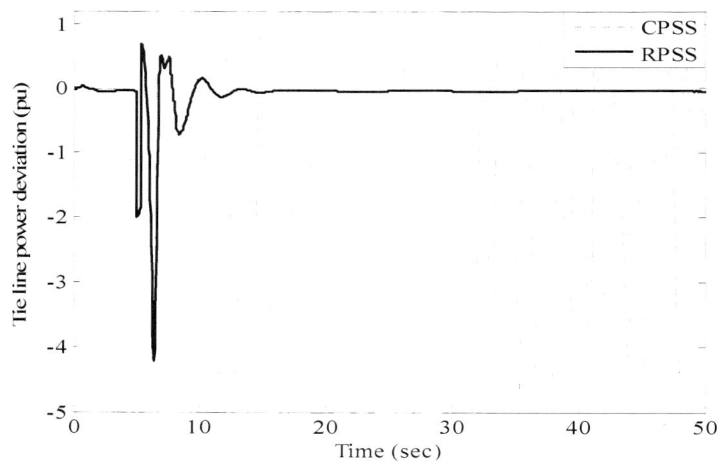

Fig. 12. System responses in case 4 (heavy load & weak line).

Finally, in case 4, it is assumed that one line of 3-101 is in service. A 3ø fault occurs at line 3-101 at 5s and the fault is cleared after 150 ms. Simulation results in Fig. 12 show that CPSS completely loses its control effect. On the other hand, the RPSS still retains system stability successfully. This explicitly shows the superior robustness of RPSS beyond CPSS.

7. Conclusion

Robust PSS design based on inverse additive perturbation in a power system with wind farms has been proposed in this work. The parameters optimization of PSS is formulated based on an enhancement of system robust stability margin. Solving the problem by GA, PSS parameters are automatically obtained. The designed PSS is based on the conventional 2nd-order lead-lag compensator. Accordingly, it is easy to implement in real systems. The damping effects and robustness of the proposed PSS have been evaluated in the two areas four machines power system with wind farms. Simulation results confirm that the robustness of the proposed PSS is much superior to that of the CPSS against various uncertainties.

8. References

Elgerd O.I. & Fosha C.E. (1970), Optimum Megawatt Frequency - control of Multi-area Electric Energy Systems, *IEEE Transactions on Power Systems*, Vol. 89, pp. 556–63.

deMello F. & Concordia C. (1969), Concepts of Synchronous Machine Stability as Affected by Excitation Control, *IEEE Transactions on Power Apparatus and Systems*, Vol. 88, pp. 316-329.

Larsen E.V. & Swann D.A. (1981), Applying Power System Stabilizers. Part I: General Concepts; Part II: Performance Objectives and Tuning Concepts; Part III: Practical Considerations, *IEEE Transactions on Power Apparatus and Systems*, Vol. 100, pp. 3017-3046.

Murdoh A.; Sanderson H.C. & Lawson R. (2000), Excitation System Performance Specification to Meet Interconnection Requirements, *Panel Session for the 2000 IEEE Power and Energy System Winter Meeting*, Singapore, Vol. 1, pp.597- 602.

Klein M.; Rogers G.J. & Kundur P. (1991), A Fundamental Study of Inter-area Oscillations in Power Systems, *IEEE Transactions on Power Systems*, Vol. 6, No. 3, pp. 914-921.

Klein M.; Rogers G.J.; Moorty S. & Kundur P. (1992), Analytical Investigation of Factors Influencing Power System Stabilizers Performance, *IEEE Transactions on Energy Conversion*, Vol. 7, No. 3, pp. 382-390.

Abdalla O.H.; Hassan S.A. & Tweig N.T. (1984), Co-ordinated Stabilization of a Multi-machine Power System, *IEEE Transactions on Power Apparatus and Systems*, Vol. 103, pp.483-494.

DeMello F.P.; Nolan P.J.; Laskowski T.F. & Undrill J.M. (1980), Co-ordinated Application of Stabilizers in Multi-machine Power Systems, *IEEE Transactions on Power Apparatus and Systems*, Vol. 99, pp. 892-901.

Tse G.T. & Tso S.K. (1993), Refinement of Conventional PSS Design in Multimachine System by Modal Analysis, *IEEE Transactions on Power Systems*, Vol. 8, No. 2, pp. 598-605.

Hsu Y.Y. & Hsu C.Y. (1986), Design of a Proportional-Integral Power System Stabilizer, *IEEE Transactions on Power Systems*, Vol. 1, No. 2, pp. 46-53.

Hsu Y.Y. & Liou K.L. (1987), Design of Self-Tuning PID Power System Stabilizers for Synchronous Generators, IEEE Transactions on Energy Conversion, Vol. 2, No. 3, pp. 343-348.

Yu Y.N & Siggers C. (1971), Stabilization and Optimal Control Signals for a Power System, *IEEE Transactions on Power Apparatus and Systems*, Vol.90, pp. 1469-1481.

Moussa H.A.M. & Yu Y.N. (1972), Optimal Power System Stabilization through Excitation and/or Governor Control, *IEEE Transactions on Power Apparatus and Systems*, Vol.91, pp.1166-1174.

Fleming R.J.; Mohan M.A. & Parvatisam K. (1981), Selection of Parameters of Stabilizers in Multi-machine Power Systems, *IEEE Transactions on Power Apparatus and Systems*, Vol.100, pp. 2329-2333.

Zhou E.Z.; Malik O.P. & Hope G.S. (1992), Design of Stabilisers for a Multimachine Power System Based on the Sensitivity of PSS Effect, *IEEE Transaction on Energy Conversion*, Vol. 7, pp.606-613.

Aldeen M. & Crusca F. (1995), Multimachine Power System Stabilisers Design Based on New LQR Approach, *Proceeding of IEE Proceeding Generation Transmission and Distribution*, Vol. 142, pp. 494-502.

Cheng S.J.; Chow Y.S.; Malik O.P. & Hope G.S. (1986), An Adaptive Synchronous Machine Stabilizer, *IEEE Transactions on Power Systems*, Vol. 1, pp.101-109.

Lim C.M (1989), A Self-tuning Stabilizer for Excitation or Governor Control of Power Systems, *IEEE Transactions on Energy Conversion*, Vol. 4, pp. 152-159, June 1989.

Hsu Y.Y. & Cheng C.H. (1990), Design of Fuzzy Power System Stabilisers for Multimachine Power Systems, *Proceeding of IEEE*, Vol. 137, No. 3.

Hoang P. & Tomsovic K. (1996), Design and Analysis of an Adaptive Fuzzy Power System Stabilizer, *IEEE Transactions on Energy Conversion*, Vol. 11, pp. 455 - 461.

Zhang Y.; Chen G.P ; Malik O.P. & G.S. Hope (1993), An Artificial Neural Network Based Adaptive Power System Stabilizer, *IEEE Transactions on Energy Conversion*, Vol. 8, No.1.

Segal R,; Kothari M.L & Madnani S. (2000), Radial Basis Function (RBF) Network Adaptive Power System Stabilizer, *IEEE Transactions on Power Systems*, Vol. 15, pp. 722-727.

Abido M.A & Abdel-Magid Y.L (1998), A Hybrid Neuro-fuzzy Power System Stabilizer for Multimachine Power Systems, *IEEE Transactions on Power Systems*, Vol. 13, No. 4, November 1998.

Chen S. & Malik O.P (1995), H_∞ Optimisation-based Power System Stabilizer Design, *Proc. of IEE Generation Transmission and Distribution*, Vol. 142, pp.179-184.

Yan T.C (1997), Applying Optimisation Method to Power System Stabiliser Design -Parts 1 & 2, *Transactions on Electrical Power and Energy System*, Vol. 19, pp. 29-43.

Abdel-Magid Y.L; Abido M.A & Mantawy A.H (2001), Robust Tuning of Power System Stabilizers in Multimachine Power Systems, *IEEE Transactions on Power Systems*, Vol. 15, pp.735-740.

Abdel-Magid Y.L; Abido, M.A; Al-Baiyat S. & Mantawy A.H (1999), Simultaneous Stabilization of Multimachine Power Systems via Genetic Algorithm, *IEEE Transactions on Power Systems*, Vol. 14, No. 4, pp. 1428-1439.

Abido M.A (2000), Robust Design of Multi-machine Power System Stabilisers using Simulated Annealing, IEEE Transactions on Energy Conversion, Vol. 15, pp. 297-304.

Cuk Supriyadi A.N, Ngamroo I.; Kaitwanidvilai S.; Kunakorn A.; Hashiguchi T. & Goda T. (2008), Design of Robust Power System Stabilizer using Genetic Algorithm-based Fixed-Structure H_∞ Loop Shaping Control, *Proceeding of the 17th World Congress The International Federation of Automatic Control Seoul (IFAC)*, Korea, pp.11086- 11091.

Dong-Jiang & Li Wang (2008), Small-signal stability analysis of an autonomous hybrid renewable energy power generation/energy storage system part I : time-domain simulations, *IEEE Transactions on Energy Conversion* ,Vol. 23, No.1, pp. 311-320.

Gu P.; Petkov Hr. & Konstantinov M.M. (2005) , *Robust control design withMATLAB*, Springer, London.

GAOT (1995), *A Genetic Algorithm for Function Optimization: A Matlab Implementation*, [Online] Available: http://www.ie.ncsu.edu/mirage/GAToolBox/gaot/

17

Wind Power Impact on Power System Dynamic Performance

Emmanuel S. Karapidakis
Technological Educational Institute of Crete
Greece

1. Introduction

It is well known that wind power is one of the major sources of renewable energy with a remarkable contribution to the installed capacity of electrical power systems. In this chapter, the impact of large scale wind power generation on the dynamic performance especially of weak and/or islands power systems are presented.

Weak or Autonomous power systems, like the ones operating in idiomorphic areas or in islands, face increased problems related to their operation and control, [Hatziargyriou, N. & Papadopoulos, M. (1997)]. In most of these systems, the real cost of electricity production is much higher than in interconnected systems due to the high operating costs of their thermal generating units, mainly diesel and gas turbines, and the import and transportation costs of the fuel used. Security is also a major concern, since mismatches in generation and load and/or unstable system frequency control might lead to system failures, easier than in interconnected systems, [Hatziargyriou, N. et al. (1997)].

Although under ongoing energy policies wind power exploitation appear particularly attractive, the integration of a substantial amount of wind power in isolated electrical systems needs careful consideration, so as to maintain a high degree of reliability and security of the system operation. The main problems identified concern operational scheduling (mainly unit commitment) due to high production forecasting uncertainties, as well as steady state and dynamic operation disorder, [Dialynas, E. et al. (1998)]. These problems may considerably limit the amount of wind generation that can be connected to the power systems, increasing the complexity of their operation. Thus, next to the more common angle and voltage stability concerns, frequency stability must be ensured, [Hatziargyriou, N. et al. (1998)]. This depends on the ability of the system to restore balance between generation and load in case of a severe system upset with minimum loss of load.

Dynamic simulation studies are the first step in determining the level of wind power penetration in power systems. Analytical studies are required in order to derive security rules and guidelines for the optimal operation of each system [Arrilaga, J. & Arnold, C.P. (1993)]. Simulations of a power system dynamic performance mainly cover voltage and frequency calculations under several abnormal operating conditions, start-up or sudden disconnection of wind generation, wind fluctuations and short circuits on the transmission and distribution network. In order to ensure the maximum exploitation of the available renewable power sources and to operate systems with increased wind power penetration in

the most economic and secure way, advanced energy management systems (EMS) are needed, [Nogaret, E. et al. (1997)].
Moreover, in order to operate optimally within the new market conditions, the price of providing a given level of security has to be accounted for. This is directly linked to the provision of remedial actions, in case of insecure situations, [La Scala, M. et al. (1998)]. Especially, for Dynamic Security, unlike Steady State security, remedial actions can only be preventive leading to load shedding or generation rescheduling. Consequently, the cost of load shedding has to be balanced with the cost of providing adequate spinning reserves in order to avoid it.

2. Power system dynamic performance

Power system's stability has been recognized as an important problem for secure system operation since the 1920s in [Steinmetz, C. P. (1920)] and in [AIEE Subcommittee on Interconnections and Stability Factors]. Generally, transient stability is the main concern on the majority of the power systems. As power systems have evolved through continuing growth, new operation technologies and controls in highly stressed conditions have emerged. More precisely, voltage stability and frequency stability have become greater concerns than in the past. A clear understanding of different types of stability and how they are interrelated is essential for the satisfactory design and operation of power systems, [Kundur, P. & Morison, G.K. (1997)].

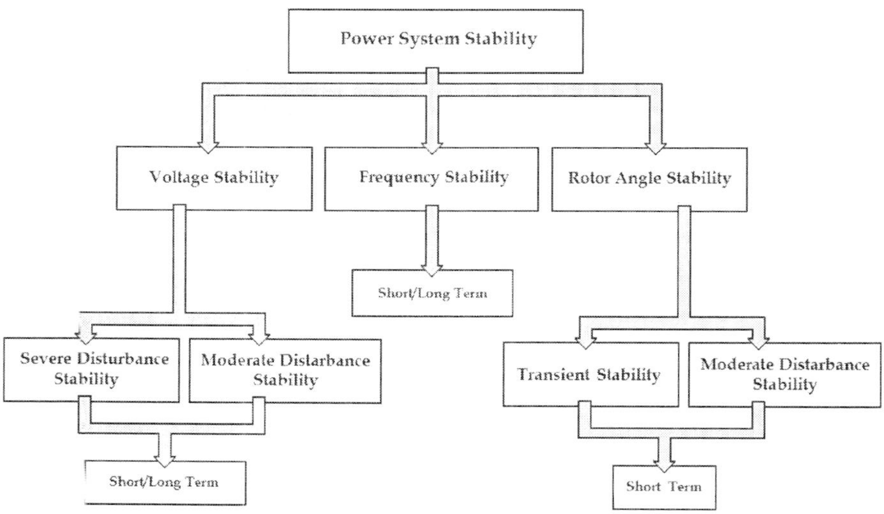

Fig. 1. Categories of power system stability

Power system stability is similar to the stability of any dynamic system, and has fundamental mathematical underpinnings. Precise definitions of stability can be found in the literature dealing with the rigorous mathematical theory of stability of dynamic systems. A circumstantial definition of Power System Dynamic Security and the corresponding types of Power System Stability is provided in [Kundur Prabha, et al. (2004)]. In Fig.1 a concise classification of power system stability is provided. Addionally, another significant issue is the relationship between the concepts of power system reliability, security, and stability of a power system, [Crary, S. B., Herlitz, I., Favez, B. (1948)].

- Power System Reliability: it refers to the probability of a required operation condition achievement, with few interruptions over an extended time period.
- Power System Security: it refers to the risk assessment of system ability to survive disturbances (taking into account the probability of these contingencies) without any power supply interruption.
- Power System Stability: it refers to the dynamic operation after a severe or moderate disturbance, consequently to an initial steady state operating condition.

As it is clearly mentioned in [Kundur Prabha, et al. (2004)], the analysis of security relates to the determination of the power system robustness to imminent disturbances. Assuming that a power system subjected to changes, it is important that when the changes are completed, the system settles to a new operating state where no technical constraints are violated. This implies that, in addition to the next operating conditions being acceptable, the system should survive the transition to these conditions. The above characterization of system security clearly highlights two aspects of its analysis:

- Static Security Analysis: It involves steady-state analysis of post-disturbance system conditions to verify that no voltage constraints are violated.
- Dynamic Security Analysis: It involves examining different categories of system stability described in Section III.

The general practice for dynamic security assessment has been to use a deterministic approach. The power system is designed and operated to withstand a set of contingencies selected on the basis that they have a significant possibility of occurrence. In practice, they are usually defined as the loss of any single element in a power system either spontaneously or preceded by a single, double, or three phase fault. This method is generally called as the N-1 criterion because it examines the behavior of an N-component network following the loss of any one of its components. In addition, emergency controls, such as generation tripping, load shedding, and controlled islanding, may be used to withstand such events and prevent widespread blackouts.

Ongoing power systems under deregulated energy markets with a diversity of participants, the deterministic approach may not be appropriate. There is a need to account for the probabilistic nature of system conditions and events, and to quantify and manage risk. The trend will be to expand the use of risk-based security assessment. In this approach, the probability of the system becoming unstable and its consequences are examined, and the degree of exposure to system failure is estimated. This approach is computationally intensive but is possible with today's computing and analysis tools.

3. Wind power penetration

As it is well known, wind is a diffused source of energy and wind parks are mainly a dispersed generation for conventional power system. Thus wind generation is often considered as embedded generation in the power systems with a variable penetration rate. Conventional power systems were designed and developed as centralized networks, where generated power flows unidirectional from higher to lower voltage levels and from large power plants to diverse loads through transmission and distribution systems. Distributed generation and especially dispersed generation of renewable energy sources can provide significant benefits, such as improved system reliability and enhanced power quality, [CIGRE Study Committee (1998)]. Additionally, dispersed generation could increase system efficiency; while under specific conditions could even decrease network operational cost. Generally, the dispersed generation changes distribution networks from passive networks, with power flows from higher to lower voltage levels, into active networks with multi-directional power flows, [Strbac, G. (2002)]. Furthermore, transmission and distribution infrastructures require specific economic regulations, [Stoft, S. (2002)].

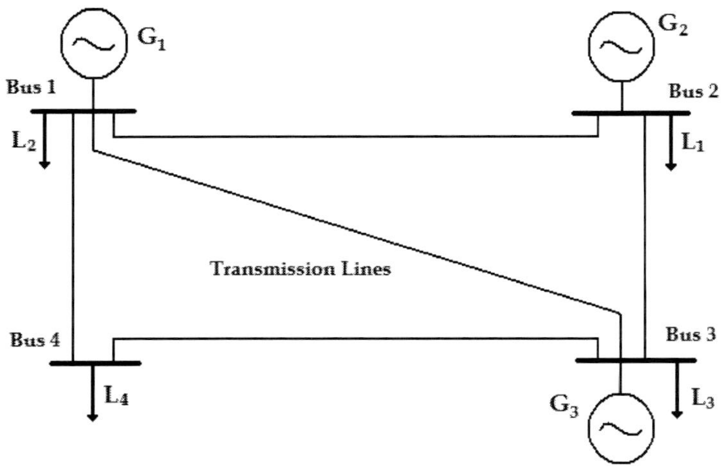

Fig. 2. Power system diagram

In Fig.2 a diagrammatic representation of a common power system is presented. Generating units could be conventional (fossil-fuel, nuclear or hydro installations) or renewables energy sources (wind parks, photovoltaics, biomass or geothermal installations). Generating units are connected to diverse loads mainly through high voltage transmission lines and distribution systems subsequently.

In case of small scale wind power installations where the wind power supplies local loads then the inpact on the grid voltage, frequency and losses is likely to be beneficial, [Burton Tony, et al. (2001)]. However, in case of large scale integrations where significant power should be flow through the trasmiision and/or distribution networks then voltage variations, frequency deviations and losses could became excessive.

For example if a wind generator operates at a power factor equal to one ($\cos\varphi=1$), then the voltage difference, in a lightly-loaded radial line, is given approximately by:

$$\Delta V = V_E - V_S = \frac{P \cdot R}{V_S} \tag{1}$$

Operating the generator at a leading power factor (absorbing reactive power) acts to reduce the voltage rise but at the expense of increased network losses. In this case the voltage rise is given by:

$$\Delta V = V_E - V_S = \frac{P \cdot R - X \cdot Q}{V_S} \tag{2}$$

Assuming a typical ratio of inductive reactive X to resistance R equal to 2 and a power factor of 0.89 for a typical uncompensated induction generator at rated output, the apparent voltage difference could be consider as zero. However, real power loss PLosses is given approximately by:

$$P_{Losses} = \left(\frac{P^2 + Q^2}{V_S} \right) \cdot R \tag{3}$$

Though the reactive power absorbed by the generator eliminates the voltage ascent, however higher real power losses are emerged in the line.

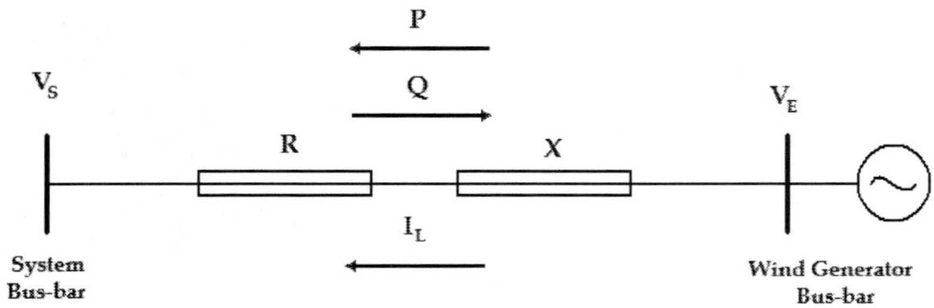

Fig. 3. Wind generator connected to a radial line

Previous equations (1) - (3) are approximate only and they are notaccurate to heavily-loaded systems, where the power factor is rarely unity. A simple but precise calculation for voltage increment in any radial circuit may be carried out using iterative techniques.

$$S_S = P - jQ \tag{4}$$

By definition S = V.I*, where * indicates the complex conjugate. Consequently, the line current is given by:

$$I_L = \left(\frac{S_E^*}{V_E^*} \right) = \left(\frac{P + jQ}{V_E^*} \right) \tag{5}$$

The voltage ascent through the line is given by I_L.Z and the V_E is equal to:

$$V_E = V_S + I_L \cdot Z = V_S + \frac{(R + jX) \cdot (P + jQ)}{V_E^*} \tag{6}$$

Finally, voltage V_E can be defined using a simple iterative expression:

$$V_E^{n+1} = V_S + \frac{(R + jX) \cdot (P + jQ)}{V_E^{n*}} \tag{7}$$

where n is the iteration number.

The previous technique is a simple approach of the conventional Gauss-Seidel load flow algorithm [Weely & Copy, (1998)]. Once the calculation converges an accurate solution is obtained. Additionaly, more complex load flow calculations could be carried out using ad hoc and advanced computer programs, [PowerWorld, (2007)]

4. Wind generation in autonomous power systems

Autonomous or isolated power systems are all the small and medium size power systems where no interconnection exists with conterminous and/or continental systems. These power systems, like the ones operating in large islands, face increased problems related to their operation and control, [Smith, P. (2006)]. In most of these systems, dynamic performance is a major concern, since mismatches in generation and load and/or unstable system frequency control might lead to system failures, easier than in interconnected systems.

Renewable sources and especially wind power exploitation appear particularly attractive, [Doherty, R. & O'Malley, M.J. (2006)]. However, the integration of a substantial amount of wind power in isolated systems needs careful consideration, so as to maintain a high degree of reliability and security of the system operation. The main problems identified concern operational scheduling (mainly unit commitment) due to high production forecasting uncertainties, as well as steady state and dynamic operating problems. These problems may considerably limit the amount of wind generation that can be connected to the island systems, increasing the complexity of their operation. Thus, next to the more common angle and voltage stability concerns, frequency stability [Karapidakis, E.S. & Thalassinakis, M. (2006)] must be ensured. This depends on the ability of the system to restore balance between generation and load following a severe system upset with minimum loss of load.

4.1 Autonomous power system of Crete

Grete is the largest Greek island with approximately 8.500 Km^2 and one of the largest in Mediterranean region. Its population is more than 600,000 inhabitants that triple in summer period. As well, it features a considerable annual increase of electricity demand approaching the 7% during the last decade. As a result, the annual energy consumption during 2008 surpassed the 3TWh in comparison with the modest 280 GWh of year 1975. Additionally, comparing the mean hourly load demand variation all year round, there is a considerable electricity generation diversification between months and seasons, as it is shown in Fig. 4.

However, even during the low consumption periods, minimum load demand is greater than current system technical minimum (approximately 100 MW). Island's electricity generation system is based mainly on three (3) oil-fired thermal power units, located as it is shown in Fig. 5. The official capacity of the local power plants is 742.9 MW, although the real power of the system is 693 MW for winter and 652 MW for summer operation. Additionally, there are 25 wind parks installed with nominal power of 124,85MW in appropriate regions of the island. These WPs are connected to the grid through MV/HV substations of 20kV/150kV. The annual peak load demand occurs on a winter day and overnight loads can be assumed to be approximately equal to 25% of the corresponding daily peak loads. The steam and diesel units mainly supply the base-load. The Gas turbines normally supply the daily peak load or the load that cannot be supplied by the other units in outage conditions. These units have a high running cost that increases significantly the average cost of the electricity being supplied.

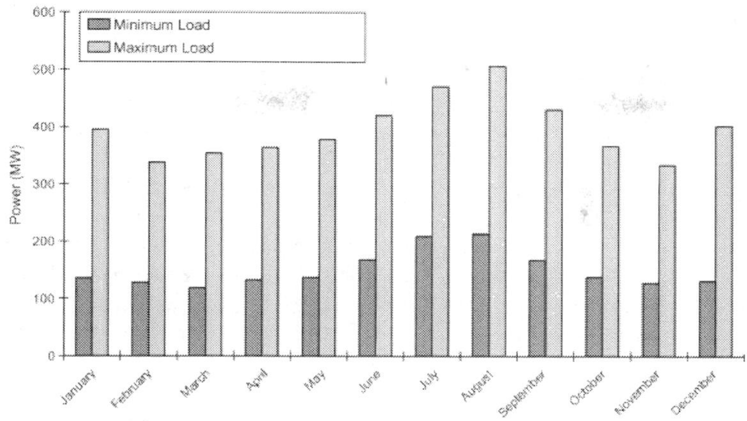

Fig. 4. Monthly variation of min and max load demand

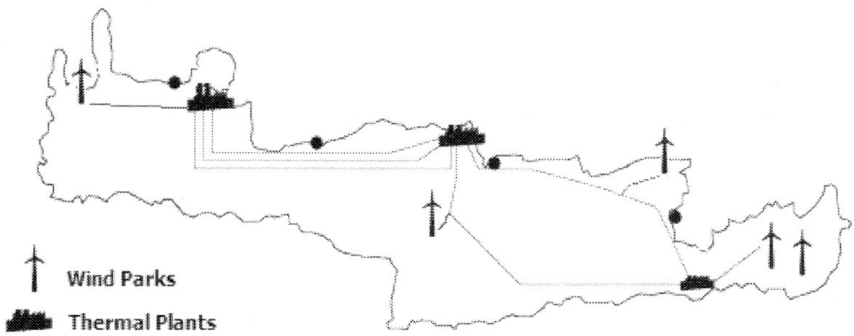

Fig. 5. Power plants and wind parks locations

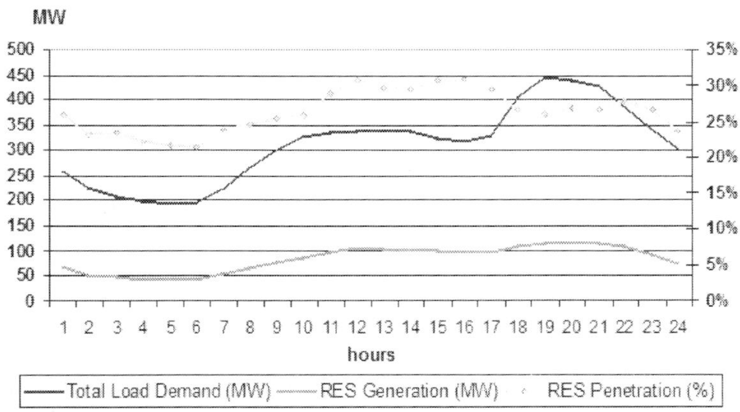

Fig. 6. Wind power penetration in power system of Crete

In Fig.6 the wind power production in parallel with the overall production in a specific day within 2008 is presented. In this case the portion of the corresponding wind generation varies between 22% and 32% of the total power supply that is considered as a significant high penetration for an autonomous system such as Crete's network. Wind energy exploitation activities in Crete started since mid eighties. As a result, a remarkable wind park installation activity has started since 1992, leading by 2008 to the existence of 25 wind power stations of rated power 124.85 MW. While a licensed capacity of additional 95 MW nominal power is planned to be installed till 2012.

Assuming the current wind parks and the prospect of many PV installations (104 MW till 2012) Crete deals even now with a significant dispersed generation and high RES penetration. Thus autonomous power system of Crete is an excellent represenative for dynamic performance estimations.

4.2 Power system dynamic performance

Regarding security, a large number of wind parks are installed at the eastern part of the island, which presents the most favorable wind conditions. As a result, in case of faults on some particular lines, these wind parks are disconnected. Frequency oscillations might easily trigger the under-frequency protection relays of the wind parks, thus causing further imbalance in the system. Fig. 7 depicts a real situation when short-circuit occured at 12:10 p.m. leading the frequency, in spite of the fast load shedding to drop till 49 Hz. The frequency oscillations are the main characteristics of an autonomous power system as Crete's.

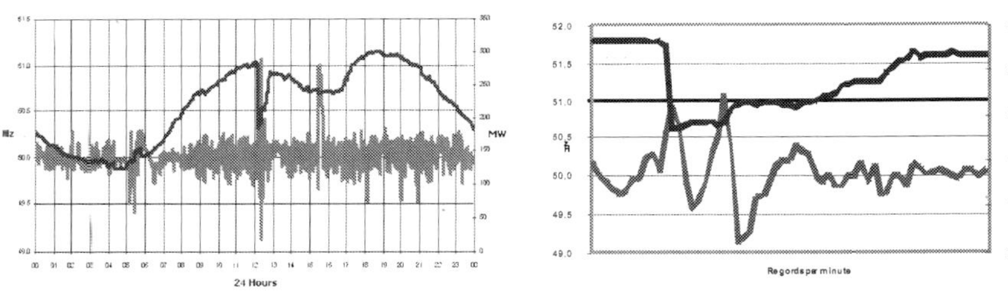

Fig. 7. Frequency fluctuations and load disconnection

In order to investigate and assess the dynamic behaviour of an examined power system, accurate simulationa are needed. So that the models, which are used for the presentation of the system components were chosen taking into account that the duration of the transient phenomena under study are between 0.1 and 10 sec approximately. The formulas for the main components of the system are as follow:

a. Diesel motors and Gas turbines

The generic model, illustrated in Fig. 8, is used for the simulation of the diesel engines and the gas turbines speed governors.

Input is the frequency change and output is the produced mechanical power, while the diesel engine or the gas turbine is represented by a first order lag with a time constant T_D. T_G is the time constant of the hydraulic actuator of the governor mechanism.

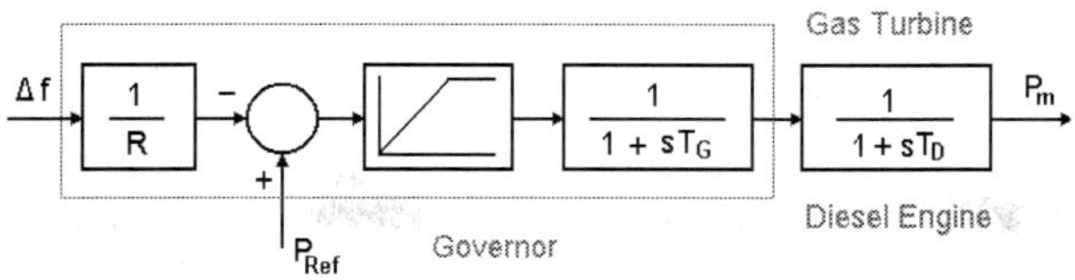

Fig. 8. Diesel-Gas speed control system

Δf, is the per unit frequency change (f_o=50Hz).

P_m, is the mechanical power of the diesel motor.

R, is the droop of the speed governor.

b. Steam unit

The block diagram of Fig. 9 represents the speed governor system considered for each steam unit.

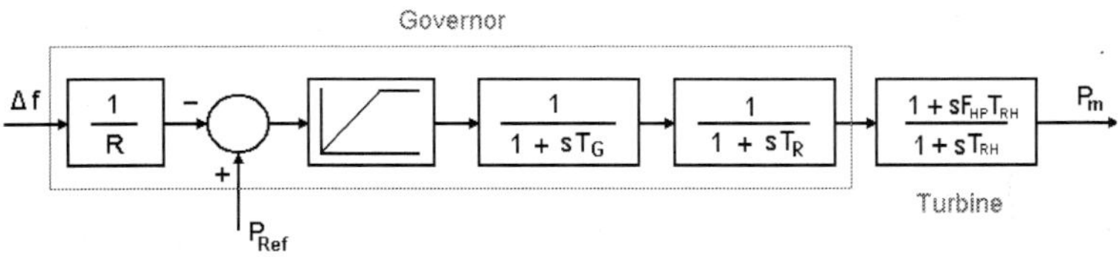

Fig. 9. Steam speed control system

Δf, is the per unit frequency change (f_o=50Hz)

P_m, is the mechanical power of the steam turbine.

The transfer function for the governor includes speed relay and transient droop. The steam turbine is represented as single reheat type whose transfer function is:

$$G_T(s) = \frac{1+F_{HP}.T_{RH}.s}{1+T_{RH}.s} \tag{8}$$

F_{HP}, is the fraction of total turbine power generated.

T_{RH}, is the time constant of reheater.

For the simulation procedure, an integrating control parallel to machine droop is added to the speed controllers of Fig. 8 and Fig. 9, as shown in next Fig. 10.

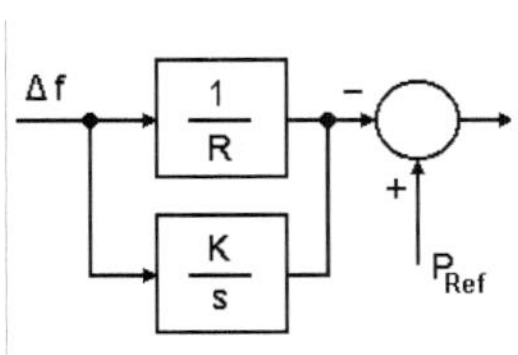

Fig. 10. Addition of integrating control block

c. Voltage regulator

The standard DC1 model of IEEE, is considered for the voltage regulator of each generator of the system.

d. Asynchronous generator equations

Wind generators are simulated mainly as induction machines with a sort-circuited double cage rotor. These induction machines are derived from synchronous machines, with the excitation winding short-circuited. Besides this, the machines are assumed to be perfectly symmetrical. The initial slip corresponds to the intersection of the electrical torque curve and the opposing mechanical torque, as shown in Fig. 11. The mechanical power is a linear function of the asynchronous wind generator speed:

$$P_m = T_m.\omega_r \tag{9}$$

In steady state conditions and in case of a disturbance where the wind remains stable, the mechanical power is assumed to be constant, therefore:

$$T_m.\omega_r = \text{constant} \tag{10}$$

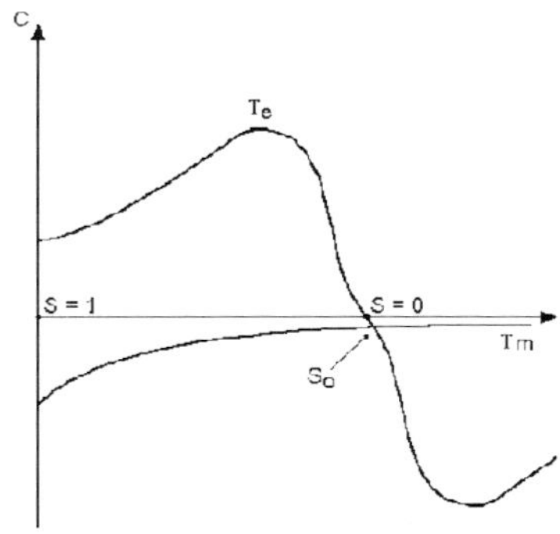

Fig. 11. Intersection of electrical and mechanical torque

e. Load equations

In general, power system loads are composed of a variety of electrical devices. For resistive loads, such as lighting and heating loads, the electrical power is independent of frequency. In case of motor loads, the electrical power changes with the frequency due to changes in motor speed. The overall frequency dependent characteristic of a composite load may be expressed as:

$$\Delta P_e = \Delta P_L + D\Delta f \tag{11}$$

ΔP_L, is the non frequency sensitive load change

$D\Delta f$, is the frequency sensitive load change

D, is the load damping constant

In the absence of a speed governor, the system response to a load change is determined by the inertia constant and the damping constant. The steady state speed deviation is such that the change in load is compensated by the variation in load due to frequency sensitivity.

4.3 Wind measurements

Although, many wind farms are under operation in the island of Crete, there is a lack of sufficient data from different wind farms. The data (time series of 10 minute time step) that are used in this study are derived by a wind farm of 20MW located in the east part of the island. The capacity factor of the wind farm is defined:

$$C.F = \frac{\overline{P}}{P_R} = \frac{E}{8760 \cdot P_R} \tag{12}$$

where, $\overline{P}$ is the mean power of a measured power time series, P_R the rated power of the wind farm, E is the annual energy production and 8760 the hours of a year.
The calculated annual capacity factor of the wind farm is 41.5%. The annual mean wind speed of the wind farm is defined by the annual series of data:

$$\overline{v} = \frac{1}{N} \cdot \sum_{1}^{N} v_n \tag{13}$$

where, v_n is the wind speed at data point n, and n=1,2,…,N is the number of measurement data.
The calculated annual mean wind speed is approximately 9.0 m/sec. The main wind direction is the North West. Then, standard deviation σ is calculated:

$$\sigma = \sqrt{\frac{1}{N-1} \cdot \sum_{1}^{N} \left(v_n - \overline{v}\right)^2} \tag{14}$$

The calculated value of standard deviation is 4.58. The power curve of the wind farm for the sixteen various wind directions was formulated given the collected data. The power curve of one wind direction is presented in Fig. 12.

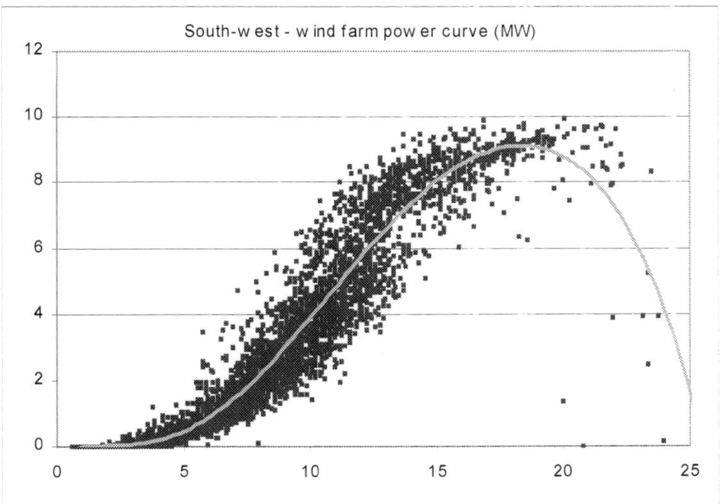

Fig. 12. Wind farm power curve and measures

The aim of this analysis was to record any sudden variation of the wind speed and of the power production of the wind farm. Two kinds of sudden variations were distinguished: "sudden loss" and "sudden blow" of the wind.

In the Fig. 13 and Fig. 14, the variation of the wind speed causes different variation of the produced power (case of power rejection and case of a sudden wind increase).

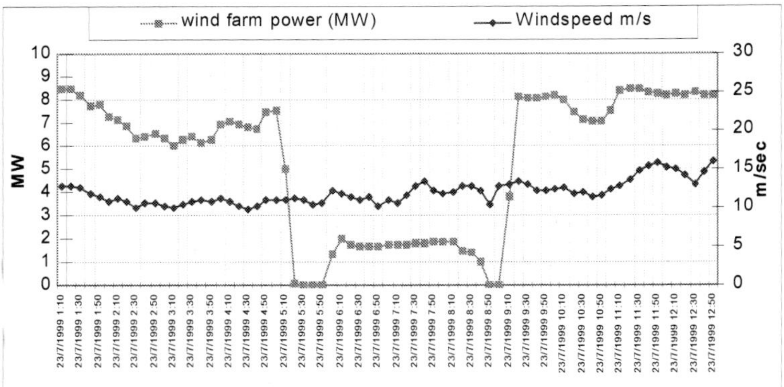

Fig. 13. Sudden power rejection

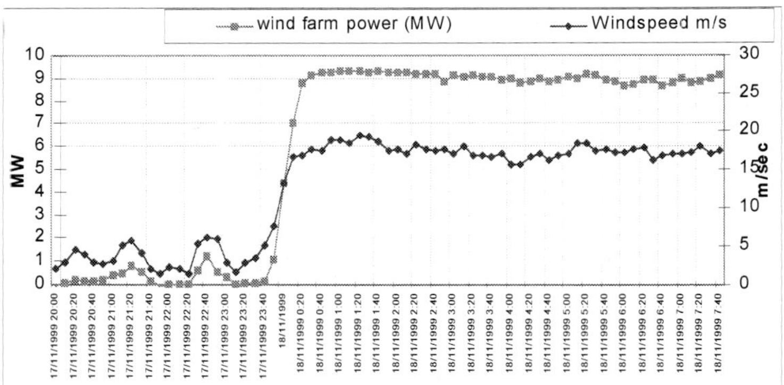

Fig. 14. Sudden increase of the produced wind power

It is obvious, that we are interested in variations of the produced power, which are caused by variations of the wind speed. The moving average of the wind farm power and the wind speed were calculated for short term (3 data points - half an hour) and medium term (12 data points - 2 hours) and then compared. When a significant deviation between the short term and the medium term moving average of the power was recorded and caused by a deviation of the wind speed, a "sudden variation" is occurred. During a "sudden blow" of the wind the short term moving average is bigger than the medium term, since the short term follows the wind speed closely. During a "sudden loss" of the wind the moving average of the short term is smaller than the medium term.

4.4 Dynamic security assessment

EUROSTAG program [Meyer; B & Stubbe, M. (1992)], PowerWorld Simulator [PowerWorld, (2007)] and Matlab [Power System Toolbox, (2006)] have been used for the simulation of the transient operation of the examined power system, under several operating conditions. Disconnection of conventional machines and wind generators as well as wind velocity fluctuations are the main disturbances under investigation. Especially, the following cases are presented:

a. Generator Trip

The system was examined for a case of power unit disconnection (Gas Turbine), which was producing 20MW. In Fig. 15 the change of the frequency and the diesel machine power in three different operating conditions, are shown. At first, the system is considered to operate without wind turbines and it seems to be quite stable. Secondly, the system is considered to operate with 28% of wind power, equal to 46MW and with the fast conventional units such as diesel machines and gas turbines to be in operation (fast spinning reserve). In this case, the system seems to be stable again. The lower value of the frequency is almost the same as in the previous case.

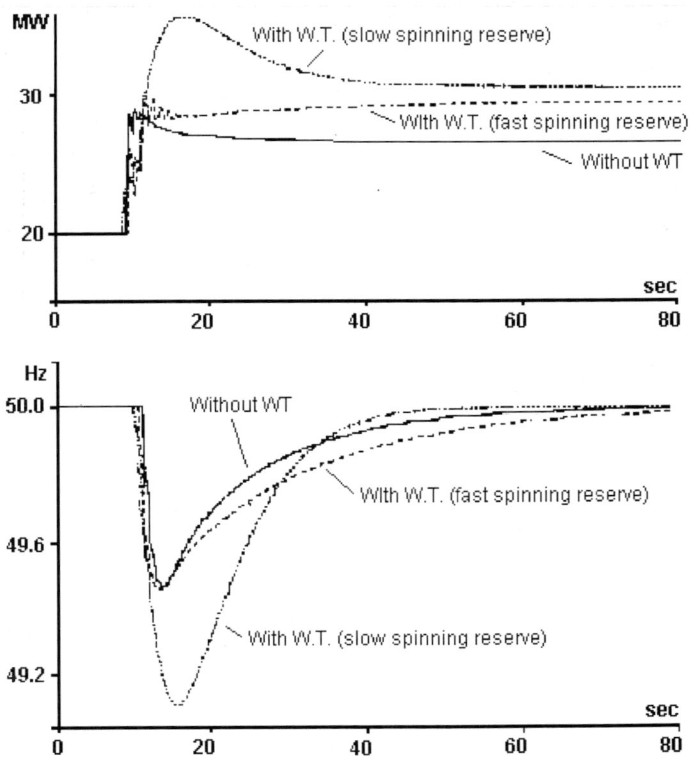

Fig. 15. Frequency and power change

Thirdly, the system is again considered to operate with the same high percent of wind power but with the slow machines, such as steam turbines, to cover the main spinning reserve (slow spinning reserve). In this case, the lower frequency value, which is equal to 49.14Hz, surely causes the operation of wind parks protection devices, leading the system to collapse after the total wind power disconnection. Therefore, it is obvious that in case of large wind power penetration, the operation of the diesel machines and the gas turbines is necessary for the dynamic security of the system.

b. Wind Power Change

In Fig. 16 the variation of the frequency and the voltage at the main wind park substation, are shown. The frequency follows the wind power changes, while the voltage profile follows an opposite trend. It can be seen that in case of normal wind power fluctuation, when the wind parks are not suddenly disconnected, and with sufficient spinning reserve, the power system remains satisfactorily stable.

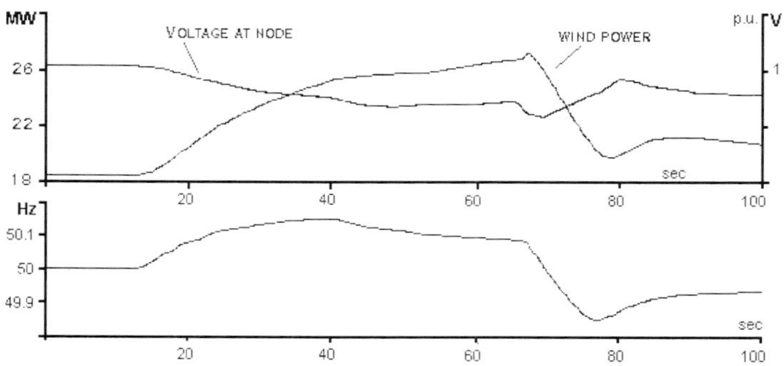

Fig. 16. Frequency and voltage variation

c. Unit Commitment Change

A maximum wind power penetration of 30% has been used by the system operators as the respective security margin. However, extensive transient analysis studies are conducted in order to assess the dynamic behavior of the system under various disturbances. Different combinations of the generating units have shown that a fixed security margin does not guarantee the system security and it distorts its economical operation. Thus, under the same contingency the system is shown to collapse with lower than 30% of wind power penetration, while survives with higher penetrations.

Fig. 17 depicts the change of frequency caused by the outage of a Gas turbine, providing 23 MW under two different operating conditions. Case 1 corresponds to a total load of 207.2 MW supplied as follows: 27 MW by Combined Cycle (18 MW of spinning reserve), 56.8 MW by the new Steam turbines (18.2 MW spinning reserve), 21.3 MW by Diesel (27.9 MW spinning reserve), 10.1 MW by the remaining Gas turbine (6.1 MW spinning reserve of maximum 16.2 MW), while the Wind power is 69 MW, corresponding to 33.3% penetration. It can be seen that the frequency undergoes a severe transient reaching a lowest value of 49.1 Hz, however the system restores its balance in about 50 seconds. Case 2 corresponds to a lower load of 199 MW supplied by 27.57 MW of Combined Cycle (17.43 MW spinning reserve), 69.3 MW of new Steam Turbines (5.7 MW of spinning reserve), 23.4 MW of Diesel (25.8 MW of spinning reserve), and 55.73 MW of Wind corresponding to 28% penetration.

Although the wind power penetration is lower than the security margin adopted the system does not manage to regain its stability and is led to frequency collapse. The difference is attributed to the fact that in the first case the spinning reserve is higher (70.2MW) and provided by faster units (Gas Turbines), while in the second case by slower units (48.93MW). The need for spinning reserve optimization can be clearly seen.

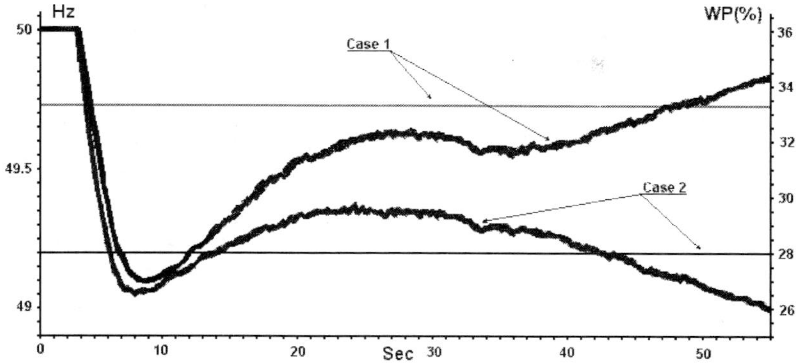

Fig. 17. Simulation results of Crete power system

5. Preventive dynamic security

In this paragraph a method for on-line preventive dynamic security of isolated power systems is presented, [Karapidakis, E.S. & Hatziargyriou, N.D. (2001)]. The method is based on Decision Trees which provide the necessary computational speed for on-line performance and the flexibility of providing preventive control. Emphasis is placed on the on-line use of the method to test the dynamic security of each generation dispatch scenario and thus to provide corrective advice via generation re-dispatch. Moreover, the algorithm implemented provides the flexibility of displaying the cost of each re-dispatch. In this way, the method can help in objective decision-making. Results from the application of the system on actual load series from the island of Crete, where the proposed system is in trial operation, are presented.

A dispatch algorithm approximating actual operating practices followed in the Control system of Crete is applied next in order to complete the pre-disturbance Operating Points (OPs). For a given load demand P_L and wind power P_W, the total conventional generation P_C is equal to:

$$P_C = P_L + P_{Losses} - P_W \quad (15)$$

P_C is dispatched to the units in operation, depending on their type and their nominal power. The various thermal units are grouped according to their type. The attributes characterizing each Operating Point comprise the active power and spinning reserve of all conventional power units. Ten variables are selected as initial attributes. Five attributes correspond to the active production of the conventional unit groups and five attributes to the spinning reserves, respectively. For each of the Operating Points produced, two characteristic disturbances have been simulated using:

- Outage of a major gas turbine
- Three-phases short-circuit at a critical bus near the Wind Parks.

The first of these disturbances happens very frequently, while the second is particularly severe leading to the disconnection of most wind parks. For each Operating Point the maximum frequency deviation and the rate of change of frequency are recorded. Both of these parameters are checked against the values activating the under-frequency relays used for load shedding and the OPs are labeled accordingly. The security criteria were:

If fmin < 49 Hz **And** df/dt > 0.4 **then**
The system is insecure **else** is secure

5.1 Secure economic dispatch

Economic dispatch analysis determines the power setpoints of the online generating units (15), so as to meet the system load and losses at least cost.

$$P_C = P_1 + P_2 + ... + P_i + ... + P_n \tag{16}$$

where P_C is the total conventional generation,
P_i is the generation of the i-th unit.
n is the number of units

Traditional dispatch algorithms tackle this problem as a constrained optimization problem and base its solution on the concept of equal incremental cost, also known as the Lambda Iteration algorithm: The total production cost of a set of generators is minimized, when all the units operate at the same incremental cost. In order to ensure that the operating setpoints proposed by the Economic Dispatch algorithm will provide a dynamically secure operating state of the system following pre-specified disturbances, the rules extracted by the relevant Decision Trees (if-then-else rules) can be used as additional constraints in the above constrained optimization problem.

5.2 Cost analysis

The presented approach provides the flexibility of displaying the cost of security, i.e. the cost associated with each re-dispatch. This is easily provided as the difference between the operating cost of the original dispatch and the operating cost of the secure re-dispatch. These costs can be calculated from the cost functions of the generating units, once the unit productions have been determined.

In addition, the security cost can be compared to the cost of load shedding. The unsupplied electric energy can be easily calculated from the operating settings of the under-frequency relays and the load forecasted at each bus affected. Alternatively, it can be estimated from the pre-disturbance load and the forecasted load as a whole, however its cost is more difficult to determine. For the dispatcher the cost of load shedding can be the price the regulator imposes for energy not served. In the traditional monopoly operation this cost can be the revenue lost due to the unsupplied electric energy, although this by no means reflects the true cost of load shedding. In any case, the total cost can be calculated from:

$$S_L = C * \int_{t=0}^{T} P_L(t)dt \tag{17}$$

where: P_L is the load shed.

C is the cost of kWh in Euros (€).

T is the time of load disconnection.

5.3 Cost of security

In this paragraph results from the application of the secure economic dispatch algorithm on actual load series of Crete are presented. In Fig. 19, the total load, the corresponding security classification (1 for secure and 0 for insecure) for the machine outage contingency and the operating cost in Euros of a characteristic day are plotted. In the upper diagram, it is shown that, approximately between 9:00 and 10:30, the system is insecure, i.e. at least a significant load shedding will take place. In the lower diagram, the effects of the secure economic dispatch algorithm on the security classification and the system operating costs are shown. The increase of costs during the previously insecure period, provided by the increased and probably faster (more expensive) spinning reserve, is notable. The effect of the two dispatch scenaria on the system frequency deviation, in the case of the machine outage, as obtained by simulation programs, is shown in Fig. 20. It is clearly shown that the proposed re-dispatch will not cause load-shedding. The probability of the contingency occurrence however is not considered in this study.

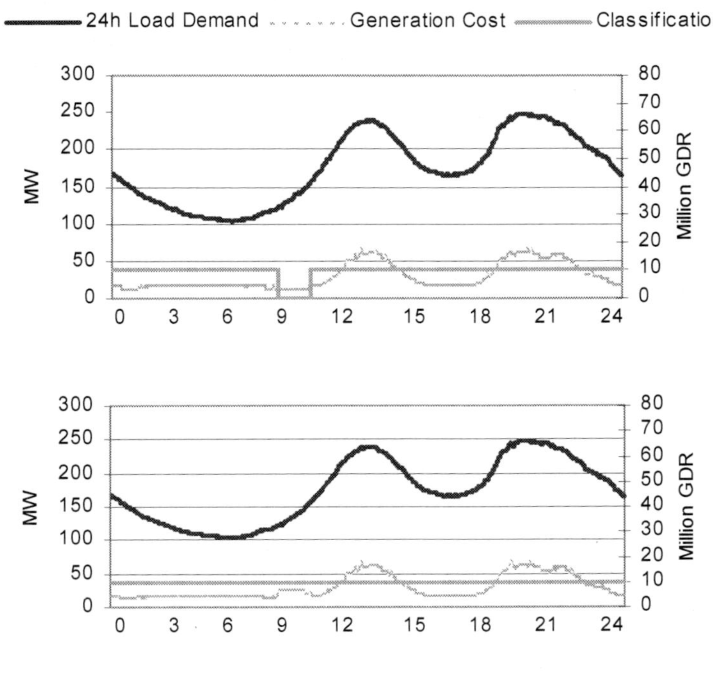

Fig. 19. 24-hour diagrams illustrating load, security classification and operating cost

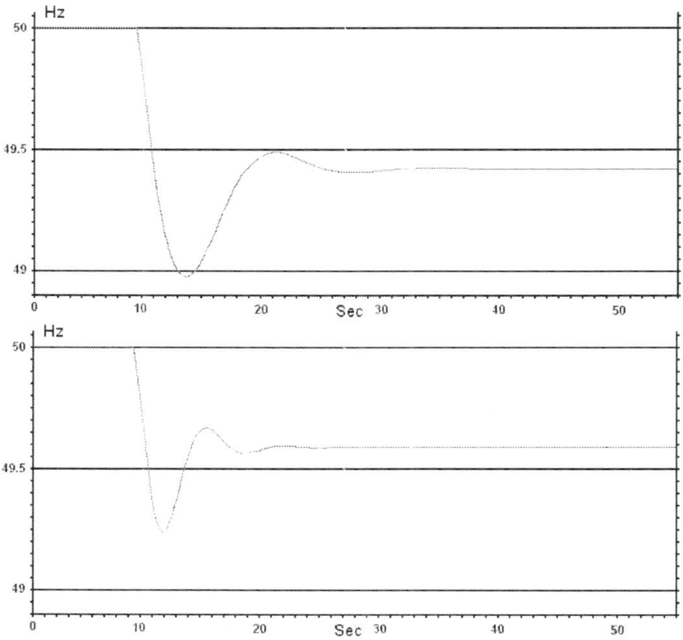

Fig. 20. Effect of dispatch on system frequency deviation

7. Conclusion

In this chapter the dynamic behavior of a power system with high percentage of wind power penetration (up to 40%) was studied with emphasis given to the modeling of the system, in order to examine the probable impacts. More precisely, several simulations were performed to study the impact of the wind park on the dynamic behavior of a representative autonomous power system as Crete's power system. The most considerable disturbances that were invastigated are the short circuit, the sudden disconnection of conventional power units as well as wind parks and the strong wind velocity fluctuations. Simulations have shown that the deviations of the power system voltage and frequency remain acceptable under most examined perturbations. However, the situation depends on the scheduling of the power units and the amount of allocated spinning reserve.

Cause to significant replacement of conventional power generation that was supplied by synchronous generators, with wind turbines that operate either asynchronous or variable-speed generators, the dynamic performance of the power system will indeed be affected. Thus, although wind turbines affect the transient stability of a power system, they are not a principal obstacle to an adequate secure and reliable operation. The stability of a power system can be maintained even if high penetration of wind power exist by additional system measures, control enhancement and preventive actions.

Finally, a method for on-line preventive dynamic security is proposed, in order to determine optimal reserves and to provide corrective advice considering dynamic security. Based on

the Decision Trees classification new unit dispatch is calculated on-line, until a dynamically secure operating state is reached. This technique provides the flexibility of displaying the cost of each proposed solution weighted against the cost of load shedding; it forms therefore the basis for valuable decision-making aid. Results from the application of the method on actual load series from the island of Crete show the accuracy and versatility of the method. Moreover, the fast execution times required for on-line classification of the current operating state make the method suitable for large systems, as well.
Therefore, there is a considerabe impact of wind parks generation to the power system that they are embeded. This impact is generally propotional to the wind power penetration percentage (running active power injection and/or reactive power absorbedness). Furthermore most of the perturbations exclusively due to the operation of the wind generators do not affect significantly the operation of the power system. Concluded it should be noted that it is possible to operate a power system with a high level of wind penetration maintaining a high level of security. This is possible, if adequate spinning reserve of the conventional units is available. The issue of spinning reserve is particularly important; therefore it must be further investigated.

8. References

AIEE Subcommittee on Interconnections and Stability Factors, (1926). First report of power system stability, AIEE Transactions, 1926, pp. 51-80.

Arrilaga; J. & Arnold, C.P. (1993), Computer Modeling of Electrical Power Systems, John Wiley & Sons, 1993.

Burton; Tony, Sharpe David, Jenkins Nick, Bossanyi Ervin (2001). Wind Energy Handbook, John Wiley & Sons, Chichester, UK, 2001.

CIGRE Study Committee (1998), Impact of Increasing Contribution of Dispersed Generation on the Power System, Final Report WG 37-23, N.37, 1998.

Crary; S. B., Herlitz, I., Favez, B. (1948). CIGRE SC32 Report: System stability and voltage, power and frequency control, CIGRE, Appendix 1, Rep. 347, 1948.

Dialynas; E.N., Hatziargyriou, N.D., Koskolos, N.C., Karapidakis, E.S. (1998). Effect of high wind power penetration on the reliability and security of isolated power systems, CIGRE Session, Paris, 30 August 1998.

Doherty; R. and O'Malley, M.J. (2006). Establishing the role that wind generation may have in future generation portfolios, IEEE Transactions on Power Systems, Vol. 21, 2006, pp. 1415 - 1422.

Hatziargyriou; N. & Papadopoulos, M. (1997). Consequences of High Wind Power Penetration in Large Autonomous Power Systems, CIGRE Symposium, Neptun, Romania, 18-19 September 1997.

Hatziargyriou; N., Karapidakis, E., Hatzifotis, D. (1998). Frequency Stability of Power Systems in large Islands with high Wind Power Penetration, Bulk Power Systems Dynamics and Control Symposium - IV Restructuring, Santorini, August 24-28 1998.

Hatziargyriou; N.D., Papadopoulos, M., Tentzerakis, S., (1997). Control requirements for optimal operation of large isolated systems with increased wind power penetration, EWEC, Dublin, Ireland, 6-9 October, 1997.

Karapidakis; E.S. & Hatziargyriou, N.D. (2001). On-Line Preventive Dynamic Security of Isolated Power Systems Using Decision Trees, IEEE Transactions on Power Systems, Vol. 17, No. 2, May 2002.

Karapidakis; E.S. & Thalassinakis, M. (2006). Analysis of Wind Energy Effects in Crete's Island Power System, 6th International World Energy System Conference, Turin, Italy, July 2006.

Kundur; P. & Morison, G.K. (1997). A Review of Definitions and Classification of Stability Problems in Today's Power Systems, Panel Session on Stability Terms and Definitions IEEE PES Meeting, New York, February 2-6, 1997.

Kundur; Prabha, Paserba John, Ajjarapu Venkat, Andersson Göran, Bose Anjan, Canizares Claudio, Hatziargyriou Nikos, Hill David, Stankovic Alex, Taylor Carson, Thierry Van Cutsem, and Vittal Vijay (2004). IEEE Transactions on Power Systems, Vol. 19, No. 2, May 2004, pp.1387-1401.

La Scala; M., Trovato, M., Antonelli, C. (1998). On-line Dynamic Preventive Control: An Algorithm for Transient Security Dispatch, IEEE Trans. on PWRS, Vol. 13, No. 2, May 1998, pp. 601-610.

Meyer; B., Stubbe, M. (1992). EUROSTAG: A Single Tool for Power System Simulation, Transmission and Distribution International, March 1992.

Nogaret; E., Stavrakakis, G., Kariniotakis, G. (1997). An advanced control system for the optimal operation and management of medium size power systems with a large penetration form renewable power sources, Renewable Energy, Vol. 12, No. 2, Elsevier Science, November 1997, pp. 137-149.

Power System Toolbox; (2006). User's Guide, MATLAB 7 Package.

PowerWorld; (2007). PowerWorld User's Guide, PowerWorld Corporation, Simulator Version 13, 2001 South First Street Champaign, IL 61820.

Smith; P., O'Malley, M., Mullane, A., Bryans, L., Nedic, D. P., Bell, K., Meibom, P., Barth, R., Hasche, B., Brand, H., Swider, D. J., Burges, K., Nabe, C., (2006). Technical and Economic Impact of High Penetration of Renewables in an Island Power System, CIGRE Session 2006, Paper C6-102.

Steinmetz, C. P. (1920). Power control and stability of electric generating stations, AIEE Transactions, vol. XXXIX, Part II, July 1920, pp. 1215–1287.

Stoft; S. (2002). Power System Economics, IEEE, Wiley Interscience Publication, Piscataway NJ, 2002.

Strbac; G. (2002). Impact of dispersed generation on distribution systems: a European perspective, Power Engineering Society Winter Meeting, vol.1, 2002, pp. 118-120.

Weedy; B. & Cory, B. (1998). Electric Power Systems, John Wiley & Sons, Chichester, UK, 1998.

18

Wind Power: Integrating Wind Turbine Generators (WTG's) with Energy Storage

Septimus van der Linden
BRULIN Associates, LLC
U.S.A.

1. Introduction

Energy Storage is the missing link between wind driven power generation and delivering power in a sustainable manner that can be dispatched at times of high demand from the grid. Transmission systems that cover large territories such as in North America are particularly vulnerable, requires additional dedicated transmission and readily dispatcheable backup power systems.

The installed capacity of Wind Turbine Generators (WTG's) in the US and worldwide, while impressive, suffers from a low capacity factor of 30% or less due to the variability and intermittency of wind as the motive force. In 2007 the global installed capacity was 94 GW with a predicted capacity of 136 GW by 2010, 55% would be installed in Europe and 23 % (31 GW) in North America, these numbers could be exceeded, as the US already has over 29 GW installed capacity with 99 GW in planning in the next 10 years.

The demand for electricity has considerable daily and seasonal variations and the maximum demand may only last for a few hours each year. As a result, some power plants are required to operate for short periods each year - an inefficient use of expensive plants. Without any additional storage above the present 2.5%, mainly PHS, of the installed base load in the USA, base loaded plants are being detrimentally cycled at higher frequency and the situation is further exacerbated by the latest growing demand for renewable energy such as wind energy. In the US, this capacity has now reached in excess of 29,000 MW [Fig 1] summarized by the American Wind Energy Association (AWEA) projects; in Canada the current 2800 MW projects under consideration or contract will grow to 7400 MW to meet energy objectives set for 2015.

Installing larger wind farms, to cover the deficiency of a higher capacity factor, results in high costs per delivered kW/hr. This requires continued tax incentives to deliver "green" energy to the consumers. The full capability of the WTG is never realized, as at high wind speeds, some of the wind energy has to be "spilled" to maintain a smooth delivery profile.

Technology improvements have not overcome the "wasted" capacity of these modern marvels except where Hydro or Pumped Hydro Storage (PHS) facilities are utilized. The Hydro power station can compensate for wind variability and intermittency while PHS provides energy storage and delivers power during high demand periods. Wind Energy Storage results in a much higher capacity factor, in effect reducing the cost of delivered kW/hrs., PHS amounts to less than 2.3 % of the current installed 1000 GW generating capacity and will decrease with the increasing addition of wind generation.

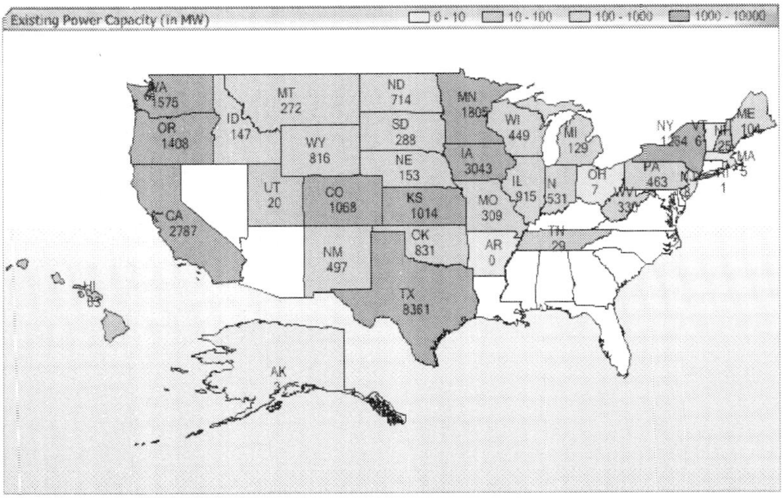

Fig. 1. Installed US Capacity 2009 by State (AWEA)

2. Decoupling energy production from supply

Storage allows energy production to be de-coupled from its supply, self-generated or purchased. WTG's can only receive energy payments for delivered power, requiring the installation of Gas Turbines or cycling of thermal plants to provide capacity that cannot be delivered by wind. The wind generation variation vs. daily demand requirement is illustrated in Fig. 2.

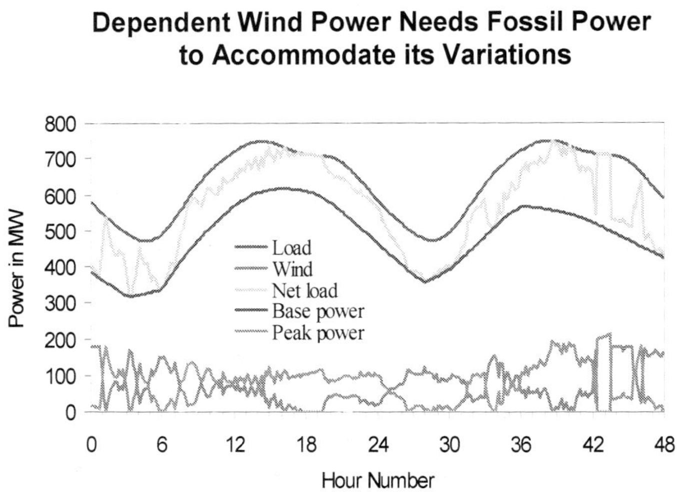

Fig. 2. Wind Energy not available during peak

The problem with the proven bulk energy PHS solution is that the USA or the worldwide installation of WTG's do not have such facilities readily available (some exceptions in Europe), are expensive to construct and difficult to permit in the USA.

A readily available, cost effective alternative bulk-energy storage technology is ready for deployment. The Gas Turbine-Compressed Air Energy Storage (GT-CAES) concept incorporates a standard production GT with CAES technology and so covers a wide range of power production that can be matched to specific storage sites. During excess wind power production or nighttime wind, this power is used to drive air compressors to pump up or pressurize storage facilities such as salt caverns, deep aquifers (depleted natural gas wells) or above ground storage tanks (Pipelines). The stored compressed air is released to an air expander to recover the stored energy. The air to the expansion turbine is pre-heated to 510 oC to 565 oC using the Gas Turbine exhaust energy recovered in a Heat Recovery Unit (HRU). The Gas Turbine low exhaust emissions are reduced further with Selective Catalytic Reduction (SCR) in the HRU. Adiabatic expansion without pre-heating the air before expansion is another possibility.

The electric motors driving the air compressors are large for Bulk Energy Storage facilities, and can absorb large and varying quantities of wind generated power and thus regulate the delivered kW/hrs delivered during peak demand, or store the excess power during low grid demand. Wind as a renewable resource would be able to deliver a larger percentage of "green" capacity with the ancillary power benefits of Storage such as Voltage Regulation, load following, spinning reserve, etc., not a feature of WTG's. Smaller capacity systems of 3 to 30 MW/hrs serve a different purpose for smaller wind farms, primarily in a "smoothing" function of decoupling for power delivery and meeting short duration peak hour generation Fig 3 illustrates the basic concept and motivation for Energy Storage.

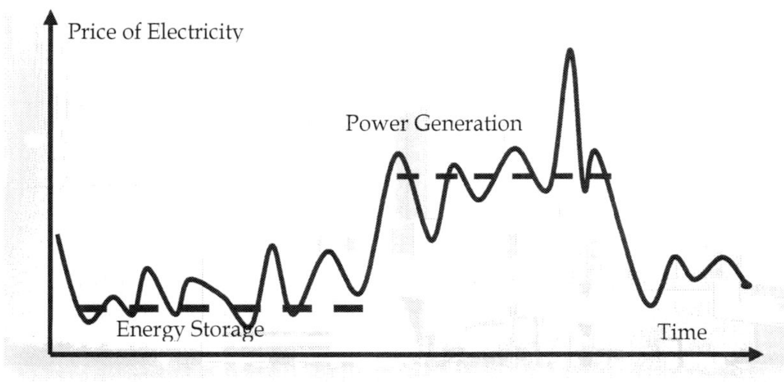

Fig. 3. Motivation for electrical energy storage

The different storage technologies illustrated [Fig 4], can be used in different combinations, to suit the specific needs of the power delivery system, not only in plant output capacity but in response times as well. Response systems such as Flywheels or Flow Batteries (seconds or milliseconds) can be combined with larger bulk systems (minutes and hours) such as CAES or with SSCAES (small surface storage), 60MW/hr systems or larger 135 MW units in several configurations up to 1000 MW or more, depending on storage cavern volume.

By having large-scale electricity storage capacity available over any time, system planners would need to build only sufficient generating capacity to meet average electrical demand rather than peak demands. Fig. 4 shows a multitude of smaller short duration devices with quick response discharge times, this is where a lot of the research and development was focused and does not accommodate large wind contribution. The emphasis today has to be on large scale systems such as pumped hydro and compressed air energy storage to fully integrate the growing installed wind capacity.

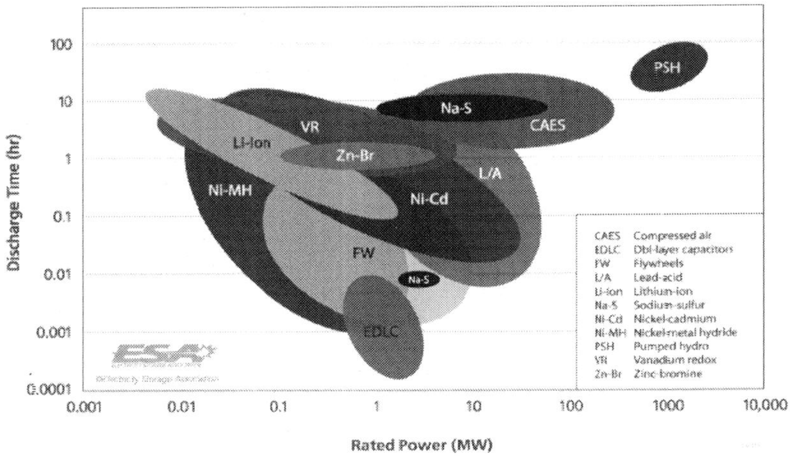

Fig. 4. Operating Regimes for Several Energy Storage Systems (ESA)

In theory, a typical power plant could operate with 40% less generating capacity than would otherwise be required when supported by Energy Storage. This represents considerable financial savings in peaking and intermediate plants. Additional reductions in emissions and capital investment can occur due to the base load generators operating more efficiently at steady state output. The wind energy can be stabilized as well as increased in capacity toward the nameplate rating. Grid instability does lead to regional blackouts. This does open the door for more consideration of Energy Storage. While this is encouraging, there are institutional hurdles to overcome, one of which is the lack of understanding of the value and benefits of Bulk Energy Storage as well as some perceived concepts that simply adding more new power plants and transmission capability will cure blackout problems experienced in recent times in the USA. Storage is probably the better solution!

Storage of electricity (energy) will significantly change the Power Industry for the better: better utilization of resources, better system efficiency, lower emissions, better reliability and security. Geologically suitable identified sites for bulk energy storage using salt domes, hard rock or aquifers can be readily exploited for 20/30 GW capability by 2020 or sooner, a fact not fully recognized by power entities. (van der Linden, Septimus, 2006)

3. How does a CAES system work?

The fundamentals of a Gas Turbine are well understood: atmospheric air is compressed to a higher pressure, fuel is added in a combustion chamber and the hot, high pressure combustion gas expands through a turbine that provides both the motive power for the compressor (60% or more) and the balance of the power (40% or less) as mechanical energy to drive an electric generator.

In a CAES cycle variation of a standard gas turbine, the compression cycle is separated from the combustion and generation cycle; by using low cost, off-peak or excess electricity, motor driven inter-cooled compressors provide the compressed air held in storage to be released from storage to the modified gas turbine for power generation on demand. In this process, some dramatic changes in the power and economic cycles have occurred. The gas turbine expander absent of its large parasitic load delivers approximately two thirds more power with no increase in fuel consumption. The required compressed air comes at a much lower cost thus enabling lower cost of electricity generation during high demand cycles from other intermediate load systems, in particular the increasing renewable energy mandates and others such as Gas Fired Thermal or Combined Cycle power plants, or even the lower cost Simple Cycle gas turbine power plants. The illustration Fig 5 below will help clarify the CAES concept.

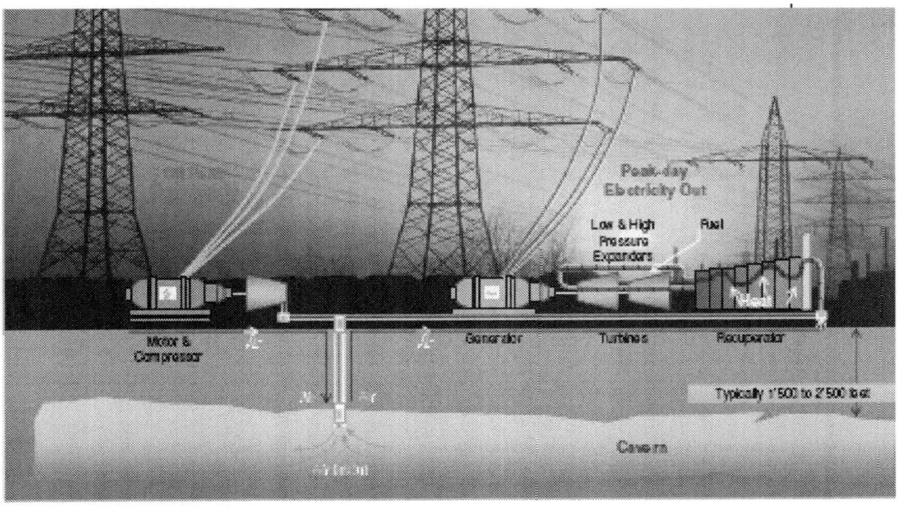

Fig. 5. CAES Concept

The Compressors utilize off peak wind energy to store high pressure air in the storage cavern, which is expanded to generate power when there is a demand during the day; this diurnal wind energy as depicted in Fig. 6 brings maximum wind capability to the grid.

4. CAES technology: storage concepts

Decoupling the Compressor trains from the generating train allows for more flexibility in compression optimization and utilization. Motor driven compressors in 50 MW or lesser increments allow sites and storage volume to best serve the transmission grid needs as well as act as load sinks of 100/200 MW or 300 MW to avoid unnecessary cycling at base loaded plants. The illustration Fig. 7 below captures the decoupling of compression from the power

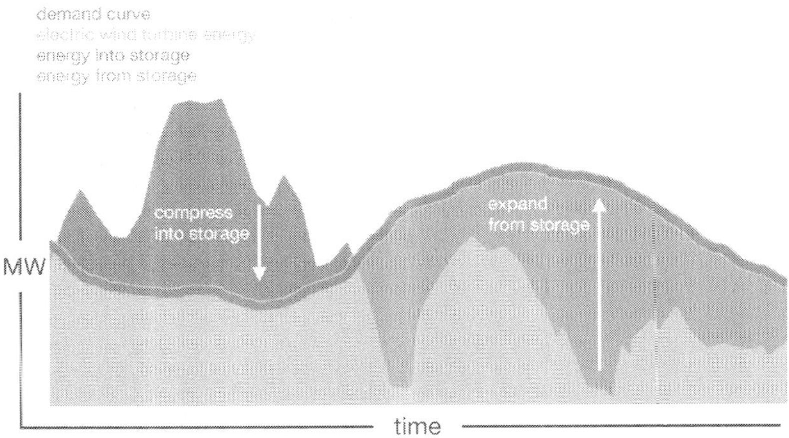

Fig. 6. Meeting the demand curve with stored wind energy (General Compression)

cycle. The Huntorf, Germany and the McIntosh CAES plant compressors are driven by the Generator which operates as a motor during the compression cycle. This limits the flexibility of the plant which today allows compression power to be designed with multiple units to accommodate different rated compression units to balance the wind variability. This optimization allows a larger range of options for different geographical regions with wind variances.

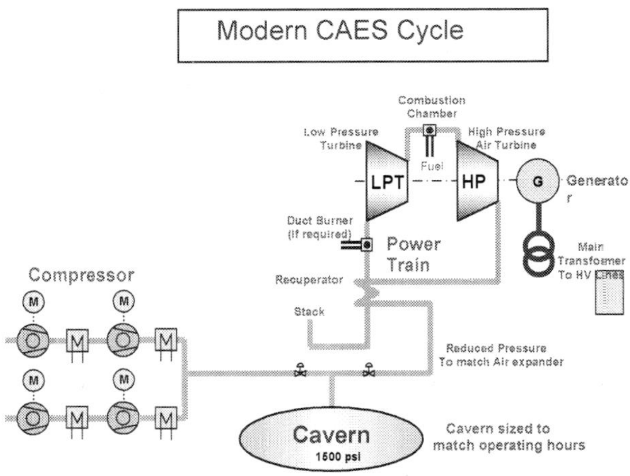

Fig. 7. Decoupling the Compression from Power (Alstom Power)

5. Applications

Stored energy integration into the generation-grid system is best illustrated in (Fig. 8) "Energy Storage Applications on the Grid". This covers a wide field in every aspect of generation-transmission and distribution. The ability of the various technologies to react quickly, converting the stored energy back to electricity readily provides three primary functions: Energy Management (hours of duration) load leveling or peak period needs; Bridging Power (seconds or minutes duration) assuring continuity of service, contingency reserves or UPS (Uninterruptible Power Supply); and Power Quality & Reliability (milliseconds or seconds duration) in support of manufacturing facilities, voltage and frequency controls.

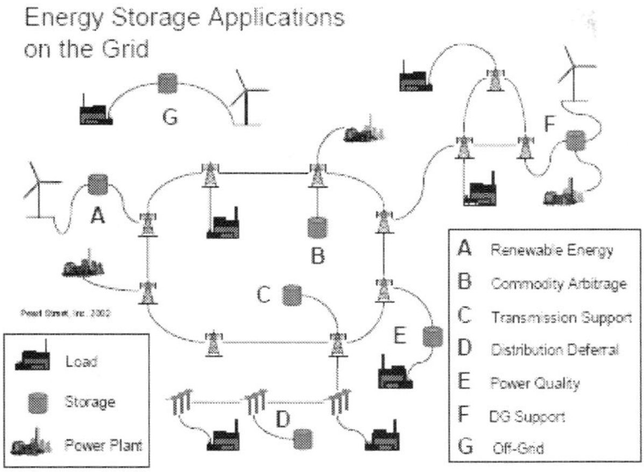

Fig. 8. Energy Storage Applications on the Grid (Pearl Street Inc)

This storage pipe concept could be applied to existing GT/CC plants. Increasing the hot day output 20/25% by injecting the stored air into the combustors with or without humidification (Fig. 9). By applying the humidification concept, the air supply in a CAES plant could reduce the required storage volume by 30% or more, or increase the operating hours by 30% of the specific cavern storage volume. (Nakhamkin, Michael. et al, 2004)

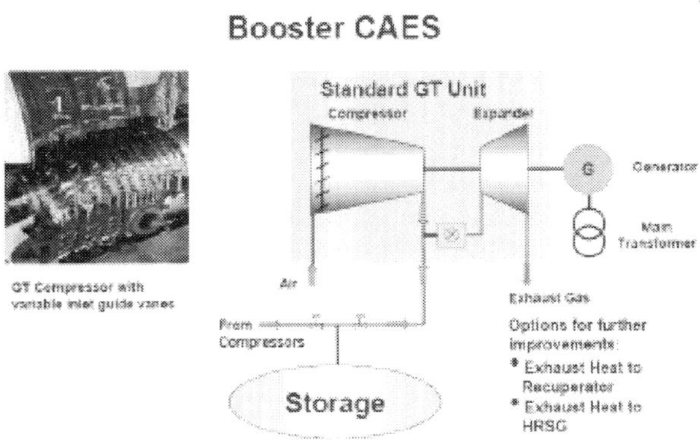

Fig. 9. Air Injection (AI) or booster CAES (Alstom Power)

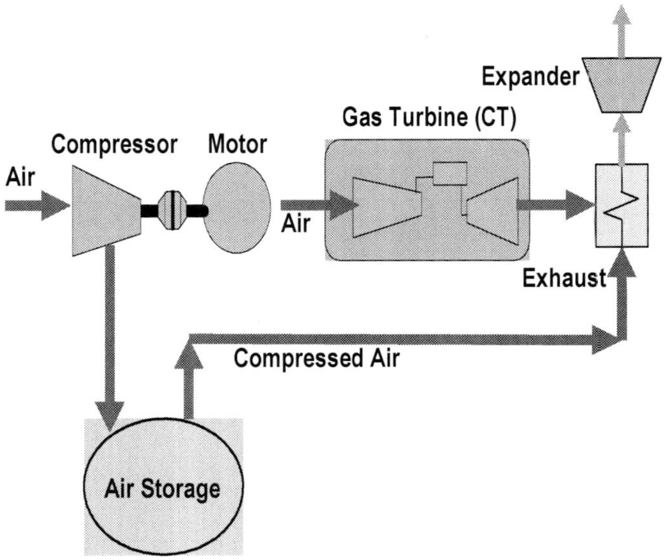

Fig. 10. GT-CAES with production GT (ES & Power LLC.)

In another Hybrid proposed concept, a conventional gas turbine is coupled with storage and a separate unfired air expander for increased flexibility of operation. Using a 180 MW gas turbine, the plant output would exceed 400 MW. (Fig.10). The advanced technology gas turbine with 38% efficiency can be operated independently when the cavern air supply has been drawn down. (Nakhamkin, Michael .et al, 2000)

The separate expander (bottoming cycle) allows stored wind "green" energy to remain clean without products of combustion. Systems of 100 MW supported by a 45 MW Gas Turbine is another of several size options using available production gas turbines rather than specially designed Combustion expanders. The first unit in Huntorf Germany (290 MW) and the first unit in McIntosh Alabama (110MW) have high and low pressure sequential combustors, and inline motor/generator driven compressors.

Advanced concepts of **Adiabatic** Compression & Expansion, requiring Thermal Energy Storage (TES) have been studied in the US, but more recently in Europe. Such systems would ideally benefit renewable energy systems such as wind, solar and biomass, adding capacity with no premium fuel consumption.

Diabatic CAES plant loses heat from the compression cycle which must be re-generated or added to the compressed air before entering the turbine expansion cycle. Adiabatic CAES will benefit from the thermal energy storage to preheat the stored air which will expand adiabatically through a sliding pressure air turbine, with the added benefit that no CO2 is generated in the process. Such studies have been completed in Europe with 19 different partners with support and involvement of the European Commission through a research contract. (Bulloch, Chris. et al, 2004)

Thermal storage devices such as the "Cowper" heat storage devices in glass and metallurgical industries were investigated for suitable thermal storage solutions. The study detailed concept sizes of 30, 150 and 300 MW respectively. (Fig.11) From the study it is concluded that the Dutch electricity market is the most promising for mass storage whereas, less promising, but not ruled out, are Italy, Norway, Sweden France, the Alpine countries with PHS will be exceptions. A series of economic calculations using the national spot market prices for the years 2001 and 2002, and a range of storage capacities, demonstrate that the opportunities are greatest on the Dutch market, with a plant storage capacity of about 3000MWh. The increased fuel costs in 2009 and higher equipment costs for WTG's will change the dynamics for AA-CAES storage/generation.

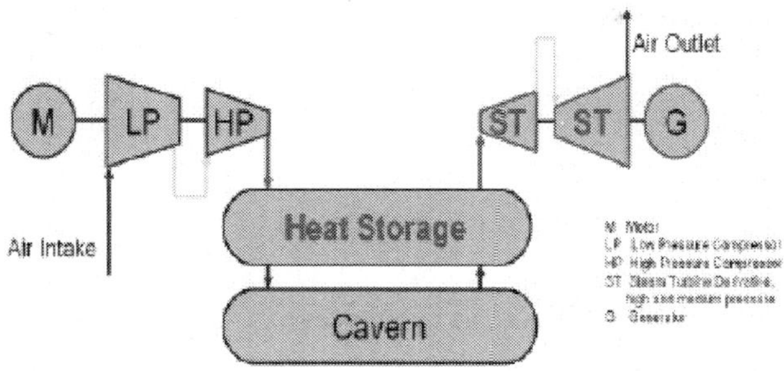

Fig. 11. Advanced Adiabatic CAES Concept (Alstom Power)

Smaller adiabatic systems suitable for isolated Wind Turbine systems where no fuel is added are under development, with utilization of the cold exhaust air to be used for cold storage systems or advanced concepts of "freeze" desalination. Such units of 500 kW and larger are ideal for wind power "smoothing" and distributed generation. The T-CAES 500kW system can produce 3600 liters/hr of fresh water from seawater or saline/brackish water. (Dr. Ben Enis et al., 2006)

The European wind resources and potential are substantial and cover a large area for suitable wind /storage integration, other than the current Hydro plant and PHS already in operation (Fig. 12).

6. Benefits from energy storage

One of the first benefits would be to fully utilize capital assets, considering that the national average for generation capacity factor is 58/60% and transmission 50/52%. Bulk Energy Storage will allow the most efficient units to be fully utilized and allow optimization of the generation mix. The integration of ever increasing renewable sources such as wind with energy storage will bring a larger contribution to the Generation mix. Furthermore, it will avoid the use of inefficient units using premium fuels during peak periods. Needle peaks can be readily met with storage as the distribution level or with current installed "peaker" unit capacity.

The market or economic benefits from Energy Storage can be quantified in four major areas of the electricity supply chain, namely: generation, transmission & distribution, energy services, and renewable energy storage. Projected benefits over a 15 year period for the USA Generation and T&D system could exceed $100 Billion. Other benefits of Wind Storage are reducing water consumption, CO2 reduction, Ancillary Service Value and Transmission Value as part of the value chain illustrated in Fig. 13.

Fig. 12. West Europe wind speeds Salt deposits and active salt facilities (Courtesy KBB)

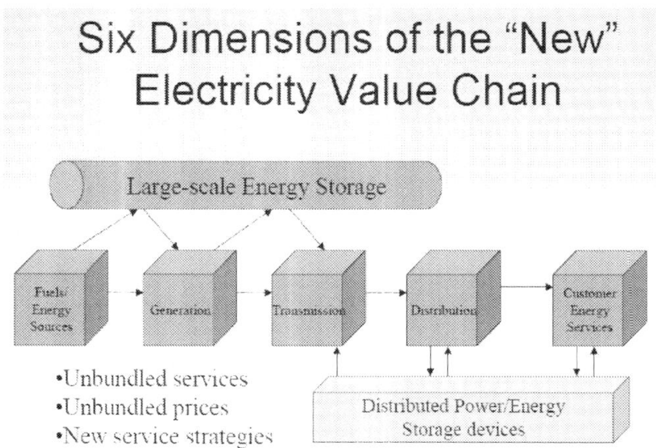

Fig. 13. Large-scale energy storage the "new" Electricity Value Chain 6th dimension (Pearl Street Inc)

Close to 90% of all new U.S. generation capacity added since 2005 has been a combination of natural gas and wind power. The U.S. electric industry faces dramatic transformations as it wrestles with the challenges of the 21st century. The capacity factor of wind requires that 3 MW is installed to displace 1.0 MW of base load coal power, and subsequently backed by gas fired power plant. This is a clarion call for integrating wind with storage technologies increasing the clean mix of renewable and flexible technologies.

Large-scale storage is the 6th dimension in the Electricity value chain which can bring new possibilities to the Utility industry with the growing mandates for 20/30%power generation from renewable energy in particular wind energy. Note that this does not exclude Distributed Power Energy Storage devices as illustrated in Fig. 13.

7. Water consumption

CAES can increase clean water and more of it: Wind power does not require water, nor does CAES plant. If wind were to provide 20% of our base load electricity by 2030, using Energy Storage technologies, water use by the electricity sector would be cut by 17% in that year. Water is a precious commodity which current fossil plants demand in high quantities for cooling. Many new power plants are using air cooled condensers to conserve water especially in California and Nevada. The Western United States face critical water issues today and renewable energy sources such as concentrated solar power (CSP) now in construction in the high solar radiation areas will have water curtailment for cooling, requiring dry cooling with performance degradation. CAES avoids water use entirely.

8. CO2 reduction

CAES contributes to increasing the CO2 reduction contributed by wind energy displacing fossil power generation, assuming zero CO2 emissions for wind whereas Coal would produce 974 Tonnes CO2/GWe/h and gas fired plant 464 Tonnes CO2/GWe/h. Increasing wind energy contribution from variable and unpredictable to a dispatcheable base load contribution from a capacity factor of 30% to 55 % or higher would significantly mitigate CO2 issues the power sector faces. The GT/CAES has a heat rate of 4010 kJ/kWh and that of a CC power plant 6500 kJ/kWh (lower heating value) LHV for comparative purposes a 40% improvement over CC and a 64 % improvement an efficient open cycle fast start and load aero-derivative gas turbine used to supplement short coming of wind generation due to variability of wind. The CO2 reduction of CAES/Wind integration is significant factor in the overall economics considering the CO2 sequestration programs being promoted.

9. Ancillary service value

The rapid response of a CAES plant capability allows it to provide automatic generation control/regulation in both generation and compression modes. CAES plant provides spinning reserve; a CAES plant with independent compression/generation trains can bring the full generation capacity online in less than 10 minutes. CAES plant can also be considered to be providing quick start or operating reserve with the ability to rapidly shed load while in compression mode, or ramp up while in generation mode. CAES can also provide other ancillary services such as balancing energy and voltage/VAR support.

10. Transmission value

Strategic location of a CAES plant may be valuable from a transmission perspective. When located in an area where wind production is congesting the local transmission lines the CAES plant has the potential to significantly reduce or eliminate congestion and increase grid efficiency by storing the wind energy and releasing it when the wind plants are at a low output and more transmission line capacity is available. The CAES plant has reactive capability in both the generation and compression modes, and when configured with

clutches it can operate in the synchronous condenser mode, to provide and absorb reactive power as needed.
A CAES plant can provide voltage reliability benefit to the transmission grid. Furthermore CAES plant reactive support can be particularly useful when combined with wind plant operation, especially as wind plants have been manufactured with limited reactive capability and installed on weak areas of the grid.
Conventional power plants have more reactive power capability and voltage stability, so when displaced by wind power these abilities are substantially diminished , the negative effects of wind on the transmission grid can be avoided with CAES plant acting as a dynamic Reactive System .

11. Future prospects (developments)

Pumped Hydro has clearly demonstrated the value of Bulk Energy Storage. While these benefits are recognized and utilized, new facilities have languished; projects in development do show promise and opportunities for implementation. The requirement for efficient Clean Coal concepts such as IGCC (gasification) can be enhanced with storage systems to keep the plant at an 80% or better load factor during the off-peak demand periods and deliver the added stored capacity during high demand.
New concepts are being proposed especially with the growing capacity of wind energy, currently backed by tax incentives. However, at 29 GW and projected substantial growth, energy storage and wind energy integration using CAES or Flow Batteries or ganged Flywheels could lead to better economic utilization of a substantial resource operating at below 30% capacity factor - storage could drive this capacity factor to 65% or higher.
Concepts outlined in a paper presented at EESAT 2003 Conference (van der Linden, S., 2003) suggested sub-surface storage using large diameter pipes such as typically used for natural gas transportation. Using a storage complex of 2000 meters of pipe, a system that will provide 60 MW/hrs (15 MW x 4hrs) could enhance power supply at remote wind farms. Introducing such smaller systems will help the industry gain confidence in the value of energy storage, gain operational experience without large expenditure. Smaller capacity systems of 3 to 30 MW/hrs serve a different purpose for smaller wind farms, primarily in a "smoothing" function of decoupling for power delivery and meeting short duration peak hour generation. (Dr. Ben Enis et al., 2006) Other concepts proposed would transmit stored air by pipeline to industrial areas or to bring power to where existing transmission is available. Permitting for new power lines is constantly challenged by environmental groups. Pipelines to transmit the stored energy would help bring more renewable energy into the demand cycle.
The overall principle of operation of the T-CAES system is depicted simply in Figure 14. Note that if the system efficiency is about 50%, then the excess power versus time area is able to deliver either the same excess power in half the time (First part of Figure 14, early time history) or half the excess power for twice the time (Second part of Figure 14, later in the time history).
In order to discuss the performance of the T-CAES system, it is necessary to consider several simplifying assumptions to readily demonstrate trends. First, wind history has 24 identical successive 1-hour periods for each 24-hour day. This assumption produces the most effective volume tank. Second, average power of 46.9% was used compared to the peak power in selecting the shape of the power history sinusoid. Third, the nighttime power

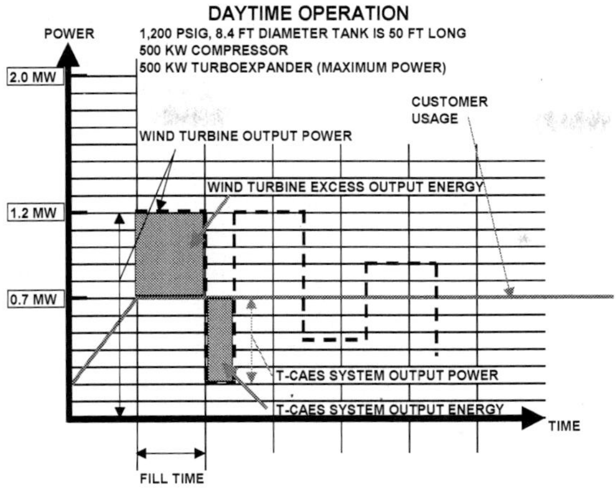

Fig. 14. Overall principal of T-CAES system (Enis Windgen)

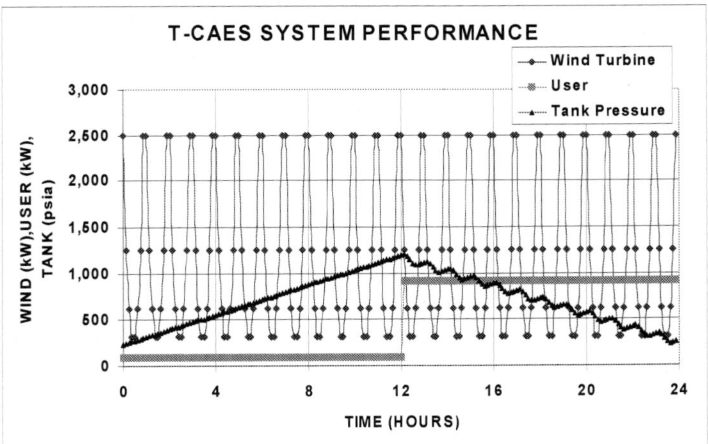

Fig. 15. Electric Power and Storage Histories (Enis Windgen)

requirement is a fraction of the daytime power requirement. Fourth, when daytime and nighttime power usage was excessive for the T-CAES system, a diesel backup was used. (Dr. Ben Enis et al., 2006). Figure 13 shows the wind turbine 2,500-kW power dropping to 312.5-kW and then rising back to 2,500-kW during each 1-hour successive period. The System is delivering 90-kW to the user during the 12-hour period at night and 910-kW to the user during the 12-hour daytime period. The 15.24 m long cylindrical storage tank that is 2.56 m in diameter permits the tank pressure to rise from 16.3 bar at the start and to reach 82.75 bar at the peak, and to return to 17.3 bar-psia at the end of the day. This cycle is therefore ready to repeat itself continuously. If the storage tank is made twice as long, say 30.48 m, then the power lulls can be extended from 1-hour to 2 hours. If there is a set of two 30.48 m long tanks, then there can be a four-hour lull. If there is a wind speed lull that extends

continuously for several days, then neither the T-CAES system nor the underground cavern CAES system can support the user. The diesel system or Gas Turbine is required.
The T-CAES system is amenable to all geological and geographical locations. It falls in the power versus duration region where other energy systems do not apply (Fig. 4). It operates at high power levels (0.5 MW to 10MW) and over many hours. It operates in three modes for different daily scenarios at the same facility: (1) Electrical power mode, (2) Chilled air co-generation and (3) Drives pneumatic equipment and pneumatic tools.
The T-CAES system provides electrical power history "smoothing" so that even smaller wind turbines can provide steady power histories to the user. When there is a differential in price of electricity during the summer daytime and the summer evening, the T-CAES system provides "peak shaving" advantages. The T-CAES system provides backup power when the electrical grid is down or when the wind turbine is idle. The Wind Turbine Generator would be remote at a suitable location, except in some rural area with co-ops or small town Municipalities. Where there are additional services from T-CAES system, such as pneumatic power or chilling then the storage and generation system are co-located with buildings, as the T-CAES system does not burn fossil fuel.

12. US storage geology and +4 wind resources and population density

The two maps Fig 16 & 17 of the US show an interesting perspective where the high wind resources are located and the population density; basically many wind farms are distant from end users with high energy demands, resulting in transmission constraints as well as time of day generation, generally not co-incident with demand. This is also illustrated in Fig 18 as experienced in Europe; imbalances between production and demand. Many small rural towns who depend on their own power generation have opted for wind energy to

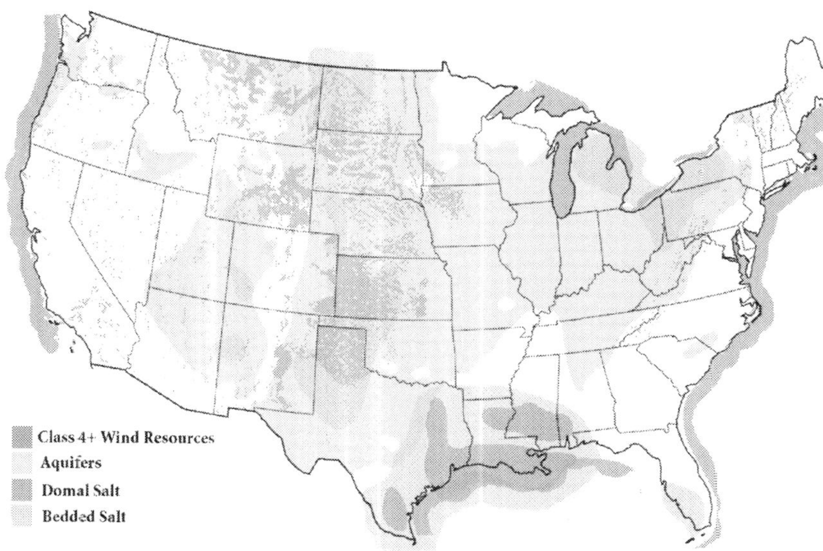

Fig. 16. Areas with geologies favorable for CAES and class 4+ winds

offset increasing fuel costs. These smaller systems may be from one to three WTG's, while helpful do not realize the full benefits of the investment. This also applies to some large schools and Colleges who support the emissions reduction effort in their communities and subsurface storage would be the right answer. The map also indicates that the favorable wind resources in the Midwest have domal and bedded salt as well as aquifers suitable for energy storage. The Iowa Storage Project using a deep saline aquifer is a good example of an Association of Midwest municipalities taking the initiative to collectively harvest their wind resources.

13. Wind resources and population

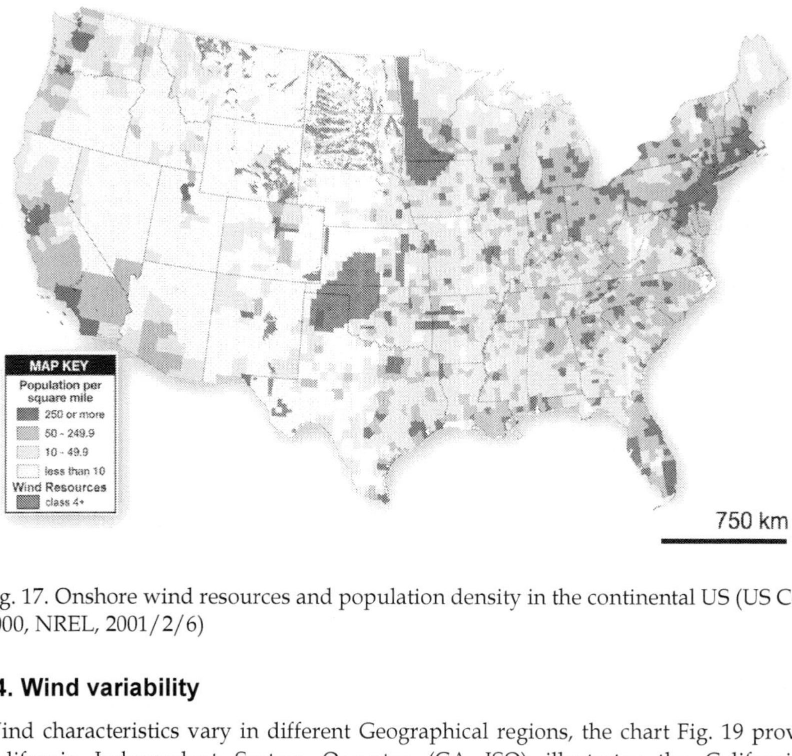

Fig. 17. Onshore wind resources and population density in the continental US (US Census 2000, NREL, 2001/2/6)

14. Wind variability

Wind characteristics vary in different Geographical regions, the chart Fig. 19 provided by California Independent System Operator (CA ISO) illustrates the California Wind Generation in January of 2005, clearly the best generation is at night not serving the high load demand during the day. Storing the wind energy for day time demand would greatly enhance the value of renewable energy with the added benefit of low CO2 emissions. Other high density wind areas such as in Texas exhibit similar characteristics. The Bonneville Power Authority in the North West does have the benefit of Hydro Power to compensate for the wind variations.

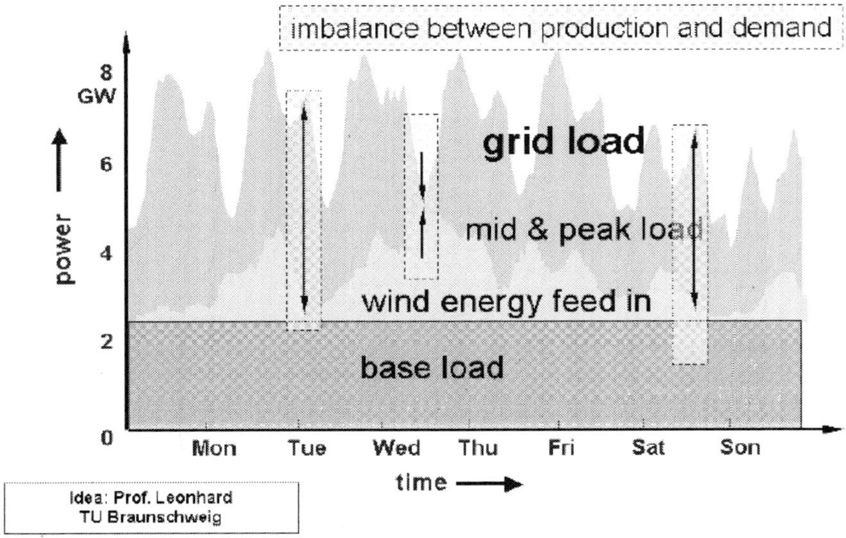

Fig. 18. Imbalances between Wind production and Power demand

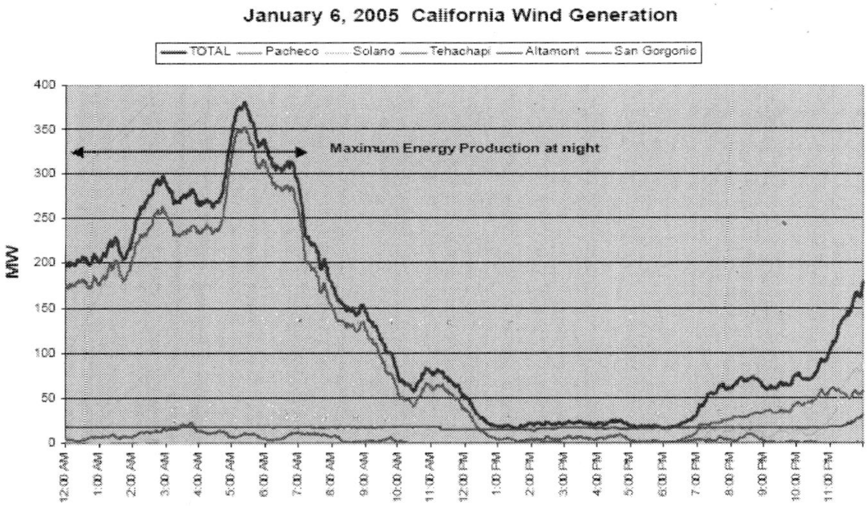

Fig. 19. California Wind Generation Jan. 2005

15. Projects in development

Several large CAES projects with different storage media are in development. Two are fully permitted and of particular note, even when the financial climate for new projects requiring major investments has slowed for such innovative concepts.

Storage Technology of Renewables and Green Energy Act of 2009 is funded by the US Department of Energy (DOE) to assist getting storage projects launched with Industry partnership. Load shifting for Wind Farm diurnal operations and ramping control to demonstrate CAES projects of 10 to 50 MW /2 to 5 hour storage will benefit from $50/60 Million cost share.

Both above ground and below ground CAES projects will be considered for demonstration. It is expected that at least 35% lower CO2 emissions than simple-cycle will be achieved, with a predicted economic payback based on the 24 months of project data. These projects must be ready for operation within 4 years of project award. Some of the larger projects in development could benefit from the US DOE serious consideration for energy storage based on increasing wind power generation.

15.1 Iowa Stored Energy Project (ISEP)

This project under development by Iowa Association of Municipal Utilities promises to be exciting and innovative. The compressed air will be stored in an underground aquifer and wind energy will be used to compress air in addition to available off-peak power. A separate section of the underground aquifer will be utilized for the storage of natural gas, allowing the CAES facility and other utilities to purchase gas when prices are lower.

The plant configuration is for 200 MW of CAES generating capacity, with 100 MW of wind energy. While wind might be the lowest cost generation system, it is variable and not reliable as a constant source. CAES provides the 'battery' storage for wind energy and makes wind energy a dispatch resource. CAES will expand the role of wind energy in the region generation mix and will operate to follow loads and provide capacity when other generation is unavailable or non-economic. The underground aquifer near Fort Dodge has the ideal dome structure allowing large volumes of air storage at 80 bar pressure or more with injection depth of 3000 feet (Fig. 20). With recent funding from the US DOE several exploratory wells have been drilled under the guidance of The Hydrodynamics Group team led by Michael King a natural gas engineer with 30 years of hydrogeology experience. The aquifer has been defined and additional test wells will need to be drilled prior to injection tests to determine the permeability of the sand stone. The results are promising thus far and further progress towards the first aquifer storage for air is well on the way.

Other states such as Illinois also have this potential for Wind & Storage. "Energy Storage Options for Central Illinois", (Makansi et al., 2003) but Iowa is in the forefront, possessing a site ideal for a CAES power plant and wind farm. These development plans have a future vision for the value of carbon reduction: adding reliable renewable resources with storage concepts such as CAES. In reality, there is no shortage of potential projects and suitable sites (van der Linden & Septimus, 2006) for Bulk Energy Storage development; there is no energy policy or incentive to implement the advantages and benefits demonstrated by NG Storage or the Pumped Hydro storage now serving the nation's power system. This long term lack of support is now getting some in-depth consideration from the US DOE driven by growing wind farm developments.

15.2 Project Markham, Texas

This 540 MW project in Matagorda County, Texas, developed by Ridge Energy Services, will consist of four 135 MW CAES units with separate LP and HP motor driven compression trains. The smaller 135 MW units in this project provide a very wide load range from 10 MW

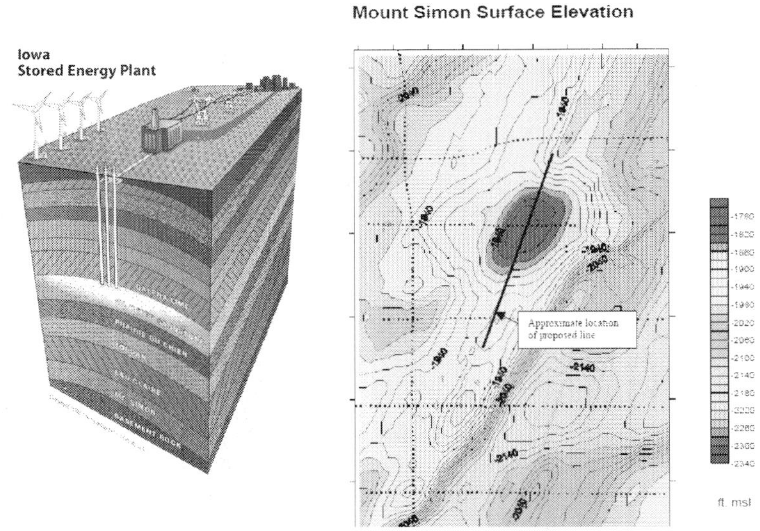

Fig. 20. Aquifer Air Storage Concept and the aquifer near Fort Dodge (Hydrodynamics Group)

minimum per unit and incremental output until all four units provide the system 540 MW. The full 540 MW can be delivered in less than 15 minutes. This is a tremendous value to the grid, providing reserve capacity, before cycling of base-loaded plant is required. The variable capacity range would be 840 MW (300 MW Compressor + 540 MW Generator). Nox emissions will be controlled to 5.0 vppm or lower with SCR in the HRU.

This site has Salt Dome cavern storage suitable for high pressure air storage and is unique in that natural gas storage is available on the site as well. This is ideal as energy can be arbitraged either as electrons (electricity) or Btu's (natural gas), or a combination of both.

Compression trains totaling 300 MW, for the required shorter off-peak charging period, will also act as a very large load sink on the system. This project stalled some years ago when funding for the project ceased due to financial circumstances, the same group contributed to a study commissioned by the Texas State Energy Conservation office, to look at the impact of CAES on Wind in Texas, Oklahoma and New Mexico. (Desai N, et al 2005)

The study results were positive in spite of very little day time to off-peak spark spread: "We have been working on combining compressed air energy storage (CAES) with wind generation in West Texas, and we have shown that storage can actually reduce the burden on the transmission system. In fact, storage can allow additional wind generation capacity to be served within an existing transmission plan. Our estimates suggest that a combined wind and storage project would be able to produce shaped, dispatchable energy for less than 5 cents per kWh with the capacity benefits of traditional thermal generation, with over 90 percent renewable content. "The conditions are different in 2009 and new developments are expected.

15.3 Norton Energy Storage, Ohio

One of the first potential CAES projects in the USA, developed by Haddington Ventures, Inc., is the huge facility at Norton in Ohio which is permitted for 2700 MW of capacity and as a commercial project when completed will be one of the largest Bulk Energy Storage facilities, including PHS, to be built in the USA. As originally planned, this will consist of 9 x 300 MW (or larger) nominally rated CAES units supported by an underground storage cavern volume of 120 million cubic meters, 722 meters below the surface, originally mined in a limestone formation. Fig 21 attests to the dry cavern walls and height of the pillar and post mining creating the large storage capability.

Fig. 21. Norton Limestone Cavern (Hydrodynamics Group)

Using 200 MW (4 x 50 MW) compression trains for each 300 MW power train will allow for 16 hours generation by day for 5 days a week. Four units producing 1200 MW could operate for 4x16 hour days without requiring recharging of the cavern. With more available surface space, cavern volume could support 5400 MW or more for 8 to 10 hours operation, 5 days a week. This cavern was originally permitted for a PHS that would only support a small fraction of that capacity. The project has had many stops and starts, and is now once again being developed by Haddington using the McIntosh Turbo machinery arrangement with an up rated version of 135 MW modules; while this changes the dynamics of the original planning, it allows for lower initial investment. Decoupling the compressors from the power train will allow the compressors to be located away from the storage cavern, allowing more generating units to be located on the site, and so reach the full potential of this storage facility. With this modular approach, the capacity could be added over 5 years allowing full integration in Ohio and the East Central Area Reliability (ECAR) region.

16. What are the economics of CAES systems?

The best proof of the economics are to look at what Alabama Electric Cooperative (McIntosh CAES Plant 2600 MW /hr storage) are achieving as well as many different studies that have been conducted comparing CAES with current CCPP as well as IGCC and PC coal-fired power plants. Renewable energy such as Wind also demonstrates lower grid costs when integrated with CAES. Site and location specifics will obviously indicate different values and a comparison will have to be made for different regions considering "off-peak" power prices or "spilled" wind energy costs as well as the optimized benefits such as capacity value, transmission value, dispatch value, firming value, shaping value (wind), etc. The Total Operating costs for the CAES plant which includes charging costs as well as natural gas prices at $6/MMBtu as provided by AEC and Electric Power Research Institute (EPRI) show a clear advantage for CAES with $40/MWh when compared to Battery, Pumped Hydro and Gas Turbine plant; CAES power plants like Pumped Hydro have to consider the storage volume and type of storage, such as solution mined caverns in salt domes (as done for NG), hard rock caverns, and aquifers or depleted gas wells. Solution mined caverns are the least costly and would add from \$50 to \$65/kW depending on the volume required (kW/hrs) to the overall installed cost (\$700/750/kW) which would be comparable or lower to a CCPP installation of equal size. In Texas, the economic impact of CAES could realize approx. \$10 Million per annum, integrating an additional 500MW of Wind just in one region. From a Pearl Street Executive Briefing Report, *"Energy Storage, the Sixth Dimension of the Electricity Value Chain,"* the following excerpts are provided, indicating the possible economic impact utilizing Bulk Energy Storage: (www.energystoragecouncil.org).

According to a 1993 DOE Study, the direct impact of energy storage on the US power Industry is estimated at \$ 57.1 Billion based on wide spread us of "high density" storage devices. The potential value today with the increased generation and load requirements is estimated at \$174.4 Billion. Improving the operational efficiency of the generation segment holds a potential of \$10.5 Billion worth of positive economic impact over the next 15 years. By improving the operational efficiency of the transmission segment another \$ 29.9 Billion potential positive economic gain, is achievable over the next 15 years. The economic screening Analysis of Energy storage options with Gas Turbines and Combined Cycle plant by Electric Power Research Institute (EPRI) indicates that CAES plant is less expensive in levelized cost in \$/kW-yr as well as Batteries and Pumped Hydro with capacity factors higher than 3% indicated by the chart Fig. 22 below.

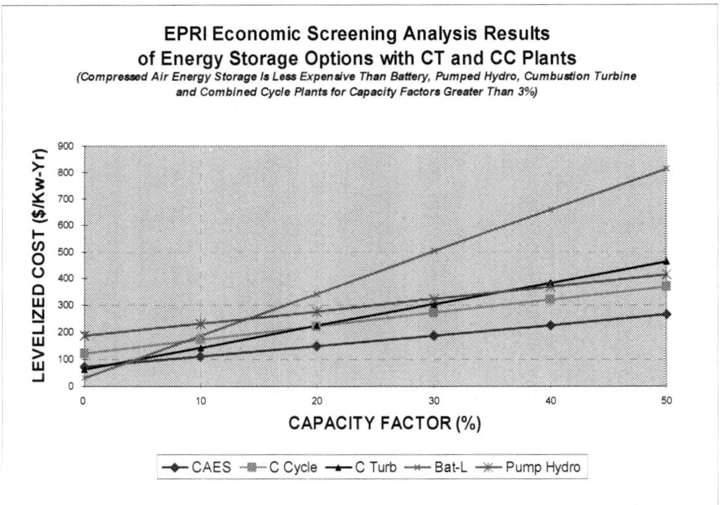

Fig. 22. EPRI Economic Screening Analysis of Energy Storage Options

The additional costs of energy storage when integrated with wind, often cited as being non-economic, is not born out by reality of the facts. Wind power for example by nameplate ratings are at \$2000/kW (or higher) for onshore installations, however when based on deliverable power capacity factors of 30% the equivalent cost to a base load plant now increases to \$4500/kW. CAES plant would add \$750/kW with the same fuel consumption as a Gas Turbine at \$350/kW. The CAES plant can readily improve the wind deliverable power to 45% to 55% or higher reducing the basis of $4500/kW to $3000 and $2500/kW respectively, the delta difference pays for the integrated CAES plant. This is a rough analysis as many other factors enter the actual economic screening.

17. Conclusions and recommendations

The current storage concepts are ready for deployment. Storage needs to be implemented in particular for Wind Energy, not just here in the US but in all developing countries. The biggest impact is probably the flexibility of operation. Economic dispatch to meet market needs, absorb excess capacity or large load swings with compression are powerful market tools. It is possible to improve energy management and obtain better value from bulk power purchase and sales; reduce risks and vulnerabilities from fuel price shocks. In particular, volatility in the US will always be a factor; long term projections show that natural gas prices will continue to rise with increased demand which cannot readily be met from new sources other than LNG imports.

The trend of increased harvesting of wind energy will put further stress on the grid reliability. This is already manifested in Europe where a far greater percentage of its generating base is committed to the variances of wind power production.

Most importantly, Bulk Energy Storage will "buffer" utilities from the lack of spinning reserve and load following capability, a result of many independent Wind Generating Farms installed in the last 5 years and substantial planned capacity. It will remove concerns about power quality and new threats to reliability. CAES as a generating asset has capacity value, as if it were a thermal asset, fully dispatchable, and a low emissions profile.

Using liquid air as an energy storage medium could be a potential solution to being able to locate storage for wind energy closer to load centers. This concept is proposed by Expansion Energy LLC. The Cryogenic tanks store liquid air are at relatively low pressure, which when required can be pumped at 42 bar to suit the prime mover, the liquid air is preheated and vaporized with the GT exhaust to feed an expander generating unit. Like many innovative concepts, ideas such as this will need further development ad investigation for wind integration.

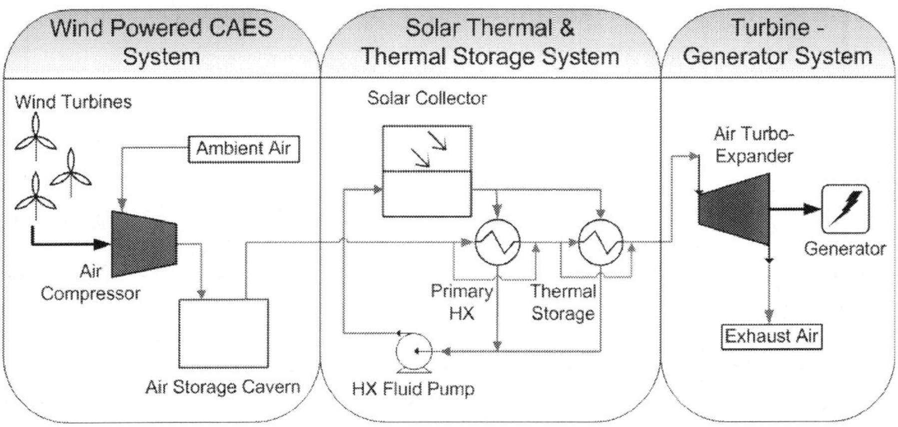

Fig. 23. Combined wind/solar thermal, using CAES & Thermal Storage (Courtesy Mark Kapner P.E. Austin Energy)

Eliminating the use of fossil fuel such as in adiabatic compression and expansion as discussed should get further attention and development funds for demonstration. Austin Energy in Texas has taken a step in this direction proposing the incorporation of Solar Energy in the thermal storage integrated with Wind and CAES as a Dispatcheable Hybrid Wind / Solar Power Plant (Fig. 23); the study is being cooperatively supported by the University of Austin.

Energy Storage provides security, reduces transmission constraints, importantly extends (optimizes) the capabilities of efficient clean coal plants, reduces emissions, and primarily enhances and integrates wind energy as a valuable renewable resource. It provides load management, (rapid response) frequency and voltage control, spinning reserve, black start capabilities and supports distributed generation. Energy Storage and Wind integration will be a Paradigm shift in the entire Utility System

18. References

Bulloch, Chris; Gatzen, Christoph; et al. (November 2004). "Advanced Adiabatic Compressed Air Energy Storage for the Integration of Wind Energy." Wind Energy Conference, EWEC 2004, London U.K.

Enis, Dr. Ben; Lieberman, Dr. Paul; Rubin, Irving; van der Linden, Septimus. (June 2007). "Wind Turbine Generator and Compressed Air Energy Storage System for Production of Electrical Power and Co-generation of Chilled Air for Desalination." ECOS 2007, Padua, Italy.

Desai, N; Gonzalez, S; Pemberton, D.J; and Rathjen, T.W. (June 2005) "The Economic Impact of CAES on Wind in TX, OK, and NM." Ridge Energy Storage & Grid Services L.P., Texas State Energy Conservation Office.

Makansi, Jason; van der Linden, Septimus; Schien, Kent. (October 2003). "Energy Storage Options for Central Illinois."

Electrical Energy Storage Applications & Technology (EESAT) Conference, San Francisco, CA.

Nakhamkin, Michael; van der Linden, Septimus. (May 2000). "Integration of a Gas Turbine (GT) with Compressed Air Energy Storage (CAES) provides the best alternative for Mid Range and Daily Cyclic Generation Needs."ASME/IGTI Congress, Munich, Germany. Paper 2000-GT-82 Novel Concept ASME.

Nakhamkin, Michael; van der Linden, Septimus; Wolk, Ronald; Patel, Manu. (June 2004). "New Compressed Air Energy Storage Concept Can Improve the Profitability of Existing Simple Cycle, Combined Cycle, Wind Energy, and Landfill Gas Combustion Turbine-based Power Plants." ASME /IGTI Congress, Vienna, Austria. Paper GT2004-54278.

Van der Linden S. (April 2002). "CAES for Today's Market." EESAT Conference, San Francisco, CA.

Van der Linden, Septimus. (2006). "Bulk Energy Storage potential in the USA, current developments and future prospects." Energy 31, 3446-3457, Science Direct.

19

Optimization of Spinning Reserve in Stand-alone Wind-Diesel Power Systems

Fernando Olsina[1] and Carlos Larisson[2]
[1]*Institute of Electrical Energy, CONICET*
[2]*Inar Technologies, Inc.*
Argentina

1. Introduction

Spinning reserve carried on synchronized units is the most effective resource available to the system operators for managing unforeseen power unbalances, such as demand fluctuations and the sudden loss of generation equipment. The amount of reserve and the speed that it can effectively be deployed determine the supply reliability that the generation system can achieve.

Carrying more spinning reserve reduces the probability that the generation system become unable to preserve the momentary power balance and costly remedial actions, such as involuntary load shedding, turns unavoidable to prevent a system collapse. Nevertheless, providing spinning reserve on a continuous basis is expensive. Indeed, the provision of spinning reserve entails incurring in startup costs to commit generating units in excess of the forecasted load, which consequently have to be dispatched at less efficient operating points.

The problem of keeping the power balance is still more difficult in stand-alone wind-diesel power systems, since these systems are additionally subjected to random power fluctuations originated in the uncertain and intermittent nature of the wind resource. Furthermore, autonomous power system cannot rely on power imported from interconnections for preserving the power balance. The inherent characteristics of these systems require scheduling more reserve on synchronized units for ensuring adequate security and reliability levels. The higher reserve requirements may substantially deteriorate the economy of these supply systems.

The costs of keeping spinning reserve must be compared with the benefits that it provides in terms of lower expected costs of interruptions. In essence, the optimal reserve level can be set so that the marginal cost of carrying an additional MW equals the marginal reduction of the expected load curtailment costs.

Despite the apparent simplicity of this optimality condition, determining the optimal amount of spinning reserve in a practical setting presents substantial modelling complexities and computational challenges. Given the random nature of the disturbances and contingencies that may face a generation system, assessing the benefits of carrying a certain amount of spinning reserve involves quantifying the occurrence probability, duration, extent and costs of load loss events. Such evaluation entails modelling the stochastic behaviour of system operation by considering the random failure of system components and the stochastic fluctuations of load and wind generation. The problem is probabilistic in its very nature and thus it may be appropriately treated by applying stochastic modelling techniques. Only with the advent of more powerful computing hardware, the problem of optimizing the spinning reserve has attracted the interest of researchers and its solution is currently deemed practicable.

This chapter proposes a novel method for determining the optimal amount of spinning reserve that should be carried in autonomous hybrid wind-diesel generation systems. The optimal spinning reserve is determined by comparing the cost of its provision with the economic benefits it delivers in terms of supply reliability. The proposed approach is still general and can be applied in straightforward manner to establish the optimal reserve level in large interconnected systems.

The presented methodology considers with accuracy the probabilistic features of the load and the wind generation, as well as the random outages of the conventional generating units. By applying high-resolution chronological simulation techniques, the stochastic features of the integrated operation of the diesel units and the wind turbine can be detailed replicated. The mathematical model appropriately considers all relevant characteristics and operational constraints of the generating units, e.g. non-linear heat rate curve, maximum and minimum output, startup and synchronization time, minimum down and uptime, ramping, etc. Massive stochastic simulation methods allow assessing the system reliability and valuing the economic costs of loss load events.

Global search methods like particle swarm optimization (PSO) are proposed for finding the optimal scheduling policy and spinning reserve requirement that minimizes the sum of the expected operation costs and the expected costs of the energy not served.

The remaining of this chapter is organized as follows. Section 2 is devoted to revisit the conventional Unit Commitment problem and presents a new stochastic formulation for coping with uncertainties affecting renewable-integrated systems. In Section 3, a number of models for simulating the chronological operation of wind-diesel systems under stochastic conditions are described. Section 4 provides some exemplary high-resolution simulations of the integrated operation of the wind generator and the diesel generating units. Additionally, results of the optimization procedure are given. Conclusions and suggestions on further research work are drawn on Section 6.

2. Mathematical formulation of the reserve optimization problem

2.1 The deterministic thermal UC problem

The single-bus Unit Commitment (UC) problem consists on scheduling available generating units and setting their respective generation outputs in order to meet a forecasted load sequence, so that all relevant unit specific and system-wide constraints are satisfied while a performance measure is optimized, e.g. minimum production costs, maximum social welfare, etc. Mathematically, solving the UC problem entails the formulation of a complex optimization problem, which is stochastic, non-linear and mixed-integer in its very nature.

Let consider a thermal-only generation system with I generating units. In discrete time, the objective function of the standard deterministic reserve-constrained UC problem for T time stages of duration Δt can be mathematically formulated as the minimization of the sum of unit start-ups and generation costs over the considered time span as follows:

$$\min_{u_i^t, P_i^t} \left[\sum_{t=1}^{T} \sum_{i=1}^{I} C_i(u_i^t, P_i^t) + S_i^t(u_i^t) \right] \tag{1}$$

where $u_i^t \in \{0,1\}$, $i = 1,2,...,I$ and $t = 1,2,...,T$ are binary decision variables indicating whether unit i is scheduled to generate on time period t (0: stand-by, 1: synchronized); P_i^t is the output generation level of generator i during period t; C_i and S_i^t are the generation and time-dependent start-up costs of unit i respectively.

Solving the UC problem involves choosing a set of decision variables so that the objective function in (1) is minimized subjected to a number of constraints:

System constraints

System demand: the total system generation must meet the forecasted power demand L^t at each time period

$$L^t - \sum_{i=1}^{I} u_i^t P_i^t = 0 \quad \forall t \in T \tag{2}$$

System reserve: the scheduled spinning reserve on the committed units must satisfy the exogenous reserve requirement R^t set by the system operator based on deterministic or probabilistic criteria

$$L^t + R^t - \sum_{i=1}^{I} u_i^t P_i^{\max} \leq 0 \quad \forall t \in T \tag{2}$$

Technical constraints on the operation of generating units

Typically, generating units impose some strict operating limits in order to ensure a secure operation and safeguard their lifetime.

Generation limits: the power generation of each scheduled unit i at any time t should be within its lower and upper rated output capabilities, $P_i^{\min}$ and $P_i^{\max}$ respectively

$$u_i^t P_i^{\min} \leq P_i^t \leq u_i^t P_i^{\max} \quad \forall i \in I,\ \forall t \in T \tag{3}$$

Ramping limits: In addition, important intertemporal constraints on the operation of the generating units must be accounted for in the problem formulation. The change in generation output between adjacent time intervals should observe units ramping capabilities

$$r_i^{\min} \Delta t \leq \left(P_i^t - P_i^{t-1}\right) \leq r_i^{\max} \Delta t \quad \forall i \in I,\ \forall t \in T \tag{4}$$

where $r_i^{\min}$ and $r_i^{\max}$ are respectively the minimum and maximum permissible change rate per unit time of the generation output, expressed for example in kW/s.

Minimum up/down time: the scheduling decisions must also comply with the minimum time in standby T_i^{off} between consecutive shut down/start-up decisions and minimum operating time T_i^{on} between consecutive start-up/shut down decisions

$$\begin{aligned} \left(X_{i,t-1}^{on} - T_i^{on}\right)\left(u_i^{t-1} - u_i^t\right) \geq 0 \quad \forall i \in I,\ \forall t \in T \\ \left(X_{i,t-1}^{off} - T_i^{off}\right)\left(u_i^t - u_i^{t-1}\right) \geq 0 \quad \forall i \in I,\ \forall t \in T \end{aligned} \tag{5}$$

where $X_{i,t}^{on}$ and $X_{i,t}^{off}$ are the time durations the unit i has been on and off at time stage t from the last start-up and shut down decision respectively.

2.2 The stochastic wind-diesel UC problem

The conventional deterministic UC problem formulated in Section 2.1 needs some important modifications and extensions in order to consider the costs of scheduling decisions under uncertain future operating conditions due to fluctuating wind generation and in order to accommodate particular constraints of diesel gensets.

Unlike the reserve-constrained UC problem described in the previous section, in the proposed formulation the reserve requirement is endogenously determined for each time period being itself a result of the optimization procedure. By introducing in the objective function the expected damage costs $E[C_E]$ associated to supply interruptions, the spinning reserve requirement may be optimized by trading off its economic benefits with the cost of its provision.

The objective function of the stochastic wind-diesel UC problem can be formulated in terms of the mathematical expectation of the overall system costs as follows

$$\min_{u_i^t, P_i^t} E\left[\sum_{t=1}^{T}\sum_{i=1}^{I} C_i(u_i^t, P_i^t) + S_i^t(u_i^t) + C_E(u_i^t, P_i^t)\right] \tag{6}$$

The proposed formulation does not require imposing system-wide constraints since the optimization procedure determine the optimal load demand to be met as well as the optimal spinning reserve held on committed units.

It is important to mention that in small autonomous systems the power needed for serving the unit-related auxiliary loads (e.g. fans, pumps, heaters, etc.) $u_i^t L_i^{aux}$ are often relevant in relation with the system demand, and hence, their serving costs must also taken into consideration. The amount of parasitic loads to be served mainly depends on the number of the committed units, and thereby is a result of the scheduling decisions.

In addition to the unit specific constraints stated in (3) to (5), further operational limits of diesel units have to be introduced in order to find feasible solutions.

Generation limits

It is important to distinguish the various rating limits of diesel gensets. The *continuous rating* is the maximum power that the diesel generator can delivered to a constant load for unlimited time, i.e. load factor of 100%. The *prime rating* refers to the peak power that can be delivered to a time-varying load for unlimited time. Typical load factors are 60% to 70%. The *emergency rating* is the genset overload capability for a time-constrained emergency use. Typically the overload capacity is 10% above the prime rating for a maximum duration of 1 h, maximum frequency of 1/12 and cumulated overload operating hours not exceeding 24 h/yr. The continuous, prime and emergency ratings involve the consideration of integral constraints over the optimization period, which must be appropriately handled.

Start up and synchronization time

After receiving the starting signal, diesel generators require some time before they can effectively deliver electrical power. This time is needed for cranking the diesel engine, accelerate to rated speed, warm-up and synchronize with the system. This time depends on the size of the genset and the prevailing ambient conditions. Typically, diesel generators can accept load from 10 s to few minutes after the start signal.

2.3 Solution techniques

The UC problem is probably one of the most investigated scheduling problems, for which a wide variety of approaches has been proposed along the years. Most notably, Dynamic Programming (DP), Lagrangian Relaxation (LR), Linear Programming (LP), Quadratic Programming (QP), Mixed-Integer Programming (MIP) as well as Artificial Intelligence based algorithms like Genetic Algorithm (GA), Artificial Neural Networks (ANN), Tabu Search (TS), Simulated Annealing (SA), Ant Colony Systems (ACS) have been proposed for solving the underlying optimization problem (Padhy, 2004; Yamin, 2004; Sen & Kothari, 1998; Sheble *et al.*, 1994).

Most recently, emerging techniques from Swarm Intelligence are being investigated for treating complex optimization problems (Kennedy & Eberhart, 1995; Kennedy & Eberhart, 1995). Particle Swarm Optimization (PSO) is currently considered a suitable derivative-free search method for dealing with many optimization problems present in the planning and operation of power systems, e.g. reserve scheduling, reactive power dispatching, power system control (AlRashidi & El-Hawary, 2009; del Valle *et al.*, 2008). PSO-based algorithms are also being increasingly considered suitable for treating the UC problem (Lee & Chen, 2007; Zhao *et al.*, 2006; Ting *et al.*, 2006). Moreover, the optimal scheduling of wind-integrated power systems solved with PSO-based techniques has recently been investigated (Swaroop *et al.*, 2009).

In this chapter, a hybrid variant of the conventional PSO algorithm referred as EPSO (Miranda & Fonseca, 2006, 2002a, 2002b), which incorporates elements of evolutionary programming (i.e. mutation, reproduction and selection), is applied for finding the optimal schedule of diesel units for each 5-min time interval over a 24-h planning horizon in order to minimize total expected system costs.

In the PSO terminology, a particle p is a matrix of scheduling decisions for each generating unit i and for each time stage t. For each particle p at the j-th iteration, the fitness of the proposed scheduling decisions is assessed by evaluating the objective function stated in (6). The expected costs are estimated by simulating the system operation under the proposed commitment decisions for a large number of possible realizations of the uncertain variables, i.e. random unit outages and stochastic fluctuations of the power demand and wind generation. In order to capture the influence of ramping constraints on accessing the spinning reserve for matching fast wind power fluctuations, the operation of the system is simulated with a time resolution of 10 seconds.

It is noteworthy to mention that the high time resolution required for simulating relevant operational features of these systems together with the computationally intensive nature of Monte Carlo and PSO methods impose a rather big computational challenge. However, the coarse-grained nature of the problem makes it amenable to be solved in a distributed computing environment.

3. The simulation model

3.1 The exemplary system

A real stand-alone hybrid wind-diesel system comprising 10 thermal units and a 2-MW wind turbine has been selected for illustrating the applicability of the proposed optimization framework. The total installed capacity of this exemplary generation system is 15.4 MW. The time horizon of the simulation model spans 24 h in order to account for the daily seasonality of the load demand and the wind resource. The time resolution for simulating the operation of the system and the load dispatch is 10 s, which allows capturing ramping constraints of the diesel units when managing power fluctuations. The unit commitment is set for each 5 minutes with a time horizon of 24 h. In the following sections, details of various models necessary for describing the stochastic behaviour of operating conditions of the autonomous wind-diesel system are presented. Emphasis on stochastic models describing the random nature of unit outages and wind power fluctuations is given.

3.2 Modeling the conventional diesel generation system

The considered thermal generation system encompasses ten identical diesel gensets with prime power rated capacity at site elevation of 1339 ekW. Further relevant technical specifications of diesel gensets are summarized in Table 1.

Specification	Parameter value
Caterpillar CAT3516B (4-stroke bi-turbo V16)	50 Hz/1500 rpm/6.6 kV
Fuel	LFO/Cmin
Gross engine capacity (at site elevation)	1415 bkW
Generator efficiency	0.95
Prime rating	1339 ekW
Continuous rating	972 ekW
Minimum operable output	280 ekW
Unit-related auxiliary load consumption	209 kW
Upward (downward) ramping capacity	50 (15) kW/s
Starting and synchronization time	120 s
Minimum up (down) time	300 (300) s

Table 1. Technical data of the considered diesel units

The operating cost of diesel generating units can be distinguished in start-up and variable hourly production costs.

According to experimental data shown in Fig. 1, the hourly fuel consumption c_F of a typical diesel genset within its operating limits is nearly linear with the delivered power output P

$$c_F = c_1 P + c_0 \tag{7}$$

where $c_0 = 0$ is a constant term and $c_1 = 0.225$ [l/ekWh] is the unit's average specific fuel consumption. The fuel consumption at idle is about 32 l/h. The lube oil consumption c_{oil} has been estimated in about 0.00106 [l/ekWh]. Assuming a fuel price $p_F = 0.811$ [US\$/l] and lube oil price $p_{oil} = 1.81$ [US\$/l], the total hourly generation costs C_G^T [US\$/h] can be computed in terms of the hourly fuel C_F costs and the lube oil costs C_{oil}

$$C_G^T = C_F + C_{oil} = \left(p_F c_1 + p_{oil} c_{oil}\right) P = 0.18438P \text{ [US\$/h]} \tag{8}$$

Start-up costs are incurred mainly in the fuel and oil consumption during the phase of engine start, warm-up, acceleration to rated speed and synchronization before it delivers electrical power to the supply system. These costs are modeled as a constant value

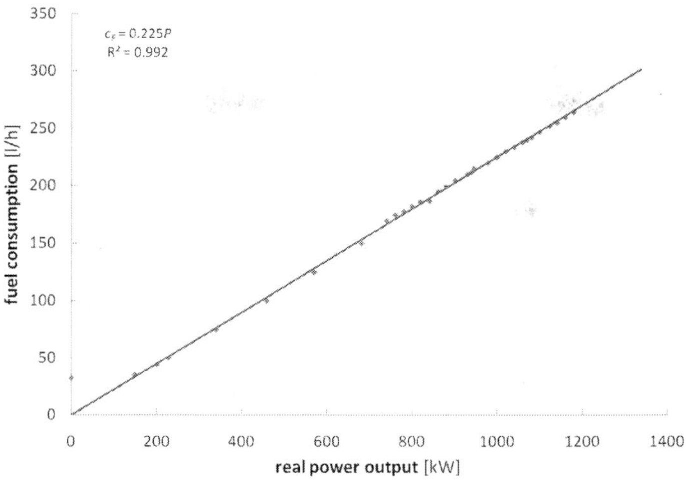

Fig. 1. Measured hourly fuel consumption of a diesel-fuelled generating unit

irrespective of the time the unit has been is standby. Considering a synchronization time of 120 s after receiving the start signal, the startup costs can be estimated in 1.07 US$. In addition, incremental parasitic loads due to startup decisions are also taken into consideration, but they are summed to the system load. Power consumption of unit-related auxiliary loads is estimated in 209 kW. The cost of supplying auxiliary loads cannot be neglected as they represent about 21.5% of the unit's continuous rate capacity[1]. Economic costs related to reduction of engine lifetime and incremental need for unit maintenance with the number of cold starts are here not considered.

The system must hold spinning reserve for managing power imbalances resulting of the sudden loss of generation equipment. Therefore, an appropriate reliability model for describing the stochastic behavior of unit failures and repair process is needed. Presently, it is well known the fact that the two-state unit model is inadequate for cycling units (IEEE Task Group, 1972). As the probability of a failure when the unit is down is typically very low compared to failure probability when the unit is operating, a simple four-state unit reliability model has been proven adequate for describing the interaction with the operating pattern of cycling units (Billinton & Jingdong, 2004). Neglecting the possibility of failure during the time the unit is unsynchronized, the simplest 4-state reliability model of a diesel unit is illustrated in Fig. 2, where transition rates λ and μ are the mean failure and repair rate respectively.

If the failure and repair rate are assumed time-invariant, the time between failures t_O and the repair time t_F are exponentially distributed and the model hold the Markov properties. This simple model does not consider failure to synchronize, postponable outages and failures leading to derate the unit capacity. Typical reliability parameters for diesel units

[1] Power consumption of plant-related auxiliary loads, i.e. loads that do not depend on UC decisions, such as lights, fuel pump and heaters, etc., must be simply added to the system demand. For the considered facilities, the fixed plant consumption is estimated in 274 kW.

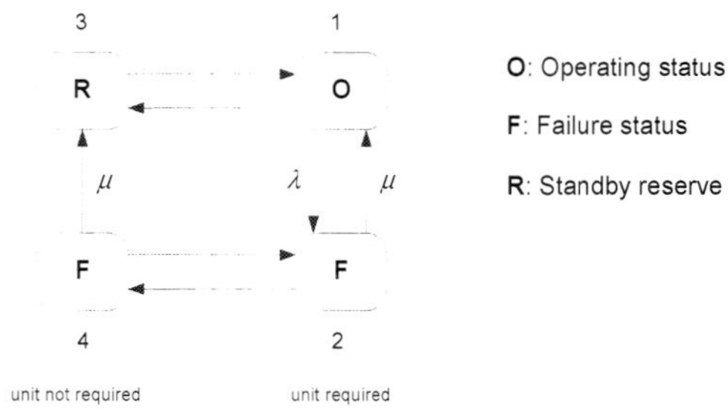

Fig. 2. Four-state Markov reliability model for a cycling diesel generating unit

FOR	λ	μ	MTTF	MTTR
0.02	0.00102041 h^{-1}	0.05 h^{-1}	980 h	50 h

FOR: Forced Outage Rate, **MTTF:** Mean Time to Failure, **MTTR:** Mean Time to Repair

Table 2. Typical reliability parameters of diesel units

have been adopted form the literature (NERC, 2006) and are summarized in Table 2. Relationships between these reliability parameters are given below

$$\begin{aligned} \text{FOR} &= \Pr(\text{F}) = \lambda(\lambda+\mu)^{-1} \\ \text{MTTF} &= \text{E}[t_{\text{O}}] = \lambda^{-1} \\ \text{MTTR} &= \text{E}[t_{\text{F}}] = \mu^{-1} \end{aligned} \tag{9}$$

where FOR is the forced outage rate representing the unit's failure probability, MTTF is the mean time to failure or the expected operating time $\overline{t_{\text{O}}}$ between two consecutive failures and MTTR is the mean time to repair or expected time $\overline{t_{\text{F}}}$ the unit resides in the failed state.
Under the Markov hypothesis, simulations of operation and repair times, t_{O} and t_{F} respectively, of generating units can be obtained by taking i.i.d. random samples from an exponential distribution with parameters λ and μ respectively (Billinton & Allan, 1996):

$$t_{\text{O}} = -\frac{1}{\lambda} ln\left(\text{U}[0,1]\right) \; ; \; t_{\text{F}} = -\frac{1}{\mu} ln\left(\text{U}[0,1]\right) \tag{10}$$

where U [0,1] are uniform i.i.d. samples over the interval [0,1].
It is important to mention that in addition to random failures, deterministic unit unavailability periods due to planned maintenance activities must also be taken into consideration in the scheduling algorithm.

3.2 Modeling the wind generator

Wind turbines exhibit highly non-linear generation characteristics. A typical wind speed - power curve is illustrated in Fig. 3. Four well-defined operating zones of the wind generator can be distinguished.

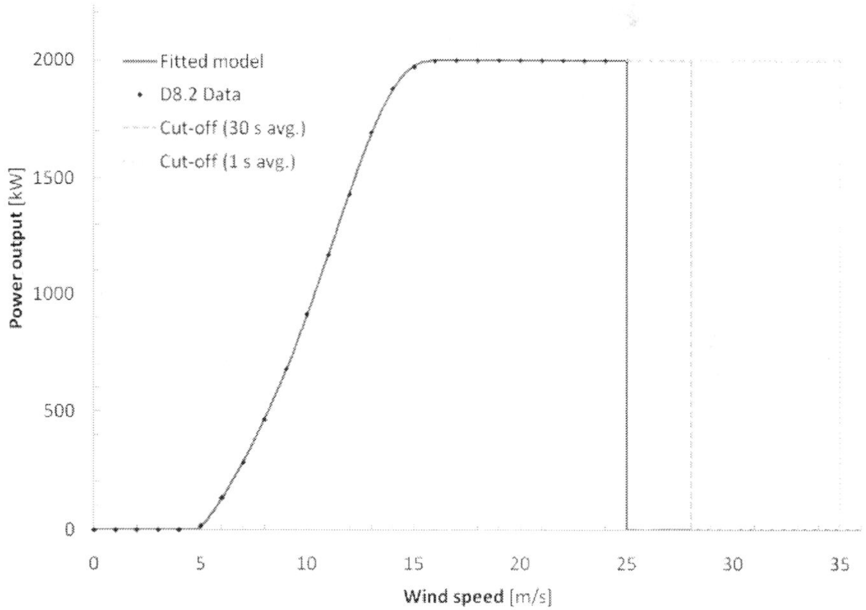

Fig. 3. Characteristic wind speed - power output of the DEWind D8.2 wind turbine

Generation is zero if prevailing wind speeds is lower than the cut-in wind velocity v_{in}. Wind power output rapidly increases from this point to the rated wind speed v_r at which the wind generator delivers its rated power capacity P_W^{max}. Fluctuations of the wind speed between these operating limits leads to large power fluctuations, which have to be balanced by the available spinning reserve carried on diesel units. If wind speed exceeds the rated velocity, the pitch control keeps the power output at the rated generation capacity. In order to safeguard the equipment, the turbine is shut-down if wind speeds exceed some predefined thresholds during certain time period. The cut-off wind speeds v_{off} are typically defined in term of time moving averages for various window widths.

A piecewise non-linear function describing the wind-power characteristic curve for the 2-MW wind turbine integrated to the considered supply system is given by the following expression:

$$P_W(v) = \begin{cases} 0 & 0 \le v \le v_{in} \\ \sum_{i=0}^{6} a_i v^i & v_{in} \le v < v_r \\ P_W^{\max} & v_r \le v < v_{off} \\ 0 & v_{off} \le v < \infty \end{cases} \tag{11}$$

where

$$\begin{aligned} P_W^{\max} &= 2000 \text{ kW} \\ v_r &= 16 \text{ m/s} \\ v_{in} &= \frac{1}{T}\sum_{t=\tau}^{\tau+T} v_t = 4 \text{ m/s} \quad T = 600 \text{ s} \\ v_{off} &= \begin{cases} \frac{1}{T}\sum_{t=\tau}^{\tau+T} v_t = 35 \text{ m/s} & T = 1 \text{ s} \\ \frac{1}{T}\sum_{t=\tau}^{\tau+T} v_t = 28 \text{ m/s} & T = 30 \text{ s} \\ \frac{1}{T}\sum_{t=\tau}^{\tau+T} v_t = 25 \text{ m/s} & T = 600 \text{ s} \end{cases} \end{aligned} \tag{12}$$

It is noteworthy to mention that once some of the cut-off conditions is reached, the turbine cannot be restarted while the 10-min average wind speed do not fall below 22 m/s. The synchronization time of the wind generator is 300 s. The ramping rate from 0 kW to the power output corresponding to the prevailing wind speed conditions is 33 kW/s. The down-ramping of the wind generator after a cut-off event is 200 kW/s, what imposes a considerable burden to the response capability of the thermal generation system.
For describing the stochastic behavior of turbine outages, a 3-state reliability model is proposed and illustrated in Fig. 4.

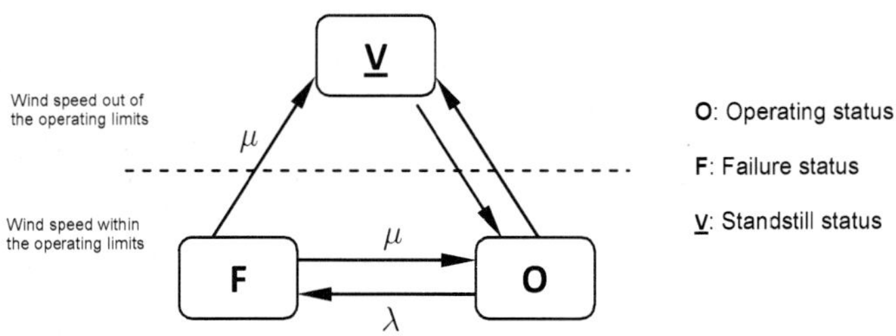

Fig. 4. Proposed Markov reliability model for the wind generator

The proposed reliability model assumes that wind turbine can fail only when it is synchronized and generating. Transitions from the operating status to idle and vice versa occur when the moving-averaged wind speeds are either lower than v_{in} or higher than v_{off}. If the turbine is repaired, depending of the prevailing wind conditions, transitions either to the operating status or standstill are possible.
Assumed reliability parameters of the wind generator for this study are given in Table 3. These values are consistent with a turbine sited in a remote location and subjected to extreme weather conditions (Castro Sayas & Allan, 1996).

FOR	λ	μ	MTTF	MTTR
0.05	0.0004 h^{-1}	0.0076 h^{-1}	2500 h	131.6 h

FOR: Forced Outage Rate, **MTTF:** Mean Time to Failure, **MTTR:** Mean Time to Repair

Table 3. Reliability parameters of a wind turbine

3.3 Modeling wind power fluctuations

The rapid fluctuations introduced by wind power generation are a major source of variability and uncertainty in the short-term operation planning of small autonomous supply systems. Because of the unpredictable nature of fast wind speed changes, additional spinning reserve must be carried to ensure that the power balance is kept at any time instant. In order to compute the spinning reserve requirement for balancing the fluctuations of the wind generation, a model accurately reproducing the severity and occurrence probability of the possible wind speed excursions is therefore needed.
By applying such a stochastic model, the system operation can sampled for a large number of possible chronological realizations of the wind speed. This allows exploring the rare occurrence of severe operating conditions, under which the system find exhausted its balancing resources and load shedding actions are needed.
In this section, results from a developed algorithm for simulating the stochastic dynamics of horizontal wind velocities are presented. The mathematical modeling details of the developed stochastic wind model are extensively treated in (Olsina & Larisson, 2008a, 2008b). High-resolution wind time series are generated in two sequential stages, i.e. 10-min average wind speeds and, based on this information, the non-stationary wind turbulence.
The proposed methodology rely on frequency-domain techniques, namely the well-known spectral representation theorem, for synthesizing random fluctuations of wind speeds over the various time scales according to the probabilistic and spectral properties observed in wind data gathered at the turbine site.
The simulation algorithm is able to accurately reproduce the remarkable non-Gaussian and non-stationary features of wind speeds. An iterative procedure and a non-linear memoryless transformation are applied to simultaneously match the observed evolutionary spectral content and the marginal non-Gaussian probability density function (PDF) of the random wind speed fluctuations. In addition, the proposed method is non-parametric, i.e there is not model parameters to be calibrated. Therefore, the proposed model does not require neither assumptions on the dataset nor the expedient postulation of a model structure or model equations to represent the wind speed variability. This is in fact an important advantage, as the very general nature of the non-parametric modeling framework allows applying the

wind stochastic model to sites with very different wind characteristics. The main parts of the developed wind speed simulation procedure are illustrated in Fig. 5.

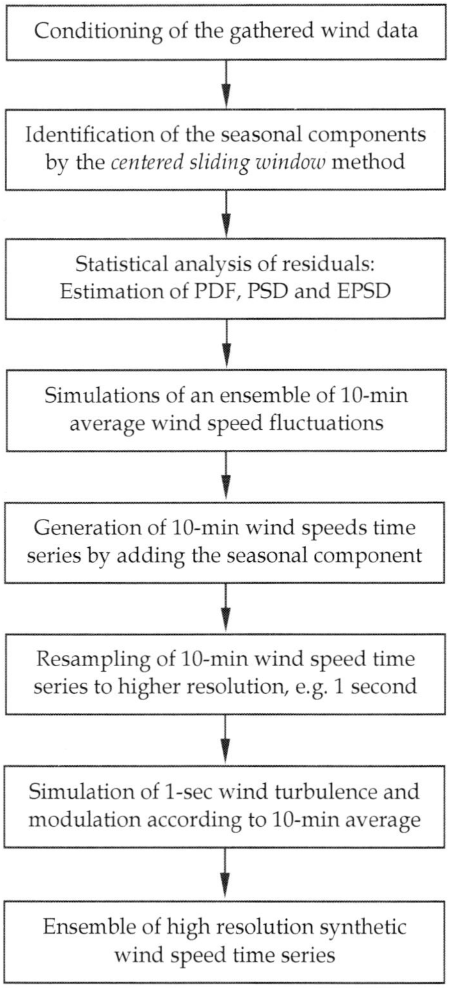

Fig. 5. Chart flow of the stochastic wind speed simulation algorithm

Before proceeding with further analysis, the gathered 10-min wind speed data, measured at 20 m, must conditioned for possible missing entries and outliers. Correction of wind data to the hub height (in this case 60 m) is typically also required at this stage. For this purpose, either the power or logarithmic law for the vertical wind speed profile can be used.

Changes in the prevailing wind conditions can be distinguished in deterministic and stochastic variability. The wind dataset v_t can be decomposed as the sum of deterministic regularities m_t, which can exactly be predicted, and random fluctuations z_t:

$$v_t = m_t + z_t \tag{10}$$

Regular deterministic patterns at daily and seasonal scales are identified by averaging wind data of the same hour over a *sliding window* centered at the time instant being estimated. A window width of 30 days has been applied to the available dataset. As an example, Fig. 6 shows the strong daily pattern identified for a week in January (summer).

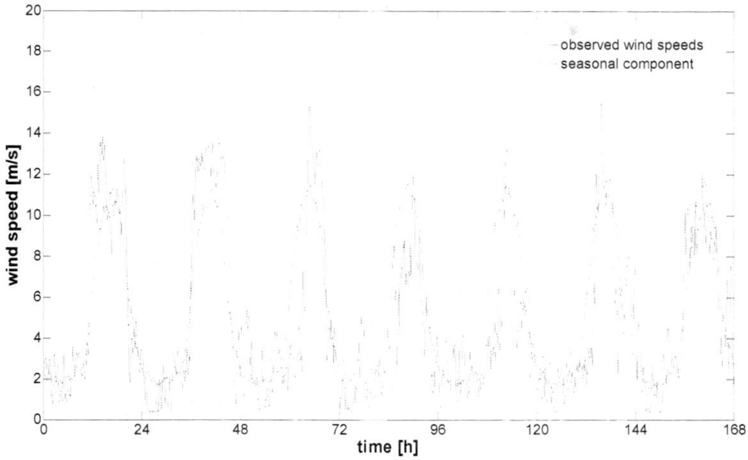

Fig. 6. Estimated deterministic component of wind speeds by using a 30-day sliding window

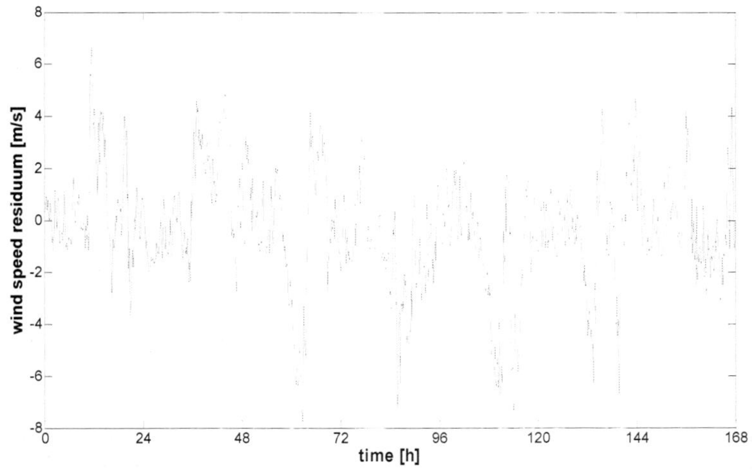

Fig. 7. Deseasonalized residuum of the 10-min average wind speed time series

After identifying the deterministic component, it can be subtracted from the dataset leaving only the stochastic part of the wind speed fluctuations. Fig. 7 shows the residuals z_t for the same January week after removing the seasonal pattern.

Fig. 7 clearly reveals the non-stationary behavior of the wind variance along the day. In fact, wind speed is much more variable an uncertain on daylight hours than on night hours. Furthermore, large deviations from the time-varying mean occurs frequently at about 10:00 AM and at 8:00 PM primarily as consequence of the uncertainty on the time the wind arrive and cease to blow.

Once the random part of the wind speed fluctuations has been isolated, a statistical analysis for characterizing the probabilistic and spectral properties of the time series is required, as simulation were generated according to this information.

The probabilistic characteristics of the wind speed time series are obtained by computing the empirical probability density function (PDF) by means of the non-parametric Kaplan-Meier method.

The stochastic dynamics of the wind fluctuations over the various time scales (seconds to months) is characterized by the (stationary) power spectral density function (PSD), which is also computed by non-parametric techniques. The Welch method provides a parameter-free, accurate and smoothed estimate of the PSD of wind residuals.

The non-stationary features of wind speeds over the year are adequately captured by computing the evolutionary power spectral density function (EPSD), which represents a complete time-frequency description of the time- varying statistical properties of the wind dataset. Here, it is assumed that non-stationary characteristic of the wind data are well represented by an intensity (uniformly) modulated stochastic process. The modulating function is therefore only function of time and depends on the local process variance. Similarly to the local mean identification, the time-varying local variance of the process is also computed with the sliding window method as referred above.

The first step in the synthesis of 10-min wind speeds is the generation of a zero-mean stationary Gaussian ensemble, i.e. a set of independently generated random processes, according to the empirical PSD by means of the spectral representation method (Shinozuka & Deodatis, 1991). Before proceeding to the next step, the generated ensemble is modulated with an envelope function in order to account for non-stationary characteristics. By means of a memoryless non-linear transformation the observed non-Gaussian PDF is satisfied. The non-stationary ensemble is iteratively corrected according tho the target PSD. The reader interested in details of the mathematical formulation of the iterative spectral correction procedures is further referred to (Deodatis & Micaletti, 2001). It is important to mention that the same spectral-based algorithm has been applied to the stochastic simulation of electricity prices, which are random processes well-known for their stochastic complexities and the challenging modeling difficulties they present (Olsina & Weber, 2008).

Fig. 8 illustrates a single sample of the generated stochastic ensemble of 10-min wind speed fluctuations. The statistical properties of the simulated samples are compared with the observed characteristics in Table 4. It can be concluded that synthetic wind fluctuations accurately replicate observed statistical features of the wind dataset.

In Fig. 9, the probability density functions for the observed wind speed residuals as well as for the simulated sample are compared by potting them together. This plot confirms that the probabilistic properties of simulated wind speed fluctuations are nearly identical to those observed at the wind turbine site. It must be noted the excellent agreement at the tails of the distribution, as the probability of large wind speed excursions largely determines the spinning reserve requirement.

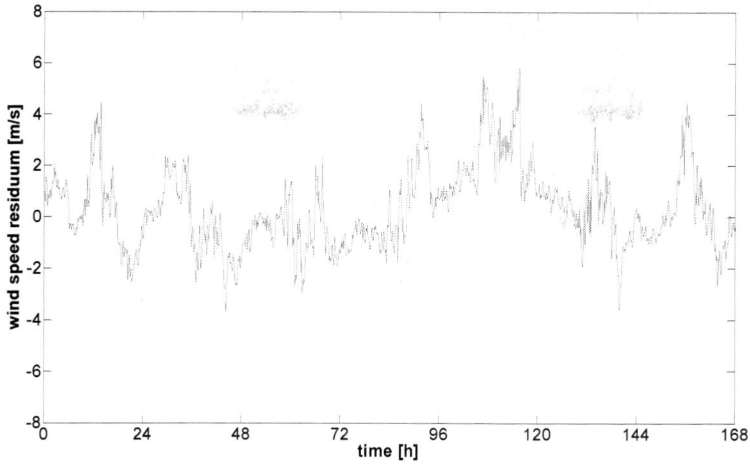

Fig. 8. Non-Gaussian non-stationary simulated sample of 10-min wind speed changes

Parameter	Observations	Simulations
Minimum [m/s]	-10.29	-10.2967
Maximum [m/s]	19.66	19.6647
Mean [m/s]	0	0.0382
Std. Dev [m/s]	3.95	3.9050
Skewness	0.5274	0.6204
Kurtosis	3.4044	3.8279

Table 4. Descriptive first-order statistics of simulated random wind speed residuals

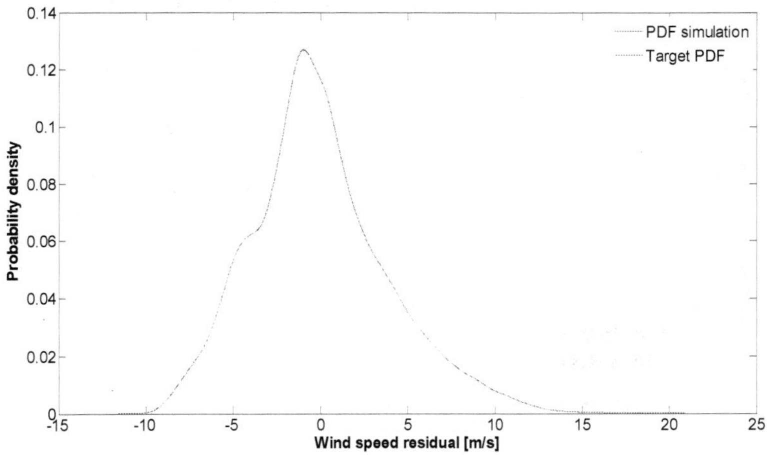

Fig. 9. PDFs of a simulated annual sample and the wind speed fluctuations over a year

In addition to the first-order statistical properties, second-order statistics are compared with the aim at checking for similarity of the stochastic dynamics of simulated samples to the observed random fluctuations. Fig. 10 depicts the frequency content of wind speed fluctuations for both the observed time series and synthetically generated ensemble. The figure shows that the ensemble-average PSD is practically identical to the observed PSD over all frequency bands.

The non-stationary behavior of the synthetically generated ensemble of wind speed residuals is tested by computing the time-varying variance. The local variance computed across the simulated ensemble matches with very high accuracy the observed local variance of the random wind fluctuations, as it can be observed in Fig 11.

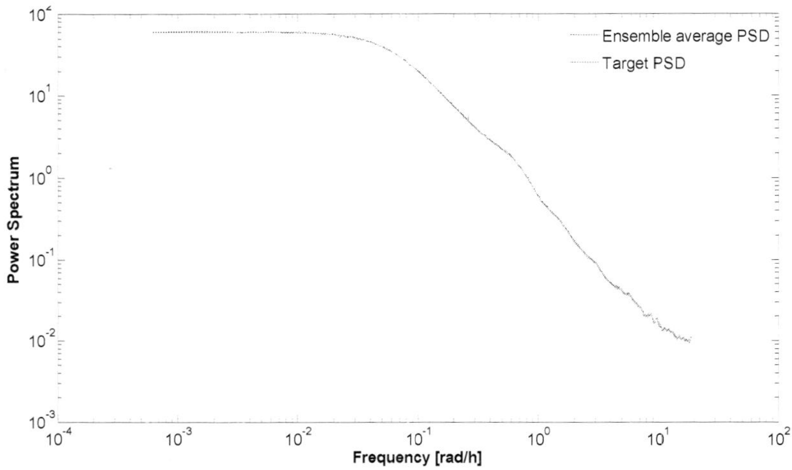

Fig. 10. Power spectra for the simulated ensemble and for observed wind speed fluctuations

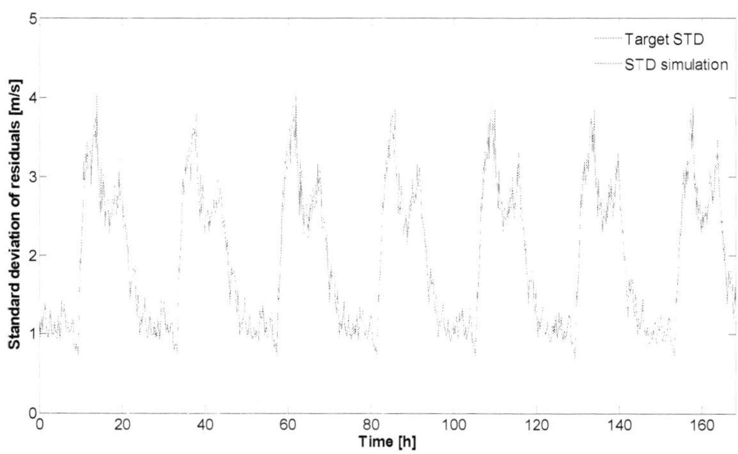

Fig. 11. Local non-stationary variance of simulated samples and observed fluctuations

By adding the deterministic component to the simulated random fluctuations, an ensemble of 10-min average wind speeds can be generated, as shown in Fig. 12. By comparing this plot with Fig. 7, we can see that the most important features of wind are captured by the proposed algorithm. This is also confirmed by comparing the spectral content of wind fluctuations in Fig. 13.
Likewise, first-order statistics provided in Table 5, as well as the probability and occurrence frequency of wind speeds exceeding the cut-off, computed from observations and the simulated ensemble are very similar.

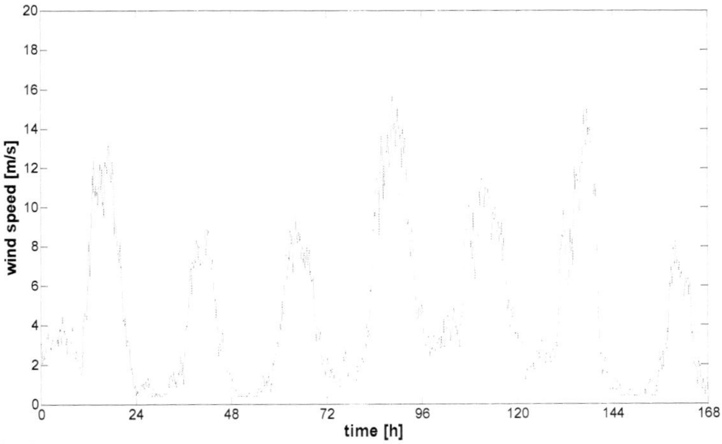

Fig. 12. Simulated sample of 10-min average wind speeds for week in January

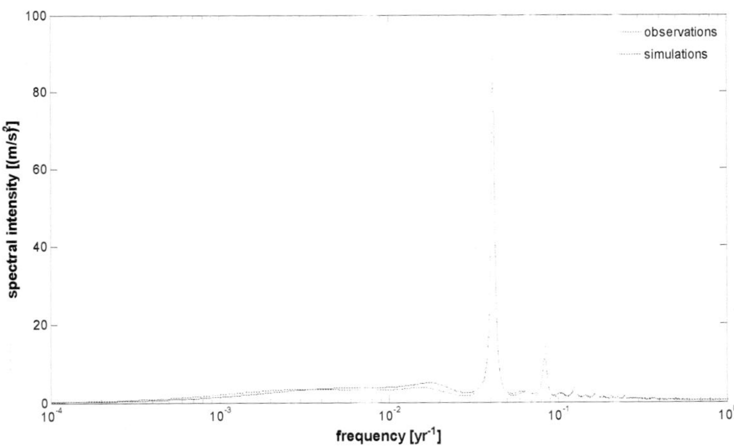

Fig. 13. Mean spectral intensity of observed and simulated 10-min wind speeds

Parameter	Observed	Simulated
Minimum [m/s]	0.4	0.4
Maximum [m/s]	26.42	32.2096
Mean [m/s]	7.0554	7.1879
Std. dev. [m/s]	5.0641	4.9590
Skewness	0.4361	1.6327
Kurtosis	2.1028	3.9796
Pr(v>25 m/s)	0.0003044	0.00024507
Up-crossing freq. [yr^{-1}]	4.00	6.023

Table 5. Descriptive first-order statistics of simulated 10-min average wind speeds

For simulating the fast power fluctuations introduced by the wind generator, simulations must incorporate the rapid wind speed changes due to the turbulence phenomena. Wind turbulence is typically regarded a Gaussian frequency-modulated non-stationary stochastic process. In order to apply the same simulation algorithm, we assume that the turbulence can still be modeled as an intensity-modulated process. For this purpose, an intermediate velocity (v = 10 m/s) for the Kaimal spectrum of wind turbulence has been selected (Kaimal *et al.*, 1972). The envelope function modulating the local variance of the turbulence is determined by imposing the condition that the turbulence intensity remains constant. The mean turbulence intensity of simulation and observations is 0.1558 and 0.1613 respectively. Fig. 14 illustrates the excellent agreement of the probabilistic and spectral properties of simulated turbulence samples when compared to observed data.

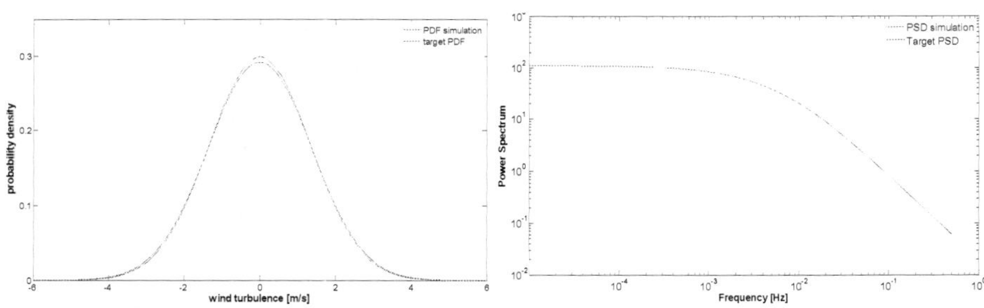

Fig. 14. PDF (left) and PSD (right) of the simulated and observed wind turbulence

The non-stationary (local-mean dependent) wind turbulence is added to the 10-min average wind speed time series in order to obtain wind speed samples at very high resolution (1 sec). For doing this, the 10-min wind speed time series are resampled at the desired temporal resolution by means of filtering techniques. A generated 24-h sample of wind speeds resulting from summing the deterministic component, the random fluctuations of 10-min wind speed averages and the wind turbulence simulated at 1 Hz is depicted in Fig. 15. The plot also shows in red the resampled 10-min average wind speed. We can observe that wind turbulence depends on the prevailing mean wind speed. The observed fast excursions of the wind speeds require scheduling significant balancing resources in order to compensate for the related wind power fluctuations.

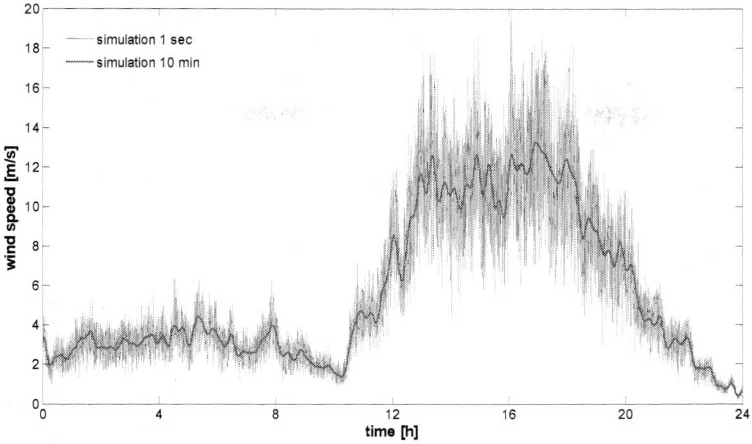

Fig. 15. Wind speed sample simulated at high resolution (1 second) for a day in January

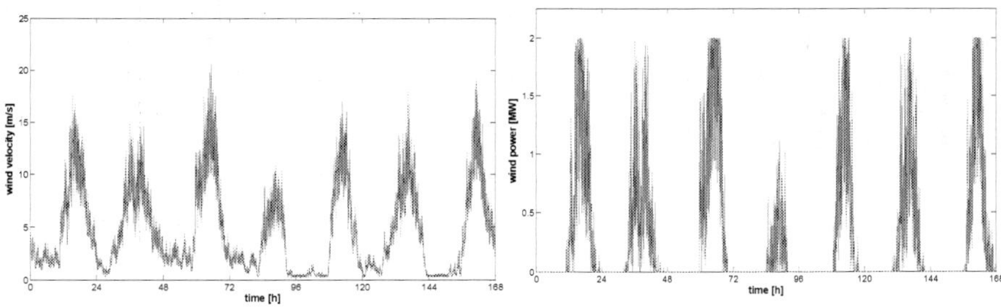

Fig. 16. Simulated 10-sec wind speeds and wind power samples for a week in January

Generated wind samples are averaged on a 10-sec-width moving window for matching the time resolution of simulations of the system operating conditions. By applying to the wind speed samples the non-linear transformation given by the wind speed - power curve of the turbine (see eqs. 11 and 12, and Fig. 3), wind power time series can be generated. Fig. 16 illustrates the resulting wind power time series from a given wind speed sample.

3.4 Modeling load demand fluctuations

The developed spectral-based simulation algorithm described in the previous section has been applied without modifications to the simulation of random fluctuations of the power demand. Load samples are simulated according to observed spectral and probabilistic properties of load measurements. Fig. 17 shows a weekly sample of the load demand simulated at very high resolution and averaged on a moving window spanning 10 s.

3.4 Prediction of the residual load demand

The residual load L_R is the power demand that must effectively supplied by the conventional generation system. As the variable costs of wind generation are considered

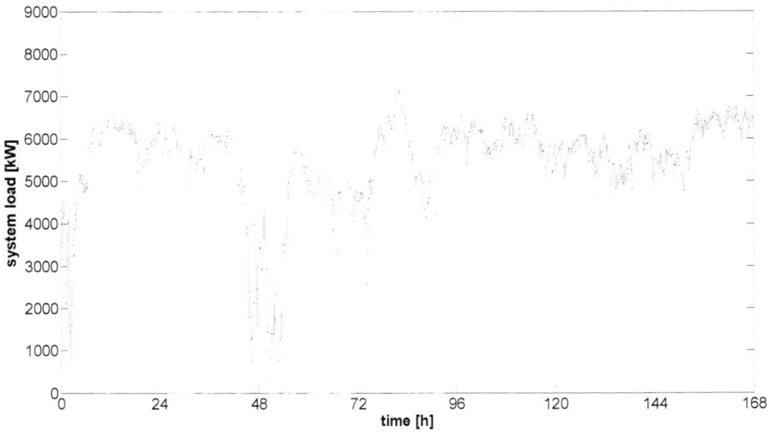

Fig. 17. Simulated 10-sec mean of the system power demand over a week

negligible, wind is dispatched first if available. Therefore, the residual load can be computed by discounting wind generation P_{WV} from the power demand L. For optimally scheduling the diesel units, the residual load must be predicted for each time interval t of the optimization horizon. Here, we assume that the prediction of the 5-min mean of the residual demand $\overline{L}_R^t$ coincides with its expected value, i.e. the forecast is unbiased. Mathematically, the residual load prediction is given by:

$$\chi_{L_R}^t = \mathrm{E}\left[\overline{L}_R^t\right] = \mathrm{E}\left[\overline{L}^t - \overline{P}_{WV}^t\right] = \mathrm{E}\left[\frac{1}{T}\sum_{t=kT}^{(k+1)T}\left(L^t - P_{WV}^t\right)\right];\ T = 300\text{ s};\ k = 0,1,2,\ldots,287 \qquad (13)$$

The forecast of the 5-min average residual load computed for each five minutes of the considered 24 h optimization horizon is depicted in Fig. 18.

3.5 Simulating the integrated operation of the hybrid generation system

For each set (particle) of generation and reserve scheduling decisions proposed by the optimization algorithm, the expected operating and interruption costs are accurately estimated by chronologically simulating the integrated operation of the diesel units and the wind generator over a large number of possible realizations. The software package Stochastic Wind Diesel Simulator (SWDS®) has been applied for this purpose (Olsina & Larisson, 2008c). The optimal load sharing (referred also as economic dispatch, ED) among the synchronized units is run each 10 s of the 24-h optimization horizon and under 100 independent daily realizations of the uncertain variables. Therefore, the evaluation of the fitness of each particle position entails solving 864000 time-coupled ED problems.

Minimum and maximum unit power output, up and down ramping constraints, as well as synchronization times, shut-down times, minimum up and down times are strictly enforced by the simulation algorithm. The processes of starting, synchronizing and ramping a committed unit as well as downramping a decommited unit are detailed modeled. Power output of units ramping-up after synchronization or ramping-down for being decommited are considered as must-run power for purpose of solving the ED problem.

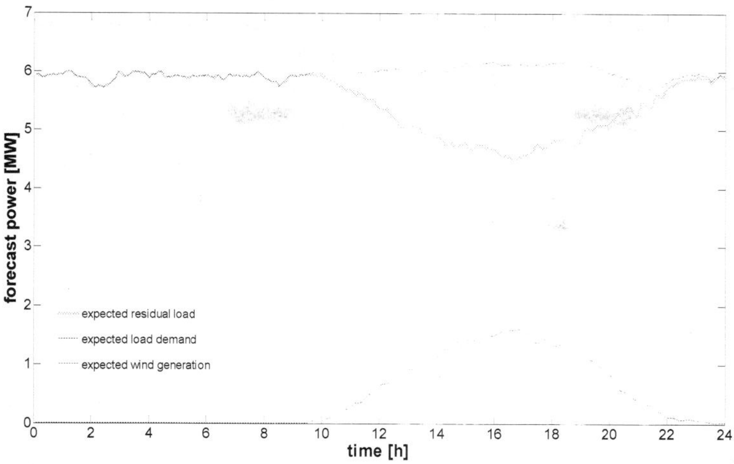

Fig. 18. Expected value of 5-min residual load average for a time horizon of 24 hours

The simulation model also takes into consideration all possible situations under which the system is unable to keep the momentary power balance. For example, a large upward excursion of the demand may lead to exhaust the spinning reserve carried on synchronized generators. Under such circumstances, load shedding is needed in order to prevent a system collapse. Likewise, rapid wind power fluctuations may exceed the aggregate ramping capacity available on synchronized units. Even though the amount of the spinning reserve may be sufficient to balance the magnitude of such fluctuations, ramping constraints impose severe limits to the speed it can effectively be deployed and again load curtailments may turn necessary.

The considered exemplary hybrid stand-alone generation system lacks of automatic load disconnection devices in order to protect the system from collapsing due to frequency instability. Therefore, if for any reason the conventional diesel system is unable to meet the instantaneous residual load a system collapse follows. The system load cannot begin to be restored before 10 minutes after collapse but the beginning of the restoration does not exceed 15 minutes. After a system collapse, the load is restored as quickly as possible by starting and synchronizing all available diesel units. The time elapsed between the generator receives the starting signal and that it delivers the prime power rating is 150 s. The wind generator does not participate during the load restoration period. The wind turbine can only be synchronized 5 minutes after the load restoration is concluded. A battery system supports all auxiliary loads while the system runs on black start until the diesel units are synchronized. The deficit power is computed as the unserved load of each time interval. The interruption costs expressed as per unserved energy unit are estimated in 9 US$/kWh.

4. Numerical results

4.1 Stochastic simulation

In this section, results from high-resolution simulations of the integrated operation of the autonomous wind-diesel generation system under a number of stochastic realizations of the exogenous variables, i.e. load demand, wind resource and random generator outages, are shown. Simulations are run on a single PC with a 2.33 GHz Intel Core 2 Quad processor and 4 GB of RAM. The CPU time required for computing a set of 100 independent daily simulations of the system operation at 10 second resolution is about 180 seconds.

The chronological simulation of the system operation for a large number of possible scenarios allows exploring and analyzing events of low probability under which the system operates near its limits or in emergency conditions. The expected operating costs (fuel costs and startup costs) as well as the expected interruption costs can be accurately assessed. Fig. 19 illustrates the chronological simulation of the operating conditions of the hybrid generation system at 10 second time resolution.

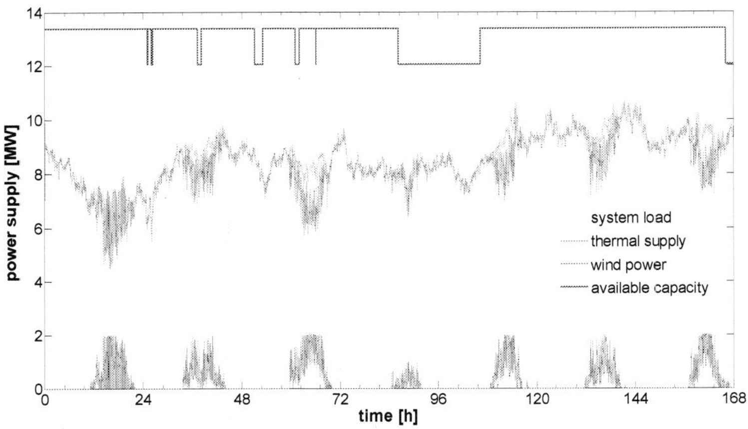

Fig. 19. Simulated chronological operation of the wind-diesel system for a week in January

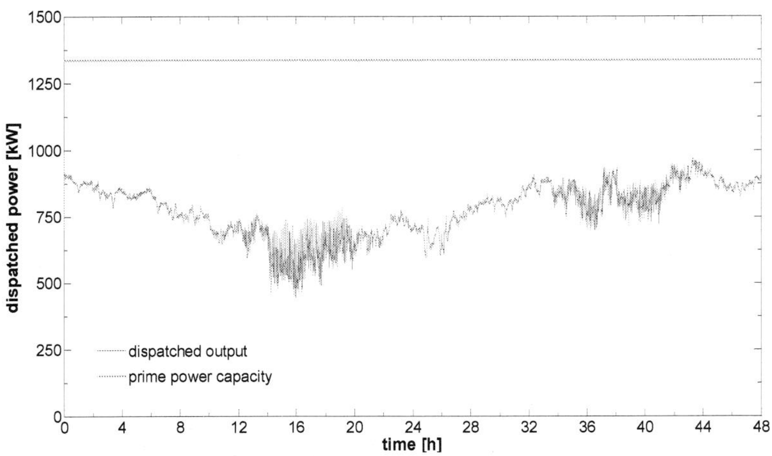

Fig. 20. 10-sec simulation of the dispatched power of a diesel generator over a 48-h period

It can be clearly observed that operating conditions are much more volatile and irregular during the time periods the wind turbine is generating. This can be better analyzed by plotting the power output of a diesel generator for two days as shown in Fig. 20.

4.2 Stochastic optimization

Solving the optimization problem involves finding the combination of scheduled units that minimizes the expected system costs over the day. Considering a system of 10 diesel gensets and 288 scheduling intervals, the total number of binary control variables is 2880. The size of the search space is given by the number of all possible permutations P_r^n of $r = 2880$ elements taken from a binary set with size $n = 2 \cdot 2880 = 5760$:

$$P_r^n = \frac{n!}{(n-r)!} = \frac{5760!}{2880!} = 4.3249 \cdot 10^{10722} \tag{14}$$

The massive dimension of the problem turns it intractable for conventional optimization methods. Thus, the EPSO algorithm is applied for solving the combinatorial stochastic optimization problem, as stated in eq. 6.

Particles are binary coded in order to represent the schedule status of diesel generators for each five-minute time interval over the 24-h optimization horizon. For reducing wasted CPU time, all proposed generation schedules are checked for infeasibility before simulating the system operation. Specific knowledge on the scheduling problem has been exploited, in order to reduce the problem dimension to 864 binary variables. Accordingly, the search space is reduced to $1.7691 \cdot 10^{2682}$, which is still a very huge number. In addition, some known good schedules are provided as initial solutions to the algorithm for speeding up convergence.

A population of 85 particles has been selected for finding the best solution. The swarm stops evaluating the search hyperspace after 20 iterations. By applying evolutionary techniques, the EPSO self-adapts progressively the coefficients and weights of the searching algorithm. Solving these stochastic optimization problem entail running the SWDS simulation engine 420 times for evaluating the fitness of the proposed solution, which requires about 21 CPU hours on 5 desktop PC. The expected total costs of the best generation scheduling found after 20 iterations is 33241.26 US$/day. The unit scheduling that minimizes the expected generation and production costs over the optimization horizon along with the expected residual load is shown in Fig. 21. The available power capacity on the synchronized units in excess of the forecasted residual load is the scheduled spinning reserve.

It is interesting to note that despite the expected residual load is lower than at other time periods, an additional diesel unit is needed for about 50 minutes at about 8:00 PM. This occurs because of the high uncertainty and variability of wind at these moments.

A breakdown of the expected system costs evaluated at the best found scheduling policy is provided in Table 6.

The total system cost is itself a random variable as it depends on stochastic factors affecting the system operation. The probability density function of the total system costs determined at the best solution found from a sample of 100 daily realizations of the operating conditions is depicted in Fig. 22.

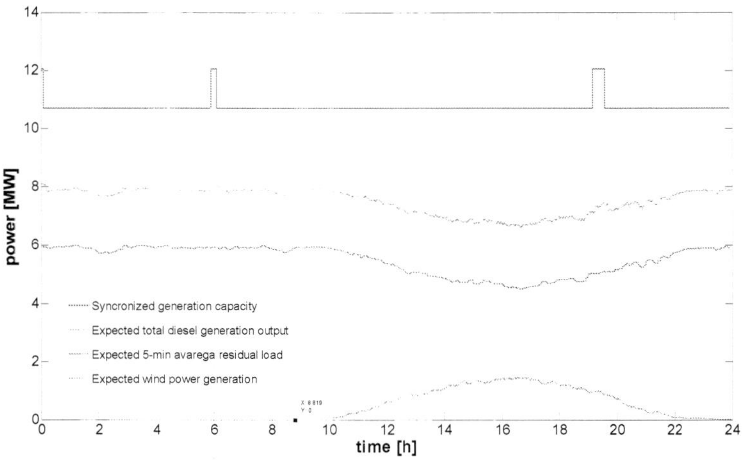

Fig. 21. Optimal scheduling of diesel units over the 24-h optimization horizon

System cost statistics	**Value** [US$/day]
Expected fuel costs	32893.31
Expected lube oil costs	335.94
Expected total generation costs	33229.25
Expected annual starts costs	1 2.01
Expected total operating costs	33241.26
Expected unserved energy costs	0.00*
Expected total system costs	**33241.26**

* The actual expected interruption costs must always be higher than zero. The relatively low sample size of 100 realizations does not allow a better estimation of the expected costs of the energy not served.

Table 6. Cost statistics at optimum

Since operating conditions are typically time-varying over the day, the expected total operating costs are not constant and vary with the time interval considered. The chronological simulation of the expected system costs for each time interval evaluated with the optimal scheduling solution found is plotted in Fig. 23. Clearly, it can be identified the fuel cost reduction during the time periods the wind turbine supply the system.

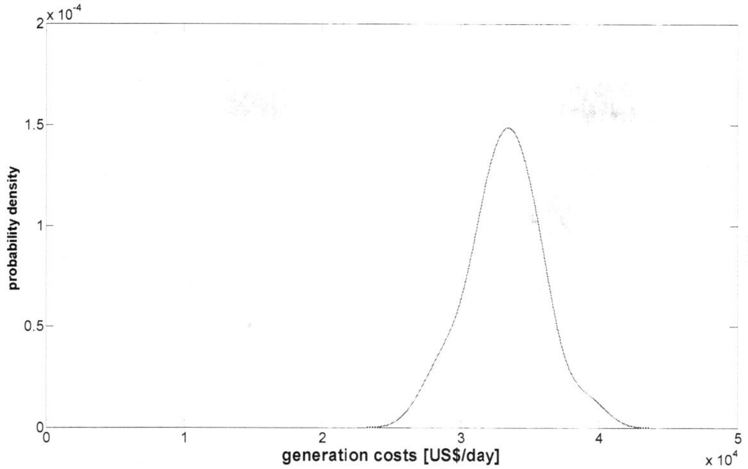

Fig. 22. Probability distribution of the daily total system costs

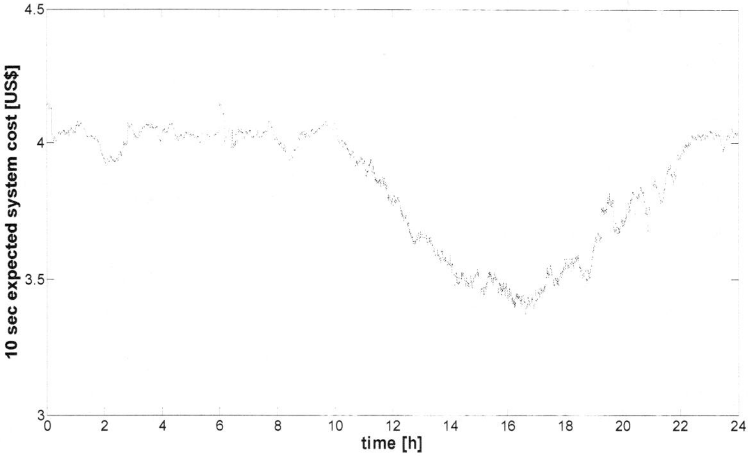

Fig. 23. Time-varying expected total system costs over the 24-h optimization horizon

5. Conclusion

The economy and the supply reliability level that an autonomous wind-diesel generation system can achieve largely depend on the UC decisions. The scheduling optimization of diesel units is a large-scale time-coupled stochastic non-linear mixed-integer problem. This chapter proposes a new methodology for solving this complex combinatorial optimization problem based on stochastic simulation and hybrid metaheuristic optimization methods.

A number of models accurately representing the several parts of the generation system, its operation as well as models for simulating the exogenous random factors affecting the operating conditions have been developed and described along this chapter. Chronological simulations of the system operation are carried out at a very high temporal resolution for capturing the effect of ramping constraints on the speed the reserve is effectively activated. A hybrid PSO algorithm for finding five-minute generation schedule that minimizes the sum of the expected operating and interruption costs over a 24-h horizon is applied. The spinning reserve is scheduled based on a cost-benefit analysis. The resulting optimal spinning reserve scheduling is uneven over the day, which reflects the changing conditions over the day and the trade-off between the costs of its provision and the benefits it provides in terms of lower interruption costs.

The proposed methodology, though very accurate, is costly in terms of computational effort. Nevertheless, PSO-based optimization techniques are naturally suited to distributed computation. Distributed-based PSO is currently an open avenue for further investigation which would enable algorithms of this kind to be implemented in real time applications, such as in closed-loop unmanned dispatch centers for stand-alone wind-diesel systems. For the purpose of further reducing the computation burden, more expert knowledge on the problem needs also to be introduced.

6. References

AlRashidi, M. R. & El-Hawary, M. E. (2009). A survey of particle swarm optimization applications in electric power systems, *IEEE Transactions on Evolutionary Computation*, 2009 (accepted for publication).

Billinton, R. & Allan, R.N. (1996). *Reliability Evaluation of Power Systems*. Plenum Press, NY.

Billinton, R. & Jingdong, G. (2004). A comparison of four-state generating unit reliability models for peaking units, *IEEE Transactions on Power Systems*, Vol. 19, No. 2, pp. 763–768, 2004.

Castro Sayas, F. & Allan, R.N. (1996). Generation availability assessment of wind farms, *IEE Proceedings on Generation, Transmission & Distribution*, Vol. 143, pp. 507 - 518, 1996.

del Valle, Y.; Venayagamoorthy, G.K.; Mohagheghi, S.; Hernandez, J.-C. & Harley, R.G. (2008). Particle swarm optimization: Basic concepts, variants and applications in power systems, *IEEE Transactions on Evolutionary Computation*, Vol. 12, No. 2, pp. 171-195, 2008.

Deodatis, G. & Micaletti, R. (2001). Simulation of highly skewed non-Gaussian stochastic processes, Journal of Engineering Mechanics, Vol. 127, pp. 1284-1295, 2001.

Eberhart, R. & Kennedy, J. (1995). A new optimizer using particle swarm theory, *Proceedings of 6th International Symposyum of Micro Machine Human Science*, pp. 39–43, 1995.

IEEE Task Group on Peaking Service Units, A four state model for estimation of outage risk for units in peaking service, *IEEE Transactions on Power Apparatus & Systems*, Vol. PAS-91, pp. 618 - 627, 1972.

Kaimal, J.C.; Wyngaard, J.C.; Izumi, Y. & Coté, O.R. (1972). Spectral characteristics of surface-layer turbulence, *Quarterly Journal of Royal Meteorological Society*, Vol. 98, pp. 563-589, 1972.

Kennedy, J. & Eberhart, R. (1995). Particle swarm optimization, *Proceedings of IEEE International Conference Neural Network*, Vol. 4, pp. 1942–1948, 1995.

Lee, T.-Y. & Chen, C.-L. (2007). Unit commitment with probabilistic reserve: An IPSO approach, *Energy Conversion and Management*, Vol. 48, No. 2, pp. 486-493, 2007.

Miranda, V. & Fonseca, N. (2002a). EPSO best of two worlds meta-heuristic applied to power system problems, *Proceedings of IEEE Congress on Evolutionary Computation*, Vol. 2, pp. 1080-1085, May 2002.

Miranda, V. & Fonseca, N. (2002b) EPSO - Evolutionary Particle Swarm Optimization, a new algorithm with applications in power systems, *Proceedings of the IEEE Power Engineering Society Transmission and Distribution Conference*, Vol. 2, pp. 745-750, Asia Pacific, Oct. 2002.

Miranda, M. & Win-Oo, N. (2006). New experiments with EPSO - Evolutionary particle swarm optimization, *Proceedings of the IEEE Swarm Intelligence Symposium*, pp. 162-169, Indianapolis, Indiana, USA, May 2006.

NERC (2006). Generating Availability Report (GAR), *North American Reliability Corporation*, November 2006. Available Online: http://www.nerc.com.

Olsina, F. & Larisson, C. (2008a). Iterative procedure for massive simulation of non-stationary, non-Gaussian, 10- min mean wind speed samples by means of spectral representation, *INAR Technical Report TR-003-2008*, San Juan, Argentina, 2008.

Olsina, F. & Larisson, C. (2008b). Two-stage simulation of non-stationary, uniformly modulated wind speed turbulence, *INAR Technical Report TR-004-2008*, San Juan, Argentina, 2008.

Olsina, F. & Larisson, C. (2008c). SWDS® – Stochastic Wind-Diesel Simulator – Methodology Manual, *INAR Technical Report TR-002-2008*, San Juan, Argentina, 2008.

Olsina, F. & Weber, C. (2009). Stochastic simulation of spot power prices by spectral representation, *IEEE Transactions on Power Systems* (accepted for published).

Padhy, N.P. (2004). Unit commitment - A bibliographical survey, *IEEE Transactions on Power Systems*, Vol. 19, No. 2, pp. 1196-1205, 2004.

Sen, S. & Kothari, D.P. (1998). Optimal thermal generating unit commitment: A review, *International Journal of Electrical Power and Energy Systems*, Vol. 20, No. 7, pp. 443-451, 1998.

Sheble, G. B. & Fahd, G. N. (1994). Unit commitment literature synopsis, *IEEE Transactions on Power Systems*, Vol. 9, No. 1, pp. 128-135, 1994.

Shinozuka, M. & Deodatis, G. (1991). Simulation of stochastic processes by spectral representation, *Applied Mechanical Review*, Vol. 44, pp. 191-203, 1991.

Swaroop, P.V.; Erlich, I.; Rohrig, K. & Dobschinski, J. (2009). A stochastic model for the optimal operation of a wind-thermal power system, *IEEE Transactions on Power Systems*, Vol. 24, No. 2, pp. 940-950, 2009.

Ting, T.O.; Rao, M.V.C. & Loo, C.K. (2006). A novel approach for unit commitment problem via an effective hybrid particle swarm optimization, *IEEE Transactions on Power Systems*, Vol. 21, No. 1, pp. 411-418, 2006.

Yamin, H.Y. (2004). Review on methods of generation scheduling in electric power systems, *Electric Power Systems Research*, Vol. 69, No. 2-3, pp. 227-248, 2004.

Zhao, B.; Guo, C.X.; Bai, B.R. & Cao, Y.J. (2006). An improved particle swarm optimization algorithm for unit commitment, *International Journal of Electrical Power & Energy Systems*, Vol. 28, No. 7, pp. 482-490, 2006.

20

Power Characteristics of Compound Microgrid Composed from PEFC and Wind Power Generation

Shin'ya Obara
Dep. of Electrical and Electric Eng., Power Eng. Lab., Kitami Institute of Technology
Japan

1. Introduction

It is predicted that a micro-grid technique is effective about a backup power supply in an emergency, a peak cut of power plants, and exhaust heat utilization. Furthermore, when renewable energy is connected to a micro-grid, there is potential to reduce the amount of greenhouse gas discharge (Abu-Sharkh *et al.*, 2006, Carlos & Hernandez, 2005, Robert, 2004). A micro-grid has an interconnection system with commercial power etc., and the independence supplying system of the power. The micro-grid with an interconnection system outputs and inputs the power between other grids. Therefore, the dynamic characteristic of the grid is influenced by the grid of a connection destination. When a micro-grid and a large-scale grid such as a commercial power system are interconnected, the dynamic characteristics of the power depend on the commercial power system. For this reason, in the micro-grid of the interconnection type, the option of the equipment to connect is wide. On the other hand, since micro-grid can reduce transportation loss of power and heat, this technique may become the major energy supply. The method of connecting two or more small-scale fuel cells and renewable energy equipment by a micro-grid, and supplying power to the demand side is effective in respect of environmental problems. So, this paper examines the independent micro-grid that connects fuel cells and wind power generation. In order to follow load fluctuation with an independent grid system, there are a method of installing a battery and a method of controlling the output of power generators. Since the battery is expensive, in this paper, it corresponds to load fluctuation by controlling the power output of the fuel cell. The output adjustment of the fuel cell has the method of controlling the production of electricity of each fuel cell, and the method of controlling the number of operations of the fuel cell. However, adjustment of the production of electricity of each fuel cell connected to the micro-grid may operate some fuel cell with a partial load with low efficiency. So, in this paper, the number of operations of fuel cells is controlled to follow fluctuations in the electricity demand.

In an independent micro-grid, a certain fuel cell connected to the micro-grid is chosen, and it is considered as a power basis. The power (voltage and frequency) of the other fuel cells is controlled to synchronize with this base power. Therefore, if the fuel cell that outputs base power is unstable, the power quality of the whole grid will deteriorate. Fuel cells other than base load operation are controlled to synchronize with the base power. The power quality

(voltage and frequency) of the micro-grid depends on the difference in the demand-and-supply balance.
A 2.5 kW fuel cell is installed in one house of the micro-grid formed from ten houses. This fuel cell is operated corresponding to a base load. A 1 kW fuel cell is installed in seven houses, and a 1.5 kW wind power generator is connected to the micro-grid. According to the difference in electricity demand of the grid and power produced by the wind power generator, the number of operations of 1 kW fuel cells is controlled. A city gas reformer is installed in houses in which fuel cells are installed, and hydrogen is produced by city gas reforming. By adding random fluctuation to an average power load pattern, the power demand of a general residence is simulated and it uses for analysis. The dynamic characteristics of the micro-grid and the efficiency of the system that are assumed in this paper are investigated by numerical analysis.

2. Micro-grid model

2.1 System scheme

Figure 1 shows the fuel cell independent micro-grid model investigated in this paper. There is a network of the power and city gas in this micro-grid. Although a power network connects all houses, a city gas network connects houses in which a fuel cell is installed. The fuel cell installed in each house is a proton exchange membrane type (PEM-FC). The output of a 2.5kW fuel cell is decided to be a base power of the micro-grid. Moreover, PEM-FC of 1 kW power is installed in seven houses. However, the fundamental dynamic characteristics of all the fuel cells are the same, and a fuel cell and a city gas reformer are installed as a pair. One set of wind power generator is installed, and the power produced by wind force is supplied to a micro-grid through an inverter and an interconnection device. The power supply of the micro-grid assumes 50-Hz of the single-phase 200 V.

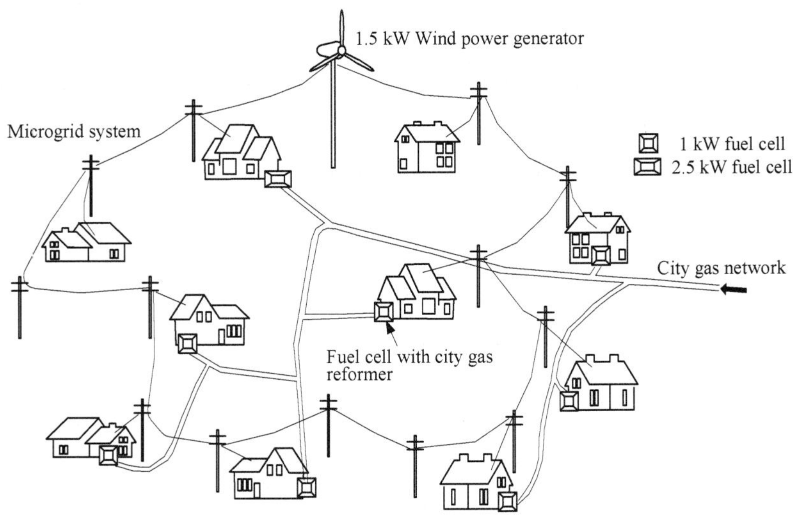

Fig. 1. Fuel cell micro-grid system with wind power generator

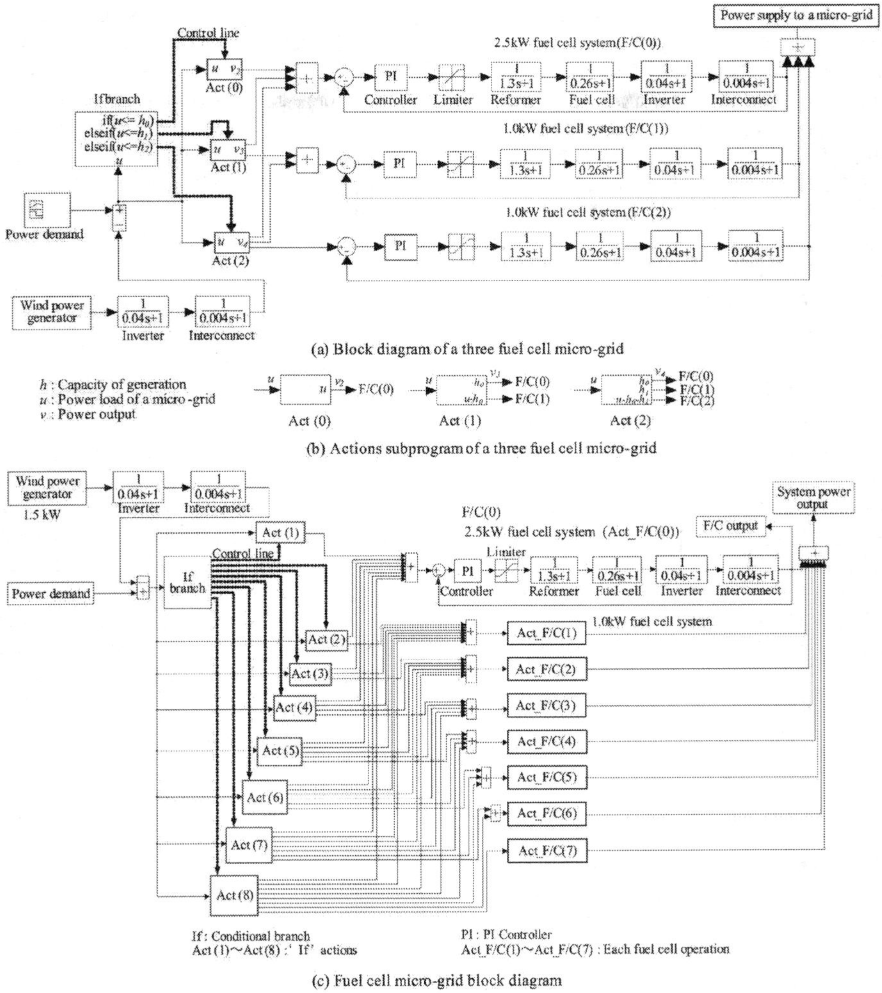

Fig. 2. System block diagram

2.2 System control

Figure 2 (a) is a block diagram of the micro-grid formed from three sets of fuel cell systems of F/C(0) to F/C(2) and one wind power generator. A fuel cell system consists of a controller, a power limitation device (Limiter), a reformer, a fuel cell, an inverter, and a system interconnection device. F/C(0) is a fuel cell corresponding to a base load, and operates F/C(1) and F/C(2) with the magnitude of load. The production of electricity required for F/C(2) from F/C(0) is taken as the value excluding the electric energy produced by wind power generation from the amount of electricity demand. The power of a

wind power generator is supplied to the grid through an inverter and a system interconnection device. Section 3.5 describes the dynamic characteristics of an inverter and a system interconnection device. The power generated by each fuel cell is decided by "If branch" and Act(0) to Act(2) in Fig. 2(a). Figure 2(b) shows the input and output of each block of Act(0) to Act(2). u expresses the power load and v_2 to v_4 expresses the output power in the block (from Act(0) to Act(2)) that branches in the magnitude of u. Moreover, h_0 and h_1 express the power generation capacity of the fuel cell of F/C(0) and F/C(1), respectively. In this system, when the value of u exceeds capacity h_0 of F/C(0), F/C(1) is operated first. F/C (2) is operated when the production of electricity is still less than the value of u. Thus, the number of operations of a fuel cell is controlled by the magnitude of the load added to the grid. The value except the power produced by wind power generation from electricity demand is the production of electricity required of fuel cell systems. Act(0) to Act(2) is chosen from magnitude (u) of the load, and the capacities of the fuel cells under IF conditions. In Act (0) to Act (2), as Fig. 2 (b) shows, the production of electricity of each fuel cell is calculated and outputted. Controlling each fuel cell by PI controller, a limiter limits the production of electricity of a fuel cell. The next section describes each dynamic characteristic of a reformer, a fuel cell, an inverter, and a system interconnection device.
Figure 2 (c) is a block diagram of the system installed in the micro-grid shown in Fig. 1. This system extends the system shown in Fig. 2 (a). F/C (0) is a 2.5-kW fuel cell corresponding to a base load, and F/C (1) to F/C (7) is a 1-kW fuel cell. Moreover, a wind power generator of 1.5-kW is connected to the grid. The dynamic characteristics of a fuel cell system are decided using the dynamic characteristics of a reformer, a fuel cell, and an inverter, and the control variables of a controller and a limiter. This paper shows the dynamic characteristics of each device with the transfer function of a primary delay system, described in the following section. Each parameter of PI control (proportional control (P) and integral control (I)) is given to the controller of a fuel cell system beforehand, and each fuel cell system is controlled.

3. Response characteristic of system configuration equipment

3.1 Power generation characteristic of fuel cell

Figure 3 (a) shows the result of measurement when inputting a load of 45 W into the testing equipment of PEM-FC (maximum output 100 W) stepwise. In the test, the ambient temperature was set to 293 K, and reformed gas and air were supplied to an anode and a cathode, respectively. An approximated curve is prepared from the result of the measurement in Figure 3 (a), and the transfer function of a primary delay is obtained. Strictly, although a transfer function is considered depending on the load factor, it is not taken into consideration because this difference is small by test results.

3.2 Output characteristics of city gas reformer

Figure 3 (b) shows the output model that inputted a load of 100% load factor into the city gas reformer stepwise (Nagano, 2002, Obara & Kudo, 2005, Lindstrom & Petterson, 2003, Oda. 1999, Takeda. 2004, Ibe. 2002). An approximated curve is prepared from the result of the measurement, and the transfer function of the primary delay of the city gas reformer is obtained. As a fuel cell, although the transfer function of a city gas reformer influences the magnitude of the load significantly, since there is no large difference, the result of Figure 3

(b) is used. Compared with the condition of the steady operation of the reformer, the characteristics of a startup and a shutdown differ greatly. Cold start operation and shutdown operation require about 20 minutes, respectively. In the analysis of this paper, it is assumed that the startup of the methanol reformer is always a hot start.

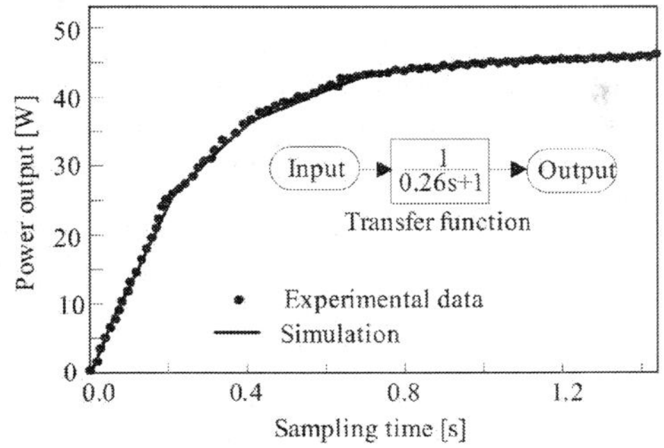

(a) Characteristics of transient response for a PEM fuel cell

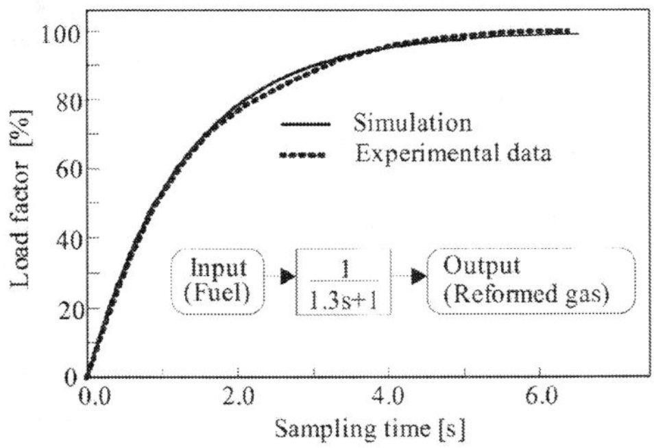

(b) Characteristics of transient response of a reformer

Fig. 3. Response characteristics of system configuration equipment (Oda. 1999, Takeda. 2004, Ibe. 2002)

3.3 Power generation characteristics of wind power generation

The model of power obtained by wind power generation is decided at random between 0 to 1.5 kW for every sampling time, as shown in Figure 4 (a). The power of wind power generator is supplied to a micro-grid through an inverter and a system interconnection device. Figure 4 (b) shows the output model of the wind power generator through an inverter and a system-interconnection device. Because influence is taken in the dynamic characteristic of an inverter and a system-interconnection device, the output of wind power generation is settled on a width of 0.75 kW ±0.25 kW range, as shown in Figure 4 (b). The details of the transfer function of an inverter and a system interconnection device are given with Section 3.5. The dynamic characteristics of the inverter and system interconnection device significantly influence the power output characteristics of wind power generation.

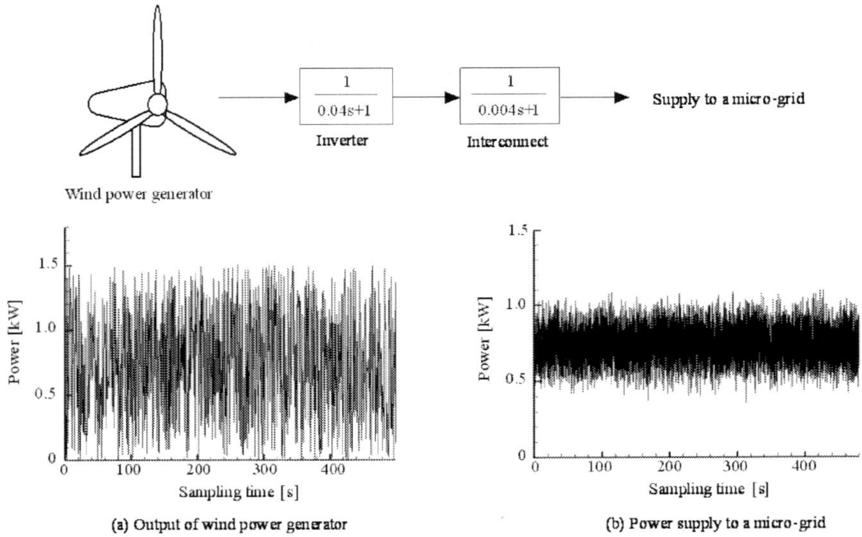

Fig. 4. Output model of wind power generator

3.4 Generation efficiency of the fuel cell system

Figure 5 shows a model of the relation between the load factor of a fuel cell, and generation efficiency (Obara & Kudo, 2005, 2005). Power-generation efficiency is obtained by dividing "the power output of the fuel cell system" by "the city gas calorific power supplied to the system." This model was prepared from the results of the power output when attaching the fuel cell show in Figure 3 (a) to the city gas reformer show in Figure 3 (b). If the load of a fuel cell is given to Figure 5, power generation efficiency is calculable. The maximum efficiency of one set of a fuel cell system is 32%.

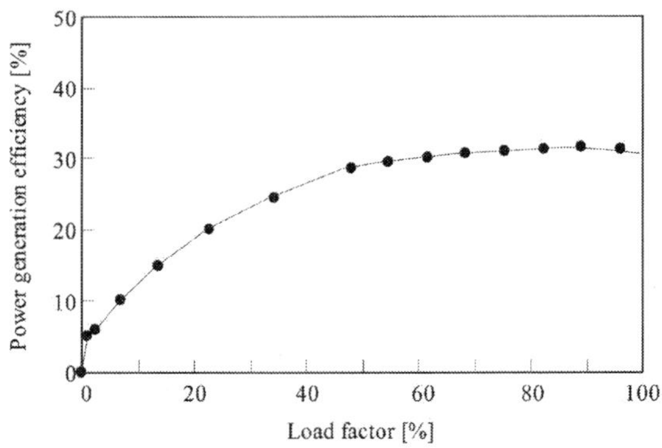

Fig. 5. Output characteristics of a PEM-FC with city-gas reformer

3.5 Inverter and system interconnection device

It is assumed that an inverter of a voltage control type is used, and 120 ms is required to output power on regular voltage and frequency (in this paper, it is less than 95%) (Kyoto Denkiki Co., Ltd. 2001). Figure 6 (a) expresses the transfer function of such an inverter with primary delay.

When changing power with a system interconnection device, the change takes about 10μs (Kyoto Denkiki Co., Ltd. 2001). However, there is the operation of taking the synchronism of the frequency between systems, and the model of the system interconnection device sets the change time to 12 ms. As a result, the transfer function of the system interconnection device by primary delay is shown Figure 6 (b).

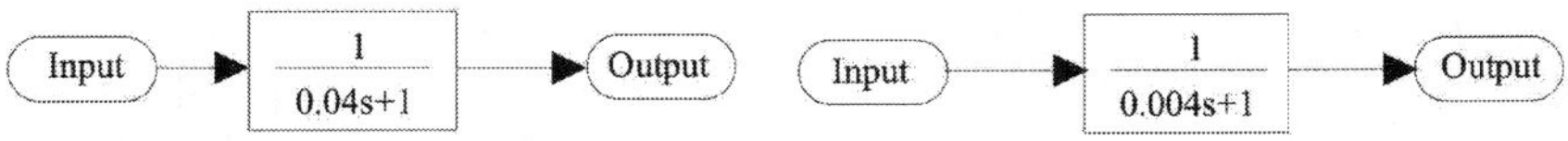

(a) Transfer function of an inverter (b) Transfer function of an interconnection device

Fig. 6. Transfer function of an inverter and interconnection device

4. Control parameters and analysis method

The response characteristics of the 1 kW fuel cell system when inputting 0.2, 0.6, and a 1.0 kW load stepwise is shown in Figure 7. The response characteristics of a fuel cell system changes by the control parameters set up with the controller. As shown in Figure 7 (c), in 1 kW step input, the rising time and settling time (time to converge on ±5% of the target output) are not based on control parameters. In 0.2 kW step input, the rise time of "P = 12.0,

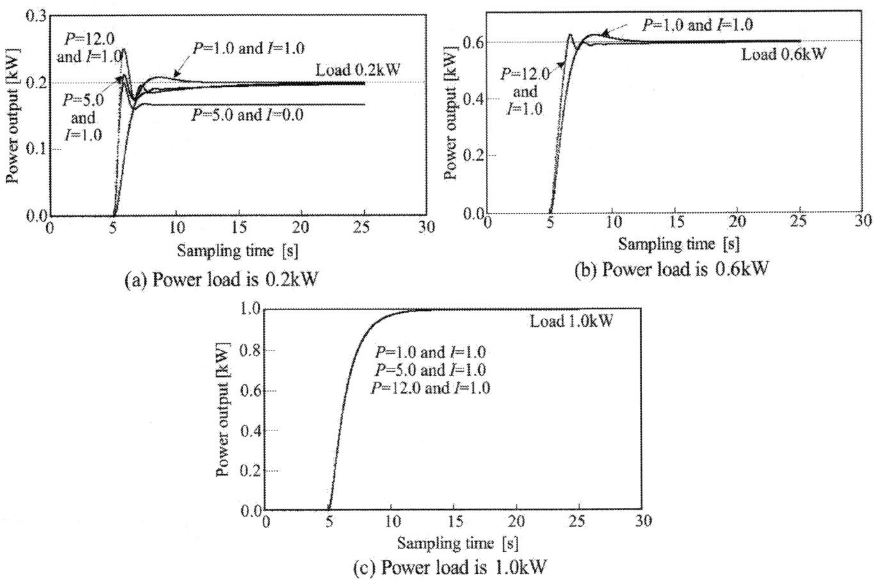

(a) Power load is 0.2kW (b) Power load is 0.6kW (c) Power load is 1.0kW

Fig. 7. Characteristics of electric power output of the system (Obara. 2005)

I = 1.0" is short, and the settling time of "P = 1.0, I = 1.0" is short. In 0.6 kW step input, "P = 12.0, I = 1.0", and "P = 1.0, I = 1.0" have almost the same settling time. Moreover, overshooting is large although the rise time of "P = 12.0, I = 1.0" is short. Considering the following load fluctuations, the control parameters of the fuel cell are analyzed by "P = 12.0, I = 1.0." The dynamic characteristics of a micro-grid are analyzed using MATLAB (Ver.7.0) and Simulink (Ver.6.0) of Math Work Corporation. However, in analysis, the solver to be used is the positive Runge-Kutta system, and this determines the sampling time from calculation converged to less than 0.01% by error.

5. Control parameters and analysis method

5.1 Step response

The response results when applying the stepwise input of 2, 4, 6 or 8 kW to the micro-grid at intervals of 30 seconds are shown in Figure 8 (a). The left-hand side in Figure 8 (a) shows the result of not installing a wind power generator. The right-hand side of the figure shows the result of a installing wind power generator. The maximum power by a overshooting and settling time (time to converge on ±5% of the target output) are described on the left-hand of Figure 8 (a). Moreover, the maximum power due to over shoot is described in the right-hand side figure. The settling time when not installing a wind power generator has the longest

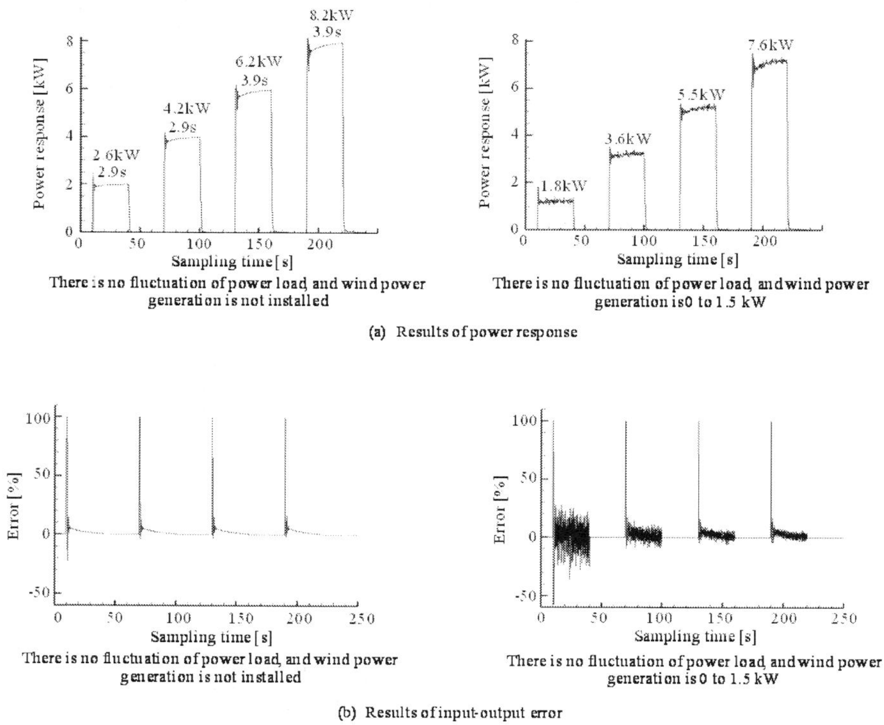

Fig. 8. Results of step response

period of step input of 6 kW and 8 kW for 3.9 seconds. If a wind power generator is connected to the micro-grid, many fluctuations in the system response characteristics will occur in a short period. If the power produced by wind power generation is supplied to the micro-grid, the dynamic characteristics of power of the micro-grid will be influenced. Figure 8 (b) shows the analysis result of the response error corresponding to Figure 8 (a). If wind power generator is connected to the grid, the response error will become large as the load of the grid becomes small. It is expected that the power range of the fluctuation of the micro-grid will increase as the output of the wind power generation grows. Therefore, when the load of a micro-grid is small compared with the output of wind power generator, the power supply of the independent micro-grid becomes unstable.

5.2 Load response characteristics of cold region houses

Figure 9 (a) shows the power demand pattern of a micro-grid formed from ten individual houses in Sapporo in Japan, and assumes a representative day in February (Narita, 1996). This power demand pattern is the average value of each hour, and the sampling time of analyses and the assumption time are written together on the horizontal axis. As a base load of the power demand pattern shown in Fig. 9 (a), F/C (0) is considered as operation of 2.5 kW constant load. Figure 9 (b) and (c) are the power demand patterns when adding load fluctuations (±1 kW and ±3 kW) to Fig. 9 (a) at random. The variation of the load was decided at random within the limits of the range of fluctuation for every sampling time.

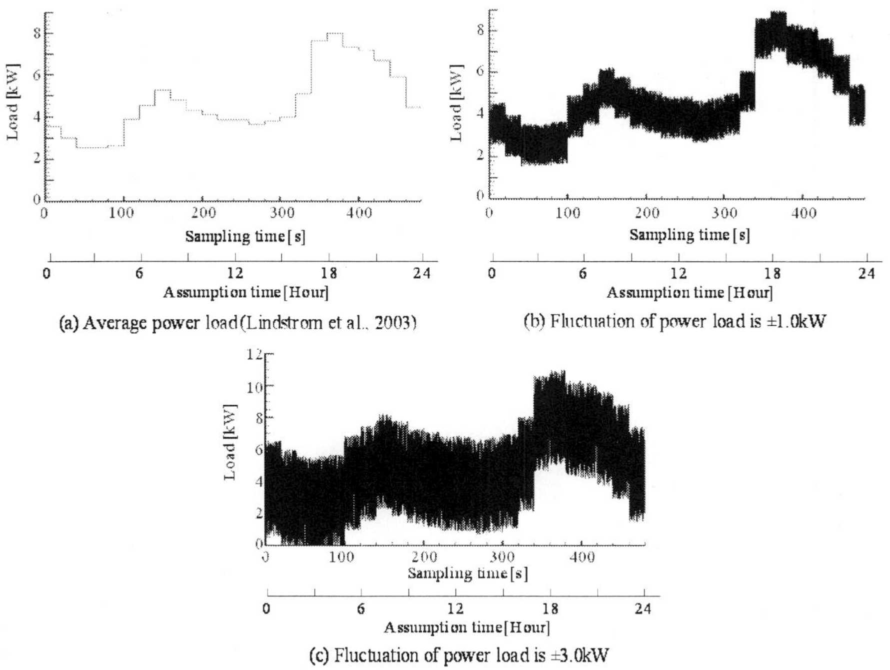

(a) Average power load(Lindstrom et al., 2003)

(b) Fluctuation of power load is ±1.0kW

(c) Fluctuation of power load is ±3.0kW

Fig. 9. 480s demand model for 10 houses in February in Sapporo

Figure 10 shows the response results of F/C (0) to F/C (6) when wind power generation is connected to the micro-grid and the power load has ±1 kW fluctuations. F/C (0) assumed operation with 2.5 kW constant output, with the result that the response of F/C (0) is much less than 2.5 kW in less than the sampling time of 100 s as shown in Figure 10 (a). This reason is because F/C (0) was less than 2.5 kW with the power of wind power generation. Although the micro-grid assumed in this paper controlled the number of operations of F/C (1) to F/C (7) depending on the magnitude of the load, since the power supply of wind power generation existed, there was no operating time of F/C (7).

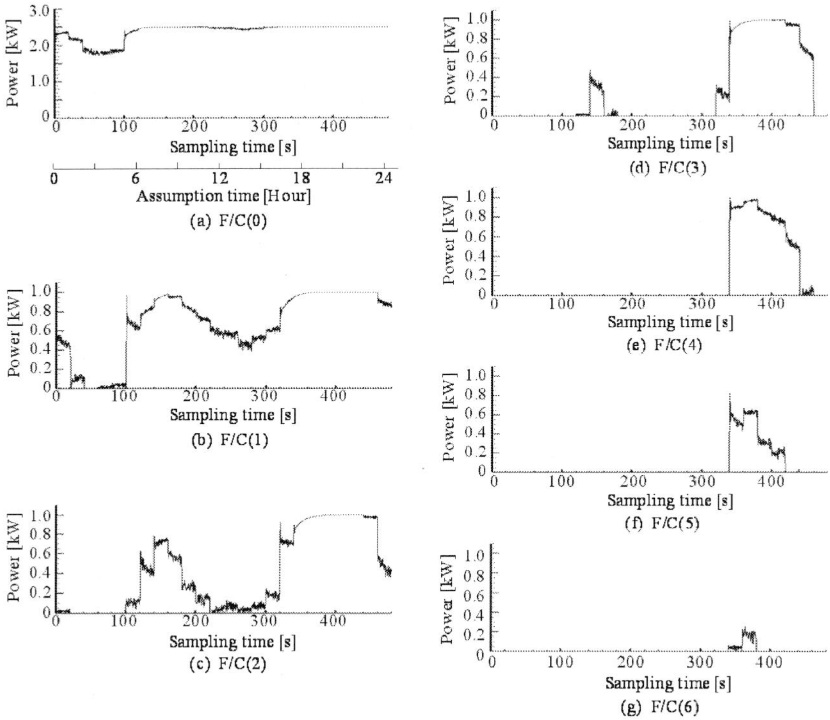

Fig. 10. Response results of each fuel cell

5.3 Power generation efficiency

Figure 11 shows the analysis results of the average power generation efficiency of fuel cell systems for every sampling time. The average efficiency of a fuel cell system is the value averaging the efficiency of F/C (0) to F/C (7) operated at each sampling time. However, the fuel cell system to stop is not included in average power generation efficiency. The average power generation efficiency of Figure 11 (a) is 13.4%, and Figure 10 (b) shows 14.3%. The difference in average efficiency occurs in the operating point of a fuel cell system shifting to the efficient side, when load fluctuations are added to the micro-grid. Thus, if load fluctuations are added to the micro-grid, compared with no load fluctuations, the load factor of the fuel cell system shown in Figure 4 will increase.

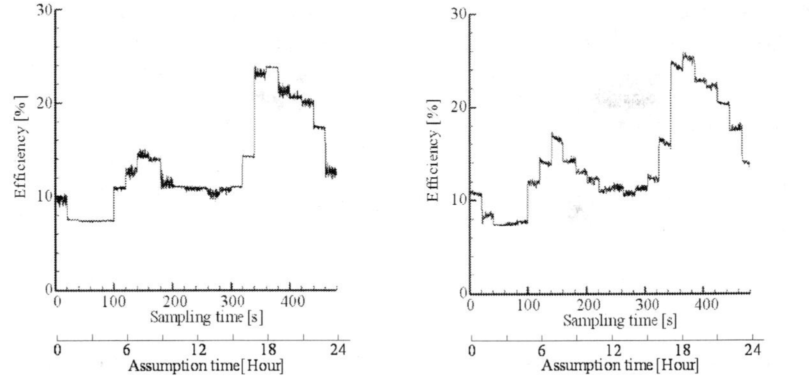

(a) There is no fluctuation of power load and wind power generation is 0 to 1.5 kW

(b) Fluctuation of power load is ±3.0kW and wind power generation is 0 to 1.5 kW

Fig. 11. Results of micro-grid average efficiency

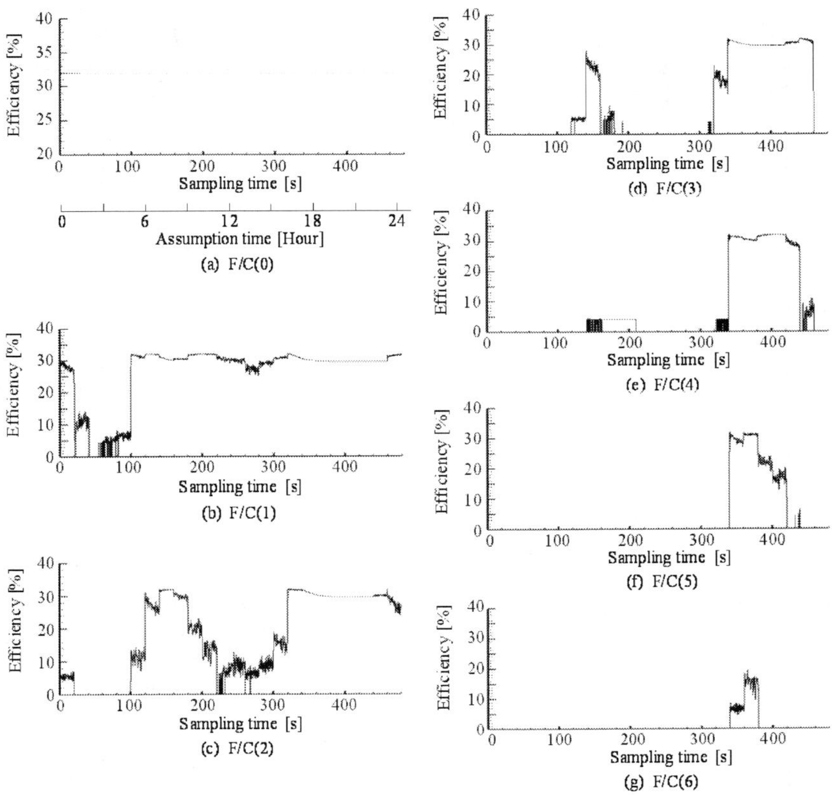

(a) F/C(0)

(b) F/C(1)

(c) F/C(2)

(d) F/C(3)

(e) F/C(4)

(f) F/C(5)

(g) F/C(6)

Fig. 12. Results of efficiency for each fuel cell

Figure 12 shows the power generation efficiency of each fuel cell in the case of connecting wind power generation to the micro-grid of ±1.0kW of load fluctuation. F/C (0) operated corresponding to a base load has maximum power generation efficiency at all sampling times. Since the number of operations of a fuel cell is controlled by the magnitude of the load added to the micro-grid, the operating time falls in the order of F/C (1) to F/C (6). Moreover, there is no time to operate F/C (7) in this operating condition.

The relation between the range of fluctuation of the power load and the existence of wind power generation, and the amount of electricity demand of a representative day is shown Fig. 13. When the load fluctuation of the power is large, although the power demand amount of the micro-grid on a representative day increases slightly, it is less than 2%. Moreover, when installing wind power generation, the power demand amount of the micro-grid of a representative day decreases compared with the case of not installing. This decrement is almost equal to the value that integrated the power (average of 0.75 kW) supplied to a grid by the wind power generation of Fig. 4 (b).

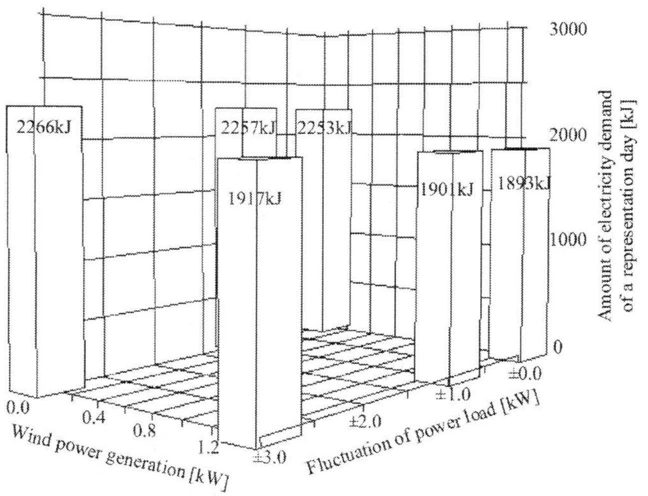

Fig. 13. Amount of 480s demand model with power fluctuation and wind power generation

Figure 14 shows the range of fluctuation of power load and the existence of wind power generation, and the relation to city gas consumption on a representative day of the micro-grid. If the range of fluctuation of the power load becomes large, city gas consumption will decrease. This is because electric power supply cannot follow the load fluctuations of the micro-grid if the range of fluctuation of the power load is large. Moreover, in ±3 kW of load fluctuation, some loads become zero (it sees from 20s to 100s of sampling times) and city gas consumption lowers. In ±3 kW of load fluctuation of the power, it is expected that the power of a micro-grid is unstable and introduction to a real system is not suitable.

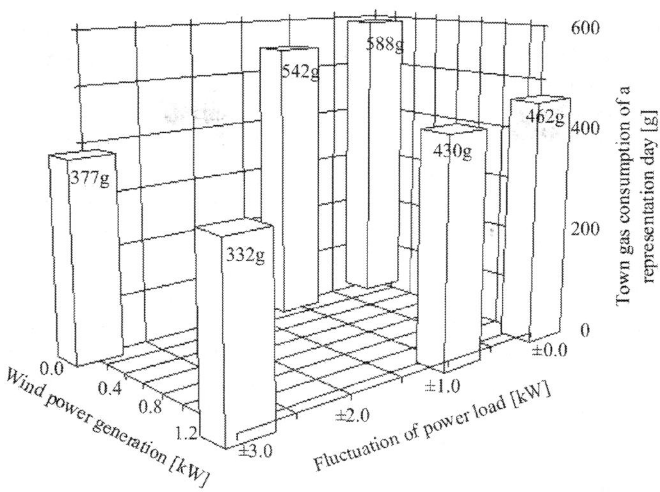

Fig. 14. Analysis result of town gas consumption for 480s demand model with power fluctuation and wind power generation

6. Conclusions

A 2.5 kW fuel cell was installed in a house linked to a micro-grid, operation corresponding to a base load was conducted, and the dynamic characteristics of the grid when installing a 1 kW fuel cell system in seven houses were investigated by numerical analysis. A wind power generator outputted to a micro-grid at random within 1.5 kW was installed, and the following conclusions were obtained.

1. Although the settling time (time to converge on ±5% of the target output) of the micro-grid differs with the magnitude of the load, and the parameters of the controller, it is about 4 seconds.
2. When connecting a wind power generator to the micro-grid, the instability of the power of the grid due to supply-and-demand difference is an issue. This issue is remarkable when the load of an independent micro-grid is small compared to the production of electricity of unstable wind power generation.
3. When wind power equipment is connected to the micro-grid with load fluctuation, the operating point of the fuel cell system may shift and power generation efficiency may improve.

7. Acknowledgements

This work was partially supported by a Grant-in-Aid for Scientific Research(C) from the JSPS.KAKENHI (17510078).

8. Nomenclature

Act : " If " action

Act_FC	:	Each fuel cell operation
F/C	:	Fuel cell
h	:	Capacity of generation W
I	:	Integral parameter
P	:	Proportionality parameter
PI	:	Proportion integration control
u	:	Power load of a micro-grid W
v	:	Power output W

9. References

Abu-Sharkh, S.; Arnold, R. J.; Kohler, J.; Li, R.; Markvart, T.; Ross, J. N.; Steemers, K.; Wilson, P. & Yao, R. (2006). Can microgrids make a major contribution to UK energy supply?. *Renewable and Sustainable Energy Reviews*, Vol. 10, No. 2, pp. 78-127.

Carlos, A. & Hernandez, A. (2005). Fuel consumption minimization of a microgrid. *IEEE Transactions on Industry Applications*, Vol. 41, No. 3, pp. 673- 681.

Ibe, S.; Shinke, N.; Takami, S.; Yasuda, Y.; Asatsu, H. & Echigo, M. (2002). Development of Fuel Processor for Residential Fuel Cell Cogeneration System, *Proc. 21th Annual Meeting of Japan Society of Energy and Resources*, pp. 493-496, Osaka, June 12-13, ed., Abe, K. (in Japanese)

Kyoto Denkiki Co., Ltd. *A system connection inverter catalog and an examination data sheet*, 2001.

Lindstrom, B. & Petterson, L. (2003). Development of a methanol fuelled reformer for fuel cell applications, *J. Power Source*, Vol. 118, pp. 71-78.

Nagano, S. (2002). Plate-Type Methanol Steam Reformer Using New Catalytic Combustion for a Fuel Cell. *Proceedings of SAE Technical Paper Series, Automotive Eng*. pp. 10.

Narita, K. (1996). The Research on Unused Energy of the Cold Region City and Utilization for the District Heat and Cooling. *Ph.D. thesis*, Hokkaido University, Sapporo. (in Japanese)

Obara, S. & Kudo, K. (2005). Installation Planning of Small-Scale Fuel Cell Cogeneration in Consideration of Load Response Characteristics (Load Response Characteristics of Electric Power Output). *Transactions of the Japan Society of Mechanical Engineers*, Series B; Vol. 71, No.706, pp. 1678-1685. (in Japanese)

Obara, S. & Kudo, K. (2005). Study on Small-Scale Fuel Cell Cogeneration System with Methanol Steam Reforming Considering Partial Load and Load Fluctuation. *Transactions of the ASME, Journal of Energy Resources Technology*, Vol. 127, pp. 265-271.

Oda, K.; Sakamoto, S.; Ueda, M.; Fuji, A. & Ouki, T. (1999). A Small-Scale Reformer for Fuel Cell Application. *Sanyo Technical Review*, Vol. 31, No. 2, pp. 99-106, Sanyo Electric Co., Ltd., Tokyo, Japan. (in Japanese)

Robert, H. (2004). Microgrid: A conceptual solution. *Proceedings of the 35th Annual IEEE Power Electronics Specialists Conference*, Vol. 6, pp. 4285-4290.

Takeda, Y.; Iwasaki, Y.; Imada, N. & Miyata, T. (2004). Development of Fuel Processor for Rapid Start-up, *Proc. 20th Energy System Economic and Environment Conference*, Tokyo, January 29-30, ed., K. Kimura, pp. 343-344. (in Japanese)

21

Large Scale Integration of Wind Power in Thermal Power Systems

Lisa Göransson and Filip Johnsson
Chalmers University of Technology
Sweden

1. Introduction

This chapter discusses and compares different modifications of wind-thermal electricity generation systems, which have been suggested for the purpose of handling variations in wind power generation. Wind power is integrated into our electricity generation systems to decrease the amount of carbon dioxide emissions associated with the generation of electricity as well as to enhance security of supply. However, the electricity generated by wind varies over time whereas thermal units are most efficient if run continuously at rated power. Thus, depending on the characteristics of the wind-thermal system, part of the decrease in emissions realized by wind power is offset by a reduced efficiency in operation of the thermal units as a result of the variations in generation from wind. This chapter discusses the extent to which it is possible to improve the ability of a wind-thermal system to manage such variations.

The first part of the chapter deals with the nature of the variations present in a wind-thermal power system, i.e. variations in load and wind power generation, and the impact of these variations on the thermal units in the system. The second part of the chapter investigates and evaluates options to moderate variations from wind power by integrating different types of storage such as pumped hydro power, compressed air energy storage, flow batteries and sodium sulphur batteries. In addition, the option of interconnecting power systems in a so called "supergrid" is discussed as well as to moderate wind power variations by managing the load on the thermal units through charging and discharging of plug-in hybrid electric vehicles.

Data from the power system of western Denmark is used to illustrate various aspects influencing the ability of a power system to accommodate wind power. Western Denmark was chosen primarily due to its current high wind power grid penetration level (24% in 2005 (Ravn 2001; Eltra 2005)) and that data from western Denmark is easily accessible through Energinet (2006).

2. Impact of wind power variations on thermal plants

The power output of a single wind turbine can vary rapidly between zero and full production. However, since the power generated by one turbine is small relative to the capacity of a thermal unit, such fluctuations have negligible impact on the generation pattern of the thermal units in the overall system. With several wind farms in a power system, the total possible variation in power output can add up to capacities corresponding to the thermal units and influence the overall generation pattern. At times of low wind speeds, some thermal unit might for example need to be started. The power output of the aggregated wind power is, however, quite different from the power output of a single turbine. Wind speeds depend on weather patterns as well as the landscape around the wind turbines (i.e. roughness of the ground, sea breeze etc.). Thus, the greater the difference in weather patterns and environmental conditions between the locations of the wind turbines, the lower the risk of correlation in power output. In a power system with geographically dispersed wind farms, the effect of local environmental conditions on power output will be reduced. Since it takes some time for a weather front to pass a region, the effect of weather patterns will be delayed from one farm to another, and the alteration in aggregated power output thus takes place over a couple of hours rather than instantaneously. This effect is referred to as power smoothing (Manwell et al. 2005). Western Denmark is a typical example of a region with dispersed wind power generation. The aggregated wind power output for this region during one week in January can be found in Figure 1. As seen in Figure 1, variations in the range of the capacity of thermal units do occur (e.g. between 90 hours and 100 hours the wind power generation decreases with 1 000MW), but the increase or decrease in power over such range takes at least some hours (e.g. approximately 10 hours for the referred to example).

2.1 Variations in load and wind power generation

Figure 2 illustrates the variations in total load (electricity consumption) in western Denmark during the same week as shown in Figure 1. As seen, the amplitude of the wind power variations at current wind power grid penetration (i.e. 24%) and the variations in load are not much different. However, there are two aspects of wind power variations which make these more complicated to manage than fluctuations in load; the unpredictability and the irregularity. Since the total load variations are predictable, it is possible to plan the scheduling of the thermal units to compensate for the load variations. The unpredictability of wind power makes it difficult to accurately schedule units with long start-up times. Variations in a system dominated by base load units create a need for what is here referred to as moderator which is a unit in the power system with the ability to reallocate power in time, such as a storage unit or import/export capacity. Since the total load variations are regular, to manage these a moderator would only need to have "storage" capacity which can displace one such variation at a time (i.e. absorb power for a maximum of 12 hours and then deliver this power to the system). Due to the irregularity of wind power variations "storage" capacities of a moderator for this application need to be more extensive than if variations were regular.

For the thermal units it is obviously the aggregated impact of the wind power and the total load which is of importance. The load on the thermal units (i.e. the total load reduced by the wind power generation) will become both less predictable and less regular as wind power is introduced to the system. In the Nordic countries, there is some correlation between wind speeds and electric load in the summer, but no correlation of significance in winter time (Holttinen 2005). However, a decrease/increase in wind power output might obviously coincide with an increase/decrease in demand at any time of the year, resulting in large variations in load on the thermal units. At times when wind power output is high and demand is low, systems with wind power in the range of 20% grid penetration or higher might face situations where power generation exceeds demand (although this obviously depends on the extent of the variations in load). Without a moderator in the system, which can displace the excess power in time, some of the wind power generated will have to be curtailed in such situations. With base load capacity in the system which has to run continuously, situations where curtailment cannot be avoided will arise more frequently[1].

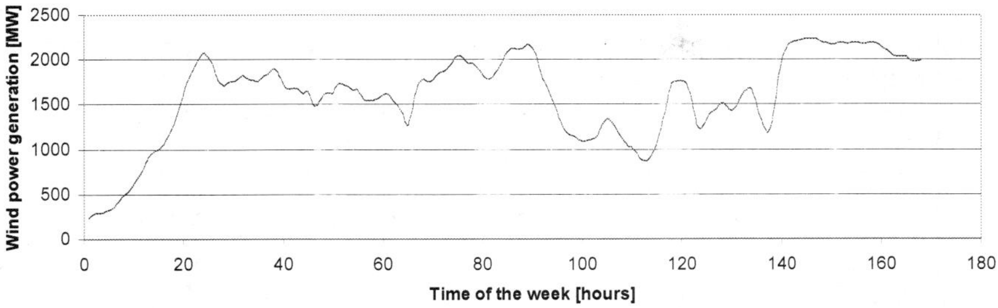

Fig. 1. Wind power generation in western Denmark during the first week in January 2005. Source (Energinet 2006)

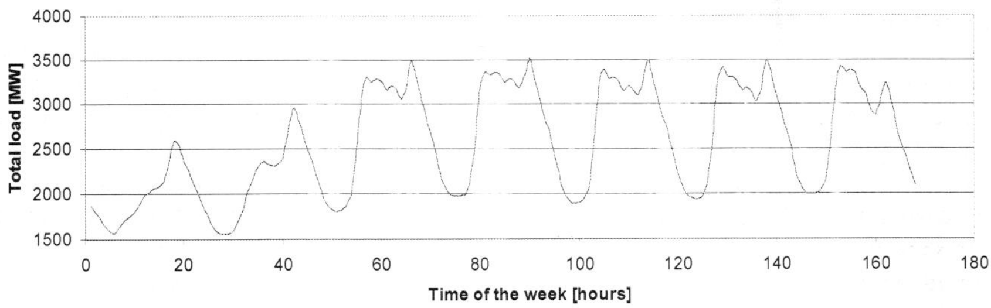

Fig. 2. Total load in western Denmark the first week in January 2005. Source (Energinet 2006)

2.2 Response to variations in wind power generation and electricity consumption

Variations in load in a wind-thermal power system that uses no active strategy for variation management can be managed in three different ways;

- by part load operation of thermal units,
- by starting/stopping thermal units or
- by curtailing wind power.

The choice of variation management strategy depends on the properties of the thermal units which are available for management (e.g. in order to choose to stop a unit it obviously has to

[1] It should be pointed out that the Nordic system (Nordpool electricity market) of which western Denmark is part, is special in the context of wind power integration, since variations in wind power can, to a certain extent, be managed by hydropower (with large reservoirs).

be running) and the duration of the variation. In a power system where cost is minimized, the variation management strategy associated with the lowest cost is obviously chosen. If, for example, the output of wind power and some large base load unit exceeds demand for an hour, curtailment of wind power (or possibly some curtailment in combination with part load of the thermal unit) might be the solution associated with the lowest total system cost. If the same situation lasts for half a day, stopping the thermal unit might be preferable from a cost minimizing perspective. To be able to take variation management decisions into account in the dispatch of units, knowledge of the start-up and part load properties of the thermal units is necessary.

Two aspects of the start-up of thermal units will have an immediate impact on the scheduling of the units; the start-up time and the start-up cost. The start-up time is either measured as the time it takes to warm up a unit before it reaches such a state that electricity can be delivered to the grid (time for synchronization) or as the time before it delivers at rated power (time until full production). In both cases, the start-up time ultimately depends on the capacity of the unit, the power plant technology and the time during which the unit has been idle. Small gas turbines have relatively short start-up times, in the range of 15 minutes, and large steam turbines have long start-up times, in the range of several hours. If a large unit has been idle for a few hours, materials might still be warm and the start-up time can be reduced. Table 1 presents the required start-up times of units in the Danish power system.

The costs associated with starting a thermal unit are a result of the cost of the fuel required during the warm-up phase and the accelerated component aging due to the stresses on the plant from temperature changes. Lefton et al. (1995) have shown that the combined effect of creep, due to base load operation, and fatigue, due to cycling (start-up/shutdown and load following operation), can significantly reduce the lifetime of materials commonly used in fossil fuel power plants in comparison to creep alone. They estimate the cycling costs (the cost to stop and then restart a unit) of a conventional fossil power plant to $1 500-$500 000 per cycle (around EUR 1 170-400 000) with the range corresponding to differences in cycling ability of different technologies and the duration of the stop. These costs include the cost of increased maintenance, as well as an increase in total system costs due to lower availability of cycled units, and an increase in engineering costs to adapt units to the new situation (i.e. improve the cycling ability).

Time since last stop	**Starting time for synchronisation**	**Starting time until full production**
	[min.]	[min.]
Immediately after stop	120	210
Up to 8 hours	180	300
Between 8 and 36 hours	300	480
Over 36 hours (cold start)	600	840

Table 1. Maximum allowed starting time for power plants in the Danish power system with nominal maximum power above 25 MW. Source: (Energinet 2007).

One alternative to shutting down and restarting a thermal unit is to reduce the load in one or several units. The load reduction in each unit is restricted by the maximum load turn-down ratio. The minimum load level of a thermal unit depends on the power plant technology and the fuel used in combustion units. The minimum load level on the Danish units range from 20% of rated power for gas- and oil-fired steam power plants to 70% of rated power for waste power plants (Energinet 2007). Minimum load level of coal fired power plants range from 35% to 50% of rated power depending on technology (Energinet 2007).

Running thermal units at part load is associated with an increase in costs and emissions per unit of energy generated (i.e. per MWh), since the efficiency decreases with the load level. The rate of the decrease in efficiency depends on the power plant technology and the level to which the load is reduced. Figure 3 illustrates the relation between efficiency and load level for three different thermal units. As shown in Figure 3, the rate of decrease in efficiency is lower at high load levels than at low load levels. It is also shown that the rate of decrease in efficiency is higher in the combined cycle plant (CC) than in the steam plant (since gas turbines are sensitive to part load operation).

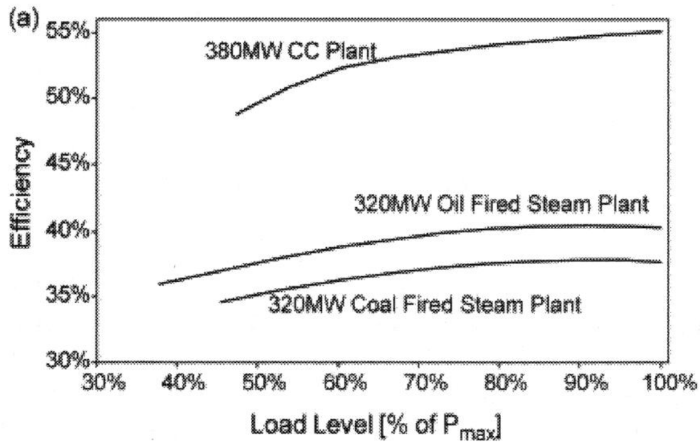

Fig. 3. Typical electric efficiency versus load level curves of different power plants. Source: (Carraretto 2006)

Work with models of the power system of western Denmark suggests that wind power variations introduce aspects that influence the competitiveness of the thermal units in the power system relative to one another (Göransson & Johnsson 2009a). In general, simulations show that an increase in the amount of wind power reduces the periods of constant production and the duration of these periods. The capacity factor of units with low start-up and turn down performance and high minimum load level (i.e. base load units) will decrease more than the capacity factor of units with high start-up and turn down performance and/or low minimum load level. This result might seem trivial. However, low start-up and turn down performance and high minimum load levels are common properties of units with low running costs designed for base load production. Thus, low running costs compete against flexibility and in a system with significant wind power capacity, the unit with the lowest running costs is not necessarily the unit which is run the most.

Figure 4 shows the capacity factors of the thermal units in the power system of western Denmark at three different levels of wind power capacity ("without wind", "current wind" corresponding to around 20% wind power grid penetration and "34% wind" with 34% wind power grid penetration) from simulations of three weeks in July 2005 (Göransson & Johnsson 2009a). As can be seen in Figure 4, the dominating trend is a decrease in import and an increase in export as the wind power capacity in the system increases.

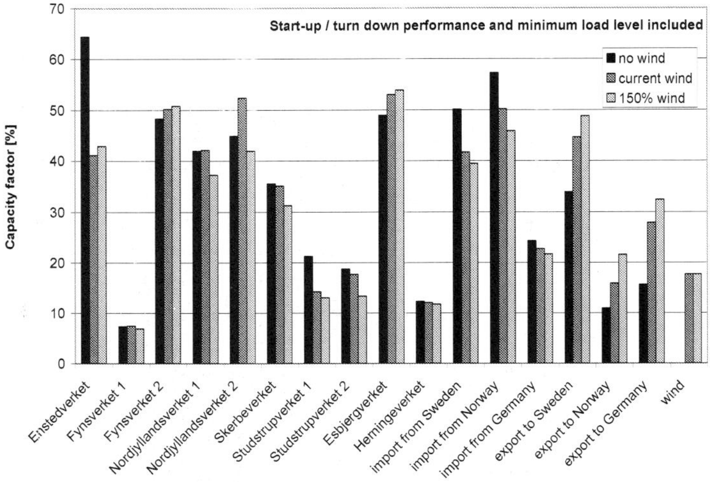

Fig. 4. Simulated impact of variations in wind power generation on the capacity factor of thermal units in western Denmark. For further details see (Göransson & Johnsson 2009a).

Enstedtsvaerket B3 also experiences a significant decrease in its capacity factor with increased wind power capacity. Enstedtsvaerket is the least flexible unit in the system (most expensive start-up and highest minimum load level), and it has a lower capacity factor than several other units in the current wind and 34% wind case despite that it has the lowest running costs of the system. The variations in wind power production have thus altered the dispatch order of the units in these two cases, favouring units with more flexible properties to the unit with the lowest running costs.

The effect of a shift from base load generation to generation in more flexible units on total system emissions depends on the specific technologies in question. A small increase in magnitude of the variations may boost the capacity factors of units with low emissions (e.g. gas-fired peak load units), whereas a large increase in magnitude of the variations may be followed by an increase in capacity factor of units with high emissions (e.g. oil-fired back-up units). The impact of the change in capacity factors on system emissions thus depends both on the power system configuration and the amount of wind power which is integrated.

3. Moderation strategies

The purpose of a moderation strategy is to improve the efficiency of the wind-thermal system by reducing the variations in the load on the thermal units, thus avoiding thermal plant cycling and part load operation. Moderation strategies reduce variations either by displacing power over time or by displacing load over time. Traditional storage forms displace power in time. A grid solution, where power is imported to and exported from a system, works according to the same principle from a power generation perspective. Strategies where the load is displaced over time are generally referred to as demand side management strategies. As an example, the charging of plug-in hybrid electric vehicles can be used for demand side management.

3.1 Storage technologies and grid strategies

Thermal units run at maximum efficiency if they generate power continuously at or near rated power whereas the demand for electricity varies in time. To avoid inefficient operation of the thermal units, the variations in load on the power system are conventionally managed by some unit which consumes some of the excess power generated (i.e. to keep the thermal units at rated power) at times of low load, to return this power to the system at times of high load levels. Storage technologies, such as pumped hydro storage and compressed air energy storage (CAES) operate in this manner. Pumped hydro has been applied for decades, while CAES is hardly a commercial alternative under present conditions. Nourai (2002) gives a thorough evaluation of storage technologies for energy management. Different types of storage technologies all have the same effect on the system, i.e. they shift some of the generated power in time. Using the grid and connections to other regions, where power is exported at times of low load and imported at times of high load levels, has the same impact on the thermal units in the system.

Shifting power in time is obviously useful also when managing wind power. The storage would then consume some of the excess wind power generated at times of high wind power generation levels and return this power to the system at times of low wind power generation levels. Literature presents thorough evaluations on the interaction between wind power (i.e. a wind farm) and one storage unit. Particularly well covered is the interaction between wind power and a (pumped) hydro power plant (Castronuovo & Lopes 2004; Jaramillo et al. 2004) and the interaction between wind power and a CAES unit (Cavallo 2007; Greenblatt et al. 2007). In such studies, the wind farm is combined with storage so that the total output resembles a conventional power plant, i.e. closer to base load (Jaramillo et al., 2004; Greenblatt et al., 2007) or maximizes return according to a given price signal (Castronuovo and Lopes, 2004). If instead the storage is a common resource which manages the total power generation level in the system, i.e. the sum of generation in thermal units and wind power plants, variations in wind power generation are allowed to compensate for variations in electric load on the power system and the benefit of the storage for the thermal units is maximized. Storage as a common resource to the system is the focus of this chapter.

3.1.1 Impact on a wind-thermal system

As the storage or transmission capacity is introduced to the power system the system emissions can be influenced in four different ways; start-up emissions decrease, part load emissions decrease, wind power curtailment decrease and the capacity factors of typical base load units increase. An example of the impact of a general moderator (i.e. a lossless storage or lossless transmission capacity) on power system emissions and wind power curtailment is illustrated in Figures 5a-c. The power system used as an example here is an isolated system containing the thermal units of western Denmark and two levels of wind power (2 374 MW, generating 20% of the total electricity demand, and 4 748 MW, generating 40% of the total electricity demand if no wind is curtailed). Details are given by Göransson and Johnsson (2009b). The ability of a general moderator to displace power in time depends on the power rating and the storage capacity of the moderator. In figures 5a-c, emissions and wind power curtailment are investigated at five different moderator power ratings (0, 500, 1000, 1500, 2000 MW) and at two different storage capacities; daily and weekly, where the charging and discharging of the storage is balanced on a daily and weekly basis, respectively.

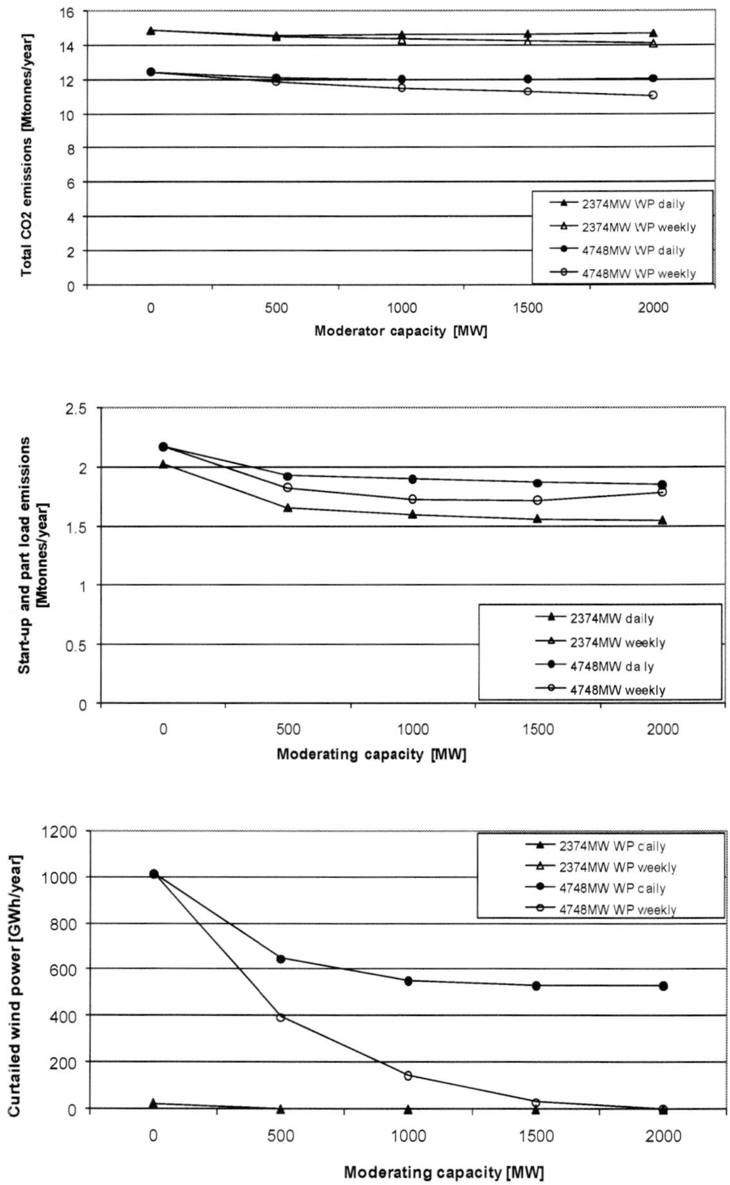

Fig. 5. Impact of moderator power rating and capacity on a: total system emissions, b: start-up and part load emissions and c: wind power curtailment. Source: (Göransson & Johnsson 2009b).

A weekly balanced moderator is obviously at least as qualified at reducing emissions as a daily balanced moderator (since the weekly balanced unit can also be balanced over each day). Figure 5a shows that the advantage of a weekly balanced moderator, compared to a daily balanced moderator, is more significant in the power system with 4 748 MW wind than in the power system with 2 374 MW wind. With a weekly balanced moderator emissions are reduced as the power rating of the moderator increases, whereas the emission reduction from applying 500 MW moderator capacity is just as large as the emission reduction from applying 2 000 MW moderator capacity if it is daily balanced. The largest emission reduction is attained in the wind-thermal power system with 4 748 MW wind, in which a 2 000 MW moderator capacity that is balanced on a weekly basis can reduce emissions with 11% (Göransson & Johnsson 2009b).

Figure 5b shows the start-up and part load emissions of the power systems. The start-up and part load emissions are higher in the system with 4 748 MW wind power capacity than in the system with 2 374 MW wind power capacity due to the greater system variations in the 4 748 MW wind system compared to the 2 374 MW wind system. The major part of the reduction is realised by the first 500 MW of moderating capacity and is mainly due to load variation management. Since variations in load occur with a daily frequency, the storage capacity of a daily balanced moderator is sufficient to manage the variations. Thus, for the start-up and part load emissions of the system, it is of little or no importance whether the moderating capacity is daily or weekly balanced.

Figure 5c displays the relation between wind power curtailment and moderator power rating. By shifting the wind power generation in time so that the correlation between load and wind power generation is improved, the moderator enables a shift from thermal power to wind power. Avoiding 1 000 GWh of wind curtailment per year corresponds to a decrease in system emissions with 0.60 Mtonnes/year[2]. A decrease of this magnitude is realised in the 4 748 MW wind system by a 2 000 MW weekly balanced moderator. In this case the avoidance of wind power curtailment is the most important factor which contributes to reduction in emissions. The daily balanced moderator does not provide the same possibility to avoid wind power curtailment as a weekly balanced moderator.

3.1.2 The choice of variation moderator

There are many technologies for storing power. Figure 6 illustrates how different storage technologies are suitable for different applications. The focus of this chapter is to discuss the ability of a moderator to allow thermal units to run continuously, despite variations in wind power generation and load. This requires significant power ratings and charge/discharge times in the scale of hours, i.e. technologies for energy management. As shown in Figure 6 pumped hydro power, compressed air energy storage (CAES), flow batteries and sodium sulphur (NaS) batteries are moderators suitable for such a purpose.

From Figure 5 the following choice of moderator properties seem sensible for the system investigated; a daily balanced moderator (3 GWh storage) of 500 MW for wind-thermal systems with around 20% wind power grid penetration, and a weekly balanced moderator (33 GWh storage) of 2 000 MW for wind-thermal systems with around 40% wind power grid penetration. From Figure 6 it can be seen that pumped hydro stations, CAES units, flow batteries and NaS batteries have discharge times in the range of hours and are thus all

[2] The average emissions of the thermal units are approximately 600kg CO_2/MWh.

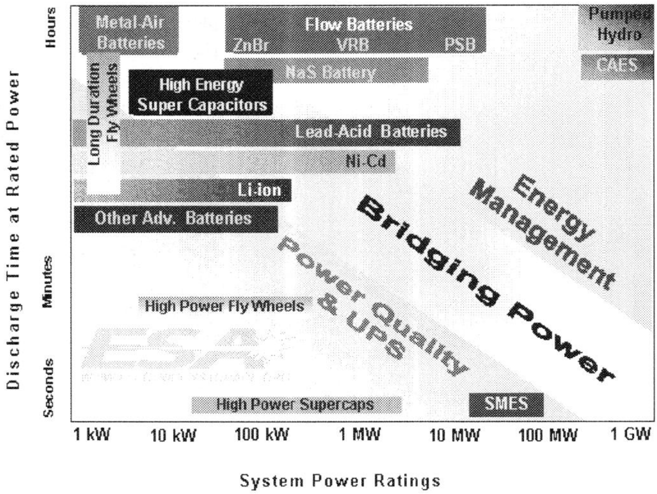

Fig. 6. Typical power ratings and discharge times of storage technologies. Source: (ElectricityStorageAssociation).

candidates to serve as daily moderation. While there are pumped hydro stations fulfilling the requirements stated (the Dinorwig pumped hydro power station in Wales has for example a power rating of 1 700 MW and is able to store 8 GWh of energy) and CAES units of this magnitude are under consideration (for example the project concerning a 2 700 MW CAES in Norton, Ohio), flow batteries and sodium sulphur batteries have only been evaluated on a smaller scale. Pumped hydro is the only technology which has been applied to storage schemes anywhere near the range required for the weekly balanced moderation of this work (the Guangzhou pumped hydro station, China, has a capacity of 2 400 MW and can store 14.4 GWh energy). Reaching a power rating of 2 000 MW with CAES or battery solutions should not pose a problem since it is merely a matter of adding a sufficient number of identical units. The problem lies in the ability to store the volumes required when reallocating power from one week to another. When it comes to the CAES technology, storage capacities are restricted by the volume of the cavern and the maximum pressure that can be applied to the air without loosing too much energy as heat. As mentioned previously, an additional alternative to moderate variations is to displace power through import and export over the system boundary. This is the main way in which western Denmark manages its variations today and the possibility to use this method on European scale is being discussed (sometimes referred to as "supergrid"). Trade over transmission lines could of course be balanced both on a daily and a weekly basis.

Figure 7 compares the reduction in emissions and costs due to the introduction of a weekly balanced moderator in the system with 40% wind power and the total LCA costs and emissions of possible moderators. Applying existing moderator technology, a net reduction in emissions of 7.5 to 10.3% is possible (Göransson & Johnsson 2009b). However, if assuming a cost of 20 EUR for emitting one tonne of carbon dioxide (corresponding to the solid line in Figure 7), overhead transmission lines is the only moderator which can lower the system

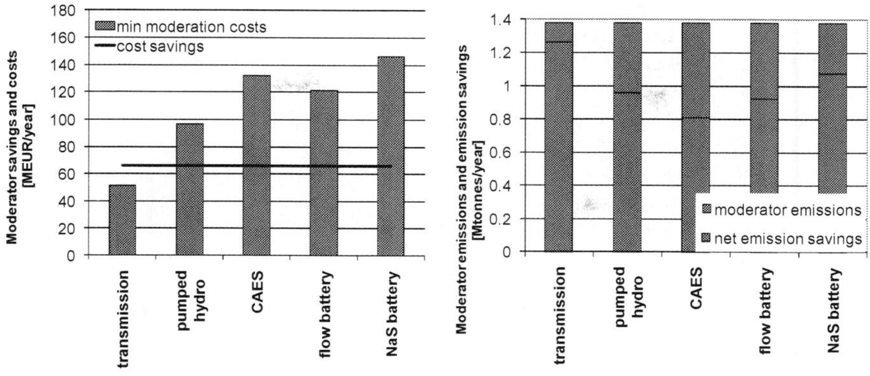

Fig. 7. Costs for different moderators and their emissions (weekly balanced moderator as example). a) Cost savings (solid line) compared with costs for moderation (bars). b) Emission savings reduced by emissions associated with moderation. Sources: (Rydh 1999; Nourai 2002; EC 2003; Denholm & Holloway 2005; IEA 2005; Kuntz 2005; Rydh & Sandén 2005; Ravenmark & Normark 2006; Göransson & Johnsson 2009b).

costs. With overhead lines, system costs can be decreased if the imported power can be bought at prices which do not exceed the yield from exported power by more than about 4 EUR/MWh. However, as noted earlier, using transmission as moderator requires either transmission lines to a region with excess flexible capacity or to a region sufficiently far away to make wind speeds and/or demand uncorrelated. Transmission lines to such a region would in many cases have to cover some distance and pass several other regions. The profitability and acceptance of building such transmission lines would improve if all regions within some large geographical scope share a system of lines for cooperative variation management. Also, the risk of correlated variations is generally smaller (i.e. the moderation of variations is more efficient) over a wider geographical scope. A system of transmission lines of such a kind, often referred to as a "supergrid", has been proposed (Airtricity 2007) to handle wind power variations in Europe. The results from the work by Göransson and Johnsson (2009b) indicate that investments in transmission lines is generally attractive since costs and emissions associated with transmission lines are lower than those of other moderator options (*cf.* Figure 7). This, provided that it is sufficient for each country to invest in 1 000 km of line (i.e. the distance assumed necessary to provide moderation in the calculations presented here). However, since the reduction in system costs from moderation only just compensates for the cost to install overhead lines (Figures 7a), the cost for underground lines and cables at sea (which are likely to make up a significant part of a "supergrid") will probably not be compensated for at a cost of 20 EUR per tonne of carbon dioxide emitted.

Although more expensive than overhead lines, underground cables are associated with a cost lower than the other moderator technologies in Figure 7 (Göransson & Johnsson 2009b). Thus, transmission in general seems to be a good option with regard to both costs and emissions compared to alternative moderation. However, at the moment construction of local storage seems to be closer to implementation than transmission lines for variation management. There are at least two factors which steer development in this direction. To start with, the EU renewable energy targets are translated into national goals, stimulating national rather than international solutions. Using transmission as wind power moderator, part of the green electricity is exported and there may be uncertainties regarding how this should be accounted for until the system of guarantees of origin is properly in place. See European Commission (2008) for details regarding such a system. Thus, even though the reduction in emissions would be maximized on an EU level with transmission as moderator, storage technologies might be favoured since they retain the green electricity within the national boundaries. Another factor counteracting the supergrid is the desire to protect the local power market.

Finally, it should be noted that when comparing different moderator technologies, the order of preference of moderator technologies depends on the average emissions of the power system in which the moderator will be integrated. This is also exemplified in the work with the western Denmark model. In the system with up to 20% wind power, average emissions are still higher than emissions from generating electricity from the combustion of natural gas. In this situation the CAES technology is the storage solution which reduces emissions the most amongst the energy management technologies in Figure 7 (Göransson & Johnsson 2009b). For a system with up to 40% wind power, and lower average emissions associated with the power generation, the CAES technology is less efficient at reducing emissions than the other alternatives (see Figure 7). Since major rearrangements of present power systems are under consideration, it is important to take future development of the system into account when choosing moderator technology.

3.2 Demand side management

Another way to keep the generation in thermal units constant at desired level is to displace some of the load, rather than some of the generated power. This strategy is referred to as demand side management. Demand side management can be exercised by any load which can be allocated to any point in time. Demand for electricity for heating purposes is suitable for demand side management since heat is relatively easy to store. One example of this is given by Stadler (2008), who shows how storage heaters have decreased the variations in electric load in Germany. Stadler further suggests a joint strategy combining CHP systems and heat pumps to improve the system ability to integrate renewable generation. This strategy, where heat pumps generate heat while consuming electricity at times of excess electricity generation and the CHP unit generates heat and power at times when the system experiences electricity shortage, is also suggested by Blarke and Lund (2007). However, there are many ways in which heat can be produced resulting in lower losses with regard to exergy (waste heat, CHP, solar collectors). Demand side management strategies where electricity is difficult to replace or where electricity replaces more emissive alternatives should obviously be prioritised to minimize overall system emissions. Stadler (2008) give several examples of loads where electricity is difficult to replace; ventilation systems, refrigeration and pumps in hot water heating systems to mention a few. An example where electricity could potentially replace more emissive fuels is the electrification of the transportation sector. Below follows an example of how demand side management could be exercised by means of choosing appropriate charging strategies for Plug-in Hybrid Electric Vehicles (PHEV).

PHEV:s have the potential to reduce the effects of variations in demand and wind power generation (i.e. to reduce the variations in load on the thermal units in the case of a wind-thermal system) through two different mechanisms, here referred to as the correlation mechanism and the flexibility mechanism. The correlation mechanism is a result of the change in load profile, which arises as soon as the non-PHEV load (dominated by the electricity consumption of households and industry) is not perfectly correlated with the demand for electricity to charge the vehicles. The flexibility mechanism is due to some active control (built-in intelligence) of charging and discharging of the PHEV:s which offers the possibility to adjust the charging and discharging of the PHEV:s to fit the generation and load pattern. Obviously, vehicles will typically be charged when the non-PHEV load is low or when wind power generation is high in order to minimize changes in load on the thermal units in the system. With a flexible load in the system, reserve requirements can be relaxed since variations in generation or non-PHEV load partly can be managed by starting/stopping charging of the vehicles.

Both the correlation mechanism and the flexibility mechanism reduce variations in the total load. A difference between the mechanisms is that the flexibility mechanism can manage irregular variations whereas the correlation mechanism only manages regular trends. The flexibility mechanism can thus manage variations in intermittent generation (such as wind power generation) and reduces the need of reserve capacity of the system whereas the correlation mechanism will not directly benefit the system with respect to influence from intermittent generation.

Kempton and Tomic (2005) propose the use of large parts of the light vehicle fleet as operational back-up and storage of wind power. They have calculated that if 50% of the US electricity would be provided by wind power, some 34% of the light vehicle fleet can provide the back-up and storage required if operated as PHEV:s under a V2G contract. Kempton and Dhanju (2006) have looked further into the ability to handle wind power variations with PHEV:s and conclude that if the light vehicle fleet was entirely made up of PHEV:s, the power rating of the batteries in the vehicles if connected at 15kW, would significantly exceed the average national load on the power system in most OECD countries. Göransson et al. (2009) found that if the electricity consumption of PHEV:s correspond to some 12 % of the electricity consumption, wind power curtailment in a system with 20% wind power can be completely avoided.

3.2.1 Impact on a wind-thermal system

The load managing ability of the PHEV:s depends on the PHEV share of total electricity consumption as well as limitations under which the power system is free to allocate the PHEV load. Denholm and Short (2006) found that if charging of the PHEV:s is optimally dispatched from a power system perspective, the PHEV:s will decrease the cycling of the power plants and increase the load factor of the base load plants. Hadley and Tsvetkova (2008) on the other hand found that, with a fixed PHEV load starting at 5 p.m., the evening peak in load will be augmented and the use of peak load units increase. A comparison of the PHEV impact on power system emissions at four different integration strategies is presented in Figure 8, taken from Göransson et al. (2009). The example in Figure 8 is a wind-thermal system (thermal units and total wind power generation from western Denmark in 2005) to which PHEV:s have been integrated at three implementation levels (3%, 12% and 20% of the total electricity consumption). The four integration strategies have the following characteristics;

- S-DIR where the charging time of the PHEV:s occurs immediately after driving and the PHEV:s are charged as soon as they return home (it is assumed that the PHEV:s will always be recharged due to the relatively low cost of driving on electricity compared to gasoline).
- S-DELAY where the charging time of the PHEV:s is delayed (i.e. with a timer) to minimize average correlations with demand,
- S-FLEX where the charging of PHEV:s can take place when it is most favourable from a power system perspective, but the entire PHEV fleet has to be charged during the night and a part of it during the workday,
- S-V2G where the power system is free to dispatch the PHEV load and to discharge PHEV:s as desired. However, charging and discharging is restricted to the PHEV capacity available to the grid and the power level of the batteries. The power level of the batteries depends on charging and discharging history and the daily driving pattern for which electricity could have been used.

As shown in Figure 8, the lowest emissions are obtained for the S-V2G strategy at 20% PHEV share of the total electricity consumption and a 4.7% reduction in power system emissions is obtained. On the other hand, when the charging time of the PHEV:s occurs immediately after driving and the PHEV:s are charged as soon as they return home (S-DIR) there is a clear increase in CO2-emissions from the power system as the share of PHEV electricity consumption increase. The other integration strategies produce emissions between the S-V2G and the S-DIR cases. As seen in Figure 8, strategies where the flexibility mechanism is present (S-FLEX, S-V2G) have lower emissions than the strategies which are limited to the correlation mechanism at high PHEV shares of consumption (i.e. 20%).

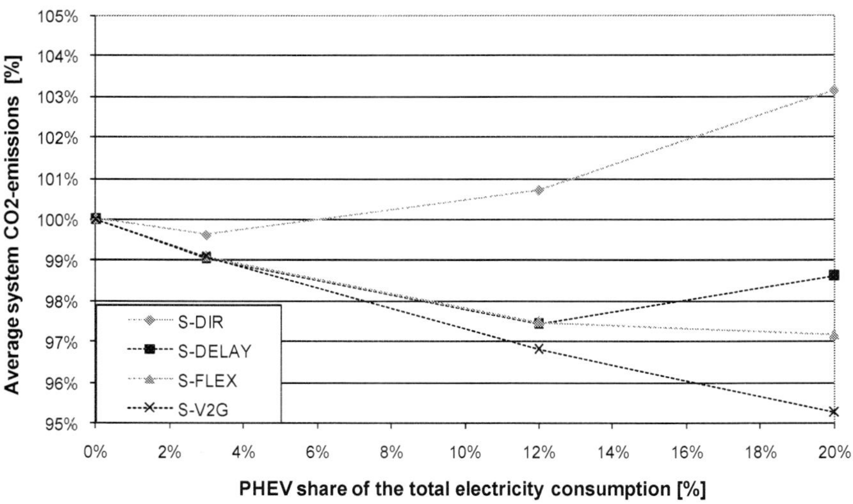

Fig. 8. Impact on CO2-emissions due to PHEV integration as obtained from the simulations of an isolated wind-thermal power system (vehicle emissions not included). Average system emissions in the system without PHEV:s are 649kgCO2/MWh, thus, in the plot 1% is equivalent to 6.49kg/MWh Source: (Göransson et al. 2009)

Which are the mechanisms behind the increase/decrease in emissions under the four different integration strategies? Figure 9 shows a weekly time series of the total consumption of electricity divided into consumption of household and industry (white) and

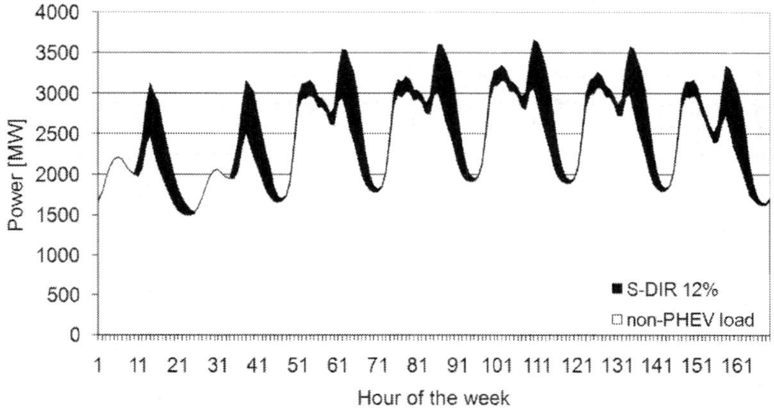

a.

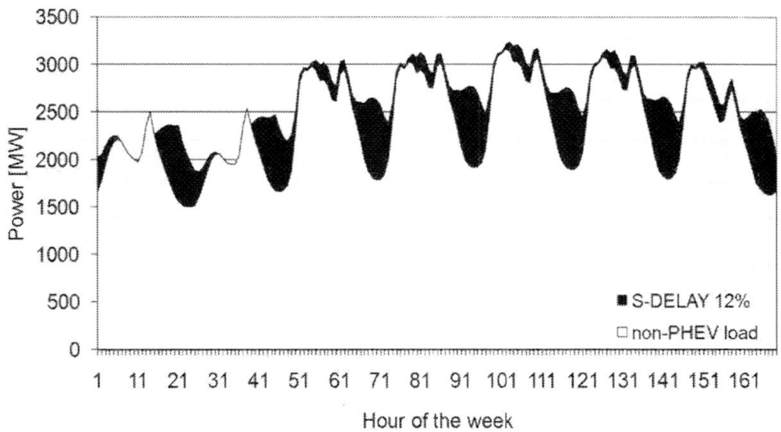

b.

Fig. 9. Total electricity consumption in the system modelled by Göransson et al. (2009) divided into consumption of household and industry (white) and consumption of vehicles (black). The example shown is for 12% PHEV share of electricity consumption. a: S-DIR integration strategy where consumption of households and industry is strongly correlated with the consumption of vehicles. b: S-DELAY strategy where a shift in charging start time decreases the correlation and evens out overall electricity consumption. This smoothening of electricity consumption through a decrease in correlation is, in this work, referred to as the correlation mechanism. Source: (Göransson et al. 2009) consumption of PHEV vehicles (black). Data of the household and industry consumption was obtained from Energinet (Energinet 2006) and PHEV consumption was taken from (Göransson et al. 2009). In Figure 9, the PHEV consumption is 12% of the total electricity consumption, and the household and industry consumption is scaled down to 88%. As can be seen from Figure 9a, in the S-DIR strategy (i.e. vehicles are charged as soon as they return home), the PHEV integration in the system does not imply a smoothening of the total load, but rather an accentuation of the peaks. As PHEV:s are integrated under the S-DIR strategy, there is a decrease in the amount of thermal units which can run continuously and most units also have to cover peak load. The result is an increase in emissions from the power generation system compared to the reference case without PHEV:s (cf. Figure 8).

Applying the S-DELAY strategy (i.e. where vehicle charging is delayed with a timer), the PHEV consumption is shifted so that it occurs at times of low non-PHEV load, and the overall load is evened out as shown in Figure 9b. This simple adjustment proves to be an efficient way to smoothen the overall load, and the integration of PHEV:s will reduce average system emissions under this strategy (cf. Figure 8). However, a large PHEV share of consumption would create new peaks in the total load at times when the PHEV load is at maximum. These new peaks would increase part load emissions of the system and the total reduction in system emissions is counteracted (cf. Figure 8 at a 20% PHEV share).

Under the S-FLEX strategy a moderate PHEV share (i.e. 12%) is sufficient to avoid situations where wind power generation competes with the generation in base load units with low running costs and high start-up costs. Start-up emissions and wind power curtailment are thus minimized already at a moderate level of integration. If the PHEV share increases, the capacity which has to be charged is of such magnitude that it creates new variations. However, due to the flexible distribution of the charging, these new variations can be allocated so that they can be met by units which are already running. Changes in capacity factors of these units cause a decrease in emissions (cf. Figure 9).

Under the S-V2G strategy the system ability to accommodate variations of both short and long duration increases with the PHEV load share, since charging is optional at all times and any increase in PHEV capacity in the system thus improves the system flexibility. However, wind power curtailment is lowest at a 12% PHEV share. This is due to the car-owners' great willingness to pay for the electricity in this example. In a system where the willingness to pay for PHEV charging is small, vehicles would always be charged so that the load would suit the generation under the V2G strategy. However, when the willingness to pay for charging is great, as in the system considered in Figure 8, vehicles are charged as much as the battery capacity and availability allows and the load variations due to PHEV charging will increase. In such situations, a higher PHEV share of consumption does not imply a greater ability of the system to accommodate wind power.

3.2.2 The choice of integration strategy

The choice of PHEV integration strategy obviously depends on the cost to implement the strategies. If the majority of the charging of the vehicles takes place at home, there is an implementation cost associated with each vehicle. The implementation cost then simply corresponds to the cost of the device for connecting and controlling PHEV:s at the charging point (e.g. the garage). There is a significant difference in implementation cost between the strategies, where the cost for sophisticated controlling (i.e. S-V2G) is particularly high. However, under a sophisticated controlling mechanism, the fleet of PHEV:s is able to improve the power system efficiency (and thus reduce costs) more than under a less sophisticated controlling mechanism. Table 2 compares the costs of implementing PHEV:s with the change in cost to supply the electricity generation system with power as PHEV:s are integrated for the western Denmark example. As shown in Table 2, the reduction in costs is always smaller than the implementation cost for the S-V2G strategy, whereas the implementation costs of the S-FLEX and S-DELAY strategies are compensated for at a 3% and 12% PHEV share.

Thus, from a maximum CO2 reduction perspective, the S-V2G strategy is the preferable integration alternative. However, as indicated above (the rightmost column in Table 2) the implementation cost of the S-V2G strategy is higher than the implementation cost of the other strategies. Also, it might be difficult to reach agreement for a strategy for which the transmission system operator has full control of the charging and discharging of the vehicle and the car owner has no say in the state in which he/she will find the car (charged/discharged). Under the S-FLEX and S-DELAY strategies, the car owner will always find the car charged at a specified/contracted time, so these strategies would probably be more convenient to implement in reality.

[EUR/vehicle and year]	Reduction in cost 20% PHEV	Reduction in cost 12% PHEV	Reduction in cost 3% PHEV	Implemen-tation cost [3]
S-DIR (fixed load –no control)	-17.16	-11.58	-4.00	0
S-DELAY (fixed load -timer)	1.54	11.23	20.85	4
S-FLEX (free load distribution)	6.29	15.25	28.76	14
S-V2G (free load, V2G allowed)	12.57	19.39	32.07	52

Table 2. Reduction in total system costs (as compared to the case without PHEV integration) per vehicle compared with implementation cost (rightmost column) under different PHEV integration strategies and implementation levels. Negative numbers imply an increase in system costs due to PHEV integration. From Göransson et al. (2009).

4. Summary

Emission savings due to wind power integration in a thermal power system are partly offset by an increase in emissions due to inefficiencies in operation of the thermal units caused by the variations in wind power generation. To reduce the variations a moderator or some demand side management strategy, i.e. a fleet of PHEV:s, can be integrated in the wind-thermal system. A reduction in variations (in load and/or wind power generation) will be

[3] (Capital costs*r/(1-(1+r)^-lifetime))
10 years' life time assumed. r =0.05 as in one of the IEA cases IEA (2005). Projected Costs of Generating Electricity, OECD/IEA.. Costs for S-FLEX US$150 and S-V2G US$550 from Tomic and Kempton (2007) Cost for S-DELAY 298SEK at standard hardware store. 2007 average exchange rate from the Swedish central bank.

reflected in the generation pattern of the electricity generating units in the system in one or several of the following ways:

- Reduction in number of start-ups
- Reduction in part load operation hours
- Reduction in wind power curtailment
- Shift from peak load to base load generation

All of the above alterations in production pattern will decrease the system generation costs. The first three effects also imply a decrease in system emissions and an improvement of system efficiency, whereas the consequences of the fourth effect depend on the specific peak load and base load technologies. By using the moderator or the fleet of PHEV:s as a common resource of the system (i.e. managing the aggregated variations of load and wind power generation), the operation of the thermal units will be more efficient after the implementation of variation management than prior the wind power integration.

Examples from results from a simulation model of the power system of western Denmark in isolation shows that a daily balanced moderator with modest power rating (i.e. 500 MW) is sufficient to reduce a significant share of the emissions due to start-ups and part load operation, whereas higher power ratings and storage capacities are required to avoid wind power curtailment. In a wind-thermal system with up to 20% wind power (i.e. 2 374 MW), wind power curtailment is modest and the advantage of a weekly balanced moderator with high power rating (i.e. 2 000 MW) compared to a daily balanced moderator with low power rating (i.e. 500 MW) are small. In a system with up to 40% wind power (i.e. 4 748 MW), however, wind power curtailment is substantial and the avoidance of curtailment is the heaviest post in the reduction of emissions through moderation. A comparison between the costs and emission savings due to moderation to the costs and emissions associated with five available moderation technologies (transmission, pumped hydro, compressed air energy storage, sodium sulphur batteries and flow batteries) indicate that all these moderators are able to decrease system emissions but only transmission lines can decrease the total system costs at a cost of 20EUR/tonne for emitting CO2 (i.e. higher CO2 prices are required to make the other moderators profitable for the system exemplified).

The chapter looks closer at Plug-in Hybrid Electric Vehicles as moderating wind power and it is shown that the ability of a fleet of PHEV:s to reduce emissions depend on integration strategy and the PHEV share of the total electricity consumption. An active integration strategy (rather than charging vehicles as they return home in the evening) is desirable already at moderate shares of consumption (i.e. 12%). An integration strategy which gives the power system full flexibility in the distribution of the charging (i.e. S-V2G) is particularly desirable at high PHEV shares (i.e. 20%). However, such a strategy is perceived as difficult to implement for two reasons; the high implementation cost relative to the system savings from moderation and the uncertainty of the car owner with respect to the state in which he/she will find the battery.

Finally, there is obviously no difference from a wind power integration perspective if variations are managed by shifting power in time compared to if they are met by shifting load in time. This, since the objective is to match load with power generation. Yet, what seems to be of importance is the time span over which the shift can be implemented. Demand side management in general implies a shift in load within a 24 hour time span since most loads are recurrent on a daily basis. This corresponds to a daily balanced storage. By shifting power or load over the day it is possible to avoid competition between wind power and base load units and thus the efficiency in generation will be improved (by a decrease in start-ups, part load operation and/or wind power curtailment). Also, the daytime peak will be reduced and some associated start-ups avoided (although start-up avoidance is of secondary importance, since the peak load units generally have good cycling ability). Results from simulation of the western Denmark system indicate that it is sufficient to manage the variations in load over the day (by shifting power or load) to efficiently accommodate wind power generation corresponding to 20% of the total demand.

It should be noted that, just as in the case of any daily balanced demand side management strategy, it is possible to avoid competition between wind power and base load units through night time charging of PHEV:s. However, unless V2G is applied, there still has to be sufficient thermal capacity in the system to supply the peaks in demand of household and industry at times of low wind speeds. Implementing PHEV:s under a V2G strategy the batteries of the PHEV:s serve as storage. It seems reasonable to assume that the PHEV battery is (at the most) sized to cover the average daily distance driven (typically to and back from work). Thus, the electricity which is stored in the battery as the vehicle leaves home in the morning corresponds to the demand of the vehicle throughout the day and any electricity which the vehicle is to deliver to the grid during the day has to be delivered to the vehicle during that same day. The V2G ability of the PHEV:s thus corresponds to storage balanced over the day (i.e. from the time people leave home in the morning until they return in the evening).

With wind power generation in the range of 40% of the total demand, the variations in wind power exceed the variations in load and, since the variations in wind power often are of longer duration (i.e. there can be strong winds affecting a region for more than 12 hours), power or load has to be shifted over longer time spans. As mentioned above, a weekly balanced moderator (typically pumped hydro or transmission) would be suitable for a wind-thermal system in this case. Some flexible generation such as hydro power or co-generation might also be applicable. However, since it is difficult to find a demand for electricity which can be delayed with a week, demand side management is difficult to apply for wind power variation management at these grid penetration levels.

For the future it seems crucial to evaluate the potential of matching wind power generation and electricity consumption on a European level. Thus, also on a European level, it is of interest to investigate the interaction between wind power variations and load variations. It is also perceived as important to evaluate the correlation between variations in wind power and other renewable power sources. The aggregated effects of large-scale wind power and solar power is of particular interest.

5. Acknowledgement

The work presented in this chapter was financed by the AGS project Pathways to Sustainable European Energy Systems.

6. References

Airtricity (2007). Building a more powerful Europe. www.airtricity.com.

Blarke, M. B. and H. Lund (2007). "Large-scale heat pumps in sustainable energy systems: system and project perspectives." *Thermal Science* 2(3): 143-152.

Carraretto, C. (2006). "Power plant operation and management in a deregulated market." *Energy* 31: 1000-1016.

Castronuovo, E. and J. P. Lopes (2004). "Optimal operation and hydro storage sizing of a wind-hydro power plant." *Electrical Power and Energy Systems* 26: 771-778.

Cavallo, A. (2007). "Controllable and affordable utility-scale electricity from intermittent wind resources and compressed air energy storage (CAES)." *Energy* 32(2): 120-127.

Denholm, P. and T. Holloway (2005). "Improved accounting of emissions from utility energy storage system operation." *Environmental Science and Technology* 39(23): 9016-9022.

EC (2003). Undergrounding of Electricity Lines in Europe, Commission of the European Communities.

ElectricityStorageAssociation. Retrieved 2008-07-15, from www.electricitystorage.org.

Eltra (2005). PUDDEL projektet slutrapport.

Energinet. (2006). Retrieved 2006-10-15, from www.energinet.dk.

Energinet (2007). Technical Regulations for Thermal Power Station Units of 1.5 MW and higher.

European Comission (2008). COM 2008, 19 Final, Proposal for a DIRECTIVE OF THE EUROPEAN PARLIAMENT AND OF THE COUNCIL on the promotion of the use of energy from renewable sources. Brussels.

Greenblatt, J., S. Succar, et al. (2007). "Base load wind energy: modeling the competition between gas turbines and compressed air energy storage for supplemental generation." *Energy Policy* 35: 1474-1492.

Göransson, L. and F. Johnsson (2009a). "Dispatch modeling of a regional power generation system - Integrating wind power." *Renewable Energy* 34(4): 1040-1049.

Göransson, L. and F. Johnsson (2009b). "Moderating power plant cycling in wind-thermal power systems." *Submitted.*

Göransson, L., F. Johnsson, et al. (2009). "Integration of plug-in hybrid electric vehicles in a regional wind-thermal power system." *Submitted.*

Holttinen, H. (2005). "Impact of hourly wind power variations on the system operation in the Nordic countries." *Wind Energy* 8(2): 197-218.

IEA (2005). *Projected Costs of Generating Electricity*, OECD/IEA.

Jaramillo, O. A., M. A. Borja, et al. (2004). "Using hydropower to complement wind energy: A hybrid system to provide firm power." *Renewable Energy* 29(11): 1887-1909.

Kuntz, M. T. (2005). 2-MWh Flow Battery Application by PacifiCorp in Utah. VRB.

Lefton, S. A., P. M. Besuner, et al. (1995). Managing utility power plant assets to economically optimize power plant cycling costs, life, and reliability. Sunnyvale, California, Aptech Engineering Services, Inc.

Manwell, J. F., J. G. McGowan, et al. (2005). *Wind energy explained*, Wiley.

Nourai, A. (2002). Large-scale electricity storage technologies for energy management. *Proceedings of the IEEE Power Engineering Society Transmission and Distribution Conference*, Chicago, IL.

Ravenmark, D. and B. Normark (2006) "Light and invisible." ABB Review 4, 25-29.

Ravn, H. (2001). BALMOREL: A Model for Analyses of the Eletcricity and CHP Markets in the Baltic Sea Region.

Rydh, C. J. (1999). "Environmental assessment of vanadium redox and lead-acid batteries for stationary energy storage." *Journal of Power Sources* 80(1): 21-29.

Rydh, C. J. and B. A. Sandén (2005). "Energy analysis of batteries in photovoltaic systems. Part I: Performance and energy requirements." *Energy Conversion and Management* 46(11-12): 1957-1979.

Stadler, I. (2008). "Power grid balancing of energy systems with high renewable energy penetration by demand response." *Utilities Policy* 16: 90-98.

Tomic, J. and W. Kempton (2007). "Using fleets of electric-drive vehicles for grid support." *Journal of Power Sources* 168(2): 459-468.

22

The Future Energy Mix Paradigm: How to Embed Large Amounts of Wind Generation While Preserving the Robustness and Quality of the Power Systems?

Ana Estanqueiro
Laboratório Nacional deEnergia e Geologia, I.P.
Portugal

1. Introduction

The 2001/77/CE Renewable Energies European Directive together with Kyoto Protocol ratification by many countries, supported by some Governments vision and strong objectives on the reduction of external oil dependence, put Europe and other developed economies in the front line to achieve a remarkable wind energy penetration within ten years time. These goals will not be achieved without technical costs and risks, but mainly, without a careful planning and assessment of the power system behaviour with large amounts of wind generation (SRA, 2008; IEAWind, 2008).

These days, one of the most relevant difficulties the wind sector faces was caused by this technology own extreme success. The high capacity installed in the last decade introduced a brand new set of power system technological concerns that recently became one of the more referenced subjects among developers, network planners and system operators.

These concerns are not anymore a negligible distribution grid integration issue that some years ago the experts tended not to give too much relevance since they were easily solved and even more easily avoided through good design and planning, but this is a real power system operation and planning challenge (Holttinen *et al*, 2009): will the power systems be capable to cope with the specificities of the wind power production in large quantities (aka "high penetration") without requiring new wind park models, system operation tools, increased performance of the wind turbines or even a change in the Transmission System Operators (TSOs) conventional mode of operation?

The recent concern of the TSOs is very legitimate, since it is their responsibility to design and manage the power system global production and its adjustment to the consumer loads as well as to assure the technical quality of the overall service, both in steady-state and under transient occurrences.

The wind power capacity reached such a dimension in some European power systems that obliged the TSOs not to neglect the typical behaviour of these spatially distributed renewable power plants, that being a situation that must be addressed by the wind park developers, the wind manufacturers, the TSO planners and regulators together with the experts in this technology grid integration behaviour.

Notwithstanding these reasonable concerns, the current trend in this R&D area is already that wind generation can be embedded in the system in large amounts and these resources managed through adequate interconnection, holistic transmission planning and system operation adaptation.
The fact that large wind parks started to be seen as "normal power plants" that have to behave as any other generating unit in the system is also a very positive sign of the wind technology maturity. This recent maturity brought a few obligations related to this technology "adult age":

- Wind park models have to be developed and to allow the TSO to simulate, at least, the large wind parks connected to the transmission network in order to study their grid integration, address their behaviour and assess their stability under transient perturbations of the system.
- Part of the already planned/existing wind capacity has to be selected or adapted to remain in parallel after the occurrence of identified perturbations that produce serious voltage dips (or at least the most common ones).
- The "tools" to address and enable to cope with both the spatial and the time variability of the wind production need to be developed. That includes the necessity of accurate wind forecast models together with spatial correlation assessment.
- In extreme cases the "Wind Power Plant" must act as a contributor to the power system regulation (e.g. frequency control by request of the TSO ...).

This chapter presents the new existing technological capabilities that should equip any wind turbine and wind power plant installed in a modern power system facing high to very high wind penetration, as well as it identifies the new wind power plants aggregation and clustering principles that are already being implemented in countries as Spain and Portugal. Moreover, the changes in strategies and methodologies of planning and operation of power systems required to implement (with minimal investments and risks) the paradigm of the future energy mix with a high amount of time-dependent renewable generation are also addressed.

2. Technical barriers to high wind penetration

A fact that should be acknowledged is that several countries and regions in Europe already have a very high penetration[1] of wind generation. Among others, one should mention Denmark, whose wind capacity provides typically 20% of the annual consumption, but also Spain, Portugal and Ireland, these later all above 10% and growing steadily every year (IEAWind, 2009).
There has always been some general concerns associated with the particularities of the wind generation in the power sector. Among others, the fact that wind power is highly variable in time and space and it doesn't offer guarantee of power. Another concern is that high (>10%) penetration requires added reserves and costs. Recently, IEA Wind Implementing Agreement R&D Task 25 report (Holttinen al, 2009) compared the costs computed for the additional reserves motivated by wind power concluding that, in the worst case scenario,

[1] several definitions of wind penetration exist, being the most common the percentage of the yearly consumption provided by the wind and used in this text. It is also used, but less common the definition based on the ratio between the wind capacity and the peak load of the power system.

these costs are always bellow 4 cent.Euro/MWh what constitutes less than 10% of the wind energy value.

Another preoccupation within the power sector is that the operation strategies to cope with wind generation and its characteristic fluctuations under very high penetration scenarios are still being developed: there are solutions being identified and some already in use for the most common grid and system transient constraints, but neither all the possible probable occurrences are addressed nor detailed adequate tools to characterize them are already fully available.

2.1 Transmission limited capacity

The first historical reason normally invoked to limit the amount of wind generation embedded in the grid is the grid limited capacity. That limitation of capacity usually refers only to the transmission capacity, once in most countries the developers of a new wind park are asked to invest themselves on the distribution grid reinforcement and even pay the totality of the cost to build the interconnection lines to the already existing network. In European countries this limitation is being addressed in different ways, but the vast majority of countries are dealing with this classic barrier and nowadays are starting to include renewable energy in general and wind energy in particular in their transmission system development plans (DENA study, 2005; REN 2008).

But constructing new transmission lines is a long and difficult path for all developed countries where environmental and social impacts prevent and delay the installation of new electric lines. In realistic terms, with the existing constraints to reinforce the transmission network, and on a "business as usual" scenario, it could take several decades to reach 20% distributed renewable penetration on a European scale.

2.2 Security of supply. Power unit scheduling

a. Balancing Power.

Being a time depended and highly variable energy source, wind power gives no guarantee of firm power generation at all or, in the limit, gives a quite reduced one at a very short production forecasting time scale. It is a commonly accepted fact that there is a threshold, above which, increasing the wind power penetration also increases the power reserve requirements of a system (Holttinen *et al*, 2009). This has been addressed in detail for some power systems or control areas, e.g. Nordpool (Holttinen, 2004) and the results are quite encouraging: the associated costs are much lower than expected up to a certain upper limit (typically 10%) and are only representative for very high penetrations above 20%. The increase level is strongly depending, as expected, on the system generation mix.

b. Wind Power Time and Space Variability

It was back in the early 1980' that some R&D groups started to address the problematic issue of the excessive "wind variability" and typical fluctuations (Lipman *et al*, 1980) and, at that time, the almost impossible task of forecasting the wind production within time intervals useful for power system operation (Troen & Landberg, 1990).

Another issue strongly related to the wind generation used to be the high frequency content of the power delivered to the system, mainly in the range of flicker emission (from 0.1 to 20 Hz). Those fluctuations could degrade the quality of the service in the surroundings of wind parks (Sorensen, 2007; IEA, 2005; Estanqueiro, 2007) and limits were successfully defined through international standards in order to guarantee an acceptable level of quality (e.g. IEC 61400-21, 2001).

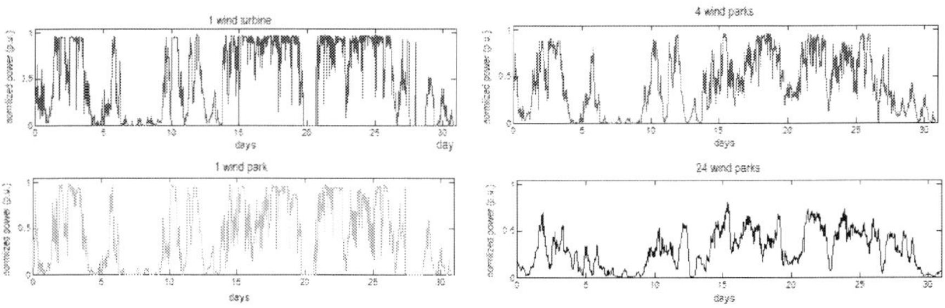

Fig. 1. Wind Power variability and aggregation smoothing effect

c. Wind Generation Technical Reliability

The main concern of every TSO with a large wind capacity in the grid is the sudden disconnection from the grid of all or most of the wind generation as a response to a fast grid perturbation, normally referred as a "voltage dip". Low voltages or dips are usually originated by short circuits and may lead to the islanding of some parts of the network including some conventional generating units. For the wind generation capacity to remain connected to the grid under such circumstances, it is necessary that the wind turbine generators can withstand these voltage dips, a characteristic known as the "ride through fault -RTF" capability (or LVRTF - low voltage ride through fault) which is nowadays requested by most grid codes and national or local regulations.

2.3 Operational energy congestion. Surplus management

In power systems where the energy mix is flexible in terms of regulation (e.g. high penetration of hydro plants with storage capacity) and has a "portfolio approach" with complementary regulation capabilities, the cost with added reserves associated with the large integration of wind in the system is normally lower than in rigid, inflexible power systems.

An issue that is commonly raised when the integration of large amounts of wind power is addressed is: what if the situation of excess of renewable penetration (e.g. wind + hydro) occurs? Should the wind parks be disconnected? would the hydro be reduced?... what is the most important value to preserve, the volatile energy that, if nor extracted from wind will be lost, or the sensible "business as usual" approach "if the hydro is historically in the system, it is a reliable and a unexpensive renewable source", therefore it should never be disconnected...

This situation, commonly referred as surplus of renewable generation raises the uncomfortable issue of either disconnecting wind generators or spilling water which would be turbined in the absence of wind. This issue is again more economical than technical, but a regulated market approach recognizing the benefit of all renewable generation has the ability to overcome these difficulties.

More straightforward approaches - although not necessarily simpler to deploy - consist on having added interconnection with neighbor power systems and use the available ancillary services on larger scales as a contribution to overcome this problem.

These barriers will be addressed in the subsequent sections, together with the possible solutions to overcome the wind integration limitation imposed by them.

3. Technical solutions for large integration: wind power plants innovative concepts

3.1 Innovative characteristics of the wind systems

a. **Low Voltage Ride Through Fault**

A matter of great concern for the TSO, confronted with the large expansion of wind generation, is the reduced capability of some wind turbines to stay connected to the grid, in the event of faults which give rise to voltage dips.

Therefore, and recognizing the large potential of wind energy, but also revealing an extreme concern towards its growth and future development and in a very acceptable form, almost all TSOs with an already representative wind energy penetration have issued grid codes requiring the wind turbines and power plants to contribute with some basic - but slightly "anti-natural" for the wind technology - power system operation functionalities, a feature which considerably increases the stability margin of the power system under transient perturbations. The more publicized one is the LVRTF - low voltage ride through fault capability, whose characteristics for several grid codes (e.g. the German, the Spanish and the US) are depicted in Fig. 2.

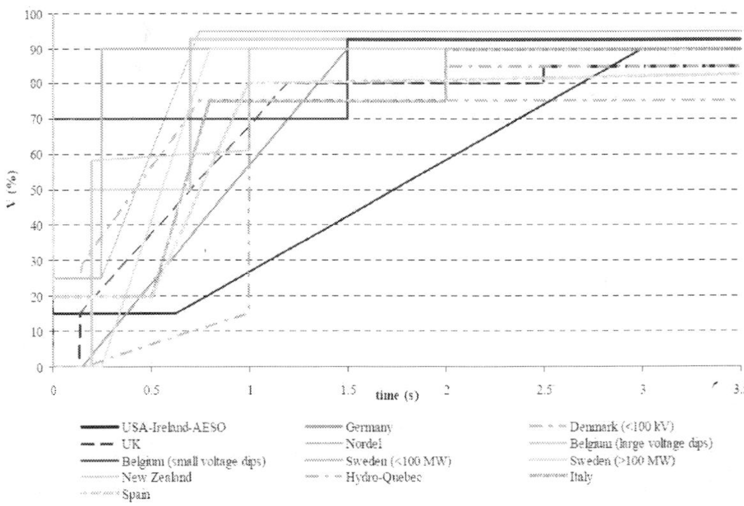

Fig. 2. LVRTF requirements for various grid codes (Tsiliet *al*, 2009)

Most wind turbine manufacturers nowadays offer this capability at an additional cost (usually 5% approx.), which allows the wind generators to withstand a wider range of voltage variations, for longer periods, without disconnection. It should be noted that a power system equipped with less modern wind technology (e.g. without RTF capability) does not have an intrinsic limitation regarding the behavior of the older wind parks under the occurrence of voltage dips. The large electrical industry has already developed RTF systems specifically for the wind industry that, when installed on a wind park without this capability are able to control its response under faults and emulate this new capability of modern wind turbines.

The wind technology RTF capability was one of the most relevant steps this industry has taken once it enabled to put it at a response level similar to the conventional generation in the occurrence of transient events and thus, enabling the TSOs to maintain or in some areas of the network even increase the power quality offered to the consumers.

b. **Participation in the primary frequency control. Low Frequency Ride through Fault**

Large scale recent events (e.g. 4th November 2006) that were propagated to almost all the European Network (1st UCTE synchronous area) and affected even some North-African countries raised the issue of wind turbine response to extreme low frequency occurrences as the one depicted in Fig.3.

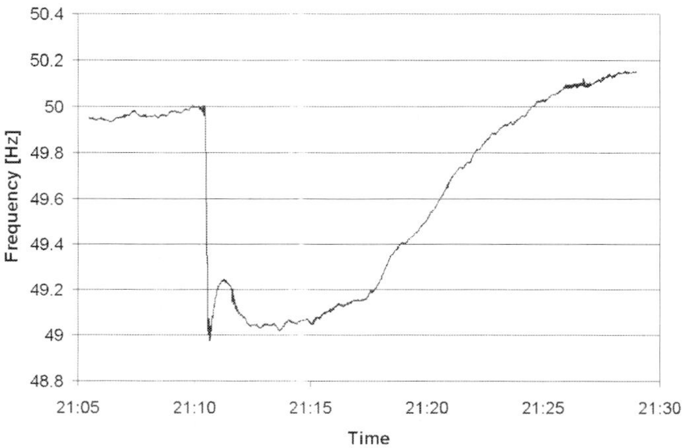

Fig. 3. Frequency dip in the European network on the 4th November 2006

If wind generators with primary frequency regulation capabilities are used, which means adopting a specific primary frequency control and a deload operation strategy - below the maximum extraction power curve (95% for example) a considerable contribution can be obtained from these units to reduce the impact of this frequency dip (Almeida & Peças Lopes, 2005). Such control strategy may provide a considerable contribution for the frequency regulation, especially in windy regions and power systems with reduced flexibility, e.g. without hydro power or reduced regulation capability.

The use of "frequency flexible" power electronics will definitely provide a relevant contribution for the power system robustness, by avoiding grid electronic interfaced wind generators disconnection from the grid when these system disturbances take place: the 2006 event shown in Fig. 3 was extremely useful to show that different wind turbine manufacturers show completely different capabilities, and moreover, that technical solutions for this concern already exist in some wind turbine manufacturers.

Isolated windy power systems with traditional frequency control problems are the typical example for the privileged application of this recent functionality of the wind turbines.

3.2 Wind power control, curtailment and overcapacity

The replacement of large conventional power plants by hundreds of wind generation units spread over the transmission and distribution system requires the development of new concepts for monitoring, controlling and managing these generation resources having in mind network operational restrictions and also market procedures.

Innovative strategies and equipments are already in operation in some European countries. The capacity of a wind park is usually limited by the capacity of the interconnecting grid. However, in wind generation most of the time wind turbines are operated far from their nominal ratings (see Fig. 4). Therefore, in order to optimise the grid connection costs, some agencies authorize the so-called "over capacity" installation in wind parks provided that a control of production is performed to avoid the injection of power larger than the initially defined by grid technical constraints. Since monitoring and control of this generation can be performed using the wind power dispatch centres, this limit can be adapted to the network operating conditions without compromising network security operational levels.

An economically effective tool is to draw wind power purchase agreements that safeguard the possibility to interrupt (curtail) the wind generation in cases technically documented and justified. This possibility is already being used in some countries together with overcapacity. This is a legal innovative approach in Europe where the permanent access of renewable sources to the system was normally widely accepted.

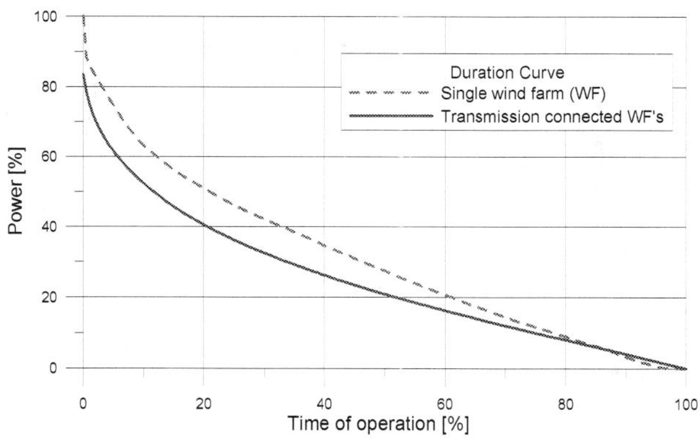

Fig. 4. Comparison of wind power duration curves for a single wind park and the all the wind farms connected to the transmission network

Fig. 4 also highlights the fact that it may be economically interesting and very relevant for low wind regions where the wind park nameplate power is never or very seldom achieved (areas with a wind Weibull distribution with almost "no tail") to reduce the nominal power of the local transformers and the dedicated interconnection line to values around 80 to 90% of the nominal capacity of the wind power plant. This is due to the fact that the investment costs associated with the remaining 10 to 20% of the grid capacity (and equipments) are high, but the value of the energy generated in these maximum operation conditions of a wind power plant is rather low, typically below 5% of the annual profits. This approach should be handled with care in turbulent windy areas where the high resource regimes may bring added control problems for the wind power plant.

The uncorrelated fluctuations of the power output of an aggregate of wind power plants allow to take that effect into the design of the electric infrastructure and sub-sizing both the transmission line and the transformer. On a power system/control area scale this has a huge impact (~10% connected capacity)

3.3 Wind generation aggregation. Virtual wind power plants

Wind power has developed in varied forms in different countries: while in some regions remains an essentially distributed electrical energy source (e.g. Denmark, Netherlands and some areas of Germany) connected to the medium voltage distribution grid, and sometimes even to the low voltage; in others as Spain, Portugal and also the United States this topology is being overcome by the installation of extremely large wind parks (with several hundreds of MW) connected to high (or even very high) voltage transmission lines.

This recent and innovative tendency of the wind industry required the operation of these power plants to be adapted to the new configuration and dimension of the wind plants. In Spain the generation of large transmission connected wind parks is already being aggregated and centrally managed by clusters that constituted a "local wind power dispatch center" and adopt an hierarchical control architecture as depicted in Fig. 5. A similar approach is already defined for the Portuguese power system for the latest generation of wind parks and will be the technical basis for the future development of the remaining sustainable wind energy potential (Estanqueiro, 2007).

This aggregation of the wind generation has several positive side effects as it enables to take advantage of one of the most basic characteristics of the wind resource: its spatial lack of correlation in what concerns the fast wind fluctuations (Estanqueiro, 2008). Other wider studies (Holttinen *et al*, 2009) have shown that a part of this smoothing effect may extend to the spatial scale of one control area, but a deep knowledge of the frequency of the fluctuations involved in the cancellation effects is still not available. Nevertheless, what could be, at a first glance, a negative characteristic may turn, in fact, to be extremely beneficial for the power system operation, since the most hazardous oscillations induced by wind tend naturally to cancel themselves. In order to profit from that effect, it is required the share of common grid interconnection, otherwise large power fluctuation may not be felt by central dispatches, while they are affecting local or regional parts of the transmission network. The smoothing effect is also not present when a whole country (or power system) is immersed in high (or low) pressure atmospheric circulations or passed by large frontal areas.

The need to monitor remotely the state and level of generation of wind power plants was recognized both by the manufacturers of wind turbines and the International Electrotechnical Commission (IEC) several years ago. The IEC Technical Committee 88 - Wind Turbines started the development of a new international set of standards on communications (see IEC 61400-25-1, 2006) and is currently updating the power quality Standard IEC 61400-21 (2001).

But the possibilities offered by the aggregation of hundreds of wind generation units spread over the transmission and distribution system largely exceed the static information contained in the simple monitoring of the wind power plant production with dispatching purposes. After implementing this type of tools, and benefiting from the natural behaviour of wind turbines (cancellation of fluctuations, modular generation, high inertia, among others) it is possible to operate these large clusters of generating groups as a Virtual Wind Power Plant and thus managing these generation sources having in mind network operational restrictions and market procedures.

Regarding wind parks, the characteristic nature of the installed energy conversion systems usually requires specific applications to be installed at the wind park managing system level. Under the typical architecture proposed in Fig. 5, such applications should be able to "dispatch" some active and reactive generation, when the system/grid operator set points are sent to the wind park, thus contributing (till a certain extent) to the frequency and voltage regulation, what reinforces their perception effectively as VWPP - Virtual Wind Power Plants.

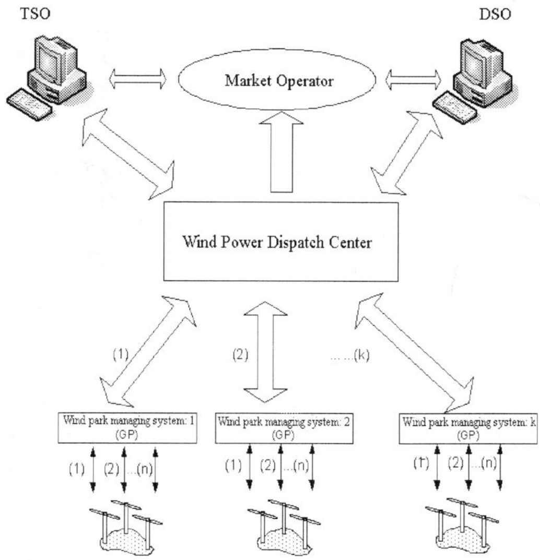

Fig. 5. A possible architecture for the management of the power system with wind aggregation agents (Estanqueiro et al, 2007).

The operation of these local dispatching centres at distribution level requires also the availability of new managing tools, one of the most relevant being the wind generation forecast. Wind forecasts are improving every day, being used by all TSOs in Europe with acceptable deviations within the useful time ranges for power system operation. These forecasting tools provide information about the wind generation within acceptable error margins with time horizons of, at least, 48 hours ahead and the larger the control system, the lower the wind correlation and the smoother the wind power output and the forecast quality.
The best existing tools use Global Numerical Weather Prediction (NWP) models results that are afterwards combined with online data through assimilation techniques, using mesoscale climatic models together with physical or statistical adaptive tools (Tambke *et al*, 2006).
The installation of wind power control at a distribution system operator level and the introduction of wind generation aggregation agents is already enabling to develop and implement the concept of Virtual Wind Power Plants. It should be noted, however, this concept is much more powerful than just the aggregation of wind generation, this later almost a logical procedure having into consideration its spatial distribution and the cancellation of the fluctuation produced at large geographic scales. Therefore a new wider concept is emerging and deals with Virtual Renewable Power Plants (VRPP) that may benefit from the generation aggregation of the natural complementary of several renewable resources as generation of electricity in PV solar power plants, that may be associated to wind power plants with generation profiles where the night periods are dominant but also

with biomass or thermo solar power plants, both having some capacity of production regulation. These new concepts enable to the:

a. Clustering of wind generation (onshore and offshore) for power output smoothing, power control and partial curtailment in large wind power plants.
b. Enhancement of distributed generation systems (DGS) use by regional/local treatment of biomass for electricity generation integrated with wind and PV applications.
c. Correlation of renewable distributed resources, assessment of the excess of renewable energy generation and need for added large/local energy storage capacity (e.g. pumped hydro, FC/H2, VRB batteries and plug-in vehicles)

3.4 Additional remote reactive power control

In order to assure the wind power plants have the capability to deliver reactive power during voltage dips, thus providing support for the network voltage, some TSOs are requiring reactive voltage support similar to the one presented in Fig. 6.

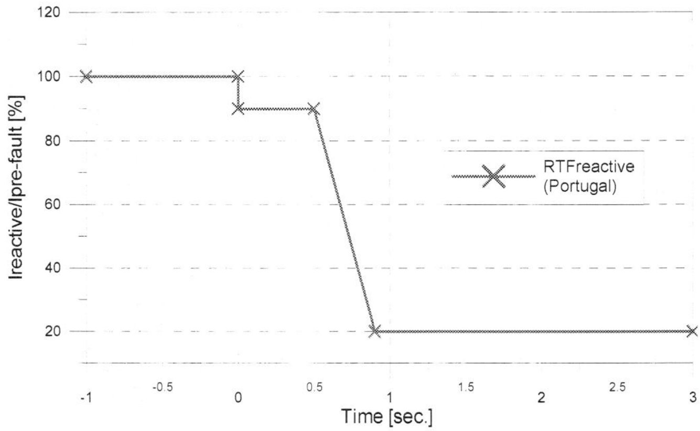

Fig. 6. Characteristic curve of reactive power delivery by wind power plants during/after voltage dips (at t=0).

This capability is also required to enable the adjustment, by request of the TSO, of the reactive power injected in the network in predefined ranges, that in some countries assume values within the interval $tg\varphi \in [0,0.2]$.

3.5 Wind power security of supply

One of the main negative characteristics related to the wind generation usually pointed out by the power systems planners and operators is its non contribution for the security of supply, due to its intrinsic time dependency and variability. Although the wind power variability and the reduced contribution to the capacity credit of a power system is a well know characteristic of the wind generation, one should be also aware of its trustful contribution for the system security of supply in time scales larger than a few days. In Fig. 7 the energy generation contribution of several renewable sources is compared in terms of annual equivalent number of rated hours, being the wind energy surprisingly more reliable than hydro power on a yearly basis.

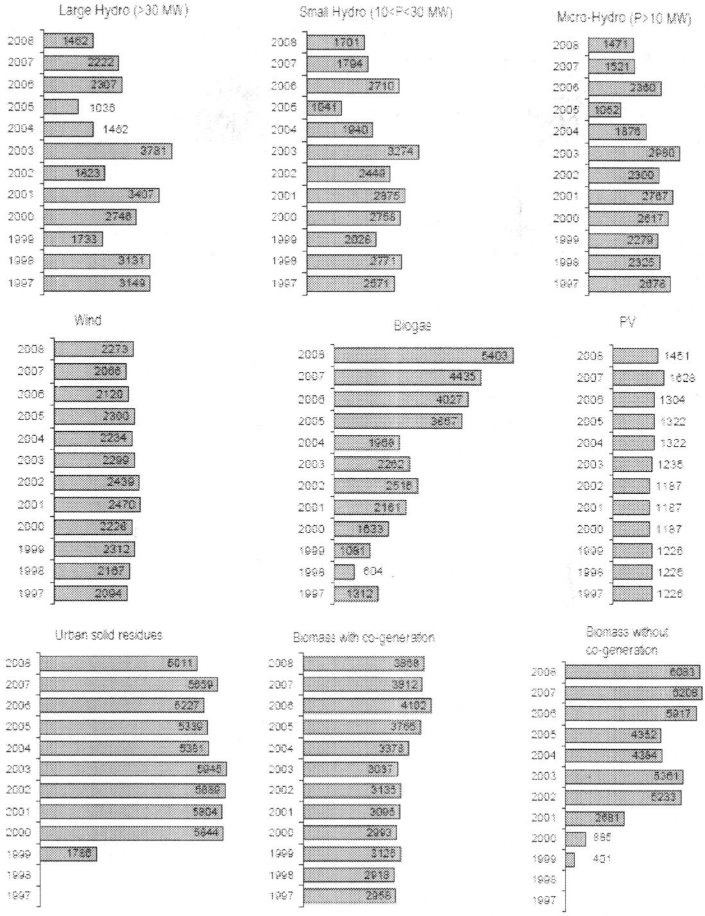

Fig. 7. Comparison of annual equivalent production hours of different renewable sources (source: DGEG, 2009)

4. Large integration of wind generation: the power system challenges

The imposing question for the large integration of wind generation in power system is then: what are the real challenges and how to address them?

Mainly, the characteristics of the wind generation that clearly differentiate it from the conventional generation and have a higher impact on its large integration in the power system are within the list below:

- Wind is regarded has not offering security of supply, may require significant added reserves and also impacts on conventional power unit scheduling;
- There is a limited capacity on the grid to embed this spatial distributed generation;

- Wind is (totally) time dependent and gives (almost) no guarantee of firm power... there are added costs for wind integration in some power systems, specially for penetration >10%;
- There are also operation and management great "challenges": in power systems with significant amounts of rigid generation (either non-dispatchable renewable or nuclear, for example), to foresee large integration of wind may produce Energy Congestion and a difficult Surplus Management;
- Large wind integration affects the robustness of the system operation.

The only possible answer being "one by one..." and using the scientific and technical tools already available, when that is the case, and develop new ones, in the areas still unaddressed...

It is commonly accept in the field that the key to overcome these issues are to add flexibility to the power system, and to study/simulate all possible occurrences using comprehensive, inclusive models.

4.1 A new holistic approach for the transmission power system

The most classical "technical" barrier for the large penetration of wind energy is the limited capacity of the transmission grid. One should clarify this is really an economic, environmental and social barrier, not a technical limitation that, is common to all new power plants, based on spatially distributed renewable energy system (RES) or not. Moreover, large power plants currently being interconnected to the power systems usually benefit from no (physical) grid integration costs at all.

The common approach with large conventional power plants is for the Transmission System Operator (TSO) or the Government of the control zone where a new conventional power plant will be interconnected to provide direct access to the transmission network, being the reinforced costs taken by the operator of the transmission network and distributed by the final consumers. Some associations of wind power developers tried, in the recent past to adopt a similar approach for the grid connection of the wind power plants, but few countries have pursued this path with the relevant exception of Germany.

More relevant for the wind sector large deployment than the distribution grid costs is the non inclusion of the wind energy and other renewable capacity goals in the transmission network development plans. One of the few countries where a holistic approach to the power system planning was implemented was Portugal. There, all forecasted power plants and power sources have been systematically included in the TSO recent development plans (REN, 2005; REN, 2008), having as direct result the fact that Portugal presents the lowest grid integration costs reported by the IEA Wind Task 25. Fig. 8 highlights (with dashed lines, including existing transmission lines to be upgraded) the main investment projects in the Portuguese transmission network until 2010 that are totally or in part induced by the Governmental renewable goals.

In Portugal the methodology followed by the TSO was based on wind resource scenarios that identified the value and location of wind resources and took into consideration the wind power already in service. With those inputs, the TSO was able to define reasonable targets for the wind generation (and ranges of uncertainty) for the different areas of the country. Adding other renewable objectives such as new large hydro, the network planning division performed a transmission network development plan. This task should not be overlooked, since most of the several thousand MW of new wind and hydro power plants

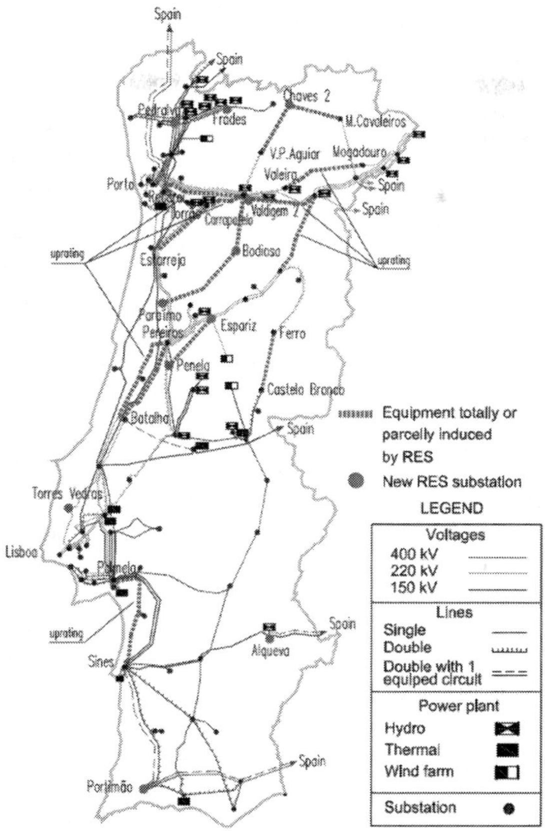

Fig. 8. Example of the Portuguese new transmission lines induced by renewable energy sources (source: REN, 2008)

will, in the future, be located in inner rural areas of the country with very small demand, that implied a large increase of the regional power surplus to be transported to the large load centers, so dictating the need for a non-negligible increase in transmission capacity.

The Portuguese plan also pursued some other investment goals such as the overall system adequacy and security as well as the quality of supply for clients and other users of the transmission and distribution network.

Throughout a 6 year period 2006-2011, 190 million Euro were allocated by this TSO for the investment directly related to the integration and transmission of renewable generation, including 4500 MW of wind power (Fig. 9). This value represents one fifth of the total TSO investment in the transmission network. The investment costs on new lines induced by the renewable generation were allocated according to their relative use of transmission capacity. It should be referred these investment values do not include the costs of wind power plant interconnection to the existing grid in form of dedicated lines, once these are not supported by the TSO in Portugal, but rather by each wind park developer.

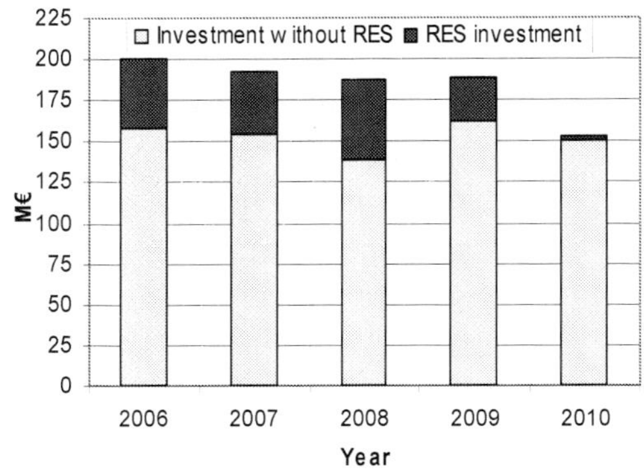

Fig. 9. TSO's total investment costs (MEuro) and reweable energy sources (RES) associated share (source: REM, 2008).

4.2 The power system contribution for large wind integration

Taking into consideration the enormous difficulties felt by all TSOs for the construction of new transmission lines, it is rather surprising to conclude that little has been done to improve the existing network efficiency and utilization, for the benefit of the smooth grid integration of wind power and other distributed renewable, but also to lighten the pressure from the difficult construction of new transmission lines. Nowadays, the environmental and social impacts of new electrical lines turns into "mandatory" measures as:

- online monitoring of transmission lines (temperature, wind, loads, etc);
- introduction of new network components (e.g. phase shift transformers);
- use of Flexible AC Transmission Systems (FACTS) devices;
- upgrading degraded components as cables, lines, protections and transformers.

All of these urgent measures to be implemented by the transmission grid operators. Notwithstanding the measures just presented that enable to operate existing lines with higher efficiency levels, with the existing steady increase in consumption, the construction of new transmission lines and the reinforcement of the existing ones will be needed. In order to optimize the integration costs of the wind power, it would be desirable that the wind deployment official national objectives would be included in the medium to long term development plans of the power systems. However this situation has seldom been reported. Some innovative strategies and equipments are already in operation in some countries where wind penetration is growing very fast: a new not very common transmission network element was recently included and is being suggested for many power systems; the phase-shifter (transformer). These electrical machines can "force" the wind power flow, injected in the high voltage levels of 60, 150 or 220 kV in some specific geographic areas to enter the 400 kV grid, using the available higher voltage lines capacity and thus avoiding the construction of new high voltage lines.

4.3 Improving the power system dynamic behaviour with FACTS

The loss of large amounts of wind generation may lead to system instability problems or to overload of interconnection lines. The capability of survival of these generation units to voltage dips that follow a short circuit in the grid is thus becoming a mandatory requirement in several Grid Codes. However, some of the already installed wind turbines are not capable of withstanding such grid disturbances, which may require the adoption of external measures like the installation of Flexible AC Transmission Systems (FACTS) devices. These devices are capable of providing a support to voltage profiles, limiting the voltage dip during the short-circuit duration time. Some studies (Franco Marques & Peças Lopes, 2006) strongly suggest that it would be beneficial to install FACTS in strategic buses of the transmission network in order to mitigate the impact of short circuits that may occur in the grid and that may lead to the disconnection of large amounts of wind generation due the tripping of their under voltage protection relays.

Apart from avoiding the tripping of some generation units, also a good damping of the oscillations can be obtained as depicted in Fig. 10 that describes the behaviour of the sum of the power flow in all the interconnection lines between Portugal and Spain following a short-circuit in an important transmission bus in the scenarios without and with FACTS installed.

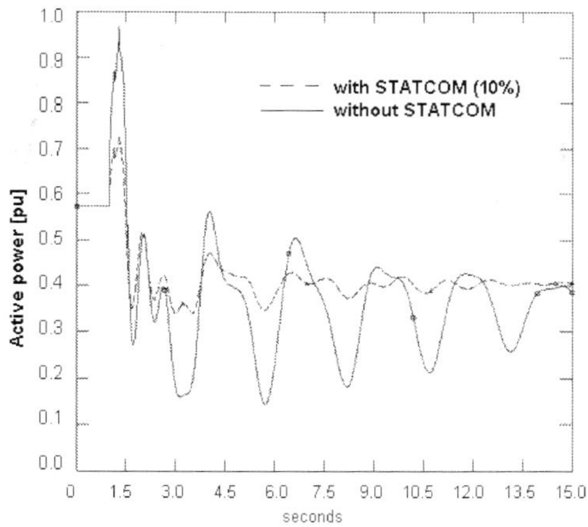

Fig. 10. Active power flow in the interconnections, following a short circuit in a relevant transmission bus - with and without STATCOM (Franco Marques & Peças Lopes, 2006).

4.4 Adding flexibility to the power system: storage and transmission reinforcement

Different generation mixes face different challenges when integrating large amounts of wind power. However, it is commonly acknowledged in the area of wind grid integration that adding flexibility to the power system easies its operation under high penetration of fluctuating renewable sources as the energy (Chandler, 2008; Holttinen *et al*, 2009). There are several ways to add flexibility to the system, being the simplest to handle to add storage hydro capacity in the geographical areas where this resource exist and is possible to deploy in a sustainable way. Another possibility is the added flexibility obtained through the interconnection with neighbouring countries and the reinforcement of the transmission network (Ackermman (Ed.), 2005).

The example of the Portuguese hydrologic plan (PNBEPH, 2007) presented below and currently underway was not only to increase the renewable generation penetration, but also to promote a smoother operation of the power system with the forecasted very high wind penetration (above 20% after 2015).

In some generation mixes the main power system constraint may end up being excess of renewable generation (e.g. wind + run-of-river hydro) during the no-load hours. Due to this fact it was recently introduced in some countries the concept of *wind energy storage* - and other highly variable time-dependent renewable primary sources - through the establishment of bilateral contracts between the wind park owners and the operators of reversible hydro power stations.

The identification of the optimised daily operation strategy for the mix of renewables can be determined by solving a linear hourly-discretized optimisation problem where the economic benefits of such strategy are driving an objective function (Leite da Silva *et al*, 2007).

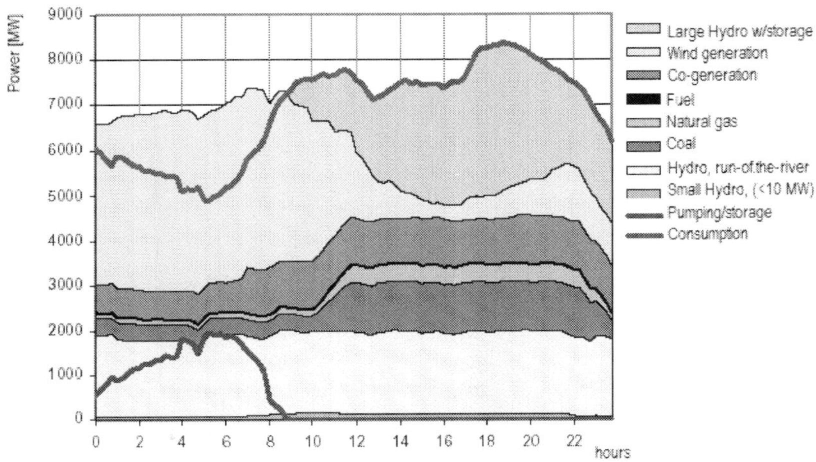

Fig. 11. Scenario of a power system generation profile for a wet windy day in 2011 (PNBEPH, 2007).

This form of *wind energy storage* enables to optimise the daily operation strategy of the power system and allows to:

- Minimize deviations to participate in structured markets;
- Contribute to the secondary and tertiary power reserves;
- Increase of wind contribution for the regulation capacity

When hydro pumping storage is available, the existing methodologies able to identify the best combined wind/hydro pumping storage strategies should be used. In the absence of hydro energy resources, other storage techniques may also be helpful and should be investigated (e.g. H2/Fuel Cells, compressed air/gas, flywheels, etc).

5. Power system studies for wind generation safe integration

5.1 Transient stability assessment

In order to ensure that planned wind generation for the near future will be managed within safety ranges, TSOs are assessing the response of their control area and the impact of the committed wind capacity on the transient stability of the transmission network (e.g. Sucena Paiva *et al*, 2005; GE Energy, 2005).

The main objective of these transient stability studies is to evaluate the percentage of the wind generation that will be disconnected due to voltage dips produced by drastic events as three-phase short-circuits in the network, for different scenarios of conventional generation / demand, and wind power penetration, as well as the spatial distribution of the wind power.

As far as wind generation is concerned, several scenarios may be considered: (i) Uniformly distributed wind generation, with all wind generators injecting a similar percentage of their rated power (80%); and (ii) the most realistic situation where the wind generation is uncorrelated and distributed, in a form that reproduces the passage of the large air masses that pass through control zone several hundreds or thousands of kilometres wide.

In the transient stability studies the most disseminated wind turbine technologies should be modelled: (i) Classical wind turbines equipped with induction generators; (ii) Wind turbines with double fed induction generator (DFIG); (iii) Wind turbines equipped with variable speed synchronous generator, connected to the grid through a rectifier/inverter system.

The main conclusions that may be achieved at the end of such transient studies normally address the following subjects:

- For some faults in the transmission grid, when wind turbines are equipped with conventional technologies (non-ride through fault), it may occur an almost complete loss of wind power. This, in some rare situations, may also originate a loss of synchronism in some parts of the power systems.
- The loss of wind power in a country or control zone has an impact on their neighbours through existing interconnections. Substantial loss of wind power in some areas can give rise to overloads, creating the risk of electrical separation of some power systems, a situation that can lead to local blackouts.
- The addition of control equipment to ensure that the wind turbines remain connected during most short-circuit situations (ride through fault capability) results in a significant reduction in the loss of wind power under faults and transient events, thus largely increasing the stability margin of a power system.

It should be referred that, the relative immature phase of grid codes development specifically to cope with the large and increasing penetration of wind energy have generated a wide range of requirements, typically one by each TSO, without any concern about the need to standardize the industrial production of wind turbines (see Fig. 2).

The need to standardize the fault ride through requirements has recently been a subject of concern of the European Wind Energy Association (EWEA) which currently runs a working group on harmonization of grid codes (EWEA, 2008).

5.2 Power reserves and security of supply.

In a near future scenario characterized by such large amount of wind power integration, system planners are under a huge pressure to come out with solutions for the determination of the required amount of system capacity to guarantee an adequate supply.

Traditionally the power reserve requirements have been based on criteria that protect against the loss of the largest power group delivering to the system. These deterministic criteria do not take into account neither the accuracy of the demand and wind power

forecasts, nor the probability of the largest generator or interconnection outages, and the consequences of such contingencies.
For this purpose, probabilistic nature studies started to be used for assessing the performance of the power system regarding this issue and are still being further developed to address the wind generation forecast errors (Ortega-Vazquez & Kirschen, 2009). The objective is to investigate the behaviour of the common reliability indices, like LOLP = loss of load probability; LOLE = loss of load expectation; EPNS = expected power not supplied; EENS = expected energy not supplied; LOLF = loss of load frequency; LOLD = loss of load duration; LOLC = loss of load cost, as well as well-being indices.
Chronological Monte Carlo simulation may be used to evaluate the reserve requirements of the future expansion plans of the generation system, to be defined by a TSO considering a large penetration of wind generation (Leite da Silva *et al*, 2007).
This analysis requires a proper modelling of system components regarding their reliability, which involves the treatment of the primary energy resource availability (hydro, wind, cogeneration, etc.), maintenance policies and specific forced outage rates.

6. Wind power plant models for the 21st century power system

One of the difficulties faced by power system planners and operators is there are not enough accurate off-the shelf tools to describe the dynamic, non-linear behavior of Wind Power Plants and, by doing that, increasing the degree of confidence of the TSOs in this form of power generation. To allow for a smooth and safe integration of large amounts of wind generation the power system needs:

a. Simulation platforms with distributed Renewable Energy Sources (RES) and non-linear system devices;
b. To model the behavior of the power system/grid with large scale integration of renewable generation on large/European scale using the classical power system approaches.

Although several models were published in the latest years (e.g. Estanqueiro, 2007; Perdana, 2007 among others) and a few European projects as Tradewind (2009) and EWIS (2008) have given excellent 1st steps in the recent past, new wind power dynamic models for power system stability studies including aggregation and clustering of wind turbines are still requested.
The relevant role of wind turbine dynamic and transient models in the large integration of wind generation planned for the next decade is due to:

- the 20% renewable penetration (mainly wind) forecasted for 2020 in Europe, may reveal itself a too risky task in technical terms if the European TSOs do not possess the requested tools to simulate the power system under extreme occurrences and carefully plan for secure operation under those circumstances.
- It should not be asked to the system operators to (apparently) reduce the robustness of their systems without providing them with the simulation models that enable to characterize the wind power plants response to every possible occurrence in the system.
- It is questionable if the European Power Systems will be manageable with a 20% penetration of variable generation (on an yearly basis) without more sophisticated simulation tools and a more detailed knowledge of the renewable power plants transient behavior.

In the future it is expectable that the network areas are classified in terms of transient response capability of wind turbines to survive low voltage ride through faults; the voltage and frequency regulation will, most probably, be asked for some pre-identified grid areas. Wide wind power control and, in certain conditions, curtailment of wind generation will be a reality. This allows to conclude that the better the available dynamic wind power plant models are in the future - and the larger their capacity to perform under transient behavior of the network - and the more effective is the wind power control capability, the more reduced will be the need to curtail this renewable electrical energy source.

7. Conclusions

Increasing the penetration of wind energy for high levels, typically near 20% on an annual basis, requires an articulated common effort of the TSOs, regulatory official agencies and wind park developers to use and require the most recent and high performing wind and power system technologies in order to guarantee the overall power quality and security of supply, and thus enabling to maximize the wind and other renewables embedded capacity. The main concepts that need to be addressed in the near future are:

- real-time assessment of transmission capacity.
- use of DGS as grid active voltage controllers;
- coordination of ancillary services on a European scale;
- integration of balancing markets and coordination of reserves within EU grids/control areas;
- implementation of solutions to allow for efficient and robust system operation with significant amounts of highly variable generation and storage;
- full deployment of the VRPP - Virtual Renewable Power Plants Concept;
- use of DSM - Demand side management for system added flexibility;
- Fuel Cells/hydrogen generation for regulation of highly variable renewable sources;
- Inclusion of plug-in vehicles as distribution storage units in the distribution network planning;

so as to become feasible the management of the power systems while preserving the global quality characteristics and security of operation under a 20% penetration of sources of electrical energy as variable as the wind.

8. References

Ackermann, T. (2005). (Ed.) "*Wind Power in Power Systems*". Wiley & Sons, Chichester, UK, January 2005, pp 691.

Almeida, R.G. and J. A. Peças Lopes (2005). "Primary Frequency Control Participation provided by Doubly Fed Induction Wind Generators", in *Proc. 15thPower System Computation Conference*, Liège, Belgium, Aug. 2005.

Almeida, R. G., E. D. Castronuovo, and J. A. Peças Lopes (2006)."Optimum Generation Control in Wind Parks When Carrying Out System Operator Requests", *IEEE Trans. Power Systems*, vol. 21, No. 2, pp. 718-725, May. 2006.

Chandler, H. (2008). Empowering Variable Renewables Options for Flexible Electricity Systems, IEA, Ed. by OECD/IEA, Paris. pp 35

CAPRPE (2005) *Concurso para Atribuição de Pontos de Recepção de Energia Eléctrica para Ligação à Rede do SEP de Parques Eólicos - Programa de Concurso, DGGE – Direcção Geral de Geologia e Energia*, Lisbon, July 2005 (*in Portuguese*).

DENA Grid study (2005). Planning of the grid integration of wind energy in Germany onshore and offshore up to the year 2020. Deutsche Energie-Agentur, Ed.s. (Dena Grid study). Deutsche Energie-Agentur Dena, March 2005. English summary and full German version available at URL: http://www.dena.de/themen/thema-reg/projektarchiv/., last accessed 15.11.2009).

Estanqueiro, A. (2008) "Impact of Wind Generation Fluctuations the Design and Operation of Power System"7th International Workshop on Large Scale. In Proceedings of the *Integration of Wind Power and on TN for Offshore Wind Farms Workshop. Madrid*, Spain. June 2008. Paper 85.

Estanqueiro, A., R. Castro, P. Flores, J. Ricardo, Medeiros Pinto, Reis Rodrigues, J. Peças Lopes. (2007) "How to prepare a power system for 15% wind energy penetration: the Portuguese case study" *Wind Energy*, Vol. 11 Issue 1, Pages 75 - 84.

Estanqueiro, A., (2007) "A Dynamic Wind Generation Model for Power Systems Studies," *Power Systems, IEEE Transactions on* , vol.22, no.3, pp.920-928, Aug. 2007

EWEA (2008), EWEA Working Group on Grid Code Requirements - Position Paper European Grid Code Requirements for Wind Power Generation (URL, available at http://www.ewea.org/fileadmin/ewea_documents/documents/publications/position_papers/080307_WGGCR_final.pdf, last accessed 15.11.2009). February 2008. pp.12.

EWIS European Wind Integration Study (2008). "Wind Turbine Model Validation Report", (URL, Available at http://www.wind-integration.eu/ downloads/library/EWIS-Wind_Turbine_Model_Validation_Report.pdf, last accessed 15.11.2009). Dec. 2008.

Franco Marques, P. J. and J. A. Peças Lopes (2006), "Impact of the Use of FACTS to Increase Robustness of Operation in Grids with Large Scale Wind Generation", *Proc. Nordic Wind Power Conference*, May 2006, Espoo, Finland.

Tradewind (2009). Frans van Hulle (Ed.). "Integrating Wind. Developing Europe's power market for the large-scale integration of wind power, Tradewind Final Report, (URL, available at http://www.tradewind.eu/fileadmin/documents/publications/Final_Report.pdf, last accessed 15.11.2009). Feb. 2009.

GE Energy (2005). *The Effects of Integrating Wind Power on Transmission System Planning, Reliability, and Operations. Report on Phase 2*, Prepared for The New York State Energy Research and Development Authority, City, State, Mar. 2005. (URL, available at: http://www.nyserda.org/publications/wind_integration_report.pdf, last accessed 15.11.2009).

H. Holttinen, P. Meibom, A. Orths, F.van Hulle, B.Lange, M. O'Malley, J. Pierik, B. Ummels, J.O. Tande, A. Estanqueiro, M. Matos, E. Gomez, L. Söder, G. Strbac, A. Shakoor, J.Ricardo, J. C. Smith, M.Milligan, E. Ela.(2009) "Design and operation of power systems with large amounts of wind power" Final report, IEA WIND Task 25, Phase one 2006-2008. Espoo, VTT. 200 p. + app. 29 p. VTT Tiedotteita - Research Notes 2493. ISBN 978-951-38-7308-0. (URL, Available at http://www.vtt.fi/inf/pdf/tiedotteet/2009/ T2493.pdf). VTT/Julkaisiga (Ed.). Helsinky.

Holttinen, H (2004). The impact of large scale wind power production on the Nordic electricity system. VTT Publications 554. Espoo, VTT Processes, 2004. 82 p. + app. 111 p. Available: http://www.vtt.fi/inf/pdf/publications/2004/P554.pdf (URL, accessed 30.11.2006).

IEA Wind (2008). IEA Wind Strategic Plan for 2009-2013 and EOT report (URL, available at http://www.ieawind.org/strategy_plan.html, last accessed 15.11.2009)

IEA Wind (2009). 2008 IEA Wind Energy Annual Report. (URL available at http://www.ieawind.org, last accessed 15.11.2009). Published by IEA Wind, ISBN 0-9786383-3-6. Boulder. pp 315.

IEA (2005). Variability of wind power and other renewables. Management options and strategies. (URL, Available: http://www.iea.org/Textbase/publications/free_new_ Desc.asp?PUBS_ID=1572, last accessed 30.11.2006).

IEC 61400-21 (2001)."Wind turbine - Part 21: Measurement and assessment of power quality characteristics of grid connected wind turbines", IEC Standard.

IEC 61400-25-1 (2006) Wind turbines - Part 25-1: Communications for monitoring and control of wind power plants - Overall description of principles and models", IEC Standard.

Leite da Silva, A.M.,L.F.Manso, W.S. Sales, L.C. Resende, M. Aguiar, M.A.Matos, J.A.Peças Lopes and Miranda, V.(2007). Application of Monte Carlo simulation to generating system well-being analysis considering renewable sources in *European Transactions on Electrical Power*, Vol. 17, Issue 4, John Wiley and Sons (Ed.), pages 387 to 400.

N. H. Lipman, E. A. Bossanyi, P. D. Dunn, P. J. Musgrove, G. E. Whittle, and C. Maclean. (1980). "Fluctuations in the output from wind turbine clusters", *Wind Engineering*, vol. 4, n° 1, pp.1-7, 1980.

Ortega-Vazquez, M. A., and D. S. Kirschen (2009), "Estimating the Spinning Reserve Requirements in Systems With Significant Wind Power Generation Penetration" in IEEE Transactions on Power Systems, Vol. 24, no. 1, February 2009, pages 114 to 124.

Perdana, A. S.U. Joutsenvuo, O. Carlson, B. Lemström. (2007) "Comparison of an aggregated model of a wind farm consisting of fixed-speed wind turbines with field measurement" in Wind Energy, Vol. 11 Issue 1, pages 13 - 27.

REN (2005). Plano de Investimentos da Rede Nacional de Transporte 2004-2009, REN, S.A., Lisbon, November 2005 (*in Portuguese*). (URL, available at http://www.ren.pt)

REN (2008). Plano de Investimentos da Rede Nacional de Transporte 2009-2011, REN, S.A., Lisbon, November 2005 (*in Portuguese*). (URL, available at http://www.ren.pt)

PNBEPH (2007). Hydrologic National Plan for Dams with high Potential (*in Portuguese*). (URL, available at http://pnbeph.inag.pt/np4/documentos.html, last accessed 15.11.2009).

Soerensen, P., N.A. Cutululis, A. Vigueras-Rodriguez, L.E. Jensen, J. Hjerrild, M.H. Donavan and, H. Madsen(2007)."Power Fluctuations from Large Wind Parks". *Power Systems, IEEE Transactions on*, vol.22, no.3, pp.920-928, Aug. 2007

SRA (2008). Strategic Research Agenda. Market Deployment Strategy FROM 2008 TO 2030. European Wind Energy Technological Platform (TPWind), (URL, available at http://www.windplatform.eu/fileadmin/ewetp_docs/Bibliography/SRA_MDS_July_2008.pdf), pp 52.

Sucena Paiva, J.P.; J.M. Ferreira de Jesus; Rui Castro; Pedro Correia; João Ricardo; A. Reis Rodrigues; João Moreira and Bruno Nunes (2005), "Transient stability study of the Portuguese transmission network with a high share of wind power", in Proceedings of the *XI ERIAC CIGRÉ - Undécimo Encuentro Regional Iberoamericano de Cigré*, Paraguay.

Tambke, J., C. Poppinga, L.V. Bremen, L. Claveri, M. Lange, U. Focken, U., J.A.T.Bye, and J.O. Wolff (2006), "Advanced Forecast Systems for the Grid Integration of 25GW Offshore Wind Power in Germany", Proceedings of the Scientific Track of the European Wind Energy Conference EWEC, Athens, 2006

Troen, J. and L. Landberg (1990), "Short Term Prediction of Local Wind Condition", W. Palz, CEC (Ed.), Proceeding of 1990 ECWEC, H. S. Stephens and Associates, Bedford, pp. 76-78, 1990

Tsili, M. , Ch. Patsiouras and S. Papathanassiou (2008)."Grid Code Requirements for Large Wind Farms: A Review of Technical Regulations and Available Wind Turbine Technologies". In Proceedings of the *European Wind Energy Conference, 2008*. (URL available at http://www.ewec2008proceedings.info/ ewec2008/allfiles2/ 144_EWEC2008fullpaper.pdf. Madrid, April 2008.

Part D
The Environmental Issues

23

Environmental Impact of Modern Wind Power under LCA Methodology

Eduardo Martínez Cámara, Emilio Jiménez Macías, Julio Blanco Fernández and Mercedes Pérez de la Parte
University of La Rioja
Spain

1. Introduction

Renewable energy sources constitute an alternative to fossil fuels and their problems, which are, on the one hand, the pollution and CO_2 emissions that they produce and, on the other hand, the diminution of reserves, in addition to other economical and political problems, such as their increasing prices and the economic dependence of non-producers countries on those that produce fossil fuels.

At the present time, renewable energy, and particularly wind power energy, is becoming increasingly relevant in the world's electricity market, based on its advances and on the legislative support of governments in several countries (Río del & Unruh, 2007; Jager-Waldau, 2007; Karki, 2007; Breukers & Wolsink, 2007), for instance with legal frameworks presenting stable and lasting premiums. Figure 1 shows the contribution and the provisions of wind power to the electricity supply network in several countries, both at a European and world level; current forecasts predict that wind power will contribute 12% of the global demand for electricity by 2020 (GWEC, 2005). This huge boom in implementation and forecasts of this power source justify the need to increase its people's understanding (Jungbluth et al., 2005; Gurzenich et al., 1999), based on scientific studies, especially from the point of view of its environmental impact.

Wind power produces electrical energy from the kinetic energy of the wind, without directly producing any pollution or emissions during the conversion process, but this does not means that it is free of contamination or CO_2 emissions. The question is that it should be considered that there is an environmental impact due to the manufacturing process of the wind turbine and the disposal process at the end of the wind turbine life cycle. And this environmental impact should be quantified in order to compare the effects of the production of energy, and to analyse the possibilities of improvement of the process from that point of view.

Thus, the aim of this chapter is to show a methodology of analysis of the environmental impact of the wind energy technology, considering the whole life cycle of the wind power systems. The application of the ISO 14040 standard (ISO, 1998) allows us to quantify the overall impact of a wind turbine and each of its components from a Life Cycle Assessment (LCA) study. It also allows us to analyse the issues that produce more impact and the aspects that could be improved in order to reduce the effective impact. The LCA model has been developed with the purpose of determining and quantifying the related emissions and the impact of wind energy production technology; additionally, the LCA model can be used to define the energy payback time (Martínez et al., 2009; Martínez et al., 2009b; Martínez et al., 2009c).

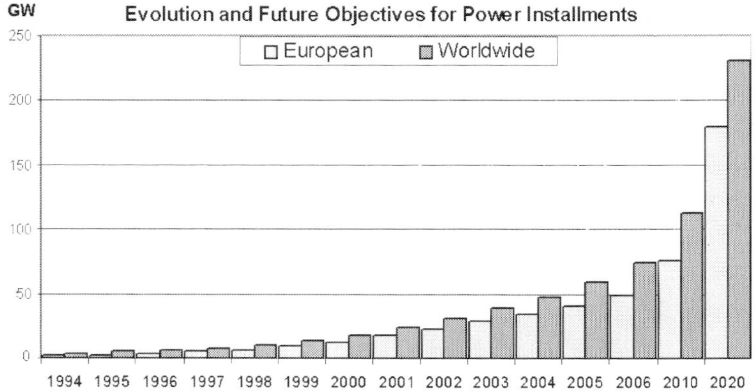

Fig. 1. Evolution and future objectives for wind power instalments

Within the existing LCA studies, there are several ones based on renewable energies in general (Gurzenich et al., 1999; Góralczyk 2003), which do not analyse in detail the LCA of a wind turbine. Reference (Gurzenich et al.,1999), for instance, shows (in its section 2) a comparison of the results of several renewable energy sources, without actually explaining in detail the LCA made in each case, and then focuses on the development of dynamic life cycle assessment as a central part of the study. There are also more specific studies on wind turbines, but they are generally based on older machines and lower rated power, less than 1 MW (Celik et al., 2007; Jungbluth et al., 2005; Ardente et al. 2008), or they refer to hybrid technologies (Khan et al., 2005). In the reference (Celik et al., 2007), micro-turbines and low power urban installation, for example, are studied. The work (Jungbluth et al., 2005) analyses the rapprochement of the database Ecoinvent to wind powers, focusing on studying wind turbines with power ranges from 30 kW to 800 kW. Reference (Ardente et al., 2008) deepens in the LCA of a wind farm with 11 turbines of 660 kW rated power. Reference (Khan et al., 2005) develops an LCA on a hybrid system of wind turbine with fuel cells, with a wind turbine of 500 kW rated power. In addition to these studies about low-power turbines, there are also other analyses focused on multi-megawatt wind turbines, as for instance references (Tryfonidou & Wagner, 2004; Douglas et al., 2008), both of which are focused on offshore wind turbines. On the other hand, there are indeed studies based on multi-megawatt wind turbines, but basically outside the LCA point of view, and focused exclusively on the potential of wind generation of certain areas or regions (Ben Amar et al., 2008; Carolin Mabel & Fernandez, 2008; Wichser & Klink, 2008).

On the other hand, such as it has been previously mentioned, Life Cycle Assessment (LCA) methodology (ISO, 2006) is useful for analysing the environmental impact occasioned by any type of product or process; however, the results obtained with LCA present some uncertainties that have to be considered and assessed in an appropriate way. In general, these LCA uncertainties can be classified into, at least, five types: parameter uncertainty, model uncertainty, spatial variability, temporal variability, and uncertainty due to choices.

For this reason, one of the purposes of this work is to assess the relevance of different choices that have been made during the development of the LCA. Five alternative scenarios have been studied. The first one (AS1) represents an increase in maintenance during the lifetime of this wind turbine. The second alternative scenario (AS2) analyses an increase in the needs of material and energy used. The third scenario (AS3) studies a change in the percentage of recycled materials during the disposal and waste treatment of the wind turbine. The fourth alternative scenario (AS4) analyses a change in the composite waste treatment of the blades at disposal time, from landfill to recycling. Finally, the fifth scenario (AS5) analyses the effect of an increase in the estimated annual production of the wind turbine (Martínez et al., 2009; Martínez et al., 2009b; Martínez et al., 2009c).

These scenarios can facilitate to assess the degree of uncertainty of the developed LCA due to choices made. But this study does not analyse the uncertainty due to imprecise knowledge of the different parameters used in the Life Cycle Inventory (LCI), the spatial and temporal variability in different parameters of the LCI, or the uncertainty due to the inaccuracy and the simplification of the environmental models used.

Finally, another aspect to consider when analysing the environmental impacts by using LCA methodology is the choice of the method used. This chosen method is rarely discussed, and although there exist several works discussing the topic (Schulze et al., 2001; Brent & Hietkamp, 2003; Dreyer et al., 2003; Pant et al., 2004; Bovea & Gallardo, 2006; Renou et al., 2008; Hung & Ma, 2009), usually they focus on specific case studies, and no one is focused on the specific case of renewable energy. Hence it is legitimate to ask whether the LCA results may be influenced by the choice of the LCIA method, between all the scientifically sound methods. This is a key issue, especially if the results of the assessment should be presented to non LCA specialist people. For that reason throughout this chapter an overview of the influence that this choice may have on the final result is provided.

All these analysis, studies and results summarize the work that has been carried by the research group in recent years and have driven to various scientific publications in several important journals related to environment and renewable energy (Martínez et al., 2009; Martínez et al., 2009b; Martínez et al., 2009c). This chapter explain these works, in a summarised and qualitative way, but all the quantitative information should be obtained from the mentioned published works of the group.

2. LCA methodology

2.1 Method and scope

For presenting the main points of the environmental impact study of the wind turbine, the method CML Leiden 2000 has been selected in order to avoid subjectivity (Guinée et al., 2001). The midpoint impact categories considered have been:

- Abiotic depletion (AD): This impact category is concerned with protection of human welfare, human health and ecosystem health, and is related to extraction of minerals and fossil fuels. The Abiotic Depletion Factor (ADF) is determined for each extraction of minerals and fossil fuels (kg antimony equivalents/kg extraction) based on concentration of reserves (Goedkoop et al., 2004).
- Climate change (GW): Climate change can result in adverse affects upon ecosystem and human health and is related to emissions of greenhouse gases to air. GW change factor is expressed as global warming potential for 100 years time horizon, in kg carbon dioxide/kg emission (Goedkoop et al., 2004).

- Stratospheric ozone depletion (OLD): This category is related to the fraction of UV-B radiation reaching the earth surface. The characterisation model defines the ozone depletion potential of different gasses (kg CFC-11 equivalent/kg emission) (Goedkoop et al., 2004).
- Human toxicity (HT): This impact category is related to exposure and effects of toxic substances for an infinite time horizon. For each toxic substance, human toxicity potential is expressed as 1,4-dichlorobenzene equivalents/kg emission (Goedkoop et al., 2004).
- Fresh-water aquatic eco-toxicity (FWAE): It is related to the impact on fresh water ecosystems, as a result of emissions of toxic substances to air, water, and soil, for an infinite time horizon. For each toxic substance, eco-toxicity potential is expressed as 1,4-dichlorobenzene equivalents/kg emission (Goedkoop et al., 2004).
- Marine eco-toxicity (MAE): This impact category is related to the impact on marine ecosystems. As in the human toxicity category, the eco-toxicity potential is expressed as 1,4-dichlorobenzene equivalents/kg emission (Goedkoop et al., 2004).
- Terrestrial eco-toxicity (TE): This impact category is related to the impact on terres-trial ecosystems. As in the human toxicity category, the eco-toxicity potential is expressed as 1,4-dichlorobenzene equivalents/kg emission (Goedkoop et al., 2004).
- Photochemical oxidation (PO): This category is related to the formation of reactive substances (mainly ozone) that are injurious to human health and ecosystems and which may also damage crops. The impact potentials are expressed as an equivalent emission of the reference substance ethylene, C2H4 (Hauschild & Wenzel, 1998).
- Acidification (AC): This category is related to the acidifying substances that cause a wide range of impacts on soil, groundwater, surface water, organisms, ecosystems, and materials. The major acidifying substances are SO2, NOX, HCl and NH3. For emissions to air, the acidification potential is defined as the number of H+ ions produced per kg substance relative to SO2 (Bauman & Tillman, 2004).
- Eutrophication (EU): This category is related to all impacts due to excessive levels of macro-nutrients in the environment caused by emissions of nutrients to air, water, and soil. Nitrogen (N) and phosphorus (P) are the two nutrients most implicated in eutrophication (Bauman & Tillman, 2004).

In addition, an energy input assessment was carried out, using cumulative energy demand to calculate the total direct and indirect amount of energy consumed throughout the life cycle (Boustead & Hancock, 2003; Pimentel, 2003). The software used in the environmental analysis was SimaPro 7.0 by Pré Consultants (SimaPro, 2006).

A LCA model of a wind turbine with Double Feed Inductor Generator (DFIG) has been developed with the object of identifying the main types of environmental impact throughout the life cycle, in order to define possible ways of achieving environmental improvements for the particular type of wind turbine analysed, or for similar ones. The wind turbine is a Gamesa onshore wind turbine, G8X model, with 2 MW rated power, and general dimensions: 80m rotor blade, 5,027m^2 sweep area, and 70m height.

The wind turbine is installed in the Munilla wind farm, in northern Spain, where it has been analysed during the different stages of its life cycle, from cradle to grave, taking into consideration the production of each of its component parts, the transport to the wind farm, the installation, the start-up, the maintenance and final decommissioning, with its subsequent disposal of waste residues. An LCA model of a wind turbine can be appreciated in (Martínez et al., 2009).

2.2 System boundary

Within the limits of the system studied fall the construction of the main components of the turbine, the transportation of the turbine to the wind farm, the assembly, the installation, and the start-up, as well as the process of dismantling the wind turbine and the subsequent treatment of generated waste. A graphical representation of the limits of the system can be seen in (Martínez et al. 2009c).

Outside the limits of the system under study fall the system of distribution of the electricity generated by the wind turbine; that is, the medium-voltage wiring, the transformer substation, and the national electrical power network.

2.3 Functional unit

The aim of the work is to know the environmental impact of wind power, and to quantify it, but it is necessary to relate this impact to the electricity generated, in order to be able to make a posterior comparative study with regard to other types of energy producing technology. Thus, the functional unit has been defined as the production of 1 kWh of electricity.

2.4 Data collection

A wind turbine consists of many components, which also comprise many sub-components, of different nature and eventually with mechanical, electrical, and electronic parts; so, it is difficult to gather from suppliers the information on all the parts that compose the turbine. We have focused on compiling the life cycle inventory (LCI) data on the most important components, specifically the foundation, the tower, the nacelle, and the rotor. In the few cases in which the data found have not been sufficiently reliable and proven, quasi-process information from commercial Ecoinvent database of SimaPro software has been used.

For instance, the materials and energy used in the diverse components have been incorporated into the model using data provided by Gamesa. The distances of transport have been calculated from specific maps as far as the real emplacement of the Munilla wind farm. The main materials that constitute the most important components of the turbine, as well as the selected reference database Ecoinvent, can be seen in (Martínez et al. 2009), specifically the Inventory per component and the Ecoinvent process selected per material.

The owner company of the wind farm performs the maintenance operations, and the information about them is recorded in its environmental management system according to the ISO 14001 standard (ISO, 2004). Based on this important information, all the maintenance operations have been taken into account during the operational phase, such as quantities of oil and grease used or replacement of filters, and transport, among others. Transport processes include the impact of emissions caused by the extraction and production of fuel and the generation of energy from that fuel during transport (Spielmann & Scholz, 2005).

2.5 Key assumptions

As previously mentioned, the LCA model developed includes both the turbine and the foundations that support it, but not the system for connection to the grid (medium voltage lines and transformer substation).

A series of cut-off criteria have been established in order to develop the study, by defining the maximum level of detail in the gathering of data for the different components of the wind turbine. The main cut-off criterion chosen is the weight of each element in relation to the total weight. This limitation in data collection does not mean a significant weakening of the final results obtained, but allows us to streamline, facilitate, and adjust the LCA study to make it more flexible.

The characterisation of each component has been obtained from the most important basic data of the manufacture, which are: the raw material required, the direct consumption of energy involved in the manufacturing processes, and the information of transport used. The information published by Riso National Laboratory has been used when it has not been possible to obtain the energy cost of the manufacturing process directly. This information for specific substances includes the primary energy consumption use related to the production, transportation, and manufacture of 1 kg of material (Etxeberria et al., 2007).

Thus, this LCA has been performed under the following conditions, due to limitations of time and cost:

- The cut-off criterion used has been the weight of the components. The elements that have been taken into account, altogether, make up 95% of the foundations, 95% of the tower, and 85% of the nacelle and rotors.
- All data on electricity has been obtained from the SimaPro database (Frischknecht & Rebitzer, 2005; Frischknecht et al., 2005).
- The wind turbine lifetime is 20 years.
- The assumed current recycling rate of waste wind turbine has been estimated based on the wind farm decommissioning projects prepared by the company (GER, 2004). (Martínez et al., 2009) presents a table with the type of dismantling of the different materials.
- The production is 4 GWh per wind turbine and year.
- One replacement generator has been estimated during the complete lifetime of the wind turbine.

According to the requirements of the standard ISO14044 (ISO, 2006), allocation has been avoided, since in this study only the production of electrical power is considered as the function of the system and, therefore, allocation has not been considered in any component or process.

2.6 Analysed scenarios

The LCA above mentioned contains several uncertain parameters, and therefore, a variability analysis has been developed in order to find the impact of variations in the most significant of these parameters. A series of variables on which to focus the research have been selected, in order to develop the variability analysis of the results of the LCA. These selected variables are presented in the following scenarios, explained in detail in (Martínez et al., 2009c):

- AS1: It is focused on increasing corrective maintenance throughout the life of the turbine. This aspect of the increased requirements for maintenance is of vital importance in the world of wind power, and it is the reason why predictive maintenance systems reducing these major corrective to the minimum are being investigated nowadays. With the aim of considering the various possible alternatives, three alternative scenarios have been considered.

  AS2: It has been established, considering an increase in energy and in materials, in order to compensate for the effect of possible elements that have not been included in

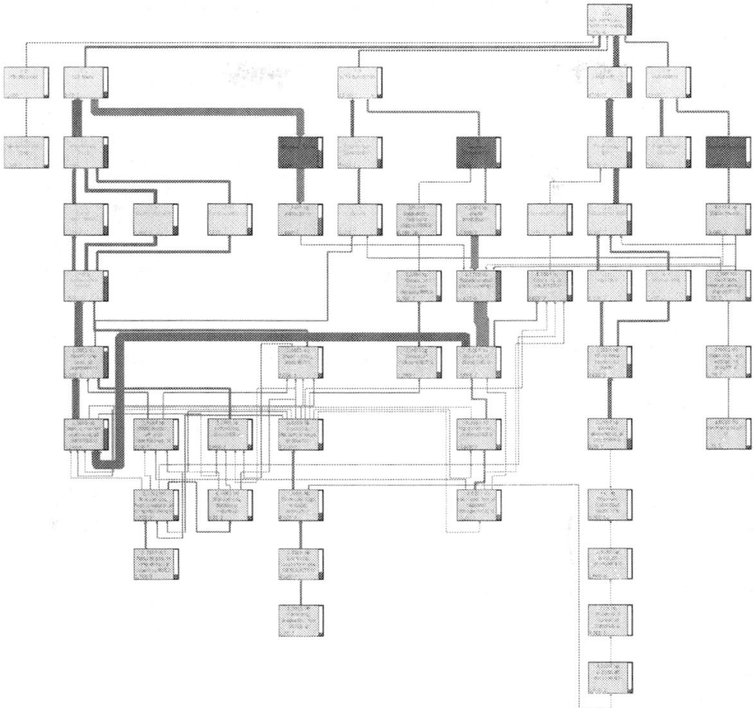

Fig. 1. ACV structure of the basic scenario

the LCA because of the use of the cut-off criteria. Moreover, each increase has been analysed separately in order to better assess the impact of each deviation relative to the basic scenario, according to the following scenarios:

- AS21: In this scenario only the increases corresponding to the energy and transportation required in the LCA are applied.
- AS22: In this scenario an increase in the consumption of different materials used throughout the lifetime of the wind turbine is contemplated.

- AS3: It has been established in order to assess the impact of reducing the criteria when the recycling process of dismantling and disposal is carried out in practice.
- AS4: A recycling of part of the composite material of the blades has been considered in this alternative scenario. This tries to assess the trend and the future changes of composite materials recycling, since the current industrial regulations begins to consider unfavourably sending composites to landfill.

Figures 1 to 7 represent the ACV of the seven previous scenarios.

The Disposal considerations of the basic scenario are published in a table in (Martínez et al. 2009c); in that work another Scenario is also analysed, AS5, associated to 3000 equivalent hours, i.e. 6 GWh annual production.

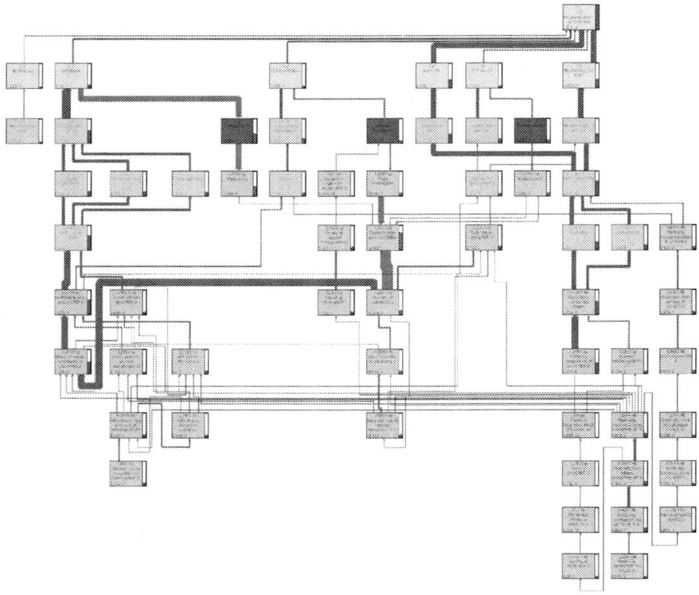

Fig. 2. ACV structure of the alternative scenario AS11

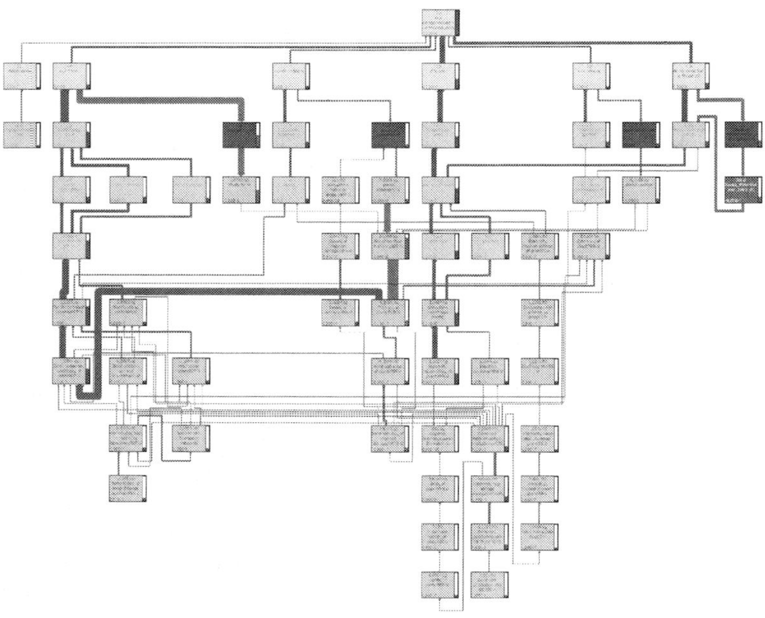

Fig. 3. ACV structure of the alternative scenario AS12

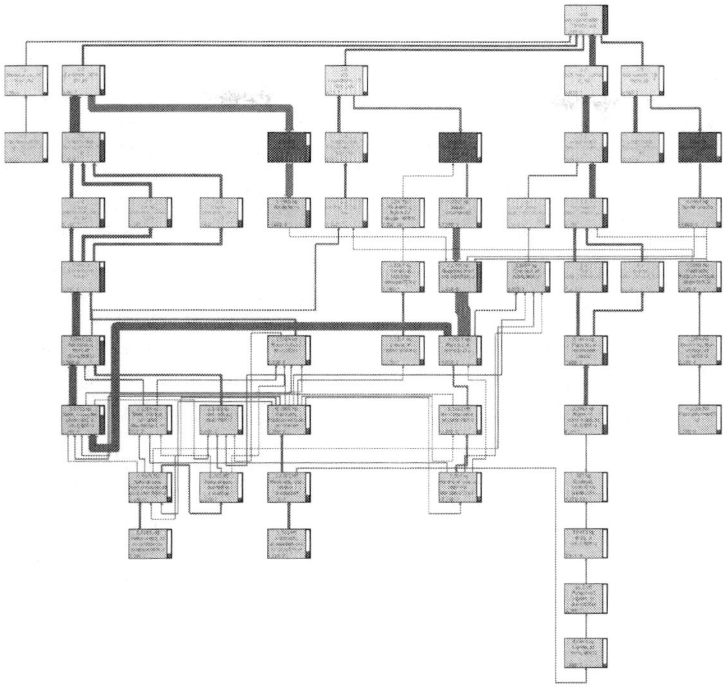

Fig. 4. ACV structure of the alternative scenario AS21

2.7 Comparative of LCIA methods

Seven methods have been selected in order to develop the life cycle impact assessment for the comparative analysis.

- CML 2 baseline 2000 V2.03 / World, 1990
- Eco-indicator 99 (E) V2.03 / Europe EI 99 E/E
- Ecopoints 97 (CH) V2.03 / Ecopoints
- EDIP/UMIP 97 V2.03 / EDIP World/Dk
- EPS 2000 V2.02 / EPS
- IMPACT 2002+ V2.02 / IMPACT 2002+
- TRACI V2.00

In the comparative results obtained with each method, the references have been, on the one hand the impact categories related to energy, and, secondly, those related to toxicity. The first ones are summarized in acidification, nutrient enrichment (eutrophication), global warning (climate change), abiotic depletion and ozone layer depletion. And those relative to toxicity are concentred in ecotoxicity and human toxicity. As expected not all impact methods present these categories or other directly comparable. In these cases the most suitable approximation has been searched, within the different impact categories available in each method, or when it has not been possible to find one or more categories of comparable impact, they have been eliminated from the comparative LCIA

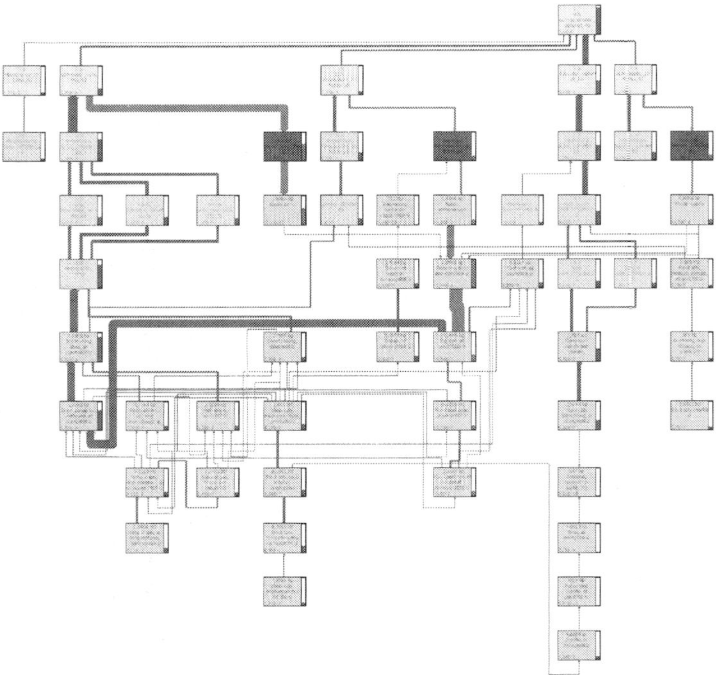

Fig. 5. ACV structure of the alternative scenario AS22

3. Results

3.1 Environmental impact

The results obtained per impact category are shown in (Martínez et al. 2009), especially in the tables Characterization results, Percentage reduction of environmental impacts of wind turbine versus the electricity mix of Spain, and Environmental impact prevented by recycling, as well as in their following analysis of results of that reference (Martínez et al. 2009).

3.2 Cumulative Energy Demand

The Cumulative Energy Demand (CED) is calculated for five classes of primary energy carriers: fossil, nuclear, hydro, biomass and others (wind, solar, geothermal). Differences for different types of cumulative energy demands are mainly due to the consideration of location-specific electricity mixes. The preponderance of non-renewable energies in Spain, especially energy from fossil fuels, is clearly demonstrated (see Table of Cumulative Energy Demand results, in Martínez et al. 2009).

3.2.1 Energy payback time

Another important aspect is to evaluate the Energy Payback and Energy Yield Ratio. The definition of both terms is as follows:

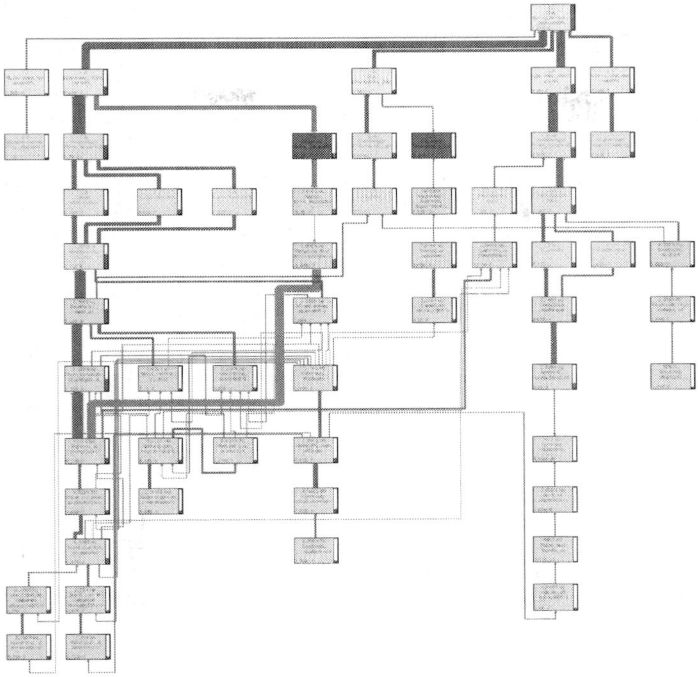

Fig. 6. ACV structure of the alternative scenario AS3

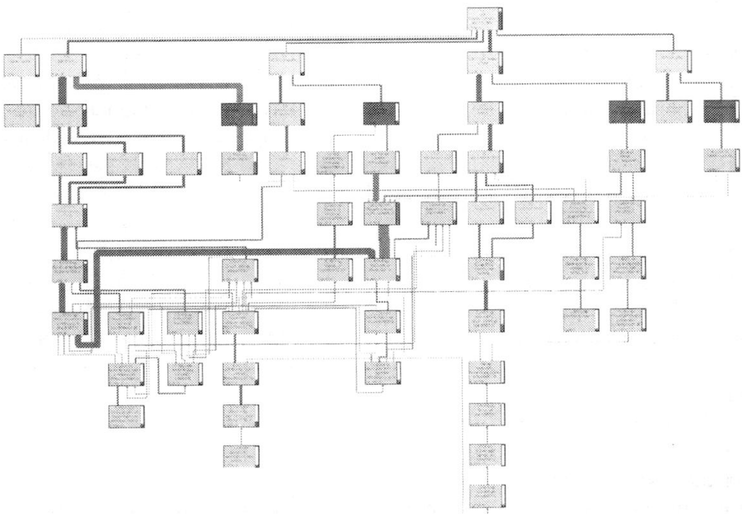

Fig. 7. ACV structure of the alternative scenario AS4

- Energy payback time: this term indicates the years that the system under study must be operating to return the amount of energy that has been needed for their manufacture, start-up, and operation throughout its lifespan.
- Energy yield ratio: This term represents the relationship between the energy generated by the system throughout its lifetime and the energy consumed by the system (CED).

(Martínez et al. 2009) presents a table with the CED value of the wind turbine. From this basis and with the average annual production of wind turbine (Troen & Petersem, 1991), the energy payback time and the energy yield ratio are obtained (Martínez et al. 2009). In addition, the time needed to compensate for the environmental impact generated by manufacturing, launching and operating of wind turbines, by the reduction of requirements for conventional electric energy generation has been calculated (Martínez et al. 2009). This study has considered again the electricity mix in Spain from the database Ecoinvent.

3.3 Results of the variability analysis

The results of the variability analysis are described, depending on the different scenarios and their characteristics (Additional maintenance, Additional inputs of materials and energy, Reduction by half of the recycling, Inclusion of blade recycling, and Increased power generation), in (Martínez et al., 2009c), where some interesting tables present the results, which are then analysed and explained. The most interesting tables are: LCA results of the alternative scenarios AS11, AS12 and AS13 and the basic scenario, Percentage of variation of the alternative scenarios AS11, AS12 and AS13 relative to the basic scenario, Percentage of variation of the alternative scenarios AS21 and AS22 relative to the basic scenario, Percentage of variation of the alternative scenario AS3 relative to the basic scenario, LCA results of the alternative scenario AS4 and percentage of variation relative to the basic scenario.

3.4 Results of the methods comparative

3.4.1 Environmental impact

Of the different impact categories studied, we can highlight the results, among others, of the impact categories of acidification and Human Toxicity. The results obtained after comparing the level of acidification in different LCIA under study are presented in (Martínez et al., 2009c). The results are very similar although a small difference can be appreciated between two groups, with results very closed between the methods of every group, the first one made up of CML 2000, EDIP 96, EPS, TRACI and Ecopoints methods, and the second one which comprises Eco Indicator 99 and Impact2002 methods. This can be explained by several reasons:

- Eco Indicator 99 combines in one single LCIA impact, acidification and eutrophication.
- In the case of Impact2002 the characterization factors for the cathegory TAN are taken directly from Eco Indicator 99 (Goedkoop & Spriensma, 2000).
- Eco Indicator 99 and Impact2002 seems to give a higher importance to nitrogen oxides in acidification phenomena compared with the rest of the methods analysed in the work.

3.4.2 Recycling

Following, the effect of using different LCIA in the results of the positive effect of recycling in the performed LCA is examined. As expected the same relationships found in the specific analysis of each category are held when analysing only recycling, although it could be noted, for example, the results obtained in the case of some impact category, with almost complete unanimity on the importance of recycling of the metallic material of the turbine tower.

3.4.3 Sensitivity analysis

Likewise, when studing the results for each of these alternative scenarios in the sensitivity analysis performed, it can be observed that the choice of the LCIA method leads to emphasize, to a certain or lesser extent, various alternative scenarios. Generally, it can be deduced that in all impact categories studied, there are one or more LCIA's alternative scenarios that provide major increase in that category for the base case.

4. Conclusions

Throughout this chapter, a methodology of analysis of the environmental impact generated by a wind turbine has been presented, based on previous works of the research group composed of the authors. From the results obtained, an important conclusion is the significant impact generated by the turbine blades and, especially, their non-recycling status. Here is found a need for further research into recycling processes of this type of material (Pickering, 2006; Cunliffe et al., 2003; Marco de et al., 1997; Torres et al., 2000; Williams et al., 2005; Vallee et al., 2004; Perrin et al., 2006), as well as for their practical application in the final dismantling and waste treatment phases of wind turbines. Another material that presents a significant impact within the study is the copper (Lunt et al., 2002; Norgate & Rankin, 2000) present in the nacelle of the turbine, but in this case with the advantage of being a recyclable material (Norgate et al., 2007).

In any case, although there are components with a significant environmental impact within the turbine, it has also been verified that these impacts are much smaller than those generated by conventional power plants in operation, with reductions in the impact about 95%, depending on the category. In addition, the energy payback time (time regarding the energy required to produce and implement a turbine) is less than one year, much smaller than the useful lifetime of the system, which is at least 20 years.

Moreover, the different uncertainties arising from the options given during the development of the LCA of a wind turbine have been analysed throughout this work. Five different scenarios within the LCA of a multimegawatt wind turbine have been analysed. In addition, the impact that these scenarios may present on the final LCA has also been assessed. From the results can be clearly emphasized the specific case of large corrections in the maintenance phase. Undoubtedly, the choices made at the turbine maintenance stage have an important effect on the results of the LCA. Therefore it is necessary to analyse and define more precisely the average of major corrections that may experience a model of wind turbine along its 20 years of life. Another issue that significantly influences the final results of the LCA study of the multi-megawatt wind generator in question is the considerations made about recycling and reuse of components and materials. A clear example is the impact of materials such as the fibreglass of the blades of the wind turbine when they are not recycled but sent directly to landfill.

Finally, seven LCIA methods have also been compared in relation to different impact categories, and significant discrepancies have been found in the results. Furthermore, some of these methods may not consider impact categories that are currently debated, which may be relevant to the LCA study being conducted.

On many occasions, a LCA practitioner may simply select a LCIA methodology provided as part of a LCA software tool. In these cases, the impact category, indicator and model selection, and classification have been preselected for the user. This is appealing from the practitioner's point of view since it is faster and less costly. However, it must be noted and cautioned that depending on the methodology chosen and the impact categories of interest, the user may obtain qualitatively different results.

5. References

Ardente, F.; Beccali, M.; Cellura, M. & Lo Brano, V. (2008). Energy performances and life cycle assessment of an Italian wind farm. *Renewable and Sustainable Energy Reviews*, 12, 1, (200-217).

Banister, D.; Stead, D.; Steen, P.; Åkerman, J.; Dreborg, K.; Nijkamp, P. & Schleicher-Tappeser, R. (2000). *European Transport Policy and Sustainable Mobility*, Spon Press, Suffolk (UK) 255 pp.

Bauman & Tillman (2004). *The Hitch Hiker's Guide to Live Cycle Assessment*, Studentlitteratur AB, Sweden

Ben Amar, F.; Elamouri, M. & Dhifaoui, R. (2008). Energy assessment of the first wind farm section of Sidi Daoud, Tunisia. *Renewable Energy*, 33, 10, (2311-2321).

Boustead, I. & Hancock, G.F. (2003). Handbook of industrial energy analysis. In: *Implementation of life cycle assessment methods*. Frischknecht & Jungbluth (eds) ecoinvent report n. 3. Swiss Centre for Life Cycle Inventories, Dübendorf, pp. 22–28

Bovea, M.D. & Gallardo, A. (2006). The influence of impact assessment methods on materials selection for eco-design. *Materials and Design*, 27, 3, (209-215).

Brent, A.C. & Hietkamp, S. (2003). Comparative evaluation of life cycle impact assessment methods with a South African case study. *International Journal of Life Cycle Assessment*, 8, 1, (27-38).

Breukers, S. & Wolsink, M. (2007). Wind power implementation in changing institutional landscapes: An international comparison. *Energy Policy*, 35, 5, (2737-2750).

Carolin Mabel, M. & Fernandez, E. (2008). Growth and future trends of wind energy in India. *Renewable and Sustainable Energy Reviews*, 12, 6, (1745-1757).

Celik, A.N.; Muneer, T. & Clarke, P. (2007). An investigation into micro wind energy systems for their utilization in urban areas and their life cycle assessment. *Proceedings of the Institution of Mechanical Engineers, Part A: Journal of Power and Energy*, 221, 8, (1107-1117).

Cunliffe, A.M.; Jones, N. and Williams, P.T. (2003). Pyrolysis of Composite Plastic Waste. *Environmental Technology*, 24, (653-663).

Douglas, C.A.; Harrison, G.P. & Chick, J.P. (2008). Life cycle assessment of the Seagen marine current turbine. *Proceedings of the Institution of Mechanical Engineers Part M: Journal of Engineering for the Maritime Environment* 222, 1, (1-12).

Dreyer, L.C.; Niemann, A.L. & Hauschild, M.Z. (2003). Comparison of three different LCIA methods: EDIP97, CML2001 and eco-indicator 99: Does it matter which one you choose. *International Journal of Life Cycle Assessment*; 8, 4, (191-200).

Etxeberria, M.; Vázquez, E.; Marí, A. & Barra, M. (2007). Influence of amount of recycled coarse aggregates and production process on properties of recycled aggregate concrete. *Cement and Concrete Research* 37, 5, (735-742).

Frischknecht, R.; Jungbluth, N.; Althaus, H.-J.; Doka, G.; Dones, R. & Heck, T. (2005). The ecoinvent database: Overview and methodological framework. *International Journal of Life Cycle Assessment*, 10, 1, (3-9).

Frischknecht, R. & Rebitzer, G. (2005). The ecoinvent database system: A comprehensive web-based LCA database. *Journal of Cleaner Production*, 13, 13-14, (1337-1343).

GER (2004) Proyecto de desmantelamiento del parque eólico de Munilla - Lasanta. Proyecto interno

Goedkoop, M.; Goedkoop, M. & Effting, S. (2004). SimaPro Database Manual Methods library, PRé Consultants, Netherlands.

Goralczyk, M. (2003). Life-cycle assessment in the renewable energy sector. *Applied Energy* 75, 3-4, (205-211).

Guinée, J.B.; Gorree, M.; Heijungs, R.; Huppes, G.; Kleijn, R.; van Oers, L.; Wegener Sleeswijk, A.; Suh, S.; Udo de Haes, H.A.; de Bruijn, J.A.; van Duin, R. & Huijbregts, M.A.J. (2001). *Life cycle assessment: An operational guide to the ISO standards*. Kluwer Publishers Amsterdam, Netherlands

Gurzenich, D.; Mathur, J.; Bansal, N.K. & Wagner, H.-J. (1999). Cumulative energy demand for selected renewable energy technologies. *International Journal of Life Cycle Assessment*, 4, 3, (143-149).

GWEC Global Wind Energy Council (2005) Wind Force 12. A blueprint to achieve 12% of the world's electricity from wind power by 2020.

Hauschild, M. & Wenzel, H. (1998). *Environmental Assessment of Products*, Scientific background vol. 2, Chapman & Hall, UK.

Hung, M.-L. & Ma, H.-W. (2009) Quantifying system uncertainty of life cycle assessment based on Monte Carlo simulation. *International Journal of Life Cycle Assessment*, 14, 1, (19-27).

ISO (1998) ISO 14040. Environmental management -life cycle assessment- principles and framework. International Standard Organization, Geneva, Switzerland.

ISO (2004) ISO 14001 Environmental Management Systems - Requirements with guidance for use. International Standard Organization, Geneva, Switzerland.

ISO (2006) ISO 14044: Environmental management - Life cycle assessment - Requirements and guidelines. International Standard Organization, Geneva, Switzerland.

Jager-Waldau (2007). Photovoltaics and renewable energies in Europe. *Renewable and Sustainable Energy Reviews*, 11, 7, (1414-1437).

Jungbluth, N.; Bauer, C.; Dones, R. & Frischknecht, R. (2005). Life cycle assessment for emerging technologies: Case studies for photovoltaic and wind power. *International Journal of Life Cycle Assessment*, 10, 1, (24-34).

Karki, R. (2007). Renewable energy credit driven wind power growth for system reliability. *Electric Power Systems Research*, 77, 7, (797-803).

Khan, F.I.; Hawboldt, K. & Iqbal, M.T. (2005). Life cycle analysis of wind–fuel cell integrated system. *Renewable Energy*, 30, 2, (157–77).

Lunt, D.; Zhuang, Y. & La Brooy, S. (2002). Life Cycle Assessment of Process Options for Copper Production. Green Processing 2002 - *Proceedings: International Conference on the Sustainable Processing of Minerals*, pp. 185-193.

Marco de, I. et al. (1997). Recycling of the Products Obtained in the Pyrolysis of Fibre-Glass Polyester SMC. *Journal of chemical technology and biotechnology*, 69, (187-192).

Martínez, E.; Jiménez, E.; Blanco, J. & Sanz, F. (2009c). 2-MW Rated Power Wind Turbine: Variability Analysis. *International Journal of Life Cycle Assessment*. In press.

Martínez, E.; Sanz, F.; Pellegrini, S.; Jiménez, E. & Blanco, J. (2009). Life cycle assessment of a 2-MW rated power wind turbine. *International Journal of Life Cycle Assessment* 14, 1, (52–63).

Martínez, E.; Sanz, F.; Pellegrini, S.; Jiménez, E. & Blanco, J. (2009b). Life cycle assessment of a multi-megawatt wind turbine. *Renewable Energy*, 34 (667-673).

Norgate, T.E.; Jahanshahi, S. & Rankin, W.J. (2007). Assessing the environmental impact of metal production processes. *Journal of Cleaner Production*, 15, 8-9, (838-848).

Norgate, T.E. & Rankin, W.J. (2000). Life cycle assessment of copper and nickel production. *Proceedings of MINPREX 2000*, pp. 133-138. The Australian Institute of Mining and Metallurgy, Melbourne

Pant, R.; Van Hoof, G.; Schowanek, D.; Feijtel, T.C.J.; De Koning, A.; Hauschild, M.; Pennington, D.W.; Olsen & Rosenbaum, R. (2004). Comparison between three different LCIA methods for aquatic ecotoxicity and a product environmental risk assessment: Insights from a detergent case study within OMNIITOX. *International Journal of Life Cycle Assessment*, 9, 5, (295-306).

Pehnt, M. (2006). Dynamic life cycle assessment (LCA) of renewable energy technologies. *Renewable Energy*, 31, 1, (55-71).

Perrin, D.M; Guillermain, C.; Bergeret, A.; Lopez-Cuesta, J.M. & Tersac, G. (2006) SMC Composites Waste Management as Reinforcing Fillers in Polypropylene by Combination of Mechanical and Chemical Recycling Processes. *Journal of Materials Science*, 41, 12, (3593-3602).

Pickering, S.J. (2006) Recycling Technologies for Thermoset Composite Materials - Current Status. *Composites: Part A*, 37, (1206-1215).

Pimentel, D. (2003). Food production and energy crisis. In: *Implementation of life cycle assessment methods*. Frischknecht & Jungbluth (eds). Ecoinvent report no. 3. Swiss Centre for Life Cycle Inventories, Dübendorf, pp. 22–28

Renou, S.; Thomas, J.S.; Agustin, E. & Pons, M.N. (2008). Influence of impact assessment methods in wastewater treatment LCA. *Journal of Cleaner Production*, 16, 10, (1098-1105).

Rieradevall, J.; Domenech, X. & Fullana, P. (1997). Application of life cycle assessment to landfilling. *International Journal of Life Cycle Assessment*, 2, 3, (141-144).

Río del, P. & Unruh, G. (2007). Overcoming the lock-out of renewable energy technologies in Spain: The cases of wind and solar electricity. *Renewable and Sustainable Energy Reviews*, 11, 7, (1498-1513).

Schulze, C.; Jödicke, A.; Scheringer, M.; Margni, M.; Jolliet, O.; Hungerbühler, K. & Matthiest, M. Comparison of different life-cycle impact assessment methods for aquatic ecotoxicity (2001) *Environmental Toxicology and Chemistry*, 20, 9, (2122-2132).

SimaPro 7 (2006) Ecoinvent data base. PRé Consultants. Amersfoort, The Netherlands.

Spielmann, M. & Scholz, R.W. (2005). Life cycle inventories of transport services: Background data for freight transport. *International Journal of Life Cycle Assessment*, 10, 1, (85-94).

Torres, A. et al. (2000) Recycling by Pyrolysis of Thermoset Composites: Characteristics of the Liquid and Gaseous Fuels Obtained. *Fuel*, 79, (897-902).

Troen, I. & Petersem, E.L. (1991). European Wind Atlas, Risø National Laboratory, Roskilde, Denmark, ISBN 87-550-1482-8. pp. 656

Tryfonidou, R. & Wagner, H.-J. (2004). Multi-megawatt wind turbines for offshore use: Aspects of Life Cycle Assessment. *International Journal of Global Energy*, 21, 3, (255-262).

Vallee, M.; Tersac, G.; Destais-Orvoen, N. & Durand, G. (2004). Chemical Recycling of Class A Surface Quality Sheet-Molding Composites. *Industrial & engineering chemistry research*, 43, 20 (6317-6324).

Wichser, C. & Klink, K. (2008). Low wind speed turbines and wind power potential in Minnesota, USA. *Renewable Energy*, 33, 8, (1749-1758).

Williams, P.T.; Cunliffe, A. & Jones, N. (2005). Recovery of Value-added Products from the Pyrolytic Recycling of Glass Fibre-Reinforced Composite Plastic Waste. *Journal of the Energy Institute*, 78, 2, (51-61).

Young, S. & Dubreuil, M. (2006). Qualitative Maps of Metal Recycling, Draft Report prepared for the International Council on Mining & Metals, London, UK.

24

Wind-Solar Driven Natural Electric Hybrid Ventilators

N.A.Ahmed
Associate Professor
Aerospace Engineering
School of Mechanical and Manufacturing Engineering
University of New South Wales, NSW 2052,
Australia

1. Introduction

The new millennium has commenced with great concern arising from the unnatural contingencies of energy utilization that have the potential to destroy the very environment which sustains life. In recent times, such environmental awareness is rapidly resulting in favourable realignment of peoples' attitudes towards lifestyles and products whose production, usage and maintenance are less harmful to nature. It is in this context that we will discuss Wind Driven Natural-Solar/Electric Hybrid ventilator.

Ventilation of domestic, commercial and industrial building is necessary to provide an optimum or at least a satisfactory environment. Such requirements are detailed in publications by organisations such as standards Australia [1] Current ventilation devices are using wind or solar energy are configured to rely exclusively on one of these natural sources of power [2,3]. Proper ventilation requires also that there be a movement or circulation of the air within the space and that the temperature and humidity be maintained within a range that allows adequate evaporation of perspiration from the skin. It was formerly believed that the discomfort, headache, and lethargy were caused entirely by the increase in the amount of carbon dioxide and the decrease in the oxygen content of the air. There is now evidence to show [4] that the deleterious effects may also result largely from interference with the heat-regulating mechanism of the body. Despite the wide distribution of air pollutant sources, the concentration of indoor pollutants may be the dominant risk factor in relation to personal exposure, as most people spend an average of 87% and 6% of their time within buildings and enclosed vehicles, respectively [5]. Indoor exposure may pose more harmful health effects, as the indoor concentrations of many pollutants are often higher than those typically encountered outside [6]. To maintain a safe working environment, many dwellings and factories now need adequate fresh air exchange to remove gaseous, process emissions and/or heat build up.

The high priority placed on indoor air quality from health considerations has prompted New York as the first state in the USA to pass legislation effective from December, 2008 to require landlords to notify tenants, prospective tenants and building occupants of indoor air test results [7]. This is a ground breaking requirement that will no doubt be followed by other places in the world. Although conventional air conditioning (AC) is thought to improve air quality by lowering ambient outdoor allergens, reports of adverse health associations of AC, such as cold and flu, various respiratory problems, sick building syndrome, rising levels of eye, nose and throat irritation, headache, allergic irritations, and so forth are long standing [8]. Proper AC maintenance and other interventions such as the use of ultraviolet irradiation of cooling coils are often suggested to reduce these effects [8]. However, these solutions are usually very expensive.

Under these circumstances, wind driven rotating ventilators, which use natural energy and are cheap to manufacture, install and maintain, can become very cost effective non-polluting means of alleviating the problems by (a) lowering temperatures in homes and buildings during summer by cooling roof spaces and even removing trapped ceiling heat through ceiling grilles and (b) improving air quality through reduction of impurity levels caused by human respiration and chemical emissions - mainly volatile organic compounds, from carpets, furniture, paints, cleaning products and the like.

The author has been pioneering roof top ventilator research, design and performance studies for over a decade at the University of New South Wales in Australia [2-3, 9-11] for over a decade. Work on wind driven ventilators started at the University of New South Wales when Edmonds Products Australia Pty Ltd required testing of their ventilators designed to operate on rooftops of dwellings to meet a single regulatory requirement, that is can the ventilators withstand hurricane wind forces, that is wind forces of a minimum velocity of 215 km/ hr without being ripped off from so that damage to the roof and its surroundings could be avoided. The experiments were essentially in the form of 'destructive testings'. These studies lead to substantially strengthening of the spindle rods and rotors used so that they could withstand the loads generated by high wind.

Subsequent studies lead to the aerodynamic investigation of the performance of these ventilators. The major issue that were investigated involved the efficient operation of these ventilators in low speeds. The studies lead to modifications to the rotors that lead to substantial performance of the Edmonds ventilators to the tune of 15% increase in air extraction efficiency.

Another problem that required attention was the operation of these ventilators in adverse weather conditions of rain. The objective was to ensure that rain did not sip through into roof cavities and buildings causing dampness and the associated health problems. A purpose built mobile rain chamber that could be joined to an existing open circuit wind tunnel was built to investigate the effect of rain on the operation of these ventilators. These studies have lead to specific modifications to the ventilator base holder to redirect the rain particles and obstruct their entry to the inside of dwellings without any penalties to the performance of these ventilators.

However, the dependency on wind depends that the operation of ventilation system based purely on the availability of wind makes its operation twenty four hours a day and whenever required a difficult proposition. A hybrid Wind Solar Driven Natural Electric ventilation system, with solar system complementing primarily as an alternative natural energy source, appears to offer a viable engineering solution that will also help reduce electricity usage compared to conventional air-condition systems thereby contributing significantly towards lower depletion of resources such as oil, coal or gas. This consideration has lead to the novel concept of a hybrid ventilator the 'ECOPOWER' that won the 2008

AIRAH award of Excellence in the HVAC-Achiever category. The award recognizes a distinguished Australian product, invention or innovation in heating, ventilation and air conditioning.

Most of the above works were experimentally based and conducted primarily on real and industrial models with substantial funding from the Australian Research Council in collaboration with industry funding from Edmonds Products Australia Pty Ltd (now part of CSR as CSR Edmonds Australia Pty Ltd) at the Aerodynamics Laboratory of the University of New South Wales. The present paper is basically organised around these activities of the author and his team starting from purely wind driven ventilator to the hybrid ventilation concept to contribute towards the cause of providing higher air quality and comfort with minimum impact to environment.

2. A brief look at various ventilation devices

Ventilation devices powered by energy sources that are renewable non-renewable or a combination of both, may be discussed under the following four headings:

i. ***Ventilation through natural convection:*** This form of ventilation is the simplest means of ventilation and relies upon natural convection. The roof a building contains hot air due to heating caused by the sun. The areas of occupation generally contain air that is much cooler. To allow natural convection to occur, provision is made for a vent at the top of the roof cavity. This allows the hot air to rise out of the roof cavity and escape. Opening windows and doors in the areas of occupation allows cooler air to be drawn into the building by the hot, rising air in the roof cavity.
   This cool air will itself be drawn into the roof cavity and heated, causing it to rise and escape through the roof vent, and perpetuate the cycle. The simplicity and elegance of this method of ventilation is thus readily apparent (Fig. 1).

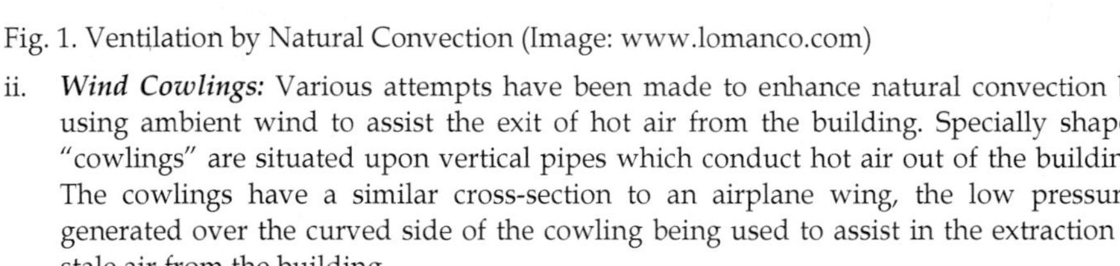

Fig. 1. Ventilation by Natural Convection (Image: www.lomanco.com)

ii. ***Wind Cowlings:*** Various attempts have been made to enhance natural convection by using ambient wind to assist the exit of hot air from the building. Specially shaped "cowlings" are situated upon vertical pipes which conduct hot air out of the building. The cowlings have a similar cross-section to an airplane wing, the low pressures generated over the curved side of the cowling being used to assist in the extraction of stale air from the building.
   The use of cowlings to enhance air extraction is applicable for domestic and commercial uses. When used for commercial purposes, the size of the cowlings can become appreciable (Fig 2). To ensure that the cowlings are orientated correctly with respect to the prevailing wind, a facility for swivelling is provided.

Fig. 2. Air extraction cowlings, University of Nottingham
(Image: Designing with Solar Power, D. Prasad & M. Snow, 2005)

iii. ***Turbine Ventilator:*** The third type of ventilation method uses wind powered centrifugal pumps. These are the "whirlybirds" (Fig 3) with which many people are familiar. Most of these simple devices combine the function of turbine and pump into one, single rotating assembly. That is, one set of blades is used to both spin the device and extract the stale air.

Fig. 3. Cylindrical centrifugal ventilator at the University of New South Wales

Some designs are slightly different in that they have separate blade groups for spinning the device and extracting the air. These blade groups are stacked vertically, but are still part of a single rotating assembly.

Fig. 4. Spherical centrifugal ventilators at the University of New South Wales

These ventilators operate as drag devices, and hence, the angular velocity of the device cannot exceed the ambient wind speed. The difference in the coefficient of drag on the convex and concave sides of the blades causes the device to rotate (Fig 4). This rotation forces exhaust air to be drawn into the centre of the pump, where it is subsequently centrifuged out of the device.

iv. ***Solar Ventilator:*** The fourth type of ventilation method uses solar power exclusively to operate an air extraction fan (Fig.5). This fan is usually of the axial type.

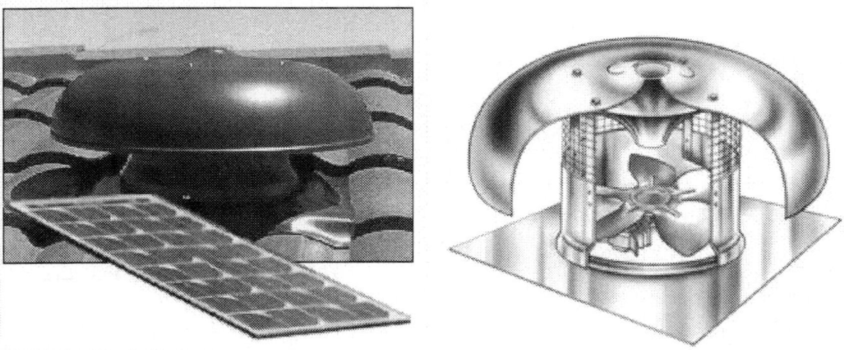

Fig. 5. External and internal views of solar powered ventilator (Image; www.edmonds.com.au)

For many commercial buildings, Australian Standards demand a minimum flow rate of fresh air, and a minimum number of air-changes per hour. Such requirements are usually met with mains-powered air extraction fans.

Such fans are usually axial types, and are very similar in concept and construction to the solar ventilator of figure 4. The main difference between the two lies in the absence of the solar panel and the requirement for hard-wiring to a mains power source.

v. ***Mains powered ventilators:*** They include various forms of ventilation devices that are powered by mains electricity supply. These are essentially dependent on the non-renewable powered systems. Although they are the most reliable systems, they come at a cost to environment and hence the push to seek greener alternatives.

vi. ***Hybrid ventilators:*** If human comfort, convenience and reliability are sought with equal concern for environmental impact, it appears that a compromise solution may be the most effective. Thus some form of hybrid solution may be explored that will provide air extraction capacity at all times and operate in all conditions. This would make it useful for applications that require a continuous flow rate of air.
The hybrid solution will have high initial costs, but these can be offset by designing the device such that the use of mains electricity is minimised. The use of solar power may be promoted by sizing the solar array such that sufficient power is available for good ventilator performance during marginal light conditions. Using, for example, a standard "whirlybird" as the basis of the hybrid ventilator may also allow the wind to power the device.
Attempts can be made to improve the performance of the device with respect to the ability to extract energy from the wind. The current wind driven device uses a single element to act simultaneously as both a turbine and pump. Due to this compromise, neither the tasks of spinning the ventilator nor extracting air is performed in an optimum manner.
The use of wind power can be promoted by physically separating the turbine and pump. The turbine can then be optimised to extract energy from low speed wind more effectively.

Each of the methods of air extraction discussed above has its own advantages and disadvantages. These are summarised in Table 1 to highlight where a hybrid solution will be useful.

3. Towards hybrid ventilation solution

In this section, various attempts made by the author and his team at the University of New South Wales leading towards the development of concepts in favour of hybrid ventilation systems are described.

The most important parameter by which ventilation device is sold is by air extraction or volumetric flow rate. The experimental procedure used was formulated after considering testing procedures outlined in Australian Standards on the classification and performance testing of natural ventilators [12], and the measurement of fluid flow in closed conduits [13] Consideration was also given to general wind tunnel testing procedures at low wind speeds [14] The aim of the project was to discover performance benefits on a comparative basis. The procedures outlined in Australian Standards are designed to produce exact quantitative values for the purposes of classification and calibration.

General scientific testing methods were more appropriate for this situation. This included such procedures as keeping external variables constant whilst a given variable of interest was tested, taking measurement values as a mean over a given time interval, and the use of due care when instruments were set up and calibrated. Adhering to standard scientific

	Advantages	Disadvantages
Natural convection	- Uses renewable energy - Simple - Cheap - Very Reliable.	- Marginal flow rate - Cannot guarantee flow rates required for occupational health and safety - Less effective at night
Wind cowlings	- Uses renewable energy - Improved air Extraction rate compared with natural convection - Relatively simple - Reliable	- Low flow rate - Flow rate not guaranteed. - Less effective at night - Cowling may impede natural convection
Wind powered centrifugal ventilator	- Uses renewable energy - Flow rate good - Simple - Reliable	- Cannot guarantee flow rate - Relies exclusively on wind energy for operation - Flow rate depends on wind strength - Combined pump and turbine design a compromise - Can be expensive
Solar powered axial ventilator	- Uses renewable energy - Flow rate good - Relatively simple - Reliable	- Cannot guarantee flow rate - Relies exclusively on solar Energy for operation - Flow rate depends on light levels - High initial cost
Mains powered ventilator	- Flow rate excellent - Flow rate continuous - Operates at all times and in all conditions	- Relies completely on mains power (non –renewable energy) - High initial cost
Hybrid solution	- Flow rate very good - Flow rate continuous - Powered mainly by renewable energy - Operates at all times and in all conditions	- May sometimes rely on mains power - High initial cost - Complex - May be less reliable mechanically

Table 1. Advantages and disadvantages of current ventilation technologies

testing protocols provided rapid evidence of performance trends, and confidence in the trends being genuine. The purpose of the project was to discover these performance trends, which had a higher priority than obtaining extremely precise measurements.

Experimental set-up

The testing of various ventilation devices was undertaken in the aerodynamics laboratory at the University of New South Wales. The ventilators were screwed to the top of an inlet tube fitting and placed at the exit of a low speed jet wind tunnel [15]

The inlet tube fitting was used to stabilize the airflow to the ventilation devices while they were under test. The inlet tube arrangement had a total centreline length of approximately 2660 mm, which was measured from the inlet plane between the top of the vertical tube and ventilator mounting flange. The 900 elbow had seven turning vanes mounted internally to reduce losses as the air negotiated the bend. The air entering the inlet cone was under the ambient conditions of the laboratory, and was not conditioned in any way.

A precision anemometer equipped with a long sensor probe was used to take velocity measurements across the flow profile in the inlet tube fitting (Fig 6). The anemometer sensor probe was used to take individual air velocity readings as averages over a one minute interval. The anemometer probe was traversed using a laboratory stand equipped with a precision vertical screw adjustment.

Readings were taken at each centimetre across the central 12 cm of the inlet tube (Fig 7), which had an overall internal diameter of 14.625 cm. The velocity at the inside tube walls was assumed to be zero. These 15 air velocity measurements (13 measured plus 2 assumed) were averaged to get the mean velocity of the flow profile in the tube. This velocity was then used with the internal tube diameter to determine the volumetric flow rate.

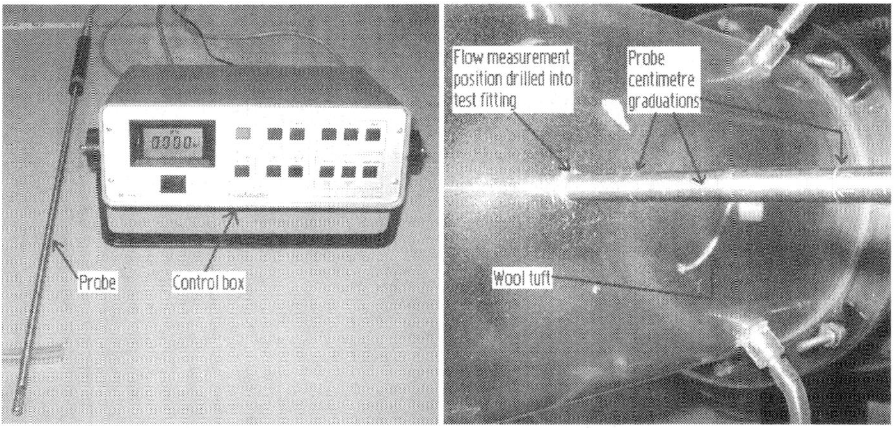

Fig. 6. Precision anemometer / centimetre graduations on probe

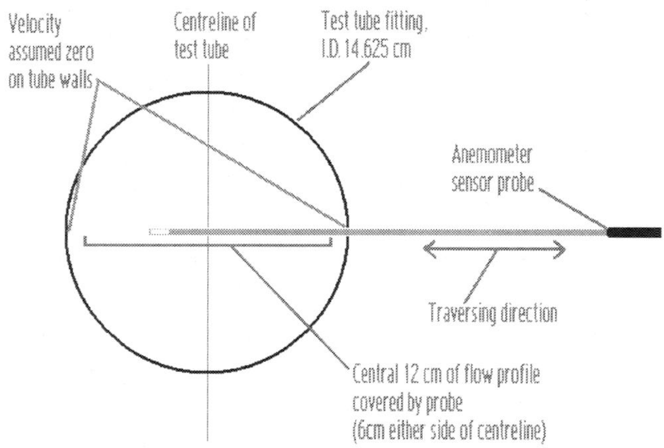

Fig. 7. Velocity measurements across flow profile in test tube fitting

Tests on Standard turbine ventilator

The testing of the standard turbine ventilator (Fig 8) is a commercially available turbine ventilator manufactured by CSR Edmonds Pty Australia Ltd was carried out to serve as a benchmark. The rotating element was 200 mm in diameter whilst the blades had a height of 47.5 mm. The device operated by using a small portion of the blades to extract energy from the incident wind. This energy spun the device which extracted stale air by centrifugal action.

Fig. 8. Standard turbine ventilator by Edmonds
(Image: www.edmonds.com.au)

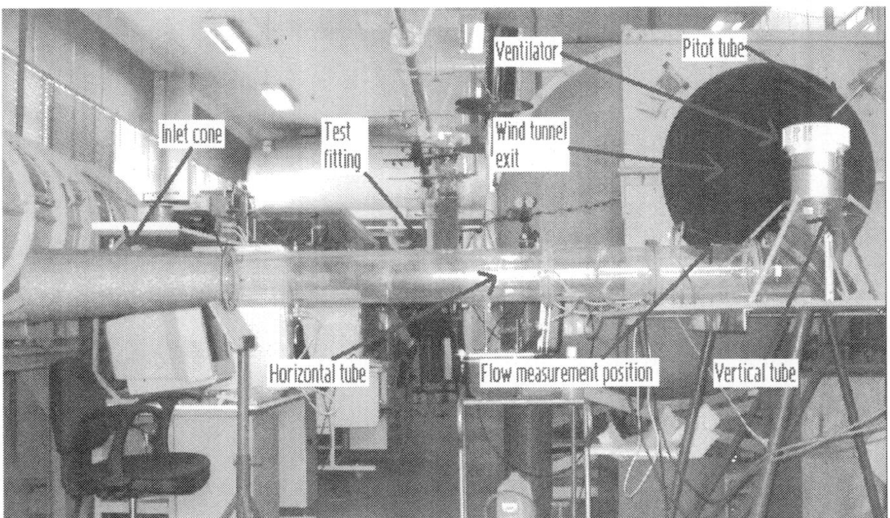

Fig. 9. Ventilator test fitting located at wind tunnel exit

The three graphs (Graphs 1-3) were established as the benchmark for comparison with other modes of ventilation. The only feature worth noting is the linear relationship that exists between wind speed and volumetric flow that is the higher the wind speed the higher is the volume flow rate.

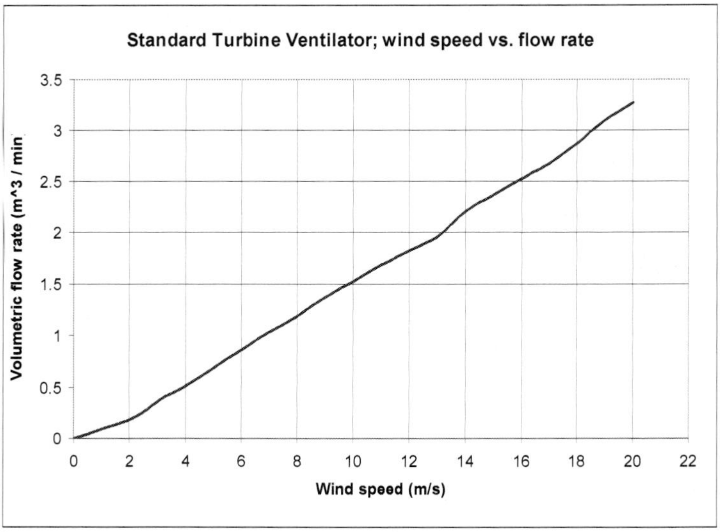

Graph 1. Standard turbine ventilator; wind speed vs. volumetric flow rate

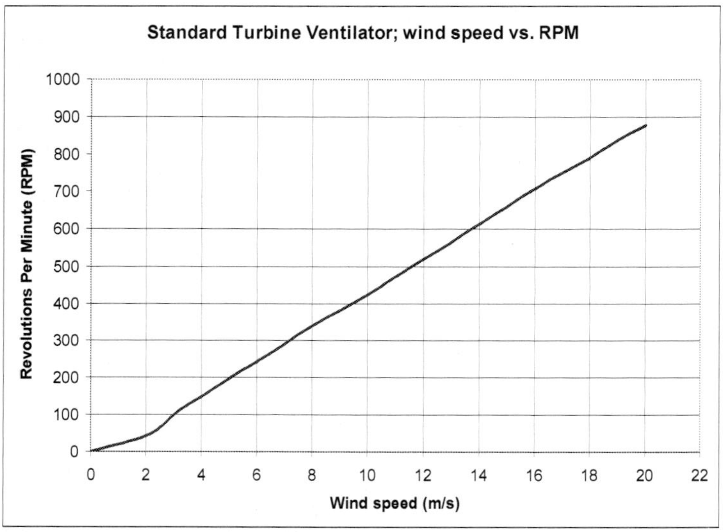

Graph 2. Standard turbine ventilator: Wind speed vs. RPM

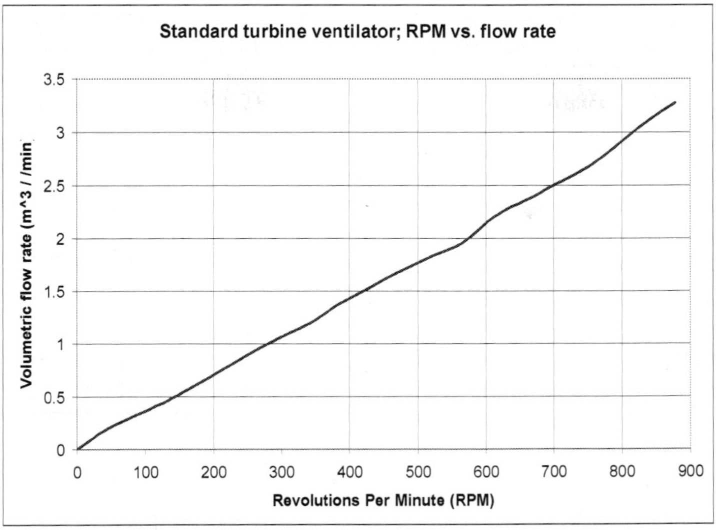

Graph 3. Standard turbine ventilator; RPM vs. volumetric flow rate

Tests on Standard Solar powered ventilator

The solar powered ventilator used in this study was a single unit that contained the solar cell, motor, and fan (Fig 10).

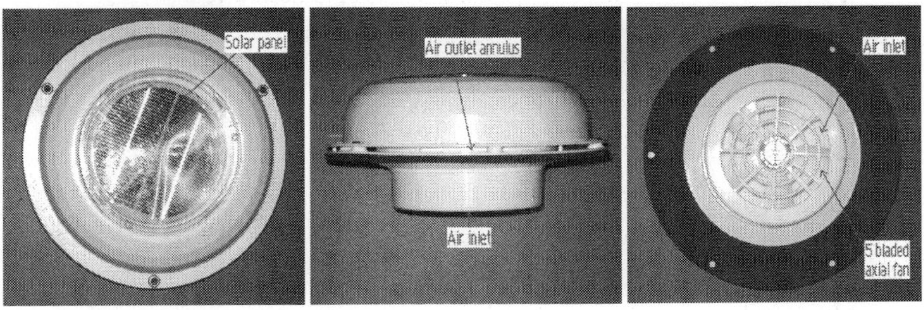

Fig. 10. Solar powered ventilator

The Solar Ventilator was a commercial ventilator intended for use on water vessels, trailers and camper vans. The device uses an axial fan to extract stale air. The stale air is drawn into the inlet situated on the bottom of the device by the 5-bladed axial fan. This air then travels through internal passages where it is subsequently expelled through the annular outlet. The diameter of the propeller and inlet is approximately 98mm. The diameter of the outlet annulus is approximately 246 mm with a height of 5mm. The device was chosen as it was intended for the same applications as the turbine ventilator, and it had the same overall physical size, which facilitated testing on the same apparatus.

The performance of the solar ventilator was severely hampered by the small size of the fan, the tortuous internal flow path and the very small height (and subsequent area) of the exit

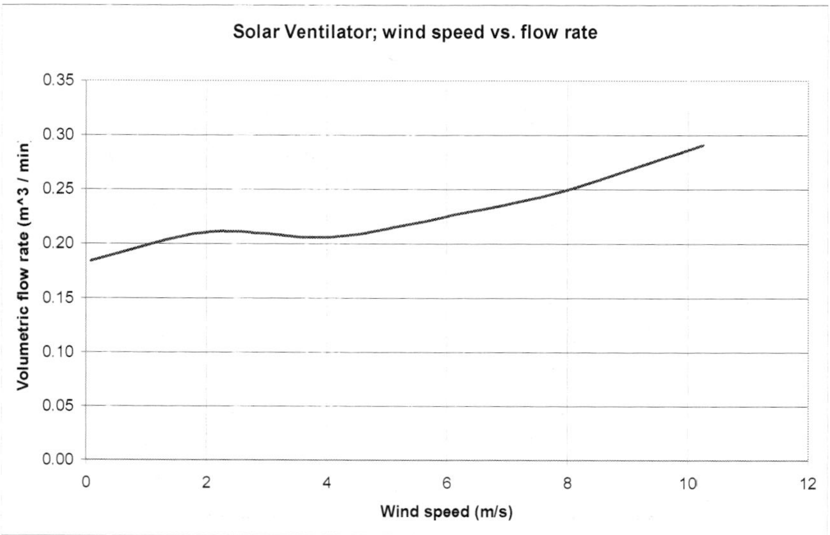

Graph 4. Solar ventilator; wind speed vs. volumetric flow rate

annulus. One of the intended applications of the solar ventilator is the ventilation of boat cabins. As a consequence, the ventilator is designed to keep water out, and the ventilation ability of the device suffers.

Due to the poor flow characteristics of the device, the only useful data was collected when the device was operating at full voltage conditions. This voltage was an average of 1.0195 volts, which was close to the figure collected from the outside sun survey. Under full voltage conditions, the volumetric flow rate was about 0.194 m^3/ min at zero wind tunnel speed.

A cross wind of 10 m/s gave a flow rate approaching 0.3 m^3/ min. The inclusion of cross wind in the air extraction capability of the solar ventilator seemed to be the intent of the manufacturer, as they quoted a flow rate of 680 ft^3/hr (0.3209 m^3/min) under normal conditions. The physical arrangement of the solar ventilator made it impossible to get RPM readings whilst the device was mounted on the test tube fitting.

Graph 4 shows the relationship between wind speed and volumetric flow rate for a variety of cell voltages. As seen in this Graph 4, the advantage of the solar ventilator was lost regardless of the cell voltage at wind speeds above 10 m/s. The advantage of higher cell voltages was most apparent at zero and low wind speeds, which was the most important consideration for the project. Both graphs indicated the performance benefit of the design at zero and low wind speed when a reasonable amount of sunlight was present.

Tests on Hybrid Ventilator: Standard ventilator with solar ventilator on top

A solution to the problem of zero wind speed operation was conceived to be a ventilator that could be powered by the wind and the sun. The hybrid device was constructed from the two ventilators both powered by renewable energy.

Fig. 11. Constituent parts of wind-solar hybrid ventilator

Fig. 12. Test set-up of wind-solar hybrid ventilator

The solar cell and motor from the solar ventilator was combined with the Edmonds turbine ventilator to produce the Solar-Wind Hybrid design (Fig 11). The test set-up is shown in Fig 12.

The feasibility and shortcomings of the initial hybrid design were confirmed by comparing the performance characteristics of the three devices. The solar ventilator was compared with the hybrid device at the same voltage levels whilst the turbine ventilator was compared to the hybrid device at the same wind speeds.

Graph 5 represents the rotational speed of the hybrid ventilator under various wind speeds and cell voltages. The performance chart shows the convergence of the RPM under various cell voltages above 10 m/s. The important characteristic for the project was the RPM advantage enjoyed at zero and low wind speeds (below 4m/s wind speed) when cell voltages were at 0.409 V and above. The cell voltage of 1.03625V was slightly less than the cell voltage achieved under ideal conditions during the sun survey. Despite the low power output of the cell, there was enough energy to spin the turbine ventilator at approximately 140 RPM under ideal sun conditions with no wind. Part power of 0.409V was able to spin the ventilator at around 43 RPM. This would certainly give some ventilation capacity at zero wind speed.

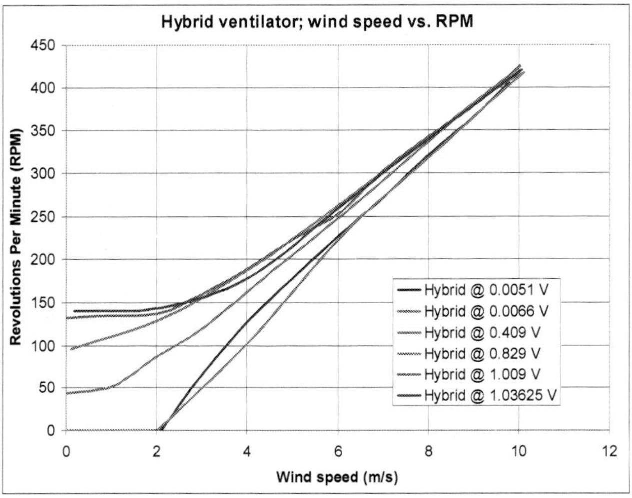

Graph 5. Solar-Wind Hybrid ventilator; wind speed vs. RPM for various cell voltages

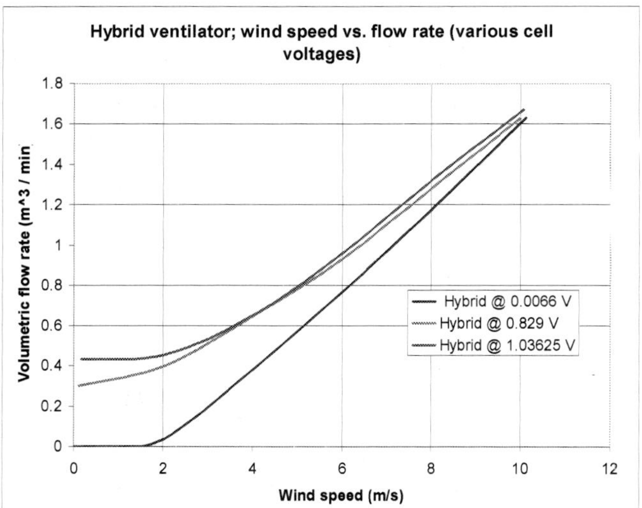

Graph 6. Solar-Wind Hybrid ventilator; wind speed vs. flow rate for various cell voltages

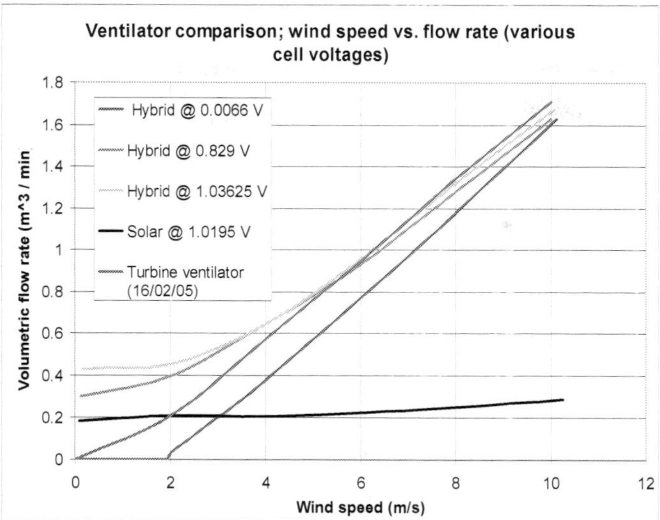

Graph 7. Ventilator comparison; wind speed vs. volume flow rate

Graph 6 shows the relationship between wind speed and volumetric flow rate for a variety of cell voltages. As with Graph 5, the advantage of the Wind-Solar Hybrid ventilator was lost regardless of the cell voltage at wind speeds above 10 m/s. The advantage of higher cell voltages was most apparent at zero and low wind speeds, which was the most important consideration for the project. Both graphs indicated the performance benefit of the design at zero and low wind speed when a reasonable amount of sunlight was present.

Graph 7 reveals the performance of the ventilators under different wind and sun conditions. The first point of interest was the vastly superior performance of the hybrid device compared to the solar ventilator. The performance curve for the solar ventilator was taken under full cell voltage conditions. When compared to the hybrid ventilator under the same power level, the hybrid ventilator had much better volume flow rate.

Even under zero wind conditions, the hybrid ventilator had a higher flow rate than the solar ventilator subjected to 10 m/s wind speed. This advantage was enjoyed even when the hybrid ventilator was subjected to less than full power.

When compared at 10 m/s wind speed, the hybrid ventilator had a flow rate more than 5 times greater than the solar ventilator. The performance curves starkly illustrated the higher efficiency of the hybrid ventilator compared to the standard solar ventilator. Such a performance advantage added to the weight behind the feasibility of the hybrid device.

The Wind-Solar Hybrid device also compared well with the turbine ventilator. Graphs 7 and 8 showed that the performance advantage of the Solar-Wind Hybrid ventilator under full power was not lost to the turbine ventilator until the wind speed was above 6.5 m/s (Graph 7). Even under part power conditions of 0.409V, the hybrid device had an advantage of up to around 5 m/s wind speed (Graph 8). For the zero to low wind speed regime (less than 4m/s), the hybrid device enjoyed an advantage even under less than ideal sun conditions.

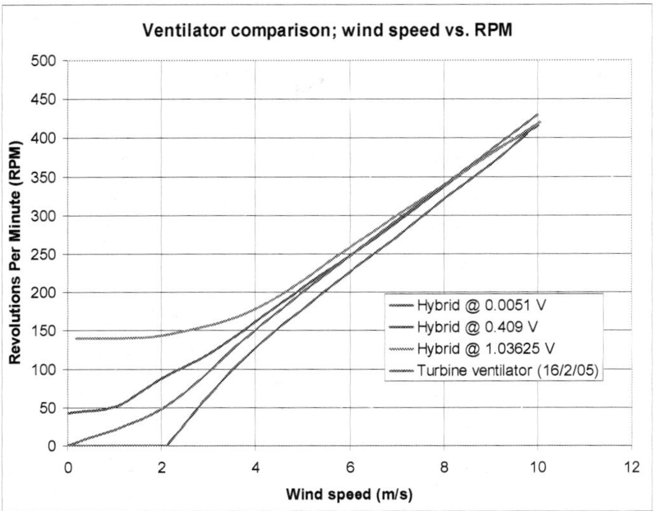

Graph 8. Comparison of Solar-Wind Ventilator with Standard Wind or Turbine Ventilator; wind speed vs. RPM

The most important finding was that the hybrid ventilator enjoyed a performance advantage above both the turbine and solar ventilators at the zero to low wind speed regime (0-4 m/s). This advantage was apparent even under less than ideal sun conditions.

The major shortcoming of the hybrid device was operations under wind power alone (zero cell voltage). The performance of the hybrid device under such conditions lagged behind the turbine ventilator for all wind speeds. The performance of the hybrid device under such conditions also lagged behind the solar ventilator below a wind speed of 3 m/s. This performance deficit under zero cell voltage was attributable to the wind having to back-drive the electric motor, which acted as a generator under such

Tests on Hybrid Ventilator with a horizontal axis wind turbine

The test fitting was modified to accommodate the horizontal axis configuration and the use of an additional test stand containing the propeller and bearing housing was required (Fig 13). The combined test set-up with standard ventilator is shown in Fig 14.

Graph 9 is a performance plot of wind speed vs. RPM, which is a measure of the effectiveness at which energy is extracted from the wind. The numbers refer to the blade pitch angles of the propeller.

The horizontal axis design exhibited superior performance to the turbine ventilator (dark blue line) at blade pitch angles above 37.5°. The blade pitch angle of 75° (purple line) gave the best performance. For any given wind speed, the horizontal axis ventilator with a 75° blade pitch angle managed to extract enough energy to spin at 2.5 times the rotational velocity of the standard turbine ventilator.

Beyond an angle of 75°, the performance of the horizontal axis ventilator dropped off, as the blade chord was becoming perpendicular to the incident wind.

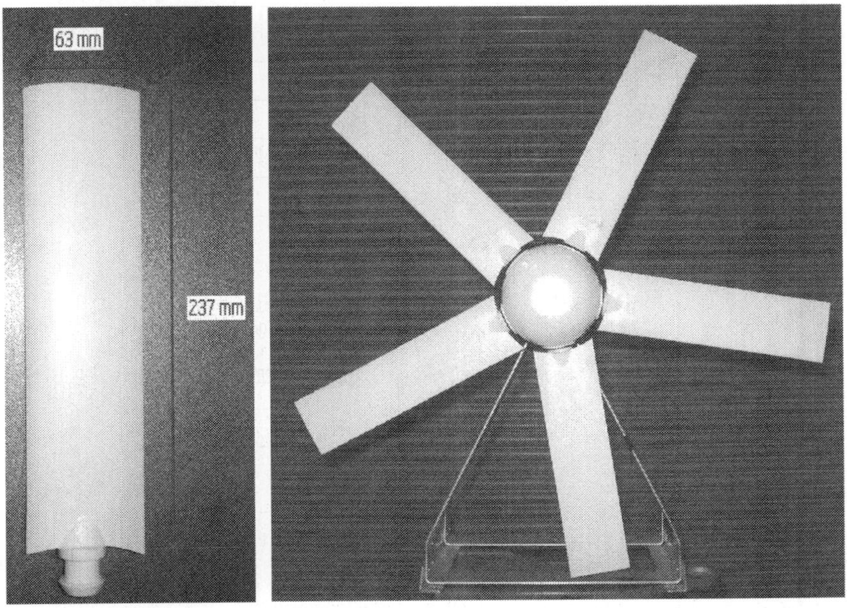

Fig. 13. Individual blade / complete propeller

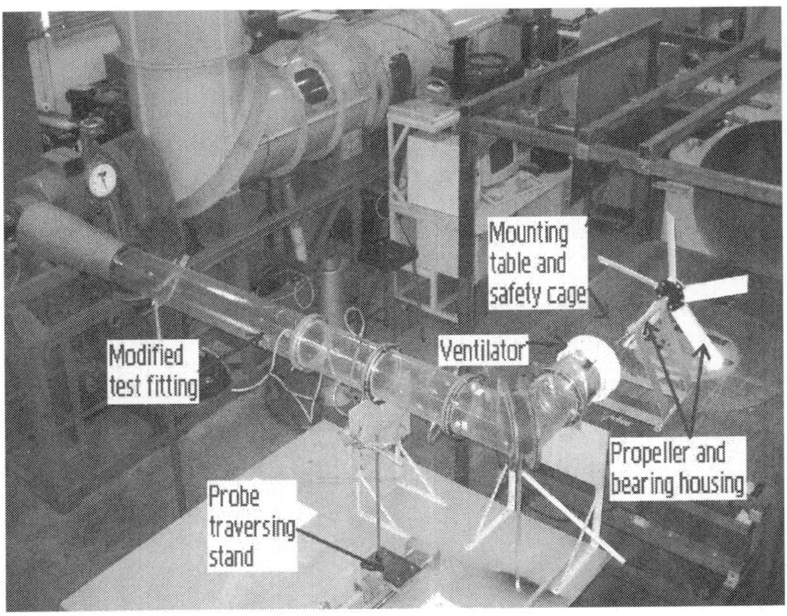

Fig. 14. Horizontal axis ventilator test set-up

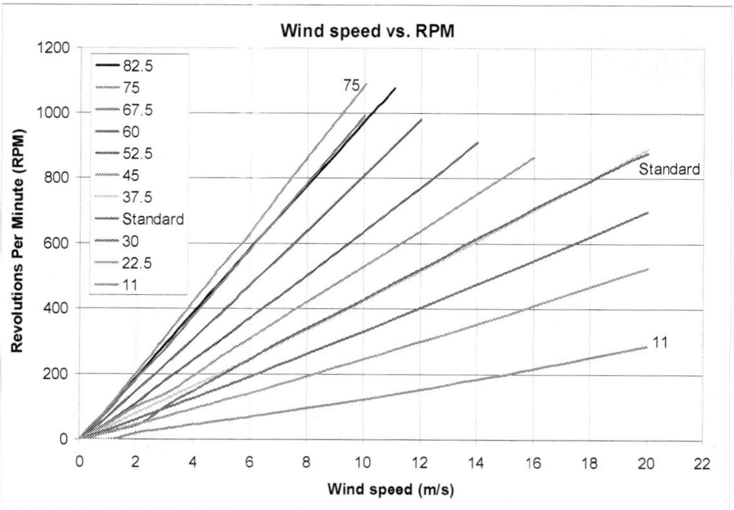

Graph 9. Wind speed vs. RPM

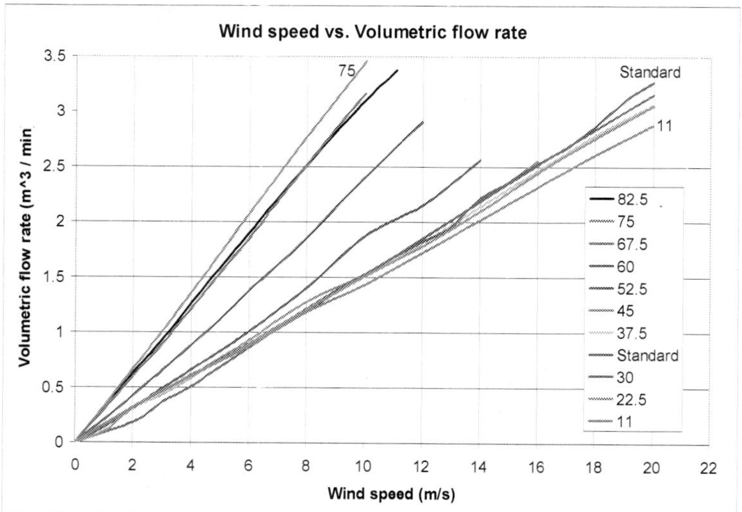

Graph 10. Wind speed vs. volumetric flow rate

Graph 10 is a performance plot of wind speed vs. volume flow rate. Again, the 75° pitch angle (purple line) proved to have the best performance. For any given wind speed, the horizontal axis ventilator with a 75° blade pitch angle managed to create an air flow that was more than 2 times greater compared with the standard turbine ventilator (dark blue line).

An interesting observation was the performance of the horizontal axis ventilator with blade angles below 37.5°. Compared with graph 9, the volume flow rate did not drop off as dramatically as RPM for the shallower pitch angles (blade chord approaching parallel with incident wind). Such an interesting result was accounted for by the cross-flow of incident wind across the ventilator (pump) due to the horizontal axis configuration. The following performance graphs quantify the phenomenon.

Graph 11 is the performance plot of RPM vs. volumetric flow rate, which indicates the effectiveness of the pump with respect to rotational velocity. The 11° pitch angle proved to have the best pump performance with respect to RPM. It was somewhat unfortunate that this shallow blade pitch angle never produced enough RPM to exploit the advantage. The standard turbine ventilator proved to have slightly better performance than the horizontal axis ventilator at a blade pitch of 75°.

A surprising result was that a blade pitch angle of 52.5° produced the worst pump performance with respect to RPM. This may be accounted for by the combined swirl and axial velocity of the incident wind after it has passed through the propeller disc. This particular combination of swirl and axial velocities seemed to minimize the beneficial cross-flow effect. The actual flow rates induced by the cross-flow appear in the following performance chart.

Graph 12 gives an indication of the volume flow rate induced by cross flow across the ventilator (pump). This data was taken by restraining the propeller, and gives a rough indication of the significance of cross flow.

A blade pitch angle of 11° gave the most amount of induced flow rate, with a blade angle of 45° giving the least amount. This data confirms the results plotted on performance Graph 11. As the incident wind passes through the propeller disc, energy is extracted which rotates the device. The propeller induces a residual swirl on the incident wind as it leaves the propeller disc. The results indicate that at blade angles around 45°, the residual swirl was of such a magnitude and direction as to significantly reduce the amount of cross-wind induced flow.

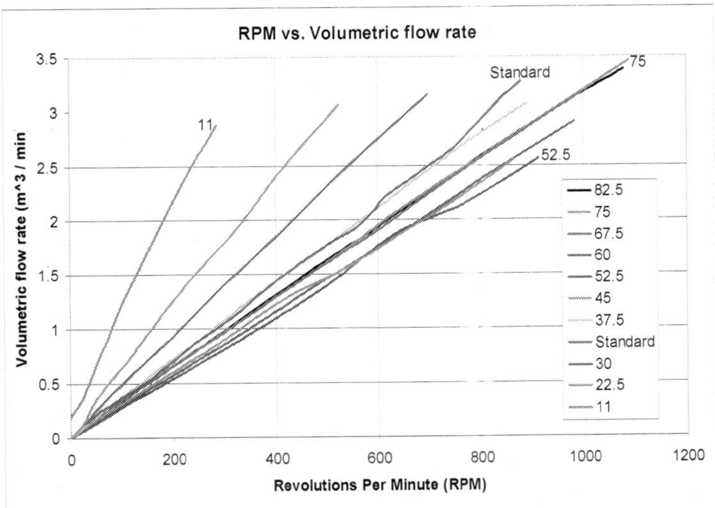

Graph 11. RPM vs. volume flow rate

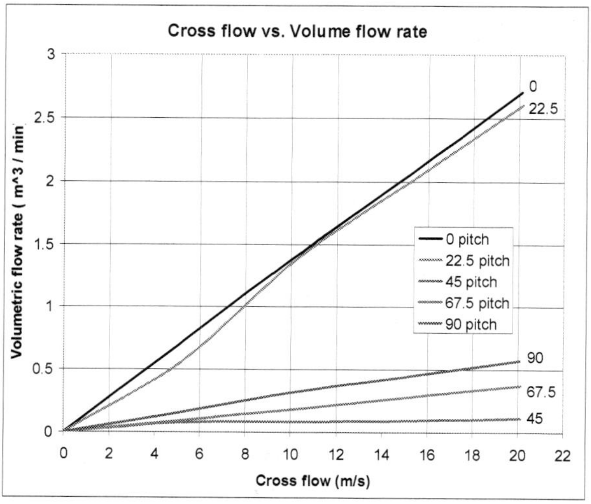

Graph 12. Cross flow vs. volume flow rate

4. Conclusions

Current building ventilators individually rely upon a single source of energy for operation. The turbine ventilator relies entirely of the prevailing wind conditions with no facility to extract energy from the sun. The solar ventilator is at the complete mercy of ambient solar radiation conditions and cannot extract energy from the wind.

The initial Wind-Solar hybrid ventilator was considered a solution to the problem of turbine ventilator operation at zero wind speeds. Air extraction capability at zero wind speed was provided by using an electric motor and solar cell to power the turbine ventilator. The significant findings upon testing of this hybrid design were the vastly improved flow rate performance compared with a purely solar powered ventilator; comparable performance with the standard turbine ventilator, and the vastly improved operational flexibility of the device. The standard turbine ventilator acting as a centrifugal pump provided much better air flow compared to an axial propeller subjected to the same power input. The hybrid design had slightly less performance than the turbine ventilator alone. This was mainly due to the back-driving of the electric motor under zero solar radiation conditions, and the crudity of the device.

The performance level of the hybrid device was vastly improved by removing the solar cell from atop the rotating ventilator and decoupling the electric motor on overrun with a one way bearing. The combination of the turbine ventilator and solar powered ventilator provided a hybrid design that had vastly improved flexibility of operation compared to the individual constituent components.

The horizontal axis ventilator was a solution to the marginal performance of a turbine ventilator at low wind speeds. Testing of the horizontal axis ventilator found significantly improved performance at low wind speed conditions. The device extracted more than double the volume flow rate of air and spun at more than twice the RPM for any given wind speed condition.

The overall conclusion is that a continuous pre-determined volume air-extraction ventilator that relies predominantly on renewable energy is entirely possible.

5. Future possibilities

With environmental issues taking centre stage and government and private funding forthcoming, future possibilities may result in completely different philosophies and different models of energy usage and human life style. The performance criteria of high volume air extraction rate of natural ventilators that rely on wind and sun may be replaced by the philosophy of providing an optimum temperature, humidity and air circulation levels. From a consideration of this philosophy the concept of the Wind-Electric Hybrid ventilator, the 'ECO-POWER' was conceived with the collaboration of CSR Edmonds Australia Pty Ltd as an alternative to the conventional air conditioning units. The electric power currently is drawn from the mains power supply. Various improvements are still needed to make this type of ventilator a commercial reality for both domestic and industrial applications. A computer aided drawing of the ventilator is shown in figure 15.

Fig. 15. A Computer aided image of Wind-Electric ECO-POWER

From the studies presented in this chapter at least, a system is entirely feasible that involves the convergence of the hybrid ventilation of standard wind powered design with possibly horizontal axis design and solar powered models. This with further improvements in electricity storage capabilities and efficient electronic control module, a vastly improved single cost effective ventilation system is just around the corner.

With rapid improvements in the performance of solar cells, electronics and power storage systems and continuous drop in costs of their production, together with the emergence of new technologies, it is not unrealistic to expect future ventilators to evolve with many innovative concepts and ideas currently unheard of.

6. Acknowledgements

The author is heavily indebted to his student Simon Shun for his unselfish contribution in wind tunnel testing and in the preparation of the graphs and figures and manuscript of this chapter. Thanks are also due to Jim Beck and Terry Flynn, the Technical Officers of the Aerodynamic Laboratory at the University of the University of New South Wales and Allan Ramsay, Derek Munn and Tarek Alfakhrany of CSR Edmonds Australia for their continuous collaboration and enthusiastic support. Thanks are also due to CSR Edmonds and Australian Research Council for providing funding to various aspects of investigations associated with wind driven ventilation over the years.

7. References

[1] Standards Australia, AS 1668.2 - 2002:
Part 2, Ventilation design for indoor contaminant control.
Section 4, Mechanical ventilation - supply systems.
Section 5, Mechanical ventilation - exhaust systems.
Section 6, Mechanical ventilation of enclosures used for particular health care functions

[2] Rashid, D.H., Ahmed, N.A. and Archer, R.D., 'Study of aerodynamic forces on a Rotating wind driven ventilator. Wind Engineering, vol. 27, no.1, pp 63-72, 2003

[3] Shun, S., and Ahmed, N.A., 'Utilising wind and solar energy as power sources For a hybrid building ventilation device', Renewable Energy, vol. 33, pp 1392- 1397, 2008

[4] Kreichelt, T.E., Kern, G.R., 'Natural ventilation in hot process buildings in the steel Industry', Journal of Iron and Steel Engineering, December, 1976, pp 39-46.

[5] W.Yang, et al, 'IAQ investigation according to school buildings in Korea', Environ Managem, 90, 348-354, 2009

[6] A.P. Jones, 'IAQ and health', Atmospheric Environ., 33, 4535-2464, 1999

[7] A.C. Biblow, 'NY to require landlords to notify tenants of IAQ results', Real Estate Finance, pp29-31, Feb, 2009

[8] Sahakian, N., et al, 'Respiratory morbidity from dampness and AC in Offices/homes', Indoor Air, 19, 58-67, 2009

[9] N.A.Ahmed and J., Back, 'Destructive wind tunnel tests', UNSW Unisearch Rep. no. 23214-10, 1996.

[10] N.A.Ahmed and J.,Back, 'Wind tunnel tests on ventilators', UNSW Unisearch Rep. no. 29295-01, 1997

[11] T.G.Flynn and N.A.Ahmed, 'Investigation of Rotating Ventilator using Smoke Flow Visualisation and Hot-wire anemometer', Proc. of 5th Pacific Symposium on Flow Visualisation and Image Processing, 27-29 September, 2005, Whitsundays, Australia, Paper No. PSFVIP-5-214

[12] Standards Australia,
AS / NZS 4740:2000, Natural ventilators - Classification and performance

[13] Standards Australia,
AS 2360.0 - 1993, Measurement of fluid flow in closed conduits, Part 0:
AS 2360.1.1 - 1993, Measurement of fluid flow in closed conduits, Part 1.1;
AS 2360.1.3 - 1993, Measurement of fluid flow in closed conduits, Part 1.3;
AS 2360.1.4 - 1993, Measurement of fluid flow in closed conduits, Part 1.4;
AS 2360.7.1 - 2001, Measurement of fluid flow in closed conduits, Part 7.1:
AS 2360.7.2 - 1993, Measurement of fluid flow in closed conduits, Part 7.2.

[14] Barlow J.B, Rae, Jr., and Pope, W.H., 'Low Speed Wind Tunnel Testing', 3rd edition, New York, Wiley, 1999

[15] Ahmed, N.A. and Archer, R.D., 'Performance improvement of bi-plane with endplates', AIAA Journal of Aircraft, vol. 38, no. 2, pp 398-400, 2001